Riemann Solvers and Numerical Methods for Fluid Dynamics

A Practical Introduction

Springer

Berlin
Heidelberg
New York
Barcelona
Hong Kong
London
Milan
Paris
Singapore
Tokyo

Eleuterio F. Toro

Riemann Solvers and Numerical Methods for Fluid Dynamics

A Practical Introduction

2 nd Edition
with 247 Figures and 26 Tables

 Springer

Prof. Dr. Eleuterio F. Toro

Manchester Metropolitan University
Department of Computing and Mathematics
Chester Street
M1 5GD Manchester
U. K.

ISBN 3-540-65966-8 2nd Edition Springer-Verlag Berlin Heidelberg New York

Cataloging–in–Publication Data applied for
Die Deutsche Bibliothek – CIP-Einheitsaufnahme
Toro, Eleuterio F.:
Riemann solvers and numerical methods for fluid dynamics: a practical introduction/
Eleuterio F. Toro.–2. ed.–Berlin; Heidelberg; New York; Barcelona; Hong Kong;
London; Milan; Paris; Singapore; Tokyo: Springer, 1999
      ISBN 3-540-65966-8

Production: ProduServ GmbH Verlagsservice, Berlin
Typesetting: Camera-ready by author
Cover design: MEDIO GmbH, Berlin
SPIN:10706153   62/3020-5 4 3 2 1 0 - Printed on acid-free paper

Preface

In 1917, the British scientist L. F. Richardson made the first reported attempt to predict the weather by solving partial differential equations numerically, by hand! It is generally accepted that Richardson's work, though unsuccessful, marked the beginning of Computational Fluid Dynamics (CFD), a large branch of Scientific Computing today. His work had the four distinguishing characteristics of CFD: a PRACTICAL PROBLEM to solve, a MATHEMATICAL MODEL to represent the problem in the form of a set of partial differential equations, a NUMERICAL METHOD and a COMPUTER, human beings in Richardson's case. Eighty years on and these four elements remain the pillars of modern CFD. It is therefore not surprising that the generally accepted definition of CFD as *the science of computing numerical solutions to Partial Differential or Integral Equations that are models for fluid flow phenomena*, closely embodies Richardson's work.

COMPUTERS have, since Richardson's era, developed to unprecedented levels and at an ever decreasing cost. The range of application areas giving rise to PRACTICAL PROBLEMS to solved numerically has increased dramatically. In addition to the traditional demands from Meteorology, Oceanography, some branches of Physics and from a range of Engineering Disciplines, there are at present fresh demands from a dynamic and fast–moving manufacturing industry, whose traditional build–test–fix approach is rapidly being replaced by the use of quantitative methods, at all levels. The need for new materials and for decision–making under environmental constraints are increasing sources of demands for mathematical modelling, numerical algorithms and high–performance computing. MATHEMATICAL MODELS have improved, though the basic equations of continuum mechanics, already available more than a century before Richardson's first attempts at CFD, are still the bases for modelling fluid flow processes. Progress is required at the level of Thermodynamics, equations of state, and advances into the modelling of non–equilibrium and multiphase flow phenomena. NUMERICAL METHODS are perhaps the success story of the last eighty years, the last twenty being perhaps the most productive. This success is firmly based on the pioneering works of scientists such as von Neumann, whose research on *stability* explained and resolved the difficulties experienced by Richardson. This success would have been impossible without the contributions from Courant, Friedrichs,

Richtmyer, Lax, Oleinik, Wendroff, Godunov, Rusanov, van Leer, Harten, Roe, Osher, Colella, Yee, and many others. The net result is: more accurate, more efficient, more robust and more sophisticated numerical methods are available for ambitious practical applications today.

Due to the massive demands on CFD and the level of sophistication of numerical methods, new demands on *education and training* of the scientists and engineers of the present and the future have arisen. This book is an attempt to contribute to the training and education in numerical methods for fluid dynamics and related disciplines.

The contents of this book were developed over a period of many years of involvement in research on numerical methods, application of the methods to solve practical problems and teaching scientist and engineers at the post–graduate level. The starting point was a module for a Masters Course in Computational Fluid Dynamics at the College of Aeronautics, Cranfield, UK. The material was also part of short courses and lectures given at Cranfield UK; the Ernst Mach Institute, Freiburg, Germany; the Shock Wave Research Centre, Tohoku University, Sendai, Japan; the Department of Mathematics and the Department of Civil and Environmental Engineering, University of Trento, Italy; the Department of Mathematics, Technical University Federico Santa Maria, Chile; the Department of Mechanics, Technical University of Aachen, Germany; and the Manchester Metropolitan University (MMU), Manchester, UK.

This book is about modern shock–capturing numerical methods for solving time–dependent hyperbolic conservation laws, with smooth and discontinuous solutions, in general multidimensional geometries. The approach is comprehensive, practical and, in the main, informal. All necessary items of information for the practical implementation of all methods studied here, are provided in detail. All methods studied are illustrated through practical numerical examples; numerical results are compared with exact solutions and in some cases with reliable experimental data.

Most of the book is devoted to a coherent presentation of *Godunov methods*. The developments of Godunov's approach over the last twenty years have led to a mature numerical technology, that can be utilised with confidence to solve practical problems in established, as well as new, areas of application. Godunov methods rely on the solution of the Riemann problem. The exact solution is presented in detail, so as to aid the reader in applying the solution methodology to other hyperbolic systems. We also present a variety of approximate Riemann solvers; again, the amount of detail supplied will hopefully aid the reader who might want to apply the methodologies to solve other problems. Other related methods such as the Random Choice Method and the Flux Vector Splitting Method are also included. In addition, we study *centred* (non–upwind) shock–capturing methods. These schemes are much less sophisticated than Godunov methods, and offer a cheap and simple alternative. High–order extensions of these methods are constructed for scalar

PDEs, along with their Total Variation Diminishing (TVD) versions. Most of these TVD methods are then extended to one–dimensional non–linear systems. Techniques to deal with PDEs with source terms are also studied, as are techniques for multidimensional systems in general geometries.

The presentation of the schemes for non–linear systems is carried out through the time–dependent Euler equations of Gas Dynamics. Having read the relevant chapters/sections, the reader will be sufficiently well equipped to extend the techniques to other hyperbolic systems, and to advection–reaction–diffusion PDEs.

There are at least two ways of utilising this book. First, it can be used as a means for *self–study*. In the presentation of the concepts, the emphasis has been placed on clarity, sometimes sacrificing mathematical rigour. The typical reader in mind is a graduate student in a Department of Engineering, Physics, Applied Mathematics or Computer Science, embarking on a research topic that involves the implementation of numerical methods, from first principles, to solve advection–reaction–diffusion problems. The contents of this book may also be useful to numerical analysts beginning their research on algorithms, as elementary background reading. Such users may benefit from a comprehensive self–study of all the contents of the book, in a period of about two months, perhaps including the practical implementation and testing of most numerical methods presented. Another class of readers who may benefit from self–studying this book are scientists and engineers in industry and research laboratories. At the cost of some repetitiveness, each chapter is almost self–contained and has plenty of cross-referencing, so that the reader may decide to start reading this book in the middle or jump to the last chapter.

This book can also be used as a *teaching aid*. Academics involved in the teaching of numerical methods may find this work a useful reference book. Selected chapters or sections may well form the bases for a final year undergraduate course on numerical methods for PDEs. In a Mathematics or Computer Science Department, the contents may include: some sections of Chap. 1, Chaps. 2, 5, 13, some sections of Chap. 14, Chap. 15 and some sections of Chap. 16. In a Department of Engineering or Physics, one may include Chaps. 3, 4, 6, 7, 8, 9, 10, 11, 12, 15, 16 and 17. A postgraduate course may involve most of the contents of this book, assuming perhaps a working knowledge of compressible fluid dynamics. Short courses for training engineers and scientists in industry and research laboratories can also be based on most of the contents of this book.

Tito Toro
Manchester, UK
March 1997.

ACKNOWLEDGEMENTS

The author is indebted to many colleagues who over the years have kindly arranged short and extended visits to their establishments, have organised seminars, workshops and short courses. These events have strongly influenced the development of this book. Special thanks are due to Dr. W. Heilig (Freiburg, Germany); Professors V. Casulli and A. Armanini (Trento, Italy); Professor K. Takayama (Sendai, Japan); Professor J. Figueroa (Valparaiso, Chile) and Professor J. Ballmann (Aachen, Germany). The final stages of the typing of the material were carried out while the author was at MMU, Manchester, UK; the support provided is gratefully acknowledged. Thanks are also due to my former and current PhD students, for their comments on my lectures, on the contents of this book and for their contribution to research in relevant areas. Special thanks are due to Stephen Billett (Centre for Atmospheric Science, University of Cambridge, UK), Caroline Lowe (Department of Aerospace Science, Cranfield University, UK), Nikolaos Nikiforakis (DAMTP, University of Cambridge, UK), Ed Boden (Cranfield and MMU), Ms Wei Hu, Mathew Ivings and Richard Millington (MMU). Thanks are due to my son Marcelo, who helped with the type–setting, and to David Ingram and John Nuttall for their help with the intricacies of Latex. The author thanks Springer–Verlag, and specially Miss Erdmuthe Raufelder in Heidelberg, for their professional assistance.

The author gratefully acknowledges useful feedback on the manuscript from Stephen Billett (University of Cambridge, UK), John Clarke (Cranfield University, UK), Jean–Marc Moschetta (SUPAERO, Toulouse, France), Claus–Dieter Munz (University of Stuttgart, Germany), Jack Pike (Bedford, UK), Ning Qin (Cranfield University, UK), Tsutomu Saito (Cray Japan), Charles Saurel (University of Provence, France), Trevor Randall (MMU, Manchester, UK), Peter Sweby (University of Reading, UK), Marcelo Toro (IBM Glasgow, UK), Helen Yee (NASA Ames, USA) and Clive Woodley (DERA, Fort Halstead, UK).

The permanent encouragement from Brigitte has been an immensely valuable support during the writing of this book. Thank you Brigitte!

To my children

MARCELO

CARLA

VIOLETA

and

EVITA

Preface to the Second Edition

This edition includes a number of minor changes. First, I have corrected all the errors found so far and, hopefully, I have not made new ones in the process. Thanks are due to many people who, mostly using the wonders of the email age, have sent their encouraging words as well as their suggested corrections. I have attempted some improvements to the presentation of certain concepts; questions posed by readers have been helpful in this respect. Having used the book for lectures, since the publication of the first edition in 1997, has been useful in re–assessing some of the material and its mode of presentation.

Suggestions have been made by some readers to the effect of including a list of exercises, with solutions, at the end of each chapter. I have, for the moment, conveniently ignored such valuable suggestions. I have kept the exercises, some of them with solutions, interspersed in the text, as their contents become relevant to the presentation of new concepts. I have added new references in appropriate sections, in an attempt to partially correct the unavoidable injustice of excluding reference to many valuable contributions to the themes of interest of this book.

I have also added a new test problem to the existing suite of tests. Each test problem of the suite is designed to expose particular, potential limitations of numerical methods for hyperbolic conservation laws. The tests have not been chosen with the purpose of producing fanciful results to entice the casual reader. The new test problem added has the purpose of illustrating via practical numerical results the main differences between *upwind* and *centred* methods and between *Godunov* and most *Flux Vector Splitting* methods. Finally, I also admit to having re–plotted all the numerical results of the first edition that had a position error of Δx!

A third edition, if is to happen at all, will include a much needed *Introduction*, conceived as a guiding tour to the general area of high–resolution numerical methods for hyperbolic conservation laws and to the contents of this book.

Tito Toro
Manchester, UK
March 1999

Table of Contents

1. The Equations of Fluid Dynamics

In this chapter we present the governing equations for the dynamics of a compressible material, such as a gas, along with closure conditions in the form of equations of state. Equations of state are statements about the nature of the material in question and require some notions from Thermodynamics. There is no attempt to provide an exhaustive and rigourous derivation of the equations of continuum mechanics; such a task is beyond the scope of this book. Instead, we give a fairly self–contained summary of the equations and the Thermodynamics in a manner that is immediately useful to the main purpose of this book, namely the detailed treatment of Riemann solvers and numerical methods.

The presentation of the equations is unconventional. We first introduce the differential form of the Euler equations along with basic physical quantities and thermodynamic relations leading to equations of state. Then the effects of viscous diffusion and heat transfer are added to the Euler equations. After this, the fundamental integral form of the equations is introduced; conventionally, this is the starting point for presenting the governing equations. This chapter contains virtually all of the necessary background on Fluid Dynamics that is required for a fruitful study of the rest of the book. It also contains useful information for those wishing to embark on complex practical applications. A hierarchy of submodels is also presented. This covers four systems of hyperbolic conservation laws for which Riemann solvers and upwind methods are directly applicable, namely (i) the time–dependent Euler equations, (ii) the steady supersonic Euler equations, (iii) the shallow water equations and (iv) the artificial compressibility equations associated with the incompressible Navier–Stokes equations. Included in the hierarchy are also some simpler models such as linear systems and scalar conservation laws.

Some remarks on notation are in order. A Cartesian frame of reference (x, y, z) is chosen and the time variable is denoted by t. Transformation to other coordinate systems is carried out using the chain rule in the usual way, see Sect. 16.7.2 of Chap. 16. Any quantity ϕ that depends on space and time will be written as $\phi(x, y, z, t)$. In most situations the governing equations will be partial differential equations (PDEs). Naturally, these will involve partial derivatives for which we use the notation

$$\phi_t \equiv \frac{\partial \phi}{\partial t} \; , \; \phi_x \equiv \frac{\partial \phi}{\partial x} \; , \; \phi_y \equiv \frac{\partial \phi}{\partial y} \; , \; \phi_z \equiv \frac{\partial \phi}{\partial z} \; .$$

We also recall some basic notation involving scalars and vectors. The *dot product* of two vectors $\mathbf{A} = (a_1, a_2, a_3)$ and $\mathbf{B} = (b_1, b_2, b_3)$ is the *scalar* quantity

$$\mathbf{A} \cdot \mathbf{B} = a_1 b_1 + a_2 b_2 + a_2 b_3 \; .$$

Given a scalar quantity ϕ that depends on the spatial variables x, y, z the *gradient operator* ∇ as applied to ϕ is the vector

$$\mathrm{grad} \, \phi \equiv \nabla \phi \equiv (\frac{\partial \phi}{\partial x}, \frac{\partial \phi}{\partial y}, \frac{\partial \phi}{\partial z}) \; .$$

The *divergence operator* applies to vectors and the result is a scalar quantity; for a vector $\mathbf{A}$, the divergence of $\mathbf{A}$ is

$$\mathrm{div} \, \mathbf{A} \equiv \nabla \cdot \mathbf{A} \equiv \frac{\partial a_1}{\partial x} + \frac{\partial a_2}{\partial y} + \frac{\partial a_3}{\partial z} \; .$$

1.1 The Euler Equations

In this section we consider the time–dependent Euler equations. These are a system of non–linear hyperbolic conservation laws that govern the dynamics of a compressible material, such as gases or liquids at high pressures, for which the effects of body forces, viscous stresses and heat flux are neglected.

There is some freedom in choosing a set of variables to describe the flow under consideration. A possible choice is the so called *primitive variables* or *physical variables*, namely, $\rho(x, y, z, t)$ = density or mass density, $p(x, y, z, t)$ = pressure, $u(x, y, z, t)$ = x–component of velocity, $v(x, y, z, t)$ = y–component of velocity, $w(x, y, z, t)$ = z–component of velocity. The velocity vector is $\mathbf{V} = (u, v, w)$. An alternative choice is provided by the so called *conserved variables*. These are the mass density ρ, the x–momentum component ρu, the y–momentum component ρv, the z–momentum component ρw and the total energy per unit mass E. Physically, these conserved quantities result naturally from the application of the fundamental laws of conservation of mass, Newton's Second Law and the law of conservation of energy. Computationally, there are some advantages in expressing the governing equations in terms of the conserved variables. This gives rise to a large class of numerical methods called *conservative methods*, which will be studied later in this book. We next state the equations in terms of the conserved variables under the assumption that the quantities involved are sufficiently smooth to allow for the operation of differentiation to be defined. Later we remove this smoothness constraint to allow for solutions containing discontinuities, such as shock waves.

1.1.1 Conservation–Law Form

The five governing conservation laws are

$$\rho_t + (\rho u)_x + (\rho v)_y + (\rho w)_z = 0 \ , \tag{1.1}$$

$$(\rho u)_t + (\rho u^2 + p)_x + (\rho uv)_y + (\rho uw)_z = 0 \ , \tag{1.2}$$

$$(\rho v)_t + (\rho uv)_x + (\rho v^2 + p)_y + (\rho vw)_z = 0 \ , \tag{1.3}$$

$$(\rho w)_t + (\rho uw)_x + (\rho vw)_y + (\rho w^2 + p)_z = 0 \ , \tag{1.4}$$

$$E_t + [u(E + p)]_x + [v(E + p)]_y + [w(E + p)]_z = 0 \ . \tag{1.5}$$

Here E is the total energy per unit volume

$$E = \rho \left(\frac{1}{2} \mathbf{V}^2 + e \right) , \tag{1.6}$$

where

$$\frac{1}{2} \mathbf{V}^2 = \frac{1}{2} \mathbf{V} \cdot \mathbf{V} = \frac{1}{2}(u^2 + v^2 + w^2)$$

is the *specific kinetic energy* and e is the *specific internal energy*. One generally refers to the full system (1.1)–(1.5) as the Euler equations, although strictly speaking the Euler equations are just (1.2)–(1.4).

The conservation laws (1.1)–(1.5) can be expressed in a very compact notation by defining a column vector $\mathbf{U}$ of conserved variables and flux vectors $\mathbf{F}(\mathbf{U})$, $\mathbf{G}(\mathbf{U})$, $\mathbf{H}(\mathbf{U})$ in the x, y and z directions, respectively. The equations now read

$$\mathbf{U}_t + \mathbf{F}(\mathbf{U})_x + \mathbf{G}(\mathbf{U})_y + \mathbf{H}(\mathbf{U})_z = \mathbf{0} \ , \tag{1.7}$$

with

$$\left. \begin{array}{c}
\mathbf{U} = \begin{bmatrix} \rho \\ \rho u \\ \rho v \\ \rho w \\ E \end{bmatrix} , \quad
\mathbf{F} = \begin{bmatrix} \rho u \\ \rho u^2 + p \\ \rho uv \\ \rho uw \\ u(E + p) \end{bmatrix} , \\[4em]
\mathbf{G} = \begin{bmatrix} \rho v \\ \rho uv \\ \rho v^2 + p \\ \rho vw \\ v(E + p) \end{bmatrix} , \quad
\mathbf{H} = \begin{bmatrix} \rho w \\ \rho uw \\ \rho vw \\ \rho w^2 + p \\ w(E + p) \end{bmatrix} .
\end{array} \right\} \tag{1.8}$$

It is important to note that $\mathbf{F} = \mathbf{F}(\mathbf{U})$, $\mathbf{G} = \mathbf{G}(\mathbf{U})$, $\mathbf{H} = \mathbf{H}(\mathbf{U})$; that is, the flux vectors are to be regarded as functions of the conserved variable vector $\mathbf{U}$. Any set of PDEs written in the form (1.7) is called a system of *conservation laws*. As partial derivatives are involved we say that (1.7) is a system of conservation laws in differential form. The differential formulation assumes smooth solutions, that is, partial derivatives are assumed to exist. There are other ways of expressing conservation laws in which the smoothness assumption is relaxed to include discontinuous solutions.

1.1.2 Other Compact Forms

An even more compact way of expressing equations (1.1)–(1.5) involves tensors. First note that the spatial derivatives in (1.1) can be expressed in terms of the divergence operator e.g.

$$\text{div}(\rho \mathbf{V}) = \nabla \cdot (\rho \mathbf{V}) = (\rho u)_x + (\rho v)_y + (\rho w)_z .$$

Thus equation (1.1) for conservation of mass can be written as

$$\rho_t + \nabla \cdot (\rho \mathbf{V}) = 0 . \tag{1.9}$$

As the divergence operator may also be applied to tensors, the three momentum equations for conservation of momentum can be written in compact form as

$$(\rho \mathbf{V})_t + \nabla \cdot (\rho \mathbf{V} \otimes \mathbf{V} + p\mathbf{I}) = 0 , \tag{1.10}$$

where $\mathbf{V} \otimes \mathbf{V}$ is the tensor product and $\mathbf{I}$ is the unit tensor. These are given respectively by

$$\mathbf{V} \otimes \mathbf{V} = \begin{pmatrix} u^2 & uv & uw \\ vu & v^2 & vw \\ wu & wv & w^2 \end{pmatrix} , \quad \mathbf{I} = \begin{pmatrix} 1 & 0 & 0 \\ 0 & 1 & 0 \\ 0 & 0 & 1 \end{pmatrix} .$$

The conservation of energy equation can be written as

$$E_t + \nabla \cdot [(E + p)\mathbf{V}] = 0 . \tag{1.11}$$

In fact the complete system of equations (1.9)–(1.11) can be written in divergence form as

$$\mathbf{U}_t + \nabla \cdot \mathbf{H} = \mathbf{0} , \tag{1.12}$$

where $\mathbf{H}$ is the tensor

$$\mathbf{H} = \begin{bmatrix} \rho u & \rho u^2 + p & \rho v u & \rho w u & u(E + p) \\ \rho v & \rho u v & \rho v^2 + p & \rho w v & v(E + p) \\ \rho w & \rho u w & \rho v w & \rho w^2 + p & w(E + p) \end{bmatrix} . \tag{1.13}$$

Note that the rows of the tensor $\mathbf{H}$ are the flux vectors $\mathbf{F}$, $\mathbf{G}$ and $\mathbf{H}$, understood as row vectors. For computational purposes it is the compact conservative form (1.7)–(1.8) of equations (1.1)–(1.5) that is most useful. In Chap. 3 we study some mathematical properties of the Euler equations and in Chap. 4 we solve exactly the Riemann problem for the one–dimensional Euler equations for ideal and covolume gases. Numerical methods for the Euler equations are discussed in Chaps. 6–12, 14 and 16.

1.2 Thermodynamic Considerations

The stated governing partial differential equations (1.1)–(1.5) for the dynamics of a compressible material are insufficient to completely describe the physical processes involved. There are more unknowns than equations and thus closure conditions are required. Physically, such conditions are statements related to the nature of the material or medium. Relation (1.6) defines the total energy E in terms of the velocity vector $\mathbf{V}$ involved in equations (1.1)–(1.5) and a new variable e, the specific internal energy. One therefore requires another relation defining e in terms of quantities already given, such as pressure and density, as a closure condition. For some applications, or when more physical effects are added to the basic equations (1.1)–(1.5), other variables, such as temperature for instance, may need to be introduced.

Central to providing closure conditions is a discussion of the fundamentals of Thermodynamics. This introduces new physical variables and provides relations between variables. Under certain conditions the governing equations may be approximated so as to make such discussion of Thermodynamics unnecessary; two important examples are incompressible flows and isentropic flows [75]. The specific internal energy e has an important role in the *First Law of Thermodynamics*, while the entropy s is intimately involved with the *Second Law of Thermodynamics*. Entropy plays a fundamental role not just in establishing the governing equations but also at the level of their mathematical properties and the designing of numerical methods to solve them. In addition to the basic thermodynamic variables density ρ, pressure p, temperature T, specific internal energy e and entropy s, one may define other new variables that are combinations of these.

1.2.1 Units of Measure

A brief discussion of physical quantities and their units of measure is essential. We consistently adopt, unless otherwise stated, the *International System of Units* or *SI Units*. Three basic quantities are length (l), mass (m) and time (t). The unit of measure of length is: one metre $= 1$ m. Submultiples are: one decimetre $= 10^{-1}$ m, one centimetre $= 10^{-2}$ m, one millimetre $= 10^{-3}$ m. Multiples are: 10^1 m, 10^2 m and 10^3 m $=$ one kilometre. From length one can establish the units of measure of area: one square metre $= 1$ m^2 and the units of measure of volume: one cubic metre $= 1$ m^3. The unit of measure of mass is: one kilogram $= 1$ kg. A useful submultiple is: one gram $= 1$ g $= 10^{-3}$ kg. As density is $\rho = m/V$, where V is the total volume of the system, the unit of measure of density is: one kilogram per cubic metre $= 1$ kg/m^3 $= 1$ kg m^{-3}. The unit of measure of time is: one second $= 1$ s. The unit of measure of speed is: one metre per second $= 1$ m/s$= 1$ m s^{-1}. The unit of measure of acceleration is: one metre per second per second $= 1$ m/s$^2 = 1$ m s^{-2}. The unit of measure of force is: one Newton $= 1$ N. The Newton N is defined as the force required to give a mass of 1 kg an acceleration of 1 m s^{-2}. Newton's Second Law states that *force = constant × mass × acceleration*. The value of the unit of force is then chosen so as to make *force* = 1 when *constant* = 1. Therefore the unit of force (N) is: 1 kg m s^{-2}. We now give the unit of measure of pressure. Pressure p is the magnitude of force per unit area and therefore its unit of measure is: one Newton per square metre $= 1 \text{ N m}^{-2} \equiv 1$ Pa : one Pascal. Two common units of pressure are 1 bar$=10^5$ Pa and 1 atm (atmosphere) $= 101\,325$ Pa. An important rule in manipulating physical quantities is *dimensional consistency*. For example, in the expression $\rho u^2 + p$ in the momentum equation (1.2), the dimensions of ρu^2 must be the same as those of (pressure) p. This is easily verified.

To introduce the unit of measure of energy we first recall the concept of *Work*. Work (W) is done when a force produces a motion and is measured as $W = $ *force × distance moved in the direction of the force*. The unit of measure of work is: one Joule = 1 J. One Joule is the work done when the point of application of a force of 1 N moves through a distance of 1 m in the direction of the force. As energy is the capacity to perform work, the unit of measure of energy is also one Joule. The temperature T will be measured in terms of the *Thermodynamic Scale* or the *Absolute Scale*, in which the unit of measure is: one kelvin = 1 K.

Thermodynamic properties of a system that are proportional to the mass m of the system are called *extensive* properties. Examples are the total energy E and the total volume V of a system. Properties that are independent of m are called *intensive* properties; examples are temperature T and pressure p. Extensive properties may be converted to their *specific* (intensive) values by dividing that property by its mass m. For instance, from the total volume V we obtain the *specific volume* $v = V/m$ (the reader is warned that v is also used for velocity component). As $\rho = m/V$, the specific volume is the reciprocal of density. The units of measure of other quantities will be given as they are introduced.

1.2.2 Equations of State (EOS)

A system in thermodynamic equilibrium can be completely described by the basic thermodynamic variables pressure p and specific volume v. A family of states in thermodynamic equilibrium may be described by a curve in the p–v plane, each characterised by a particular value of a variable temperature T. Systems described by the p–v–T variables are usually called p–v–T systems. There are physical situations that require additional variables. Here we are only interested in p–v–T systems. In these, one can relate the variables via the *thermal equation of state*

$$T = T(p, v) . \tag{1.14}$$

Two more possible relations are

$$p = p(T, v) , \quad v = v(T, p) .$$

The p–v–T relationship changes from substance to substance. For *thermally ideal gases* one has the simple expression

$$T = \frac{pv}{R} \, ,$$ (1.15)

where R is a constant which depends on the particular gas under consideration.

The First Law of Thermodynamics states that for a non–adiabatic system the change Δe in internal energy e in a process will be given by $\Delta e = \Delta W + \Delta Q$, where ΔW is the work done on the system and ΔQ is the heat transmitted to the system. Taking the work done as $dW = -pdv$ one may write

$$dQ = de + pdv \, .$$ (1.16)

The internal energy e can also be related to p and v via a *caloric equation of state*

$$e = e(p, v) \, .$$ (1.17)

Two more possible ways of expressing the p–v–e relationship are

$$p = p(v, e) \, , \; v = v(e, p) \, .$$

For a *calorically ideal gas* one has the simple expression

$$e = \frac{pv}{\gamma - 1} = \frac{p}{\rho(\gamma - 1)} \, ,$$ (1.18)

where γ is a constant that depends on the particular gas under consideration.

The thermal and caloric equations of state for a given material are closely related. Both are necessary for a complete description of the thermodynamics of a system. Choosing a thermal EOS does restrict the choice of a caloric EOS but does not determine it. Note that for the Euler equations (1.1)–(1.5) one only requires a caloric EOS, e.g. $p = p(\rho, e)$, unless temperature T is needed for some other purpose, in which case a thermal EOS needs to be given explicitly.

1.2.3 Other Variables and Relations

The *entropy s* results as follows. We first introduce an integrating factor $1/T$ so that the expression

$$de + pdv = \left(\frac{\partial e}{\partial v} + p \right) dv + \frac{\partial e}{\partial p} dp$$

in (1.16) becomes an exact differential. Then the Second Law of Thermodynamics introduces a new variable s, called entropy, via the relation

$$Tds = de + pdv \, .$$ (1.19)

For any process the change in entropy is $\Delta s = \Delta s_0 + \Delta s_i$, where Δs_0 is the entropy carried into the system through the boundaries of the system and Δs_i is the entropy generated in the system during the process. Examples of entropy–generating mechanisms are heat transfer and viscosity, such as may operate within the internal structure of shock waves. The Second Law of Thermodynamics states that $\Delta s_i > 0$ in any irreversible process. Only in a reversible process is $\Delta s_i = 0$.

Another variable of interest is the *specific enthalpy h*. This is defined in terms of other thermodynamic variables, namely

$$h = e + pv \ . \tag{1.20}$$

One can also establish various relationships amongst the basic thermodynamic variables already defined. For instance from (1.19)

$$de = Tds - pdv \ , \tag{1.21}$$

that is to say, one may choose to express the internal energy e in terms of the variables appearing in the differentials, i.e.

$$e = e(s, v) \ . \tag{1.22}$$

Also, taking the differential of (1.20) we have $dh = de + pdv + vdp$, which by virtue of (1.21) becomes

$$dh = Tds + vdp \ , \tag{1.23}$$

and thus we can choose to define h in terms of s and p, i.e.

$$h = h(s, p) \ . \tag{1.24}$$

Relations (1.22) and (1.24) are called *canonical equations of state* and, unlike the thermal and caloric equations of state (1.14) and (1.17), each of these provides a complete description of the Thermodynamics. For instance, given (1.22) in which e is a function of s and v (independent variables) the pressure p and temperature T follow as

$$p = - \left(\frac{\partial e}{\partial v} \right)_s \ , \ T = \left(\frac{\partial e}{\partial s} \right)_v \ . \tag{1.25}$$

Relations (1.25) follow from comparing

$$de = \left(\frac{\partial e}{\partial s} \right)_v ds + \left(\frac{\partial e}{\partial v} \right)_s dv$$

with equation (1.21). It is conventional in Thermodynamics to specify clearly the independent variables in partial differentiation, as changes of variables often take place. In (1.25), obviously the independent variables are s and v, as is also indicated in (1.22). For instance, the first partial derivative in (1.25) means differentiation of e with respect to v while holding s constant;

the second partial derivative in (1.25) means differentiation of e with respect to s while holding v constant. In a similar manner, equation (1.24) (where s and p are the independent variables) produces T and v from relation (1.23) and

$$dh = \left(\frac{\partial h}{\partial s}\right)_p ds + \left(\frac{\partial h}{\partial p}\right)_s dp \, .$$

Hence,

$$T = \left(\frac{\partial h}{\partial s}\right)_p \, , \, v = \left(\frac{\partial h}{\partial p}\right)_s \, . \tag{1.26}$$

The Helmholtz *free energy* f is defined as

$$f = e - Ts \, . \tag{1.27}$$

A corresponding canonical EOS is

$$f = f(v, T) \, ,$$

from which one can obtain

$$s = -\left(\frac{\partial f}{\partial T}\right)_v \, , \, p = -\left(\frac{\partial f}{\partial v}\right)_T \, . \tag{1.28}$$

Two more quantities can be defined if a thermal EOS $v = v(p, T)$ is given. These are the *volume expansivity* α (or expansion coefficient) and the *isothermal compressibility* β, namely

$$\alpha = \frac{1}{v}\left(\frac{\partial v}{\partial T}\right)_p \, , \, \beta = -\frac{1}{v}\left(\frac{\partial v}{\partial p}\right)_T \, . \tag{1.29}$$

Using equations (1.28) and (1.27) we obtain

$$\left(\frac{\partial s}{\partial v}\right)_T = \left(\frac{\partial p}{\partial T}\right)_v = \frac{\alpha}{\beta} \, ,$$

from which it can be shown that

$$\left(\frac{\partial e}{\partial v}\right)_T = \frac{\alpha T - \beta p}{\beta} \, . \tag{1.30}$$

The *heat capacity at constant pressure* c_p and the *heat capacity at constant volume* c_v (specific heat capacities) are now introduced. In general, when an addition of heat dQ changes the temperature by dT the ratio $c = dQ/dT$ is called the *heat capacity* of the system. For a process at constant pressure relation (1.16) becomes

$$dQ = de + d(pv) = dh \, ,$$

where definition (1.20) has been used. The heat capacity c_p at constant pressure becomes $c_p = dQ/dT = dh/dT$. From (1.23), since $dp = 0$, $dh = Tds$. Assuming $h = h(T, p)$ we obtain

$$c_p = \left(\frac{\partial h}{\partial T}\right)_p = T \left(\frac{\partial s}{\partial T}\right)_p . \tag{1.31}$$

The heat capacity c_v at constant volume may be written, following a similar argument, as

$$c_v = \left(\frac{\partial e}{\partial T}\right)_v = T \left(\frac{\partial s}{\partial T}\right)_v . \tag{1.32}$$

The speed of sound is another variable of fundamental interest. For flows in which particles undergo unconstrained thermodynamic equilibrium one defines a new state variable a, called the equilibrium speed of sound or just speed of sound. Given a caloric equation of state

$$p = p(\rho, s) , \tag{1.33}$$

one defines the speed of sound a as

$$a = \sqrt{\left(\frac{\partial p}{\partial \rho}\right)_s} . \tag{1.34}$$

This basic definition can be transformed in various ways using established thermodynamic relations. For instance, given a caloric EOS in the form $h = h(p, \rho)$, from (1.23) we can write

$$\left(\frac{\partial h}{\partial p}\right)_\rho dp + \left(\frac{\partial h}{\partial \rho}\right)_p d\rho = Tds + \frac{1}{\rho}\left[\left(\frac{\partial p}{\partial \rho}\right)_s d\rho + \left(\frac{\partial p}{\partial s}\right)_\rho ds\right] .$$

Setting $ds = 0$ and using definition (1.34) we obtain

$$a^2 = -\frac{\left(\frac{\partial h}{\partial \rho}\right)_p}{\left(\frac{\partial h}{\partial p}\right)_\rho - \frac{1}{\rho}} .$$

For a thermally ideal gas $h = h(T)$ and thus

$$a^2 = \frac{\left(\frac{\partial h}{\partial T}\right)_p \left(\frac{\partial T}{\partial \rho}\right)_p}{\left(\frac{\partial h}{\partial T}\right)_p \left(\frac{\partial T}{\partial p}\right)_\rho - \frac{1}{\rho}} .$$

From (1.31) $\left(\frac{\partial h}{\partial T}\right)_p = c_p$ and, if the thermal EOS (1.15) is acceptable, we obtain

$$a = \sqrt{\gamma(T)RT} = \sqrt{\frac{\gamma p}{\rho}} . \tag{1.35}$$

For a general material the caloric EOS is a functional relationship involving the variables p–ρ–e. One may also use the specific volume $v = 1/\rho$ instead of density ρ. The derived expression for the speed of sound a depends on the choice of independent variables. Two possible choices and their respective expressions for a are

$$\left. \begin{array}{l} p = p(\rho, e) \ , \ a = \sqrt{\frac{p}{\rho^2} p_e + p_\rho} \ , \\[2mm] e = e(\rho, p) \ , \ a = \sqrt{\frac{p}{\rho^2 e_p} - \frac{e_\rho}{e_p}} \ , \end{array} \right\} \tag{1.36}$$

where subscripts denote partial derivatives.

1.2.4 Ideal Gases

We consider gases obeying the ideal thermal EOS

$$pV = n\mathcal{R}T \ , \tag{1.37}$$

where V is the volume, $\mathcal{R} = 8.134 \times 10^3$ J kilomole^{-1}K^{-1}, called the *Universal Gas Constant*, and T is the temperature measured in degrees kelvin (K). Two more universal constants are now introduced. Recall that a mole of a substance is numerically equal to ω gram and contains 6.02×10^{23} particles of that substance, where ω is the relative atomic mass (RAM) or relative molecular mass (RMM); 1 kilomole $= \omega$ kg. One kilomole of a substance contains $N_A = 6.02 \times 10^{26}$ particles of that substance. The constant N_A is called the *Avogadro Number*. Sometimes this number is given in terms of one mole. The *Boltzmann Constant* k is now defined as $k = \mathcal{R}N_A$; n in (1.37) is the number of kilomoles in volume V, that is $n = N/N_A$ and N is the number of molecules. On division by the mass $m = n\omega$ we have

$$pv = RT \ , \ R = \frac{\mathcal{R}}{\omega} \ , \tag{1.38}$$

where R is called the *Specific Gas Constant* or simply *Gas Constant*. Solving for v we write the ideal gas thermal equation of state as

$$v = v(T, p) = \frac{RT}{p} \ . \tag{1.39}$$

The volume expansivity α and the isothermal compressibility β defined by (1.29) become

$$\alpha = \frac{1}{T} \ , \ \beta = \frac{1}{p} \ . \tag{1.40}$$

Substitution of these into (1.30) gives

$$\left(\frac{\partial e}{\partial v} \right)_T = 0 \ .$$

This means that if the ideal thermal EOS (1.39) is assumed, then it follows that the internal energy e is a function of temperature alone, that is $e = e(T)$. In the particular case in which

$$e = c_v T ,\tag{1.41}$$

where the specific heat capacity c_v is a constant, one speaks of a *calorically ideal* gas, or a *polytropic gas*.

It is possible to relate c_p and c_v via the general expression

$$c_p = c_v + \frac{\alpha^2 T v}{\beta} .\tag{1.42}$$

For a thermally ideal gas equations (1.39) and (1.40) apply and thus

$$c_p - c_v = R .\tag{1.43}$$

A necessary condition for thermal stability is $c_v > 0$ and for mechanical stability $\beta > 0$ [21]. From (1.42) the following inequalities result

$$c_p > c_v > 0 .\tag{1.44}$$

The *ratio of specific heats* γ, or *adiabatic exponent*, is defined as

$$\gamma = \frac{c_p}{c_v} ,\tag{1.45}$$

which if used in conjunction with (1.43) gives

$$c_p = \frac{\gamma R}{\gamma - 1} , \ c_v = \frac{R}{\gamma - 1} .\tag{1.46}$$

For a calorically ideal gas (polytropic gas) γ is a constant and for a thermally ideal gas γ is a function of temperature, i.e. $\gamma = \gamma(T)$.

In order to determine the caloric EOS (1.41) we need to determine the specific heat capacities, c_v in particular. Molecular Theory and the principle of equipartition of energy [304] can also provide an expression for the specific internal energy of a molecule. In general a molecule, however complex, has M degrees of freedom, of which three are translational. Other possible degrees of freedom are rotational and vibrational. From Molecular Theory it can be shown that if the energy associated with any degree of freedom is a quadratic function in the appropriate variable expressing that degree of freedom, then the mean value of the energy is $\frac{1}{2}kT$ where k is the *Boltzmann constant*. Moreover, from the principle of equipartition of energy this is the same for each degree of freedom. Therefore, the mean total energy of a molecule is $\bar{e} = \frac{1}{2}MkT$, and for N molecules we have

$$N\bar{e} = \frac{1}{2}MNkT ,$$

from which the specific internal energy is

$$e = \frac{1}{2}MRT .$$

Use of (1.32) gives directly

$$c_v = \left(\frac{\partial e}{\partial T}\right)_v = \frac{1}{2}MR ,$$

and thus we obtain

$$c_p = \frac{M+2}{2}R .$$

The ratio of specific heats becomes

$$\gamma = \frac{M+2}{M} . \tag{1.47}$$

From the thermal EOS for ideal gases (1.39) we have

$$e = \frac{1}{2}Mpv .$$

But from (1.43) and (1.46)

$$M = \frac{2}{\gamma - 1}$$

and hence

$$e = \frac{pv}{(\gamma - 1)} = \frac{p}{(\gamma - 1)\rho} , \tag{1.48}$$

which is the expression for the specific internal energy advanced in (1.18).

The theoretical expressions for c_p, c_v and γ in terms of R and M are found to be very accurate for monatomic gases, for which $M = 3$ (three translational degrees of freedom). For polyatomic gases rotational and vibrational degrees of freedom contribute to M but now the expressions might be rather inaccurate when compared with experimental data. A strong dependence on T is observed. However, the inequality $1 < \gamma < \frac{5}{3}$, predicted from (1.47) for the limiting values $M = 3$ and $M = \infty$, holds true.

1.2.5 Covolume and van der Waal Gases

A very simple generalisation of the ideal–gas thermal EOS, $pv = RT$, is the so–called *covolume equation of state*

$$p(v - b) = RT , \tag{1.49}$$

where b is called the *covolume* and in SI units has dimensions of m^3kg^{-1}. This EOS applies to dense gases at high pressure for which the volume occupied by the molecules themselves is no longer negligible. There is therefore a reduction in the volume available to molecular motion. This type of correction to the ideal gas EOS is said to have first been suggested by Clausius. Hirn is credited with first having written down EOS (1.49). Sometimes, this equation is also called the Noble–Abel EOS. In the study of propulsion systems, gaseous combustion products at very high densities are reasonably well described by the covolume EOS. In its simplest version the covolume b is a constant and is determined experimentally or from equilibrium thermochemical calculations. Corner [95] reports on good experimental results for a range of solid propellants and observes that b changes very little, usually in the range $0.9 \times 10^{-3} \leq b \leq 1.1 \times 10^{-3}$. The best values of b lead to errors of no more than 2% and thus there is some justification in using (1.49) with b constant. A more accurate covolume EOS defines b as a function of ρ, i.e. $b = b(\rho)$. Such dependence of b on ρ can be given in either tabular or algebraic form. A simple example of an algebraic form is $b(\rho) = \exp^{-0.4\rho}$, for $\rho < 2 \, \text{g cm}^{-3}$. The thermal covolume EOS (1.49) leads to a caloric covolume EOS $e = e(p, \rho)$ with a corresponding sound speed a. These are given by

$$e = \frac{p(1 - b\rho)}{\rho(\gamma - 1)} \ , \quad a = \left[\frac{\gamma p}{(1 - b\rho)\rho} \right]^{\frac{1}{2}} \ , \tag{1.50}$$

where γ is the ratio of specific heats as before.

The covolume EOS (1.49) can be further corrected to account for the forces of attraction between molecules, the van der Waal forces. These are neglected in both the ideal and covolume equations of state. Accounting for such forces results in a reduction of the pressure by an amount c/v^2, where c is a quantity that depends on the particular gas under consideration. Thus from (1.49) the pressure is corrected as

$$p = \frac{RT}{v - b} - \frac{c}{v^2} \ .$$

Then we can write

$$(p + \frac{c}{v^2})(v - b) = RT \ . \tag{1.51}$$

This is generally known as the *van der Waal's equation of state* for real gases.

General background on Thermodynamics and equations of state can be found in virtually any textbook on Thermodynamics or Gas Dynamics. We particularly recommend the book by Sears and Salinger [304], Chap. 1 of the book on Gas Dynamics by Becker [21] and the book by Clarke and McChesney [81]. The review paper by Menikoff and Plohr [240] is highly relevant to the themes of this book. A useful reference on equations of state for combustion problems is [403].

So far, we have presented the Euler equations for the dynamics of a compressible medium along with some elementary notions on Thermodynamics so that a closed system is obtained. Given initial and boundary conditions the conservation equations can be solved. In this book we are interested in numerical methods to solve the governing equations.

1.3 Viscous Stresses

Here we augment the Euler equations (1.7) by adding the physical effects of viscosity. Strictly speaking it is only the momentum equations in (1.7) that are modified. The stresses in a fluid, given by a tensor S, are due to the effects of the thermodynamic pressure p and the viscous stresses. Thus the stress tensor can be written as

$$S = -p\mathsf{I} + \mathit{\Pi} \ , \tag{1.52}$$

where $p\mathsf{I}$ is the spherically symmetric tensor due to p, I is the unit tensor as in (1.10) and $\mathit{\Pi}$ is the viscous stress tensor. It is desirable to express S in terms of flow variables already defined. For the pressure contribution this has already been achieved by defining p in terms of other thermodynamic variables via an equation of state. Recall that equations of state are approximate statements about the nature of a material. In defining the viscous stress contribution $\mathit{\Pi}$ one may resort to the *Newtonian approximation*, whereby $\mathit{\Pi}$ is related to the derivatives of the velocity field $\mathbf{V} = (u, v, w)$ via the *deformation tensor*

$$\mathsf{D} = \begin{bmatrix} u_x & \frac{1}{2}(v_x + u_y) & \frac{1}{2}(w_x + u_z) \\ \frac{1}{2}(u_y + v_x) & v_y & \frac{1}{2}(w_y + v_z) \\ \frac{1}{2}(u_z + w_x) & \frac{1}{2}(v_z + w_y) & w_z \end{bmatrix} \ .$$

The Newtonian assumption is an idealisation in which the relationship between $\mathit{\Pi}$ and D is linear and homogeneous, that is $\mathit{\Pi}$ will vanish only if D vanishes, and the medium is isotropic with respect to this relation; an isotropic medium is that in which there are no preferred directions. By denoting the stress tensor by

$$\mathit{\Pi} = \begin{bmatrix} \tau^{xx} & \tau^{xy} & \tau^{xz} \\ \tau^{yx} & \tau^{yy} & \tau^{yz} \\ \tau^{zx} & \tau^{zy} & \tau^{zz} \end{bmatrix} \ , \tag{1.53}$$

the Newtonian approximation becomes

$$\mathit{\Pi} = 2\eta\mathsf{D} + (\eta_b - \frac{2}{3}\eta)(\mathrm{div}\mathbf{V})\mathsf{I} \ , \tag{1.54}$$

or in full

$$\tau^{xx} = \tfrac{4}{3}\eta u_x - \tfrac{2}{3}\eta(v_y + w_z) + \eta_b \operatorname{div}\mathbf{V} \;,$$

$$\tau^{yy} = \tfrac{4}{3}\eta v_y - \tfrac{2}{3}\eta(w_z + u_x) + \eta_b \operatorname{div}\mathbf{V} \;,$$

$$\tau^{zz} = \tfrac{4}{3}\eta w_z - \tfrac{2}{3}\eta(u_x + v_y) + \eta_b \operatorname{div}\mathbf{V} \;,$$

$$\tau^{xy} = \tau^{yx} = \eta(u_y + v_x) \;,$$

$$\tau^{yz} = \tau^{zy} = \eta(v_z + w_y) \;,$$

$$\tau^{zx} = \tau^{xz} = \eta(w_x + u_z) \;. \tag{1.55}$$

In the Newtonian relationship (1.54) there are two scalar quantities that are still undetermined, these are the *coefficient of shear viscosity* η and the *coefficient of bulk viscosity* η_b. Approximate expressions for these are obtained from experimentation and results from Molecular Theory. In particular, for monatomic gases Molecular Theory based on the hard sphere assumption gives $\eta_b = 0$, which is found to agree well with experiment. For polyatomic gases $\eta_b \neq 0$ and appropriate values for η_b are to be given experimentally. Concerning the coefficient of shear viscosity η, it is observed that, as long as temperatures are not too high, η depends strongly on temperature and only slightly on pressure. Again, Molecular Theory and experimentation suggest that η be proportional to T^n; in fact $n = \tfrac{1}{2}$ in Molecular Theory. A relatively accurate relation between η and T is the Sutherland formula

$$\eta = C_1 \left[1 + \frac{C_2}{T} \right]^{-1} \sqrt{T} \;, \tag{1.56}$$

where C_1 and C_2 are two experimentally adjustable constants. When T is measured in kelvin, η has the units of $\mathrm{kg\,m^{-1}s^{-1}}$. For the case of air one has

$$C_1 = 1.46 \times 10^{-6} \;,\; C_2 = 112 \text{ K} \;.$$

Sutherland's formula describes the dependence of η on T rather well for a wide range of temperatures, provided no dissociation or ionisation take place. These phenomena occur at very high temperatures where the dependence of η on pressure p, in addition to temperature T, cannot be neglected. Useful background on high temperature gas dynamics is found in the book by Anderson [7] and in the book by Clarke and McChesney [81].

In summary, the Navier–Stokes equations (momentum equations) can now be written in differential conservation law form as

$$(\rho\mathbf{V})_t + \nabla \cdot (\rho\mathbf{V} \otimes \mathbf{V} + p\mathsf{I} - \Pi) = \mathbf{0} \;, \tag{1.57}$$

where Π is given by (1.55) with $\eta_b = 0$. Compare with the Euler equations (1.10).

1.4 Heat Conduction

Influx of energy contributes to the rate of change of total energy E. We denote by $\mathbf{Q} = (q_1, q_2, q_3)^T$ the energy flux vector, which results from (i) heat flow due to temperature gradients, (ii) diffusion processes in gas mixtures and (iii) radiation. Here we only consider effect (i) above. $\mathbf{Q}$ is identical to the heat flux vector caused by temperature gradients. In a similar manner to that in which viscous stresses were related to gradients of the velocity vector $\mathbf{V}$, one can relate $\mathbf{Q}$ to gradients of temperature T via Fourier's heat conduction law

$$\mathbf{Q} = -\kappa \nabla T , \tag{1.58}$$

where κ is a positive scalar quantity called the *coefficient of thermal conductivity* or just *thermal conductivity*, and is yet to be determined. Note the analogy between η and κ. This analogy between η and κ goes further in that κ, just as η, depends on T but only slightly on pressure p. In fact, Molecular Theory says that κ is directly proportional to η. Under the assumption that the specific heat at constant pressure c_p is constant, the dimensionless quantity

$$P_\mathrm{r} \equiv \frac{c_p \eta}{\kappa} \tag{1.59}$$

is a constant, and is called the Prandtl number. For monatomic gases P_r is very nearly constant. For air in the temperature range $200 \text{ K} \leq T \leq 1000 \text{ K}$ P_r differs only slightly from its mean value of 0.7. A formula attributed to Eucken [21] relates P_r to the ratio of specific heats γ via

$$P_\mathrm{r} = \frac{4\gamma}{9\gamma - 5} , \tag{1.60}$$

to account for departures from calorically ideal gas behaviour.

When the effects of viscosity and heat conduction are added to the basic Euler equations (1.7) one has the Navier–Stokes equations with heat conduction

$$\mathbf{U}_t + \mathbf{F}_x^\mathrm{a} + \mathbf{G}_y^\mathrm{a} + \mathbf{H}_z^\mathrm{a} = \mathbf{F}_x^\mathrm{d} + \mathbf{G}_y^\mathrm{d} + \mathbf{H}_z^\mathrm{d} , \tag{1.61}$$

where $\mathbf{U}$ is the vector of conserved variables, the flux vectors $\mathbf{F}^\mathrm{a}$, $\mathbf{G}^\mathrm{a}$ and $\mathbf{H}^\mathrm{a}$ are the inviscid fluxes (a stands for advection) for the Euler equations as given by (1.8) and the respective flux vectors $\mathbf{F}^\mathrm{d}$, $\mathbf{G}^\mathrm{d}$ and $\mathbf{H}^\mathrm{d}$ (d stands for diffusion) due to viscosity and heat conduction are

$$\mathbf{F^d} = \begin{bmatrix} 0 \\ \tau^{xx} \\ \tau^{xy} \\ \tau^{xz} \\ u\tau^{xx} + v\tau^{xy} + w\tau^{xz} - q_1 \end{bmatrix} \,,$$

$$\left. \mathbf{G^d} = \begin{bmatrix} 0 \\ \tau^{yx} \\ \tau^{yy} \\ \tau^{yz} \\ u\tau^{yx} + v\tau^{yy} + w\tau^{yz} - q_2 \end{bmatrix} \,, \right\} \qquad (1.62)$$

$$\mathbf{H^d} = \begin{bmatrix} 0 \\ \tau^{zx} \\ \tau^{zy} \\ \tau^{zz} \\ u\tau^{zx} + v\tau^{zy} + w\tau^{zz} - q_3 \end{bmatrix} \,.$$

The form of the equations given by (1.61) splits the effect of advection on the left–hand side from those of viscous diffusion and heat conduction on the right–hand side. For numerical purposes, the particular form of the equations adopted depends largely on the numerical technique to be used to solve the equations. One possible approach is to split the advection effects from those of viscous diffusion and heat conduction during a small time interval Δt, in which case form (1.61) is perfectly adequate. An alternative form is obtained by combining the fluxes due to advection, viscous diffusion and heat conduction into new fluxes so that the governing equations look formally like a homogeneous system (zero right–hand side) of conservation laws

$$\left. \begin{aligned} \mathbf{U}_t + \mathbf{F}_x + \mathbf{G}_y + \mathbf{H}_z = \mathbf{0}\,, \\ \mathbf{F} = \mathbf{F^a} - \mathbf{F^d}\,,\ \mathbf{G} = \mathbf{G^a} - \mathbf{G^d}\,,\ \mathbf{H} = \mathbf{H^a} - \mathbf{H^d}\,. \end{aligned} \right\} \qquad (1.63)$$

This form is only justified if the numerical method employed actually exploits the coupling of advection, viscosity and heat conduction when defining numerical approximations to the flux vectors $\mathbf{F}$, $\mathbf{G}$, and $\mathbf{H}$ in (1.63).

1.5 Integral Form of the Equations

The actual derivation of the governing equations, such as the Euler and Navier–Stokes equations stated earlier, is based on integral relations on control volumes and their boundaries. The differential form of the equations results from further assumptions on the flow variables (smoothness). In the absence of viscous diffusion and heat conduction one obtains the Euler equations. These admit discontinuous solutions and the smoothness assumption that leads to the differential form no longer holds true. Thus one must return to the more fundamental integral form involving integrals over control volumes and their boundaries. From a computational point of view there is another good reason for returning to the integral form of the equations. Discretised domains result naturally in finite control volumes or computational cells. Local enforcement of the fundamental equations in these volumes lead to *Finite Volume* numerical methods.

1.5.1 Time Derivatives

Before proceeding to the derivation of the equations in integral form we review some preliminary concepts that are needed. Consider a scalar field function $\phi(x, y, z, t)$, then the time rate of change of ϕ as registered by an observer moving with the fluid velocity $\mathbf{V} = (u, v, w)$ is given by

$$\frac{D\phi}{Dt} = \frac{\partial \phi}{\partial t} + \mathbf{V} \cdot \text{grad}\phi \, . \tag{1.64}$$

This time derivative D/Dt following a particle is usually called the *substantial derivative* or material derivative. The first term $\partial \phi / \partial t$ in (1.64) denotes the partial derivative of ϕ with respect to time and represents the *local rate of change of ϕ*; the second term in (1.64) is the *convective rate of change*. The operator D/Dt can also be applied to vectors in a component–wise manner, in which case equation (1.64) is to be interpreted as a vector equation. In particular, we can obtain the substantial derivative of $\mathbf{V} = (u, v, w)$, that is

$$\frac{D\mathbf{V}}{Dt} = \frac{\partial \mathbf{V}}{\partial t} + \mathbf{V} \cdot \text{grad}\mathbf{V} \, , \tag{1.65}$$

which in full becomes

$$\begin{pmatrix} \frac{Du}{Dt} \\ \frac{Dv}{Dt} \\ \frac{Dw}{Dt} \end{pmatrix}^{T} = \begin{pmatrix} \frac{\partial u}{\partial t} \\ \frac{\partial v}{\partial t} \\ \frac{\partial w}{\partial t} \end{pmatrix}^{T} + (u, v, w) \cdot \begin{bmatrix} \frac{\partial u}{\partial x} & \frac{\partial v}{\partial x} & \frac{\partial w}{\partial x} \\ \frac{\partial u}{\partial y} & \frac{\partial v}{\partial y} & \frac{\partial w}{\partial y} \\ \frac{\partial u}{\partial z} & \frac{\partial v}{\partial z} & \frac{\partial w}{\partial z} \end{bmatrix} \, .$$

The symbol $(\,)^{T}$ denotes transpose of $(\,)$. Actually, equation (1.65) is the acceleration vector of an element of a moving fluid. Let us now consider

$$\Psi(t) = \int \int \int_{V} \phi(x, y, z, t) \, \mathrm{d}V \, , \tag{1.66}$$

where the integrand ϕ is any scalar field function and the volume of integration V is enclosed by a piece–wise smooth boundary surface A that moves with the material under consideration. It can be shown that the material derivative of Ψ is given by

$$\frac{D\Psi}{Dt} = \int\int\int_V \frac{\partial\phi}{\partial t}\,\mathrm{d}V + \int\int_A (\mathbf{n}\cdot\phi\mathbf{V})\,\mathrm{d}A\ , \tag{1.67}$$

where $\mathbf{n} = (n_1, n_2, n_3)$ is the outward pointing unit vector normal to the surface A. The proof of statement (1.67) is based on a three dimensional generalisation of the following result:

$$\frac{\mathrm{d}}{\mathrm{d}\alpha}\int_{\xi_1(\alpha)}^{\xi_2(\alpha)} f(\xi,\alpha)\,\mathrm{d}\xi = \int_{\xi_1(\alpha)}^{\xi_2(\alpha)} \frac{\partial f}{\partial\alpha}\,\mathrm{d}\xi + f(\xi_2,\alpha)\frac{\mathrm{d}\xi_2}{\mathrm{d}\alpha} - f(\xi_1,\alpha)\frac{\mathrm{d}\xi_1}{\mathrm{d}\alpha}\ . \tag{1.68}$$

Expression (1.67) can be generalised to vectors $\Psi(x,y,z,t)$ as follows:

$$\frac{D\Psi}{Dt} = \int\int\int_V \frac{\partial\Phi}{\partial t}\,\mathrm{d}V + \int\int_A \Phi(\mathbf{n}\cdot\mathbf{V})\,\mathrm{d}A\ . \tag{1.69}$$

The first term on the right hand side of (1.67) represents the local contribution of the field ϕ to the time rate of change of $\Psi(t)$. The second term is the contribution due to the motion of the surface moving at the fluid velocity $\mathbf{V}$. The surface integral may be transformed to a volume integral by virtue of Gauss's theorem. This states that for any differentiable vector field $\Phi = (\phi_1, \phi_2, \phi_3)$ and a volume V with smooth bounding surface A the following identity holds

$$\int\int_A (\mathbf{n}\cdot\Phi)\,\mathrm{d}A = \int\int\int_V \mathrm{div}\Phi\,\mathrm{d}V\ . \tag{1.70}$$

Gauss's theorem also applies to differentiable scalar fields and tensor fields.

1.5.2 Conservation of Mass

The law of conservation of mass can now be stated in integral form by identifying the scalar ϕ in (1.66) and (1.67) as the density ρ. In this case $\Psi(t)$ in (1.66) becomes the total mass in the volume V. By assuming that *no mass is generated or annihilated within* V and recalling that the surface A moves with the fluid velocity, which means that no mass flows across the surface, we have $D\Psi/Dt = 0$, or from (1.67)

$$\int\int\int_V \frac{\partial\rho}{\partial t}\,\mathrm{d}V + \int\int_A \mathbf{n}\cdot(\rho\mathbf{V})\,\mathrm{d}A = 0\ .$$

This is the integral form of the law of conservation of mass corresponding to the differential form (1.1). This integral conservation law may be generalised to include sources of mass, which will then appear as additional integral terms. A useful reinterpretation of the integral form results if we rewrite it as

$$\int\int\int_V \frac{\partial\rho}{\partial t}\,\mathrm{d}V = -\int\int_A \mathbf{n}\cdot(\rho\mathbf{V})\,\mathrm{d}A\ . \tag{1.71}$$

If now V is a fixed control volume independent of time t then the left hand side of (1.71) becomes

$$\int \int \int_V \frac{\partial \rho}{\partial t} \, dV = \frac{d}{dt} \int \int \int_V \rho \, dV \ ,$$

and thus is the time–rate of change of the mass enclosed by the volume V. The right hand side of (1.71) is the net mass inflow, per unit time, over the mass outflow. That is, the mass enclosed by the control volume V, in the absence of sources or sinks, can only change by virtue of mass flow through the boundary of the control volume V. Thus we rewrite (1.71) as

$$\frac{d}{dt} \int \int \int_V \rho \, dV = - \int \int_A \mathbf{n} \cdot (\rho \mathbf{V}) \, dA \ . \tag{1.72}$$

Here $\rho \mathbf{V}$ is the mass flow vector and $\mathbf{n} \cdot (\rho \mathbf{V})$ is its normal component through the surface A with outward unit normal vector $\mathbf{n}$. For computational purposes this is the formulation of the integral form of the law of conservation of mass that is most useful.

The integral form is actually equivalent to the differential form (1.1) of the law conservation of mass if we assume sufficient smoothness of the flow variables in (1.72). Then we can apply Gauss's theorem and write

$$\int \int_A \mathbf{n} \cdot (\rho \mathbf{V}) \, dA = \int \int \int_V \operatorname{div}(\rho \mathbf{V}) \, dV \ .$$

Then (1.71) becomes

$$\int \int \int_V \left[\frac{\partial \rho}{\partial t} + \operatorname{div}(\rho \mathbf{V}) \right] dV = 0 \ .$$

As V is arbitrary it follows that the integrand must vanish, that is

$$\rho_t + (\rho u)_x + (\rho v)_y + (\rho w)_z = 0 \ ,$$

which is (1.1). As pointed out earlier, the Euler equations (1.7) admit discontinuous solutions, such as shock waves and contact surfaces. Hence the differential form (1.7) is not valid in general. The integral form (1.72), however, remains valid.

1.5.3 Conservation of Momentum

The differential form of the law of conservation of momentum for the inviscid case was stated as equations (1.2)–(1.5), or in more compact form involving tensors as equation (1.10), the Euler equations. Equations (1.61) contain the momentum equations augmented by the effects of viscosity, which gives the Navier–Stokes equations, and heat conduction. As done for the mass equation, we now provide the foundations for the law of conservation of momentum, derive its integral form in quite general terms and show that under appropriate smoothness assumptions the differential form is implied by the integral form. A control volume V with bounding surface A is chosen and the total momentum in V is given by

$$\Psi(t) = \int\int\int_V \rho \mathbf{V}\, dV \,. \tag{1.73}$$

The law of conservation of momentum results from the direct application of Newton's law: *the time rate of change of the momentum in V is equal to the total force acting on the volume V*. The total force is divided into surface forces f_S and volume forces f_V given by

$$f_S = \int\int_A \mathbf{S}\, dA\,, \quad f_V = \int\int\int_V \rho \mathbf{g}\, dV \,. \tag{1.74}$$

Here $\mathbf{g}$ is the specific volume–force vector and may account for inertial forces, gravitational forces, electromagnetic forces and so on. $\mathbf{S}$ is the stress vector, which is given in terms of a stress tensor S as $\mathbf{S} = \mathbf{n} \cdot \mathsf{S}$. The stress tensor S can be split into a spherically symmetric part $-p\mathsf{I}$ due to pressure p, and a viscous part Π given by (1.52)–(1.53). Application of Newton's Law gives

$$\frac{D\Psi}{Dt} = f_S + f_V \,,$$

which by virtue of (1.69) with $\Phi \equiv \rho \mathbf{V}$ gives

$$\int\int\int_V \frac{\partial}{\partial t}(\rho \mathbf{V})\, dV = -\int\int_A \mathbf{V}(\mathbf{n} \cdot \rho \mathbf{V})\, dA + f_S + f_V \,.$$

Regarding V as a fixed volume in space, independent of time, we write

$$\frac{d}{dt}\int\int\int_V (\rho \mathbf{V})\, dV = -\int\int_A \mathbf{V}(\mathbf{n} \cdot \rho \mathbf{V})\, dA + f_S + f_V \,, \tag{1.75}$$

which may be interpreted as saying that the *time rate of change of momentum within the fixed control volume V is due to the net momentum inflow over momentum outflow, given by the first term in (1.75), plus surface and volume forces.* Substituting S from (1.52) into (1.74) and writing all surface terms into a single integral we have

$$\left.\begin{aligned} \tfrac{d}{dt}\int\int\int_V (\rho \mathbf{V})\, dV \;=\;& -\int\int_A [\mathbf{V}(\mathbf{n} \cdot \rho \mathbf{V}) + p\mathbf{n} - \mathbf{n} \cdot \Pi]\, dA \\ & + \int\int\int_V \rho \mathbf{g}\, dV \,. \end{aligned}\right\} \tag{1.76}$$

This general statement is valid even for the case of discontinuous solutions. The differential conservation law form can now be derived from (1.76) under the assumption that the integrand in the surface integral is sufficiently smooth so that Gauss's theorem may be invoked. Consider the first term of the integrand of the surface integral in (1.76)

$$\mathbf{V}(\mathbf{n} \cdot \rho \mathbf{V}) = \mathbf{n} \cdot \rho \mathbf{V} \otimes \mathbf{V} \,,$$

where $\mathbf{V} \otimes \mathbf{V}$ is the tensor in (1.10). The three columns of the left hand side are

$$
\begin{aligned}
u \cdot (\mathbf{n} \cdot \rho \mathbf{V}) &= \mathbf{n} \cdot \left[\rho u^2, \rho uv, \rho uw\right]^T , \\
v \cdot (\mathbf{n} \cdot \rho \mathbf{V}) &= \mathbf{n} \cdot \left[\rho uv, \rho v^2, \rho vw\right]^T , \\
w \cdot (\mathbf{n} \cdot \rho \mathbf{V}) &= \mathbf{n} \cdot \left[\rho uw, \rho vw, \rho w^2\right]^T .
\end{aligned}
$$

Application of Gauss's theorem to each of the surface terms gives

$$
\iiint_V \frac{\partial}{\partial t}(\rho \mathbf{V})\,\mathrm{d}V = -\iiint_V \left[\mathrm{div}(\rho \mathbf{V} \otimes \mathbf{V}) + \mathrm{grad}\,p - \mathrm{div}\,\Pi\right]\,\mathrm{d}V
$$
$$
+ \iiint_V \rho \mathbf{g}\,\mathrm{d}V .
$$

As this is valid for any arbitrary volume V the integrand must vanish, i.e.

$$
\frac{\partial}{\partial t}(\rho \mathbf{V}) + \mathrm{div}\left[\rho \mathbf{V} \otimes \mathbf{V} + p\mathsf{I} - \Pi\right] = \rho \mathbf{g} . \tag{1.77}
$$

This is the differential form of the momentum equation, including a source term due to volume forces. When the viscous stresses are identically zero, $\Pi \equiv \mathbf{0}$ and the volume forces are neglected, we obtain the Euler equations (1.10). If the viscous stresses are given by (1.55) under the Newtonian assumption we obtain the Navier–Stokes equations (1.57) in differential conservation law form.

1.5.4 Conservation of Energy

As done for mass and momentum we now consider the total energy $\Psi(t)$ in a control volume V, that is

$$
\Psi(t) = \iiint_V E\,\mathrm{d}V . \tag{1.78}
$$

The time rate of change of total energy $\Psi(t)$ is equal to the work done, per unit time, by all the forces acting on the volume plus the influx of energy per unit time into the volume. Recall that a force $\mathbf{f}$ acting on a point moving with velocity $\mathbf{V}$ produces the work $\mathbf{V} \cdot \mathbf{f}$ per unit time. The surface and volume forces in (1.74) respectively give rise to the following terms:

$$
E_{\mathrm{surf}} = -\iint_A p(\mathbf{V} \cdot \mathbf{n})\,\mathrm{d}A + \iint_A \mathbf{V} \cdot (\mathbf{n} \cdot \Pi)\,\mathrm{d}A , \tag{1.79}
$$

$$
E_{\mathrm{volu}} = \iiint_V \rho(\mathbf{V} \cdot \mathbf{g})\,\mathrm{d}V . \tag{1.80}
$$

The first term in (1.79) corresponds to the work done by the pressure while the second term corresponds to the work done by the viscous stresses. E_{volu} in (1.80) is the work done by the volume force $\mathbf{g}$. To account for the influx of energy into the volume we denote the energy flow vector by $\mathbf{Q} = (q_1, q_2, q_3)$; the flow of energy per unit time across a surface element dA is given by $-(\mathbf{n} \cdot \mathbf{Q})\,dA$. This gives another term,

$$E_{\text{infl}} = -\int\int_A (\mathbf{n} \cdot \mathbf{Q})\,dA \,, \tag{1.81}$$

to be included in the equation of balance of energy, which now reads

$$\frac{D\Psi}{Dt} = E_{\text{surf}} + E_{\text{volu}} + E_{\text{infl}} \,. \tag{1.82}$$

The left hand side of (1.82) can be transformed via (1.67) with the definition $\phi \equiv E$ and the result is

$$\frac{D\Psi(t)}{Dt} = \int\int\int_V \frac{\partial}{\partial t} E\,dV + \int\int_A (\mathbf{n} \cdot E\mathbf{V})\,dA \,. \tag{1.83}$$

As done for the laws of conservation of mass and momentum we now reinterpret the volume V as fixed in space and independent of time and rewrite (1.82)–(1.83) as

$$\frac{d}{dt}\int\int\int_V E\,dV = -\int\int_A (\mathbf{n} \cdot E\mathbf{V})\,dA + E_{\text{surf}} + E_{\text{volu}} + E_{\text{infl}} \,,$$

which in full becomes

$$\left.\begin{aligned} \frac{d}{dt}\int\int\int_V E\,dV =\ & -\int\int_A [\mathbf{n} \cdot (E\mathbf{V} + p\mathbf{V} + \mathbf{Q}) - \mathbf{V} \cdot (\mathbf{n} \cdot \mathit{\Pi})]\,dA \\ & + \int\int\int_V \rho(\mathbf{V} \cdot \mathbf{g})\,dV \,. \end{aligned}\right\} \tag{1.84}$$

Thus the time rate of change of total energy enclosed in the volume V equals the net flow of energy through the boundary surface A plus the forces E_{surf}, E_{volu} and E_{infl} as given by (1.79)–(1.81).

The differential form of the conservation of energy law (1.84) can now be derived by assuming sufficient smoothness and applying Gauss's theorem to all surface integrals. Direct application of Gauss's theorem to the first term of (1.79) and to (1.81) gives

$$-\int\int_A \mathbf{n} \cdot (p\mathbf{V})\,dA = -\int\int\int_V \operatorname{div}(p\mathbf{V})\,dV \,,$$

$$-\int\int_A \mathbf{n} \cdot \mathbf{Q}\,dA = -\int\int\int_V \operatorname{div}\mathbf{Q}\,dV \,.$$

The second term of (1.79) can be transformed via Gauss's theorem by first observing that $\mathbf{V} \cdot (\mathbf{n} \cdot \mathit{\Pi}) = \mathbf{n} \cdot (\mathbf{V} \cdot \mathit{\Pi})$. This follows from the symmetry of the viscous stress tensor $\mathit{\Pi}$, see (1.53), (1.55). Hence

$$\int\int_A \mathbf{V} \cdot (\mathbf{n} \cdot \mathit{\Pi})\,dA = \int\int\int_V \operatorname{div}(\mathbf{V} \cdot \mathit{\Pi})\,dV \,.$$

Substitution of these volume integrals into the integral form of the law of conservation of energy (1.84) gives

$$\int \int \int_V \{E_t + \operatorname{div}[(E+p)\mathbf{V} - \mathbf{V} \cdot \Pi + \mathbf{Q}] - \rho(\mathbf{V} \cdot \mathbf{g})\}\, dV = 0 \ .$$

Since V is arbitrary the integrand must vanish identically, that is

$$E_t + \operatorname{div}[(E+p)\mathbf{V} - \mathbf{V} \cdot \Pi + \mathbf{Q}] = \rho(\mathbf{V} \cdot \mathbf{g}) \ . \tag{1.85}$$

This is the differential form of the law of conservation of energy with a source term accounting for the effect of body forces; if these are neglected we obtain the homogeneous energy equation contained in (1.61). When viscous and heat conduction effects are neglected we obtain the energy equation (1.5) or (1.11) corresponding to the compressible Euler equations.

1.6 Submodels

In this section we consider simplified versions, or submodels, of the governing equations and their closure conditions. Compressible submodels will include flows with area variation; flows with axial symmetry; flows with cylindrical and spherical symmetry; plane one–dimensional flow and further simplifications of this to include linearised and scalar submodels; the one–dimensional version of the Navier–Stokes equations. Incompressible submodels will include free–surface gravity flows and the derivation of the shallow water equations as a special case; we also study various formulations of the incompressible Navier–Stokes equations.

1.6.1 Summary of the Equations

Before proceeding with the study of particular situations we summarise the general laws of conservation of mass, momentum and total energy. In differential conservation law form these read

$$\rho_t + \nabla \cdot (\rho \mathbf{V}) = 0 \ , \tag{1.86}$$

$$\frac{\partial}{\partial t}(\rho \mathbf{V}) + \nabla \cdot [\rho \mathbf{V} \otimes \mathbf{V} + pI - \Pi] = \rho \mathbf{g} \ , \tag{1.87}$$

$$E_t + \nabla \cdot [(E+p)\mathbf{V} - \mathbf{V} \cdot \Pi + \mathbf{Q}] = \rho(\mathbf{V} \cdot \mathbf{g}) \ . \tag{1.88}$$

where $\mathbf{g} = (g_1, g_2, g_3)$ is a body force vector. The integral form of the conservation laws is given by

$$\frac{\mathrm{d}}{\mathrm{d}t} \int \int \int_V \rho\, \mathrm{d}V = -\int \int_A \mathbf{n} \cdot (\rho \mathbf{V})\, \mathrm{d}A \ , \tag{1.89}$$

$$\frac{\mathrm{d}}{\mathrm{d}t} \int \int \int_V (\rho \mathbf{V}) \, \mathrm{d}V = -\int \int_A [\mathbf{V}(\mathbf{n} \cdot \rho \mathbf{V}) + p\mathbf{n} - \mathbf{n} \cdot \varPi] \, \mathrm{d}A \left.\vphantom{\int}\right\}$$

$$+ \int \int \int_V \rho \mathbf{g} \, \mathrm{d}V \,, \tag{1.90}$$

$$\frac{\mathrm{d}}{\mathrm{d}t} \int \int \int_V E \, \mathrm{d}V = -\int \int_A [\mathbf{n} \cdot (E\mathbf{V} + p\mathbf{V} + \mathbf{Q}) - \mathbf{V} \cdot (\mathbf{n} \cdot \varPi)] \, \mathrm{d}A \left.\vphantom{\int}\right\}$$

$$+ \int \int \int_V \rho(\mathbf{V} \cdot \mathbf{g}) \, \mathrm{d}V \,, \tag{1.91}$$

where V is the total volume of an element of fluid and A is its boundary. Computationally, V will be a finite volume or computational cell. When body forces are included via a source term vector but viscous and heat conduction effects are neglected we have the Euler equations

$$\mathbf{U}_t + \mathbf{F}(\mathbf{U})_x + \mathbf{G}(\mathbf{U})_y + \mathbf{H}(\mathbf{U})_z = \mathbf{S}(\mathbf{U}) \,, \tag{1.92}$$

$$\mathbf{U} = \begin{bmatrix} \rho \\ \rho u \\ \rho v \\ \rho w \\ E \end{bmatrix} \,, \quad \mathbf{F} = \begin{bmatrix} \rho u \\ \rho u^2 + p \\ \rho uv \\ \rho uw \\ u(E + p) \end{bmatrix} \,, \left.\vphantom{\begin{bmatrix} \rho \\ \rho u \\ \rho v \\ \rho w \\ E \end{bmatrix}}\right\}$$

$$\mathbf{G} = \begin{bmatrix} \rho v \\ \rho uv \\ \rho v^2 + p \\ \rho vw \\ v(E + p) \end{bmatrix} \,, \quad \mathbf{H} = \begin{bmatrix} \rho w \\ \rho uw \\ \rho vw \\ \rho w^2 + p \\ w(E + p) \end{bmatrix} \,. \tag{1.93}$$

Here $\mathbf{S} = \mathbf{S}(\mathbf{U})$ is a *source* or *forcing term*. Body forces such as gravity may be represented in $\mathbf{S}$. Injection of mass, momentum or energy may also be included in $\mathbf{S}$. Usually, $\mathbf{S}(\mathbf{U})$ is a prescribed *algebraic* function of the flow variables and does not involve derivatives of these, but there are exceptions. Equations (1.92) are said to be *inhomogeneous*. When $\mathbf{S}(\mathbf{U}) \equiv \mathbf{0}$ one speaks of *homogeneous* equations. There are other situations in which source terms $\mathbf{S}(\mathbf{U})$ arise as a consequence of approximating the homogeneous equations in (1.92) to model situations with particular geometric features. In this case the source term is of geometric character, but we shall still call it a source term.

Sometimes it is convenient to express the equations in terms of the *primitive* or *physical* variables ρ, u, v, w and p. By expanding derivatives in (1.92), using the mass equation into the momentum equations and in turn using these into the energy equation one can re-write the three-dimensional Euler equations for ideal gases with a body-force source term as

$$\left.\begin{array}{c} \rho_t + u\rho_x + v\rho_y + w\rho_z + \rho(u_x + v_y + w_z) = 0 \;, \\[2mm] u_t + uu_x + vu_y + wu_z + \frac{1}{\rho}p_x = g_1 \;, \\[2mm] v_t + uv_x + vv_y + wv_z + \frac{1}{\rho}p_y = g_2 \;, \\[2mm] w_t + uw_x + vw_y + ww_z + \frac{1}{\rho}p_z = g_3 \;, \\[2mm] p_t + up_x + vp_y + wp_z + \gamma p(u_x + v_y + w_z) = 0 \;. \end{array}\right\} \tag{1.94}$$

Next we consider simplifications of the Euler and Navier–Stokes equations augmented by source terms to account for additional flow physics.

1.6.2 Flow with Area Variation

Flows with area variation arise naturally in the study of fluid flow phenomena in ducts, pipes, shock tubes and nozzles. One may start from the three dimensional homogeneous version of (1.92) to produce, under the assumption of smooth area variations, a one–dimensional system with geometric source terms. Denote the cross–sectional area of the nozzle by $A = A(x,t)$, where x denotes distance along the nozzle and t denotes time. It is assumed that the area varies smoothly with space and time and its variation is due to both *translation* and *deformation*. Assume that the speed of translation is given by $c(x,t)$.

Most of the presentation that follows is motivated by some useful remarks by Professor Tim Swafford (Swafford, 1998, private communication) on the first edition of this book. He pointed out the correct derivation of the equations for the case in which the area depends on time, and also provided the following key references: Varner *et. al.* [399], Chessor [70] and Warsi [405].

The governing equations read

$$\mathbf{U}_t + \mathbf{F}(\mathbf{U})_x = \mathbf{S}(\mathbf{U}) \;, \tag{1.95}$$

where

$$\mathbf{U} = \begin{bmatrix} A\rho \\ A\rho u \\ AE \end{bmatrix} \;, \quad \mathbf{F} = \begin{bmatrix} A\rho(u-c) \\ A[\rho u(u-c)+p] \\ A[(u-c)E+up] \end{bmatrix} \;, \quad \mathbf{S} = \begin{bmatrix} 0 \\ pA_x \\ -pA_t \end{bmatrix} \;. \tag{1.96}$$

In what follows we assume $c = 0$. Manipulation of equations (1.95)–(1.96) lead to the following alternative form

$$\mathbf{U}_t + \mathbf{F}(\mathbf{U})_x = \mathbf{S}(\mathbf{U}) \;, \tag{1.97}$$

where now

$$\mathbf{U} = \begin{bmatrix} \rho \\ \rho u \\ E \end{bmatrix} , \quad \mathbf{F} = \begin{bmatrix} \rho u \\ \rho u^2 + p \\ u(E + p) \end{bmatrix} , \quad \mathbf{S} = -\frac{1}{A}\frac{dA}{dt}\begin{bmatrix} \rho \\ \rho u \\ (E + p) \end{bmatrix} . \quad (1.98)$$

Here the source term vector contains the differential operator

$$\frac{dA}{dt} = A_t + u A_x , \qquad (1.99)$$

which expresses the time variation of the area $A(x,t)$ along particle paths $dx/dt = u$. The upwind nature of the coefficient $\frac{dA}{dt}$ in the source term vector suggests possible discretisation procedures.

Yet another form of equations (1.95)–(1.96), for the case in which the area is independent of time and $c = 0$, is the following

$$(A\mathbf{U})_t + [A\mathbf{F}(\mathbf{U})]_x = -A\mathbf{S}(\mathbf{U})_x , \qquad (1.100)$$

where

$$\mathbf{U} = \begin{bmatrix} \rho \\ \rho u \\ E \end{bmatrix} , \quad \mathbf{F} = \begin{bmatrix} \rho u \\ \rho u^2 \\ u(E + p) \end{bmatrix} , \quad \mathbf{S} = \begin{bmatrix} 0 \\ p \\ 0 \end{bmatrix} . \qquad (1.101)$$

Note that the momentum flux does not include the pressure term p. For details see Ben–Artzi and Falcovitz [24].

1.6.3 Axi–Symmetric Flows

Here we consider domains that are symmetric around a coordinate direction. We choose this coordinate to be the z–axis and is called the axial direction. The second coordinate is r, which measures distance from the axis of symmetry z and is called the radial direction. There are two components of velocity, namely $u(r, z)$ and $v(r, z)$. These are respectively the radial (r) and axial (z) components of velocity. Then the three dimensional (inhomogeneous) conservation laws (1.92) are approximated by a two dimensional problem with geometric source terms $\mathbf{S}(\mathbf{U})$, namely

$$\mathbf{U}_t + \mathbf{F}(\mathbf{U})_r + \mathbf{G}(\mathbf{U})_z = \mathbf{S}(\mathbf{U}) , \qquad (1.102)$$

where

$$\mathbf{U} = \begin{bmatrix} \rho \\ \rho u \\ \rho v \\ E \end{bmatrix} , \quad \mathbf{F} = \begin{bmatrix} \rho u \\ \rho u^2 + p \\ \rho uv \\ u(E + p) \end{bmatrix} ,$$

$$\mathbf{G} = \begin{bmatrix} \rho v \\ \rho uv \\ \rho v^2 + p \\ v(E + p) \end{bmatrix} , \quad \mathbf{S} = -\frac{1}{r}\begin{bmatrix} \rho u \\ \rho u^2 \\ \rho uv \\ u(E + p) \end{bmatrix} . \qquad (1.103)$$

An alternative form of (1.102)–(1.103) is

$$\bar{\mathbf{U}}_t + \bar{\mathbf{F}}(\bar{\mathbf{U}})_r + \bar{\mathbf{G}}(\bar{\mathbf{U}})_z = \bar{\mathbf{S}}(\bar{\mathbf{U}}) ,\qquad(1.104)$$

where

$$\left.\begin{array}{c} \bar{\mathbf{U}} = r\mathbf{U} ,\ \bar{\mathbf{F}} = r\mathbf{F} ,\ \bar{\mathbf{G}} = r\mathbf{G} , \\[2mm] \bar{\mathbf{S}} = (0,\ -p,\ 0,\ 0)^T . \end{array}\right\}\qquad(1.105)$$

From a numerical point of view this form of the equations has its attractions (Pike and Roe, 1989, private communication).

1.6.4 Cylindrical and Spherical Symmetry

Cylindrical and spherically symmetric wave motion arises naturally in the theory of explosion waves in water, air and other media. In these situations the multidimensional equations may be reduced to essentially one–dimensional equations with a geometric source term vector $\mathbf{S}(\mathbf{U})$ to account for the second and third spatial dimensions. We write

$$\mathbf{U}_t + \mathbf{F}(\mathbf{U})_r = \mathbf{S}(\mathbf{U}) ,\qquad(1.106)$$

where

$$\mathbf{U} = \left[\begin{array}{c} \rho \\ \rho u \\ E \end{array}\right] ,\quad \mathbf{F} = \left[\begin{array}{c} \rho u \\ \rho u^2 + p \\ u(E + p) \end{array}\right] ,\quad \mathbf{S} = -\frac{\alpha}{r}\left[\begin{array}{c} \rho u \\ \rho u^2 \\ u(E + p) \end{array}\right] .\qquad(1.107)$$

Here r is the radial distance from the origin and u is the radial velocity. When $\alpha = 0$ we have plane one–dimensional flow; when $\alpha = 1$ we have cylindrically symmetric flow, an approximation to two–dimensional flow. This is a special case of equations (1.102)–(1.103) when no axial variations are present ($v = 0$). For $\alpha = 2$ we have spherically symmetric flow, an approximation to three–dimensional flow. Approximations (1.106)–(1.107) can easily be solved numerically to a high degree of accuracy by a good one–dimensional numerical method. These accurate one–dimensional solutions can then be very useful in partially validating two and three dimensional numerical solutions of the full models, see Sect. 17.1 of Chap. 17.

1.6.5 Plain One–Dimensional Flow

We first consider the one–dimensional time dependent case

$$\mathbf{U}_t + \mathbf{F}(\mathbf{U})_x = \mathbf{0} ,\qquad(1.108)$$

where

$$\mathbf{U} = \left[\begin{array}{c} \rho \\ \rho u \\ E \end{array}\right] ,\quad \mathbf{F} = \left[\begin{array}{c} \rho u \\ \rho u^2 + p \\ u(E + p) \end{array}\right] .\qquad(1.109)$$

These equations also result from equations (1.106)–(1.107) with $\alpha \equiv 0$ and r replaced by x. Under suitable physical assumptions they produce even simpler mathematical models. In all the submodels studied so far we have assumed some thermodynamic closure condition given by an Equation of State (EOS).

The **isentropic equations** result under the assumption that the entropy s is constant everywhere, which is a simplification of the thermodynamics. Now the EOS is

$$p = p(\rho) \equiv C\rho^\gamma , \quad C=\text{constant} . \tag{1.110}$$

This makes the energy equation redundant and we have the 2×2 system

$$\left.\begin{array}{c} \mathbf{U}_t + \mathbf{F}(\mathbf{U})_x = 0 , \\[4mm] \mathbf{U} = \left[\begin{array}{c} \rho \\ \rho u \end{array} \right] , \quad \mathbf{F} = \left[\begin{array}{c} \rho u \\ \rho u^2 + p \end{array} \right] , \end{array}\right\} \tag{1.111}$$

with the pressure p given by the simple EOS (1.110).

The **isothermal equations** are even a simpler model than the isentropic equations, still non–linear. These may be viewed as resulting from the isentropic equations (1.111) with the EOS (1.110) further simplified to

$$p = p(\rho) \equiv a^2\rho , \tag{1.112}$$

where a is a non–zero constant propagation speed of sound. The isothermal equations are

$$\left.\begin{array}{c} \mathbf{U}_t + \mathbf{F}(\mathbf{U})_x = 0 , \\[4mm] \mathbf{U} = \left[\begin{array}{c} \rho \\ \rho u \end{array} \right] , \quad \mathbf{F} = \left[\begin{array}{c} \rho u \\ \rho u^2 + \rho a^2 \end{array} \right] . \end{array}\right\} \tag{1.113}$$

More submodels may be obtained by writing the isentropic equations as

$$\rho_t + \rho u_x + u\rho_x = 0 , \tag{1.114}$$

$$u_t + uu_x = -\frac{1}{\rho}p_x . \tag{1.115}$$

These result from (1.94) with the appropriate simplifications.

The **inviscid Burgers equation** is a scalar (single equation) non–linear equation given by

$$u_t + uu_x = 0 , \tag{1.116}$$

and can be obtained from the momentum equation (1.115) by neglecting density, and thus pressure, variations. In conservative form equation (1.116) reads

$$u_t + \left(\frac{u^2}{2} \right)_x = 0 . \tag{1.117}$$

The **Linearised Equations of Gas Dynamics** are obtained from (1.114)–(1.115) by considering small disturbances $\bar{u}$, $\bar{\rho}$ to a motionless gas.

Set $u = \bar{u}$ and $\rho = \bar{\rho} + \rho_0$, where ρ_0 is a constant density value. Recall that $p = p(\rho)$ and neglecting products of small quantities we have

$$p = p(\rho_0) + \bar{\rho}\frac{\partial p}{\partial \rho}(\rho_0) ,$$

that is, $p = p_0 + a^2\bar{\rho}$ with $p_0 = p(\rho_0)$, and

$$a^2 = \frac{\partial p}{\partial \rho}(\rho_0) = \text{constant} . \tag{1.118}$$

Substituting into (1.114)–(1.115) and neglecting squares of small quantities we obtain the linear equations

$$\bar{\rho}_t + \rho_0\bar{u}_x = 0 , \tag{1.119}$$

$$\bar{u}_t + \frac{a^2}{\rho_0}\bar{\rho}_x = 0 . \tag{1.120}$$

Elimination of $\bar{u}$ gives

$$\bar{\rho}_{tt} = a^2\bar{\rho}_{xx} , \tag{1.121}$$

which is the linear second–order wave equation for $\bar{\rho}(x,t)$. In matrix form system (1.119)–(1.120) reads

$$\mathbf{W}_t + \mathbf{A}\mathbf{W}_x = \mathbf{0} , \tag{1.122}$$

$$\mathbf{W} = \begin{bmatrix} \rho \\ u \end{bmatrix} , \quad \mathbf{A} = \begin{bmatrix} 0 & \rho_0 \\ a^2/\rho_0 & 0 \end{bmatrix} , \tag{1.123}$$

where bars have been dropped. The coefficient matrix $\mathbf{A}$ is now constant and thus the system (1.122) is a linear system with constant coefficients, the linearised equations of gas dynamics.

The **linear advection**, sometimes called linear convection, equation

$$u_t + au_x = 0 , \tag{1.124}$$

where a is a constant speed of wave propagation, is a further simplification to (1.121). This is also known as the *one–way wave equation* and plays a major role in the designing, analysing and testing of numerical methods for wave propagation problems.

1.6.6 Steady Compressible Flow

The steady, or time independent, homogeneous three dimensional Euler equations (1.92) are

$$\mathbf{F(U)}_x + \mathbf{G(U)}_y + \mathbf{H(U)}_z = \mathbf{0} . \tag{1.125}$$

In the steady regime it is important to identify *subsonic* and *supersonic flow*. To this end we recall the definition of Mach number M

$$M = \left[\frac{(u^2 + v^2 + w^2)}{a^2} \right]^{\frac{1}{2}} , \tag{1.126}$$

where a is the speed of sound; for ideal gases $a = \sqrt{(\gamma p/\rho)}$. Supersonic flow requires $M > 1$, while for subsonic flow we have $M < 1$. For sonic flow $M = 1$. Computationally, the three–dimensional equations may be treated via the method of dimensional splitting, which in essence reduces the three–dimensional problem to a sequence of *augmented* two–dimensional problems, see Chap. 16. The basic approach therefore relies on the two–dimensional case. In differential conservation form we have

$$\mathbf{F(U)}_x + \mathbf{G(U)}_y = \mathbf{0} , \tag{1.127}$$

$$\mathbf{F} = \begin{bmatrix} \rho u \\ \rho u^2 + p \\ \rho u v \\ u(E + p) \end{bmatrix} , \quad \mathbf{G} = \begin{bmatrix} \rho v \\ \rho u v \\ \rho v^2 + p \\ v(E + p) \end{bmatrix} . \tag{1.128}$$

As discontinuous solutions such as shock waves and slip surfaces are to be admitted, we replace the differential form (1.127)–(1.128) by the more general integral conservation form

$$\oint (\mathbf{F} \, dy - \mathbf{G} \, dx) = \mathbf{0} . \tag{1.129}$$

The integral is to be evaluated over the boundary of the appropriate control volume. In numerical methods this will be a computational cell.

   Steady linearised models can be obtained from the steady Euler equations (1.127)–(1.128). An interesting submodel is the small perturbation, two–dimensional steady supersonic equations

$$u_x - a^2 v_y = 0 , \quad v_x - u_y = 0 , \tag{1.130}$$

with

$$a^2 = \frac{1}{M_\infty^2 - 1} . \tag{1.131}$$

$M_\infty = $ constant denotes the free–stream Mach number and $u(x, y)$, $v(x, y)$ are small perturbations of the x and y velocity components respectively. In matrix form these equations read

$$\mathbf{W}_x + \mathbf{A}\mathbf{W}_y = \mathbf{0} \,, \tag{1.132}$$

with

$$\mathbf{W} = \begin{bmatrix} u \\ v \end{bmatrix} \,, \quad \mathbf{A} = \begin{bmatrix} 0 & -a^2 \\ -1 & 0 \end{bmatrix} \,. \tag{1.133}$$

See Sect. 18.2.2 of Chapter 18 for more information about steady supersonic flow.

1.6.7 Viscous Compressible Flow

The one–dimensional version of the compressible Navier–Stokes equations with heat conduction can be obtained from (1.61) by setting $v = w = 0$. The result is

$$\mathbf{U}_t + \mathbf{F}(\mathbf{U})_x = \mathbf{S} \,, \tag{1.134}$$

where

$$\mathbf{U} = \begin{bmatrix} \rho \\ \rho u \\ E \end{bmatrix} \,, \quad \mathbf{F} = \begin{bmatrix} \rho u \\ \rho u^2 + p \\ u(E + p) \end{bmatrix} \,, \quad \mathbf{S} = \begin{bmatrix} 0 \\ \frac{4}{3}(\eta u_x)_x \\ \frac{4}{3}(\eta u u_x)_x - (\kappa T_x)_x \end{bmatrix} \,. \tag{1.135}$$

Burgers's equation is the viscous version of (1.117) corresponding to a scalar non–linear simplification of (1.135), namely

$$u_t + \left(\frac{u^2}{2} \right)_x = \alpha u_{xx} \,, \tag{1.136}$$

where α is a coefficient of viscosity. A linearised form of this is

$$u_t + a u_x = \alpha u_{xx} \,, \tag{1.137}$$

which is the viscous version of the linear advection equation (1.124).

Next we consider two examples of incompressible flow. The first concerns inviscid incompressible flow with body forces, with special reference to free–surface gravity flow, as in oceans and rivers for example. The second example concerns the incompressible viscous equations with body forces and heat conduction neglected.

1.6.8 Free–Surface Gravity Flow

Consider the flow of water in a channel and assume the water to be in-compressible, non–viscous, non–heat conducting and subject to gravitational forces. We adopt the convention that the horizontal plane is given by the coordinates x and z and that the vertical direction is given by y. Denote the body force vector by $\mathbf{g} = (g_1, g_2, g_3) \equiv (0, -g, 0)$ where g is the acceleration due to gravity, assumed constant. Two important boundaries of the three dimensional domain are the bottom of the channel, denoted by $y = -h(x, z)$

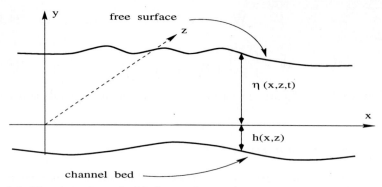

Fig. 1.1. Flow in a channel with free surface under gravity

and assumed fixed in time, and the free surface under gravity $y = \eta(x, z, t)$, which depends on space and time. Fig. 1.1 illustrates the situation for a vertical plane $z = $ constant. The state $y = 0$, $u = 0$, $w = 0$ corresponds to the rest position of equilibrium.

For an incompressible fluid the law of conservation of mass can be shown [75] to produce

$$\mathrm{div}\mathbf{V} = u_x + v_y + w_z = 0 \,. \tag{1.138}$$

From the mass equation in (1.86) we write

$$\frac{D\rho}{Dt} + \rho(\mathrm{div}\mathbf{V}) = 0 \,,$$

and from (1.138) it follows that

$$\frac{D\rho}{Dt} = \rho_t + \mathbf{V} \cdot \mathrm{grad}\rho = 0 \,.$$

That is, the mass density following the fluid is constant. If in addition we assume the water to be homogeneous, no variations in space, then it follows that ρ does not change with time. Practical examples of incompressible non–homogeneous fluids arise in Oceanography in the study of stratified flows. From the momentum equations in (1.94) we write

$$\left.\begin{aligned}
\frac{Du}{Dt} &\equiv u_t + uu_x + vu_y + wu_z = -\tfrac{1}{\rho}p_x \,, \\[2mm]
\frac{Dv}{Dt} &\equiv v_t + uv_x + vv_y + wv_z = -\tfrac{1}{\rho}p_y - g \,, \\[2mm]
\frac{Dw}{Dt} &\equiv w_t + uw_x + vw_y + ww_z = -\tfrac{1}{\rho}p_z \,.
\end{aligned}\right\} \tag{1.139}$$

There are now four equations and four unknowns, namely u, v, w and p. In principle, given initial and boundary conditions one should be able to solve (1.138)–(1.139) for u, v, w and p as functions of space x, y, z and time t. Boundary conditions are required at the bottom $y = -h(x, z)$ and on the free surface under gravity $y = \eta(x, z, t)$. Two boundary conditions are given for the free surface

$$\left.\begin{array}{c} \frac{D}{Dt}(\eta - y) = 0 \\[2ex] p = p_{atm} \end{array}\right\} \quad y = \eta(x, z, t) \,. \tag{1.140}$$

For the bottom boundary one takes

$$\frac{D}{Dt}(h + y) = 0 \,, \; y = -h(x, z) \,. \tag{1.141}$$

In equation (1.140), p_{atm} is the atmospheric pressure, which for convenience is assumed to be zero. Equation (1.141) states that the normal component of velocity vanishes, i.e. there is no flow through the bottom of the channel. In spite of the strong physical assumptions made, the free–surface problem as stated remains a very difficult problem to solve numerically. Analytical solutions are impossible to obtain. Further approximations result in more tractable mathematical models. For general background on the topic see the book on water waves by Stoker [328].

1.6.9 The Shallow Water Equations

The shallow water equations are an approximation to the full free–surface problem and result from the assumption that the vertical component of the acceleration Dv/Dt can be neglected. Inserting $Dv/Dt = 0$ in the second of equations (1.139) gives

$$p = \rho g(\eta - y) \,. \tag{1.142}$$

This is called the *hydrostatic pressure relation*. Differentiation of p with respect to x and z gives

$$p_x = \rho g \eta_x \,, \tag{1.143}$$

$$p_z = \rho g \eta_z \,, \tag{1.144}$$

i.e. both p_x and p_z are independent of y and so the x and z components of the acceleration Du/Dt and Dw/Dt in (1.139) are independent of y. Thus, the x and z components of velocity are also independent of y for all t if they were at a given time, $t = 0$, say. Hence, the first and third equations in (1.139), making use of (1.143)–(1.144), become

$$u_t + u u_x + w u_z = -g \eta_x \,, \tag{1.145}$$

$$w_t + u w_x + w w_z = -g \eta_z \,. \tag{1.146}$$

An important step in deriving the shallow water equations now follows; we integrate the continuity equation (1.138) with respect to y (vertical direction) to obtain

$$\int_{-h}^{\eta} u_x \, dy + \int_{-h}^{\eta} w_z \, dy + v|_{-h}^{\eta} = 0 \,. \tag{1.147}$$

By expanding the boundary conditions (1.140) and (1.141) according to the definition of total derivative D/Dt, see (1.64), we obtain

$$(\eta_t + u\eta_x + w\eta_z - v)|_{y=\eta} = 0 , \qquad (1.148)$$

$$(uh_x + wh_z + v)|_{y=-h} = 0 , \qquad (1.149)$$

which, if used in (1.147), give

$$\left.\begin{aligned}
&\int_{-h}^{\eta} u_x \, dy + \int_{-h}^{\eta} w_z \, dy + \eta_t \\
&+(u|_{y=\eta})\eta_x + (w|_{y=\eta})\eta_z + (u|_{y=-h})h_x + (w|_{y=-h})h_z = 0 .
\end{aligned}\right\} \qquad (1.150)$$

Equation (1.150) can finally be expressed as

$$\eta_t + \frac{\partial}{\partial x} \int_{-h}^{\eta} u \, dy + \frac{\partial}{\partial z} \int_{-h}^{\eta} w \, dy = 0 . \qquad (1.151)$$

This follows by using the relations

$$\frac{\partial}{\partial x} \int_{-h(x,z)}^{\eta(x,z,t)} u \, dy = \int_{-h}^{\eta} u_x \, dy + (u|_{y=\eta})\eta_x + (u|_{y=-h})h_x ,$$

$$\frac{\partial}{\partial z} \int_{-h(x,z)}^{\eta(x,z,t)} w \, dy = \int_{-h}^{\eta} w_z \, dy + (w|_{y=\eta})\eta_z + (w|_{y=-h})h_z .$$

Equation (1.151) can be simplified further. This follows from the observation that both u and w are independent of y and so equation (1.151) becomes

$$\eta_t + [u(\eta + h)]_x + [w(\eta + h)]_z = 0 . \qquad (1.152)$$

The governing two–dimensional shallow water equations are (1.145), (1.146) and (1.152). We now express these equations in conservation law form. Since $h(x, z)$ is independent of time t we have $h_t = 0$ and so equation (1.152) can be re–written as

$$(\eta + h)_t + [u(\eta + h)]_x + [w(\eta + h)]_z = 0 , \qquad (1.153)$$

which if multiplied by u and added to (1.145), premultiplied by $\eta + h$, gives

$$[u(\eta + h)]_t + [u^2(\eta + h)]_x + [uw(\eta + h)]_z = -g(\eta + h)\eta_x . \qquad (1.154)$$

In an analogous way equations (1.146) and (1.153) give

$$[w(\eta + h)]_t + [uw(\eta + h)]_x + [w^2(\eta + h)]_z = -g(\eta + h)\eta_z . \qquad (1.155)$$

The right–hand side term of (1.154) can be re–written as

$$- g(\eta + h)\eta_x = g(\eta + h)h_x - \frac{1}{2}g[(\eta + h)^2]_x \qquad (1.156)$$

and so (1.154) becomes

$$(\phi u)_t + (\phi u^2 + \frac{1}{2}\phi^2)_x + (\phi uw)_z = g\phi h_x \;. \tag{1.157}$$

Similarly, equation (1.155) becomes

$$(\phi w)_t + (\phi uw)_x + (\phi w^2 + \frac{1}{2}\phi^2)_z = g\phi h_z \;, \tag{1.158}$$

where

$$\phi = gH \;,\; H = \eta + h \;. \tag{1.159}$$

H is the total depth of the fluid and ϕ is called the *geopotential*. The two–dimensional shallow water equations are now written in compact conservation form with source terms, i.e. equations (1.153), (1.157) and (1.158) can be expressed as

$$\mathbf{U}_t + \mathbf{F}(\mathbf{U})_x + \mathbf{G}(\mathbf{U})_z = \mathbf{S}(\mathbf{U}) \;, \tag{1.160}$$

with

$$\mathbf{U} = \begin{bmatrix} \phi \\ \phi u \\ \phi w \end{bmatrix} \;,\; \mathbf{F} = \begin{bmatrix} \phi u \\ \phi u^2 + \frac{1}{2}\phi^2 \\ \phi uw \end{bmatrix} \;,$$

$$\mathbf{G} = \begin{bmatrix} \phi w \\ \phi uw \\ \phi w^2 + \frac{1}{2}\phi^2 \end{bmatrix} \;,\; \mathbf{S} = \begin{bmatrix} 0 \\ g\phi h_x \\ g\phi h_z \end{bmatrix} \;. \tag{1.161}$$

In equations (1.160) $\mathbf{U}$ is the vector of conserved variables, $\mathbf{F}(\mathbf{U})$ and $\mathbf{G}(\mathbf{U})$ are flux vectors and $\mathbf{S}(\mathbf{U})$ is the source term vector. For many applications there will be additional terms in the vector $\mathbf{S}(\mathbf{U})$ to account for Coriolis forces, wind forces, bottom friction, etc. Numerical solution procedures will deal essentially with the homogeneous part of (1.160). The numerical treatment of the source terms is a relatively standard process as is the treatment of the two–dimensional homogeneous problem. Both can be dealt with via splitting schemes; for details see Chaps. 15 and 16. From a computational point of view, most of the effort goes into devising schemes for the basic homogeneous one–dimensional system

$$\mathbf{U}_t + \mathbf{F}(\mathbf{U})_x = \mathbf{0} \;, \tag{1.162}$$

with $\mathbf{U}$ and $\mathbf{F}(\mathbf{U})$ given by

$$\mathbf{U} = \begin{bmatrix} \phi \\ \phi u \end{bmatrix} \;,\; \mathbf{F} = \begin{bmatrix} \phi u \\ \phi u^2 + \frac{1}{2}\phi^2 \end{bmatrix} \;. \tag{1.163}$$

The conservation laws (1.162)–(1.163) can be written in integral form as

$$\oint (\mathbf{U} \, dx - \mathbf{F}(\mathbf{U}) \, dt) = \mathbf{0} \;. \tag{1.164}$$

Equations (1.164) admit discontinuous solutions while equations (1.162)–(1.163) do not.

See Sect. 18.2.1 of Chapter 18 for more information about the shallow water equations. Se also the textbook [370].

1.6.10 Incompressible Viscous Flow

We assume the fluid to be incompressible, homogeneous, non–heat conducting and viscous, with constant coefficient of viscosity η. Body forces are also neglected. We study three mathematical formulations of the governing equations in Cartesian coordinates and restrict our attention to the two–dimensional case.

The **primitive variable formulation** of the incompressible two dimensional Navier–Stokes equations is given by

$$u_x + v_y = 0 , \tag{1.165}$$

$$u_t + uu_x + vu_y + \frac{1}{\rho}p_x = \nu\left[u_{xx} + u_{yy}\right] , \tag{1.166}$$

$$v_t + uv_x + vv_y + \frac{1}{\rho}p_y = \nu\left[v_{xx} + v_{yy}\right] , \tag{1.167}$$

where

$$\nu = \frac{\eta}{\rho} \tag{1.168}$$

is the *kinematic viscosity* and η is the coefficient of *shear viscosity*. We have a set of three equations (1.165)–(1.167) for the three unknowns u, v, p, the primitive variables. In principle, given a domain along with initial and boundary conditions for the equations one should be able to solve this problem using the primitive variable formulation.

The **stream–function vorticity formulation** is another way of expressing the incompressible Navier–Stokes equations. This formulation is attractive for the two–dimensional case but not so much in three dimensions, in which the role of a stream function is replaced by that of a *vector potential*. The magnitude of the vorticity vector can be written as

$$\zeta = v_x - u_y . \tag{1.169}$$

Introducing a stream function ψ we have

$$u = \psi_y , \quad v = -\psi_x . \tag{1.170}$$

By combining equations (1.166) and (1.167), so as to eliminate pressure p, and using (1.169) we obtain

$$\zeta_t + u\zeta_x + v\zeta_y = \nu\left[\zeta_{xx} + \zeta_{yy}\right] , \tag{1.171}$$

which is called the *vorticity transport equation*. This is an advection–diffusion equation of *parabolic* type. In order to solve (1.171) one requires the solution for the stream function ψ, which is in turn related to the vorticity ζ via

$$\psi_{xx} + \psi_{yy} = -\zeta . \tag{1.172}$$

This is called the *Poisson equation* and is of *elliptic type*. There are numerical schemes to solve (1.171)–(1.172) using the *apparent* decoupling of the parabolic–elliptic problem (1.165)–(1.167) to transform it into the parabolic equation (1.171) and the elliptic equation (1.172). A relevant observation, from the numerical point of view, is that the advection terms of the left hand side of equation (1.171) can be written in conservative form and hence we have

$$\zeta_t + (u\zeta)_x + (v\zeta)_y = \nu \left[\zeta_{xx} + \zeta_{yy}\right] . \tag{1.173}$$

This follows from the fact that $u_x + v_y = 0$, which was also used to obtain (1.171) from (1.166)–(1.167).

1.6.11 The Artificial Compressibility Equations

The artificial compressibility formulation is yet another approach to formulate the incompressible Navier–Stokes equations and was originally put forward by Chorin [72] for the steady case. Consider the two–dimensional equations (1.165)–(1.167) in non–dimensional form

$$u_x + v_y = 0 , \tag{1.174}$$

$$u_t + uu_x + vu_y + p_x = \alpha \left[u_{xx} + u_{yy}\right] , \tag{1.175}$$

$$v_t + uv_x + vv_y + p_y = \alpha \left[v_{xx} + v_{yy}\right] , \tag{1.176}$$

where the following non–dimensionalisation has been used

$$u \leftarrow u/V_\infty , \; v \leftarrow v/V_\infty , p \leftarrow \frac{p}{\rho_\infty V_\infty^2} ,$$

$$x \leftarrow x/L , \; y \leftarrow y/L , \; t \leftarrow tV_\infty/L ,$$

$$\alpha = 1/R_{\mathrm{eL}} , \; R_{\mathrm{eL}} = \frac{V_\infty L}{\nu_\infty} .$$

Multiplying (1.174) by the non–zero parameter c^2 and adding an *artificial compressibility* term p_t the first equation reads

$$p_t + (uc^2)_x + (vc^2)_y = 0 .$$

By using equation (1.174) the advective terms in (1.175)–(1.176) can be written in *conservative form*, so that the modified system becomes

$$p_t + (uc^2)_x + (vc^2)_y = 0 , \tag{1.177}$$

$$u_t + (u^2 + p)_x + (uv)_y = \alpha \left[u_{xx} + u_{yy}\right] , \tag{1.178}$$

$$v_t + (uv)_x + (v^2 + p)_y = \alpha \left[v_{xx} + v_{yy}\right] . \tag{1.179}$$

These equations can be written in compact form as

$$\mathbf{U}_t + \mathbf{F}(\mathbf{U})_x + \mathbf{G}(\mathbf{U})_y = \mathbf{S}(\mathbf{U}) , \tag{1.180}$$

where

$$
\mathbf{U} = \begin{bmatrix} p \\ u \\ v \end{bmatrix} , \quad \mathbf{F} = \begin{bmatrix} c^2 u \\ u^2 + p \\ uv \end{bmatrix} ,
$$

$$
\mathbf{G} = \begin{bmatrix} c^2 v \\ uv \\ v^2 + p \end{bmatrix} , \quad \mathbf{S} = \begin{bmatrix} 0 \\ \alpha \left(u_{xx} + u_{yy} \right) \\ \alpha \left(v_{xx} + v_{yy} \right) \end{bmatrix} .
$$

$$(1.181)$$

Equations (1.180)–(1.181) are called the *artificial compressibility equations*. Here c^2 is the *artificial compressibility factor*, a constant parameter. The source term vector in this case is a function of second derivatives. Note that the modified equations are equivalent to the original equations in the *steady state* limit.

The left–hand side of the artificial compressibility equations form a non–linear hyperbolic system. The Riemann problem can be defined and solved exactly or approximately. See Chaps. 4, 8, 9, 10, 11 and 12 on approaches to solve the Riemann problem. Once a Riemann solver is available one can deploy Godunov–type numerical methods to solve the equations with general initial data. See Chaps. 6, 7, 13, 14 and 16 for possible numerical methods. The topic of numerical methods for the artificial compressibility equations is not pursued in this textbook; the interested reader is referred to [366] for details on exact and approximate Riemann solvers for the artificial compressibility equations. Further information about the artificial compressibility equations is found in Sect. 18.2.3 of Chapter 18.

2. Notions on Hyperbolic Partial Differential Equations

In this chapter we study some elementary properties of a class of hyperbolic Partial Differential Equations (PDEs). The selected aspects of the equations are those thought to be essential for the analysis of the equations of fluid flow and the implementation of numerical methods. For general background on PDEs we recommend the book by John [185] and particularly the one by Zachmanoglou and Thoe [425]. The discretisation techniques studied in this book are strongly based on the underlying Physics and mathematical properties of PDEs. It is therefore justified to devote some effort to some fundamentals on PDEs. Here we deal almost exclusively with *hyperbolic* PDEs and *hyperbolic conservation laws* in particular. There are three main reasons for this: (i) The equations of compressible fluid flow reduce to hyperbolic systems, the Euler equations, when the effects of viscosity and heat conduction are neglected. (ii) Numerically, it is generally accepted that the hyperbolic terms of the PDEs of fluid flow are the terms that pose the most stringent requirements on the discretisation techniques. (iii) The theory of hyperbolic systems is much more advanced than that for more complete mathematical models, such as the Navier–Stokes equations. In addition, there has in recent years been a noticeable increase in research and development activities centred on the theme of hyperbolic problems, as these cover a wide range of areas of scientific and technological interest. A good source of up–to–date work in this field is found in the proceedings of the series of meetings on Hyperbolic Problems, see for instance [59], [119], [142]. See also [219]. Other relevant publications are those of Godlewski and Raviart [144], Hörmander [176] and Tveito and Winther [385].

We restrict ourselves to some of the basics on hyperbolic PDEs and choose an informal way of presentation. The selected topics and approach are almost exclusively motivated by the theme of the book, namely the Riemann problem and high–resolution upwind and centred numerical methods.

2.1 Quasi–Linear Equations: Basic Concepts

In this section we study systems of first–order partial differential equations of the form

$$\frac{\partial u_i}{\partial t} + \sum_{j=1}^{m} a_{ij}(x, t, u_1, \ldots, u_m)\frac{\partial u_j}{\partial x} + b_i(x, t, u_1, \ldots, u_m) = 0 \,, \qquad (2.1)$$

for $i = 1, \ldots, m$. This is a system of m equations in m unknowns u_i that depend on space x and a time–like variable t. Here u_i are the *dependent variables* and x, t are the *independent variables*; this is expressed via the notation $u_i = u_i(x, t)$; $\partial u_i/\partial t$ denotes the partial derivative of $u_i(x, t)$ with respect to t; similarly $\partial u_i/\partial x$ denotes the partial derivative of $u_i(x, t)$ with respect to x. We also make use of subscripts to denote partial derivatives. System (2.1) can also be written in matrix form as

$$\mathbf{U}_t + \mathbf{A}\mathbf{U}_x + \mathbf{B} = \mathbf{0} \,, \qquad (2.2)$$

with

$$\mathbf{U} = \begin{bmatrix} u_1 \\ u_2 \\ \vdots \\ u_m \end{bmatrix}, \ \mathbf{B} = \begin{bmatrix} b_1 \\ b_2 \\ \vdots \\ b_m \end{bmatrix}, \ \mathbf{A} = \begin{bmatrix} a_{11} & \cdots & a_{1m} \\ a_{21} & \cdots & a_{2m} \\ \vdots & \vdots & \vdots \\ a_{m1} & \cdots & a_{mm} \end{bmatrix} . \qquad (2.3)$$

If the entries a_{ij} of the matrix $\mathbf{A}$ are all constant and the components b_j of the vector $\mathbf{B}$ are also constant then system (2.2) is *linear with constant coefficients*. If $a_{ij} = a_{ij}(x, t)$ and $b_j = b_j(x, t)$ the system is *linear with variable coefficients*. The system is still linear if $\mathbf{B}$ depends linearly on $\mathbf{U}$ and is called *quasi–linear* if the coefficient matrix $\mathbf{A}$ is a function of the vector $\mathbf{U}$, that is $\mathbf{A} = \mathbf{A}(\mathbf{U})$. Note that quasi–linear systems are in general systems of non–linear equations. System (2.2) is called *homogeneous* if $\mathbf{B} = \mathbf{0}$. For a set of PDEs of the form (2.2) one needs to specify the range of variation of the independent variables x and t. Usually x lies in a subinterval of the real line, namely $x_l < x < x_r$; this subinterval is called the *spatial domain* of the PDEs, or just *domain*. At the values x_l, x_r one also needs to specify *Boundary Conditions (BCs)*. In this Chapter we assume the domain is the full real line, $-\infty < x < \infty$, and thus no boundary conditions need to be specified. As to variations of time t we assume $t_0 < t < \infty$. An *Initial Condition (IC)* needs to be specified at the initial time, which is usually chosen to be $t_0 = 0$.

Two scalar ($m = 1$) examples of PDEs of the form (2.1) are the linear advection equation

$$\frac{\partial u}{\partial t} + a\frac{\partial u}{\partial x} = 0 \qquad (2.4)$$

and the inviscid Burgers equation

$$\frac{\partial u}{\partial t} + u\frac{\partial u}{\partial x} = 0 \,, \qquad (2.5)$$

both introduced in Sect. 1.6.2 of Chap. 1. In the linear advection equation (2.4) the coefficient a (a constant) is the wave propagation speed. In the Burgers equation $a = a(u) = u$.

Definition 2.1.1 (Conservation Laws). *Conservation laws are systems of partial differential equations that can be written in the form*

$$\mathbf{U}_t + \mathbf{F}(\mathbf{U})_x = \mathbf{0} \,, \tag{2.6}$$

where

$$\mathbf{U} = \begin{bmatrix} u_1 \\ u_2 \\ \vdots \\ u_m \end{bmatrix} \,, \quad \mathbf{F}(\mathbf{U}) = \begin{bmatrix} f_1 \\ f_2 \\ \vdots \\ f_m \end{bmatrix} \,. \tag{2.7}$$

$\mathbf{U}$ *is called the vector of conserved variables,* $\mathbf{F} = \mathbf{F}(\mathbf{U})$ *is the vector of fluxes and each of its components* f_i *is a function of the components* u_j *of* $\mathbf{U}$.

Definition 2.1.2 (Jacobian Matrix). *The Jacobian of the flux function* $\mathbf{F}(\mathbf{U})$ *in (2.6) is the matrix*

$$\mathbf{A}(\mathbf{U}) = \partial \mathbf{F}/\partial \mathbf{U} = \begin{bmatrix} \partial f_1/\partial u_1 & \cdots & \partial f_1/\partial u_m \\ \partial f_2/\partial u_1 & \cdots & \partial f_2/\partial u_m \\ \vdots & \vdots & \vdots \\ \partial f_m/\partial u_1 & \cdots & \partial f_m/\partial u_m \end{bmatrix} \,. \tag{2.8}$$

The entries a_{ij} *of* $\mathbf{A}(\mathbf{U})$ *are partial derivatives of the components* f_i *of the vector* $\mathbf{F}$ *with respect to the components* u_j *of the vector of conserved variables* $\mathbf{U}$, *that is* $a_{ij} = \partial f_i/\partial u_j$.

Note that conservation laws of the form (2.6)–(2.7) can also be written in quasi–linear form (2.2), with $\mathbf{B} \equiv \mathbf{0}$, by applying the chain rule to the second term in (2.6), namely

$$\frac{\partial \mathbf{F}(\mathbf{U})}{\partial x} = \frac{\partial \mathbf{F}}{\partial \mathbf{U}} \frac{\partial \mathbf{U}}{\partial x} \,.$$

Hence (2.6) becomes

$$\mathbf{U}_t + \mathbf{A}(\mathbf{U})\mathbf{U}_x = \mathbf{0} \,,$$

which is a special case of (2.2). The scalar PDEs (2.4) and (2.5) can be expressed as conservation laws, namely

$$\frac{\partial u}{\partial t} + \frac{\partial f(u)}{\partial x} = 0 \,, \ f(u) = au \,, \tag{2.9}$$

$$\frac{\partial u}{\partial t} + \frac{\partial f(u)}{\partial x} = 0 \,, \ f(u) = \frac{1}{2}u^2 \,. \tag{2.10}$$

Definition 2.1.3 (Eigenvalues). *The eigenvalues* λ_i *of a matrix* $\mathbf{A}$ *are the solutions of the characteristic polynomial*

$$|\mathbf{A} - \lambda \mathbf{I}| = \det(\mathbf{A} - \lambda \mathbf{I}) = 0 \,, \tag{2.11}$$

where $\mathbf{I}$ *is the identity matrix. The eigenvalues of the coefficient matrix* $\mathbf{A}$ *of a system of the form (2.2) are called the eigenvalues of the system.*

Physically, eigenvalues represent speeds of propagation of information. Speeds will be measured positive in the direction of increasing x and negative otherwise.

Definition 2.1.4 (Eigenvectors). *A right eigenvector of a matrix $\mathbf{A}$ corresponding to an eigenvalue λ_i of $\mathbf{A}$ is a vector $\mathbf{K}^{(i)} = [k_1^{(i)}, k_2^{(i)}, \ldots, k_m^{(i)}]^T$ satisfying $\mathbf{A}\mathbf{K}^{(i)} = \lambda_i \mathbf{K}^{(i)}$. Similarly, a left eigenvector of a matrix $\mathbf{A}$ corresponding to an eigenvalue λ_i of $\mathbf{A}$ is a vector $\mathbf{L}^{(i)} = [l_1^{(i)}, l_2^{(i)}, \ldots, l_m^{(i)}]$ such that $\mathbf{L}^{(i)}\mathbf{A} = \lambda_i \mathbf{L}^{(i)}$.*

For the scalar examples (2.9)–(2.10) the eigenvalues are trivially found to be $\lambda = a$ and $\lambda = u$ respectively. Next we find eigenvalues and eigenvectors for a system of PDEs.

Example 2.1.1 (Linearised Gas Dynamics). The linearised equations of Gas Dynamics, derived in Sect. 1.6.2 of Chap. 1, are the 2×2 linear system

$$\left. \begin{array}{l} \frac{\partial \rho}{\partial t} + \rho_0 \frac{\partial u}{\partial x} = 0 \, , \\[2mm] \frac{\partial u}{\partial t} + \frac{a^2}{\rho_0} \frac{\partial \rho}{\partial x} = 0 \, , \end{array} \right\} \tag{2.12}$$

where the unknowns are the density $u_1 = \rho(x, t)$ and the speed $u_2 = u(x, t)$; ρ_0 is a constant reference density and a is the sound speed, a positive constant. When written in the matrix form (2.2) this system reads

$$\mathbf{U}_t + \mathbf{A}\mathbf{U}_x = \mathbf{0} \, , \tag{2.13}$$

with

$$\mathbf{U} = \begin{bmatrix} u_1 \\ u_2 \end{bmatrix} \equiv \begin{bmatrix} \rho \\ u \end{bmatrix} \, , \quad \mathbf{A} = \begin{bmatrix} 0 & \rho_0 \\ a^2/\rho_0 & 0 \end{bmatrix} . \tag{2.14}$$

The eigenvalues of the system are the zeros of the characteristic polynomial

$$|\mathbf{A} - \lambda\mathbf{I}| = \det \begin{bmatrix} 0 - \lambda & \rho_0 \\ a^2/\rho_0 & 0 - \lambda \end{bmatrix} = 0 \, .$$

That is, $\lambda^2 = a^2$, which has two real and distinct solutions, namely

$$\lambda_1 = -a \, , \quad \lambda_2 = +a \, . \tag{2.15}$$

We now find the right eigenvectors $\mathbf{K}^{(1)}$, $\mathbf{K}^{(2)}$ corresponding to the eigenvalues λ_1 and λ_2.

The eigenvector $\mathbf{K}^{(1)}$ for eigenvalue $\lambda = \lambda_1 = -a$ is found as follows: we look for a vector $\mathbf{K}^{(1)} = [k_1, k_2]^T$ such that $\mathbf{K}^{(1)}$ is a right eigenvector of $\mathbf{A}$, that is $\mathbf{A}\mathbf{K}^{(1)} = \lambda_1 \mathbf{K}^{(1)}$. Writing this in full gives

$$\begin{bmatrix} 0 & \rho_0 \\ a^2/\rho_0 & 0 \end{bmatrix} \begin{bmatrix} k_1 \\ k_2 \end{bmatrix} = \begin{bmatrix} -ak_1 \\ -ak_2 \end{bmatrix} \, ,$$

which produces two linear algebraic equations for the unknowns k_1 and k_2

$$\rho_0 k_2 = -ak_1 \; , \quad \frac{a^2}{\rho_0} k_1 = -ak_2 \; . \tag{2.16}$$

The reader will realise that in fact these two equations are equivalent and so effectively we have a single linear algebraic equation in two unknowns. This gives a one–parameter family of solutions. Thus we select an arbitrary non–zero parameter α_1, a scaling factor, and set $k_1 = \alpha_1$ in any of the equations to obtain $k_2 = -\alpha_1 a/\rho_0$ for the second component and hence the first right eigenvector becomes

$$\mathbf{K}^{(1)} = \alpha_1 \begin{bmatrix} 1 \\ -a/\rho_0 \end{bmatrix} \; . \tag{2.17}$$

The eigenvector $\mathbf{K}^{(2)}$ for eigenvalue $\lambda = \lambda_2 = +a$ is found in a similar manner. The resulting algebraic equations for $\mathbf{K}^{(2)}$ corresponding to the eigenvalue $\lambda_2 = +a$ are

$$\rho_0 k_2 = ak_1 \; , \quad \frac{a^2}{\rho_0} k_1 = ak_2 \; . \tag{2.18}$$

By denoting the second scaling factor by α_2 and setting $k_1 = \alpha_2$ we obtain

$$\mathbf{K}^{(2)} = \alpha_2 \begin{bmatrix} 1 \\ a/\rho_0 \end{bmatrix} \; . \tag{2.19}$$

Taking the scaling factors to be $\alpha_1 = \rho_0$ and $\alpha_2 = \rho_0$ gives the right eigenvectors

$$\mathbf{K}^{(1)} = \begin{bmatrix} \rho_0 \\ -a \end{bmatrix} \; , \quad \mathbf{K}^{(2)} = \begin{bmatrix} \rho_0 \\ a \end{bmatrix} \; . \tag{2.20}$$

Definition 2.1.5 (Hyperbolic System). *A system (2.2) is said to be hyperbolic at a point (x, t) if $\mathbf{A}$ has m real eigenvalues $\lambda_1, \ldots, \lambda_m$ and a corresponding set of m linearly independent right eigenvectors $\mathbf{K}^{(1)}, \ldots, \mathbf{K}^{(m)}$. The system is said to be strictly hyperbolic if the eigenvalues λ_i are all distinct.*

Note that strict hyperbolicity implies hyperbolicity, because real and distinct eigenvalues ensure the existence of a set of linearly independent eigenvectors. The system (2.2) is said to be *elliptic* at a point (x, t) if none of the eigenvalues λ_i of $\mathbf{A}$ are real. Both scalar examples (2.9)–(2.10) are trivially hyperbolic. The linearised gas dynamic equations (2.12) are also hyperbolic, since λ_1 and λ_2 are both real at any point (x, t). Moreover, as the eigenvalues are also distinct this system is strictly hyperbolic.

Example 2.1.2 (The Cauchy–Riemann Equations). An example of a first–order system of the form (2.2) with t replaced by x and x replaced by y is the Cauchy–Riemann equations

$$\frac{\partial u}{\partial x} - \frac{\partial v}{\partial y} = 0 \; , \quad \frac{\partial v}{\partial x} + \frac{\partial u}{\partial y} = 0 \; , \tag{2.21}$$

where $u_1 = u(x,y)$ and $u_2 = v(x,y)$. These equations arise in the study of *analytic functions* in Complex Analysis [260]. When written in matrix notation (2.2) equations (2.21) become

$$\mathbf{U}_x + \mathbf{A}\mathbf{U}_y = 0 , \tag{2.22}$$

with

$$\mathbf{U} = \begin{bmatrix} u \\ v \end{bmatrix} , \quad \mathbf{A} = \begin{bmatrix} 0 & -1 \\ 1 & 0 \end{bmatrix} . \tag{2.23}$$

The characteristic polynomial $|\mathbf{A} - \lambda\mathbf{I}| = 0$ gives $\lambda^2 + 1 = 0$, which has no real solutions for λ and thus the system is *elliptic*.

Example 2.1.3 (The Small Perturbation Equations). In Sect. 1.6.2 of Chap. 1, the small perturbation steady equations were introduced

$$u_x - a^2 v_y = 0 , \quad v_x - u_y = 0 , \tag{2.24}$$

with

$$a^2 = \frac{1}{M_\infty^2 - 1} . \tag{2.25}$$

$M_\infty = constant$ denotes the free–stream Mach number and $u(x,y)$, $v(x,y)$ are small perturbations of the x and y velocity components respectively. In matrix notation these equations read

$$\mathbf{U}_x + \mathbf{A}\mathbf{U}_y = 0 , \tag{2.26}$$

with

$$\mathbf{U} = \begin{bmatrix} u \\ v \end{bmatrix} , \quad \mathbf{A} = \begin{bmatrix} 0 & -a^2 \\ -1 & 0 \end{bmatrix} . \tag{2.27}$$

The character of these equations depends entirely on the value of the Mach number M_∞. For subsonic flow $M_\infty < 1$ the characteristic polynomial has complex solutions and thus the equations are of elliptic type. For supersonic flow $M_\infty > 1$ and the system is strictly hyperbolic, with eigenvalues

$$\lambda_1 = -a , \quad \lambda_2 = +a . \tag{2.28}$$

It is left to the reader to check that the corresponding right eigenvectors are

$$\mathbf{K}^{(1)} = \alpha_1 \begin{bmatrix} 1 \\ 1/a \end{bmatrix} , \quad \mathbf{K}^{(2)} = \alpha_2 \begin{bmatrix} 1 \\ -1/a \end{bmatrix} , \tag{2.29}$$

where α_1 and α_2 are two non–zero scaling factors. By taking the values $\alpha_1 = \alpha_2 = a$ we obtain the following expressions for the right eigenvectors

$$\mathbf{K}^{(1)} = \begin{bmatrix} a \\ 1 \end{bmatrix} , \quad \mathbf{K}^{(2)} = \begin{bmatrix} a \\ -1 \end{bmatrix} ,$$

2.2 The Linear Advection Equation

A general, time–dependent linear advection equation in three space dimensions reads

$$u_t + a(x, y, z, t)u_x + b(x, y, z, t)u_y + c(x, y, z, t)u_z = 0 , \qquad (2.30)$$

where the unknown is $u = u(x, y, z, t)$ and a, b, c are variable coefficients. If the coefficients are sufficiently smooth one can express (2.30) as a conservation law with source terms, namely

$$u_t + (au)_x + (bu)_y + (cu)_z = u(a_x + b_y + c_z) . \qquad (2.31)$$

In this section we study in detail the initial–value problem (IVP) for the special case of the linear advection equation, namely

$$\left. \begin{array}{ll} \text{PDE:} & u_t + au_x = 0 , \ -\infty < x < \infty , \ t > 0 . \\[2mm] \text{IC:} & u(x, 0) = u_0(x) , \end{array} \right\} \qquad (2.32)$$

where a is a constant wave propagation speed. The initial data at time $t = 0$ is a function of x alone and is denoted by $u_0(x)$. We warn the reader that for systems we shall use a different notation for the initial data. Generally, we shall not be explicit about the conditions $-\infty < x < \infty$; $t > 0$ on the independent variables when stating an initial–value problem. The PDE in (2.32) is the simplest hyperbolic PDE and in view of (2.9) is also the simplest hyperbolic conservation law. It is a very useful model equation for the purpose of studying numerical methods for hyperbolic conservation laws, in the same way as the linear, first–order ordinary differential equation

$$\frac{\mathrm{d}x}{\mathrm{d}t} = \beta , \ x = x(t) , \ \beta = \text{constant} , \qquad (2.33)$$

is a popular model equation for analysing numerical methods for Ordinary Differential Equation (ODEs). Two useful references on ordinary differential equations are Brown [55] and Lambert [199]. In Sect. 15.4 of Chap. 15 we study numerical methods for ODEs in connection with source terms in inhomogeneous PDEs.

2.2.1 Characteristics and the General Solution

We recall the definition of *characteristics* or *characteristic curves* in the context of a scalar equation such as that in (2.32). Characteristics may be defined as curves $x = x(t)$ in the t–x plane along which the PDE becomes an ODE. Consider $x = x(t)$ and regard u as a function of t, that is $u = u(x(t), t)$. The rate of change of u along $x = x(t)$ is

$$\frac{\mathrm{d}u}{\mathrm{d}t} = \frac{\partial u}{\partial t} + \frac{\mathrm{d}x}{\mathrm{d}t} \frac{\partial u}{\partial x} . \qquad (2.34)$$

If the characteristic curve $x = x(t)$ satisfies the ODE

$$\frac{\mathrm{d}x}{\mathrm{d}t} = a \,, \qquad (2.35)$$

then the PDE in (2.32), together with (2.34) and (2.35), gives

$$\frac{\mathrm{d}u}{\mathrm{d}t} = \frac{\partial u}{\partial t} + a\frac{\partial u}{\partial x} = 0 \,. \qquad (2.36)$$

Therefore the rate of change of u along the characteristic curve $x = x(t)$ satisfying (2.35) is zero, that is, u is *constant along the curve* $x = x(t)$. The speed a in (2.35) is called the *characteristic speed* and according to (2.35) it is the slope of the curve $x = x(t)$ in the t–x plane. In practice it is more common to use the x–t plane to sketch the characteristics, in which case the slope of the curves in question is $1/a$. The family of characteristic curves

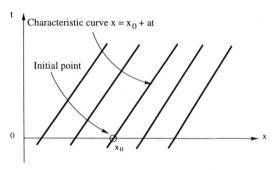

Fig. 2.1. Picture of characteristics for the linear advection equation for positive characteristic speed a. Initial condition at time $t = 0$ fixes the initial position x_0

$x = x(t)$ given by the ODE (2.35) are illustrated in Fig. 2.1 for $a > 0$ and are a one–parameter family of curves. A particular member of this family is determined when an *initial condition* (IC) at time $t = 0$ for the ODE (2.35) is added. Suppose we set

$$x(0) = x_0 \,, \qquad (2.37)$$

then the single characteristic curve that passes through the point $(x_0, 0)$, according to (2.35) is

$$x = x_0 + at \,. \qquad (2.38)$$

This is also illustrated in Fig. 2.1. Now we may regard the initial position x_0 as a parameter and in this way we reproduce the full one–parameter family of characteristics. The fact that the curves are *parallel* is typical of linear hyperbolic PDEs with constant coefficients.

Recall the conclusion from (2.36) that u remains constant along characteristics. Thus, if u is given the initial value $u(x, 0) = u_0(x)$ at time $t = 0$, then along the whole characteristic curve $x(t) = x_0 + at$ that passes through the initial point x_0 on the x–axis, the solution is

$$u(x, t) = u_0(x_0) = u_0(x - at) . \qquad (2.39)$$

The second equality follows from (2.38). The interpretation of the solution (2.39) of the PDE in (2.32) is this: *given an initial profile $u_0(x)$, the PDE will simply translate this profile with velocity a to the right if $a > 0$ and to the left if $a < 0$*. The shape of the initial profile *remains unchanged*. The model equation in (2.32) under study contains some of the basic features of wave propagation phenomena, where a wave is understood as some recognisable feature of a disturbance that travels at a finite speed.

2.2.2 The Riemann Problem

By using geometric arguments we have constructed the analytical solution of the general IVP (2.32) for the linear advection equation. This is given by (2.39) in terms of the initial data $u_0(x)$. Now we study a special IVP called the Riemann problem

$$
\left.
\begin{array}{ll}
\text{PDE:} & u_t + au_x = 0 . \\[2mm]
\text{IC:} \quad u(x, 0) = u_0(x) = \left\{ \begin{array}{ll} u_L & \text{if } x < 0 , \\ u_R & \text{if } x > 0 , \end{array} \right.
\end{array}
\right\} \qquad (2.40)
$$

where u_L (left) and u_R (right) are two constant values, as shown in Fig. 2.2. Note that the initial data has a discontinuity at $x = 0$. IVP (2.40) is the simplest initial–value problem one can pose. The trivial case would result when $u_L = u_R$. From the previous discussion on the solution of the general

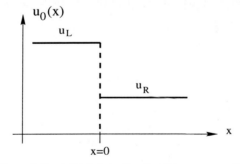

Fig. 2.2. Illustration of the initial data for the Riemann problem. At the initial time the data consists of two constant states separated by a discontinuity at $x = 0$

IVP (2.32) we expect any point on the initial profile to propagate a distance $d = at$ in time t. In particular, we expect the initial discontinuity at $x = 0$

to propagate a distance $d = at$ in time t. This particular characteristic curve $x = at$ will then separate those characteristic curves to the left, on which the solution takes on the value u_L, from those curves to the right, on which the solution takes on the value u_R; see Fig. 2.3. So the solution of the Riemann problem (2.40) is simply

$$u(x,t) = u_0(x - at) = \begin{cases} u_L & \text{if } x - at < 0 , \\ u_R & \text{if } x - at > 0 . \end{cases} \qquad (2.41)$$

Solution (2.41) also follows directly from the general solution (2.39), namely $u(x,t) = u_0(x-at)$. From (2.40), $u_0(x-at) = u_L$ if $x-at < 0$ and $u_0(x-at) = u_R$ if $x - at > 0$. The solution of the Riemann problem can be represented in the x–t plane, as shown in Fig. 2.3. Through any point x_0 on the x–axis one can draw a characteristic. As a is constant these are all parallel to each other. For the solution of the Riemann problem the characteristic that passes through $x = 0$ is significant. This is the only one across which the solution changes.

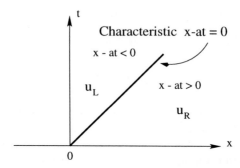

Fig. 2.3. Illustration of the solution of the Riemann problem in the x–t plane for the linear advection equation with positive characteristic speed a

2.3 Linear Hyperbolic Systems

In the previous section we studied in detail the behaviour and the general solution of the simplest PDE of hyperbolic type, namely the linear advection equation with constant wave propagation speed. Here we extend the analysis to sets of m hyperbolic PDEs of the form

$$\mathbf{U}_t + \mathbf{A}\mathbf{U}_x = \mathbf{0} , \qquad (2.42)$$

where the coefficient matrix $\mathbf{A}$ is constant. From the assumption of hyperbolicity $\mathbf{A}$ has m real eigenvalues λ_i and m linearly independent eigenvectors $\mathbf{K}^{(i)}$, $i = 1, \ldots, m$.

2.3.1 Diagonalisation and Characteristic Variables

In order to analyse and solve the general IVP for (2.42) it is found useful to transform the dependent variables $\mathbf{U}(x,t)$ to a new set of dependent variables $\mathbf{W}(x,t)$. To this end we recall the following definition

Definition 2.3.1 (Diagonalisable System). *A matrix* $\mathbf{A}$ *is said to be diagonalisable if* $\mathbf{A}$ *can be expressed as*

$$\mathbf{A} = \mathbf{K}\Lambda\mathbf{K}^{-1} \text{ or } \Lambda = \mathbf{K}^{-1}\mathbf{A}\mathbf{K}, \tag{2.43}$$

in terms of a diagonal matrix Λ *and a matrix* $\mathbf{K}$. *The diagonal elements of* Λ *are the eigenvalues* λ_i *of* $\mathbf{A}$ *and the columns* $\mathbf{K}^{(i)}$ *of* $\mathbf{K}$ *are the right eigenvectors of* $\mathbf{A}$ *corresponding to the eigenvalues* λ_i, *that is*

$$\Lambda = \begin{bmatrix} \lambda_1 & \cdots & 0 \\ 0 & \cdots & 0 \\ \vdots & \vdots & \vdots \\ 0 & \cdots & \lambda_m \end{bmatrix}, \quad \mathbf{K} = [\mathbf{K}^{(1)}, \ldots, \mathbf{K}^{(m)}], \quad \mathbf{A}\mathbf{K}^{(i)} = \lambda_i \mathbf{K}^{(i)}. \tag{2.44}$$

A system (2.42) is said to be *diagonalisable* if the coefficient matrix $\mathbf{A}$ is diagonalisable. Based on the concept of diagonalisation one often defines a hyperbolic system (2.42) as a system with real eigenvalues and diagonalisable coefficient matrix.

Characteristic variables. The existence of the inverse matrix $\mathbf{K}^{-1}$ makes it possible to define a new set of dependent variables $\mathbf{W} = (w_1, w_2, \ldots, w_m)^T$ via the transformation

$$\mathbf{W} = \mathbf{K}^{-1}\mathbf{U} \text{ or } \mathbf{U} = \mathbf{K}\mathbf{W}, \tag{2.45}$$

so that the linear system (2.42), when expressed in terms of $\mathbf{W}$, becomes *completely decoupled*, in a sense to be defined. The new variables $\mathbf{W}$ are called *characteristic variables*. Next we derive the governing PDEs in terms of the characteristic variables, for which we need the partial derivatives $\mathbf{U}_t$ and $\mathbf{U}_x$ in equations (2.42). Since $\mathbf{A}$ is constant, $\mathbf{K}$ is also constant and therefore these derivatives are

$$\mathbf{U}_t = \mathbf{K}\mathbf{W}_t, \quad \mathbf{U}_x = \mathbf{K}\mathbf{W}_x.$$

Direct substitution of these expressions into equation (2.42) gives

$$\mathbf{K}\mathbf{W}_t + \mathbf{A}\mathbf{K}\mathbf{W}_x = \mathbf{0}.$$

Multiplication of this equation from the left by $\mathbf{K}^{-1}$ and use of (2.43) gives

$$\mathbf{W}_t + \Lambda\mathbf{W}_x = \mathbf{0}. \tag{2.46}$$

This is is called the *canonical form* or *characteristic form* of system (2.42). When written in full this system becomes

$$
\begin{bmatrix} w_1 \\ w_2 \\ \vdots \\ w_m \end{bmatrix}_t + \begin{bmatrix} \lambda_1 & \cdots & 0 \\ 0 & \cdots & 0 \\ \vdots & \vdots & \vdots \\ 0 & \cdots & \lambda_m \end{bmatrix} \begin{bmatrix} w_1 \\ w_2 \\ \vdots \\ w_m \end{bmatrix}_x = 0 . \tag{2.47}
$$

Clearly the i–th PDE of this system is

$$
\frac{\partial w_i}{\partial t} + \lambda_i \frac{\partial w_i}{\partial x} = 0 , \; i = 1, \ldots, m \tag{2.48}
$$

and involves the *single unknown* $w_i(x, t)$; the system is therefore *decoupled* and is identical to the linear advection equation in (2.32); now the characteristic speed is λ_i and there are m characteristic curves satisfying m ODEs

$$
\frac{dx}{dt} = \lambda_i , \quad \text{for } i = 1, \ldots, m . \tag{2.49}
$$

2.3.2 The General Initial–Value Problem

We now study the IVP for the PDEs (2.42). The initial condition is now denoted by superscript (0), namely

$$
\mathbf{U}^{(0)} = (u_1^{(0)}, \; \ldots, u_m^{(0)})^T ,
$$

rather than by subscript 0, as done for the scalar case. We find the general solution of the IVP by first solving the corresponding IVP for the canonical system (2.46) or (2.47) in terms of the characteristic variables $\mathbf{W}$ and initial condition $\mathbf{W}^{(0)} = (w_1^{(0)}, \ldots, w_m^{(0)})^T$ such that

$$
\mathbf{W}^{(0)} = \mathbf{K}^{-1} \mathbf{U}^{(0)} \text{ or } \mathbf{U}^{(0)} = \mathbf{K} \mathbf{W}^{(0)} .
$$

The solution of the IVP for (2.46) is direct. By considering each unknown $w_i(x, t)$ satisfying (2.48) and its corresponding initial data $w_i^{(0)}$ we write its solution immediately as

$$
w_i(x, t) = w_i^{(0)}(x - \lambda_i t) , \text{ for } i = 1, \ldots, m . \tag{2.50}
$$

Compare with solution (2.39) for the scalar case. The solution of the general IVP in terms of the original variables $\mathbf{U}$ is now obtained by transforming back according to (2.45), namely $\mathbf{U} = \mathbf{K} \mathbf{W}$.

Example 2.3.1 (Linearised Gas Dynamics Revisited). As a simple example we now study the general IVP for the linearised equations of Gas Dynamics (2.12), namely

$$
\begin{bmatrix} u_1 \\ u_2 \end{bmatrix}_t + \begin{bmatrix} 0 & \rho_0 \\ a^2/\rho_0 & 0 \end{bmatrix} \begin{bmatrix} u_1 \\ u_2 \end{bmatrix}_x = 0 , \; u_1 \equiv \rho , \; u_2 \equiv u ,
$$

with initial condition

$$\left[\begin{array}{c} u_1(x,0) \\ u_2(x,0) \end{array} \right] = \left[\begin{array}{c} u_1^{(0)}(x) \\ u_2^{(0)}(x) \end{array} \right] .$$

We define characteristic variables

$$\mathbf{W} = (w_1, w_2)^T = \mathbf{K}^{-1}\mathbf{U} ,$$

where $\mathbf{K}$ is the matrix of right eigenvectors and $\mathbf{K}^{-1}$ is its inverse, both given by

$$\mathbf{K} = \left[\begin{array}{cc} \rho_0 & \rho_0 \\ -a & a \end{array} \right] , \quad \mathbf{K}^{-1} = \frac{1}{2a\rho_0} \left[\begin{array}{cc} a & -\rho_0 \\ a & \rho_0 \end{array} \right] .$$

Since $\lambda_1 = -a$ and $\lambda_2 = a$, in terms of the characteristic variables we may write

$$\left[\begin{array}{c} w_1 \\ w_2 \end{array} \right]_t + \left[\begin{array}{cc} -a & 0 \\ 0 & a \end{array} \right] \left[\begin{array}{c} w_1 \\ w_2 \end{array} \right]_x = \mathbf{0} ,$$

or in full

$$\frac{\partial w_1}{\partial t} - a\frac{\partial w_1}{\partial x} = 0 , \quad \frac{\partial w_2}{\partial t} + a\frac{\partial w_2}{\partial x} = 0 .$$

The initial condition satisfies

$$\left[\begin{array}{c} w_1^{(0)} \\ w_2^{(0)} \end{array} \right] = \mathbf{K}^{-1} \left[\begin{array}{c} u_1^{(0)} \\ u_2^{(0)} \end{array} \right] ,$$

or in full

$$w_1^{(0)}(x) = \frac{1}{2a\rho_0} \left[au_1^{(0)}(x) - \rho_0 u_2^{(0)}(x) \right] ,$$

$$w_2^{(0)}(x) = \frac{1}{2a\rho_0} \left[au_1^{(0)}(x) + \rho_0 u_2^{(0)}(x) \right] .$$

Each equation involves a single independent variable and is a linear advection equation of the form (2.48). The solution for w_1 and w_2 in terms of their initial data $w_1^{(0)}$, $w_2^{(0)}$, according to (2.50) is

$$w_1(x,t) = w_1^{(0)}(x + at) , \quad w_2(x,t) = w_2^{(0)}(x - at) ,$$

or in full

$$w_1(x,t) = \frac{1}{2a\rho_0} \left[au_1^{(0)}(x + at) - \rho_0 u_2^{(0)}(x + at) \right] ,$$

$$w_2(x,t) = \frac{1}{2a\rho_0} \left[au_1^{(0)}(x - at) + \rho_0 u_2^{(0)}(x - at) \right] .$$

This is the solution in terms of the characteristic variables . In order to obtain the solution to the original problem we transform back using $\mathbf{U} = \mathbf{K}\mathbf{W}$. This gives the final solution as

$$u_1(x,t) = \frac{1}{2a}\left[au_1^{(0)}(x+at) - \rho_0 u_2^{(0)}(x+at)\right]$$
$$+ \frac{1}{2a}\left[au_1^{(0)}(x-at) + \rho_0 u_2^{(0)}(x-at)\right] ,$$

$$u_2(x,t) = -\frac{1}{2\rho_0}\left[au_1^{(0)}(x+at) - \rho_0 u_2^{(0)}(x+at)\right]$$
$$+ \frac{1}{2\rho_0}\left[au_1^{(0)}(x-at) + \rho_0 u_2^{(0)}(x-at)\right] .$$

Exercise 2.3.1. Find the solution of the general IVP for the Small Perturbation Equations (2.24) using the above methodology.

Solution 2.3.1. (Left to the reader).

We return to the expression $\mathbf{U} = \mathbf{KW}$ in (2.45) used to recover the solution to the original problem. When written in full this expression becomes

$$u_1 = w_1 k_1^{(1)} + w_2 k_1^{(2)} + \ldots + w_m k_1^{(m)} ,$$
$$u_i = w_1 k_i^{(1)} + w_2 k_i^{(2)} + \ldots + w_m k_i^{(m)} ,$$
$$u_m = w_1 k_m^{(1)} + w_2 k_m^{(2)} + \ldots + w_m k_m^{(m)} ,$$

or

$$\begin{bmatrix} u_1 \\ u_2 \\ \vdots \\ u_m \end{bmatrix} = w_1 \begin{bmatrix} k_1^{(1)} \\ k_2^{(1)} \\ \vdots \\ k_m^{(1)} \end{bmatrix} + w_2 \begin{bmatrix} k_1^{(2)} \\ k_2^{(2)} \\ \vdots \\ k_m^{(2)} \end{bmatrix} + \ldots + w_m \begin{bmatrix} k_1^{(m)} \\ k_2^{(m)} \\ \vdots \\ k_m^{(m)} \end{bmatrix} , \qquad (2.51)$$

or more succinctly

$$\mathbf{U}(x,t) = \sum_{i=1}^{m} w_i(x,t)\mathbf{K}^{(i)} . \qquad (2.52)$$

This means that the function $w_i(x,t)$ is the coefficient of $\mathbf{K}^{(i)}$ in an *eigenvector expansion* of the vector $\mathbf{U}$. But according to (2.50), $w_i(x,t) = w_i^{(0)}(x-\lambda_i t)$ and hence

$$\mathbf{U}(x,t) = \sum_{i=1}^{m} w_i^{(0)}(x - \lambda_i t)\mathbf{K}^{(i)} . \qquad (2.53)$$

Thus, given a point (x,t) in the x–t plane, the solution $\mathbf{U}(x,t)$ at this point depends only on the initial data at the m points $x_0^{(i)} = x - \lambda_i t$. These are the intersections of the characteristics of speed λ_i with the x-axis. The solution (2.53) for $\mathbf{U}$ can be seen as the superposition of m waves, each of which is advected independently without change in shape. The i-th wave has shape $w_i^{(0)}(x)\mathbf{K}^{(i)}$ and propagates with speed λ_i.

2.3.3 The Riemann Problem

We study the Riemann problem for the hyperbolic, constant coefficient system (2.42). This is the special IVP

$$
\left.
\begin{array}{ll}
\text{PDEs:} & \mathbf{U}_t + \mathbf{A}\mathbf{U}_x = \mathbf{0} \ , \ -\infty < x < \infty \ , \ t > 0 \ , \\[2mm]
\text{IC:} & \mathbf{U}(x,0) = \mathbf{U}^{(0)}(x) = \left\{ \begin{array}{ll} \mathbf{U}_{\mathrm{L}} & x < 0 \ , \\ \mathbf{U}_{\mathrm{R}} & x > 0 \end{array} \right.
\end{array}
\right\}
\tag{2.54}
$$

and is a generalisation of the IVP (2.32). We assume that the system is *strictly hyperbolic* and we order the real and distinct eigenvalues as

$$
\lambda_1 < \lambda_2 < \ldots < \lambda_m \ . \tag{2.55}
$$

The General Solution. The structure of the solution of the Riemann problem (2.54) in the x–t plane is depicted in Fig. 2.4. It consists of m waves emanating from the origin, one for each eigenvalue λ_i. Each wave i carries a *jump discontinuity* in $\mathbf{U}$ propagating with speed λ_i. Naturally, the solution to the left of the λ_1–wave is simply the initial data $\mathbf{U}_{\mathrm{L}}$ and to the right of the λ_m–wave is $\mathbf{U}_{\mathrm{R}}$. The task at hand is to find the solution in the wedge between the λ_1 and λ_m waves. As the eigenvectors $\mathbf{K}^{(1)}, \ldots, \mathbf{K}^{(m)}$ are

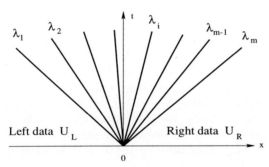

Fig. 2.4. Structure of the solution of the Riemann problem for a general $m \times m$ linear hyperbolic system with constant coefficients

linearly independent, we can expand the data $\mathbf{U}_{\mathrm{L}}$, constant left state, and $\mathbf{U}_{\mathrm{R}}$, constant right state, as linear combinations of the set $\mathbf{K}^{(1)}, \ldots, \mathbf{K}^{(m)}$, that is

$$
\mathbf{U}_{\mathrm{L}} = \sum_{i=1}^{m} \alpha_i \mathbf{K}^{(i)} \ , \quad \mathbf{U}_{\mathrm{R}} = \sum_{i=1}^{m} \beta_i \mathbf{K}^{(i)} \ , \tag{2.56}
$$

with constant coefficients α_i, β_i, for $i = 1, \ldots, m$. Formally, the solution of the IVP (2.54) is given by (2.53) in terms of the initial data $w_i^{(0)}(x)$ for the characteristic variables and the right eigenvectors $\mathbf{K}^{(i)}$. Note that each of the expansions in (2.56) is a special case of (2.53). In terms of the characteristic variables we have m scalar Riemann problems for the PDEs

$$\frac{\partial w_i}{\partial t} + \lambda_i \frac{\partial w_i}{\partial x} = 0 , \tag{2.57}$$

with initial data obtained by comparing (2.56) with (2.53), that is

$$w_i^{(0)}(x) = \begin{cases} \alpha_i & \text{if } x < 0 , \\ \beta_i & \text{if } x > 0 , \end{cases} \tag{2.58}$$

for $i = 1, \ldots, m$. From the previous results, see equation (2.50), we know that the solutions of these scalar Riemann problems are given by

$$w_i(x, t) = w_i^{(0)}(x - \lambda_i t) = \begin{cases} \alpha_i & \text{if } x - \lambda_i t < 0 , \\ \beta_i & \text{if } x - \lambda_i t > 0 . \end{cases} \tag{2.59}$$

For a given point (x, t) there is an eigenvalue λ_I such that $\lambda_I < \frac{x}{t} < \lambda_{I+1}$, that is $x - \lambda_i t > 0 \ \forall i$ such that $i \leq I$. We can thus write the final solution to the Riemann problem (2.54) in terms of the original variables as

$$\mathbf{U}(x, t) = \sum_{i=I+1}^{m} \alpha_i \mathbf{K}^{(i)} + \sum_{i=1}^{I} \beta_i \mathbf{K}^{(i)} , \tag{2.60}$$

where the integer $I = I(x, t)$ is the maximum value of the sub–index i for which $x - \lambda_i t > 0$.

The Solution for a 2 × 2 System. As an example consider the Riemann problem for a general 2 × 2 linear system. From the origin $(0, 0)$ in the (x, t) plane there will be two waves travelling with speeds that are equal to the characteristic speeds λ_1 and λ_2 ($\lambda_1 < \lambda_2$); see Fig. 2.5. The solution to the left of $dx/dt = \lambda_1$ is simply the data state $\mathbf{U}_L = \alpha_1 \mathbf{K}^{(1)} + \alpha_2 \mathbf{K}^{(2)}$ and to the right of $dx/dt = \lambda_2$ the solution is the constant data state $\mathbf{U}_R = \beta_1 \mathbf{K}^{(1)} + \beta_2 \mathbf{K}^{(2)}$. The wedge between the λ_1 and λ_2 waves is usually called

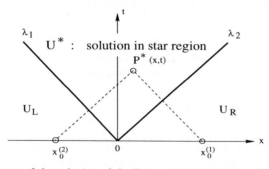

Fig. 2.5. Structure of the solution of the Riemann problem for a 2 × 2 linear system with constant coefficients

the *Star Region* and the solution there is denoted by $\mathbf{U}^*$; its value is *due to the passage of two waves emerging from the origin of the initial discontinuity.* From the point $P^*(x, t)$ we trace back the characteristics with speeds λ_1 and λ_2. These are parallel to those passing through the origin. The characteristics through P^* pass through the *initial points* $x_0^{(2)} = x - \lambda_2 t$ and $x_0^{(1)} = x - \lambda_1 t$. The coefficients in the expansion (2.60) for $\mathbf{U}(x, t)$ are thus determined. The solution at a point P^* has the form (2.60). It is a question of choosing the correct coefficients α_i or β_i. Select a time t^* and a point x_L to the left of the slowest wave so $\mathbf{U}(x_L, t^*) = \mathbf{U}_L$, see Fig. 2.6. The solution at the starting

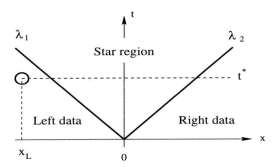

Fig. 2.6. The Riemann problem solution found by travelling along dashed horizontal line $t = t^*$

point (x_L, t^*) is obviously

$$\mathbf{U}_L = \sum_{i=1}^{2} \alpha_i \mathbf{K}_i = \alpha_1 \mathbf{K}^{(1)} + \alpha_2 \mathbf{K}^{(2)} \; ,$$

i.e. all coefficients are α's, that is, the point (x_L, t^*) lies to the left of every wave. As we move to the right of (x_L, t^*) on the horizontal line $t = t^*$ we cross the wave $dx/dt = \lambda_1$, hence $x - \lambda_1 t$ changes from negative to positive, see (2.59), and therefore the coefficient α_1 above changes to β_1. Thus the solution in the entire *Star Region*, between the λ_1 and λ_2 waves, is

$$\mathbf{U}^*(x, t) = \beta_1 \mathbf{K}^{(1)} + \alpha_2 \mathbf{K}^{(2)} \; . \tag{2.61}$$

As we continue moving right and cross the λ_2 wave the value $x - \lambda_2 t$ changes from negative to positive and hence the coefficient α_2 in (2.60) and (2.61) changes to β_2, i.e the solution to the right of the fastest wave of speed λ_2 is, trivially,

$$\mathbf{U}_R = \beta_1 \mathbf{K}^{(1)} + \beta_2 \mathbf{K}^{(2)} \; .$$

Remark 2.3.1. From equation (2.56) it is easy to see that the jump in $\mathbf{U}$ across the whole wave structure in the solution of the Riemann problem is

$$\Delta \mathbf{U} = \mathbf{U}_R - \mathbf{U}_L = (\beta_1 - \alpha_1) \mathbf{K}^{(1)} + \ldots + (\beta_m - \alpha_m) \mathbf{K}^{(m)} \; . \tag{2.62}$$

It is an eigenvector expansion with coefficients that are the strengths of the waves present in the Riemann problem. The wave strength of wave i is $\beta_i - \alpha_i$ and the jump in $\mathbf{U}$ across wave i, denoted by $(\Delta \mathbf{U})_i$, is

$$(\Delta \mathbf{U})_i = (\beta_i - \alpha_i)\mathbf{K}^{(i)} . \tag{2.63}$$

When solving the Riemann problem, sometimes it is more useful to expand the total jump $\Delta \mathbf{U} = \mathbf{U}_R - \mathbf{U}_L$ in terms of the eigenvectors and unknown wave strengths $\delta_i = \beta_i - \alpha_i$.

2.3.4 The Riemann Problem for Linearised Gas Dynamics

As an illustrative example we apply the methodology described in the previous section to solve the Riemann problem for the linearised equations of Gas Dynamics (2.12)

$$\mathbf{U}_t + \mathbf{A}\mathbf{U}_x = 0 ,$$

with

$$\mathbf{U} = \begin{bmatrix} u_1 \\ u_2 \end{bmatrix} \equiv \begin{bmatrix} \rho \\ u \end{bmatrix} , \quad \mathbf{A} = \begin{bmatrix} 0 & \rho_0 \\ a^2/\rho_0 & 0 \end{bmatrix} .$$

The eigenvalues of the system are

$$\lambda_1 = -a , \quad \lambda_2 = +a ,$$

and the corresponding right eigenvectors are

$$\mathbf{K}^{(1)} = \begin{bmatrix} \rho_0 \\ -a \end{bmatrix} , \quad \mathbf{K}^{(2)} = \begin{bmatrix} \rho_0 \\ a \end{bmatrix} .$$

First we decompose the left data state $\mathbf{U}_L = [\rho_L, u_L]^T$ in terms of the right eigenvectors according to equation (2.56), namely

$$\mathbf{U}_L = \begin{bmatrix} \rho_L \\ u_L \end{bmatrix} = \alpha_1 \begin{bmatrix} \rho_0 \\ -a \end{bmatrix} + \alpha_2 \begin{bmatrix} \rho_0 \\ a \end{bmatrix} .$$

Solving for the unknown coefficients α_1 and α_2 we obtain

$$\alpha_1 = \frac{a\rho_L - \rho_0 u_L}{2a\rho_0} , \quad \alpha_2 = \frac{a\rho_L + \rho_0 u_L}{2a\rho_0} .$$

Similarly, by expanding the right–hand data $\mathbf{U}_R = [\rho_R, u_R]^T$ in terms of the eigenvectors and solving for the coefficients β_1 and β_2 we obtain

$$\beta_1 = \frac{a\rho_R - \rho_0 u_R}{2a\rho_0} , \quad \beta_2 = \frac{a\rho_R + \rho_0 u_R}{2a\rho_0} .$$

Now by using equation (2.61) we find the solution in the star region as

$$\mathbf{U}^* = \begin{bmatrix} \rho^* \\ u^* \end{bmatrix} = \beta_1 \begin{bmatrix} \rho_0 \\ -a \end{bmatrix} + \alpha_2 \begin{bmatrix} \rho_0 \\ a \end{bmatrix} .$$

After some algebraic manipulations we obtain the solution explicitly as

$$\left.\begin{array}{l} \rho_* = \frac{1}{2}(\rho_L + \rho_R) - \frac{1}{2}(u_R - u_L)\rho_0/a \ , \\[2mm] u_* = \frac{1}{2}(u_L + u_R) - \frac{1}{2}(\rho_R - \rho_L)a/\rho_0 \ . \end{array}\right\} \tag{2.64}$$

Fig. 2.7 illustrates the solution for $\rho(x,t)$ and $u(x,t)$ at time $t = 1$ for the parameter values $\rho_0 = 1$, $a = 1$ and initial data $\rho_L = 1$, $u_L = 0$, $\rho_R = \frac{1}{2}$ and $u_R = 0$. The two symmetric waves that emerge from the initial position of the discontinuity carry a discontinuous jump in both density ρ and velocity u.

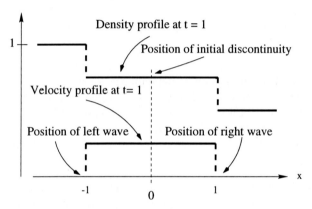

Fig. 2.7. Density and velocity solution profiles at time t=1

Remark 2.3.2. The exact solution (2.64) can be very useful in testing numerical methods for systems with discontinuous solutions.

2.3.5 Some Useful Definitions

Next we recall some standard definitions associated with hyperbolic systems.

Definition 2.3.2 (Domain of Dependence). *Recall that for the linear advection equation the solution at a given point $P = (x^*, t^*)$ depends solely on the initial data at a single point x_0 on the x-axis. This point is obtained by tracing back the characteristic passing through the point $P = (x^*, t^*)$. As a matter of fact, the solution at $P = (x^*, t^*)$ is identical to the value of the initial data $u_0(x)$ at the point x_0. One says that the domain of dependence of the point $P = (x^*, t^*)$ is the point x_0. For a 2×2 system the domain of dependence is an interval $[x_L, x_R]$ on the x-axis that is subtended by the characteristics passing through the point $P = (x^*, t^*)$.*

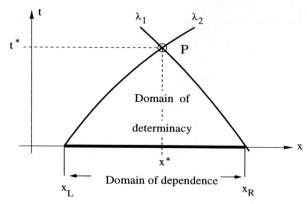

Fig. 2.8. Domain of dependence of point P and corresponding domain of determinacy, for a 2 by 2 system

Fig. 2.8 illustrates the domain of dependence for a 2×2 system with characteristic speeds λ_1 and λ_2, with $\lambda_1 < \lambda_2$. In general, the characteristics of a hyperbolic system are curved. For a larger system the domain of dependence is determined by the slowest and fastest characteristics and is always a bounded interval, as the characteristic speeds for hyperbolic systems are always finite.

Definition 2.3.3 (Domain of Determinacy). *For a given domain of dependence $[x_L, x_R]$, the domain of determinacy is the set of points (x, t), within the domain of existence of the solution $\mathbf{U}(x, t)$, in which $\mathbf{U}(x, t)$ is solely determined by initial data on $[x_L, x_R]$.*

In Fig. 2.8 we illustrate the domain of determinacy of an interval $[x_L, x_R]$ for the case of a 2×2 system with characteristic speeds λ_1 and λ_2, with $\lambda_1 < \lambda_2$.

Definition 2.3.4 (Range of Influence). *Another useful concept is that of the range of influence of a point $Q = (x_0, 0)$ on the x–axis. It is defined as the set of points (x, t) in the x–t plane in which the solution $\mathbf{U}(x, t)$ is influenced by initial data at the point $Q = (x_0, 0)$.*

Fig. 2.9 illustrates the range of influence of a point $Q = (x_0, 0)$ for the case of a 2×2 system with characteristic speeds λ_1 and λ_2, with $\lambda_1 < \lambda_2$.

2.4 Conservation Laws

The purpose of this section is to provide the reader with a succinct presentation of some mathematical properties of hyperbolic conservation laws. We restrict our attention to those properties thought to be essential to the development and application of numerical methods for conservation laws.

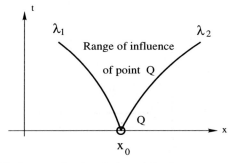

Fig. 2.9. Range of influence of point Q for a 2 by 2 system

In Chap. 1 we applied the physical principles of conservation of mass, momentum and energy to derive time–dependent, multidimensional non–linear systems of conservations laws. In this section we restrict ourselves to simple model problems. In Sect. 2.1 we advanced the formal definition of a system of m conservation laws

$$\mathbf{U}_t + \mathbf{F}(\mathbf{U})_x = \mathbf{0} , \qquad (2.65)$$

where $\mathbf{U}$ is the vector of conserved variables and $\mathbf{F}(\mathbf{U})$ is the vector of fluxes. This system is hyperbolic if the Jacobian matrix

$$\mathbf{A}(\mathbf{U}) = \frac{\partial \mathbf{F}}{\partial \mathbf{U}}$$

has real eigenvalues $\lambda_i(\mathbf{U})$ and a complete set of linearly independent eigenvectors $\mathbf{K}^{(i)}(\mathbf{U})$, $i = 1, \ldots, m$, which we assume to be ordered as

$$\lambda_1(\mathbf{U}) < \lambda_2(\mathbf{U}) <, \ \ldots, \ < \lambda_m(\mathbf{U}) ,$$

$$\mathbf{K}^{(1)}(\mathbf{U}) , \ \mathbf{K}^{(2)}(\mathbf{U}) , \ \ldots , \ \mathbf{K}^{(m)}(\mathbf{U}) .$$

It is important to note that now eigenvalues and eigenvectors depend on $\mathbf{U}$, although sometimes we shall omit the argument $\mathbf{U}$.

2.4.1 Integral Forms of Conservation Laws

As discussed in Scct. 1.5 of Chap. 1, conservation laws may be expressed in differential and integral form. There are two good reasons for considering the integral form (s) of the conservation laws: (i) the derivation of the governing equations is based on physical conservation principles expressed as integral relations on control volumes, (ii) the integral formulation requires less smoothness of the solution, which paves the way to extending the class of admissible solutions to include discontinuous solutions.

The integral form has variants that are worth studying in detail. Consider a one–dimensional time dependent system, such as the Euler equations introduced in Sect. 1.1 of Chap. 1. Choose a control volume $V = [x_L, x_R] \times [t_1, t_2]$

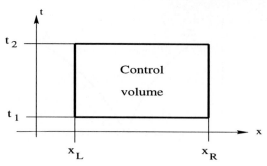

Fig. 2.10. A control volume $V = [x_L, x_R] \times [t_1, t_2]$ on x–t plane

on the x–t plane as shown in Fig. 2.10. The integral form, see Sect. 1.5, of the equation for conservation of mass in one space dimension is

$$\frac{d}{dt} \int_{x_L}^{x_R} \rho(x, t)\, dx = f(x_L, t) - f(x_R, t) \;,$$

where $f = \rho u$ is the flux. For the complete system we have

$$\frac{d}{dt} \int_{x_L}^{x_R} \mathbf{U}(x, t)\, dx = \mathbf{F}(\mathbf{U}(x_L, t)) - \mathbf{F}(\mathbf{U}(x_R, t)) \;, \tag{2.66}$$

where $\mathbf{F}(\mathbf{U})$ is the flux vector. This is one version of the integral form of the conservation laws: *Integral Form I*. The corresponding differential form reads as (2.65). Another version of the integral form of the conservation laws is obtained by integrating (2.66) in time between t_1 and t_2, with $t_1 \le t_2$. See Fig. 2.10. Clearly,

$$\int_{t_1}^{t_2} \left[\frac{d}{dt} \int_{x_L}^{x_R} \mathbf{U}(x, t)\, dx \right] dt = \int_{x_L}^{x_R} \mathbf{U}(x, t_2)\, dx - \int_{x_L}^{x_R} \mathbf{U}(x, t_1)\, dx$$

and thus (2.66) becomes

$$\left. \begin{array}{rl} \int_{x_L}^{x_R} \mathbf{U}(x, t_2)\, dx = & \int_{x_L}^{x_R} \mathbf{U}(x, t_1)\, dx + \int_{t_1}^{t_2} \mathbf{F}(\mathbf{U}(x_L, t))\, dt \\[2mm] & - \int_{t_1}^{t_2} \mathbf{F}(\mathbf{U}(x_R, t))\, dt \;, \end{array} \right\} \tag{2.67}$$

which we call: *Integral Form II* of the conservation laws.

Another version of the integral form of the conservation laws is obtained by integrating (2.65) in any domain V in x–t space and using Green's theorem. The result is

$$\oint [\mathbf{U}\, dx - \mathbf{F}(\mathbf{U})\, dt] = \mathbf{0} \;, \tag{2.68}$$

where the line integration is performed along the boundary of the domain, in an anticlock–wise manner. We call this version *Integral Form III* of the conservation laws. Note that *Integral Form II* of the conservation laws is a special case of *Integral Form III*, in which the control volume V is the rectangle $[x_L, x_R] \times [t_1, t_2]$.

A fourth integral form results from adopting a more mathematical approach for extending the concept of solution of (2.65) to include discontinuities. See Chorin and Marsden [75]. A *weak* or *generalised* solution $\mathbf{U}$ is required to satisfy the integral relation

$$\int_0^{+\infty} \int_{-\infty}^{+\infty} [\phi_t \mathbf{U} + \phi_x \mathbf{F}(\mathbf{U})] \, dx \, dt = - \int_{-\infty}^{+\infty} \phi(x, 0) \mathbf{U}(x, 0) \, dx , \qquad (2.69)$$

for all *test functions* $\phi(x, t)$ that are continuously differentiable and have *compact support*. A function $\phi(x, t)$ has compact support if it vanishes outside some bounded set. Note that in (2.69) the derivatives of $\mathbf{U}(x, t)$ and $\mathbf{F}(\mathbf{U})$ have been passed on to the test function $\phi(x, t)$, which is sufficiently smooth to admit these derivatives.

Remark 2.4.1. The integral forms (2.66)–(2.69) corresponding to (2.65) are valid for any system (2.65), not just for the Euler equations.

Examples of Conservation Laws. *Scalar* conservation laws ($m = 1$) in differential form read

$$u_t + f(u)_x = 0 , \quad f(u) : \text{flux function.} \qquad (2.70)$$

To be able to solve for the conserved variable $u(x, t)$ the flux function $f(u)$ must be a completely determined algebraic function of $u(x, t)$, and possibly some extra parameters of the problem. As seen in Sect. 2.2 the linear advection equation is the simplest example, in which the flux function is $f(u) = au$, a linear function of u.

The **inviscid Burgers's equation** has flux $f(u) = \frac{1}{2}u^2$, a quadratic function of u. Another example of a conservation law is the **traffic flow equation**

$$\rho_t + f(\rho)_x = 0 , \quad f(\rho) = u_m(1 - \frac{\rho}{\rho_m})\rho . \qquad (2.71)$$

Here the conserved variable $\rho(x, t)$ is a density function (density of motor vehicles), u_m and ρ_m are parameters of the problem, namely the maximum speed of vehicles and the maximum density, both positive constants. For details on the traffic flow equation see Whitham [411], Zachmanoglou and Thoe [425], Toro [370] and Haberman [154]. An example of practical interest in oil–reservoir simulation is the **Buckley-Leverett equation**

$$u_t + f(u)_x = 0 , \quad f(u) = \frac{u^2}{u^2 + b(1 - u)^2} , \qquad (2.72)$$

where b is a parameter of the problem. More details of this equation are found in LeVeque [207].

Systems of conservation laws are constructed, as obvious examples, from linear systems

$$\mathbf{U}_t + \mathbf{A}\mathbf{U}_x = 0 \,,$$

with constant coefficient matrix $\mathbf{A}$. The required *conservation–law form* is obtained by defining the flux function as the product of the coefficient matrix $\mathbf{A}$ and the vector $\mathbf{U}$, namely

$$\mathbf{U}_t + \mathbf{F}(\mathbf{U})_x = 0 \,, \quad \mathbf{F}(\mathbf{U}) = \mathbf{A}\mathbf{U} \,. \tag{2.73}$$

Trivially, the Jacobian matrix is $\mathbf{A}$.

Example 2.4.1 (Isothermal Gas Dynamics). The isothermal equations of Gas Dynamics, see Sect. 1.6.2 of Chap. 1, are one example of a non–linear system of conservation laws. These are

$$\left. \mathbf{U}_t + \mathbf{F}(\mathbf{U})_x = 0 \,, \right.$$

$$\left. \mathbf{U} = \left[\begin{array}{c} u_1 \\ u_2 \end{array} \right] \equiv \left[\begin{array}{c} \rho \\ \rho u \end{array} \right] \,, \quad \mathbf{F} = \left[\begin{array}{c} f_1 \\ f_2 \end{array} \right] \equiv \left[\begin{array}{c} \rho u \\ \rho u^2 + a^2 \rho \end{array} \right] \,, \right\} \tag{2.74}$$

where a is positive, constant speed of sound. The Jacobian matrix is found by first expressing $\mathbf{F}$ in terms of the components $u_1 \equiv \rho$ and $u_2 \equiv \rho u$ of the vector $\mathbf{U}$ of conserved variables, namely

$$\mathbf{F}(\mathbf{U}) = \left[\begin{array}{c} f_1 \\ f_2 \end{array} \right] \equiv \left[\begin{array}{c} u_2 \\ u_2^2/u_1 + a^2 u_1 \end{array} \right] \,.$$

According to (2.8) the Jacobian matrix is

$$\mathbf{A}(\mathbf{U}) = \frac{\partial \mathbf{F}}{\partial \mathbf{U}} = \left[\begin{array}{cc} 0 & 1 \\ -(u_2/u_1)^2 + a^2 & 2u_2/u_1 \end{array} \right] = \left[\begin{array}{cc} 0 & 1 \\ a^2 - u^2 & 2u \end{array} \right] \,.$$

It is left to the reader to verify that the eigenvalues of $\mathbf{A}$ are

$$\lambda_1 = u - a \,, \quad \lambda_2 = u + a \tag{2.75}$$

and that the right eigenvectors are

$$\mathbf{K}^{(1)} = \left[\begin{array}{c} 1 \\ u - a \end{array} \right] \,, \quad \mathbf{K}^{(2)} = \left[\begin{array}{c} 1 \\ u + a \end{array} \right] \,, \tag{2.76}$$

where the scaling factors for $\mathbf{K}^{(1)}$ and $\mathbf{K}^{(2)}$ have been taken to be unity. The isothermal equations of Gas Dynamics are thus hyperbolic.

Example 2.4.2 (Isentropic Gas Dynamics). Another non–linear example of a system of conservation laws are the isentropic equations of Gas Dynamics

$$\mathbf{U}_t + \mathbf{F}(\mathbf{U})_x = \mathbf{0} \ ,$$

$$\mathbf{U} = \left[\begin{array}{c} u_1 \\ u_2 \end{array} \right] \equiv \left[\begin{array}{c} \rho \\ \rho u \end{array} \right] \ , \quad \mathbf{F} = \left[\begin{array}{c} f_1 \\ f_2 \end{array} \right] \equiv \left[\begin{array}{c} \rho u \\ \rho u^2 + p \end{array} \right] \ , \right\} \quad (2.77)$$

together with the closure condition, or equation of state (EOS),

$$p = C\rho^\gamma \ , \quad C = \text{constant} \ . \quad (2.78)$$

See Sect. 1.6.2 of Chap. 1.

Exercise 2.4.1. (i) Find the Jacobian matrix, the eigenvalues and the right eigenvectors for the isentropic equations (2.77)–(2.78). (ii) Show that for a generalised isentropic EOS, $p = p(\rho)$, the system is hyperbolic if and only if $p'(\rho) > 0$, that is, the pressure must be a monotone increasing function of ρ. (iii) Show that the sound speed has the general form

$$a = \sqrt{p'(\rho)} \ .$$

Solution 2.4.1. The eigenvalues are

$$\lambda_1 = u - a \ , \quad \lambda_2 = u + a \ , \quad (2.79)$$

and the right eigenvectors are

$$\mathbf{K}^{(1)} = \left[\begin{array}{c} 1 \\ u - a \end{array} \right] \ , \quad \mathbf{K}^{(2)} = \left[\begin{array}{c} 1 \\ u + a \end{array} \right] \ , \quad (2.80)$$

with the sound speed a as claimed.

2.4.2 Non–Linearities and Shock Formation

Here we study some distinguishing features of non–linear hyperbolic conservation laws, such as wave steepening and shock formation. We restrict our attention to the initial–value problem for scalar non–linear conservation laws, namely

$$u_t + f(u)_x = 0 \ , \quad u(x,0) = u_0(x) \ . \quad (2.81)$$

A corresponding integral form of the conservation law is

$$\frac{\mathrm{d}}{\mathrm{d}t} \int_{x_{\mathrm{L}}}^{x_{\mathrm{R}}} u(x,t) \, \mathrm{d}x = f(u(x_{\mathrm{L}},t)) - f(u(x_{\mathrm{R}},t)) \ . \quad (2.82)$$

The flux function f is assumed to be a function of u only, which under certain circumstances is an inadequate representation of the physical problem being modelled. Relevant physical phenomena of our interest are shock waves in compressible media. These have viscous dissipation and heat conduction, in addition to pure advection. A more appropriate flux function for a model conservation law would also include a dependence on u_x, so that the modified conservation law would read

$$u_t + f(u)_x = \alpha u_{xx} \,, \tag{2.83}$$

with α a positive coefficient of viscosity. The conservation law in (2.81) may be rewritten as

$$u_t + \lambda(u)u_x = 0 \,, \tag{2.84}$$

where

$$\lambda(u) = \frac{df}{du} = f'(u) \tag{2.85}$$

is the *characteristic speed*. In the system case this corresponds to the eigenvalues of the Jacobian matrix. For the linear advection equation $\lambda(u) = a$, constant. For the inviscid Burgers equation $\lambda(u) = u$, that is, the characteristic speed depends on the solution and is in fact identical to the conserved variable. For the traffic flow equation $\lambda(u) = u_m(1 - \frac{2\rho}{\rho_m})$.

The behaviour of the flux function $f(u)$ has profound consequences on the behaviour of the solution $u(x,t)$ of the conservation law itself. A crucial property is *monotonicity* of the characteristic speed $\lambda(u)$. There are essentially three possibilities:

– $\lambda(u)$ is a monotone *increasing* function of u, i.e.

$$\frac{d\lambda(u)}{du} = \lambda'(u) = f''(u) > 0 \text{ (convex flux)}$$

– $\lambda(u)$ is a monotone *decreasing* function of u, i.e.

$$\frac{d\lambda(u)}{du} = \lambda'(u) = f''(u) < 0 \text{ (concave flux)}$$

– $\lambda(u)$ has extrema, for some u, i.e.

$$\frac{d\lambda(u)}{du} = \lambda'(u) = f''(u) = 0 \text{ (non–convex, non–concave flux)} \,.$$

In the case of non–linear systems of conservation laws the character of the flux function is determined by the Equation of State. One speaks of convex, or otherwise, equations of state. See the review paper by Menikoff and Plohr [240]. For the inviscid Burgers equation $\lambda'(u) = f''(u) = 1 > 0$, the flux is convex. For the traffic flow equation $\lambda'(u) = f''(u) = -2u_m/\rho_m < 0$, the flux is concave.

Exercise 2.4.2. Analyse the character of the flux function for the Buckley–Leverett equation and show that it is non–convex, non–concave.

Solution 2.4.2. (Left to the reader).

We study the *inviscid* IVP (2.81) and for the moment we assume that the initial data $u(x,0) = u_0(x)$ is smooth. For some finite time the solution $u(x,t)$ will remain smooth. We rewrite the IVP as

$$\left. \begin{array}{l} u_t + \lambda(u)u_x = 0 , \ \lambda(u) = f'(u) , \\[2mm] u(x,0) = u_0(x) . \end{array} \right\} \tag{2.86}$$

Note that the PDE in (2.86) is a non–linear extension of the linear advection equation in (2.32) in which the characteristic speed is $\lambda(u) = a = $ constant. We construct solutions to IVP (2.86) following characteristic curves, in much the same way as performed for the linear advection equation.

Construction of Solutions on Characteristics. Consider characteristic curves $x = x(t)$ satisfying the IVP

$$\frac{dx}{dt} = \lambda(u) , \ x(0) = x_0 . \tag{2.87}$$

Then, by regarding both u and x to be functions of t we find the total derivative of u along the curve $x(t)$, namely

$$\frac{du}{dt} = u_t + \lambda(u)u_x = 0 . \tag{2.88}$$

That is, u is constant along the characteristic curve satisfying the IVP (2.87) and therefore the slope $\lambda(u)$ is constant along the characteristic. Hence the characteristic curves are straight lines. The value of u along each curve is the value of u at the initial point $x(0) = x_0$ and we write

$$u(x,t) = u_0(x_0) . \tag{2.89}$$

Fig. 2.11 shows a typical characteristic curve emanating from the initial point x_0 on the x–axis. The slope $\lambda(u)$ of the characteristic may then be evaluated at x_0 so that the solution characteristics curves of IVP (2.87) are

$$x = x_0 + \lambda(u_0(x_0))t . \tag{2.90}$$

Relations (2.89) and (2.90) may be regarded as the analytical solution of IVP (2.86). Note that the point x_0 depends on the given point (x,t), see Fig. 2. 11, and thus $x_0 = x_0(x,t)$. The solution given by (2.89) and (2.90) is implicit, which is more apparent if we substitute x_0 from (2.90) into (2.89) to obtain

$$u(x,t) = u_0(x - \lambda(u_0(x_0))t) . \tag{2.91}$$

Note that this solution is identical in form to the solution (2.39) of the linear advection equation in (2.32).

Next we verify that relations (2.89) and (2.90) actually define the solution. From (2.89) we obtain the t and x derivatives

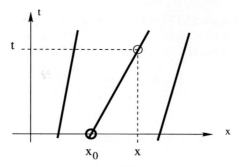

Fig. 2.11. Typical characteristic curves for a non–linear hyperbolic conservation law

$$u_t = u_0'(x_0)\frac{\partial x_0}{\partial t} \ , \quad u_x = u_0'(x_0)\frac{\partial x_0}{\partial x} \ . \tag{2.92}$$

From (2.90) the t and x derivatives are found to be

$$\left.\begin{array}{rl} \lambda(u_0(x_0)) + [1 + \lambda'(u_0(x_0))u_0'(x_0)t]\frac{\partial x_0}{\partial t} &= \ 0 \ , \\[2mm] [1 + \lambda'(u_0(x_0))u_0'(x_0)t]\frac{\partial x_0}{\partial x} &= \ 1 \ . \end{array}\right\} \tag{2.93}$$

From (2.93) we obtain

$$\frac{\partial x_0}{\partial t} = -\frac{\lambda(u_0(x_0))}{1 + \lambda'(u_0(x_0))u_0'(x_0)t} \tag{2.94}$$

and

$$\frac{\partial x_0}{\partial x} = \frac{1}{1 + \lambda'(u_0(x_0))u_0'(x_0)t} \ . \tag{2.95}$$

Substitution of (2.94)–(2.95) into (2.92) verifies that u_t and u_x satisfy the PDE in (2.86).

Wave Steepening. Recall that in the case of the linear advection equation, in which the characteristic speed is $\lambda(u) = a = $ constant, the solution consists of the initial data $u_0(x)$ translated with speed a *without distortion*. In the non–linear case the characteristic speed $\lambda(u)$ is a function of the solution itself. Distortions are therefore produced; this is a distinguishing feature of non–linear problems.

To explain the wave distortion phenomenon we consider initial data $u_0(x)$ as shown in Fig. 2.12. A smooth initial profile is shown in Fig. 2.12a along with five initial points $x_0^{(i)}$ and their corresponding initial data values $u_0^{(i)} = u_0(x_0^{(i)})$. For the moment let us assume that the flux function $f(u)$ is convex, that is $\lambda'(u) = f''(u) > 0$. In this case the characteristic speed is an increasing function of u. Fig. 2.12b shows the characteristics $x^{(i)}(t)$ emanating from the initial points $x_0^{(i)}$ and carrying the constant initial values $u_0^{(i)}$ along them. Given the assumed convex character of the flux, higher values

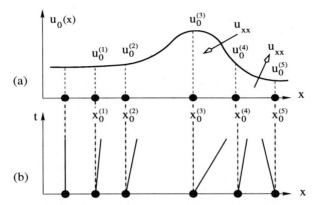

Fig. 2.12. Wave steepening in a convex, non–linear hyperbolic conservation law: (a) initial condition and (b) corresponding picture of characteristics

of $u_0(x)$ will travel faster than lower values of $u_0(x)$. There are two intervals on the x–axis where distortions are most evident. These are the intervals $I_E = [x_0^{(1)}, x_0^{(3)}]$ and $I_C = [x_0^{(3)}, x_0^{(5)}]$. In I_E the value $u_0^{(3)}$ will propagate faster than $u_0^{(2)}$ and this in turn will propagate faster that $u_0^{(1)}$. The orientation of the respective characteristics in Fig. 2.12b makes this situation clear. At a later time the initial data in I_E will have been transformed into a broader and flatter profile. We say that I_E is an *expansive* region. In the expansive region the characteristic speed increases as x increases, that is $\lambda_x > 0$. By contrast the interval I_C is *compressive* and $\lambda_x < 0$; the value $u_0^{(3)}$ will propagate faster than $u_0^{(4)}$ and this in turn will propagate faster that $u_0^{(5)}$, as shown by the orientation of the respective characteristics in Fig. 2.12b. The compressive region will tend to get steeper and narrower as time evolves. The wave steepening mechanism will eventually produce folding over of the solution profile, with corresponding crossing of characteristics, and triple–valued solutions. Note that the compressive and expansive character of the data just described reverses for the case of a concave flux, $\lambda'(u) = f''(u) < 0$. Before crossing of characteristics the single–valued solution may be found following characteristics, as described previously. When characteristics first intersect we say that the wave breaks; the derivative u_x becomes infinite and this happens at a precise breaking time t_b given by

$$t_b = \frac{-1}{\lambda_x(x_0)} . \qquad (2.96)$$

This is confirmed by equations (2.94)–(2.95). Breaking first occurs on the characteristic emanating from $x = x_0$ for which $\lambda_x(x_0)$ is negative and $|\lambda_x(x_0)|$ is a maximum. For details see Whitham [411].

This is an anomalous situation that may be rescued by going back to the physical origins of the equations and questioning the adequacy of the model furnished by (2.81). The improved model equation (2.83) says that the time rate of change of u is not just due to the advection term $f(u)_x$ but is a competing balance between advection and the diffusion term αu_{xx}. As shown in Fig. 2.12a in the interval $[x_0^{(3)}, x_0^{(4)}]$ the *wave steepening* effect of $f(u)_x$ is opposed by the *wave-easing* effect of αu_{xx}, which is negative there. In the interval $[x_0^{(4)}, x_0^{(5)}]$ the role of these contradictory effects is reversed. The more complete description of the physics does not allow folding over of the solution. But rather than working with the more complete, and therefore more complex, viscous description of the problem, it is actually possible to insist on using the inviscid model (2.81) by allowing discontinuities to be formed as a *process of increasing compression*, namely shock waves. Further details are found in Lax [203], Whitham [411] and Smoller [316].

Shock Waves. Shock waves in air are small transition layers of very rapid changes of physical quantities such as pressure, density and temperature. The transition layer for a strong shock is of the same order of magnitude as the mean–free path of the molecules, that is about 10^{-7} m. Therefore replacing these waves as mathematical discontinuities is a reasonable approximation. Very weak shock waves such as sonic booms, are an exception; the discontinuous approximation here can be very inaccurate indeed, see Whitham [411]. For a discussion on shock thickness see Landau and Lifshitz [200], pp. 337–341.

We therefore insist on using the simplified model (2.81) but in its integral form, e.g. (2.82). Consider a solution $u(x,t)$ such that $u(x,t)$, $f(u)$ and their derivatives are continuous everywhere except on a line $s = s(t)$ on the x–t plane across which $u(x,t)$ has a jump discontinuity. Select two fixed points x_L and x_R on the x–axis such that $x_L < s(t) < x_R$. Enforcing the conservation law in integral form (2.82) on the control volume $[x_L, x_R]$ leads to

$$f(u(x_L,t)) - f(u(x_R,t)) = \frac{d}{dt} \int_{x_L}^{s(t)} u(x,t)\, dx + \frac{d}{dt} \int_{s(t)}^{x_R} u(x,t)\, dx \ .$$

Direct use of formula (1.68) of Chap. 1 yields

$$\left. \begin{array}{rl} f(u(x_L,t)) - f(u(x_R,t)) = & [u(s_L,t) - u(s_R,t)]\, S \\[2mm] & + \int_{x_L}^{s(t)} u_t(x,t)\, dx + \int_{s(t)}^{x_R} u_t(x,t)\, dx \ , \end{array} \right\}$$

where $u(s_L,t)$ is the limit of $u(s(t),t)$ as x tends to $s(t)$ from the left, $u(s_R,t)$ is the limit of $u(s(t),t)$ as x tends to $s(t)$ from the right and $S = ds/dt$ is the speed of the discontinuity. As $u_t(x,t)$ is bounded the integrals vanish identically as $s(t)$ is approached from left and right and we obtain

$$f(u(x_L,t)) - f(u(x_R,t)) = [u(s_L,t) - u(s_R,t)]\, S \ . \tag{2.97}$$

This algebraic expression relating the jumps $\Delta f = f(u(x_R,t)) - f(u(x_L,t))$, $\Delta u = u(x_R,t) - u(x_L,t)$ and the speed S of the discontinuity is called the *Rankine–Hugoniot Condition* and is usually expressed as

$$\Delta f = S \Delta u . \tag{2.98}$$

For the scalar case considered here one can solve for the speed S as

$$S = \frac{\Delta f}{\Delta u} . \tag{2.99}$$

Therefore, in order to admit discontinuous solutions we may formulate the problem in terms of PDEs, which are valid in smooth parts of the solution, and the Rankine–Hugoniot Conditions across discontinuities.

Two Examples of Discontinuous Solutions. Consider the following initial–value problem for the inviscid Burgers equation

$$\left. \begin{array}{c} u_t + f(u)_x = 0 , \ f(u) = \frac{1}{2}u^2 , \\[2mm] u(x,0) = u_0(x) = \left\{ \begin{array}{l} u_L \ \text{if } x < 0 , \\ u_R \ \text{if } x > 0 . \end{array} \right. \end{array} \right\} \tag{2.100}$$

First *assume that* $u_L > u_R$. As the flux is convex $\lambda'(u) = f''(u) > 0$ the characteristic speeds on the left are greater than those on the right, that is $\lambda_L \equiv \lambda(u_L) > \lambda_R \equiv \lambda(u_R)$. Based on the discussion about Fig. 2.12 the initial data in IVP (2.100) is the extreme case of compressive data. Crossing of characteristics takes place immediately, as illustrated in Fig. 2.13b. The

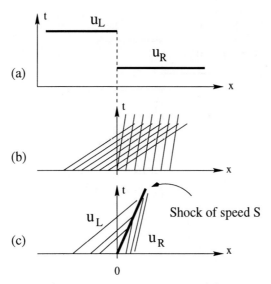

Fig. 2.13. (a) Compressive discontinuous initial data (b) picture of characteristics and (c) solution on x–t plane

discontinuous solution of the IVP is

$$u(x,t) = \begin{cases} u_L & \text{if} \quad x - St < 0 \,, \\ u_R & \text{if} \quad x - St > 0 \,, \end{cases} \tag{2.101}$$

where the speed of the discontinuity is found from (2.99) as

$$S = \frac{1}{2}(u_L + u_R) \,. \tag{2.102}$$

This discontinuous solution is a shock wave and is compressive in nature as discussed previously and as observed in Fig. 2.13a; it satisfies the following condition

$$\lambda(u_L) > S > \lambda(u_R) \,, \tag{2.103}$$

which is called the *entropy condition*. More details are found in Chorin and Marsden [75], LeVeque [207], Smoller [316], Whitham [411].

Now we *assume that* $u_L < u_R$ in the IVP (2.100). This data is the extreme case of *expansive* data, for convex $f(u)$. A possible *mathematical* solution has identical form as solution (2.101)–(2.102) for the *compressive* data case. See Fig. 2.14. However, this solution is physically incorrect. The discontinuity has not arisen as the result of compression, $\lambda_L < \lambda_R$; the characteristics diverge from the discontinuity. This solution is called a *rarefaction shock*, or *entropy–violating shock*, and does not satisfy the entropy condition (2.103); it is therefore rejected as a physical solution. Compare Figs. 2.13 and 2.14; in the compressive case characteristics run into the discontinuity. Given the

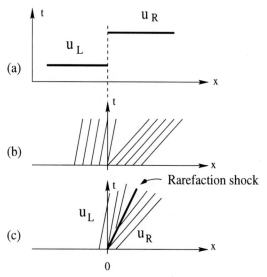

Fig. 2.14. (a) Expansive discontinuous initial data (b) picture of characteristics and (c) rarefaction shock solution on x-t plane

expansive character of the data and based on the discussion on Fig. 2.12, it would be more reasonable to expect the initial data to break up immediately and to broaden with time. This actually gives another solution to be discussed next.

Rarefaction Waves. Reconsider the IVP (2.100) with general convex flux function $f(u)$

$$u_t + f(u)_x = 0 ,$$
$$u(x,0) = u_0(x) = \left\{ \begin{array}{l} u_L \text{ if } x < 0 , \\ u_R \text{ if } x > 0 , \end{array} \right\} \tag{2.104}$$

and *expansive* initial data, $u_L < u_R$. As discussed previously, the entropy–violating solution to this problem is

$$u(x,t) = \left\{ \begin{array}{l} u_L \text{ if } x - St < 0 , \\ u_R \text{ if } x - St > 0 , \end{array} \right.$$
$$S = \frac{\Delta f}{\Delta u} . \left. \begin{array}{l} \\ \\ \\ \end{array} \right\} \tag{2.105}$$

Amongst the various other reasons for rejecting this solution as a physical solution, instability stands out as a prominent argument. By instability it is meant that *small perturbations of the initial data lead to large changes in the solution*. As a matter of fact, under small perturbations, the whole character of the solution changes completely, as we shall see.

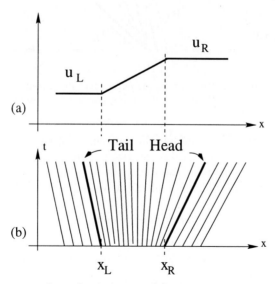

Fig. 2.15. Non–centred rarefaction wave: (a) expansive smooth initial data, (b) picture of characteristics on x–t plane

Let us modify the initial data in (2.104) by replacing the discontinuous change from u_L to u_R by a linear variation of $u_0(x)$ between two fixed points $x_L < 0$ and $x_R > 0$. Now the initial data reads

$$u_0(x) = \begin{cases} u_L & \text{if} & x \leq x_L , \\ u_L + \frac{u_R - u_L}{x_R - x_L}(x - x_L) & \text{if} & x_L < x < x_R , \\ u_R & \text{if} & x \geq x_R , \end{cases} \qquad (2.106)$$

and is illustrated in Fig. 2.15a. The corresponding picture of characteristics emanating from the initial time $t = 0$ is shown in Fig. 2.15b. The solution

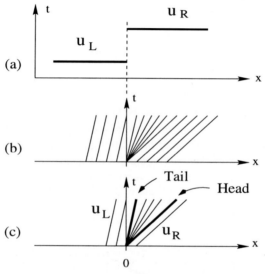

Fig. 2.16. Centred rarefaction wave: (a) expansive discontinuous initial data (b) picture of characteristics (c) entropy satisfying (rarefaction) solution on x–t plane

$u(x, t)$ to this problem is found by following characteristics, as discussed previously, and consists of two constant states, u_L and u_R, separated by a region of *smooth transition* between the data values u_L and u_R. This is called a *rarefaction wave*. The right edge of the wave is given by the characteristic emanating from x_R

$$x = x_R + \lambda(u_R)t \qquad (2.107)$$

and is called the *Head* of the rarefaction. It carries the value $u_0(x_R) = u_R$. The left edge of the wave is given by the characteristic emanating from x_L

$$x = x_L + \lambda(u_L)t \qquad (2.108)$$

and is called the *Tail* of the rarefaction. It carries the value $u_0(x_L) = u_L$.

As we assume convexity, $\lambda'(u) = f''(u) > 0$, larger values of $u_0(x)$ propagate faster than lower values and thus the wave spreads and flattens as time evolves. The spreading of waves is a typical non–linear phenomenon not seen in the study of linear hyperbolic systems with constant coefficients. The entire solution is

$$
\begin{array}{rlll}
u(x,t) & = & u_{\mathrm{L}} & \text{if} \quad \frac{x-x_{\mathrm{L}}}{t} \le \lambda_{\mathrm{L}} \,, \\
\lambda(u) & = & \frac{x-x_{\mathrm{L}}}{t} & \text{if} \quad \lambda_{\mathrm{L}} < \frac{x-x_{\mathrm{L}}}{t} < \lambda_{\mathrm{R}} \,, \\
u(x,t) & = & u_{\mathrm{R}} & \text{if} \quad \frac{x-x_{\mathrm{R}}}{t} \ge \lambda_{\mathrm{R}} \,.
\end{array}
\left.\rule{0pt}{40pt}\right\} \tag{2.109}
$$

No matter how small the size $\Delta x = x_{\mathrm{R}} - x_{\mathrm{L}}$ of the interval over which the discontinuous data in IVP (2.104) has been spread over, the structure of the above rarefaction solution remains unaltered and is entirely different from the rarefaction shock solution (2.105), for which small changes to the data lead to large changes in the solution. Thus the rarefaction shock solution is unstable. From the above construction the rarefaction solution is stable and as x_{L} and x_{R} approach zero from below and above respectively, the discontinuous data at $x = 0$ in IVP (2.104) is reproduced. Therefore, the limiting case is to be interpreted as follows: $u_0(x)$ takes on all the values between u_{L} and u_{R} at $x = 0$ and consequently $\lambda(u_0(x))$ takes on all the values between λ_{L} and λ_{R} at $x = 0$. As higher values propagate faster than lower values the initial data disintegrates immediately giving rise to a rarefaction solution. This limiting rarefaction in which all characteristics of the wave emanate from a single point is called a *centred rarefaction wave*. The solution is

$$
\begin{array}{rlll}
u(x,t) & = & u_{\mathrm{L}} & \text{if} \quad \frac{x}{t} \le \lambda_{\mathrm{L}} \,, \\
\lambda(u) & = & \frac{x}{t} & \text{if} \quad \lambda_{\mathrm{L}} < \frac{x}{t} < \lambda_{\mathrm{R}} \,, \\
u(x,t) & = & u_{\mathrm{R}} & \text{if} \quad \frac{x}{t} \ge \lambda_{\mathrm{R}} \,,
\end{array}
\left.\rule{0pt}{40pt}\right\} \tag{2.110}
$$

and is illustrated in Fig. 2.16.

Now we have at least two solutions to the IVP (2.104). Thus, having extended the concept of solution to include discontinuities, extra spurious solutions are now part of this extended class. The question is how to distinguish between a physically correct solution and a spurious solution. The anticipated answer is that a physical discontinuity, in addition to the Rankine–Hugoniot Condition (2.98), also satisfies the entropy condition (2.103).

The Riemann Problem for the Inviscid Burgers Equation. We finalise this section by giving the solution of the Riemann problem for the inviscid Burgers equation, namely

$$
\begin{array}{ll}
\text{PDE}: & u_t + \left(\frac{u^2}{2}\right)_x = 0 \,, \\
\text{IC}: & u(x,0) = \begin{cases} u_{\mathrm{L}}, & x < 0 \,, \\ u_{\mathrm{R}}, & x > 0 \,. \end{cases}
\end{array}
\left.\rule{0pt}{30pt}\right\} \tag{2.111}
$$

From the previous discussion the exact solution is a *single wave* emanating from the origin as shown in Fig. 2.17a. In view of the entropy condition this wave is either a shock wave, when $u_{\mathrm{L}} > u_{\mathrm{R}}$, or a rarefaction wave, when $u_{\mathrm{L}} \le u_{\mathrm{R}}$. The complete solution is

$$u(x,t) = \left\{ \begin{array}{ll} u_L & \text{if } x - St < 0 \\ u_R & \text{if } x - St > 0 \end{array} \right\} \text{if } u_L > u_R ,$$
$$S = \tfrac{1}{2}(u_L + u_R)$$

$$u(x,t) = \left\{ \begin{array}{ll} u_L & \text{if } \frac{x}{t} \le u_L \\ \frac{x}{t} & \text{if } u_L < x/t < u_R \\ u_R & \text{if } x/t \ge u_R \end{array} \right\} \text{if } u_L \le u_R .$$

$$(2.112)$$

Fig. 2.17 shows the solution of the Riemann problem for the inviscid Burgers equation. Fig. 2.17a depicts the structure of the general solution and consists of a single wave, Fig. 2.17b shows the case in which the solution is a shock wave and Fig. 2.17c shows the case in which it is a rarefaction wave.

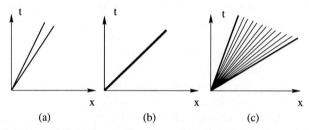

(a)　　　　　　　　(b)　　　　　　　　(c)

Fig. 2.17. Solution of the Riemann problem for the inviscid Burgers equation: (a) structure of general solution (single wave, shock or rarefaction), (b) solution is a shock wave and (c) solution is a rarefaction wave

Some of the studied notions for scalar conservations laws extend quite directly to systems of hyperbolic conservations laws, as we see in the next section.

2.4.3 Characteristic Fields

Consider a hyperbolic system of conservation laws of the form (2.65) with real eigenvalues $\lambda_i(\mathbf{U})$ and corresponding right eigenvectors $\mathbf{K}^{(i)}(\mathbf{U})$. The characteristic speed $\lambda_i(\mathbf{U})$ defines a *characteristic field*, the λ_i–field. Sometimes one also speaks of the $\mathbf{K}^{(i)}$–field, that is the characteristic field defined by the eigenvector $\mathbf{K}^{(i)}$.

Definition 2.4.1 (Linearly degenerate fields). *A λ_i–characteristic field is said to be linearly degenerate if*

$$\nabla \lambda_i(\mathbf{U}) \cdot \mathbf{K}^{(i)}(\mathbf{U}) = 0 , \quad \forall \mathbf{U} \in \Re^m , \qquad (2.113)$$

where $\Re^m$ is the set of real–valued vectors of m components.

Definition 2.4.2 (Genuinely nonlinear fields). *A λ_i–characteristic field is said to be genuinely nonlinear if*

$$\nabla \lambda_i(\mathbf{U}) \cdot \mathbf{K}^{(i)}(\mathbf{U}) \ne 0 , \quad \forall \mathbf{U} \in \Re^m . \qquad (2.114)$$

The symbol '·' denotes the dot product in *phase space*. $\nabla\lambda_i(\mathbf{U})$ is the gradient of the eigenvalue $\lambda_i(\mathbf{U})$, namely

$$\nabla\lambda_i(\mathbf{U}) = \left(\frac{\partial}{\partial u_1}\lambda_i, \frac{\partial}{\partial u_2}\lambda_i, \ldots, \frac{\partial}{\partial u_m}\lambda_i\right)^{\mathrm{T}}.$$

The *phase space* is the space of vectors $\mathbf{U} = (u_1, \ldots, u_m)$; for a 2×2 system we speak of the *phase plane* u_1–u_2. Note that for a linear system (2.42) the eigenvalues λ_i are constant and therefore $\nabla\lambda_i(\mathbf{U}) = \mathbf{0}$. Hence all characteristic fields of a linear hyperbolic system with constant coefficients are linearly degenerate.

Exercise 2.4.3. Show that both characteristic fields of the isothermal equations of Gas Dynamics (2.74) are genuinely non–linear.

Solution 2.4.3. First we write the eigenvalues (2.75) in terms of the conserved variables, namely

$$\lambda_1 = \frac{u_2}{u_1} - a, \; \lambda_2 = \frac{u_2}{u_1} + a,$$

$$\nabla\lambda_1(\mathbf{U}) = \left(-\frac{u}{\rho}, \frac{1}{\rho}\right)^{\mathrm{T}}, \; \nabla\lambda_2(\mathbf{U}) = \left(-\frac{u}{\rho}, \frac{1}{\rho}\right)^{\mathrm{T}}.$$

Therefore

$$\nabla\lambda_1(\mathbf{U}) \cdot \mathbf{K}^{(1)}(\mathbf{U}) = -\frac{a}{\rho} \neq 0,$$

$$\nabla\lambda_2(\mathbf{U}) \cdot \mathbf{K}^{(2)}(\mathbf{U}) = \frac{a}{\rho} \neq 0$$

and thus both characteristic fields are *genuinely non–linear*, as claimed.

Example 2.4.3 (Detonation Analogue). In the study of detonation waves in high energy solids it is found useful to devise mathematical objects that preserve some of the basic physical features of detonation phenomena but are simpler to analyse than more comprehensive models. Fickett [124] proposed a system that is essentially the inviscid Burgers equation plus a reaction model. He called the system *detonation analogue*. Clarke and colleagues [80] pointed out that this analogue is also exceedingly useful for numerical purposes. Writing the system in *conservation–law form* one has the *inhomogeneous system* with a source term, namely

$$\mathbf{U}_t + \mathbf{F}(\mathbf{U})_x = \mathbf{S}(\mathbf{U}), \tag{2.115}$$

$$\mathbf{U} = \begin{bmatrix} u_1 \\ u_2 \end{bmatrix} \equiv \begin{bmatrix} \rho \\ \alpha \end{bmatrix}, \; \mathbf{F} = \begin{bmatrix} \frac{1}{2}(\rho^2 + \alpha Q) \\ 0 \end{bmatrix}, \; \mathbf{S} = \begin{bmatrix} 0 \\ 2\sqrt{1-\alpha} \end{bmatrix}. \tag{2.116}$$

The parameter Q plays the role of *heats of reaction* and α is a *reaction progress variable*. The mathematical character of the system is determined solely by the homogeneous part, $\mathbf{S} = \mathbf{0}$. The Jacobian matrix is

$$\mathbf{A}(\mathbf{U}) = \frac{\partial \mathbf{F}}{\partial \mathbf{U}} = \begin{bmatrix} u_1 & \frac{1}{2}Q \\ 0 & 0 \end{bmatrix} = \begin{bmatrix} \rho & \frac{1}{2}Q \\ 0 & 0 \end{bmatrix} .$$

Simple calculations show that the eigenvalues are

$$\lambda_1 = 0 , \; \lambda_2 = \rho \tag{2.117}$$

and the right eigenvectors are

$$\mathbf{K}^{(1)} = \begin{bmatrix} 1 \\ -2\rho/Q \end{bmatrix} , \; \mathbf{K}^{(2)} = \begin{bmatrix} 1 \\ 0 \end{bmatrix} . \tag{2.118}$$

The detonation analogue is therefore hyperbolic.

Exercise 2.4.4. Check that the λ_1–field is linearly degenerate and that the λ_2–field is genuinely non–linear.

Solution 2.4.4. (Left to the reader).

Rankine-Hugoniot Conditions. Given a system of hyperbolic conservation laws

$$\mathbf{U}_t + \mathbf{F}(\mathbf{U})_x = 0 \tag{2.119}$$

and a discontinuous wave solution of speed S_i associated with the λ_i–characteristic field, the Rankine–Hugoniot Conditions state

$$\Delta\mathbf{F} = S_i\Delta\mathbf{U} , \tag{2.120}$$

with

$$\Delta\mathbf{U} \equiv \mathbf{U}_R - \mathbf{U}_L , \; \Delta\mathbf{F} \equiv \mathbf{F}_R - \mathbf{F}_L , \; \mathbf{F}_L = \mathbf{F}(\mathbf{U}_L) , \; \mathbf{F}_R = \mathbf{F}(\mathbf{U}_R) ,$$

where $\mathbf{U}_L$ and $\mathbf{U}_R$ are the respective states immediately to the left and right of the discontinuity. Fig. 2.18 illustrates the Rankine–Hugoniot Conditions. Note that unlike the scalar case, see (2.99), it is generally not possible to

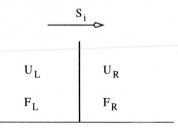

Fig. 2.18. Illustration of the Rankine–Hugoniot Conditions for a single discontinuity of speed S_i connecting two constant states $\mathbf{U}_L$ and $\mathbf{U}_R$ via a system of conservation laws

solve for the speed S_i. For a linear system with constant coefficients

$$\mathbf{U}_t + \mathbf{A}\mathbf{U}_x = 0 \, ,$$

with eigenvalues λ_i, for $i = 1, \ldots, m$, the Rankine–Hugoniot Conditions across the wave of speed $S_i \equiv \lambda_i$ read

$$\Delta\mathbf{F} = \mathbf{A}\Delta\mathbf{U} = \lambda_i(\Delta\mathbf{U})_i \, . \tag{2.121}$$

See (2.63). Actually, these conditions provide a technique for finding the solution of the Riemann problem for linear hyperbolic system with constant coefficients.

Exercise 2.4.5. Solve the Riemann problem for the linearised equations of Gas Dynamics using the Rankine–Hugoniot Conditions across each wave.

Solution 2.4.5. The structure of the solution is depicted in Fig. 2.5. The unknowns are ρ_* and u_* in the *Star Region*. Recall that the vector $\mathbf{U}$ and the coefficient matrix $\mathbf{A}$ are given by

$$\mathbf{U} = \begin{bmatrix} u_1 \\ u_2 \end{bmatrix} \equiv \begin{bmatrix} \rho \\ u \end{bmatrix} \, , \quad \mathbf{A} = \begin{bmatrix} 0 & \rho_0 \\ a^2/\rho_0 & 0 \end{bmatrix}$$

and the eigenvalues are

$$\lambda_1 = -a \, , \quad \lambda_2 = +a \, .$$

Application of the Rankine–Hugoniot Conditions across the λ_1–wave of speed $S_1 = \lambda_1$ gives

$$\begin{bmatrix} 0 & \rho_0 \\ \frac{a^2}{\rho_0} & 0 \end{bmatrix} \begin{bmatrix} \rho_* - \rho_L \\ u_* - u_L \end{bmatrix} = -a \begin{bmatrix} \rho_* - \rho_L \\ u_* - u_L \end{bmatrix} \, .$$

Expanding and solving for u_* gives

$$u_* = u_L - (\rho_* - \rho_L)\frac{a}{\rho_0} \, .$$

For the λ_2–wave of speed $S_2 = \lambda_2$ we obtain

$$u_* = u_R + (\rho_* - \rho_R)\frac{a}{\rho_0} \, .$$

The simultaneous solution of these two linear algebraic equations for the unknowns ρ_* and u_* is

$$\rho_* = \tfrac{1}{2}(\rho_L + \rho_R) - \tfrac{1}{2}(u_R - u_L)\rho_0/a \, ,$$

$$u_* = \tfrac{1}{2}(u_L + u_R) - \tfrac{1}{2}(\rho_R - \rho_L)a/\rho_0 \, ,$$

which is the solution (2.64) obtained using a different technique based on eigenvector expansion of the initial data. The technique that makes use of the Rankine–Hugoniot conditions is more direct.

Generalised Riemann Invariants. For a general quasi–linear hyperbolic system

$$\mathbf{W}_t + \mathbf{A(W)W}_x = \mathbf{0} \ , \tag{2.122}$$

with

$$\mathbf{W} = [w_1, w_2, \cdots w_m]^T \ ,$$

we consider the wave associated with the i–characteristic field with eigenvalue λ_i and corresponding right eigenvector

$$\mathbf{K}^{(i)} = \left[k_1^{(i)}, k_2^{(i)}, \cdots k_m^{(i)} \right]^T \ .$$

The vector of dependent variables $\mathbf{W}$ here is some suitable set, which may be the set conserved variables, for instance. Recall that any system of conservation laws may always be expressed in quasi–linear form via the Jacobian matrix, see (2.6) and (2.8).

The *Generalised Riemann Invariants* are relations that hold true, for certain waves, *across* the wave structure and lead the following $(m-1)$ ordinary differential equations

$$\frac{dw_1}{k_1^{(i)}} = \frac{dw_2}{k_2^{(i)}} = \frac{dw_3}{k_3^{(i)}} = \cdots = \frac{dw_m}{k_m^{(i)}} \ . \tag{2.123}$$

They relate ratios of changes dw_s of a quantity w_s to the respective component $k_s^{(i)}$ of the right eigenvector $\mathbf{K}^{(i)}$ corresponding to a λ_i–wave family . For a detailed discusssion see the book by Jeffrey [183].

Example 2.4.4 (Linearised Gas Dynamics revisited). Here we find the Generalised Riemann Invariants for the linearised equations of Gas Dynamics. The dependent variables are

$$\mathbf{W} = \begin{bmatrix} w_1 \\ w_2 \end{bmatrix} \equiv \begin{bmatrix} \rho \\ u \end{bmatrix}$$

and the right eigenvectors are

$$\mathbf{K}^{(1)} = \begin{bmatrix} \rho_0 \\ -a \end{bmatrix} \ , \ \mathbf{K}^{(2)} = \begin{bmatrix} \rho_0 \\ a \end{bmatrix} \ .$$

Across the λ_1–wave we have

$$\frac{d\rho}{\rho_0} = \frac{du}{-a} \ ,$$

which leads to

$$du + \frac{a}{\rho_0} d\rho = 0 \ .$$

After integration this produces

$$I_L(\rho, u) = u + \frac{a}{\rho_0}\rho = \text{constant} . \qquad (2.124)$$

The constant of integration is obtained by evaluating $I_L(\rho, u)$ at a reference state. Across the λ_2–wave we have

$$\frac{d\rho}{\rho_0} = \frac{du}{a} ,$$

which leads to

$$I_R(\rho, u) = u - \frac{a}{\rho_0}\rho = \text{constant} . \qquad (2.125)$$

Again the constant of integration is obtained by evaluating $I_R(\rho, u)$ at a reference state.

Exercise 2.4.6. Solve the Riemann problem for the linearised equations of Gas Dynamics using the Generalised Riemann Invariants.

Solution 2.4.6. Application of $I_L(\rho, u)$ across the left wave connecting the states $\mathbf{W}_L$ and $\mathbf{W}_*$ gives

$$u_* + \frac{a}{\rho_0}\rho_* = u_L + \frac{a}{\rho_0}\rho_L .$$

Similarly, application of $I_R(\rho, u)$ across the right wave connecting the states $\mathbf{W}_R$ and $\mathbf{W}_*$ gives

$$u_* - \frac{a}{\rho_0}\rho_* = u_R - \frac{a}{\rho_0}\rho_R$$

and the simultaneous solution for the unknowns ρ_* and u_* gives

$$\left.\begin{array}{l} \rho_* = \frac{1}{2}(\rho_L + \rho_R) - \frac{1}{2}(u_R - u_L)\rho_0/a , \\[2mm] u_* = \frac{1}{2}(u_L + u_R) - \frac{1}{2}(\rho_R - \rho_L)a/\rho_0 , \end{array}\right\}$$

which is the same solution (2.64) obtained from applying other techniques.

Exercise 2.4.7. Solve the Riemann problem for the Small Perturbation Equations (2.24) using the following techniques:

- by expanding the initial data $\mathbf{U}_L$ and $\mathbf{U}_R$ in terms of the eigenvectors, see (2.56).
- by expanding the total jump $\Delta\mathbf{U}$ in terms of the eigenvectors, see (2.62).
- by using the Rankine-Hugoniot Conditions across each wave, see (2.121).
- by applying the Generalised Riemann Invariants, see (2.123).

Solution 2.4.7. Use of any of the suggested techniques will give the general solution

$$\left.\begin{array}{l} u_* = \frac{1}{2}(u_L + u_R) + \frac{1}{2}a(v_R - v_L) , \\[2mm] v_* = \frac{1}{2}(v_L + v_R) + \frac{1}{2a}(u_R - u_L) . \end{array}\right\}$$

Example 2.4.5 (Isentropic Gas Dynamics Revisited). For this example the eigenvalues are $\lambda_1 = u - a$ and $\lambda_2 = u + a$, with $a = \sqrt{p'(\rho)} = \sqrt{\frac{\gamma p}{\rho}}$ defining the sound speed. The corresponding right eigenvectors are given by

$$\mathbf{K}^{(1)} = \begin{bmatrix} 1 \\ u - a \end{bmatrix}, \quad \mathbf{K}^{(2)} = \begin{bmatrix} 1 \\ u + a \end{bmatrix}.$$

Across the left $\lambda_1 = u - a$ wave we have

$$\frac{d\rho}{1} = \frac{d(\rho u)}{u - a},$$

which after expanding differentials yields

$$du + \frac{a}{\rho} d\rho = 0.$$

On exact integration we obtain

$$I_L(\rho, \rho u) = u + \int \frac{a}{\rho} \, d\rho = \text{constant}. \tag{2.126}$$

Across the right $\lambda_2 = u + a$ wave we obtain

$$I_R(\rho, \rho u) = u - \int \frac{a}{\rho} \, d\rho = \text{constant}. \tag{2.127}$$

As as the sound speed a is a function of ρ alone we can evaluate the integral term above exactly as

$$\int \frac{a}{\rho} \, d\rho = \frac{2a}{\gamma - 1}$$

by first noting that

$$a = \sqrt{p'(\rho)} = \sqrt{C\gamma \rho^{\gamma - 1}} = \sqrt{C\gamma} \rho^{\frac{\gamma - 1}{2}}.$$

Then the *left* and *right* Riemann Invariants become

$$\left. \begin{array}{l} I_L(\rho, \rho u) = u + \frac{2a}{\gamma - 1} = \text{constant across the } \lambda_1\text{-wave}, \\[2ex] I_R(\rho, \rho u) = u - \frac{2a}{\gamma - 1} = \text{constant across the } \lambda_2\text{-wave}. \end{array} \right\} \tag{2.128}$$

Generalised Riemann Invariants provide a powerful tool of analysis of hyperbolic conservation laws.

2.4.4 Elementary–Wave Solutions of the Riemann Problem

The Riemann problem for a general $m \times m$ non–linear hyperbolic system with data $\mathbf{U}_L$, $\mathbf{U}_R$ is the IVP

$$\mathbf{U}_t + \mathbf{F}(\mathbf{U})_x = \mathbf{0} \ ,$$
$$\mathbf{U}(x,0) = \mathbf{U}^{(0)}(x) = \left\{ \begin{array}{lll} \mathbf{U}_L & \text{if} & x < 0 \ , \\ \mathbf{U}_R & \text{if} & x > 0 \ . \end{array} \right\} \tag{2.129}$$

The similarity solution $\mathbf{U}(x/t)$ of (2.129) consists of $m+1$ constant states separated by m waves, as depicted by the x–t picture of Fig. 2.19. For each eigenvalue λ_i there is a wave family. For linear systems with constant coefficients each wave is a discontinuity of speed $S_i = \lambda_i$ and defines a linearly degenerate field.

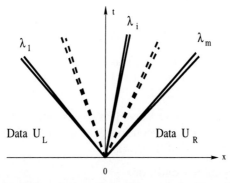

Fig. 2.19. Structure of the solution of the Riemann problem for a system of non–linear conservation laws

For non–linear systems the waves may be discontinuities such as shock waves and contact waves, or smooth transition waves such as rarefactions. The possible types of waves present in the solution of the Riemann problem depends crucially on closure conditions. For the Euler equations we shall only consider Equations of State such that the only waves present are shocks, contacts and rarefactions. Suppose that the initial data states $\mathbf{U}_L$, $\mathbf{U}_R$ are connected by a single wave, that is, the solution of the Riemann problem consists of a single non–trivial wave; all other waves have zero strength. This assumption is entirely justified as we can always solve the Riemann problem with general data and then select the constant states on either side of a particular wave as the initial data for the Riemann problem. If the wave is a discontinuity then the wave is a shock wave or a contact wave.

Shock Wave. For a shock wave the two constant states $\mathbf{U}_L$ and $\mathbf{U}_R$ are connected through a single jump discontinuity in a *genuinely non–linear field* i and the following conditions apply

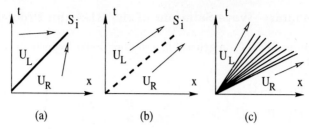

Fig. 2.20. Elementary wave solutions of the Riemann problem: (a) shock wave of speed S_i, (b) contact discontinuity of speed S_i and (c) rarefaction wave

– the Rankine–Hugoniot Conditions

$$\mathbf{F}(\mathbf{U_R}) - \mathbf{F}(\mathbf{U_L}) = S_i(\mathbf{U_R} - \mathbf{U_L}) \ . \tag{2.130}$$

– the entropy condition

$$\lambda_i(\mathbf{U_L}) > S_i > \lambda_i(\mathbf{U_R}) \ . \tag{2.131}$$

Fig. 2.20a depicts a shock wave of speed S_i. The characteristic $dx/dt = \lambda_i$ on both sides of the wave run into the shock wave, which illustrates the compressive character of a shock.

Contact Wave. For a *contact wave* the two data states $\mathbf{U_L}$ and $\mathbf{U_R}$ are connected through a single jump discontinuity of speed S_i in a *linearly degenerate field i* and the following conditions apply

– the Rankine–Hugoniot Conditions

$$\mathbf{F}(\mathbf{U_R}) - \mathbf{F}(\mathbf{U_L}) = S_i(\mathbf{U_R} - \mathbf{U_L}) \ . \tag{2.132}$$

– constancy of the Generalised Riemann Invariants across the wave

$$\frac{dw_1}{k_1^{(i)}} = \frac{dw_2}{k_2^{(i)}} = \frac{dw_3}{k_3^{(i)}} = \cdots = \frac{dw_m}{k_m^{(i)}} \ . \tag{2.133}$$

– the parallel characteristic condition

$$\lambda_i(\mathbf{U_L}) = \lambda_i(\mathbf{U_R}) = S_i \ . \tag{2.134}$$

Fig. 2.20b depicts a contact discontinuity. Characteristics on both sides of the wave run parallel to it.

Rarefaction Wave. For a rarefaction wave the two data states $\mathbf{U}_L$ and $\mathbf{U}_R$ are connected through a *smooth transition* in a *genuinely non–linear field i* and the following conditions are met

– constancy of the Generalised Riemann Invariants across the wave

$$\frac{dw_1}{k_1^{(i)}} = \frac{dw_2}{k_2^{(i)}} = \frac{dw_3}{k_3^{(i)}} = \cdots = \frac{dw_m}{k_m^{(i)}} \ . \tag{2.135}$$

– divergence of characteristics

$$\lambda_i(\mathbf{U}_L) < \lambda_i(\mathbf{U}_R) \ . \tag{2.136}$$

Fig. 2.20c depicts a rarefaction wave. Characteristics on the left and right of the wave diverge as do characteristics inside the wave.

Remark 2.4.2. The solution of the general Riemann problem contains m waves of any of the above type, namely: shock waves, contact discontinuities and rarefaction waves. In solving the general Riemann problem we shall enforce these conditions by discriminating the particular type of wave present.

For further study we recommend the following references: Lax [203], Whitham [411], Chorin and Marsden [75], Courant and Friedrichs [97], Smoller [316] and LeVeque [207]. See also the papers by Lax [204] and [202].

The introductory concepts of this chapter will we used to analyse some of the properties of the Euler equations in Chap. 3. For the time–dependent one dimensional Euler equations we solve the Riemann problem exactly in Chap. 4, while in Chaps. 9 to 12 we present approximate Riemann solvers.

3. Some Properties of the Euler Equations

In this chapter we apply the mathematical tools presented in Chap. 2 to analyse some of the basic properties of the time–dependent Euler equations. As seen in Chap. 1, the Euler equations result from neglecting the effects of viscosity, heat conduction and body forces on a compressible medium. Here we show that these equations are a system of *hyperbolic* conservations laws and study some of their mathematical properties. In particular, we study those properties that are essential for finding the solution of the Riemann problem in Chap. 4. We analyse the eigenstructure of the equations, that is, we find eigenvalues and eigenvectors; we study properties of the characteristic fields and establish basic relations across rarefactions, contacts and shock waves. It is worth remarking that the process of finding eigenvalues and eigenvectors usually involves a fair amount of algebra as well as some familiarity with basic physical quantities and their relations. For very complex systems of equations finding eigenvalues and eigenvectors may require the use of symbolic manipulators. Useful background reading for this chapter is found in Chaps. 1 and 2.

3.1 The One–Dimensional Euler Equations

Here we study the one–dimensional time–dependent Euler equations with an ideal Equation of State, using conservative and non–conservative formulations. The basic structure of the solution of the Riemann problem is outlined along with a detailed study of the elementary waves present in the solution. We provide the foundations for finding the exact solution of the Riemann problem in Chap. 4.

3.1.1 Conservative Formulation

The conservative formulation of the Euler equations, in differential form, is

$$\mathbf{U}_t + \mathbf{F}(\mathbf{U})_x = \mathbf{0} , \tag{3.1}$$

where $\mathbf{U}$ and $\mathbf{F}(\mathbf{U})$ are the vectors of conserved variables and fluxes, given respectively by

$$\mathbf{U} = \begin{bmatrix} u_1 \\ u_2 \\ u_3 \end{bmatrix} \equiv \begin{bmatrix} \rho \\ \rho u \\ E \end{bmatrix} , \quad \mathbf{F} = \begin{bmatrix} f_1 \\ f_2 \\ f_3 \end{bmatrix} \equiv \begin{bmatrix} \rho u \\ \rho u^2 + p \\ u(E + p) \end{bmatrix} . \tag{3.2}$$

Here ρ is density, p is pressure, u is particle velocity and E is total energy per unit volume

$$E = \rho(\frac{1}{2}u^2 + e) , \tag{3.3}$$

where e is the *specific internal energy* given by a caloric Equation of State (EOS)

$$e = e(\rho, p) . \tag{3.4}$$

For ideal gases one has the simple expression

$$e = e(\rho, p) = \frac{p}{(\gamma - 1)\rho} , \tag{3.5}$$

with $\gamma = c_p/c_v$ denoting the *ratio of specific heats*. From the EOS (3.5) and using equation (1.36) of Chap. 1 we write the *sound speed* a as

$$a = \sqrt{(p/\rho^2 - e_\rho)/e_p} = \sqrt{\frac{\gamma p}{\rho}} . \tag{3.6}$$

The conservation laws (3.1)–(3.2) may also be written in quasi–linear form

$$\mathbf{U}_t + \mathbf{A}(\mathbf{U})\mathbf{U}_x = \mathbf{0} , \tag{3.7}$$

where the coefficient matrix $\mathbf{A}(\mathbf{U})$ is the *Jacobian matrix*

$$\mathbf{A}(\mathbf{U}) = \frac{\partial \mathbf{F}}{\partial \mathbf{U}} = \begin{bmatrix} \partial f_1/\partial u_1 & \partial f_1/\partial u_2 & \partial f_1/\partial u_3 \\ \partial f_2/\partial u_1 & \partial f_2/\partial u_2 & \partial f_2/\partial u_3 \\ \partial f_3/\partial u_1 & \partial f_3/\partial u_2 & \partial f_3/\partial u_3 \end{bmatrix} .$$

Proposition 3.1.1 (Jacobian Matrix). *The Jacobian matrix* $\mathbf{A}$ *is*

$$\mathbf{A}(\mathbf{U}) = \begin{bmatrix} 0 & 1 & 0 \\ -\frac{1}{2}(\gamma - 3)(\frac{u_2}{u_1})^2 & (3 - \gamma)(\frac{u_2}{u_1}) & \gamma - 1 \\ -\frac{\gamma u_2 u_3}{u_1^2} + (\gamma - 1)(\frac{u_2}{u_1})^3 & \frac{\gamma u_3}{u_1} - \frac{3}{2}(\gamma - 1)(\frac{u_2}{u_1})^2 & \gamma(\frac{u_2}{u_1}) \end{bmatrix} .$$

Proof. First we express all components f_i of the flux vector $\mathbf{F}$ in terms of the components u_i of the vector $\mathbf{U}$ of conserved variables, namely $u_1 \equiv \rho$, $u_2 \equiv \rho u$, $u_3 \equiv E$. Obviously $f_1 = u_2 \equiv \rho u$. To find f_2 and f_3 we first need to express the pressure p in terms of the conserved variables. From (3.3) and (3.5) we find

$$p = (\gamma - 1)[u_3 - \frac{1}{2}(u_2^2/u_1)] .$$

Thus the flux vector can be written as

$$\mathbf{F}(\mathbf{U}) = \begin{bmatrix} f_1 \\ f_2 \\ f_3 \end{bmatrix} \equiv \begin{bmatrix} u_2 \\ \frac{1}{2}(3-\gamma)\frac{u_2^2}{u_1} + (\gamma-1)u_3 \\ \gamma\frac{u_2}{u_1}u_3 - \frac{1}{2}(\gamma-1)\frac{u_2^3}{u_1^2} \end{bmatrix} .$$

By direct evaluation of all partial derivatives we arrive at the sought result.

Exercise 3.1.1. Write the Jacobian matrix $\mathbf{A}(\mathbf{U})$ in terms of the the sound speed a and the particle velocity u.

Solution 3.1.1.

$$\mathbf{A}(\mathbf{U}) = \begin{bmatrix} 0 & 1 & 0 \\ \frac{1}{2}(\gamma-3)u^2 & (3-\gamma)u & \gamma-1 \\ \frac{1}{2}(\gamma-2)u^3 - \frac{a^2u}{\gamma-1} & \frac{3-2\gamma}{2}u^2 + \frac{a^2}{\gamma-1} & \gamma u \end{bmatrix} . \quad (3.8)$$

Often, the Jacobian matrix is also expressed in terms of the *total specific enthalpy* H, which is related to the *specific enthalpy* h and other variables, namely

$$H = (E+p)/\rho \equiv \frac{1}{2}u^2 + h , \quad h = e + p/\rho . \quad (3.9)$$

In terms of H, a and u the Jacobian matrix becomes

$$\mathbf{A}(\mathbf{U}) = \begin{bmatrix} 0 & 1 & 0 \\ \frac{1}{2}(\gamma-3)u^2 & (3-\gamma)u & \gamma-1 \\ u\left[\frac{1}{2}(\gamma-1)u^2 - H\right] & H - (\gamma-1)u^2 & \gamma u \end{bmatrix} . \quad (3.10)$$

Proposition 3.1.2 (The Homogeneity Property). *The Euler equations (3.1)–(3.2) with the ideal–gas EOS (3.5) satisfy the homogeneity property*

$$\mathbf{F}(\mathbf{U}) = \mathbf{A}(\mathbf{U})\mathbf{U} . \quad (3.11)$$

Proof. The proof of this property is immediate. By multiplying the Jacobian matrix (3.8) by the vector $\mathbf{U}$ in (3.2) we identically reproduce the vector $\mathbf{F}(\mathbf{U})$ of fluxes in (3.2).

This remarkable property of the Euler equations forms the basis for numerical schemes of the *Flux Vector Splitting* type studied in Chap. 8. Note that the relationship between the flux $\mathbf{F}$, the coefficient matrix $\mathbf{A}$ and the conserved variables $\mathbf{U}$ for the Euler equations is identical to that for linear systems with constant coefficients, see Sect. 2.4 of Chap. 2. This property is also satisfied by the Euler equations with an Equation of State that is slightly more general than (3.5). See Steger and Warming [326] for details.

Proposition 3.1.3. *The eigenvalues of the Jacobian matrix* **A** *are*

$$\lambda_1 = u - a \, , \quad \lambda_2 = u \, , \quad \lambda_3 = u + a \tag{3.12}$$

and the corresponding right eigenvectors are

$$\mathbf{K}^{(1)} = \begin{bmatrix} 1 \\ u - a \\ H - ua \end{bmatrix} \, , \quad \mathbf{K}^{(2)} = \begin{bmatrix} 1 \\ u \\ \frac{1}{2}u^2 \end{bmatrix} \, , \quad \mathbf{K}^{(3)} = \begin{bmatrix} 1 \\ u + a \\ H + ua \end{bmatrix} \, . \tag{3.13}$$

Proof. Use of the expression (3.8) for **A** and the *characteristic polynomial*

$$|\mathbf{A} - \lambda \mathbf{I}| = 0 \, ,$$

lead to

$$(\lambda - u)(\gamma u - \lambda)\left[(2u - \gamma u - \lambda] + \right. \\ (\lambda - u)\left[-a^2 - (\gamma - 1)u^2 + (\gamma - 1)\gamma u^2\right] + \Delta = 0 \, ,$$

where

$$\Delta = \frac{1}{2}(\gamma u - \lambda)(1 - \gamma)u^2 - \frac{1}{2}(\gamma - 1)u^2\left[(1 - 2\gamma)\lambda + \gamma u\right] \, .$$

Manipulations show that Δ also contains the common factor $(\lambda - u)$, which implies that $\lambda_2 = u$ is a root of the characteristic polynomial and thus an eigenvalue of **A**. After cancelling $(\lambda - u)$ the remaining terms give

$$\lambda^2 - 2u\lambda + u^2 - a^2 = 0 \, ,$$

with real roots

$$\lambda_1 = u - a \, , \quad \lambda_3 = u + a \, .$$

Therefore the eigenvalues are: $\lambda_1 = u - a$, $\lambda_2 = u$, $\lambda_3 = u + a$ as claimed. To find the right eigenvectors we look, see Sect. 2.1 of Chap. 2, for a vector $\mathbf{K} = [k_1, k_2, k_3]^T$ such that

$$\mathbf{AK} = \lambda \mathbf{K} \, .$$

By substituting $\lambda = \lambda_i$ in turn, solving for the components of the vector **K** and selecting appropriate values for the scaling factors we find the desired result.

The eigenvalues are all real and the eigenvectors $\mathbf{K}^{(1)}$, $\mathbf{K}^{(2)}$, $\mathbf{K}^{(3)}$ form a complete set of *linearly independent eigenvectors*. We have thus proved that the time–dependent, one–dimensional Euler equations for ideal gases are *hyperbolic*. In fact these equations are *strictly hyperbolic*, because the eigenvalues are all real and *distinct*, as long as the sound speed a remains positive. Hyperbolicity remains a property of the Euler equations for more general equations of state, as we shall see in Chap. 4 for covolume gases.

3.1.2 Non–Conservative Formulations

The Euler equations (3.1)–(3.2) may be formulated in terms of variables other than the conserved variables. For smooth solutions all formulations are equivalent. For solutions containing shock waves however, non–conservative formulations give incorrect shock solutions. This point is addressed via the shallow water equations and the isothermal equations in Sect. 3.3 of this chapter. In spite of this, non–conservative formulations have some advantages over their conservative counterpart, when analysing the equations, for instance. Also, from the numerical point of view, there has been a recent revival of the idea of using schemes for non–conservative formulations of the equations. See e.g. Karni [191] and Toro [360], [368].

Primitive–Variable Formulations. For smooth solutions the equations may be formulated, and solved, using variables other than the conserved variables. For the one–dimensional case one possibility is to choose a vector $\mathbf{W} = (\rho, u, p)^T$ of *primitive* or *physical* variables, with p given by the equation of state. Expanding derivatives in the first of equations (3.1)–(3.2), the mass equation, we obtain

$$\rho_t + u\rho_x + \rho u_x = 0 . \tag{3.14}$$

By expanding derivatives in the second of equations (3.1)–(3.2), the momentum equation, we obtain

$$u\left[\rho_t + u\rho_x + \rho u_x\right] + \rho\left[u_t + uu_x + \frac{1}{\rho}p_x\right] = 0 .$$

Use of (3.14) followed by division through by ρ gives

$$u_t + uu_x + \frac{1}{\rho}p_x = 0 . \tag{3.15}$$

In a similar manner, the energy equation in (3.1)–(3.2) can be rearranged so as to use (3.14) and (3.15). The result is

$$p_t + \rho a^2 u_x + up_x = 0 . \tag{3.16}$$

Thus, in quasi–linear form we have

$$\mathbf{W}_t + \mathbf{A}(\mathbf{W})\mathbf{W}_x = \mathbf{0} , \tag{3.17}$$

where

$$\mathbf{W} = \begin{bmatrix} \rho \\ u \\ p \end{bmatrix} , \quad \mathbf{A}(\mathbf{W}) = \begin{bmatrix} u & \rho & 0 \\ 0 & u & 1/\rho \\ 0 & \rho a^2 & u \end{bmatrix} . \tag{3.18}$$

Proposition 3.1.4. *The system (3.17)–(3.18) has real eigenvalues*

$$\lambda_1 = u - a \ , \ \lambda_2 = u \ , \ \lambda_3 = u + a \ , \tag{3.19}$$

with corresponding right eigenvectors

$$\mathbf{K}^{(1)} = \alpha_1 \begin{bmatrix} 1 \\ -a/\rho \\ a^2 \end{bmatrix} \ , \ \mathbf{K}^{(2)} = \alpha_2 \begin{bmatrix} 1 \\ 0 \\ 0 \end{bmatrix} \ , \ \mathbf{K}^{(3)} = \alpha_3 \begin{bmatrix} 1 \\ a/\rho \\ a^2 \end{bmatrix} \ . \tag{3.20}$$

where $\alpha_1, \alpha_2, \alpha_3$ are scaling factors, or normalisation parameters, see Sect. 2.1 of Chap. 2. The left eigenvectors are

$$\left. \begin{array}{c} \mathbf{L}^{(1)} = \beta_1 (0, 1, -\frac{1}{\rho a}) \ , \\[2mm] \mathbf{L}^{(2)} = \beta_2 (1, 0, -\frac{1}{a^2}) \ , \\[2mm] \mathbf{L}^{(3)} = \beta_3 (0, 1, \frac{1}{\rho a}) \ , \end{array} \right\} \tag{3.21}$$

where $\beta_1, \beta_2, \beta_3$ are scaling factors.

Proof. (Left to the reader).

Exercise 3.1.2. Verify that by choosing appropriate normalisation parameters $\alpha_1, \alpha_2, \alpha_3$ and $\beta_1, \beta_2, \beta_3$ in (3.20) and (3.21) respectively, the left and right eigenvectors $\mathbf{L}^{(j)}$ and $\mathbf{K}^{(j)}$ of $\mathbf{A(W)}$ are *bi–orthonormal*, that is

$$\mathbf{L}^{(j)} \cdot \mathbf{K}^{(i)} = \left\{ \begin{array}{ll} 1 & \text{if } i = j \ , \\[2mm] 0 & \text{otherwise} \ . \end{array} \right. \tag{3.22}$$

Characteristic Equations. Recall that the eigenvalues $\lambda_1 = u - a$, $\lambda_2 = u$, $\lambda_3 = u + a$ define characteristic directions $dx/dt = \lambda_i$ for $i = 1, 2, 3$. For a set of partial differential equations (3.17) a *characteristic equation* says that in a direction $dx/dt = \lambda_i$, $\mathbf{L}^{(i)} \cdot d\mathbf{W} = 0$, or in full

$$\mathbf{L}^{(i)} \cdot \begin{bmatrix} d\rho \\ du \\ dp \end{bmatrix} = 0 \ . \tag{3.23}$$

By expanding (3.23) for $\mathbf{L}^{(1)}, \mathbf{L}^{(2)}, \mathbf{L}^{(3)}$ we obtain the *characteristic equations*

$$\left. \begin{array}{c} dp - \rho a \, du = 0 \ \text{along} \ dx/dt = \lambda_1 = u - a \ , \\[2mm] dp - a^2 \, d\rho = 0 \ \text{along} \ dx/dt = \lambda_2 = u \ , \\[2mm] dp + \rho a \, du = 0 \ \text{along} \ dx/dt = \lambda_3 = u + a \ . \end{array} \right\} \tag{3.24}$$

These differential relations hold true *along characteristic* directions. For numerical purposes, linearisation of these equations provides ways of solving the Riemann problem for the Euler equations, approximately; see Sect. 9.3 of Chap. 9.

Entropy Formulation. The entropy s can be written as

$$s = c_v \ln\left(\frac{p}{\rho^\gamma}\right) + C_0 ,\qquad (3.25)$$

where C_0 is a constant. From this equation we obtain

$$p = C_1 \rho^\gamma e^{s/c_v} ,\qquad (3.26)$$

where C_1 is a constant. Now, if in the primitive–variable formulation (3.17) we use entropy s instead of pressure p we have the new vector of unknowns

$$\mathbf{W} = (\rho, u, s)^T ,\qquad (3.27)$$

and a corresponding new way of expressing the governing equations.

Proposition 3.1.5. *The entropy s satisfies the following PDE*

$$s_t + u s_x = 0 .\qquad (3.28)$$

Proof. From (3.25) and the expression (3.6) for the sound speed a we have

$$s_t = \frac{c_v}{p}\left[p_t - a^2 \rho_t\right] ,\quad s_x = \frac{c_v}{p}\left[p_x - a^2 \rho_x\right] .$$

But from (3.16) $p_t = -\rho a^2 u_x - u p_x$, and hence $s_t + u s_x = 0$, as claimed.

The significance of the result is that

$$s_t + u s_x = \frac{ds}{dt} = 0 ,\qquad (3.29)$$

and so *in regions of smooth flow*, the entropy s is constant along particle paths $dx/dt = u$. Hence, along a particle path one has the *isentropic law* given by

$$p = C\rho^\gamma ,\qquad (3.30)$$

where $C = C(s_0)$ is a function of the *initial entropy* s_0 and is constant along the path so long as the flow remains smooth; see Sect. 1.6.2 of Chap. 1. In general of course, C changes from path to path. When solving the Riemann problem the initial entropy can be computed on the initial data of the Riemann problem, which is piece–wise constant. If C is the same constant throughout the flow domain we speak of *isentropic flow*, or sometimes, *homentropic flow*. This leads to the special set of governing equations (1.109)–(1.110) presented in Chap. 1. The governing equations for the entropy formulation, written in quasi–linear form, are

$$\mathbf{W}_t + \mathbf{A}(\mathbf{W})\mathbf{W}_x = 0 ,\qquad (3.31)$$

with

$$\mathbf{A}(\mathbf{W}) = \begin{bmatrix} u & \rho & 0 \\ a^2/\rho & u & \frac{1}{\rho}\frac{\partial p}{\partial s} \\ 0 & 0 & u \end{bmatrix} .\qquad (3.32)$$

Proposition 3.1.6. *The eigenvalues of system (3.31)–(3.32) are*

$$\lambda_1 = u - a \ , \ \lambda_2 = u \ , \ \lambda_3 = u + a \tag{3.33}$$

and the corresponding right eigenvectors are

$$\mathbf{K}^{(1)} = \begin{bmatrix} 1 \\ -a/\rho \\ 0 \end{bmatrix} , \quad \mathbf{K}^{(2)} = \begin{bmatrix} -\frac{\partial p}{\partial s} \\ 0 \\ a^2 \end{bmatrix} , \quad \mathbf{K}^{(3)} = \begin{bmatrix} 1 \\ a/\rho \\ 0 \end{bmatrix} . \tag{3.34}$$

Proof. (Left to the reader).

3.1.3 Elementary Wave Solutions of the Riemann Problem

Here we describe the structure of the solution of the Riemann problem as a set of *elementary waves* such as rarefactions, contacts and shock waves, see Sect. 2.4.4 of Chapt. 4. Each of these elementary waves are studied in detail. Basic relations across these waves are established. Such relations will be used in Chap. 4 to *connect* all unknown states to the data states and thus find the complete solution of the Riemann problem.

The Riemann problem for the one–dimensional, time dependent Euler equations (3.1)–(3.2) with data $(\mathbf{U}_L, \mathbf{U}_R)$ is the IVP

$$\begin{aligned} \mathbf{U}_t + \mathbf{F}(\mathbf{U})_x &= \mathbf{0} \ , \\ \mathbf{U}(x,0) = \mathbf{U}^{(0)}(x) &= \left\{ \begin{array}{ll} \mathbf{U}_L & \text{if} \quad x < 0 \ , \\ \mathbf{U}_R & \text{if} \quad x > 0 \ . \end{array} \right. \end{aligned} \tag{3.35}$$

The physical analogue of the Riemann problem is the *shock–tube problem* in Gas Dynamics, in which the velocities u_L and u_R on either side of the diaphragm, here idealised by an initial discontinuity, are zero. Shock tubes and shock–tube problems have played, over a period of more than 100 years, a fundamental role in fluid dynamics research.

The structure of the similarity solution $\mathbf{U}(x/t)$ of (3.35) is as depicted in Fig. 3.1. There are three waves *associated* with the three characteristic fields corresponding to the eigenvectors $\mathbf{K}^{(i)}$, $i = 1, 2, 3$. We choose the convention of representing the outer waves, when their character is unknown, by a pair of rays emanating from the origin and the middle wave by a dashed line. Each wave family is shown along with the corresponding eigenvalue. The three waves separate four constant states. From left to right these are $\mathbf{U}_L$ (left data state); $\mathbf{U}_{*L}$ between the 1–wave and the 2–wave; $\mathbf{U}_{*R}$ between the 2–wave and the 3–wave and $\mathbf{U}_R$ (right data state). As we shall see the waves present in the solution are of three types: rarefaction waves, contact discontinuities and shock waves. In order to identify the types we analyse the characteristic fields for $\mathbf{K}^{(i)}$, $i = 1, 2, 3$; see Sects. 2.4.3 and 2.4.4 of Chap. 2.

Proposition 3.1.7. *The $\mathbf{K}^{(2)}$–characteristic field is linearly degenerate and the $\mathbf{K}^{(1)}$, $\mathbf{K}^{(3)}$ characteristic fields are genuinely non–linear.*

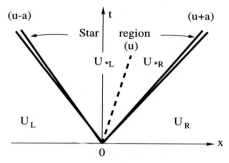

Fig. 3.1. Structure of the solution of the Riemann problem in the x–t plane for the time–dependent, one dimensional Euler equations. There are three wave families associated with the eigenvalues $u - a$, u and $u + a$

Proof. For the $\mathbf{K}^{(2)}$–characteristic field we have

$$\nabla \lambda_2(\mathbf{U}) = [\partial \lambda_2 / \partial u_1, \partial \lambda_2 / \partial u_2, \partial \lambda_2 / \partial u_3] = [-u/\rho, 1/\rho, 0] \ .$$

Hence

$$\nabla \lambda_2 \cdot \mathbf{K}^{(2)} = [-u/\rho, 1/\rho, 0] \cdot \begin{bmatrix} 1 \\ u \\ \frac{1}{2}u^2 \end{bmatrix} = 0$$

and therefore the $\mathbf{K}^{(2)}$ characteristic field is linearly degenerate as claimed. The proof that the $\mathbf{K}^{(1)}$ and $\mathbf{K}^{(3)}$ characteristic fields are *genuinely nonlinear* is left to the reader.

The wave associated with the $\mathbf{K}^{(2)}$ characteristic field is a *contact discontinuity* and those associated with the $\mathbf{K}^{(1)}$, $\mathbf{K}^{(3)}$ characteristic fields will either be rarefaction waves (smooth) or shock waves (discontinuities), see Sect. 2.4.4 of Chapt. 4. Of course one does not know in advance what types of waves will be present in the solution of the Riemann problem. The only exception is the middle wave, which is always a contact discontinuity. Fig. 3.2 shows a particular case in which the left wave is a rarefaction, the middle wave is a contact and the right wave is a shock wave. For each wave we have drawn a pair of arrows, one on each side, to indicate the characteristic directions of the corresponding eigenvalue. For the rarefaction wave we have

$$\lambda_1(\mathbf{U}_{\mathrm{L}}) \leq \lambda_1(\mathbf{U}_{*\mathrm{L}}) \ .$$

The eigenvalue $\lambda_1(\mathbf{U})$ increases monotonically as we cross the rarefaction wave from left to right and the characteristics on either side diverge from the wave; compare with Fig. 2.20 of Chap. 2. For the shock wave, characteristics run into the wave and we have

$$\lambda_3(\mathbf{U}_{*\mathrm{R}}) > S_3 > \lambda_3(\mathbf{U}_{\mathrm{R}}) \ ,$$

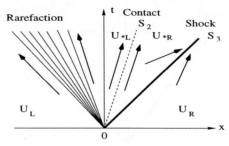

Fig. 3.2. Structure of the solution of the Riemann problem in the x–t plane for the time–dependent, one dimensional Euler equations, in which the left wave is a rarefaction, the middle wave is a contact discontinuity and the right wave is a shock wave

which is the *entropy condition*. See Sect. 2.4.4 of Chap. 2. S_3 is the speed of the 3–shock. For the contact wave we have

$$\lambda_2(\mathbf{U}_{*L}) = \lambda_2(\mathbf{U}_{*R}) = S_2 \ ,$$

where S_2 is the speed of the contact wave; the characteristics are parallel to the contact wave. Recall that this is what happens for *all characteristic fields* in linear hyperbolic systems with constant coefficients. Next we study each type of waves separately.

Contact Discontinuities. The contact discontinuity in the solution of the Riemann problem for the Euler equations can be analysed by utilising the eigenstructure of the equations. In particular the Generalised Riemann Invariants will reveal which quantities change across the wave. Recall that for a general $m \times m$ hyperbolic system, such as (3.1)–(3.2) or (3.7), with

$$\mathbf{W} = [w_1, w_2, \cdots, w_m]^T \ ,$$

and right eigenvectors

$$\mathbf{K}^{(i)} = \left[k_1^{(i)}, k_2^{(i)}, \cdots, k_m^{(i)} \right] \ ,$$

the $i-$th Generalised Riemann Invariants are the $(m-1)$ ODEs

$$\frac{dw_1}{k_1^{(i)}} = \frac{dw_2}{k_2^{(i)}} = \frac{dw_3}{k_3^{(i)}} = \cdots = \frac{dw_m}{k_m^{(i)}} \ .$$

Using the eigenstructure (3.12)–(3.13) of the conservative formulation (3.1)–(3.2), for the $\mathbf{K}^{(2)}$–wave we have

$$\frac{d\rho}{1} = \frac{d(\rho u)}{u} = \frac{dE}{\frac{1}{2}u^2} \ . \tag{3.36}$$

Manipulation of these equalities gives

$$p = \text{constant},\ u = \text{constant}$$

across the contact wave. The same result follows directly by inspection of the eigenvector $\mathbf{K}^{(2)}$ in (3.20) for the primitive–variable formulation (3.17)–(3.18): the wave jumps in ρ, u and p are proportional to the corresponding components of the eigenvector. These are zero for the velocity and pressure. The jump in ρ is in general non–trivial. To conclude: *a contact wave is a discontinuous wave across which both pressure and particle velocity are constant but density jumps discontinuously as do variables that depend on density, such as specific internal energy, temperature, sound speed, entropy, etc.*

Rarefaction Waves. Rarefaction waves in the Euler equations are associated with the $\mathbf{K}^{(1)}$ and $\mathbf{K}^{(3)}$ characteristic fields. Inspection of the eigenvectors (3.20) for the primitive–variable formulation reveals that ρ, u and p change across a rarefaction wave. We now utilise the Generalised Riemann Invariants for the eigenstructure (3.33)–(3.34) of the entropy formulation (3.31)–(3.32).

Proposition 3.1.8. *For the Euler equations the Generalised Riemann Invariants across 1 and 3 rarefactions are*

$$\left. \begin{array}{l} I_{\mathrm{L}}(u, a) = u + \frac{2a}{\gamma - 1} = \text{constant} \\ s = \text{constant} \end{array} \right\} \text{across } \lambda_1 = u - a\ , \qquad (3.37)$$

$$\left. \begin{array}{l} I_{\mathrm{R}}(u, a) = u - \frac{2a}{\gamma - 1} = \text{constant} \\ s = \text{constant} \end{array} \right\} \text{across } \lambda_3 = u + a\ . \qquad (3.38)$$

Proof. Across a wave associated with $\lambda_1 = u - a$ wave we have

$$\frac{\mathrm{d}\rho}{1} = \frac{\mathrm{d}u}{-a/\rho} = \frac{\mathrm{d}s}{0}\ .$$

Two meaningful relations are

$$u + \int \frac{a}{\rho}\,\mathrm{d}\rho = \text{constant and } s = \text{constant.} \qquad (3.39)$$

Similarly, across the $\lambda_3 = u + a$ wave we have

$$u - \int \frac{a}{\rho}\,\mathrm{d}\rho = \text{constant and } s = \text{constant.} \qquad (3.40)$$

In order to reproduce (3.37) and (3.38) we need to evaluate the integrals in (3.39) and (3.40). First we note that by inspection of the eigenvectors $\mathbf{K}^{(1)}$ and $\mathbf{K}^{(3)}$ the condition of constant entropy across the respective waves is immediate. We may therefore use the isentropic law (3.30) with the constant C evaluated at the appropriate data state (constant). Thus the integral is as found for the isentropic equations in Sect. 2.4.3 of Chap. 2, that is

$$\int \frac{a}{\rho}\,\mathrm{d}\rho = \frac{2a}{\gamma - 1}\ ,$$

and thus equations (3.37)–(3.38) are reproduced.

To summarise: *a rarefaction wave is a smooth wave associated with the 1 and 3 fields across which ρ, u and p change. The wave has a fan–type shape and is enclosed by two bounding characteristics corresponding to the Head and the Tail of the wave. Across the wave the Generalised Riemann Invariants apply.* The *solution within* the rarefaction will be given in Chap. 4, where the full solution of the Riemann problem is presented.

Shock Waves. Details on the Physics of shock waves are found in any book on Gas Dynamics. We particularly recommend Becker [21], Anderson [8], Landau and Lifshitz [200]. The specialised book by Zeldovich and Raizer [428] is highly recommended.

    In the context of the one–dimensional Euler equations, shock waves are discontinuous waves associated with the genuinely non–linear fields 1 and 3. All three quantities ρ, u and p change across a shock wave. Consider the $\mathbf{K}^{(3)}$ characteristic field and assume the corresponding wave is a right–facing shock wave travelling at the constant speed S_3; see Fig. 3.3. In terms of the primitive variables we denote the state ahead of the shock by $\mathbf{W}_R = (\rho_R, u_R, p_R)^T$ and the state behind the shock by $\mathbf{W}_* = (\rho_*, u_*, p_*)^T$. We are interested in deriving relations, across the shock wave, between the various quantities involved. Central to the analysis is the application of the Rankine–Hugoniot Conditions. It is found convenient to transform the problem to a

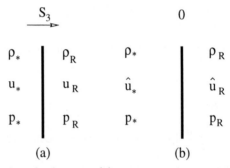

Fig. 3.3. Right–facing shock wave: (a) stationary frame of reference, shock has speed S_3; (b) frame of reference moves with speed S_3, so that the shock has zero speed

new frame of reference moving with the shock so that in the new frame the shock speed is zero. Fig. 3.3 depicts both frames of reference. In the transformed frame (b) the states ahead and behind the shock have changed by virtue of the transformation. Densities and pressures remain unaltered while velocities have changed to the *relative velocities* $\hat{u}_R$ and $\hat{u}_*$ given by

$$\hat{u}_* = u_* - S_3 \,, \; \hat{u}_R = u_R - S_3 \,. \tag{3.41}$$

Application of the Rankine–Hugoniot Conditions in the frame in which the shock speed is zero gives

$$\rho_* \hat{u}_* = \rho_R \hat{u}_R \, , \tag{3.42}$$

$$\rho_* \hat{u}_*^2 + p_* = \rho_R \hat{u}_R^2 + p_R \, , \tag{3.43}$$

$$\hat{u}_*(\hat{E}_* + p_*) = \hat{u}_R(\hat{E}_R + p_*) \, . \tag{3.44}$$

By using the definition of total energy E and introducing the *specific internal energy* e the left–hand side of (3.44) may we written as

$$\hat{u}_* \rho_* \left[\frac{1}{2} \hat{u}_*^2 + (e_* + p_*/\rho_*) \right]$$

and the right–hand side of (3.44) as

$$\hat{u}_R \rho_R \left[\frac{1}{2} \hat{u}_R^2 + (e_R + p_R/\rho_R) \right] \, .$$

Now we use the *specific enthalpy* h and write

$$h_* = e_* + p_*/\rho_* \, , \quad h_R = e_R + p_R/\rho_R \, . \tag{3.45}$$

Use of equations (3.42) and (3.44) leads to

$$\frac{1}{2} \hat{u}_*^2 + h_* = \frac{1}{2} \hat{u}_R^2 + h_R \, . \tag{3.46}$$

By using (3.42) into (3.43) we write

$$\rho_* \hat{u}_*^2 = (\rho_R \hat{u}_R) \hat{u}_R + p_R - p_* = (\rho_* \hat{u}_*) \frac{\rho_* \hat{u}_*}{\rho_R} + p_R - p_* \, .$$

After some manipulations we obtain

$$\hat{u}_*^2 = \left(\frac{\rho_R}{\rho_*} \right) \left[\frac{p_R - p_*}{\rho_R - \rho_*} \right] \, . \tag{3.47}$$

In a similar way we obtain

$$\hat{u}_R^2 = \left(\frac{\rho_*}{\rho_R} \right) \left[\frac{p_R - p_*}{\rho_R - \rho_*} \right] \, . \tag{3.48}$$

Substitution of (3.47)–(3.48) into (3.46) gives

$$h_* - h_R = \frac{1}{2}(p_* - p_R) \left[\frac{\rho_* + \rho_R}{\rho_* \rho_R} \right] \, . \tag{3.49}$$

Assuming the specific internal energy e is given by the the caloric equation of state (3.4), it is then more convenient to rewrite the energy equation (3.49) using (3.45). We obtain

$$e_* - e_R = \frac{1}{2}(p_* + p_R) \left[\frac{\rho_* - \rho_R}{\rho_* \rho_R} \right] \, . \tag{3.50}$$

Note that up to this point no assumption on the general caloric EOS (3.4) has been made. In what follows, we derive shock relations that apply to ideal gases in which the ideal caloric EOS (3.5) is assumed. By using (3.5) into (3.50) and performing some algebraic manipulations one obtains

$$\frac{\rho_*}{\rho_R} = \frac{(\frac{p_*}{p_R}) + (\frac{\gamma-1}{\gamma+1})}{(\frac{\gamma-1}{\gamma+1})(\frac{p_*}{p_R}) + 1} . \tag{3.51}$$

This establishes a useful relation between the density ratio ρ_*/ρ_R and the pressure ratio p_*/p_R across the shock wave.

We now introduce Mach numbers

$$M_R = u_R/a_R , \quad M_S = S_3/a_R , \tag{3.52}$$

where M_R is the Mach number of the flow ahead of the shock, in the original frame; M_S is the *shock Mach number*. Manipulation of equations (3.48), (3.51) and (3.52) leads to expressions for the density and pressure ratios across the shock as functions of the relative Mach number $M_R - M_S$, namely

$$\frac{\rho_*}{\rho_R} = \frac{(\gamma + 1)(M_R - M_S)^2}{(\gamma - 1)(M_R - M_S)^2 + 2} , \tag{3.53}$$

$$\frac{p_*}{p_R} = \frac{2\gamma(M_R - M_S)^2 - (\gamma - 1)}{(\gamma + 1)} . \tag{3.54}$$

The shock speed S_3 can be related to the density and pressure ratios across the shock wave. In terms of the pressure ratio (3.54) we first note the following relationship

$$M_R - M_S = -\sqrt{\left(\frac{\gamma + 1}{2\gamma}\right)\left(\frac{p_*}{p_R}\right) + \left(\frac{\gamma - 1}{2\gamma}\right)} .$$

This leads to an expression for the shock speed as a function of the pressure ratio across the shock, namely

$$S_3 = u_R + a_R \sqrt{\left(\frac{\gamma + 1}{2\gamma}\right)\left(\frac{p_*}{p_R}\right) + \left(\frac{\gamma - 1}{2\gamma}\right)} . \tag{3.55}$$

Note that as the shock strength tends to zero, the ratio p_*/p_R tends to unity and the shock speed S_3 approaches the characteristic speed $\lambda_3 = u_R + a_R$, as expected. We can also obtain an expression for the *particle velocity* u_* behind the shock wave. From (3.42) we relate u_* to the density ratio across the shock, namely

$$u_* = (1 - \rho_R/\rho_*)S_3 + u_R \rho_R/\rho_* . \tag{3.56}$$

Example 3.1.1 (Shock Wave). Consider a shock wave of shock Mach number $M_S = 3$ propagating into the atmosphere with conditions (ahead of the shock) $\rho_R = 1.225$ kg/m³, $u_R = 0$ m/s, $p_R = 101\,325$ Pa. Assume the process is suitably modelled by the ideal gas EOS (3.5) with $\gamma = 1.4$. From the definition of sound speed (3.6) we obtain $a_R = 340.294$ m/s. As the shock Mach number $M_S = 3$ is assumed (a parameter) then equation (3.52) gives the shock speed as $S = 1020.882$ m/s. From equation (3.53) we obtain $\rho_* = 4.725$ kg/m³. From equation (3.54) we obtain $p_* = 1\,047\,025$ Pa and from equation (3.56) we obtain $u_* = 756.2089$ m/s.

Remark 3.1.1. Shock relations (3.53), (3.54) and (3.56) define a state

$$(\rho_*, u_*, p_*)^T$$

behind a shock for given initial conditions $(\rho_R, u_R, p_R)^T$ ahead of the shock and a chosen shock Mach number M_S, or equivalently a shock speed S_3. The shock is associated with the 3–wave family. These relations can be useful in setting up test problems involving a single shock wave to test numerical methods.

The analysis for a 1–shock wave (left facing) travelling with velocity S_1 is entirely analogous. The state ahead of the shock (left side now) is denoted by $\mathbf{W}_L = (\rho_L, u_L, p_L)^T$ and the state behind the shock (right side) by $\mathbf{W}_* = (\rho_*, u_*, p_*)^T$. As done for the 3–shock we transform to a stationary frame of reference. The relative velocities are

$$\hat{u}_L = u_L - S_1 \ , \ \hat{u}_* = u_* - S_1 \ . \tag{3.57}$$

Mach numbers are

$$M_L = u_L/a_L \ , \ M_S = S_1/a_L \ . \tag{3.58}$$

The density and pressure ratio relationship is

$$\frac{\rho_*}{\rho_L} = \frac{\left(\frac{p_*}{p_L}\right) + \left(\frac{\gamma-1}{\gamma+1}\right)}{\left(\frac{\gamma-1}{\gamma+1}\right)\left(\frac{p_*}{p_L}\right) + 1} \ . \tag{3.59}$$

In terms of the relative Mach number $M_L - M_S$ the density and pressure ratios across the left shock can be expressed as follows

$$\frac{\rho_*}{\rho_L} = \frac{(\gamma+1)(M_L - M_S)^2}{(\gamma-1)(M_L - M_S)^2 + 2} \ , \tag{3.60}$$

$$\frac{p_*}{p_L} = \frac{2\gamma(M_L - M_S)^2 - (\gamma-1)}{(\gamma+1)} \ . \tag{3.61}$$

The shock speed S_1 can be obtained from either (3.60) or (3.61). In terms of the pressure ratio (3.61) we have

$$M_{\mathrm{L}} - M_{\mathrm{S}} = \sqrt{\left(\frac{\gamma+1}{2\gamma}\right)\left(\frac{p_*}{p_{\mathrm{L}}}\right) + \left(\frac{\gamma-1}{2\gamma}\right)} \, ,$$

which leads to

$$S_1 = u_{\mathrm{L}} - a_{\mathrm{L}}\sqrt{\left(\frac{\gamma+1}{2\gamma}\right)\left(\frac{p_*}{p_{\mathrm{L}}}\right) + \left(\frac{\gamma-1}{2\gamma}\right)} \, . \tag{3.62}$$

Note that as the shock strength tends to zero, the ratio p_*/p_{L} tends to unity and the shock speed S_1 approaches the characteristic speed $\lambda_1 = u_{\mathrm{L}} - a_{\mathrm{L}}$, as expected. The particle velocity behind the left shock is

$$u_* = (1 - \rho_{\mathrm{L}}/\rho_*)S_1 + u_{\mathrm{L}}\rho_{\mathrm{L}}/\rho_* \, . \tag{3.63}$$

Shock relations (3.60), (3.61) and (3.63) define a state $(\rho_*, u_*, p_*)^T$ behind a shock for given initial conditions $(\rho_L, u_L, p_L)^T$ ahead of the shock and a chosen shock Mach number M_S, or equivalently a shock speed S_1. The shock is associated with the 1–wave family.

3.2 Multi–Dimensional Euler Equations

In the previous section we analysed the one–dimensional, time–dependent Euler equations. Here we study a few basic properties of the two and three dimensional cases. In differential conservation–law form the three–dimensional equations are

$$\mathbf{U}_t + \mathbf{F(U)}_x + \mathbf{G(U)}_y + \mathbf{H(U)}_z = \mathbf{0} \, , \tag{3.64}$$

with

$$\mathbf{U} = \begin{bmatrix} \rho \\ \rho u \\ \rho v \\ \rho w \\ E \end{bmatrix} , \quad \mathbf{F} = \begin{bmatrix} \rho u \\ \rho u^2 + p \\ \rho u v \\ \rho u w \\ u(E+p) \end{bmatrix} ,$$

$$\mathbf{G} = \begin{bmatrix} \rho v \\ \rho u v \\ \rho v^2 + p \\ \rho v w \\ v(E+p) \end{bmatrix} , \quad \mathbf{H} = \begin{bmatrix} \rho w \\ \rho u w \\ \rho v w \\ \rho w^2 + p \\ w(E+p) \end{bmatrix} . \tag{3.65}$$

Here E is the total energy per unit volume

$$E = \rho\left(\frac{1}{2}\mathbf{V}^2 + e\right) , \tag{3.66}$$

where $\frac{1}{2}\mathbf{V}^2 = \frac{1}{2}\mathbf{V} \cdot \mathbf{V} = \frac{1}{2}(u^2 + v^2 + w^2)$ is the *specific kinetic energy* and e is *specific internal energy* given by a caloric equation of state (3.4).

The corresponding integral form of the conservation laws (3.64) is given by

$$\frac{d}{dt} \int \int \int_V \mathbf{U} \, dV + \int \int_A \mathcal{H} \cdot \mathbf{n} \, dA = \mathbf{0} \, , \tag{3.67}$$

where V is a *control volume*, A is the boundary of V, $\mathcal{H} = (\mathbf{F}, \mathbf{G}, \mathbf{H})$ is the tensor of fluxes, $\mathbf{n}$ is the outward unit vector normal to the surface A, dA is an area element and $\mathcal{H} \cdot \mathbf{n} \, dA$ is the flux component normal to the boundary A. The conservation laws (3.67) state that the time–rate of change of $\mathbf{U}$ inside volume V depends only on the *total flux* through the surface A, the boundary of the control volume V. Numerical methods of the finite volume type, see Sect. 16.7.3 of Chap. 16, are based on this formulation of the equations. For details of the derivation of integral form of the conservation laws see Sects. 1.5 and 1.6.1 of Chap. 1.

In the next section we study some properties of the two–dimensional Euler equation in conservation form

3.2.1 Two–Dimensional Equations in Conservative Form

The two–dimensional version of the Euler equations in differential conservative form is

$$\mathbf{U}_t + \mathbf{F(U)}_x + \mathbf{G(U)}_y = \mathbf{0} \, , \tag{3.68}$$

with

$$\mathbf{U} = \begin{bmatrix} \rho \\ \rho u \\ \rho v \\ E \end{bmatrix} \, , \quad \mathbf{F} = \begin{bmatrix} \rho u \\ \rho u^2 + p \\ \rho u v \\ u(E + p) \end{bmatrix} \, , \quad \mathbf{G} = \begin{bmatrix} \rho v \\ \rho u v \\ \rho v^2 + p \\ v(E + p) \end{bmatrix} \, . \tag{3.69}$$

Eigenstructure. Here we find the Jacobian matrix of the x–split equations, its eigenvalues and corresponding right eigenvectors. We also study the types of characteristic fields present.

Proposition 3.2.1. *The Jacobian matrix* $\mathbf{A(U)}$ *corresponding to the flux* $\mathbf{F(U)}$ *is given by*

$$\mathbf{A(U)} = \begin{bmatrix} 0 & 1 & 0 & 0 \\ -u^2 + \frac{1}{2}(\gamma - 1)\mathbf{V}^2 & (3 - \gamma)u & -(\gamma - 1)v & \gamma - 1 \\ -uv & v & u & 0 \\ u\left[\frac{1}{2}(\gamma - 1)\mathbf{V}^2 - H\right] & H - (\gamma - 1)u^2 & -(\gamma - 1)uv & \gamma u \end{bmatrix} \, . \tag{3.70}$$

The eigenvalues of $\mathbf{A}$ *are*

$$\lambda_1 = u - a \, , \quad \lambda_2 = \lambda_3 = u \, , \quad \lambda_4 = u + a \, , \tag{3.71}$$

with corresponding right eigenvectors

$$\mathbf{K}^{(1)} = \begin{bmatrix} 1 \\ u - a \\ v \\ H - au \end{bmatrix} , \quad \mathbf{K}^{(2)} = \begin{bmatrix} 1 \\ u \\ v \\ \frac{1}{2}\mathbf{V}^2 \end{bmatrix} , \\ \mathbf{K}^{(3)} = \begin{bmatrix} 0 \\ 0 \\ 1 \\ v \end{bmatrix} , \quad \mathbf{K}^{(4)} = \begin{bmatrix} 1 \\ u + a \\ v \\ H + ua \end{bmatrix} . \tag{3.72}$$

Proof. Exercise.

Rotational Invariance. We next prove an important property, called the *rotational invariance* of the Euler equations. The property allows the proof of hyperbolicity in time for the two–dimensional equations (3.68)–(3.69) and can also be used for computational purposes to deal with domains that are not aligned with the Cartesian directions, see Sect. 16.7.3 of Chap. 16. We first note that outward unit vector $\mathbf{n}$ normal to the surface A in the two–dimensional case is given by

$$\mathbf{n} \equiv (n_1, n_2) \equiv (\cos\theta, \sin\theta) , \tag{3.73}$$

where θ is the angle formed by x–axis and the normal vector $\mathbf{n}$; θ is measured in an anticlockwise manner and lies in the range $0 \le \theta \le 2\pi$. Fig. 3.4 depicts the situation. The integrand of the surface integral in (3.67) becomes

$$(\mathbf{F}, \mathbf{G}) \cdot \mathbf{n} = \cos\theta \mathbf{F}(\mathbf{U}) + \sin\theta \mathbf{G}(\mathbf{U}) . \tag{3.74}$$

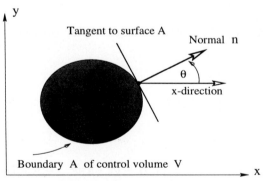

Fig. 3.4. Control volume V on x–y plane; boundary of V is A, outward unit normal vector is $\mathbf{n}$ and θ is angle between the x–direction and $\mathbf{n}$

Proposition 3.2.2 (Rotational Invariance). *The two–dimensional Euler equations (3.68)–(3.69) satisfy the rotational invariance property*

$$\cos\theta \mathbf{F}(\mathbf{U}) + \sin\theta \mathbf{G}(\mathbf{U}) = \mathbf{T}^{-1}\mathbf{F}(\mathbf{TU}) , \tag{3.75}$$

for all angles θ and vectors $\mathbf{U}$. Here $\mathbf{T} = \mathbf{T}(\theta)$ is the rotation matrix and $\mathbf{T}^{-1}(\theta)$ is its inverse, namely

$$\mathbf{T} = \begin{bmatrix} 1 & 0 & 0 & 0 \\ 0 & \cos\theta & \sin\theta & 0 \\ 0 & -\sin\theta & \cos\theta & 0 \\ 0 & 0 & 0 & 1 \end{bmatrix} , \quad \mathbf{T}^{-1} = \begin{bmatrix} 1 & 0 & 0 & 0 \\ 0 & \cos\theta & -\sin\theta & 0 \\ 0 & \sin\theta & \cos\theta & 0 \\ 0 & 0 & 0 & 1 \end{bmatrix} . \tag{3.76}$$

Proof. First we calculate $\hat{\mathbf{U}} = \mathbf{TU}$. The result is

$$\hat{\mathbf{U}} = \mathbf{TU} = [\rho, \rho\hat{u}, \rho\hat{v}, E]^T ,$$

with $\hat{u} = u\cos\theta + v\sin\theta$, $\hat{v} = -u\sin\theta + v\cos\theta$. Next we compute $\hat{\mathbf{F}} = \mathbf{F}(\hat{\mathbf{U}})$ and obtain

$$\hat{\mathbf{F}} = \mathbf{F}(\hat{\mathbf{U}}) = \left[\rho\hat{u}, \rho\hat{u}^2 + p, \rho\hat{u}\hat{v}, \hat{u}(E + p)\right]^T .$$

Now we apply $\mathbf{T}^{-1}$ to $\mathbf{F}(\hat{\mathbf{U}})$. The result is easily verified to be

$$\mathbf{T}^{-1}\hat{\mathbf{F}} = \begin{bmatrix} \rho\hat{u} \\ \cos\theta\left(\rho\hat{u}^2 + p\right) - \sin\theta\left(\rho\hat{u}\hat{v}\right) \\ \sin\theta\left(\rho\hat{u}^2 + p\right) + \cos\theta\left(\rho\hat{u}\hat{v}\right) \\ \hat{u}(E + p) \end{bmatrix} = \cos\theta\mathbf{F} + \sin\theta\mathbf{G} .$$

This is clearly satisfied for the first and fourth components. Further manipulation show that it is also satisfied for the second and third flux components and the proposition is thus proved.

Hyperbolicity in Time. Here we use the rotational invariance property of the two–dimensional time dependent Euler equations to show that the equations are *hyperbolic in time.*

Definition 3.2.1 (Hyperbolicity in time). *System (3.68)–(3.69) is hyperbolic in time if for all admissible states $\mathbf{U}$ and real angles θ, the matrix*

$$\mathbf{A}(\mathbf{U}, \theta) = \cos\theta\mathbf{A}(\mathbf{U}) + \sin\theta\mathbf{B}(\mathbf{U}) \tag{3.77}$$

is diagonalisable. Here $\mathbf{A}(\mathbf{U})$ and $\mathbf{B}(\mathbf{U})$ are respectively the Jacobian matrices of the fluxes $\mathbf{F}(\mathbf{U})$ and $\mathbf{G}(\mathbf{U})$ in (3.68).

Proposition 3.2.3. *The two–dimensional Euler equations (3.68)–(3.69) are hyperbolic in time.*

Proof. We want to prove that the matrix $\mathbf{A}(\mathbf{U}, \theta)$ in (3.77) is *diagonalisable*, see Sect. 2.3.2 of Chap. 2. That is we want to prove that there exist a diagonal matrix $\Lambda(\mathbf{U}, \theta)$ and a non–singular matrix $\mathbf{K}(\mathbf{U}, \theta)$ such that

$$\mathbf{A}(\mathbf{U}, \theta) = \mathbf{K}(\mathbf{U}, \theta)\Lambda(\mathbf{U}, \theta)\mathbf{K}^{-1}(\mathbf{U}, \theta) . \tag{3.78}$$

By differentiating (3.75) with respect to $\mathbf{U}$ we have

$$\mathbf{A}(\mathbf{U}, \theta) = \cos\theta \mathbf{A}(\mathbf{U}) + \sin\theta \mathbf{B}(\mathbf{U}) = \mathbf{T}(\theta)^{-1}\mathbf{A}\left(\mathbf{T}(\theta)\mathbf{U}\right)\mathbf{T}(\theta) .$$

But the matrix $\mathbf{A}(\mathbf{U})$ is diagonalisable, it has four linearly independent eigenvectors $\mathbf{K}^{(i)}(\mathbf{U})$ given by (3.72). Therefore we can write

$$\mathbf{A}(\mathbf{U}) = \mathbf{K}(\mathbf{U})\Lambda(\mathbf{U})\mathbf{K}^{-1}(\mathbf{U}) ,$$

where $\mathbf{K}(\mathbf{U})$ is the non–singular matrix the columns of which are the right eigenvectors $\mathbf{K}^{(i)}(\mathbf{U})$, $\mathbf{K}^{-1}(\mathbf{U})$ is its inverse and $\Lambda(\mathbf{U})$ is the diagonal matrix with the eigenvalues $\lambda_i(\mathbf{U})$ given by (3.71) as the diagonal entries. Then we have

$$\mathbf{A}(\mathbf{U}, \theta) = \mathbf{T}(\theta)^{-1}\left\{\mathbf{K}\left(\mathbf{T}(\theta)\mathbf{U}\right)\Lambda\left(\mathbf{T}(\theta)\mathbf{U}\right)\mathbf{K}^{-1}\left(\mathbf{T}(\theta)\mathbf{U}\right)\right\}\mathbf{T}(\theta)$$

$$= \left\{\mathbf{T}(\theta)^{-1}\mathbf{K}\left(\mathbf{T}(\theta)\mathbf{U}\right)\right\}\Lambda\left(\mathbf{T}(\theta)\mathbf{U}\right)\left\{\mathbf{T}(\theta)^{-1}\mathbf{K}\left(\mathbf{T}(\theta)\mathbf{U}\right)\right\}^{-1} .$$

Hence the condition for hyperbolicity holds by taking

$$\mathbf{K}(\mathbf{U}, \theta) = \mathbf{T}^{-1}(\theta)\mathbf{K}\left(\mathbf{T}(\theta)\mathbf{U}\right) , \quad \Lambda(\mathbf{U}, \theta) = \Lambda\left(\mathbf{T}(\theta)\mathbf{U}\right) .$$

We have thus proved that the time–dependent, two dimensional Euler equations are hyperbolic in time, as claimed.

Characteristic Fields. Next we analyse the characteristic fields associated with the four eigenvectors given by (3.72).

Proposition 3.2.4 (Types of Characteristic Fields). *For $i = 1$ and $i = 4$ the $\mathbf{K}^{(i)}(\mathbf{U})$ characteristic fields are genuinely non–linear, while for $i = 2$ and $i = 3$ they are linearly degenerate.*

Proof. The proof that the fields $i = 2$ and $i = 3$ are linearly degenerate is trivial. Clearly

$$\nabla\lambda_2 = \nabla\lambda_3 = (-u/\rho, 1/\rho, 0, 0) .$$

By inspecting $\mathbf{K}^{(2)}(\mathbf{U})$ and $\mathbf{K}^{(3)}(\mathbf{U})$ it is obvious that

$$\nabla\lambda_2 \cdot \mathbf{K}^{(2)}(\mathbf{U}) = \nabla\lambda_3 \cdot \mathbf{K}^{(3)}(\mathbf{U}) = 0$$

and therefore the 2 and 3 characteristic fields are linearly degenerate as claimed. The proof for $i = 1, 4$ involves some algebra. The result is

$$\nabla\lambda_1 \cdot \mathbf{K}^{(1)}(\mathbf{U}) = -\frac{(\gamma + 1)a}{2\rho} \neq 0 , \ \nabla\lambda_4 \cdot \mathbf{K}^{(4)}(\mathbf{U}) = \frac{(\gamma + 1)a}{2\rho} \neq 0$$

and thus the 1 and 4 characteristic fields are *genuinely non–linear* as claimed.

In the context of the Riemann problem we shall see that across the 2 and 3 waves both pressure p and *normal* velocity component u are constant. The 2 field is associated with a contact discontinuity, across which density jumps discontinuously. The 3 field is associated with a shear wave across which the *tangential* velocity component jumps discontinuously. The 1 and 4 characteristic fields are associated with shock waves and rarefaction waves.

3.2.2 Three–Dimensional Equations in Conservative Form

Here we extend previous results proved for the two–dimensional equations, to the time–dependent three dimensional Euler equations. Proofs are omitted, they involve elementary but tedious algebra.

Eigenstructure. The Jacobian matrix $\mathbf{A}$ corresponding to the flux $\mathbf{F}(\mathbf{U})$ in (3.64) is given by

$$
\mathbf{A} = \frac{\partial \mathbf{F}}{\partial \mathbf{U}} =
\begin{bmatrix}
0 & 1 & 0 & 0 & 0 \\
\hat{\gamma}H - u^2 - a^2 & (3 - \gamma)u & -\hat{\gamma}v & -\hat{\gamma}w & \hat{\gamma} \\
-uv & v & u & 0 & 0 \\
-uw & w & 0 & u & 0 \\
\frac{1}{2}u[(\gamma - 3)H - a^2] & H - \hat{\gamma}u^2 & -\hat{\gamma}uv & -\hat{\gamma}uw & \gamma u
\end{bmatrix},
$$
(3.79)

where

$$
H = (E + p)/\rho = \frac{1}{2}\mathbf{V}^2 + \frac{a^2}{(\gamma - 1)} \;,\; \mathbf{V}^2 = u^2 + v^2 + w^2 \;,\; \hat{\gamma} = \gamma - 1 \;. \quad (3.80)
$$

The x–split one–dimensional system is hyperbolic with real eigenvalues

$$
\lambda_1 = u - a \;,\; \lambda_2 = \lambda_3 = \lambda_4 = u \;,\; \lambda_5 = u + a \;. \quad (3.81)
$$

The matrix of corresponding right eigenvectors is

$$
\mathbf{K} =
\begin{bmatrix}
1 & 1 & 0 & 0 & 1 \\
u - a & u & 0 & 0 & u + a \\
v & v & 1 & 0 & v \\
w & w & 0 & 1 & w \\
H - ua & \frac{1}{2}\mathbf{V}^2 & v & w & H + ua
\end{bmatrix}.
$$
(3.82)

We also give the expression for the inverse matrix of $\mathbf{K}$, namely

$$
\mathbf{K}^{-1} = \frac{(\gamma - 1)}{2a^2}
\begin{bmatrix}
H + \frac{a}{\hat{\gamma}}(u - a) & -(u + \frac{a}{\hat{\gamma}}) & -v & -w & 1 \\
-2H + \frac{4}{\hat{\gamma}}a^2 & 2u & 2v & 2w & -2 \\
-\frac{2va^2}{\hat{\gamma}} & 0 & \frac{2a^2}{\hat{\gamma}} & 0 & 0 \\
-\frac{2wa^2}{\hat{\gamma}} & 0 & 0 & \frac{2a^2}{\hat{\gamma}} & 0 \\
H - \frac{a}{\hat{\gamma}}(u + a) & -u + \frac{a}{\hat{\gamma}} & -v & -w & 1
\end{bmatrix}. \quad (3.83)
$$

Rotational Invariance. We now state the rotational invariance property for the three–dimensional case.

Proposition 3.2.5. *The time–dependent three dimensional Euler equations are rotationally invariant, that is they satisfy*

$$\cos\theta^{(y)}\cos\theta^{(z)}\mathbf{F}(\mathbf{U}) + \cos\theta^{(y)}\sin\theta^{(z)}\mathbf{G}(\mathbf{U}) + \sin\theta^{(y)}\mathbf{H}(\mathbf{U}) = \mathbf{T}^{-1}\mathbf{F}\,(\mathbf{TU})\ ,$$
(3.84)

for all angles $\theta^{(y)}$, $\theta^{(z)}$ *and vectors* $\mathbf{U}$. *Here* $\mathbf{T} = \mathbf{T}(\theta^{(y)},\theta^{(z)})$ *is the rotation matrix*

$$\mathbf{T} = \begin{bmatrix} 1 & 0 & 0 & 0 & 0 \\ 0 & \cos\theta^{(y)}\cos\theta^{(z)} & \cos\theta^{(y)}\sin\theta^{(z)} & \sin\theta^{(y)} & 0 \\ 0 & -\sin\theta^{(z)} & \cos\theta^{(z)} & 0 & 0 \\ 0 & -\sin\theta^{(y)}\cos\theta^{(z)} & -\sin\theta^{(y)}\sin\theta^{(z)} & \cos\theta^{(y)} & 0 \\ 0 & 0 & 0 & 0 & 1 \end{bmatrix},$$
(3.85)

and is the product of two rotation matrices, namely

$$\mathbf{T} = \mathbf{T}(\theta^{(y)},\theta^{(z)}) = \mathbf{T}^{(y)}\mathbf{T}^{(z)}\ ,$$
(3.86)

with

$$\mathbf{T}^{(y)} \equiv \mathbf{T}^{(y)}(\theta^{(y)}) = \begin{bmatrix} 1 & 0 & 0 & 0 & 0 \\ 0 & \cos\theta^{(y)} & 0 & \sin\theta^{(y)} & 0 \\ 0 & 0 & 1 & 0 & 0 \\ 0 & -\sin\theta^{(y)} & 0 & \cos\theta^{(y)} & 0 \\ 0 & 0 & 0 & 0 & 1 \end{bmatrix},$$

$$\mathbf{T}^{(z)} \equiv \mathbf{T}^{(z)}(\theta^{(z)}) = \begin{bmatrix} 1 & 0 & 0 & 0 & 0 \\ 0 & \cos\theta^{(z)} & \sin\theta^{(z)} & 0 & 0 \\ 0 & -\sin\theta^{(z)} & \cos\theta^{(z)} & 0 & 0 \\ 0 & 0 & 0 & 1 & 0 \\ 0 & 0 & 0 & 0 & 1 \end{bmatrix}.$$
(3.87)

More details of the rotational invariance and related properties of the three–dimensional Euler equations are found in Billett and Toro [44].

3.2.3 Three–Dimensional Primitive Variable Formulation

As done for the one–dimensional Euler equations, we can express the two and three dimensional equations in terms of primitive variables.

Proposition 3.2.6. *The three–dimensional, time–dependent Euler equations can be written in terms of the primitive variables* $\mathbf{W} = (\rho, u, v, w, p)^T$ *as*

$$\rho_t + u\rho_x + v\rho_y + w\rho_z + \rho(u_x + v_y + w_z) = 0 ,$$

$$u_t + uu_x + vu_y + wu_z + \tfrac{1}{\rho}p_x = 0 ,$$

$$v_t + uv_x + vv_y + wv_z + \tfrac{1}{\rho}p_y = 0 , \qquad\qquad (3.88)$$

$$w_t + uw_x + vw_y + ww_z + \tfrac{1}{\rho}p_z = 0 ,$$

$$p_t + up_x + vp_y + wp_z + \rho a^2(u_x + v_y + w_z) = 0 .$$

Proof. To prove this result one follows the same steps as for the one–dimensional case leading to equations (3.14)–(3.16).

Equations (3.88) can be written in quasi–linear form as

$$\mathbf{W}_t + \mathbf{A}(\mathbf{W})\mathbf{W}_x + \mathbf{B}(\mathbf{W})\mathbf{W}_y + \mathbf{C}(\mathbf{W})\mathbf{W}_z = \mathbf{0} , \qquad (3.89)$$

where the coefficient matrices $\mathbf{A}(\mathbf{W})$, $\mathbf{B}(\mathbf{W})$ and $\mathbf{C}(\mathbf{W})$ are given by

$$\mathbf{A}(\mathbf{W}) = \begin{bmatrix} u & \rho & 0 & 0 & 0 \\ 0 & u & 0 & 0 & 1/\rho \\ 0 & 0 & u & 0 & 0 \\ 0 & 0 & 0 & u & 0 \\ 0 & \rho a^2 & 0 & 0 & u \end{bmatrix} , \qquad (3.90)$$

$$\mathbf{B}(\mathbf{W}) = \begin{bmatrix} v & \rho & 0 & 0 & 0 \\ 0 & v & 0 & 0 & 0 \\ 0 & 0 & v & 0 & 1/\rho \\ 0 & 0 & 0 & v & 0 \\ 0 & 0 & \rho a^2 & 0 & v \end{bmatrix} , \qquad (3.91)$$

$$\mathbf{C}(\mathbf{W}) = \begin{bmatrix} w & \rho & 0 & 0 & 0 \\ 0 & w & 0 & 0 & 0 \\ 0 & 0 & w & 0 & 0 \\ 0 & 0 & 0 & w & 1/\rho \\ 0 & 0 & 0 & \rho a^2 & w \end{bmatrix} . \qquad (3.92)$$

Proposition 3.2.7. *The eigenvalues of the coefficient matrix* $\mathbf{A}(\mathbf{W})$ *in (3.90) are given by*

$$\lambda_1 = u - a , \ \lambda_2 = \lambda_3 = \lambda_4 = u , \ \lambda_5 = u + a . \qquad (3.93)$$

with corresponding right eigenvectors

$$
\mathbf{K}^{(1)} = \begin{bmatrix} \rho \\ -a \\ 0 \\ 0 \\ \rho a^2 \end{bmatrix} , \quad \mathbf{K}^{(2)} = \begin{bmatrix} 1 \\ 0 \\ v \\ w \\ 0 \end{bmatrix} , \quad \mathbf{K}^{(3)} = \begin{bmatrix} \rho \\ 0 \\ 1 \\ w \\ 0 \end{bmatrix} ,
$$

$$
\mathbf{K}^{(4)} = \begin{bmatrix} \rho \\ 0 \\ v \\ 1 \\ 0 \end{bmatrix} , \quad \mathbf{K}^{(5)} = \begin{bmatrix} \rho \\ a \\ 0 \\ 0 \\ \rho a^2 \end{bmatrix} . \tag{3.94}
$$

Proof. The proof involves the usual algebraic steps for finding eigenvalues and eigenvectors. See Sect. 2.1 of Chap. 2.

3.2.4 The Split Three–Dimensional Riemann Problem

When solving numerically the two or three dimensional Euler equations by most methods of the upwind type in current use, one requires the solution of *split* Riemann problems. The x–split, three–dimensional Riemann problem is the IVP

$$
\mathbf{U}_t + \mathbf{F(U)}_x = \mathbf{0} ,
$$
$$
\mathbf{U}(x,0) = \mathbf{U}^{(0)}(x) = \begin{cases} \mathbf{U_L} & \text{if} \quad x < 0 , \\ \mathbf{U_R} & \text{if} \quad x > 0 , \end{cases} \tag{3.95}
$$

where

$$
\mathbf{U} = \begin{bmatrix} \rho \\ \rho u \\ \rho v \\ \rho w \\ E \end{bmatrix} , \quad \mathbf{F(U)} = \begin{bmatrix} \rho u \\ \rho u^2 + p \\ \rho u v \\ \rho u w \\ u(E + p) \end{bmatrix} . \tag{3.96}
$$

The structure of the similarity solution is shown in Fig. 3.5 and is almost identical to that for the one–dimensional case shown in Fig. 3.1. Both pressure and *normal* particle velocity u are constant in the *Star Region*, across the middle wave. There are two new characteristic fields associated with $\lambda_3 = u$ and $\lambda_4 = u$, arising from the multiplicity 3 of the eigenvalue u; these correspond to two *shear waves* across which the respective tangential velocity components v and w change discontinuously. For the two–dimensional case we proved in Sect. 3.2.1 that the λ_3–field is linearly degenerate. This result is also true for the λ_4–field in three dimensions. The 1 and 5 characteristic fields are genuinely non–linear and are associated with rarefactions or shock waves, just as in the one–dimensional case. By inspecting the eigenvectors $\mathbf{K}^{(1)}$ and $\mathbf{K}^{(5)}$ in (3.94) we see immediately that the Generalised Riemann Invariants across 1 and 5 rarefaction waves give no change in the tangential velocity components v and w across these waves, see Fig. 3.5. In fact this is also true

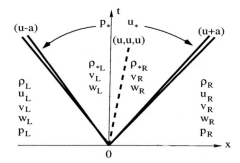

Fig. 3.5. Structure of the solution of the three–dimensional split Riemann problem

when these waves are shock waves. Consider a right shock wave of speed S associated with the 5 field. By transforming to a frame of reference in which the shock speed is zero and applying the Rankine–Hugoniot Conditions we obtain the same relations (3.42)–(3.44) as in the one–dimensional case plus two extra relations involving v and w. The three relevant relations are

$$\rho_*(u_* - S) = \rho_R(u_R - S) , \tag{3.97}$$

$$\rho_*(u_* - S)(v_* - S) = \rho_R(u_R - S)(v_R - S) , \tag{3.98}$$

$$\rho_*(u_* - S)(w_* - S) = \rho_R(u_R - S)(w_R - S) . \tag{3.99}$$

Application of the shock condition (3.97) into equations (3.98) and (3.99) gives directly $v_* = v_R$ and $w_* = w_R$. A similar analysis for a left shock wave gives an equivalent result. Hence the tangential velocity components v and w remain constant across the non–linear waves 1 and 5, irrespective of their type.

Therefore finding the solution of the Riemann problem for the split three–dimensional equations is fundamentally the same as finding the solution for the corresponding one–dimensional Riemann problem. The solution for the extra variables v and w could not be simpler: it consists of single jump discontinuities across the shear waves from the values v_L, w_L on the left data state to the values v_R, w_R on the right data state. This simple behaviour of the tangential velocity components in the solution of split Riemann problems is sometimes incorrectly modelled by some *approximate* Riemann solvers.

3.3 Conservative Versus Non–Conservative Formulations

The specific purpose of this section is first to make the point that under the assumption of smooth solutions, *conservative* and non–conservative formulations are not unique. It is vitally important to scrutinise the conservative formulations carefully, as these may be conservative purely in a mathematical sense. The key question is to see what the *conserved quantities* are in the formulation and whether the conservation statements they imply make *physical sense*. The second point of interest here is to make the reader aware of the fact that in the presence of shock waves, formulations that are *conservative purely in a mathematical sense* will produce wrong shock speeds and thus wrong solutions. We illustrate these points through the one–dimensional shallow water equations, see Sect. 1.6.3 of Chap. 1,

$$\begin{bmatrix} \phi \\ \phi u \end{bmatrix}_t + \begin{bmatrix} \phi u \\ \phi u^2 + \frac{1}{2}\phi^2 \end{bmatrix}_x = \mathbf{0} . \tag{3.100}$$

They express the physical laws of conservation of mass and momentum. Under the assumption of smooth solutions we can expand derivatives so as to write the equations in primitive–variable form

$$\phi_t + u\phi_x + \phi u_x = 0 , \tag{3.101}$$

$$u_t + uu_x + \phi_x = 0 . \tag{3.102}$$

It is tempting to derive new conservation–law forms of the shallow water equations starting from equations (3.101)–(3.102). One possibility is to keep the mass equation as in (3.101) and re–write the momentum equation (3.102) as

$$u_t + (\frac{1}{2}u^2 + \phi)_x = 0 . \tag{3.103}$$

Now we have an alternative *conservative* form of the shallow water equations, namely

$$\begin{bmatrix} \phi \\ u \end{bmatrix}_t + \begin{bmatrix} \phi u \\ \frac{1}{2}u^2 + \phi \end{bmatrix}_x = \mathbf{0} . \tag{3.104}$$

Mathematically, see Chap. 2, this is a system of conservation laws. It expresses conservation of mass, as in (3.101), and *conservation of particle speed u*. Physically, this second conservation law does not make sense. A critical question is this : *can we use the conservation–law form* (3.104) *for the shallow water equations*. The anticipated answer is : *yes we can, if and only if solutions are smooth*. In the presence of shock waves formulations (3.100) and (3.104) lead to different solutions, as we now demonstrate.

Without loss of generality we consider a right facing shock wave in which the state ahead of the shock is given by the variables ϕ_R, u_R.

Proposition 3.3.1. *A right–facing shock wave solution of (3.100) has shock speed*

$$\left. \begin{aligned} S &= u_R + Q/\phi_R , \\ Q &= \left[\frac{1}{2}\left(\phi_* + \phi_R\right)\phi_*\phi_R\right]^{\frac{1}{2}} , \end{aligned} \right\} \tag{3.105}$$

while a right–facing shock wave solution of (3.104) has speed

$$\left. \begin{aligned} \hat{S} &= u_R + \hat{Q}/\phi_R , \\ \hat{Q} &= \left[\frac{2}{\phi_*+\phi_R}\right]^{\frac{1}{2}} \phi_*\phi_R . \end{aligned} \right\} \tag{3.106}$$

Proof. This is left to the reader as an exercise. Use contents of Chap. 2 and those of Sect. 3.1.3 of Chap. 3.

Remark 3.3.1. Clearly the shock speeds S and $\hat{S}$ are equivalent only when $\phi_* \equiv \phi_R$, that is when the shock wave is trivial. In general

$$\hat{S} \leq S \tag{3.107}$$

and thus shock solutions of the *new* (incorrect) conservation laws (3.104) are slower than shock solutions of the *conventional* (correct) conservation laws (3.100). Note also that the conservative form (3.104) is non–unique.

Consider now the isothermal equations of Gas Dynamics, see Sect. 1.6.2 of Chap. 1. In conservation–law form these equations read

$$\left[\begin{array}{c} \rho \\ \rho u \end{array} \right]_t + \left[\begin{array}{c} \rho u \\ \rho u^2 + a^2 \rho \end{array} \right]_x = \mathbf{0} \,, \tag{3.108}$$

where the sound speed a is constant. These conservation laws state that mass and momentum are conserved, which is in accord with the laws of conservation of mass and momentum studied in Chap. 1. Let us now assume that solutions are sufficiently smooth so that partial derivatives exist; we expand derivatives and after some algebraic manipulations obtain the primitive–variable formulation

$$\rho_t + u\rho_x + \rho u_x = 0 \,, \tag{3.109}$$

$$u_t + uu_x + \frac{a^2}{\rho}\rho_x = 0 \,. \tag{3.110}$$

This is a perfectly acceptable formulation, valid for smooth flows.

New conservation laws can be constructed, starting from the primitive formulation (3.109)–(3.110) above. One such possible system of conservation laws is

$$\left[\begin{array}{c} \rho \\ u \end{array} \right]_t + \left[\begin{array}{c} \rho u \\ \frac{1}{2}u^2 + a^2 ln\rho \end{array} \right]_x = \mathbf{0} \,. \tag{3.111}$$

Mathematically, these equations are a set of conservation laws, see Sects. 2.1 and 2.4 of Chap. 2. Physically however, they are useless, they state that mass and *velocity* are conserved !

Exercise 3.3.1. Using the contents of Sect. 3.1.3 for isolated shock waves, compare the shock solutions of the two conservative formulations (3.108) and (3.111). Which gives the fastest shock ? Find other *conservative* formulations corresponding to (3.109)–(3.110).

4. The Riemann Problem for the Euler Equations

In his classical paper of 1959, Godunov [145] presented a conservative extension of the first–order upwind scheme of Courant, Isaacson and Rees [98] to non–linear systems of hyperbolic conservation laws. The key ingredient of the scheme is the solution of the Riemann problem. The purpose of this chapter is to provide a detailed presentation of the complete, exact solution to the Riemann problem for the one–dimensional, time–dependent Euler equations for ideal and covolume gases, including vacuum conditions. The methodology can then be applied to other hyperbolic systems.

The exact solution of the Riemann problem is useful in a number of ways. First, it represents the solution to a system of hyperbolic conservation laws subject to the simplest, non–trivial, initial conditions; in spite of the simplicity of the initial data the solution of the Riemann problem contains the fundamental physical and mathematical character of the relevant set of conservation laws. The solution of the general IVP may be seen as resulting from non–linear superposition of solutions of local Riemann problems [141]. In the case of the Euler equations the Riemann problem includes the so called *shock–tube problem*, a basic physical problem in Gas Dynamics. For a detailed discussion on the shock–tube problem the reader is referred to the book by Courant and Friedrichs [97]. The exact Riemann problem solution is also an invaluable reference solution that is useful in assessing the performance of numerical methods and to check the correctness of programs in the early stages of development. The Riemann problem solution, exact or approximate, can also be used locally in the method of Godunov and high–order extensions of it; this is the main role we assign to the Riemann problem here. A detailed knowledge of the exact solution is also fundamental when utilising, assessing and developing approximate Riemann solvers.

There is no exact closed–form solution to the Riemann problem for the Euler equations, not even for ideal gases; in fact not even for much simpler models such as the isentropic and isothermal equations. However, it is possible to devise *iterative* schemes whereby the solution can be computed numerically to any desired, practical, degree of accuracy. Key issues in designing an exact Riemann solver are: the variables selected, the equations used, the number of equations and the technique for the iterative procedure, the initial guess and the handling of unphysical iterates, such as negative pressure. Godunov is credited with the first exact Riemann solver for the Euler equations [145]. By today's standards Godunov's first Riemann solver is cumbersome and computationally inefficient. Later, Godunov [147] proposed a second exact Riemann solver. Distinct features of this solver are: the equations used are simpler, the variables selected are more convenient from the computational point of view and the iterative procedure is rather sophisticated. Much of the work that followed contains the fundamental features of Godunov's second Riemann solver. Chorin [73], independently, produced improvements to Godunov's first Riemann solver. In 1979, van Leer [392] produced another improvement to Godunov's first Riemann solver resulting in a scheme that is similar to Godunov's second solver. Smoller [316] proposed a rather different approach; later, Dutt [115] produced a practical implementation of the scheme. Gottlieb and Groth [150] presented another Riemann solver for ideal gases; of the schemes they tested, theirs is shown to be the most efficient. Toro [351] presented an exact Riemann solver for ideal and covolume gases of comparable efficiency to that of Gottlieb and Groth. More recently, Schleicher [299] and Pike [263] have also presented new exact Riemann solvers which appear to be the fastest to date. For gases obeying a general equation of state the reader is referred to the pioneering work of Colella and Glaz [89]. Other relevant publications are that of Menikoff and Plohr [240] and that of Saurel, Larini and Loraud [298].

In this chapter we present a solution procedure of the Riemann problem for the Euler equations for both ideal and covolume gases. The methodology is presented in great detail for the ideal gas case. We then address the issue of *vacuum* and provide an exact solution for the three cases that can occur. Particular emphasis is given to the *sampling* of the solution; this will be useful to provide the complete solution and to utilise it in numerical methods such as the Godunov method [145] and Glimm's method or Random Choice Method [141], [73]. The necessary background for this chapter is found in Chaps. 1, 2 and 3.

4.1 Solution Strategy

The Riemann problem for the one–dimensional time–dependent Euler equations is the Initial Value Problem (IVP) for the conservation laws

$$\left. \begin{array}{c} \mathbf{U}_t + \mathbf{F}(\mathbf{U})_x = \mathbf{0} \; , \\[2mm] \mathbf{U} = \begin{bmatrix} \rho \\ \rho u \\ E \end{bmatrix} \; , \quad \mathbf{F} = \begin{bmatrix} \rho u \\ \rho u^2 + p \\ u(E + p) \end{bmatrix} \; , \end{array} \right\} \tag{4.1}$$

with initial conditions (IC)

$$\mathbf{U}(x,0) = \mathbf{U}^{(0)}(x) = \begin{cases} \mathbf{U}_{\mathrm{L}} & \text{if} \quad x < 0 \,, \\ \mathbf{U}_{\mathrm{R}} & \text{if} \quad x > 0 \,. \end{cases} \tag{4.2}$$

The domain of interest in the x–t plane are points (x,t) with $-\infty < x < \infty$ and $t > 0$. In practice one lets x vary in a finite interval $[x_{\mathrm{L}}, x_{\mathrm{R}}]$ around the point $x = 0$. In solving the Riemann problem we shall frequently make use of the vector $\mathbf{W} = (\rho, u, p)^T$ of primitive variables, rather than the vector $\mathbf{U}$ of conserved variables, where ρ is density, u is particle velocity and p is pressure. The Riemann problem (4.1)–(4.2) is the simplest, non–trivial, IVP for (4.1). Data consists of just two constant states, which in terms of primitive variables are $\mathbf{W}_{\mathrm{L}} = (\rho_{\mathrm{L}}, u_{\mathrm{L}}, p_{\mathrm{L}})^T$ to the left of $x = 0$ and $\mathbf{W}_{\mathrm{R}} = (\rho_{\mathrm{R}}, u_{\mathrm{R}}, p_{\mathrm{R}})^T$ to the right of $x = 0$, separated by a discontinuity at $x = 0$. Physically, in the context of the Euler equations, the Riemann problem is a slight generalisation of the so called *shock–tube problem*: two stationary gases $(u_{\mathrm{L}} = u_{\mathrm{R}} = 0)$ in a tube are separated by a diaphragm. The rupture of the diaphragm generates a nearly *centred* wave system that typically consists of a rarefaction wave, a contact discontinuity and a shock wave. This physical problem is reasonably well approximated by solving the shock–tube problem for the Euler equations. In the Riemann problem the particle speeds u_{L} and u_{R} are allowed to be non–zero, but the structure of the solution is the same as that of the shock–tube problem. In general, given the conservation equations (4.1) for

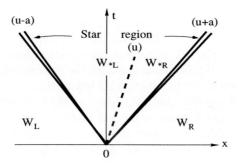

Fig. 4.1. Structure of the solution of the Riemann problem on the x-t plane for the one–dimensional time–dependent Euler equations

the dynamics, it is left to the statements about the material, the equation of state, to determine not only the structure of the solution of the Riemann problem but also the mathematical character of the equations. In this chapter we restrict our attention to *ideal gases* obeying the caloric Equation of State (EOS)

$$e = \frac{p}{(\gamma - 1)\rho} \,, \tag{4.3}$$

and *covolume gases* obeying

$$e = \frac{p(1 - b\rho)}{(\gamma - 1)\rho} ,$$ (4.4)

where γ is the ratio of specific heats, a constant, and b is the covolume, also a constant. See Sects. 1.2.4 and 1.2.5 of Chap. 1. For the case in which no

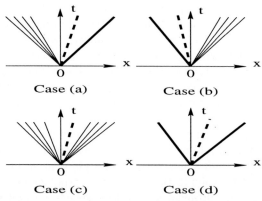

Fig. 4.2. Possible wave patterns in the solution of the Riemann problem: (a) left rarefaction, contact, right shock (b) left shock, contact, right rarefaction (c) left rarefaction, contact, right rarefaction (d) left shock, contact, right shock

vacuum is present the exact solution of the Riemann problem (4.1), (4.2) has three waves, which are *associated* with the eigenvalues $\lambda_1 = u - a$, $\lambda_2 = u$ and $\lambda_3 = u + a$; see Fig. 4.1. Note that the speeds of these waves are not, in general, the characteristics speeds given by the eigenvalues. The three waves separate four constant states, which from left to right are: $\mathbf{W}_L$ (data on the left hand side), $\mathbf{W}_{*L}$, $\mathbf{W}_{*R}$ and $\mathbf{W}_R$ (data on the right hand side).

The unknown region between the left and right waves, the *Star Region*, is divided by the middle wave into the two subregions *Star Left* ($\mathbf{W}_{*L}$) and *Star Right* ($\mathbf{W}_{*R}$). As seen in Sect. 3.1.3 of Chap. 3, the middle wave is always a contact discontinuity while the left and right (non–linear) waves are either shock or rarefaction waves. Therefore, according to the type of non–linear waves there can be four possible wave patterns, which are shown in Fig. 4.2. There are two possible variations of these, namely when the left or right non–linear wave is a *sonic rarefaction wave*; these two cases are only of interest when utilising the solution of the Riemann problem in Godunov–type methods. For the purpose of constructing a solution scheme for the Riemann problem it is sufficient to consider the four patterns of Fig. 4.2.

An analysis based on the eigenstructure of the Euler equations, Sect. 3.1.3 Chap. 3, reveals that both pressure p_* and particle velocity u_* between the left and right waves are constant, while the density takes on the two constant values ρ_{*L} and ρ_{*R}. Here we present a solution procedure which makes use of the constancy of pressure and particle velocity in the *Star Region* to derive a single, algebraic non–linear equation for pressure p_*. In summary, the main physical quantities sought are p_*, u_*, ρ_{*L} and ρ_{*R}.

4.2 Equations for Pressure and Particle Velocity

Here we establish equations and solution strategies for computing the pressure p_* and the particle velocity u_* in the *Star Region*.

Proposition 4.2.1 (solution for p_* and u_*). *The solution for pressure p_* of the Riemann problem (4.1), (4.2) with the ideal gas Equation of State (4.3) is given by the root of the algebraic equation*

$$f(p, \mathbf{W}_L, \mathbf{W}_R) \equiv f_L(p, \mathbf{W}_L) + f_R(p, \mathbf{W}_R) + \Delta u = 0 , \quad \Delta u \equiv u_R - u_L , \quad (4.5)$$

where the function f_L is given by

$$f_L(p, \mathbf{W}_L) = \begin{cases} (p - p_L) \left[\dfrac{A_L}{p + B_L} \right]^{\frac{1}{2}} & \text{if } p > p_L \text{ (shock)}, \\[3mm] \dfrac{2a_L}{(\gamma - 1)} \left[\left(\dfrac{p}{p_L} \right)^{\frac{\gamma - 1}{2\gamma}} - 1 \right] & \text{if } p \le p_L \text{ (rarefaction)}, \end{cases} \quad (4.6)$$

the function f_R is given by

$$f_R(p, \mathbf{W}_R) = \begin{cases} (p - p_R) \left[\dfrac{A_R}{p + B_R} \right]^{\frac{1}{2}} & \text{if } p > p_R \text{ (shock)}, \\[3mm] \dfrac{2a_R}{(\gamma - 1)} \left[\left(\dfrac{p}{p_R} \right)^{\frac{\gamma - 1}{2\gamma}} - 1 \right] & \text{if } p \le p_R \text{ (rarefaction)}, \end{cases} \quad (4.7)$$

and the data–dependent constants A_L, B_L, A_R, B_R are given by

$$\left. \begin{array}{ll} A_L = \dfrac{2}{(\gamma + 1)\rho_L} , & B_L = \dfrac{(\gamma - 1)}{(\gamma + 1)} p_L , \\[3mm] A_R = \dfrac{2}{(\gamma + 1)\rho_R} , & B_R = \dfrac{(\gamma - 1)}{(\gamma + 1)} p_R . \end{array} \right\} \quad (4.8)$$

The solution for the particle velocity u_* in the Star Region is

$$u_* = \frac{1}{2}(u_L + u_R) + \frac{1}{2} [f_R(p_*) - f_L(p_*)] . \quad (4.9)$$

Remark 4.2.1. Before proceeding to prove the above statements we make some useful remarks. Once (4.5) is solved for p_* the solution for u_* follows as in (4.9) and the remaining unknowns are found by using standard gas dynamics relations studied in Chap. 3. The function f_L governs relations *across* the left non–linear wave and serves to connect the unknown particle speed u_* to the known state $\mathbf{W}_L$ on the left side, see Fig. 4.3; the relations depend on the type of wave (shock or rarefaction). The arguments of f_L are the pressure p and the data state $\mathbf{W}_L$. Similarly, the function f_R governs relations across the right wave and connects the unknown u_* to the right data state $\mathbf{W}_R$; its arguments are p and $\mathbf{W}_R$. For convenience we shall often omit the data arguments of the functions f, f_L and f_R. The sought pressure p_* in the *Star Region* is the root of the algebraic equation (4.5), $f(p) = 0$. A detailed analysis of the pressure function $f(p)$ reveals a particularly simple behaviour and that for *physically relevant data* there exists a unique solution to the equation $f(p) = 0$.

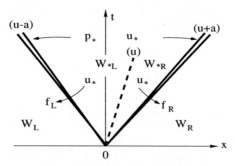

Fig. 4.3. Strategy for solving the Riemann problem via a pressure function. The particle velocity is connected to data on the left and right via functions f_L and f_R

Proof. Here we derive expressions for f_L and f_R in equation (4.5). We do this by considering each non–linear wave separately.

4.2.1 Function f_L for a Left Shock

We assume the left wave is a shock moving with speed S_L as shown in Fig. 4.4a; pre–shock values are ρ_L, u_L and p_L and post–shock values are ρ_{*L}, u_* and p_*.

As done in Sect. 3.1.3 of Chap. 3, we transform the equations to a frame of reference moving with the shock, as depicted in Fig. 4.4b. In the new frame the shock speed is zero and the *relative velocities* are

$$\hat{u}_L = u_L - S_L \ , \quad \hat{u}_* = u_* - S_L \ . \tag{4.10}$$

The Rankine–Hugoniot Conditions, see Sect. 3.1.3 of Chap. 3, give

$$\rho_L \hat{u}_L = \rho_{*L} \hat{u}_* \ , \tag{4.11}$$

$$\rho_L \hat{u}_L^2 + p_L = \rho_{*L} \hat{u}_*^2 + p_* \ , \tag{4.12}$$

$$\hat{u}_L (\hat{E}_L + p_L) = \hat{u}_* (\hat{E}_{*L} + p_*) \ . \tag{4.13}$$

We introduce the *mass flux* Q_L, which in view of (4.11) may be written as

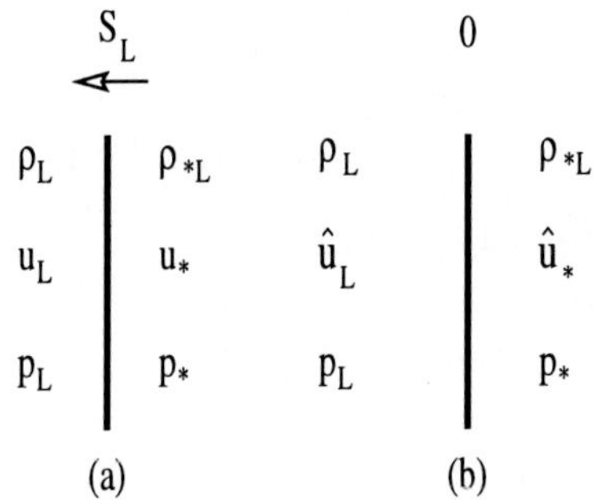

Fig. 4.4. Left wave is a shock wave of speed S_L: (a) stationary frame, shock speed is S_L (b) frame of reference moving with speed S_L, shock speed is zero

$$Q_L \equiv \rho_L \hat{u}_L = \rho_{*L} \hat{u}_* . \tag{4.14}$$

From equation (4.12)

$$(\rho_L \hat{u}_L)\hat{u}_L + p_L = (\rho_{*L} \hat{u}_*)\hat{u}_* + p_* .$$

Use of (4.14) and solving for Q_L gives

$$Q_L = -\frac{p_* - p_L}{\hat{u}_* - \hat{u}_L} . \tag{4.15}$$

But from equation (4.10) $\hat{u}_L - \hat{u}_* = u_L - u_*$ and so Q_L becomes

$$Q_L = -\frac{p_* - p_L}{u_* - u_L} , \tag{4.16}$$

from which we obtain

$$u_* = u_L - \frac{(p_* - p_L)}{Q_L} . \tag{4.17}$$

We are now close to having related u_* to data on the left hand side. We seek to express the right hand side of (4.17) purely in terms of p_* and $\mathbf{W}_L$, which means that we need to express Q_L as a function of p_* and the data on the left hand side. We substitute the relations

$$\hat{u}_L = \frac{Q_L}{\rho_L} , \quad \hat{u}_* = \frac{Q_L}{\rho_{*L}} ,$$

obtained from (4.14) into equation (4.15) to produce

$$Q_L^2 = -\frac{p_* - p_L}{\frac{1}{\rho_{*L}} - \frac{1}{\rho_L}} . \tag{4.18}$$

As seen in Sect. 3.1.3 of Chap. 3, the density ρ_{*L} is related to the pressure p_* behind the left shock via

$$\rho_{*L} = \rho_L \left[\frac{\left(\frac{\gamma-1}{\gamma+1} \right) + \left(\frac{p_*}{p_L} \right)}{\left(\frac{\gamma-1}{\gamma+1} \right) \left(\frac{p_*}{p_L} \right) + 1} \right] . \tag{4.19}$$

Substitution of ρ_{*L} into (4.18) yields

$$Q_L = \left[\frac{p_* + B_L}{A_L} \right]^{\frac{1}{2}} , \tag{4.20}$$

which in turn reduces (4.17) to

$$u_* = u_L - f_L(p_*, \mathbf{W}_L) , \tag{4.21}$$

with

$$f_L(p_*, \mathbf{W}_L) = (p_* - p_L) \left[\frac{A_L}{p_* + B_L} \right]^{\frac{1}{2}}$$

and

$$A_L = \frac{2}{(\gamma+1)\rho_L} , \quad B_L = \frac{(\gamma-1)}{(\gamma+1)} p_L .$$

Thus, the sought expression for f_L for the case in which the left wave is a shock wave has been obtained.

4.2.2 Function f_L for Left Rarefaction

Now we derive an expression for f_L for the case in which the left wave is a rarefaction wave, as shown in Fig. 4.5. The unknown state $\mathbf{W}_{*L}$ is now connected to the left data state $\mathbf{W}_L$ using the isentropic relation and the Generalised Riemann Invariants for the left wave. As seen in Sect. 3.1.2 of

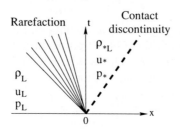

Fig. 4.5. Left wave is a rarefaction wave that connects the data state $\mathbf{W}_L$ with the unknown state $\mathbf{W}_{*L}$ in the star region to the left of the contact discontinuity

Chap. 3, the isentropic law

$$p = C\rho^\gamma , \tag{4.22}$$

where C is a constant, may be used across rarefactions. C is evaluated at the *initial* left data state by applying the isentropic law, namely

$$p_L = C\rho_L^\gamma \,,$$

and so the constant C is

$$C = p_L/\rho_L^\gamma \,,$$

from which we write

$$\rho_{*L} = \rho_L \left(\frac{p_*}{p_L}\right)^{\frac{1}{\gamma}} . \qquad (4.23)$$

In Sect. 3.1.3 of Chap. 3 we showed that across a left rarefaction the Generalised Riemann Invariant $I_L(u, a)$ is constant. By evaluating the constant on the left data state we write

$$u_L + \frac{2a_L}{\gamma - 1} = u_* + \frac{2a_{*L}}{\gamma - 1} \,, \qquad (4.24)$$

where a_L and u_{*L} denote the sound speed on the left and right states bounding the left rarefaction wave. See Fig. 4.5.

Substitution of ρ_{*L} from (4.23) into the definition of a_{*L} gives

$$a_{*L} = a_L \left(\frac{p_*}{p_L}\right)^{\frac{\gamma-1}{2\gamma}} , \qquad (4.25)$$

and equation (4.24) leads to

$$u_* = u_L - f_L(p_*, \mathbf{W}_L) \,, \qquad (4.26)$$

with

$$f_L(p_*, \mathbf{W}_L) = \frac{2a_L}{(\gamma - 1)} \left[\left(\frac{p_*}{p_L}\right)^{\frac{\gamma-1}{2\gamma}} - 1\right] .$$

This is the required expression for the function f_L for the case in which the left wave is a rarefaction wave.

4.2.3 Function f_R for a Right Shock

Here we find the expression for the function f_R for the case in which the right wave is a shock wave travelling with speed S_R. The situation is entirely analogous to the case of a left shock wave. Pre–shock values are ρ_R, u_R and p_R and post–shock values are ρ_{*R}, u_* and p_*. In the transformed frame of reference moving with the shock, the shock speed is zero and the *relative* velocities are

$$\hat{u}_R = u_R - S_R \,, \quad \hat{u}_* = u_* - S_R \,. \qquad (4.27)$$

Application of the Rankine–Hugoniot Conditions gives

$$\left. \begin{array}{l} \rho_{*R}\hat{u}_* = \rho_R\hat{u}_R \,, \\[2mm] \rho_{*R}\hat{u}_*^2 + p_* = \rho_R\hat{u}_R^2 + p_R \,, \\[2mm] \hat{u}_*(\hat{E}_{*R} + p_*) = \hat{u}_R(\hat{E}_R + p_R) \,. \end{array} \right\} \qquad (4.28)$$

Now the *mass flux* is defined as

$$Q_R \equiv -\rho_{*R}\hat{u}_* = -\rho_R\hat{u}_R .$$

(4.29)

By performing algebraic manipulations similar to those for a left shock we derive the following expression for the mass flux

$$Q_R = \left[\frac{p_* + B_R}{A_R}\right]^{\frac{1}{2}} .$$

(4.30)

Hence the particle velocity in the *Star Region* satisfies

$$u_* = u_R + f_R(p_*, \mathbf{W}_R) ,$$

(4.31)

with

$$f_R(p_*, \mathbf{W}_R) = (p_* - p_R)\left[\frac{A_R}{p_* + B_R}\right]^{\frac{1}{2}} ,$$

$$A_R = \frac{2}{(\gamma+1)\rho_R} , \quad B_R = \frac{(\gamma-1)}{(\gamma+1)}p_R .$$

This is the sought expression for f_R for the case in which the right wave is a shock wave.

4.2.4 Function f_R for a Right Rarefaction

The derivation of the function f_R for the case in which the right wave is a rarefaction wave is carried out in an entirely analogous manner to the case of a left rarefaction. The isentropic law gives

$$\rho_{*R} = \rho_R\left(\frac{p_*}{p_R}\right)^{\frac{1}{\gamma}}$$

(4.32)

and the Generalised Riemann Invariant $I_R(u, a)$ for a right rarefaction gives

$$u_* - \frac{2a_{*R}}{\gamma-1} = u_R - \frac{2a_R}{\gamma-1} .$$

(4.33)

Using (4.32) into the definition of sound speed a_{*R} gives

$$a_{*R} = a_R\left(\frac{p_*}{p_R}\right)^{\frac{\gamma-1}{2\gamma}} ,$$

(4.34)

which if substituted into (4.33) leads to

$$u_* = u_R + f_R(p_*, \mathbf{W}_R) ,$$

(4.35)

with

$$f_R(p_*, \mathbf{W}_R) = \frac{2a_R}{\gamma - 1} \left[\left(\frac{p_*}{p_R} \right)^{\frac{\gamma-1}{2\gamma}} - 1 \right] .$$

The functions f_L and f_R have now been determined for all four possible wave patterns of Fig. 4.2. Now by eliminating u_* from equations (4.21) or (4.26) and (4.31) or (4.35) we obtain a single equation

$$f(p_*, \mathbf{W}_L, \mathbf{W}_R) \equiv f_L(p_*, \mathbf{W}_L) + f_R(p_*, \mathbf{W}_R) + \Delta u = 0 , \qquad (4.36)$$

which is the required equation (4.5) for the pressure. This proves the first part of the proposition. Assuming this single non–linear algebraic equation is solved (numerically) for p_* then the solution for the particle velocity u_* can be found from equation (4.21) if the left wave is a shock $(p_* > p_L)$ or from equation (4.26) if the left wave is a rarefaction $(p_* \le p_L)$ or from equation (4.31) if the right wave is a shock $(p_* > p_R)$ or from equation (4.35) if the right wave is a rarefaction wave $(p_* \le p_R)$. It can also be found from a mean value as

$$u_* = \frac{1}{2}(u_L + u_R) + \frac{1}{2}[f_R(p_*) - f_L(p_*)] ,$$

which is equation (4.9), and the proposition has thus been proved.

4.3 Numerical Solution for Pressure

The unknown pressure p_* in the *Star Region* is found by solving the single algebraic equation (4.5), $f(p) = 0$, numerically. Any standard technique can be used. See Maron and Lopez [228] for background on numerical methods for algebraic equations. The behaviour of the pressure function $f(p)$ plays a fundamental role in finding its roots numerically.

4.3.1 Behaviour of the Pressure Function

Given data ρ_L, u_L, p_L and ρ_R, u_R, p_R the pressure function $f(p)$ behaves as shown in Fig. 4.6. It is monotone and concave down as we shall demonstrate. The first derivatives of f_K (K=L,R) with respect to p are

$$f'_K = \begin{cases} \left(\frac{A_K}{B_K + p} \right)^{1/2} \left[1 - \frac{p - p_K}{2(B_K + p)} \right] & \text{if } p > p_K \text{ (shock)} , \\ \frac{1}{\rho_K a_K} \left(\frac{p}{p_K} \right)^{-(\gamma+1)/2\gamma} & \text{if } p \le p_K \text{ (rarefaction)} . \end{cases} \qquad (4.37)$$

As $f' = f'_L + f'_R$ and by inspection $f'_K > 0$, the function $f(p)$ is monotone as claimed. The second derivatives of the functions f_K are

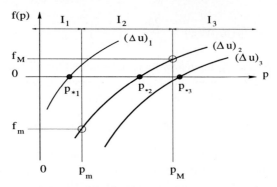

Fig. 4.6. Behaviour of the pressure function in the solution of the Riemann problem

$$f''_K = \begin{cases} -\frac{1}{4} \left(\frac{A_K}{B_K + p} \right)^{1/2} \left[\frac{4B_K + 3p + p_K}{(B_K + p)^2} \right] & \text{if } p > p_K \text{ (shock) },\\[4mm] -\frac{(\gamma+1)a_K}{2\gamma^2 p_K^2} \left(\frac{p}{p_K} \right)^{-(3\gamma+1)/2\gamma} & \text{if } p \le p_K \text{ (rarefaction) }. \end{cases} \qquad (4.38)$$

Since $f'' = f''_L + f''_R$ and $f''_K < 0$ the function $f(p)$ is concave down as anticipated. From equations (4.37) and (4.38) it can be seen that $f'_K \to 0$ as $p \to \infty$ and $f''_K \to 0$ as $p \to \infty$. This behaviour of f_K, and thus of $f(p)$, has implications when devising iteration schemes to find the zero p_* of $f(p) = 0$. The velocity difference $\Delta u = u_R - u_L$ and the pressure values p_L, p_R are the most important parameters of $f(p)$. With reference to Fig. 4. 6 we define

$$p_{\min} = \min(p_L, p_R) , \quad p_{\max} = \max(p_L, p_R) ,$$

$$f_{\min} = f(p_{\min}) , \quad f_{\max} = f(p_{\max}) .$$

For given p_L, p_R it is the velocity difference Δu which determines the value of p_*. Three intervals I_1, I_2 and I_3 can be identified:

$$\left. \begin{aligned} &p_* \text{ lies in } I_1 = (0, \, p_{\min}) && \text{if } f_{\min} > 0 \text{ and } f_{\max} > 0 ,\\[2mm] &p_* \text{ lies in } I_2 = [p_{\min}, \, p_{\max}] && \text{if } f_{\min} \le 0 \text{ and } f_{\max} \ge 0 ,\\[2mm] &p_* \text{ lies in } I_3 = (p_{\max}, \, \infty) && \text{if } f_{\min} < 0 \text{ and } f_{\max} < 0 . \end{aligned} \right\} \qquad (4.39)$$

For sufficiently large Δu, as $(\Delta u)_1$ in Fig. 4.6, the solution p_* is as p_{*1}, which lies in I_1 and thus $p_* < p_L$, $p_* < p_R$; so the two non–linear waves are rarefaction waves. For Δu as $(\Delta u)_2$ in Fig. 4.6 $p_* = p_{*2}$ lies between p_L and p_R and hence one non–linear wave is a rarefaction wave and the other is a shock wave. For sufficiently small values of Δu, as $(\Delta u)_3$ in Fig. 4.6, $p_* = p_{*3}$ lies in I_3, that is $p_* > p_L$, $p_* > p_R$, which means that both non–linear waves are shock waves. The interval where p_* lies is identified by noting the signs of $f_{\min}$ and $f_{\max}$; see 4.39.

Another observation on the behaviour of $f(p)$ is this: in I_1 both $f'(p)$ and $f''(p)$ vary rapidly; this may lead to numerical difficulties when searching for the root of $f(p) = 0$. As p increases the shape of $f(p)$ tends to resemble that of a straight line. For non–vacuum initial data $\mathbf{W}_L$, $\mathbf{W}_R$ there exists a unique positive solution p_* for pressure, provided Δu is *sufficiently small*. As a matter of fact, even for the case in which the data states are *non–vacuum states*, values of Δu larger than a critical value $(\Delta u)_{crit}$ lead to vacuum in the solution of the Riemann problem. The critical value can be found analytically in terms of the initial data. Clearly for a positive solution for pressure p_* we require $f(0) < 0$. Direct evaluation of $f(p)$ gives the *pressure positivity condition*

$$(\Delta u)_{crit} \equiv \frac{2a_L}{\gamma - 1} + \frac{2a_R}{\gamma - 1} > u_R - u_L . \tag{4.40}$$

Vacuum is created by the non–linear waves if this condition is violated. The structure of the solution in this case is different from that depicted in Fig. 4.1 and so is the method of solution, as we shall see in Sect. 4.6 of this chapter.

4.3.2 Iterative Scheme for Finding the Pressure

Given the particularly simple behaviour of the pressure function $f(p)$ and the availability of analytical expressions for the derivative of $f(p)$ we use a Newton–Raphson [228] iterative procedure to find the root of $f(p) = 0$. Suppose a guess value p_0 for the true solution p_* is available; since $f(p)$ is a smooth function we can find an approximate value of $f(p)$ at a neighbouring point $p_0 + \delta$ via a Taylor expansion

$$f(p_0 + \delta) = f(p_0) + \delta f'(p_0) + O(\delta^2) . \tag{4.41}$$

If the $p_0 + \delta$ is a solution of $f(p) = 0$ then

$$f(p_0) + \delta f'(p_0) = 0 , \tag{4.42}$$

and so the *corrected* value $p_1 = p_0 + \delta$ is

$$p_1 = p_0 - \frac{f(p_0)}{f'(p_0)} . \tag{4.43}$$

The above procedure generalises to

$$p_{(k)} = p_{(k-1)} - \frac{f(p_{(k-1)})}{f'(p_{(k-1)})} , \tag{4.44}$$

where $p_{(k)}$ is the k–th iterate. The iteration procedure is stopped whenever the relative pressure change

$$CHA = \frac{|p_{(k)} - p_{(k-1)}|}{\frac{1}{2} \left[p_{(k)} + p_{(k-1)} \right]} , \tag{4.45}$$

is less than a prescribed small tolerance TOL. Typically $TOL = 10^{-6}$.

In order to implement the iteration scheme (4.44) we need a guess value p_0 for the pressure. Given the benign behaviour of $f(p)$ the choice of p_0 is not too critical. An inadequate choice of p_0 results in a large number of iterations to achieve convergence. A difficulty that requires special handling in the Newton–Raphson method arises when the root is close to zero (strong rarefaction waves) and the guess value p_0 is too large: the next iterate for pressure can be negative. This is due to the rapid variations of the first and second derivatives of $f(p)$ near $p = 0$. We illustrate the effect of the initial guess value by considering four possible choices. Three of these are approximations to the solution p_* for pressure, see Chap. 9 for details. One such approximation is the so called *Two–Rarefaction* approximation

$$
p_{TR} = \left[\frac{a_L + a_R - \frac{1}{2}(\gamma - 1)(u_R - u_L)}{a_L/p_L^{\frac{\gamma-1}{2\gamma}} + a_R/p_R^{\frac{\gamma-1}{2\gamma}}} \right]^{\frac{2\gamma}{\gamma-1}} , \tag{4.46}
$$

and results from the exact function (4.5) for pressure under the assumption that the two non–linear waves are rarefaction waves. If the solution actually consists of two rarefactions then p_{TR} is exact and no iteration is required. A second guess value results from a linearised solution based on primitive variables. This is

$$
\left. \begin{aligned}
p_0 &= \max(TOL, p_{PV}) , \\[2mm]
p_{PV} &= \tfrac{1}{2}(p_L + p_R) - \tfrac{1}{8}(u_R - u_L)(\rho_L + \rho_R)(a_L + a_R) .
\end{aligned} \right\} \tag{4.47}
$$

A third guess value is given by a *Two–Shock* approximation

$$
\left. \begin{aligned}
p_0 &= \max(TOL, p_{TS}) , \\[2mm]
p_{TS} &= \frac{g_L(\hat{p})p_L + g_R(\hat{p})p_R - \Delta u}{g_L(\hat{p}) + g_R(\hat{p})} , \\[2mm]
g_K(p) &= \left(\frac{A_K}{p + B_K} \right)^{\frac{1}{2}} ,
\end{aligned} \right\} \tag{4.48}
$$

where A_K and B_K given by (4.8). Here $\hat{p}$ is an estimate of the solution; the value $\hat{p} = p_0$ given by (4.47) works well. Note that approximate solutions may predict, incorrectly, a negative value for pressure, even when condition (4.40) is satisfied. Thus in order to avoid negative guess values we introduce the small positive constant TOL, as used in the iteration procedure. As a fourth guess value we utilise the arithmetic mean of the data, namely

$$
p_0 = \frac{1}{2}(p_L + p_R) . \tag{4.49}
$$

Next, we carry out some tests on the effect of the various guess values for p_0 on the convergence of the Newton–Raphson iterative scheme for finding the pressure p_*.

4.3.3 Numerical Tests

Five Riemann problems are selected to test the performance of the Riemann solver and the influence of the initial guess for pressure. The tests are also used to illustrate some typical wave patterns resulting from the solution of the Riemann problem. Table 4.1 shows the data for all five tests in terms of primitive variables. In all cases the ratio of specific heats is $\gamma = 1.4$. The source code for the exact Riemann solver, called HE-E1RPEXACT, is part of the library *NUMERICA* [369]; a listing is given in Sect. 4.9.

Test 1 is the so called Sod test problem [318]; this is a very mild test and its solution consists of a left rarefaction, a contact and a right shock. Fig. 4.7 shows solution profiles for density, velocity, pressure and specific internal energy across the complete wave structure, at time $t = 0.25$ units. Test 2, called the *123 problem*, has solution consisting of two strong rarefactions and a trivial stationary contact discontinuity; the pressure p_* is very small (close to vacuum) and this can lead to difficulties in the iteration scheme to find p_* numerically. Fig. 4.8 shows solution profiles. Test 2 is also useful in assessing the performance of numerical methods for low density flows, see Einfeldt et. al. [118]. Test 3 is a very severe test problem, the solution of which contains a left rarefaction, a contact and a right shock; this test is actually the left half of the blast wave problem of Woodward and Colella [413], Fig. 4.9 shows solution profiles. Test 4 is the right half of the Woodward and Colella problem; its solution contains a left shock, a contact discontinuity and a right rarefaction, as shown in Fig. 4.10. Test 5 is made up of the right and left shocks emerging from the solution to tests 3 and 4 respectively; its solution represents the collision of these two strong shocks and consists of a left facing shock (travelling very slowly to the right), a right travelling contact discontinuity and a right travelling shock wave. Fig. 4.11 shows solution profiles for Test 5.

Test	ρ_L	u_L	p_L	ρ_R	u_R	p_R
1	1.0	0.0	1.0	0.125	0.0	0.1
2	1.0	-2.0	0.4	1.0	2.0	0.4
3	1.0	0.0	1000.0	1.0	0.0	0.01
4	1.0	0.0	0.01	1.0	0.0	100.0
5	5.99924	19.5975	460.894	5.99242	-6.19633	46.0950

Table 4.1. Data for five Riemann problem tests

Table 4.2 shows the computed values for pressure in the *Star Region* by solving the pressure equation $f(p) = 0$ (equation 4.5) by a Newton–Raphson method. This task is carried out by the subroutine STARPU, which is contained in the FORTRAN 77 program given in Sect. 4.9 of this chapter.

Test	p_*	p_{TR}	p_{PV}	p_{TS}	$\frac{1}{2}(p_L + p_R)$
1	0.30313	0.30677(3)	0.55000(5)	0.31527(3)	0.55(5)
2	0.00189	exact(1)	TOL(8)	TOL(8)	0.4(9)
3	460.894	912.449(5)	500.005(4)	464.108(3)	500.005(4)
4	46.0950	82.9831(5)	50.005(4)	46.4162(3)	50.005(4)
5	1691.64	2322.65(4)	781.353(5)	1241.21(4)	253.494(6)

Table 4.2 Guess values p_0 for iteration scheme. Next to each guess is the required number of iterations for convergence (in parentheses).

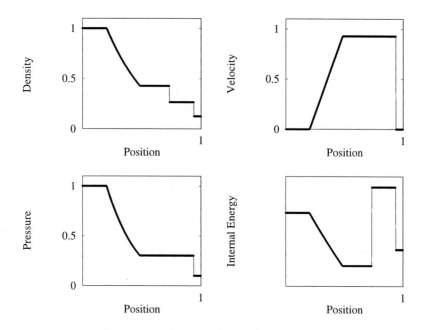

Fig. 4.7. Test 1: Exact solution for density, velocity, pressure and specific internal energy at time t = 0.25 units

The exact, converged, solution for pressure is given in column 2. Columns 3 to 6 give the guess values p_{TR}, p_{PV}, p_{TS} and the arithmetic mean value of the data. The number in parentheses next to each guess value is the number of iterations required for convergence for a tolerance $TOL = 10^{-6}$. For Test 1, p_{TR} and p_{TS} are the best guess values for p_0. For Test 2, p_{TR} is actually the exact solution (two rarefactions). By excluding Test 2, p_{TS} is the best guess overall. Experience in using hybrid schemes suggests that a combination of two or three approximations is bound to provide a suitable guess value for p_0 that is both accurate and efficient. In the FORTRAN 77 program provided in Sect. 4.9 of this chapter, the subroutine STARTE contains a hybrid scheme

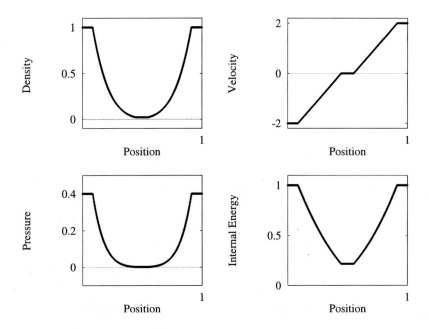

Fig. 4.8. Test 2: Exact solution for density, velocity, pressure and specific internal energy at time t = 0.15 units

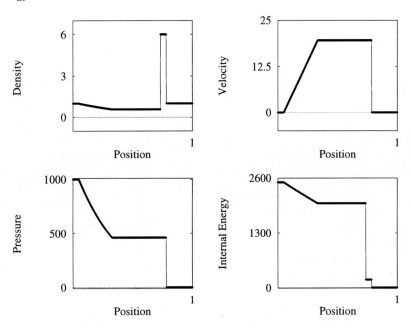

Fig. 4.9. Test 3: Exact solution for density, velocity, pressure and specific internal energy at time t = 0.012 units

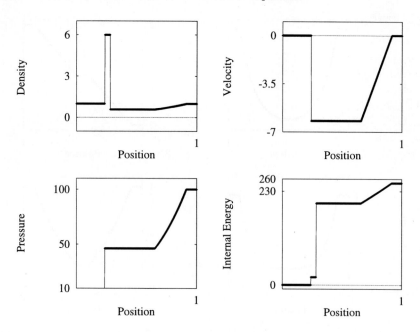

Fig. 4.10. Test 4: Exact solution for density, velocity, pressure and specific internal energy at time t = 0.035 units

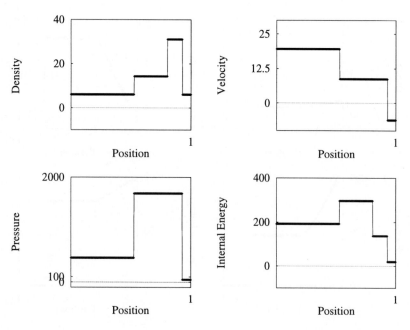

Fig. 4.11. Test 5: Exact solution for density, velocity, pressure and specific internal energy at time t = 0.035 units

involving p_{PV}, p_{TR} and p_{TS}. In a typical application of the exact Riemann solver to a numerical method, the overwhelming majority of Riemann problems will consist of *nearby states* which can be accurately approximated by the simple value p_{PV}.

Having found p_*, the solution u_* for the particle velocity follows from (4.9) and the density values ρ_{*L}, ρ_{*R} follow from appropriate wave relations, as detailed in the next section. Table 4.3 shows exact solutions for pressure p_*, speed u_*, densities ρ_{*L} and ρ_{*R} for tests 1 to 5. These quantities may prove of some use for initial testing of programs.

Test	p_*	u_*	ρ_{*L}	ρ_{*R}
1	0.30313	0.92745	0.42632	0.26557
2	0.00189	0.00000	0.02185	0.02185
3	460.894	19.5975	0.57506	5.99924
4	46.0950	-6.19633	5.99242	0.57511
5	1691.64	8.68975	14.2823	31.0426

Table 4.3. Exact solution for pressure, speed and densities for tests 1 to 5.

4.4 The Complete Solution

So far, we have an algorithm for computing the pressure p_* and particle velocity u_* in the *Star Region*. We still dot not know the sought values ρ_{*L} and ρ_{*R} for the density in this region; these are computed by identifying the types of non–linear waves, and can be done by comparing the pressure p_* to the pressures p_L and p_R, and then applying the appropriate conditions across the respective waves. Another pending task is to determine completely the left and right waves. For shock waves we only need to find the density behind the wave and the shock speed. For rarefaction waves there is more work involved: we need ρ behind the wave, equations for the *Head* and *Tail* of the wave and the full solution inside the rarefaction fan.

There are two cases. First we consider the case in which the sampling point (x, t) lies to the left of the contact discontinuity, as in Fig. 4.12. Again, there are two possibilities; these are now studied separately.

Left Shock Wave. A left shock wave, see Fig. 4.12a, is identified by the condition $p_* > p_L$. We know p_* and u_*. From the pressure ratio, see Sect. 3.1.3 of Chap. 3, we find the density according to

$$\rho_{*L} = \rho_L \left[\frac{\frac{p_*}{p_L} + \frac{\gamma-1}{\gamma+1}}{\frac{\gamma-1}{\gamma+1} \frac{p_*}{p_L} + 1} \right] . \tag{4.50}$$

The shock speed S_L is also a function of the pressure p_*. From (4.10) and (4.14) we deduce the shock speed as

$$S_{\mathrm{L}} = u_{\mathrm{L}} - Q_{\mathrm{L}}/\rho_{\mathrm{L}} \, , \tag{4.51}$$

where the mass flux Q_{L} is given by (4.20). More explicitly, see Sect. 3.1.3 of Chap. 3, one has

$$S_{\mathrm{L}} = u_{\mathrm{L}} - a_{\mathrm{L}} \left[\frac{\gamma + 1}{2\gamma} \frac{p_*}{p_{\mathrm{L}}} + \frac{\gamma - 1}{2\gamma} \right]^{\frac{1}{2}} . \tag{4.52}$$

We have therefore completely determined the solution for the entire region to the left of the contact discontinuity in the case in which the left wave is a shock wave.

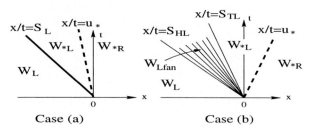

Fig. 4.12. Sampling the solution at a point to the left of the contact: (a) left wave is a shock (b) left wave is a rarefaction

Left Rarefaction Wave. A left rarefaction wave, see Fig. 4.12b, is identified by the condition $p_* \leq p_{\mathrm{L}}$. The pressure p_* and the particle velocity u_* in the *Star Region* are known. The density follows from the isentropic law as

$$\rho_{*\mathrm{L}} = \rho_{\mathrm{L}} \left(\frac{p_*}{p_{\mathrm{L}}} \right)^{\frac{1}{\gamma}} . \tag{4.53}$$

The sound speed behind the rarefaction is

$$a_{*\mathrm{L}} = a_{\mathrm{L}} \left(\frac{p_*}{p_{\mathrm{L}}} \right)^{\frac{\gamma - 1}{2\gamma}} . \tag{4.54}$$

The rarefaction wave is enclosed by the *Head* and the *Tail*, which are the characteristics of speeds given respectively by

$$S_{\mathrm{HL}} = u_{\mathrm{L}} - a_{\mathrm{L}} \, , \quad S_{\mathrm{TL}} = u_* - a_{*\mathrm{L}} \, . \tag{4.55}$$

We now find the solution for $\mathbf{W}_{\mathrm{Lfan}} = (\rho, u, p)^T$ inside the left rarefaction fan. This is easily obtained by considering the characteristic ray through the origin $(0,0)$ and a general point (x,t) inside the fan. The slope of the characteristic is

$$\frac{\mathrm{d}x}{\mathrm{d}t} = \frac{x}{t} = u - a \, ,$$

where u and a are respectively the sought particle velocity and sound speed at (x, t). Also, use of the Generalised Riemann Invariant $I_L(u, a)$ yields

$$u_L + \frac{2a_L}{\gamma - 1} = u + \frac{2a}{\gamma - 1} .$$

The simultaneous solution of these two equations for u and a, use of the definition of the sound speed a and the isentropic law give the result

$$\mathbf{W}_{Lfan} = \begin{cases} \rho = \rho_L \left[\frac{2}{(\gamma+1)} + \frac{(\gamma-1)}{(\gamma+1)a_L} \left(u_L - \frac{x}{t} \right) \right]^{\frac{2}{\gamma-1}} , \\[2mm] u = \frac{2}{(\gamma+1)} \left[a_L + \frac{(\gamma-1)}{2} u_L + \frac{x}{t} \right] , \\[2mm] p = p_L \left[\frac{2}{(\gamma+1)} + \frac{(\gamma-1)}{(\gamma+1)a_L} \left(u_L - \frac{x}{t} \right) \right]^{\frac{2\gamma}{\gamma-1}} . \end{cases} \quad (4.56)$$

Next we consider the solution at a point (x, t) to the right of the contact discontinuity for the two possible wave configurations of Fig. 4.13.

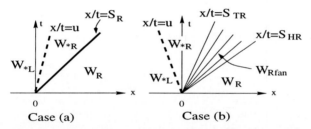

Case (a)   Case (b)

Fig. 4.13. Sampling the solution at a point to the right of the contact: (a) right wave is a shock (b) right wave is a rarefaction

Right Shock Wave. A right shock wave, see Fig. 4.13a, is identified by the condition $p_* > p_R$. We know the pressure p_* and the particle velocity u_*. The density ρ_{*R} is found to be

$$\rho_{*R} = \rho_R \left[\frac{\frac{p_*}{p_R} + \frac{\gamma-1}{\gamma+1}}{\frac{\gamma-1}{\gamma+1} \frac{p_*}{p_R} + 1} \right] \quad (4.57)$$

and the shock speed is

$$S_R = u_R + Q_R/\rho_R , \quad (4.58)$$

with the mass flux Q_R given by (4.30). More explicitly we have

$$S_R = u_R + a_R \left[\frac{(\gamma+1)}{2\gamma} \frac{p_*}{p_R} + \frac{(\gamma-1)}{2\gamma} \right]^{\frac{1}{2}} . \quad (4.59)$$

Right Rarefaction Wave. A right rarefaction wave, see Fig. 4.13b, is identified by the condition $p_* \leq p_R$. The pressure p_* and velocity u_* in the *Star Region* are known. The density is found from the isentropic law as

$$\rho_{*R} = \rho_R \left(\frac{p_*}{p_R} \right)^{\frac{1}{\gamma}} , \tag{4.60}$$

from which the sound speed follows as

$$a_{*R} = a_R \left(\frac{p_*}{p_R} \right)^{\frac{\gamma-1}{2\gamma}} . \tag{4.61}$$

The speeds for the *Head* and *Tail* are given respectively by

$$S_{HR} = u_R + a_R , \quad S_{TR} = u_* + a_{*R} . \tag{4.62}$$

The solution for $\mathbf{W}_{Rfan}$ inside a right rarefaction fan is found in an analogous manner to the case of a left rarefaction fan. The solution is

$$\mathbf{W}_{Rfan} = \begin{cases} \rho = \rho_R \left[\frac{2}{(\gamma+1)} - \frac{(\gamma-1)}{(\gamma+1)a_R} \left(u_R - \frac{x}{t} \right) \right]^{\frac{2}{\gamma-1}} , \\[2mm] u = \frac{2}{(\gamma+1)} \left[-a_R + \frac{(\gamma-1)}{2} u_R + \frac{x}{t} \right] , \\[2mm] p = p_R \left[\frac{2}{(\gamma+1)} - \frac{(\gamma-1)}{(\gamma+1)a_R} \left(u_R - \frac{x}{t} \right) \right]^{\frac{2\gamma}{\gamma-1}} . \end{cases} \tag{4.63}$$

4.5 Sampling the Solution

We have developed a solver to find the exact solution of the complete wave structure of the Riemann problem at any point (x, t) in the relevant domain of interest $x_L < x < x_R$; $t > 0$, with $x_L < 0$ and $x_R > 0$. We now provide a solution sampling procedure which, apart from being a summary of the solution, may also prove of practical use when programming the solution algorithm. Suppose we wish to evaluate the solution at a general point (x, t). We denote the solution of the Riemann problem at (x, t) in terms of the vector of primitive variables $\mathbf{W} = (\rho, u, p)^T$. As the solution $\mathbf{W}$ is a similarity solution we perform the sampling in terms of the *speed* $S = x/t$. When the solution at a specified time t is required the solution profiles are only a function of space x. In sampling the complete solution there are two cases to consider.

4.5.1 Left Side of Contact: $S = x/t \leq u_*$

As shown in Fig. 4.12 there are two possible wave configurations. Fig. 4.12a shows the case in which the left wave is a shock wave. In this case the complete solution on the left hand side of the contact wave is

$$\mathbf{W}(x,t) = \begin{cases} \mathbf{W}_{*L}^{\text{sho}} & \text{if } S_L \leq \frac{x}{t} \leq u_* , \\ \\ \mathbf{W}_L & \text{if } \frac{x}{t} \leq S_L , \end{cases} \tag{4.64}$$

where S_L is the shock speed given by (4.52), $\mathbf{W}_{*L}^{\text{sho}} = (\rho_{*L}^{\text{sho}}, u_*, p_*)^T$ with ρ_{*L}^{sho} given by (4.50) and $\mathbf{W}_L$ is the left data state. If the left wave is a rarefaction, as depicted by Fig. 4.12b, then the complete solution on the left hand side of the contact consists of three states, namely

$$\mathbf{W}(x,t) = \begin{cases} \mathbf{W}_L & \text{if } \frac{x}{t} \leq S_{\text{HL}} , \\ \\ \mathbf{W}_{\text{Lfan}} & \text{if } S_{\text{HL}} \leq \frac{x}{t} \leq S_{\text{TL}} , \\ \\ \mathbf{W}_{*L}^{\text{fan}} & \text{if } S_{\text{TL}} \leq \frac{x}{t} \leq u_* , \end{cases} \tag{4.65}$$

where S_{HL} and S_{TL} are the speeds of the head and tail of the rarefaction given by (4.55), $\mathbf{W}_{*L}^{\text{fan}} = (\rho_{*L}^{\text{fan}}, u_*, p_*)$ with ρ_{*L}^{fan} given by (4.53), $\mathbf{W}_{\text{Lfan}}$ is the state inside the rarefaction fan given by (4.56) and $\mathbf{W}_L$ is the left data state. Fig. 4.14 shows a flow chart for sampling the solution at any point (x,t) to the left of the contact discontinuity. An analogous flow chart results from the case in which the point (x,t) lies to the right of the contact discontinuity; the reader is encouraged to draw the flow chart for this case.

4.5.2 Right Side of Contact: $S = x/t \geq u_*$

As shown in Fig. 4.13 there are two possible wave configurations. Fig. 4.13a shows the case in which the right wave is a shock wave. In this case the complete solution on the right hand side of the contact wave is

$$\mathbf{W}(x,t) = \begin{cases} \mathbf{W}_{*R}^{\text{sho}} & \text{if } u_* \leq \frac{x}{t} \leq S_R , \\ \\ \mathbf{W}_R & \text{if } \frac{x}{t} \geq S_R , \end{cases} \tag{4.66}$$

where S_R is the shock speed given by (4.59), $\mathbf{W}_{*R}^{\text{sho}} = (\rho_{*R}^{\text{sho}}, u_*, p_*)^T$ with ρ_{*R}^{sho} given by (4.57) and $\mathbf{W}_R$ is the right data state. If the right wave is a rarefaction, as depicted by Fig. 4.13b, then the complete solution on the right hand side of the contact consists of three states, namely

$$\mathbf{W}(x,t) = \begin{cases} \mathbf{W}_{*R}^{\text{fan}} & \text{if } u_* \leq \frac{x}{t} \leq S_{\text{TR}} , \\ \\ \mathbf{W}_{\text{Rfan}} & \text{if } S_{\text{TR}} \leq \frac{x}{t} \leq S_{\text{HR}} , \\ \\ \mathbf{W}_R & \text{if } \frac{x}{t} \geq S_{\text{HR}} , \end{cases} \tag{4.67}$$

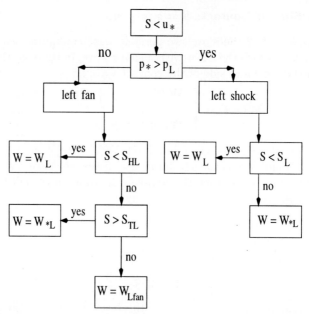

Fig. 4.14. Flow chart for sampling the solution at a point (x, t) to the left of the contact discontinuity $\frac{dx}{dt} = u_*$; $S = x/t$

where S_{HR} and S_{TR} are the speeds of the head and tail of the rarefaction given by (4.62), $\mathbf{W}_{*\text{R}}^{\text{fan}} = (\rho_{*\text{R}}^{\text{fan}}, u_*, p_*)^T$ with $\rho_{*\text{R}}^{\text{fan}}$ given by (4.60), $\mathbf{W}_{\text{Rfan}}$ is the state inside the right rarefaction fan given by (4.63) and $\mathbf{W}_{\text{R}}$ is the right data state.

Exercise 4.5.1. Write a flow chart for sampling the solution at any point (x, t) to the right of the contact discontinuity $\frac{dx}{dt} = u_*$.

Solution 4.5.1. (Left to the reader).

In Sect. 4.9 we give a program for finding the complete solution of the Riemann problem for ideal gases, excluding the cases in which vacuum is present.

4.6 The Riemann Problem in the Presence of Vacuum

The admission of flowing material adjacent to *vacuum* plays an important role in a number of practical applications. Loosely, vacuum is characterised by the condition $\rho = 0$. It follows that the total energy per unit mass also vanishes, $E = 0$. Values of pressure and particle velocity in vacuum are discussed later. Naturally, in vacuum regions the Euler equations, or any other mathematical model based on the continuum assumption, are no longer a valid description of the physics. Here we discuss solutions of the Euler equations in domains adjacent to regions of vacuum. As for the non–vacuum case described previously, the simplest problem involving the vacuum state is furnished by the Riemann problem. There are two obvious cases to consider. One is that in which the left non–vacuum state is adjacent to a right vacuum state at the initial time $t = 0$. The second case is simply the previous case reversed, the right non–vacuum state is adjacent to a left vacuum state. There is a third case, in which both left and right data states are non–vacuum states, but the vacuum state is generated in the interaction of the data states via the Riemann problem. The solution of the Riemann problem in the presence of vacuum involves the computation of free boundaries separating vacuum regions from those in which material exists.

In the presence of vacuum the structure of the solution of the Riemann problem is different from that of the conventional case shown in Fig. 4.1; the *Star Region* does no longer exist. Attempts at using the pressure equation (4.5) and an iterative scheme to solve it will fail, simply because the scheme would be assuming a solution structure that does not exist. The temptation to use *small values* of density and pressure to *simulate* vacuum with the Riemann solver for the non–vacuum case will also prove frustrating. If this is done in approximate Riemann solvers, then one is effectively changing the local wave structure of the solution.

Concerning the admissible elementary waves present in the structure of the solution of the Riemann problem including the vacuum state, an important observation is that a shock wave cannot be adjacent to a vacuum region. This is stated in the following proposition

Proposition 4.6.1. *A shock wave cannot be adjacent to a vacuum region.*

Proof. Let us consider a left non–vacuum constant state $\mathbf{W}_L = (\rho_L, u_L, p_L)^T$ adjacent to a right vacuum state $\mathbf{W}_0 \equiv (\rho_0, u_0, p_0)^T$ at the initial time $t = 0$, where $\rho_0 = 0$. Assume these states are connected by a discontinuity of speed S. Application of the Rankine–Hugoniot Conditions, see Sect. 3.1.3 of Chap. 3, gives

$$\rho_L u_L - \rho_0 u_0 = S(\rho_L - \rho_0) , \tag{4.68}$$

$$\rho_L u_L^2 + p_L - (\rho_0 u_0^2 + p_0) = S(\rho_L u_L - \rho_0 u_0) , \tag{4.69}$$

$$u_L(E_L + p_L) - u_0(E_0 + p_0) = S(E_L - E_0) . \tag{4.70}$$

As $E_0 = 0$ and assuming u_0 to be *finite*, manipulation of the equations gives

$$u_L = u_0 = S , \; p_L = p_0 . \tag{4.71}$$

It follows that a shock wave cannot be adjacent to a region of vacuum, $p_L = p_0$. The proposition is thus proved.

From the result (4.71) it also follows that a contact discontinuity can be adjacent to a region of vacuum, $u_L = u_0 = S$, which makes perfect physical sense. This wave separates a region of material from a region of no material and is therefore a boundary. The velocity u_0 of the front is also the maximum particle velocity across the wave system connecting a non–vacuum state with the vacuum state and is called the *escape velocity*. It turns out that u_0 is completely determined by the data on the non–vacuum state. As to the pressure p_0 we note that the previous result is independent of the specification of p_0. However, in order to determine the solution of the Riemann problem admitting regions of vacuum it is necessary to make some statements regarding pressure p and sound speed a. This is most conveniently done by specifying an equation of state. As the only possible waves are contacts and rarefactions, it is reasonable to adopt an isentropic–type equation of state and assume that this is valid all the way up to the boundary separating material from vacuum. We take the EOS

$$p = p(\rho) , \tag{4.72}$$

subject to the following conditions, see Liu and Smoller [256],

$$p(0) = 0 , \; p'(0) = 0, \; p'(\rho) > 0 , \; p''(\rho) > 0 . \tag{4.73}$$

Thus from now on we denote the vacuum state by $\mathbf{W}_0 = (\rho_0, u_0, p_0)^T \equiv (0, u_0, 0)^T$. From equation (4.71) the velocity u_0 is the speed of the boundary between the region of material and the region of vacuum. As pointed out in solving the Riemann problem including $\mathbf{W}_0$ there are three cases to consider. The structure of the solution of the Riemann problem for each of the three cases is depicted in Figs. 4.15, 4.16 and 4.17 respectively. We now study each case in detail.

4.6.1 Case 1: Vacuum Right State

In this case the Riemann problem has data of the type

$$\mathbf{W}(x, 0) = \begin{cases} \mathbf{W}_L \neq \mathbf{W}_0 \text{ if } x < 0 , \\ \\ \mathbf{W}_0 \text{ (vacuum) if } x > 0 , \end{cases} \tag{4.74}$$

with $\mathbf{W}_L = (\rho_L, u_L, p_L)^T$ and $\mathbf{W}_0 = (0, u_0, 0)^T$.

This problem may be seen as a mathematical model for the following physical situation: a shock tube of constant cross sectional area is filled with a gas at uniform conditions. Assume the right–hand boundary is a piston of speed $u_P = 0$ at time $t = 0$. Suppose the piston is *instantaneously accelerated* to a speed $u_P > 0$ (to the right). Clearly if $u_P > u_0$ vacuum will take place. The incipient cavitation case is $u_P = u_0$. The structure of the solution of the Riemann problem with initial data (4.74) is shown in Fig. 4.15 and consists of a left rarefaction wave and a contact wave that coalesces with the tail of the

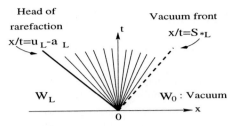

Fig. 4.15. Riemann problem solution for a right vacuum state

rarefaction. Obviously across the contact, $\Delta u = \Delta p = 0$. Note that the right non–linear wave of Fig. 4.1 is absent in Fig. 4.15. The physical interpretation of this is that there is no medium for this wave to propagate through. The exact solution follows directly from the methods of previous sections if the speed of the *contact front* is known. Application of the Generalised Riemann Invariant $I_L(u, a)$ to connect a point on the left data state with a point along the contact gives

$$u_0 + \frac{2a_0}{\gamma - 1} = u_L + \frac{2a_L}{\gamma - 1} \ . \tag{4.75}$$

From the assumptions (4.73) on the equation of state and the definition of sound speed we find that the sound speed vanishes along the contact, that is, $a_0 = 0$. Use of (4.73) with $a_0 = 0$ gives the speed of the *front* as

$$S_{*L} \equiv u_0 = u_L + \frac{2a_L}{\gamma - 1} \ . \tag{4.76}$$

It is worth remarking that the conditions $\rho_0 = 0$, $p_0 = 0$ do not, in general, imply $a_0 = 0$. The value of a_0 depends on the particular equation of state. The complete solution $\mathbf{W}(x, t)$ can now be written as

$$\mathbf{W}_{L0}(x, t) = \begin{cases} \mathbf{W}_L \text{ if } \frac{x}{t} \leq u_L - a_L \ , \\[2mm] \mathbf{W}_{\text{Lfan}} \text{ if } u_L - a_L < \frac{x}{t} < S_{*L} \ , \\[2mm] \mathbf{W}_0 \text{ if } \frac{x}{t} \geq S_{*L} \ , \end{cases} \tag{4.77}$$

where $\mathbf{W}_{\text{Lfan}}$ is the solution inside the left rarefaction fan given by (4.56).

4.6.2 Case 2: Vacuum Left State

In this case the Riemann problem has data of the type

$$\mathbf{W}(x, 0) = \begin{cases} \mathbf{W}_0 \text{ (vacuum) if } x < 0 \ , \\[2mm] \mathbf{W}_R \neq \mathbf{W}_0 \text{ if } x > 0 \ . \end{cases} \tag{4.78}$$

The structure of the solution in the x–t plane is illustrated in Fig. 4.16.

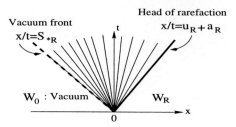

Fig. 4.16. Riemann problem solution for a left vacuum state

Compared with the regular Riemann problem case of Fig. 4.1, the left non–linear wave is missing and the contact wave separating the vacuum state from the non–vacuum state coalesces with the tail of the right rarefaction wave. The speed of the contact is found to be

$$S_{*R} = u_R - \frac{2a_R}{\gamma - 1} , \tag{4.79}$$

and thus the complete solution is given by

$$\mathbf{W}_{R0}(x, t) = \begin{cases} \mathbf{W}_0 \text{ if } \frac{x}{t} \leq S_{*R} , \\[2mm] \mathbf{W}_{Rfan} \text{ if } S_{*R} < \frac{x}{t} < u_R + a_R , \\[2mm] \mathbf{W}_R \text{ if } \frac{x}{t} \geq u_R + a_R , \end{cases} \tag{4.80}$$

where $\mathbf{W}_{Rfan}$ is the solution inside the right fan and is given by (4.63).

4.6.3 Case 3: Generation of Vacuum

This case has general data $\mathbf{W}_L = (\rho_L, u_L, p_L)^T \neq \mathbf{W}_0$, $\mathbf{W}_R = (\rho_R, u_R, p_R)^T \neq \mathbf{W}_0$ but combinations of particle and sound speeds are such that *vacuum is generated* as part of the interaction between $\mathbf{W}_L$ and $\mathbf{W}_R$. The structure of

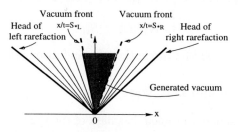

Fig. 4.17. Riemann problem solution for non–vacuum data states that do not satisfy the pressure positivity condition. Vacuum is generated in the middle of two rarefaction waves

the solution for this case is depicted in Fig. 4.17. There are two rarefaction waves and two contact waves of speeds S_{*L} and S_{*R} that enclose the generated vacuum state. The speeds S_{*L} and S_{*R} are as given by equations (4.76) and (4.79) respectively. The full solution reads

$$\mathbf{W}(x,t) = \begin{cases} \mathbf{W}_{L0}(x,t) \text{ if } \frac{x}{t} \leq S_{*L} , \\ \mathbf{W}_0 \text{ (vacuum) if } S_{*L} < \frac{x}{t} < S_{*R} , \\ \mathbf{W}_{R0}(x,t) \text{ if } \frac{x}{t} \geq S_{*R} , \end{cases} \tag{4.81}$$

where $\mathbf{W}_{L0}(x,t)$ and $\mathbf{W}_{R0}(x,t)$ are the solutions for the two previous cases given by (4.77) and (4.80) respectively. Note that for solution (4.81) to apply the condition $S_{*L} \leq S_{*R}$ must be valid. This implies

$$(\Delta u)_{\text{crit}} \equiv \frac{2a_L}{\gamma - 1} + \frac{2a_R}{\gamma - 1} \leq u_R - u_L , \tag{4.82}$$

which is consistent with the *pressure positivity condition* (4.40). The *vacuum condition* (4.82), which is stated purely in terms of the data values for particle velocity and sound speed, can be very useful in practical applications of the Riemann problem to numerical methods of the Godunov type. A note of caution is in order: it is also possible to determine a *vacuum generating condition* for approximate Riemann solvers, but this will in general be different from the exact condition (4.82). Two useful references concerning vacuum are the papers by Munz [245] and Munz et. al. [246].

4.7 The Riemann Problem for Covolume Gases

A small perturbation of the ideal gas equation of state (4.3) is the covolume EOS (4.4). For details on the Thermodynamics of covolume gases see Sect. 1.2.2 of Chap. 1. We now solve the Riemann problem for the Euler equations (4.1)-(4.2) for covolume gases. The solution methodology [351] applied is the same as for the ideal gas case and we shall therefore omit much of the detail.

The sound speed for covolume gases is

$$a = \sqrt{(p/\rho^2 - e_\rho)/e_p} = \sqrt{\frac{\gamma p}{\rho(1 - b\rho)}} . \tag{4.83}$$

The structure of the solution in the x–t plane is identical to that for the Riemann problem with the ideal gas EOS (4.3), in which $b = 0$. The solution procedure presented for the ideal gas case applies directly to the covolume case, except for finding the solution inside rarefaction waves. No closed–form solution can be found in this case and an extra iterative step is required.

4.7.1 Solution for Pressure and Particle Velocity.

In solving the Riemann problem we need to determine the flow quantities in the *Star Region*, see Fig. 4.1. The first step is to find the pressure and particle velocity.

Proposition 4.7.1 (Solution for p_* and u_*). *The solution for pressure p_* and particle velocity u_* of the Riemann problem (4.1), (4.2) with the co-volume gas Equation of State (4.4) in the unknown* Star Region *is given by the root of the algebraic equation*

$$f(p, \mathbf{W}_L, \mathbf{W}_R) \equiv f_L(p, \mathbf{W}_L) + f_R(p, \mathbf{W}_R) + \Delta u = 0 , \quad \Delta u \equiv u_R - u_L , \quad (4.84)$$

where the function f_L is given by

$$
f_L(p, \mathbf{W}_L) =
\begin{cases}
(p - p_L) \left[\dfrac{A_L}{p + B_L} \right]^{\frac{1}{2}} & \text{if } p > p_L \text{ (shock)} , \\[3mm]
\dfrac{2a_L(1 - b\rho_L)}{(\gamma - 1)} \left[\left(\dfrac{p}{p_L} \right)^{\frac{\gamma-1}{2\gamma}} - 1 \right] & \text{if } p \leq p_L \text{ (rarefaction)} ,
\end{cases}
$$

$$(4.85)$$

the function f_R is given by

$$
f_R(p, \mathbf{W}_R) =
\begin{cases}
(p - p_R) \left[\dfrac{A_R}{p + B_R} \right]^{\frac{1}{2}} & \text{if } p > p_R \text{ (shock)} , \\[3mm]
\dfrac{2a_R(1 - b\rho_R)}{(\gamma - 1)} \left[\left(\dfrac{p}{p_R} \right)^{\frac{\gamma-1}{2\gamma}} - 1 \right] & \text{if } p \leq p_R \text{ (rarefaction)} ,
\end{cases}
$$

$$(4.86)$$

and the data–dependent constants A_L, B_L, A_R, B_R are given by

$$
\left.
\begin{aligned}
A_L &= \frac{2(1 - b\rho_L)}{(\gamma+1)\rho_L} , & B_L &= \frac{(\gamma-1)}{(\gamma+1)} p_L , \\[3mm]
A_R &= \frac{2(1 - b\rho_R)}{(\gamma+1)\rho_R} , & B_R &= \frac{(\gamma-1)}{(\gamma+1)} p_R .
\end{aligned}
\right\}
$$

$$(4.87)$$

The particle velocity in the Star Region is given by

$$u_* = \frac{1}{2}(u_L + u_R) + \frac{1}{2}[f_R(p_*) - f_L(p_*)] . \qquad (4.88)$$

Proof. As for the case of ideal gases the derivation of the function f_L requires the analysis of the left non–linear wave, which can either be a shock wave or a rarefaction wave. Similarly for the function f_R.

We first consider f_L and assume the **left wave is a shock wave** of speed S_L. Application of the Rankine–Hugoniot Conditions yields the following relations of interest: The density ρ_{*L} is related to the pressure p_* behind the left shock via

$$\rho_{*L} = \rho_L \left[\frac{\frac{p_*}{p_L} + \frac{\gamma-1}{\gamma+1}}{\left(\frac{\gamma-1+2b\rho_L}{\gamma+1}\right)\frac{p_*}{p_L} + \frac{\gamma+1-2b\rho_L}{\gamma+1}} \right] \tag{4.89}$$

The *mass flux*

$$Q_L = -\frac{p_* - p_L}{u_* - u_L} \tag{4.90}$$

is worked out to be

$$Q_L = \left[\frac{p_* + B_L}{A_L} \right]^{\frac{1}{2}}, \tag{4.91}$$

with A_L and B_L as given by (4.87). From (4.90) we obtain

$$u_* = u_L - f_L(p_*, \mathbf{W}_L), \tag{4.92}$$

with

$$f_L(p_*, \mathbf{W}_L) = (p_* - p_L) \left[\frac{A_L}{p_* + B_L} \right]^{\frac{1}{2}}$$

and

$$A_L = \frac{2(1 - b\rho_L)}{(\gamma + 1)\rho_L}, \quad B_L = \frac{(\gamma - 1)}{(\gamma + 1)} p_L.$$

This is the sought expression for f_L for the case in which the left wave is a shock wave.

When the **left wave is a rarefaction** we apply the $I_L(u, a)$ Generalised Riemann Invariant to obtain

$$u_L + \frac{2a_L}{\gamma - 1}(1 - b\rho_L) = u_* + \frac{2a_{*L}}{\gamma - 1}(1 - b\rho_{*L}), \tag{4.93}$$

where a_L and a_{*L} denote the sound speed on the left and right states bounding the left rarefaction wave. Now the isentropic law gives

$$\rho_{*L} = \rho_L \left(\frac{1 - b\rho_{*L}}{1 - b\rho_L} \right) \left(\frac{p_*}{p_L} \right)^{\frac{1}{\gamma}}. \tag{4.94}$$

Substitution of ρ_{*L} from (4.94) into the definition of a_{*L} reduces equation (4.93) to

$$u_* = u_L - f_L(p_*, \mathbf{W}_L), \tag{4.95}$$

with

$$f_L(p_*, \mathbf{W}_L) = \frac{2a_L(1 - b\rho_L)}{(\gamma - 1)} \left[\left(\frac{p_*}{p_L} \right)^{\frac{\gamma-1}{2\gamma}} - 1 \right],$$

which is the required expression for the function f_L for the case in which the left wave is a rarefaction wave.

We now determine the function f_R. For a **right shock wave** of speed S_R we find

$$\rho_{*\mathrm{R}} = \rho_{\mathrm{R}} \left[\frac{\frac{p_*}{p_{\mathrm{R}}} + \frac{\gamma-1}{\gamma+1}}{\left(\frac{\gamma-1+2b\rho_{\mathrm{R}}}{\gamma+1}\right)\frac{p_*}{p_{\mathrm{R}}} + \frac{\gamma+1-2b\rho_{\mathrm{R}}}{\gamma+1}} \right]. \tag{4.96}$$

The *mass flux*

$$Q_{\mathrm{R}} = \frac{p_* - p_{\mathrm{R}}}{u_* - u_{\mathrm{R}}} \tag{4.97}$$

becomes

$$Q_{\mathrm{R}} = \left[\frac{p_* + B_{\mathrm{R}}}{A_{\mathrm{R}}}\right]^{\frac{1}{2}} \tag{4.98}$$

and leads to

$$u_* = u_{\mathrm{R}} + f_{\mathrm{R}}(p_*, \mathbf{W}_{\mathrm{R}}), \tag{4.99}$$

where

$$f_{\mathrm{R}}(p_*, \mathbf{W}_{\mathrm{R}}) = (p_* - p_{\mathrm{R}}) \left[\frac{A_{\mathrm{R}}}{p_* + B_{\mathrm{R}}}\right]^{\frac{1}{2}},$$

with A_{R} and B_{R} as given by (4.87). This is the sought expression for f_{R} for the case in which the right wave is a shock wave.

When the **right wave is a rarefaction** we apply the $I_{\mathrm{R}}(u, a)$ Generalised Riemann Invariant to obtain

$$u_{\mathrm{R}} - \frac{2a_{\mathrm{R}}}{\gamma - 1}(1 - b\rho_{\mathrm{R}}) = u_* - \frac{2a_{*\mathrm{R}}}{\gamma - 1}(1 - b\rho_{*\mathrm{R}}), \tag{4.100}$$

where $a_{*\mathrm{R}}$ and a_{R} denote the sound speed on the left and right states bounding the right rarefaction wave. Now the isentropic law gives

$$\rho_{*\mathrm{R}} = \rho_{\mathrm{R}} \left(\frac{1 - b\rho_{*\mathrm{R}}}{1 - b\rho_{\mathrm{R}}}\right)\left(\frac{p_*}{p_{\mathrm{R}}}\right)^{\frac{1}{\gamma}}. \tag{4.101}$$

Substitution of $\rho_{*\mathrm{R}}$ from (4.101) into the definition of $a_{*\mathrm{L}}$ reduces equation (4.100) to

$$u_* = u_{\mathrm{R}} + f_{\mathrm{R}}(p_*, \mathbf{W}_{\mathrm{R}}), \tag{4.102}$$

with

$$f_{\mathrm{R}}(p_*, \mathbf{W}_{\mathrm{R}}) = \frac{2a_{\mathrm{R}}(1 - b\rho_{\mathrm{R}})}{(\gamma - 1)}\left[\left(\frac{p_*}{p_{\mathrm{R}}}\right)^{\frac{\gamma-1}{2\gamma}} - 1\right],$$

which is the required expression for the function f_{R} for the case in which the right wave is a rarefaction wave.

The numerical solution of the algebraic equation $f(p) = 0$ given by (4.84) yields the pressure p_*. The particle velocity u_* may be computed from (4.92) if $p_* > p_{\mathrm{L}}$ or from (4.95) if $p_* \le p_{\mathrm{L}}$ or from (4.99) if $p_* > p_{\mathrm{R}}$ or from (4.102) if $p_* \le p_{\mathrm{R}}$. It can also be found from a mean value as

$$u_* = \frac{1}{2}(u_{\mathrm{L}} + u_{\mathrm{R}}) + \frac{1}{2}(f_{\mathrm{R}}(p_*) - f_{\mathrm{L}}(p_*))$$

and the proposition is thus proved.

4.7.2 Numerical Solution for Pressure

The numerical solution of the pressure equation is carried out by a Newton–Raphson iteration scheme, as done for the ideal gas case in Sect. 4.3. This requires the calculation of the derivative of the function $f(p)$. Details of this are found in [351]. A possible guess value p_0 for the iteration scheme is given by a *Two–Rarefaction* approximation to p_*, namely

$$
p_{\mathrm{TR}} = \left[\frac{(1 - b\rho_{\mathrm{L}})a_{\mathrm{L}} + (1 - b\rho_{\mathrm{R}})a_{\mathrm{R}} - \frac{1}{2}(\gamma - 1)(u_{\mathrm{R}} - u_{\mathrm{L}})}{(1 - b\rho_{\mathrm{L}})a_{\mathrm{L}}/p_{\mathrm{L}}^{\frac{\gamma-1}{2\gamma}} + (1 - b\rho_{\mathrm{R}})a_{\mathrm{R}}/p_{\mathrm{R}}^{\frac{\gamma-1}{2\gamma}}} \right]^{\frac{2\gamma}{\gamma-1}} , \qquad (4.103)
$$

and results from the exact function (4.84) for pressure under the assumption that the two non–linear waves are rarefaction waves. If the solution actually consists of two rarefactions then p_{TR} is exact and no iteration is required. A second guess value results from a linearised solution based on primitive variables. This is

$$
\left. \begin{aligned}
p_0 &= \max(TOL, p_{\mathrm{PV}}) , \\[6pt]
p_{\mathrm{PV}} &= \tfrac{1}{2}(p_{\mathrm{L}} + p_{\mathrm{R}}) - \tfrac{1}{8}(u_{\mathrm{R}} - u_{\mathrm{L}})(\rho_{\mathrm{L}} + \rho_{\mathrm{R}})(a_{\mathrm{L}} + a_{\mathrm{R}}) .
\end{aligned} \right\} \qquad (4.104)
$$

It is worth remarking here that the form of the p_{PV} approximation for covolume remains identical to that for the ideal gas case. The equation of state makes its input through the sound speeds a_{L} and a_{R}. The *Two–Shock* approximation is applied in exactly the same way as in the ideal gas case with the appropriate definitions for the quantities involved. See Sect. 9.4 of Chapt. 9.

4.7.3 The Complete Solution

So far, we know how to find p_* and u_*. The density values $\rho_{*\mathrm{L}}$ and $\rho_{*\mathrm{R}}$ follow from the appropriate (determined by the value of p_*) left and right wave relations. If the left wave is a shock wave, $p_* > p_{\mathrm{L}}$, then $\rho_{*\mathrm{L}}$ follows from (4.89). As in the ideal gas case the shock speed S_{L} is

$$
S_{\mathrm{L}} = u_{\mathrm{L}} - Q_{\mathrm{L}}/\rho_{\mathrm{L}} ,
$$

where the mass flux Q_{L} is given by (4.91). If the left wave is a rarefaction wave, $p_* \leq p_{\mathrm{L}}$, $\rho_{*\mathrm{L}}$ follows from (4.94). The rarefaction fan is enclosed by the characteristics of speeds

$$
S_{\mathrm{HL}} = u_{\mathrm{L}} - a_{\mathrm{L}} , \quad S_{\mathrm{TL}} = u_* - a_{*\mathrm{L}} .
$$

For a right shock wave, $p_* > p_{\mathrm{R}}$, $\rho_{*\mathrm{L}}$ is given by (4.96) and the shock speed is

$$
S_{\mathrm{R}} = u_{\mathrm{R}} + Q_{\mathrm{R}}/\rho_{\mathrm{R}} ,
$$

where the mass flux Q_R is given by (4.98). For a right rarefaction, $p_* \leq p_R$, ρ_{*L} follows from (4.101) and the wave is enclosed by the characteristics of speeds

$$S_{TR} = u_* + a_{*R} , \quad S_{HR} = u_R + a_R .$$

Solution values inside rarefaction fans cannot be obtained directly, as in the ideal gas case. For covolume gases an extra iterative procedure is needed.

4.7.4 Solution Inside Rarefactions

Unlike the ideal gas case the solution inside rarefaction waves for covolume gases is not direct. An extra iterative procedure is needed. Here we give the details for the case of a left rarefaction. Consider a general point (x, t) inside a left rarefaction fan and a characteristic ray through the origin and the point (x, t). The slope of the characteristic is

$$\frac{x}{t} = u - a . \tag{4.105}$$

Use of the $I_L(u, a)$ Generalised Riemann Invariant allows us to write

$$a \left[1 + \frac{2}{\gamma - 1}(1 - b\rho) \right] = I_L(u_L, a_L) - \frac{x}{t} , \tag{4.106}$$

where ρ, u and a are the sought solution values at (x, t) inside the left rarefaction wave. Use of the isentropic relation for the covolume EOS gives

$$p = p_L \left(\frac{1 - b\rho_L}{\rho_L} \right)^\gamma \left(\frac{\rho}{1 - b\rho} \right)^\gamma . \tag{4.107}$$

Further algebraic manipulations lead to a non–linear algebraic equation for the density ρ at the point (x, t) inside the left rarefaction fan, namely

$$Z_L(\rho) \equiv \rho^{\gamma-1}(\gamma + 1 - 2b\rho)^2 - \beta_L(1 - b\rho)^{\gamma+1} = 0 , \tag{4.108}$$

where the constant β_L is given by

$$\beta_L = \frac{\left\{ (\gamma - 1)[I_L(u_L, a_L) - \frac{x}{t}] \right\}^2}{\gamma p_L[(1 - b\rho_L)/\rho_L]^\gamma} , \tag{4.109}$$

where $I_L(u_L, a_L)$ is the left Riemann invariant evaluated on the left data state.

Equation (4.108) is solved numerically for ρ using a Newton–Raphson iteration, for which one needs the derivative

$$Z_L'(\rho) = (\gamma + 1)[b\beta_L(1 - b\rho)^\gamma + (\gamma + 1 - 2b\rho)(\gamma - 1 - 2b\rho)\rho^{\gamma-2}] . \tag{4.110}$$

Once ρ has been found to a given accuracy the pressure p follows immediately from equation (4.107). The sound speed a follows from definition (4.83) and the velocity u follows directly from (4.105).

The solution at a point (x,t) inside the right rarefaction fan is found in an entirely analogous way. In this case the density function $Z_R(\rho)$ is

$$Z_R(\rho) \equiv \rho^{\gamma-1}(\gamma + 1 - 2b\rho)^2 - \beta_R(1 - b\rho)^{\gamma+1} = 0 \,, \qquad (4.111)$$

where the constant β_R is given by

$$\beta_R = \frac{\{(\gamma - 1)[I_R(u_R, a_R) - \frac{x}{t}]\}^2}{\gamma p_R[(1 - b\rho_R)/\rho_R]^\gamma} \,, \qquad (4.112)$$

where $I_R(u_R, a_R)$ is the right Riemann invariant evaluated on the right data state.

The solution $\mathbf{W} = (\rho, u, p)^T$ of the Riemann problem for the Euler equations with the covolume equation of state is now known at any point (x,t) in the relevant domain. The solution sampling procedure is entirely analogous to the ideal gas case of Sect. 4.5 and is omitted.

Exercise 4.7.1. Write a flow chart for sampling the solution of the Riemann problem for covolume gases at any point (x,t) in a domain of interest $x_L \leq x \leq x_R$, $t > 0$, with $x_L < 0$ and $x_R > 0$.

Solution 4.7.1. (Left to the reader).

4.8 The Split Multi–Dimensional Case

For the purpose of using the Riemann problem solution in conjunction with numerical methods of the Godunov type, see Chaps. 5, 6, 14 and 16, it is useful to solve the split multi–dimensional Riemann problem. See Chap. 3. The x–split Riemann problem is the IVP for

$$\mathbf{U}_t + \mathbf{F}(\mathbf{U})_x = 0 \,,$$

$$\mathbf{U} = \begin{bmatrix} \rho \\ \rho u \\ \rho v \\ \rho w \\ E \end{bmatrix} \,, \quad \mathbf{F}(\mathbf{U}) = \begin{bmatrix} \rho u \\ \rho u^2 + p \\ \rho u v \\ \rho u w \\ u(E + p) \end{bmatrix} \,, \left.\begin{matrix} \\ \\ \\ \\ \\ \end{matrix}\right\} \qquad (4.113)$$

with initial conditions

$$\mathbf{U}(x,0) = \begin{cases} \mathbf{U}_L & \text{if } x < 0 \,, \\ \\ \mathbf{U}_R & \text{if } x > 0 \,. \end{cases} \qquad (4.114)$$

We note here that the x–direction will in general be understood as the direction *normal* to the boundary of a domain in two or three–dimensional space, see Sect. 3.2.4 of Chap. 3 and Sect. 16.7.3 of Chap. 16.

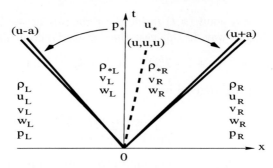

Fig. 4.18. Structure of the solution of the Riemann problem for the split three–dimensional case

As seen in Sect. 3.2.4 of Chap. 3 the exact solution to this problem is virtually identical to that of the genuine one–dimensional problem discussed previously. Fig. 4.18 shows the structure of the solution in terms of primitive variables $\mathbf{W} = (\rho, u, v, w, p)^T$. The outer non–linear waves are exactly the same as in the one–dimensional case. The multiplicity 3 of the eigenvalue $\lambda = u$ generates three, *coincident middle waves*. So effectively there are three waves that separate four constant states $\mathbf{W}_L$, $\mathbf{W}_{*L}$, $\mathbf{W}_{*R}$ and $\mathbf{W}_R$. In the *Star Region* between the left and right waves the solution for pressure, *normal velocity* and density is exactly the same as in the one–dimensional case. The two extra quantities v and w (tangential velocity components) only jump across the middle wave from their left data values to their right data values.

In summary, the solution to the x–split three–dimensional Riemann problem (4.113), (4.114) is exactly the same as that for the one–dimensional Riemann problem for the quantities ρ, u and p; for $q = v$ and $q = w$ we have, in addition,

$$
q(x, t) = \begin{cases} q_L & \text{if } \frac{x}{t} < u_* \, , \\[2mm] q_R & \text{if } \frac{x}{t} > u_* \, . \end{cases} \tag{4.115}
$$

As a matter of fact, for any passively advected quantity $q(x, y, z, t)$ by the fluid, that is

$$
q_t + u q_x + v q_y + w q_z = 0 \, , \tag{4.116}
$$

one can, by making use of the continuity equation, derive a new conservation law

$$
(\rho q)_t + (u \rho q)_x + (v \rho q)_y + (w \rho q)_z = 0 \, . \tag{4.117}
$$

When this conservation law is added to the set of one–dimensional Euler equations the x–split Riemann problem has solution for ρ, u and p as described earlier and the solution for the advected quantity q is as given by (4.115). One may have several advected quantities such as q. In chemically reactive compressible flows the quantity q may stand for the concentration of a chemical species, the progress variable of a chemical reaction or a fluid interface parameter.

Finally, we remark that although the exact solution to the split three–dimensional Riemann problem for any passively advected quantity, such as the tangential velocity components or concentration of chemical species, is so simple, approximate Riemann solvers may produce solutions for these quantities that are completely incorrect.

A highly relevant reference on the general theme of this chapter is the paper by Zhang [430]. Exact and approximate Riemann solvers for compressible liquids with various equations of state are given in [180]. For exact Riemann solvers for real gases see Colella and Glaz [89] and Menikoff and Plohr [240]. Approximate Riemann solvers for the ideal Euler equations are presented in Chaps. 9 to 12.

The methodology can be applied quite directly to solve the Riemann problem for other hyperbolic systems. For the steady supersonic Euler equations, exact Riemann solvers have been given by, amongst others, Marshall and Plohr [231]; Honma, Wada and Inomata [175]; Dawes [104]; Toro and Chou [377] and Toro and Chakravarthy [376]. Approximate Riemann solvers for the steady supersonic Euler equations have been given by Roe [281], Pandolfi [257], Toro and Chou [377] and Toro and Chakravarthy [376].

4.9 FORTRAN Program for Exact Riemann Solver

A listing of a FORTRAN 77 program to compute the exact solution to the Riemann problem for the one–dimensional time–dependent Euler equations for ideal gases is now given. The data file is called exact.ini. The main program calls two routines: STARPU and SAMPLE. The first routine solves iteratively for the pressure p_* and then computes the particle speed u_*; the guessed value for the iteration is provided by the subroutine GUESSP. The subroutine SAMPLE finds, for given p_*, u_* and $S = x/t$, the solution of the Riemann problem at the point (x, t). This routine can then be utilised in numerical methods to solve the general initial boundary value problem for the Euler equations. For the Godunov first–order method, see Chap. 6, one calls SAMPLE with $S = 0$. For Glimm's method, see Chap. 7, one calls SAMPLE with $S = \theta x/t$, where θ is a random number. The source code is also part of the library *NUMERICA* [369].

```
*
*---------------------------------------------------------------*
*                                                               *
C                  EXACT RIEMANN SOLVER                         *
C                  FOR THE EULER EQUATIONS                      *
*                                                               *
C             Name of program: HE-E1RPEXACT                     *
*                                                               *
C       Purpose: to solve the Riemann problem exactly,          *
C                for the time dependent one dimensional         *
C                Euler equations for an ideal gas               *
*                                                               *
C       Input  file: exact.ini                                  *
C       Output file: exact.out (exact solution)                 *
*                                                               *
C       Programer: E. F. Toro                                   *
*                                                               *
C       Last revision: February 1st 1999                        *
*                                                               *
C       Theory is found in Chapter 4 of Reference 1             *
*                                                               *
C       1. Toro, E. F., "Riemann Solvers and Numerical          *
C                         Methods for Fluid Dynamics"           *
C                         Springer-Verlag,                      *
C                         Second Edition, 1999                  *
*                                                               *
C       This program is part of                                 *
*                                                               *
C       NUMERICA                                                *
C       A Library of Source Codes for Teaching,                 *
C       Research and Applications,                              *
C       by E. F. Toro                                           *
C       Published by NUMERITEK LTD,                             *
C       Website: www.numeritek.com                              *
*                                                               *
*---------------------------------------------------------------*
*
        IMPLICIT NONE
*
C       Declaration of variables:
*
        INTEGER I, CELLS
*
        REAL    GAMMA, G1, G2, G3, G4, G5, G6, G7, G8,
```

```
&            DL, UL, PL, CL, DR, UR, PR, CR,
&            DIAPH, DOMLEN, DS, DX, PM, MPA, PS, S,
&            TIMEOUT, UM, US, XPOS
*
      COMMON /GAMMAS/ GAMMA, G1, G2, G3, G4, G5, G6, G7, G8
      COMMON /STATES/ DL, UL, PL, CL, DR, UR, PR, CR
*
      OPEN(UNIT = 1,FILE = 'exact.ini',STATUS = 'UNKNOWN')
*
C     Initial data and parameters are read in
*
      READ(1,*)DOMLEN      ! Domain length
      READ(1,*)DIAPH       ! Initial discontinuity position
      READ(1,*)CELLS       ! Number of computing cells
      READ(1,*)GAMMA       ! Ratio of specific heats
      READ(1,*)TIMEOUT     ! Output time
      READ(1,*)DL          ! Initial density  on left state
      READ(1,*)UL          ! Initial velocity on left state
      READ(1,*)PL          ! Initial pressure on left state
      READ(1,*)DR          ! Initial density  on right state
      READ(1,*)UR          ! Initial velocity on right state
      READ(1,*)PR          ! Initial pressure on right state
      READ(1,*)MPA         ! Normalising constant
*
      CLOSE(1)
*
C     Compute gamma related constants
*
      G1 = (GAMMA - 1.0)/(2.0*GAMMA)
      G2 = (GAMMA + 1.0)/(2.0*GAMMA)
      G3 = 2.0*GAMMA/(GAMMA - 1.0)
      G4 = 2.0/(GAMMA - 1.0)
      G5 = 2.0/(GAMMA + 1.0)
      G6 = (GAMMA - 1.0)/(GAMMA + 1.0)
      G7 = (GAMMA - 1.0)/2.0
      G8 = GAMMA - 1.0
*
C     Compute sound speeds
*
      CL = SQRT(GAMMA*PL/DL)
      CR = SQRT(GAMMA*PR/DR)
*
C     The pressure positivity condition is tested for
*
```

```
      IF(G4*(CL+CR).LE.(UR-UL))THEN
*
C         The initial data is such that vacuum is generated.
C         Program stopped.
*
          WRITE(6,*)
          WRITE(6,*)'***Vacuum is generated by data***'
          WRITE(6,*)'***Program stopped***'
          WRITE(6,*)
*
          STOP
      ENDIF
*
C     Exact solution for pressure and velocity in star
C     region is found
*
      CALL STARPU(PM, UM, MPA)
*
      DX = DOMLEN/REAL(CELLS)
*
C     Complete solution at time TIMEOUT is found
*
      OPEN(UNIT = 2,FILE = 'exact.out',STATUS = 'UNKNOWN')
*
      DO 10 I = 1, CELLS
*
         XPOS = (REAL(I) - 0.5)*DX
         S    = (XPOS - DIAPH)/TIMEOUT
*
C        Solution at point (X,T) = ( XPOS - DIAPH,TIMEOUT)
C        is found
*
         CALL SAMPLE(PM, UM, S, DS, US, PS)
*
C        Exact solution profiles are written to exact.out.
*
         WRITE(2, 20)XPOS, DS, US, PS/MPA, PS/DS/G8/MPA
*
 10   CONTINUE
*
      CLOSE(2)
*
 20   FORMAT(5(F14.6, 2X))
*
```

```
      END
*
*---------------------------------------------------------------*
*
      SUBROUTINE STARPU(P, U, MPA)
*
      IMPLICIT NONE
*
C     Purpose: to compute the solution for pressure and
C              velocity in the Star Region
*
C     Declaration of variables
*
      INTEGER I, NRITER
*
      REAL    DL, UL, PL, CL, DR, UR, PR, CR,
     &        CHANGE, FL, FLD, FR, FRD, P, POLD, PSTART,
     &        TOLPRE, U, UDIFF, MPA
*
      COMMON /STATES/ DL, UL, PL, CL, DR, UR, PR, CR
      DATA TOLPRE, NRITER/1.0E-06, 20/
*
C     Guessed value PSTART is computed
*
      CALL GUESSP(PSTART)
*
      POLD  = PSTART
      UDIFF = UR - UL
*
      WRITE(6,*)'----------------------------------------'
      WRITE(6,*)'   Iteration number        Change  '
      WRITE(6,*)'----------------------------------------'
*
      DO 10 I = 1, NRITER
*
         CALL PREFUN(FL, FLD, POLD, DL, PL, CL)
         CALL PREFUN(FR, FRD, POLD, DR, PR, CR)
         P     = POLD - (FL + FR + UDIFF)/(FLD + FRD)
         CHANGE = 2.0*ABS((P - POLD)/(P + POLD))
         WRITE(6, 30)I, CHANGE
         IF(CHANGE.LE.TOLPRE)GOTO 20
         IF(P.LT.0.0)P = TOLPRE
         POLD  = P
*
```

```fortran
   10    CONTINUE
*
         WRITE(6,*)'Divergence in Newton-Raphson iteration'
*
   20    CONTINUE
*
C        Compute velocity in Star Region
*
         U = 0.5*(UL + UR + FR - FL)
*
         WRITE(6,*)'---------------------------------------'
         WRITE(6,*)'    Pressure          Velocity'
         WRITE(6,*)'---------------------------------------'
         WRITE(6,40)P/MPA, U
         WRITE(6,*)'---------------------------------------'
*
   30    FORMAT(5X, I5,15X, F12.7)
   40    FORMAT(2(F14.6, 5X))
*
         RETURN
         END
*
*-----------------------------------------------------------*
*
         SUBROUTINE GUESSP(PM)
*
C        Purpose: to provide a guess value for pressure
C                 PM in the Star Region. The choice is made
C                 according to adaptive Riemann solver using
C                 the PVRS, TRRS and TSRS approximate
C                 Riemann solvers. See Sect. 9.5 of Chapt. 9
C                 of Ref. 1
*
         IMPLICIT NONE
*
C        Declaration of variables
*
         REAL    DL, UL, PL, CL, DR, UR, PR, CR,
     &           GAMMA, G1, G2, G3, G4, G5, G6, G7, G8,
     &           CUP, GEL, GER, PM, PMAX, PMIN, PPV, PQ,
     &           PTL, PTR, QMAX, QUSER, UM
*
         COMMON /GAMMAS/ GAMMA, G1, G2, G3, G4, G5, G6, G7, G8
         COMMON /STATES/ DL, UL, PL, CL, DR, UR, PR, CR
```

```
*
      QUSER = 2.0
*
C     Compute guess pressure from PVRS Riemann solver
*
      CUP  = 0.25*(DL + DR)*(CL + CR)
      PPV  = 0.5*(PL + PR) + 0.5*(UL - UR)*CUP
      PPV  = AMAX1(0.0, PPV)
      PMIN = AMIN1(PL,  PR)
      PMAX = AMAX1(PL,  PR)
      QMAX = PMAX/PMIN
*
      IF(QMAX.LE.QUSER.AND.
     & (PMIN.LE.PPV.AND.PPV.LE.PMAX))THEN
*
C        Select PVRS Riemann solver
*
         PM = PPV
      ELSE
         IF(PPV.LT.PMIN)THEN
*
C           Select Two-Rarefaction Riemann solver
*
            PQ  = (PL/PR)**G1
            UM  = (PQ*UL/CL + UR/CR +
     &             G4*(PQ - 1.0))/(PQ/CL + 1.0/CR)
            PTL = 1.0 + G7*(UL - UM)/CL
            PTR = 1.0 + G7*(UM - UR)/CR
            PM  = 0.5*(PL*PTL**G3 + PR*PTR**G3)
         ELSE
*
C           Select Two-Shock Riemann solver with
C           PVRS as estimate
*
            GEL = SQRT((G5/DL)/(G6*PL + PPV))
            GER = SQRT((G5/DR)/(G6*PR + PPV))
            PM  = (GEL*PL + GER*PR - (UR - UL))/(GEL + GER)
         ENDIF
      ENDIF
*
      RETURN
      END
*
*-------------------------------------------------------------*
```

```
*
        SUBROUTINE PREFUN(F, FD, P, DK, PK, CK)
*
C    Purpose: to evaluate the pressure functions
C             FL and FR in exact Riemann solver
*
      IMPLICIT NONE
*
C    Declaration of variables
*
      REAL    AK, BK, CK, DK, F, FD, P, PK, PRAT, QRT,
     &        GAMMA, G1, G2, G3, G4, G5, G6, G7, G8
*
      COMMON /GAMMAS/ GAMMA, G1, G2, G3, G4, G5, G6, G7, G8
*
      IF(P.LE.PK)THEN
*
C        Rarefaction wave
*
         PRAT = P/PK
         F    = G4*CK*(PRAT**G1 - 1.0)
         FD   = (1.0/(DK*CK))*PRAT**(-G2)
      ELSE
*
C        Shock wave
*
         AK  = G5/DK
         BK  = G6*PK
         QRT = SQRT(AK/(BK + P))
         F   = (P - PK)*QRT
         FD  = (1.0 - 0.5*(P - PK)/(BK + P))*QRT
      ENDIF
*
      RETURN
      END
*
*------------------------------------------------------------*
*
        SUBROUTINE SAMPLE(PM, UM, S, D, U, P)
*
C    Purpose: to sample the solution throughout the wave
C             pattern. Pressure PM and velocity UM in the
C             Star Region are known. Sampling is performed
C             in terms of the 'speed' S = X/T. Sampled
```

```
C                 values are D, U, P
*
C     Input variables : PM, UM, S, /GAMMAS/, /STATES/
C     Output variables: D, U, P
*
      IMPLICIT NONE
*
C     Declaration of variables
*
      REAL    DL, UL, PL, CL, DR, UR, PR, CR,
     &        GAMMA, G1, G2, G3, G4, G5, G6, G7, G8,
     &        C, CML, CMR, D, P, PM, PML, PMR,  S,
     &        SHL, SHR, SL, SR, STL, STR, U, UM
*
      COMMON /GAMMAS/ GAMMA, G1, G2, G3, G4, G5, G6, G7, G8
      COMMON /STATES/ DL, UL, PL, CL, DR, UR, PR, CR
*
      IF(S.LE.UM)THEN
*
C        Sampling point lies to the left of the contact
C        discontinuity
*
         IF(PM.LE.PL)THEN
*
C           Left rarefaction
*
            SHL = UL - CL
*
            IF(S.LE.SHL)THEN
*
C              Sampled point is left data state
*
               D = DL
               U = UL
               P = PL
            ELSE
               CML = CL*(PM/PL)**G1
               STL = UM - CML
*
               IF(S.GT.STL)THEN
*
C                 Sampled point is Star Left state
*
                  D = DL*(PM/PL)**(1.0/GAMMA)
```

```
                        U = UM
                        P = PM
                     ELSE
*
C                       Sampled point is inside left fan
*
                        U = G5*(CL + G7*UL + S)
                        C = G5*(CL + G7*(UL - S))
                        D = DL*(C/CL)**G4
                        P = PL*(C/CL)**G3
                     ENDIF
                  ENDIF
               ELSE
*
C              Left shock
*
               PML = PM/PL
               SL  = UL - CL*SQRT(G2*PML + G1)
*
               IF(S.LE.SL)THEN
*
C                 Sampled point is left data state
*
                  D = DL
                  U = UL
                  P = PL
*
               ELSE
*
C                 Sampled point is Star Left state
*
                  D = DL*(PML + G6)/(PML*G6 + 1.0)
                  U = UM
                  P = PM
               ENDIF
            ENDIF
         ELSE
*
C        Sampling point lies to the right of the contact
C        discontinuity
*
         IF(PM.GT.PR)THEN
*
C           Right shock
```

```
*
            PMR = PM/PR
            SR  = UR + CR*SQRT(G2*PMR + G1)
*
            IF(S.GE.SR)THEN
*
C              Sampled point is right data state
*
               D = DR
               U = UR
               P = PR
            ELSE
*
C              Sampled point is Star Right state
*
               D = DR*(PMR + G6)/(PMR*G6 + 1.0)
               U = UM
               P = PM
            ENDIF
         ELSE
*
C           Right rarefaction
*
            SHR = UR + CR
*
            IF(S.GE.SHR)THEN
*
C              Sampled point is right data state
*
               D = DR
               U = UR
               P = PR
            ELSE
               CMR = CR*(PM/PR)**G1
               STR = UM + CMR
*
               IF(S.LE.STR)THEN
*
C                 Sampled point is Star Right state
*
                  D = DR*(PM/PR)**(1.0/GAMMA)
                  U = UM
                  P = PM
               ELSE
```

```
*
C                         Sampled point is inside left fan
*
                    U = G5*(-CR + G7*UR + S)
                    C = G5*(CR - G7*(UR - S))
                    D = DR*(C/CR)**G4
                    P = PR*(C/CR)**G3
                ENDIF
              ENDIF
            ENDIF
        ENDIF
*
        RETURN
        END
*
*-----------------------------------------------------------*
*
```

5. Notions on Numerical Methods

We assume the reader to be familiar with some basic concepts on numerical methods for partial differential equations in general. In particular, we shall assume the concepts of truncation error, order of accuracy, consistency, modified equation, stability and convergence. For background on these concepts the reader may consult virtually any standard book on numerical methods for differential equations. As general references, useful textbooks are those of Smith [315], Anderson et. al. [5], Mitchell and Griffiths [241], Roache [279], Richtmyer and Morton [276], Hoffmann [172] and Fletcher [125]. Very relevant textbooks to the main themes of this book are Sod [319], Holt [173], Hirsch Volumes I [170] and II [171], LeVeque [207], Godlewski and Raviart [144], Kröner [196] and Thomas [344].

The contents of this chapter are designed specifically to provide the necessary background for the application of *high–resolution upwind and centred numerical methods* to hyperbolic conservation laws. Our prime objective is to present the basic Godunov method [145] in a simple setting so as to make the application of upwind methods to non–linear systems of conservation laws an easier task in the forthcoming chapters. Essential background material is given in Chap. 2. For those who are absolute beginners in the field I would recommend the following self study programme as a way of obtaining more benefit from this text book: (a) read chapters 2 to 4 of the book by Hoffmann [172] and do exercises, (b) read chapters 7 to 10 of the book by Hirsch Vol. I [170] and do exercises.

5.1 Discretisation: Introductory Concepts

Our concern is the utilisation of numerical methods for solving partial differential equations (PDEs). Numerical methods replace the *continuous* problem represented by the PDEs by a *finite* set of *discrete* values. These are obtained by first discretising the domain of the PDEs into a finite set of points or volumes via a *mesh* or *grid*. The corresponding discretisation of the PDEs on the grid results in discrete values. In the Finite Difference approach one regards these values as *point values* defined at grid points. The Finite Volume approach regards these discrete values as *averages over finite volumes*. We are mostly interested in the second approach but for the purpose of introducing some of the basic concepts in numerical methods we also consider the Finite Difference approach. For most of this chapter we restrict the discussion to model problems, such as the model PDE

$$u_t + a u_x = 0 \,,$$

with $u = u(x, t)$ and $a \neq 0$ a constant wave propagation speed. See Sect. 2.1 of Chap. 2. In this equation there are two partial derivatives, namely a time derivative u_t and a spatial derivative u_x.

5.1.1 Approximation to Derivatives

Given a sufficiently smooth function $g(x)$, by using Taylor's theorem, we can always find the value of $g(x)$ at any neighbouring point $x_0 + \Delta x$ of $x = x_0$ if we know $g(x)$ and all its derivatives $g^{(k)}(x)$ at $x = x_0$, that is

$$g(x_0 + \Delta x) = g(x_0) + \sum_k \frac{(\Delta x)^k}{k!} g^{(k)}(x_0) \,. \tag{5.1}$$

By truncating the Taylor series one can obtain approximations to $g(x_0 + \Delta x)$. Also one can obtain approximations to derivatives of $g(x)$. Consider a function $g(x)$ and three equally–spaced points $x_0 - \Delta x$, x_0 and $x_0 + \Delta x$, as shown in Fig. 5.1. As an illustrative example we shall derive three approximations to the first derivative of $g(x)$ at the point x_0. By neglecting terms of third order and higher, $O(\Delta x^3)$, we can write

$$g(x_0 + \Delta x) = g(x_0) + \Delta x g^{(1)}(x_0) + \frac{(\Delta x)^2}{2} g^{(2)}(x_0) + O(\Delta x^3) \tag{5.2}$$

and

$$g(x_0 - \Delta x) = g(x_0) - \Delta x g^{(1)}(x_0) + \frac{(\Delta x)^2}{2} g^{(2)}(x_0) + O(\Delta x^3) \,. \tag{5.3}$$

Neglecting second order terms in (5.2) leads immediately to an approximation to the first derivative $g^{(1)}(x)$ of $g(x)$, that is

$$g^{(1)}(x_0) = \frac{g(x_0 + \Delta x) - g(x_0)}{\Delta x} + O(\Delta x) \,. \tag{5.4}$$

This is a first–order approximation to $g^{(1)}(x)$ at $x = x_0$, the leading term in the remaining part contains Δx to the power unity. It is a *one–sided* approximation to the first derivative of $g(x)$, usually called a *forward* finite difference approximation; the points used by the approximation are x_0 (the point at which the derivative is sought) and its neighbour $x_0 + \Delta x$ on the right hand side. From (5.3) we can also obtain a *backward* first–order approximation

$$g^{(1)}(x_0) = \frac{g(x_0) - g(x_0 - \Delta x)}{\Delta x} + O(\Delta x) \,. \tag{5.5}$$

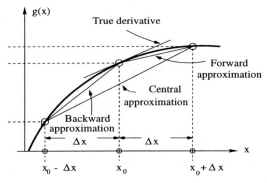

Fig. 5.1. Finite difference approximations to the first derivative of a function $g(x)$ at the point x_0: backward, centred and forward

Note that by subtracting (5.3) from (5.2) we obtain

$$g^{(1)}(x_0) = \frac{g(x_0 + \Delta x) - g(x_0 - \Delta x)}{2\Delta x} + O(\Delta x^2) \,. \qquad (5.6)$$

This is a *central* finite–difference approximation to $g^{(1)}(x_0)$ and is second-order accurate; it uses the left and right hand side neighbours of $x = x_0$. It can also be obtained by taking a mean value of the forward and backward approximations (5.4) and (5.5). For the problems of our concern here the distinction between one–sided and central approximations is significant. *Upwind Methods* may be viewed as one–sided differencing schemes. The question of which side is also of paramount importance.

5.1.2 Finite Difference Approximation to a PDE

Consider the Initial Boundary Value Problem (IBVP) for the linear advection equation in the domain $[0, L] \times [0, T]$ on the x–t plane. This consists of a PDE together with initial condition (IC) and boundary conditions (BCs), namely

$$\left. \begin{array}{lll} \text{PDE}: & u_t + a u_x = 0 \,, \\[2mm] \text{IC}: & u(x, 0) = u_0(x) \,, \\[2mm] \text{BCs}: & u(0, t) = u_l(t) \,, \quad u(L, t) = u_r(t) \,. \end{array} \right\} \qquad (5.7)$$

Solving this IBVP means evolving the solution $u(x, t)$ in time starting from the initial condition $u_0(x)$ at time $t = 0$ and subject to boundary conditions. For the moment we assume that the boundary constraints take the form of prescribed boundary functions $u_l(t)$ and $u_r(t)$. The application of boundary conditions is intimately linked to the physics of the problem at hand. A more meaningful discussion is presented in Chap. 6 in the context of physically more meaningful systems of PDEs.

A possible finite difference mesh for discretising the domain is depicted in Fig. 5.2. This is a regular grid of dimensions Δx (spacing of grid points in the x–direction) by Δt (spacing in the t–direction). In general if $[0, L]$ is discretised by M equally spaced grid points then

$$\Delta x = \frac{L}{M - 1} . \tag{5.8}$$

The points of the mesh are positioned at $(i\Delta x, n\Delta t)$ on the x–t plane, with $i = 0, \ldots, M$ and $n = 0, \ldots$. Often we shall use the notation $x_i = i\Delta x$, $t^n = n\Delta t$. The discrete values of the function $u(x, t)$ at $(i\Delta x, n\Delta t)$ will be denoted by

$$u_i^n \equiv u(i\Delta x, n\Delta t) \equiv u(x_i, t^n) .$$

The superscript n refers to the time discretisation and is called the *time level*. The subscript i refers to the space discretisation and designates the spatial position in the mesh. We shall also use the symbol u_i^n to denote an *approximation* to the exact mesh value $u(i\Delta x, n\Delta t)$. The distinction will be made at appropriate places. If the IBVP (5.7) has given data at a time level

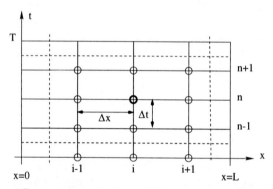

Fig. 5.2. Finite difference discretisation of domain on x–t plane. Regular mesh of dimensions $\Delta x \times \Delta t$ is assumed

n, say, this can be represented by a set of discrete values $u_i^n, i = 0, \ldots, M$. Solving (5.7) requires finding the solution at the next time level $n + 1$, that is, we want to find the set $u_i^{n+1}, i = 0, \ldots, M$. The extreme boundary points u_0^n, u_M^n are determined by the particular *boundary conditions* to be applied. For the moment we simply assume that these are prescribed for all time levels.

Consider now approximations to the time derivative u_t at the grid point (x_i, t^n). A first–order forward approximation gives

$$u_t = \frac{u_i^{n+1} - u_i^n}{\Delta t} . \tag{5.9}$$

For the spatial derivatives one could use the second–order central approximation

$$u_x = \frac{u_{i+1}^n - u_{i-1}^n}{2\Delta x} . \tag{5.10}$$

By replacing u_t and u_x in the PDE in (5.7) by their respective approximations (5.9) and (5.10) we obtain

$$\frac{u_i^{n+1} - u_i^n}{\Delta t} + a \left[\frac{u_{i+1}^n - u_{i-1}^n}{2\Delta x} \right] = 0 . \tag{5.11}$$

This is the discrete analogue of the PDE in (5.7). The differential equation has been replaced by a finite difference equation. As all values at the time level n in (5.11) are prescribed data values at the initial time, or have already been computed, we can solve for the *single unknown* u_i^{n+1} at the new time level as

$$u_i^{n+1} = u_i^n - \frac{1}{2}c \left[u_{i+1}^n - u_{i-1}^n \right] , \tag{5.12}$$

where

$$c = \frac{\Delta t a}{\Delta x} = \frac{a}{\Delta x / \Delta t} \tag{5.13}$$

is a *dimensionless* quantity known as the *Courant number*; it is also known as the Courant–Friedrichs–Lewy number, or CFL number. This quantity can be regarded as the ratio of two speeds, namely the wave propagation speed a in the partial differential equation in (5.7) and the *grid speed* $\Delta x / \Delta t$ defined by the discretisation of the domain. Formula (5.12) provides an *explicit* scheme for evolving the solution in time and has resulted from approximating the time and space derivatives of the PDE in (5.7) by first and second order finite differences respectively. This appears to be a reasonable step.

It is disappointing however, and perhaps surprising, to realise that the resulting scheme (5.12) is totally useless. It is *unconditionally unstable*. This can be seen by performing a von Neumann stability analysis. Consider the *trial solution* $u_i^n = A^n e^{I i \theta}$. A is the amplitude, $\theta = P \Delta x$ is the phase angle, P is the wave number in the x–direction, $\lambda = 2\pi / P$ is the wave length and $I = \sqrt{-1}$ is the unit complex number. Substitution of the trial solution into the scheme (5.12) gives $A = 1 - I c \sin \theta$. For stability one requires $\|A\| \leq 1$. But note that $\|A\| = 1 + c^2 \sin^2 \theta \geq 1$ and thus the scheme is unstable under all circumstances, unconditionally unstable.

5.2 Selected Difference Schemes

We study some of the most well–known finite difference schemes, of first and second order of accuracy.

5.2.1 The First Order Upwind Scheme

One way of remedying our failed attempt at devising a useful numerical method for the PDE in (5.7) is to replace the central finite difference approximation to the spatial derivative u_x by a first–order *one–sided approximation*. Two choices are

$$u_x = \frac{u_i^n - u_{i-1}^n}{\Delta x} \; , \tag{5.14}$$

$$u_x = \frac{u_{i+1}^n - u_i^n}{\Delta x} \; . \tag{5.15}$$

It turns out that only one of these yields a useful numerical scheme. The correct choice of either (5.14) or (5.15) will depend on the *sign of the wave propagation speed* a of the differential equation in (5.7). Suppose a is positive, then (5.14) together with (5.9) give the scheme

$$u_i^{n+1} = u_i^n - c(u_i^n - u_{i-1}^n) \; . \tag{5.16}$$

A von-Neumann stability analysis of (5.16) gives

$$\|A\|^2 = (1 - c)^2 + 2c(1 - c)\cos\theta + c^2 \; ,$$

from which it follows that the scheme is stable if the CFL number c lies between zero and unity; it is *conditionally stable* with *stability condition*

$$0 \le c \le 1 \; . \tag{5.17}$$

Recall that the CFL number depends on the speed a, the mesh spacing Δx and the time step size Δt. Of these, a is prescribed at the outset, Δx is chosen on desired accuracy or on computing resources available. We are left with some freedom to choose Δt, the time step size. The stability restriction (5.17) on c means a restriction on Δt and thus we are not free to choose the time step at our will, at least for the schemes under discussion.

Scheme (5.16) is called the *first–order upwind method* and is due to Courant, Isaacson and Rees [98]; we shall also call it the *CIR scheme*. The key feature of this numerical method is the fact that the discretisation has been performed according to the sign of the wave propagation speed a in the differential equation. The physics and mathematics embodied in the PDE are intimately linked to the discretisation procedure. The term *upwind*, or upstream, refers to the fact that spatial differencing is performed using mesh points on the side from which information (wind) flows; see Fig. 5.3. For positive a the upwind side is the left side and for negative a the upwind side is the right side.

Suppose that for positive a we use the *downwind* information to perform the spatial differencing. This gives the *downwind scheme*

$$u_i^{n+1} = u_i^n - c(u_{i+1}^n - u_i^n) \; , \tag{5.18}$$

which can easily be checked to be *unconditionally unstable*. Thus, in order to obtain a useful one–sided scheme one must perform the spatial differencing according to the sign of the speed a of the PDE. For negative speed a the upwind scheme is (5.18).

In order to formulate the upwind scheme, in a unified manner, for both positive and negative wave speeds a we introduce the following notation

$$a^+ = \max(a, 0) = \frac{1}{2}(a + |a|), \quad a^- = \min(a, 0) = \frac{1}{2}(a - |a|), \quad (5.19)$$

where $|a|$ denotes the absolute value of a, that is,

$$|a| = a \text{ if } a \geq 0, \quad |a| = -a \text{ if } a \leq 0.$$

It can easily be verified that for any value of the speed a the speeds a^+ and a^- satisfy

$$a^+ \geq 0, \quad a^- \leq 0.$$

Clearly, for $a \geq 0$ one has $a^+ = a$ and $a^- = 0$; for $a \leq 0$ one has $a^+ = 0$ and $a^- = a$. Based on the speeds a^+ and a^- we define Courant numbers

$$c^+ = \Delta t a^+ / \Delta x, \quad c^- = \Delta t a^- / \Delta x. \quad (5.20)$$

Using the above notation the *first–order upwind scheme* can be expressed in general form as

$$u_i^{n+1} = u_i^n - c^+ \left(u_i^n - u_{i-1}^n\right) - c^- \left(u_{i+1}^n - u_i^n\right). \quad (5.21)$$

For $a \geq 0$ the second difference term vanishes leading to (5.16); if $a \leq 0$ we obtain (5.18). The idea of *splitting the difference* into positive and negative components can be generalised to systems of conservation laws. Now the stability condition for the upwind scheme (5.21) is

$$0 \leq |c| \leq 1. \quad (5.22)$$

The *stencil* of the first–order upwind scheme for the case $a > 0$ is shown in Fig. 5.3. It has a triangular shape, the three mesh points involved define a triangle. The base of the triangle defines the *numerical domain of dependence* of the scheme. This is generally different from the (true) domain of dependence of the PDE, see Sect. 2.3.5 of Chap. 2. One can relate the meaning of the scheme and its stability condition to the exact behaviour of the differential equation in (5.7).

Consider the characteristic of speed a through the mesh point (x_i, t^{n+1}) in Fig. 5.3, at which a numerical solution u_i^{n+1} is sought. Since the exact solution of the PDE in (5.7) is constant along characteristics of speed $dx/dt = a$, see Sect. 2.2 of Chap. 2, the true solution at (x_i, t^{n+1}) is

$$u(x_i, t^{n+1}) = u(x_p, t^n), \quad (5.23)$$

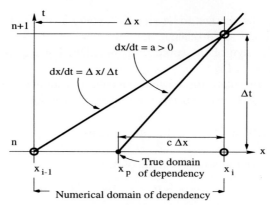

Fig. 5.3. Stencil of first–order upwind scheme (CIR) for positive speed a of wave propagation. Upwind direction lies on the left hand side.

with x_p between x_{i-1} and x_i. Unfortunately, the only values at time level n available to us are those at the grid points and thus we could not set $u_i^{n+1} = u(x_p, t^n)$, unless $c = 1$ of course.

One may however utilise the information at the mesh points x_{i-1} and x_i at the time level n to produce an estimate of the data at the point x_p. For instance we can construct a linear interpolation function $\tilde{u}(x)$ based on the two points (x_{i-1}, u_{i-1}^n) and (x_i, u_i^n). From Fig. 5.3, the distance d between x_p and x_i is

$$d = \Delta t a = \frac{\Delta t a}{\Delta x} \Delta x = c\Delta x \ ,$$

and thus

$$x_p = (i - 1)\Delta x + (1 - c)\Delta x \ .$$

The linear interpolant is then

$$\tilde{u}(x) = u_{i-1}^n + \frac{(u_i^n - u_{i-1}^n)}{\Delta x}(x - x_{i-1}) \ ,$$

which if evaluated at $x = x_p$ gives

$$\tilde{u}(x_p) = u_i^n - c(u_i^n - u_{i-1}^n) \ .$$

This is precisely the first–order upwind scheme (5.16), which can then be seen as a *linear interpolation* scheme. Fig. 5.3 is also useful for interpreting the stability condition (5.17), which states

$$0 \le \frac{a}{\Delta x/\Delta t} \le 1 \ .$$

Thus for stability, the grid speed $\Delta x/\Delta t$ must be larger than the speed a in the PDE. In the context of Fig. 5.3 this means that the characteristics

$$\frac{dx}{dt} = \Delta x/\Delta t , \quad \frac{dx}{dt} = a ,$$

must be as shown. That is *the numerical domain of dependence must contain the true domain of dependence of the PDE*, which is in fact the single point x_p.

A truncation error analysis reveals that the CIR scheme is first–order accurate in space and time. Moreover, the corresponding *modified equation* is

$$q_t + a q_x = \alpha_{\text{cir}} q_{xx} , \tag{5.24}$$

where α_{cir} is the numerical viscosity coefficient of the CIR scheme and is given by

$$\alpha_{\text{cir}} = \frac{1}{2}\Delta x a(1- | c |) . \tag{5.25}$$

The advection–diffusion equation (5.24) is the actual equation solved by the numerical scheme, to second–order accuracy in fact; see LeVeque [207]. The viscous term $\alpha_{\text{cir}} q_{xx}$ is responsible for the production of the *artificial or numerical viscosity* of the scheme. This vanishes when $\Delta x = 0$, which is impossible in practice. It also vanishes when $| c |= 1$, which is only of the academic importance, as for non–linear systems it is impossible to enforce this CFL number unity condition. The case $a = 0$ is interesting, as it allows for the perfect resolution of *stationary discontinuities*. In general $\alpha_{\text{cir}} > 0$ and one of the consequences is that discontinuities in the solution tend to be heavily *spread or smeared* and extreme values tend to be *clipped*. This is a disadvantage of the CIR scheme, which is in fact common to all first–order methods, with the exception of the Random Choice Method, see Chap. 7.

Scheme (5.21) is part of a wider class of methods that can be written as

$$\left. \begin{aligned} u_i^{n+1} &= H\left(u_{i-l_L}^n, \ldots, u_{i+l_R}^n\right) , \\ &= \sum_{k=-l_L}^{l_R} b_k u_{i+k}^n , \end{aligned} \right\} \tag{5.26}$$

with l_L and l_R two non–negative integers; b_k, $k = -l_L, \ldots, l_R$ are the coefficients of the scheme and u_{i+k}^n are data values at time level n. For the CIR scheme (5.21) the coefficients are given by

$$b_{-1} = c^+ , \quad b_0 = (1- | c |) , \quad b_1 = -c^- . \tag{5.27}$$

Under the CFL stability condition (5.22) we see that

$$b_k \geq 0 , \ \forall k , \tag{5.28}$$

i.e. all coefficients (5.27) are *positive or zero*.

Definition 5.2.1 (Monotone Schemes). *A scheme of the form (5.26) is said to be monotone if all coefficients b_k are positive or zero. An alternative definition is given in terms of the function H in (5.26); this is required to satisfy*

$$\frac{\partial H}{\partial u_k^n} \geq 0 \, , \, \forall \, k \, . \tag{5.29}$$

The class of monotone schemes form the basis of modern schemes for conservation laws. Monotone schemes are, however, at most first–order accurate; high–order extensions are studied in Chaps. 13, 14 and 16.

5.2.2 Other Well–Known Schemes

Another first–order scheme is that of **Lax and Friedrichs**. The scheme is sometimes also called the Lax Method [201], or the scheme of Keller and Lax. This does not require the differencing to be performed according to upwind directions and can be seen as a way of stabilising the unstable scheme (5.12) obtained from forward in time and central in space approximations to the partial derivatives. The Lax–Friedrichs scheme results if u_i^n in the time derivative (5.9) is replaced by

$$\frac{1}{2} \left(u_{i-1}^n + u_{i+1}^n \right) \, ,$$

a mean value of the two neighbours at time level n. Then the modified scheme becomes

$$u_i^{n+1} = \frac{1}{2} \left(u_{i-1}^n + u_{i+1}^n \right) - \frac{1}{2} c \left(u_{i+1}^n - u_{i-1}^n \right)$$

or

$$u_i^{n+1} = \frac{1}{2}(1 + c)u_{i-1}^n + \frac{1}{2}(1 - c)u_{i+1}^n \, . \tag{5.30}$$

A von Neumann stability analysis reveals that scheme (5.30) is stable under the stability condition (5.22). A truncation error analysis says that the scheme is first–order accurate in time and second–order accurate in space. The modified equation is like (5.24) with numerical viscosity coefficient given by

$$\alpha_{\text{lf}} = \frac{\Delta x a}{2c}(1 - c^2) \, . \tag{5.31}$$

By comparing α_{lf} with α_{cir} we see that the Lax–Friedrichs scheme is considerably more diffusive than the CIR scheme; in fact for $0 \leq c \leq 1$ we have

$$2 \leq \alpha_{\text{lf}}/\alpha_{\text{cir}} = \frac{1+c}{c} < \infty \, .$$

When written in the form (5.26) the coefficients of the Lax–Friedrichs scheme are found to be

$$b_{-1} = \frac{1}{2}(1 + c) \, , \quad b_0 = 0 \, , \quad b_1 = \frac{1}{2}(1 - c) \, .$$

Under the stability condition (5.22) all coefficients b_k in the Lax–Friedrich scheme (5.30) are positive or zero. Therefore the scheme is *monotone*.

A scheme of historic as well as practical importance is that of **Lax and Wendroff** [204]. For a comprehensive treatment of the family of Lax–Wendroff schemes see Hirsch [171], Chap. 17. The basic Lax–Wendroff scheme is second–order accurate in both space and time. There are several ways of deriving the scheme for the model equation in (5.7). A rather unconventional derivation is this: for the time derivative u_t insist on the first–order forward approximation (5.9); for the space derivative u_x take an average of the up-wind (stable if $a > 0$) and downwind (unstable if $a > 0$) approximations (5.14) and (5.15) respectively, that is

$$u_x = \beta_1 \frac{u_i^n - u_{i-1}^n}{\Delta x} + \beta_2 \frac{u_{i+1}^n - u_i^n}{\Delta x} . \tag{5.32}$$

If the coefficients β_1, β_2 are chosen as

$$\beta_1 = \frac{1}{2}(1 + c) , \ \beta_2 = \frac{1}{2}(1 - c) , \tag{5.33}$$

the resulting scheme is the Lax–Wendroff method

$$u_i^{n+1} = \frac{1}{2}c(1 + c)u_{i-1}^n + (1 - c^2)u_i^n - \frac{1}{2}c(1 - c)u_{i+1}^n . \tag{5.34}$$

This scheme is second–order in space and time although all finite difference approximations used to generate it are first–order accurate. Moreover, one of the terms in the spatial derivative originates from an unconditionally unstable scheme and yet the Lax–Wendroff scheme is stable with stability condition (5.22). This scheme is a good example to show that the order of accuracy of the scheme cannot in general be inferred from the order of accuracy of the finite difference approximations to the partial derivatives involved.

When written in the form (5.26) the Lax–Wendroff scheme (5.34) has coefficients

$$b_{-1} = \frac{(1 + c)c}{2} , \ b_0 = 1 - c^2 , \ b_1 = -\frac{(1 - c)c}{2} .$$

Therefore this scheme is *not monotone*. Not all coefficients in (5.34) are positive or zero. The fact that a scheme is not monotone is associated with the phenomenon of *spurious oscillations* in the numerical solution in the vicinity of sharp gradients, such as at discontinuities; see Chap. 13.

Another second–order accurate scheme for (5.7) is the upwind method of **Warming and Beam** [404]. For positive speed a it reads

$$u_i^{n+1} = \frac{1}{2}c(c - 1)u_{i-2}^n + c(2 - c)u_{i-1}^n + \frac{1}{2}(c - 1)(c - 2)u_i^n . \tag{5.35}$$

Note that the scheme is fully one–sided in the sense that all the mesh points involved, other than the centre of the stencil, are on the left hand side of the centre of the stencil. There is an equivalent scheme for negative speed a. Clearly the Warming–Beam scheme is not monotone. The stability restriction for this scheme is

$$0 \leq |c| \leq 2 . \tag{5.36}$$

The enlarged stability range means that one may advance in time with a larger time step Δt, which has a bearing on the efficiency of the scheme.

Yet another second order scheme is the **Fromm** scheme [128]. For the linear advection equation in (5.7), for $a > 0$, the scheme reads

$$\left. \begin{aligned} u_i^{n+1} = \quad & -\tfrac{1}{4}(1-c)cu_{i-2}^n + \tfrac{1}{4}(5-c)cu_{i-1}^n \\[2mm] & +\tfrac{1}{4}(1-c)(4+c)u_i^n - \tfrac{1}{4}(1-c)cu_{i+1}^n , \end{aligned} \right\} \tag{5.37}$$

which has stability restriction (5.22). Also, it can be easily verified that the Fromm scheme is not monotone.

Second–order schemes such as the Lax–Wendroff, Warming–Beam and Fromm schemes have modified equation of the form

$$q_t + aq_x = \beta q_{xxx} , \tag{5.38}$$

which is a *dispersive* equation. See LeVeque [207] for details.

5.3 Conservative Methods

Computing solutions containing discontinuities, such as shock waves, poses stringent requirements on (i) the mathematical formulation of the governing equations and (ii) the numerical schemes to solve the equations. As seen in Chaps. 2 and 3 the formulation of the equations can be differential or integral. Also, there are various choices for the set of variables to be used. One obvious choice is the set of conserved variables. In Sect. 3.3 of Chap. 3, through an example, we highlighted the fact that formulations based on variables other than the conserved variables (non–conservative variables) fail at shock waves. They give the wrong jump conditions; consequently they give the wrong shock strength, the wrong shock speed and thus the wrong shock position. Recent work by Hou and Le Floch [177] has shown that non–conservative schemes do not converge to the correct solution if a shock wave is present in the solution. The classical result of Lax and Wendroff [204], on the other hand, says that conservative numerical methods, *if convergent*, do converge to the weak solution of the conservation law. Therefore, it appears as if there is no choice but to work with conservative methods if shock waves are part of the solution. There are alternative special procedures involving shock fitting [244], [242] and adaptive primitive–conservative schemes [360], [192]. Some primitive–variable schemes are presented in Sect. 14.6 of Chap. 14.

5.3.1 Basic Definitions

Here we shall study conservative shock capturing methods. Of the class of conservative methods we are particularly interested in *upwind methods*, but not exclusively. This section is designed to introduce some basic concepts in the simple setting of model problems. Consider a scalar conservation law written in differential form

$$u_t + f(u)_x = 0 \,, \tag{5.39}$$

where $f = f(u)$ is the flux function. The choice of flux $f(u) = au$ reproduces the linear advection equation in (5.7). In order to include weak solutions of (5.39) we must use the integral form of the equations. Two possibilities are

$$\oint (u \, \mathrm{d}x - f \, \mathrm{d}t) = 0 \tag{5.40}$$

and

$$\int_{x_1}^{x_2} u(x, t_2) \, \mathrm{d}x = \int_{x_1}^{x_2} u(x, t_1) \, \mathrm{d}x + \int_{t_1}^{t_2} f(u(x_1, t)) \, \mathrm{d}t - \int_{t_1}^{t_2} f(u(x_2, t)) \, \mathrm{d}t \tag{5.41}$$

for any rectangular control volume $[x_1, x_2] \times [t_1, t_2]$. See Sect. 2.4 of Chap. 2.

Definition 5.3.1 (Conservative Method). *A* conservative scheme *for the scalar conservation law (5.39) is a numerical method of the form*

$$u_i^{n+1} = u_i^n + \frac{\Delta t}{\Delta x} \left[f_{i-\frac{1}{2}} - f_{i+\frac{1}{2}} \right] \,, \tag{5.42}$$

where

$$f_{i+\frac{1}{2}} = f_{i+\frac{1}{2}} \left(u_{i-l_L}^n, \ldots, u_{i+l_R}^n \right) \,, \tag{5.43}$$

with l_L, l_R *two non–negative integers;* $f_{i+\frac{1}{2}}$ *is called the* numerical flux, *an approximation to the* physical flux $f(u)$ *in (5.39).*

For any particular choice of numerical flux $f_{i+\frac{1}{2}}$ a corresponding conservative scheme results. A fundamental requirement on the numerical flux is the *consistency condition*

$$f_{i+\frac{1}{2}}(v, \ldots v) = f(v) \,. \tag{5.44}$$

This means that if all arguments in (5.43) are equal to v then the numerical flux is identical to the physical flux at $u = v$. See LeVeque [207].

Exercise 5.3.1. Verify that the choice of numerical flux

$$f_{i+\frac{1}{2}} = f_{i+\frac{1}{2}}(u_i^n, u_{i+1}^n) = \frac{1}{2}(f_i^n + f_{i+1}^n) \,,$$

with $f_i^n = f(u_i^n)$, $f_{i+1}^n = f(u_{i+1}^n)$, when substituted into the conservative formula (5.42), reproduces the unconditionally unstable scheme (5.12), when applied to the linear advection equation, in which $f(u) = au$.

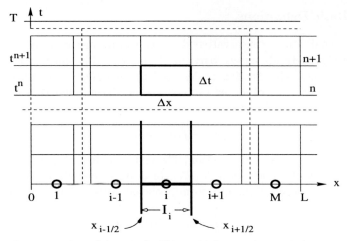

Fig. 5.4. Discretisation of domain $[0, L]$ into M finite volumes I_i (computing cells)

Solution 5.3.1. (Left to the reader).

The conservative scheme (5.42) requires a redefinition of the discretisation of the domain. Now one is concerned with cell averages defined over *finite volumes*. A domain $[0, L] \times [0, T]$ in the x–t plane is discretised as shown in Fig. 5.4. The spatial domain of length L is subdivided into M *finite volumes*, called *computing cells* or simply *cells*, given as

$$x_{i-\frac{1}{2}} = (i - 1)\Delta x \leq x \leq i\Delta x = x_{i+\frac{1}{2}} . \tag{5.45}$$

The extreme values $x_{i-\frac{1}{2}}$ and $x_{i+\frac{1}{2}}$ of cell I_i define the position of the intercell boundaries at which the corresponding intercell numerical fluxes must be specified. The size of the cell is

$$\Delta x = x_{i+\frac{1}{2}} - x_{i-\frac{1}{2}} = \frac{L}{M} . \tag{5.46}$$

Obviously one may discretise the domain into cells of irregular size. For simplicity we assume regular meshes in this chapter. The average value of $u(x, t)$ in cell i, the cell average, at a fixed time $t = t^n = n\Delta t$ is defined as

$$u_i^n = \frac{1}{\Delta x} \int_{x_{i-\frac{1}{2}}}^{x_{i+\frac{1}{2}}} u(x, t^n) \, dx . \tag{5.47}$$

Note here that although within cell i one may have spatial variations of $u(x, t)$ at time $t = t^n$, the integral average value u_i^n given above is constant. We shall assign that constant value at the centre of the cell, which gives rise to *cell–centred conservative methods*. Computationally, we shall deal with approximations to the cell averages u_i^n, which for simplicity we shall still denote as u_i^n. The set of cell averages (5.47) defines a piece–wise constant distribution of the solution at time t^n; see Fig. 5.5.

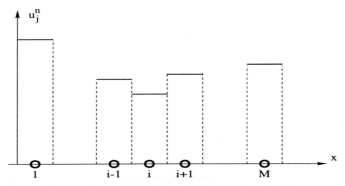

Fig. 5.5. Piece–wise constant distribution of data at time level n

5.3.2 Godunov's First Order Upwind Method

Godunov [145] is credited with the first successful conservative extension of the CIR scheme (5.21) to non–linear systems of conservation laws. When applied to the scalar conservation law (5.39) with $f(u) = au$, Godunov's scheme reduces to the CIR scheme, allowing for appropriate interpretation of the values $\{u_i^n\}$.

Godunov's first–order upwind method is a conservative method of the form (5.42), where the intercell numerical fluxes $f_{i+\frac{1}{2}}$ are computed by using solutions of local Riemann problems. A basic assumption of the method is that at a given time level n the data has a piece–wise constant distribution of the form (5.47), as depicted in Fig. 5.5. The data at time level n may be seen as pairs of constant states (u_i^n, u_{i+1}^n) separated by a discontinuity at the intercell boundary $x_{i+\frac{1}{2}}$. Then, locally, one can define a Riemann problem

$$\text{PDE}: \quad u_t + f(u)_x = 0 \,.$$

$$\text{IC}: \quad u(x,0) = u_0(x) = \begin{cases} u_i^n & \text{if } x < 0 \,, \\ u_{i+1}^n & \text{if } x > 0 \,, \end{cases}$$

This *local* Riemann problem may be solved analytically, if desired. Thus, at a given time level n, at each intercell boundary $x_{i+\frac{1}{2}}$ we have the *local* Riemann problem $RP(u_i^n, u_{i+1}^n)$ with initial data (u_i^n, u_{i+1}^n). What is then needed is a way of finding the solution of the *global* problem at a later time level $n+1$.

First Version of Godunov's Method. Godunov proposed the following scheme to update a cell value u_i^n to a new value u_i^{n+1}: *solve the two Riemann problems $RP(u_{i-1}^n, u_i^n)$ and $RP(u_i^n, u_{i+1}^n)$ for the conservation law (5.39), take an integral average in cell i of the combined solutions of these two local problems and assign the value to u_i^{n+1}.* Fig. 5.6 illustrates the situation for the special case $f(u) = au$, $a > 0$. The exact solution of $RP(u_{i-1}^n, u_i^n)$ for $a > 0$, see Sect. 2.2.2 of Chap. 2, is

$$u_{i-\frac{1}{2}}(x/t) = \begin{cases} u_{i-1}^n & \text{if } x/t < a \,, \\ u_i^n & \text{if } x/t > a \,, \end{cases} \tag{5.48}$$

where the local origin of the Riemann problem is $(0,0)$. Likewise the solution $u_{i+\frac{1}{2}}(x/t)$ of $RP(u_i^n, u_{i+1}^n)$ is given by

$$u_{i+\frac{1}{2}}(x/t) = \begin{cases} u_i^n & \text{if } x/t < a\ , \\ u_{i+1}^n & \text{if } x/t > a\ . \end{cases} \tag{5.49}$$

The Godunov scheme defines the updated solution as

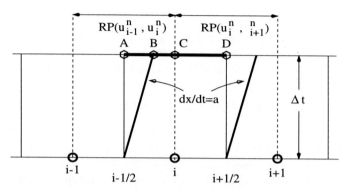

Fig. 5.6. Illustration of Godunov's method for the linear advection equation for positive speed a. Riemann problem solutions are averaged at time $t = \Delta t$ inside cell I_i

$$u_i^{n+1} = \frac{1}{\Delta x} \left[\int_0^{\frac{1}{2}\Delta x} u_{i-\frac{1}{2}}(x/\Delta t)\,\mathrm{d}x + \int_{-\frac{1}{2}\Delta x}^0 u_{i+\frac{1}{2}}(x/\Delta t)\,\mathrm{d}x \right]\ . \tag{5.50}$$

This integral is evaluated at time Δt (local time) between the points A and D in Fig. 5.6. Note that we only use the right half of the solution $u_{i-\frac{1}{2}}(x/t)$ and the left half of the solution $u_{i+\frac{1}{2}}(x/t)$. The reader should realise that *each solution has its own local frame of reference* with origin $(0,0)$ corresponding with the intercell boundaries at $x_{i-\frac{1}{2}}$ and $x_{i+\frac{1}{2}}$. In order to evaluate the integral we impose a restriction on the size of the time step Δt. We set

$$c = \frac{a\Delta t}{\Delta x} \leq \frac{1}{2}\ . \tag{5.51}$$

The first term in (5.50) involves the intervals AB and BC; these have lengths respectively given by

$$l_{AB} = c\Delta x\ , \quad l_{BC} = (\frac{1}{2} - c)\Delta x\ .$$

The integrand is give by (5.48) and (5.49); hence we have

$$\frac{1}{\Delta x} \int_0^{\frac{1}{2}\Delta x} u_{i-\frac{1}{2}}(x/\Delta t)\,\mathrm{d}x = cu_{i-1}^n + (\frac{1}{2} - c)u_i^n\ .$$

The second term gives

$$\frac{1}{\Delta x} \int_{-\frac{1}{2}\Delta x}^{0} u_{i+\frac{1}{2}}(x/\Delta t)\, \mathrm{d}x = \frac{1}{2}u_i^n \ .$$

Hence (5.50) becomes

$$u_i^{n+1} = u_i^n + c(u_{i-1}^n - u_i^n) \ , \tag{5.52}$$

which is identical to the CIR first order upwind method (5.16) for positive speed a. The conservative character of the Godunov method is self evident: the updated solution is obtained by integral averaging, a conservative process, of local exact solutions of the conservation law.

Exercise 5.3.2. Using geometric arguments, show that the Godunov approach described above also reproduces the CIR scheme for negative a.

Solution 5.3.2. (Left to the reader).

Second Version of Godunov's Method. A second interpretation of the Godunov method leads directly to the conservative formula (5.42), which is easier to apply in practice and avoids the over–restrictive CFL–like condition (5.51). The integral average (5.50) of the solution of the Riemann problems $RP(u_{i-1}^n, u_i^n)$ and $RP(u_i^n, u_{i+1}^n)$ can also be written as

$$u_i^{n+1} = \frac{1}{\Delta x} \int_{x_{i-\frac{1}{2}}}^{x_{i+\frac{1}{2}}} \tilde{u}(x, \Delta t)\, \mathrm{d}x \ , \tag{5.53}$$

where $\tilde{u}(x, t)$ is understood as the combined solution of $RP(u_{i-1}^n, u_i^n)$ and $RP(u_i^n, u_{i+1}^n)$. Since $\tilde{u}(x, t)$ is an exact solution to the original conservation law (5.39) we can make use of it in its integral form (5.41), say, in the control volume $[x_{i-\frac{1}{2}}, x_{i+\frac{1}{2}}] \times [0, \Delta t]$ to write

$$\left. \begin{aligned} \int_{x_{i-\frac{1}{2}}}^{x_{i+\frac{1}{2}}} \tilde{u}(x, \Delta t)\, \mathrm{d}x = & \int_{x_{i-\frac{1}{2}}}^{x_{i+\frac{1}{2}}} \tilde{u}(x, 0)\, \mathrm{d}x + \int_0^{\Delta t} f(\tilde{u}(x_{i-\frac{1}{2}}, t))\, \mathrm{d}t \\ & - \int_0^{\Delta t} f(\tilde{u}(x_{i+\frac{1}{2}}, t))\, \mathrm{d}t \ . \end{aligned} \right\} \tag{5.54}$$

Using the definition of cell averages (5.47) into equation (5.54) followed by division through by Δx we reproduce the conservative formula

$$u_i^{n+1} = u_i^n + \frac{\Delta t}{\Delta x}\left[f_{i-\frac{1}{2}} - f_{i+\frac{1}{2}} \right] \ , \tag{5.55}$$

with the intercell fluxes defined as time integral averages, namely

$$f_{i-\frac{1}{2}} = \frac{1}{\Delta t} \int_0^{\Delta t} f(\tilde{u}(x_{i-\frac{1}{2}}, t))\, \mathrm{d}t \ , \quad f_{i+\frac{1}{2}} = \frac{1}{\Delta t} \int_0^{\Delta t} f(\tilde{u}(x_{i+\frac{1}{2}}, t))\, \mathrm{d}t \ . \tag{5.56}$$

Thus, by invoking the integral form of the conservation laws on a control or finite volume $[x_{i-\frac{1}{2}}, x_{i+\frac{1}{2}}] \times [0, \Delta t]$ in x–t space we have arrived at the conservative formula (5.55) with intercell fluxes (5.56); these are time integral averages of the physical flux $f(u)$ evaluated at the intercell boundaries. The integrand $f(\tilde{u}(x,t))$ at each cell interface depends on the exact solution $\tilde{u}(x,t)$ of the Riemann problem along the t–axis (local coordinates); this is given by

$$\tilde{u}(x_{i-\frac{1}{2}}, t) = u_{i-\frac{1}{2}}(0) , \quad \tilde{u}(x_{i+\frac{1}{2}}, t) = u_{i+\frac{1}{2}}(0) , \qquad (5.57)$$

and the intercell fluxes $f_{i-\frac{1}{2}}$ and $f_{i+\frac{1}{2}}$ become

$$f_{i-\frac{1}{2}} = f(u_{i-\frac{1}{2}}(0)) , \quad f_{i+\frac{1}{2}} = f(u_{i+\frac{1}{2}}(0)) . \qquad (5.58)$$

In general, one expresses the Godunov intercell numerical flux as

$$f_{i+\frac{1}{2}}^{\text{god}} = f(u_{i+\frac{1}{2}}(0)) , \qquad (5.59)$$

where $u_{i+\frac{1}{2}}(0)$ denotes the exact solution $u_{i+\frac{1}{2}}(x/t)$ of the Riemann problem $RP(u_i^n, u_{i+1}^n)$ evaluated at $x/t = 0$, i.e. the solution is evaluated along the intercell boundary, which coincides with the t–axis in the local frame of the Riemann problem solution. We have thus defined the second version of Godunov's method for a general scalar conservation law (5.39), as the conservative formula (5.55) together with the intercell numerical flux (5.59). For the special conservation law with flux $f(u) = au$, $a > 0$, we have

$$f_{i-\frac{1}{2}} = au_{i-1}^n , \quad f_{i+\frac{1}{2}} = au_i^n , \qquad (5.60)$$

which if substituted in the conservative formula (5.55) reproduce the CIR scheme. The second version (5.55), (5.59) of Godunov's method is the one that is mostly used in practice.

Exercise 5.3.3. Verify that the second version of the Godunov method based on the conservative formula (5.55) and the Godunov intercell flux (5.59) also reproduces the CIR scheme when applied to (5.39) with flux $f(u) = au$ and $a < 0$.

Solution 5.3.3. (Left to the reader).

5.3.3 Godunov's Method for Burgers's Equation

As a way of illustrating Godunov's method in the context of *non–linear* PDEs we apply the scheme to the inviscid Burgers equation

$$u_t + f(u)_x = 0 , \quad f(u) = \frac{1}{2}u^2 . \qquad (5.61)$$

We adopt the second version (5.55) with numerical flux given by (5.59). We need the solution $u_{i+\frac{1}{2}}(x/t)$ of the Riemann problem $RP(u_i^n, u_{i+1}^n)$. As seen in Sect. 2.4.2 of Chap. 2, the solution is a shock wave, when $u_i^n > u_{i+1}^n$, and a rarefaction wave when $u_i^n \leq u_{i+1}^n$. The complete solution is

$$
\left.
\begin{array}{l}
u_{i+\frac{1}{2}}(x/t) = \left\{
\begin{array}{ll}
u_i^n & \text{if } S \geq x/t \\
u_{i+1}^n & \text{if } S \leq x/t
\end{array}
\right\} \text{ if } u_i^n > u_{i+1}^n , \\[2pt]
S = \frac{1}{2}(u_i^n + u_{i+1}^n) \\[12pt]
u_{i+\frac{1}{2}}(x/t) = \left\{
\begin{array}{ll}
u_i^n & \text{if } x/t \leq u_i^n \\
x/t & \text{if } u_i^n < x/t < u_{i+1}^n \\
u_{i+1}^n & \text{if } x/t \geq u_{i+1}^n
\end{array}
\right\} \text{ if } u_i^n \leq u_{i+1}^n .
\end{array}
\right\}
\tag{5.62}
$$

The Godunov's flux requires $u_{i+\frac{1}{2}}(0)$; this is the solution $u_{i+\frac{1}{2}}(x/t)$ evaluated along the intercell boundary $x_{i+\frac{1}{2}}$, that corresponds to $x/t = 0$ in the frame of the local Riemann problem. The second stage is *to identify all possible wave patterns in the solution* of the Riemann problem. For Burgers's equation there are five possibilities. These are illustrated in Fig. 5.7. If the solution is

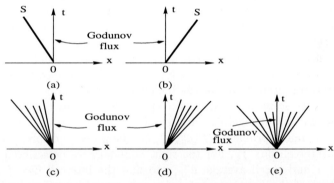

Fig. 5.7. Five possible wave patterns in the solution of the Riemann problem for the inviscid Burgers equations, when evaluating the Godunov flux

a shock wave then cases (a) and (b) can occur. The sought value $u_{i+\frac{1}{2}}(0)$, on the t–axis, depends on the sign of the shock speed S. If the solution is a rarefaction wave then the three possible cases are illustrated in Figs. 5.7c, 5.7d and 5.7e. Applying terminology from Gas Dynamics to the rarefaction cases, Fig. 5.7c is called *supersonic to the left* and that of Fig. 5.7d *supersonic to the right*. The case of Fig. 5.7e is that of a *transonic rarefaction* or *sonic rarefaction*; as the wave is crossed, there is a sign change in the characteristic speed u and thus there is one point at which $u = 0$, a *sonic point*. The complete sought solution is summarised as follows

$$u_{i+\frac{1}{2}}(0) = \begin{cases} u_i^n & \text{if } S \geq 0 \\ u_{i+1}^n & \text{if } S \leq 0 \end{cases} \Bigg\} \text{ if } u_i^n > u_{i+1}^n \,,$$
$$S = \tfrac{1}{2}(u_i^n + u_{i+1}^n)$$

$$u_{i+\frac{1}{2}}(0) = \begin{cases} u_i^n & \text{if } 0 \leq u_i^n \\ 0 & \text{if } u_i^n < 0 < u_{i+1}^n \\ u_{i+1}^n & \text{if } 0 \geq u_{i+1}^n \end{cases} \Bigg\} \text{ if } u_i^n \leq u_{i+1}^n \,. \qquad (5.63)$$

Naturally, Godunov's method can also be implemented using approximate solutions to the Riemann problem. For a review on the Godunov scheme in the context of two well–known approximations to the Riemann problem solution for Burgers's equation, the reader is referred to the paper by van Leer [395].

In applying Godunov's scheme to solve Burgers's equation there are two more issues to consider. One concerns the application of boundary conditions and the other is to do with the choice of the time step Δt.

Boundary Conditions. The conservative formula (5.55) can be applied directly to all cells i, for $i = 2, \ldots, M - 1$. The two required intercell fluxes at $x_{i-\frac{1}{2}}$ and $x_{i+\frac{1}{2}}$ are defined in terms of the corresponding Riemann problems. For each of the cells 1 and M, which are adjacent to the left and right boundaries respectively, we only have one intercell flux. Some special procedure needs to be implemented. Let us consider the left boundary $x = 0$. One possibility is to assume a boundary function $u_l(t)$ prescribed there. Then we could define an intercell flux at the boundary by setting $f_{\frac{1}{2}} = f(u_l(t))$.

A more attractive alternative is to specify a *fictitious cell* 0 to the left of the boundary $x = 0$ together with a cell average u_0^n, at each time level n, so that a Riemann problem $RP(u_0^n, u_1^n)$ can be posed and solved to find the *missing* intercell flux $f_{\frac{1}{2}}$. For the right boundary we prescribe a fictitious cell $M + 1$ and a cell average u_{M+1}^n to find the intercell flux $f_{M+\frac{1}{2}}$. The prescription of the fictitious states depends entirely on the physics of the particular problem at hand. Provisionally, for the inviscid Burgers's equation we apply the boundary conditions

$$u_0^n = u_1^n \,, \quad u_{M+1}^n = u_M^n \,. \qquad (5.64)$$

Note that the fictitious states here are given in terms of the data at the states within the computational domain adjacent to the boundaries. This particular type of boundary conditions will cause no disturbance at the boundaries; waves will hopefully go through the boundaries as if the boundaries were not there. Often one speaks of *transparent* or *transmissive* boundary conditions, in this case.

Choosing the Time Step. As seen for the linear case, the choice of the size of the time step Δt in the conservative formula is related to the stability condition of the particular scheme. For Godunov's method (the CIR scheme in this case) the choice of Δt depends on the restriction on the Courant number c. For non–linear problems, at each time level, there are multiple wave speeds and thus multiple associated Courant numbers. In deriving the second version of Godunov's method we made the implicit assumption that the value of the local Riemann problem solution along the intercell boundary is constant. This means that the fastest wave at a given time travels for at most one cell length Δx in the sought time step Δt. Denoting by S_{max}^n the maximum wave speed throughout the domain at time level n we define the *maximum Courant number* C_{cfl}

$$C_{cfl} = \Delta t S_{max}^n / \Delta x , \tag{5.65}$$

where Δt is such that

$$0 < C_{cfl} \leq 1 .$$

We shall often call C_{cfl} the *CFL coefficient* or the *Courant number coefficient*. The time step Δt follows as

$$\Delta t = C_{cfl} \Delta x / S_{max}^n . \tag{5.66}$$

A matter of crucial importance is the estimation of the maximum speed S_{max}^n. In realistic applications this can be difficult and frustrating, as inappropriate choices can cause the scheme to *crash*, no matter how sophisticated this is. For Burgers's equation one can identify wave speeds, such as those emerging from solutions of Riemann problems at the intercell boundaries, and characteristic speeds u. At any time level n, there are $M + 2$ characteristic speeds u_i^n (including the fictitious cells) and hence one possibility is to take S_{max}^n as the maximum of these, in absolute value. Another possibility is to select, at each time level, a speed $S_{i+\frac{1}{2}}^n$ from the solution Riemann problem at each cell interface; this information is available as part of the flux computation process. For the case of a shock, one obviously takes the shock speed. For the case of a rarefaction there are two characteristic speeds of significance, namely those of the head and tail of the expansion. Thus we define an intercell speed

$$S_{i+\frac{1}{2}}^n = \begin{cases} | \frac{1}{2}(u_i^n + u_{i+1}^n) | & \text{(shock)}, \\ \\ \max(| u_i^n |, | u_{i+1}^n |) & \text{(rarefaction)}. \end{cases} \tag{5.67}$$

Finally, we define a maximum wave speed at time level n as follows

$$S_{max}^n = \max_i \left\{ S_{i+\frac{1}{2}}^n \right\} \text{ for } i = 0, \ldots, M + 1 . \tag{5.68}$$

Having chosen C_{cfl}, the time step Δt to march the solution to the next time level is given by (5.66). For a scheme with linearised stability condition $|c| \leq 1$, one usually takes the empirical value $C_{cfl} = 0.9$.

In Sect. 5.6 we give a listing of a FORTRAN program for Godunov's method in conjunction with the exact Riemann solver, to solve numerically the inviscid Burgers equation.

5.3.4 Conservative Form of Difference Schemes

Conventional finite difference schemes can often be expressed in conservation form (5.42). It is a question of finding an intercell flux (5.43).

The Lax–Friedrichs Scheme. Recall that the Lax–Friedrichs scheme as applied to the linear advection equation is

$$u_i^{n+1} = \frac{(1+c)}{2} u_{i-1}^n + \frac{(1-c)}{2} u_{i+1}^n . \tag{5.69}$$

It is easy to verify that the conservative formula (5.42) together with the intercell flux

$$f_{i+\frac{1}{2}} = \frac{(1+c)}{2c} f(u_i^n) + \frac{(c-1)}{2c} f(u_{i+1}^n) \tag{5.70}$$

reproduces the Lax–Friedrichs scheme. Therefore, at least for the linear advection equation, one can recast the Lax–Friedrichs scheme in conservative form.

An interesting way of viewing the Lax–Friedrichs scheme (5.69) is as an integral average *within cell i*, namely

$$u_i^{n+1} = \frac{1}{\Delta x} \int_{x_{i-\frac{1}{2}}}^{x_{i+\frac{1}{2}}} \tilde{u}(x, \frac{1}{2}\Delta t) \, dx , \tag{5.71}$$

in which $\tilde{u}(x,t)$ is the solution of the Riemann problem $RP(u_{i-1}^n, u_{i+1}^n)$ (note subscripts), that is

$$\tilde{u}(x/t) = \begin{cases} u_{i-1}^n & \text{if } x/t < a , \\ u_{i+1}^n & \text{if } x/t > a . \end{cases} \tag{5.72}$$

Exercise 5.3.4. Verify that exact evaluation of the integral (5.71) reproduces the Lax–Friedrichs scheme (5.69).

Solution 5.3.4. (Left to the reader).

Remark 5.3.1. We note that the Lax–Friedrichs solution at cell i is a *weighted average* of the solution of the Riemann problem with the *left and right neighbour states* as data, at time $t = \frac{1}{2}\Delta t$. One could state that the Lax–Friedrichs scheme is *upwind biased,* as the upwind term has always the larger weight.

Let us now attempt to generalise interpretation (5.71) of the Lax–Friedrichs scheme to non–linear systems of conservation laws

$$\mathbf{U}_t + \mathbf{F}(\mathbf{U})_x = 0 . \tag{5.73}$$

Now (5.71) reads

$$\mathbf{U}_i^{n+1} = \frac{1}{\Delta x} \int_{x_{i-\frac{1}{2}}}^{x_{i+\frac{1}{2}}} \tilde{\mathbf{U}}(x, \frac{1}{2}\Delta t) \, dx , \tag{5.74}$$

where $\tilde{\mathbf{U}}(x, t)$ is the solution of the Riemann problem $RP(\mathbf{U}_{i-1}^n, \mathbf{U}_{i+1}^n)$. Given an exact Riemann solver, e.g. see Chap. 4, one could then implement this Riemann–problem based version of the Lax–Friedrichs scheme. The author has implemented this version of the scheme for non–linear systems. The numerical results obtained are indistinguishable from those obtained from the conventional form of the scheme. Version (5.74) offers no obvious advantages over the conventional form; in fact it is more expensive and complex and hence is of no practical use. A stochastic evaluation of the integral leads to a random choice type method; see Sect. 7.3 of Chap. 7.

If the space integral (5.74) at time $\frac{1}{2}\Delta t$ is replaced by invoking the integral form of conservation law (Chap. 2, equation (2.67)) in the control volume $[-\frac{1}{2}\Delta x, \frac{1}{2}\Delta x] \times [0, \frac{1}{2}\Delta t]$ we obtain

$$\left. \begin{array}{l} \int_{-\frac{1}{2}\Delta x}^{\frac{1}{2}\Delta x} \tilde{\mathbf{U}}(x, \tfrac{1}{2}\Delta t)\,\mathrm{d}x = \int_{-\frac{1}{2}\Delta x}^{\frac{1}{2}\Delta x} \tilde{\mathbf{U}}(x, 0)\,\mathrm{d}x + \int_0^{\frac{1}{2}\Delta t} \mathbf{F}(\tilde{\mathbf{U}}(-\tfrac{1}{2}\Delta x, t))\,\mathrm{d}t \\[2mm] \qquad\qquad - \int_0^{\frac{1}{2}\Delta t} \mathbf{F}(\tilde{\mathbf{U}}(\tfrac{1}{2}\Delta x, t))\,\mathrm{d}t \; . \end{array} \right\}$$
(5.75)

Direct evaluation of the integrals and use of the definition of integral averages (5.74) in cell i, as applied to systems, yield

$$\mathbf{U}_i^{n+1} = \frac{1}{2}(\mathbf{U}_{i-1}^n + \mathbf{U}_{i+1}^n) + \frac{1}{2}\frac{\Delta t}{\Delta x}(\mathbf{F}_{i-1}^n - \mathbf{F}_{i+1}^n) \; .$$

Simple algebraic manipulations of this expression lead to a conservative version of the Lax–Friedrichs scheme for systems

$$\mathbf{U}_i^{n+1} = \mathbf{U}_i^n + \frac{\Delta t}{\Delta x}\left[\mathbf{F}_{i-\frac{1}{2}} - \mathbf{F}_{i+\frac{1}{2}}\right] \; ,$$
(5.76)

with the Lax–Friedrichs intercell flux given by

$$\mathbf{F}_{i+\frac{1}{2}}^{LF} = \frac{1}{2}(\mathbf{F}_i^n + \mathbf{F}_{i+1}^n) + \frac{1}{2}\frac{\Delta x}{\Delta t}(\mathbf{U}_i^n - \mathbf{U}_{i+1}^n) \; .$$
(5.77)

This is the conventional numerical flux for the Lax–Friedrichs scheme when applied to systems of conservations laws (5.73). No mention of the Riemann problem is needed in this formulation. Compare the conservative formula (5.76) for systems with the conservative formula (5.55) for scalar conservation laws.

The Lax–Wendroff Scheme. The Lax–Wendroff scheme (5.34) as applied to the linear advection equation may also be written in conservation form (5.42). The intercell numerical flux is

$$f_{i+\frac{1}{2}} = \frac{(1+c)}{2}(au_i^n) + \frac{(1-c)}{2}(au_{i+1}^n) \; ,$$

which is a weighted average of fluxes on the left and right of the interface. For the linear advection equation, it is easy to check that this can also be obtained from

$$f_{i+\frac{1}{2}} = f\left(u_{i+\frac{1}{2}}^{n+\frac{1}{2}}\right) , \quad u_{i+\frac{1}{2}}^{n+\frac{1}{2}} = \frac{(1+c)}{2}u_i^n + \frac{(1-c)}{2}u_{i+1}^n .$$

For non–linear scalar conservation laws, such as Burgers's equations (5.61), this generalises to

$$f_{i+\frac{1}{2}} = f\left(u_{i+\frac{1}{2}}^{n+\frac{1}{2}}\right) ; \quad u_{i+\frac{1}{2}}^{n+\frac{1}{2}} = \frac{1}{\Delta x}\int_{-\frac{1}{2}\Delta x}^{\frac{1}{2}\Delta x} u_{i+\frac{1}{2}}\left(x, \frac{1}{2}\Delta t\right) dx , \qquad (5.78)$$

where $u_{i+\frac{1}{2}}(x,t)$ is the solution of the Riemann problem $RP(u_i^n, u_{i+1}^n)$. A straightforward Riemann–problem based generalisation of the Lax–Wendroff scheme to non–linear hyperbolic systems (5.73) ([352], [358]) reads

$$\mathbf{F}_{i+\frac{1}{2}} = \mathbf{F}(\mathbf{U}_{i+\frac{1}{2}}^{n+\frac{1}{2}}) ; \quad \mathbf{U}_{i+\frac{1}{2}}^{n+\frac{1}{2}} = \frac{1}{\Delta x}\int_{-\frac{1}{2}\Delta x}^{\frac{1}{2}\Delta x} \mathbf{U}_{i+\frac{1}{2}}\left(x, \frac{1}{2}\Delta t\right) dx ,$$

where $\mathbf{U}_{i+\frac{1}{2}}(x,t)$ is the solution of the Riemann problem $RP(\mathbf{U}_i^n, \mathbf{U}_{i+1}^n)$. This scheme is called the Weighted Average Flux (WAF) method and is studied in Chaps. 13, 14 and 16.

As done for the Lax–Friedrichs scheme one may replace the integral involving the solution of the Riemann problem by invoking the integral form of the conservation laws, see Sect. 2.4.1 of Chap. 2, in the control volume $[-\frac{1}{2}\Delta x, \frac{1}{2}\Delta x] \times [0, \frac{1}{2}\Delta t]$ to obtain

$$\mathbf{F}_{i+\frac{1}{2}} = \mathbf{F}\left(\mathbf{U}_{i+\frac{1}{2}}^{n+\frac{1}{2}}\right) ; \quad \mathbf{U}_{i+\frac{1}{2}}^{n+\frac{1}{2}} = \frac{1}{2}(\mathbf{U}_i^n + \mathbf{U}_{i+1}^n) + \frac{1}{2}\frac{\Delta t}{\Delta x}(\mathbf{F}_i^n - \mathbf{F}_{i+1}^n) . \quad (5.79)$$

This scheme is known as the two–step Richtmyer version of the Lax–Wendroff method, as applied to non–linear systems of conservation laws (5.73). No mention of the Riemann problem solution is necessary here.

Remark 5.3.2. Note the similarities between the reinterpretations and generalisations of the Lax–Friedrichs and Lax–Wendroff schemes. For both schemes one ends up with two formulations. For the Lax–Friedrichs scheme the weighted–average character leads to an integral formulation involving the solution of the Riemann problem. The second version of the scheme eliminates the role of the Riemann problem by utilising the integral form of the conservation law and leads to the conventional form of the Lax–Friedrichs scheme for non–linear systems. For the Lax–Wendroff method the procedure is entirely analogous. An integral average interpretation leads to a Riemann–problem based extension to non–linear systems [352]. Utilisation of the integral form of the conservation law eliminates the role of the Riemann problem and leads to the conventional Richtmyer version of the scheme. Both versions of the Lax–Wendroff method have actually been applied in practice.

Exercise 5.3.5. Verify that the Fromm scheme (5.37) as applied to the linear advection equation in (5.7), for positive a, can be written in conservation form (5.42) with numerical flux

$$f_{i+\frac{1}{2}}^{\text{FR}} = -\frac{1}{4}(1-c)f_{i-1} + f_i + \frac{1}{4}(1-c)f_{i+1} ,$$

where c is the Courant number.

Solution 5.3.5. (Left to the reader).

Exercise 5.3.6. Verify that the numerical flux of the scheme of Warming and Beam (5.35), as applied to (5.7) with $a > 0$, is

$$f_{i+\frac{1}{2}}^{\text{WB}} = \frac{1}{2}(c-1)f_{i-1} + \frac{1}{2}(3-c)f_i .$$

Solution 5.3.6. (Left to the reader).

Remark 5.3.3. The Warming–Beam numerical flux can be derived from (5.78) in terms of integral averages of solutions of Riemann problems under the assumption $1 \leq c \leq 2$. An extension of this interpretation to non–linear systems was proposed by Toro and Billett [373].

Exercise 5.3.7. Apply (5.78) to the linear advection equation and derive the Warming–Beam flux for negative speed a. Assume $-2 \leq c \leq -1$.

Solution 5.3.7. (Left to the reader).

5.4 Upwind Schemes for Linear Systems

Here we apply the first–order upwind scheme to hyperbolic systems with constant coefficients

$$\mathbf{U}_t + \mathbf{A}\mathbf{U}_x = \mathbf{0} . \tag{5.80}$$

For background on mathematical properties of these PDEs see Sect. 2.3 of Chap. 2. We denote the real eigenvalues of the $m \times m$ constant coefficient matrix $\mathbf{A}$ by λ_j with $j = 1, \ldots, m$ and assume they are ordered as

$$\lambda_1 < \lambda_2 < \lambda_3 \ldots < \lambda_m .$$

The corresponding right eigenvectors are denoted by $\mathbf{K}^{(1)}, \mathbf{K}^{(2)}, \ldots, \mathbf{K}^{(m)}$. Note here that in general the eigenvalues λ_j can be of any sign and thus a one–sided differencing scheme applied to (5.80) directly will only work if all the eigenvalues are of the same sign. In the general case with eigenvalues of mixed sign the particular chosen side for the differencing will be *upwind* for only some of the eigenvalues and *downwind* for the rest. The difficulty can be resolved by splitting the matrix into two matrices, one of them having positive or zero eigenvalues and the other having negative or zero eigenvalues. From the assumption of hyperbolicity, $\mathbf{A}$ may be diagonalised as

$$\mathbf{A} = \mathbf{K}\Lambda\mathbf{K}^{-1} , \qquad (5.81)$$

where $\mathbf{K}$ is the matrix whose columns are the right eigenvectors $\mathbf{K}^{(j)}$, $\mathbf{K}^{-1}$ is the inverse of $\mathbf{K}$ and Λ is the diagonal matrix formed by the eigenvalues of $\mathbf{A}$, namely

$$\Lambda = \begin{pmatrix} \lambda_1 & & 0 \\ & \ddots & \\ 0 & & \lambda_m \end{pmatrix} . \qquad (5.82)$$

See Sect. 2.3.1 of Chap. 2. In terms of the *characteristic variables*

$$\mathbf{V} = \mathbf{K}^{-1}\mathbf{U} , \qquad (5.83)$$

system (5.80) becomes the *decoupled* system

$$\mathbf{V}_t + \Lambda\mathbf{V}_x = \mathbf{0} , \qquad (5.84)$$

where the j–th equation

$$\frac{\partial}{\partial t}v_j + \lambda_j \frac{\partial}{\partial x}v_j = 0 , \text{ for } j = 1,..,m, \qquad (5.85)$$

involves only the variable $v_j(x,t)$.

5.4.1 The CIR Scheme

From a numerical point of view the decomposition of (5.80) into the decoupled set (5.84) with component equations (5.85) is very convenient. Each component equation (5.85) is a linear advection equation with characteristic speed $\lambda_j = $ constant, just as the PDE in (5.7), in which the speed is a, a constant. The CIR first–order upwind scheme (5.21) can directly be applied to each component equation (5.85). As for the scalar case we introduce the following definitions

$$\begin{aligned} \lambda_j^+ &\equiv \max(\lambda_j, 0) = \tfrac{1}{2}(\lambda_j + |\lambda_j|) , \\ \lambda_j^- &\equiv \min(\lambda_j, 0) = \tfrac{1}{2}(\lambda_j - |\lambda_j|) , \end{aligned} \right\} \qquad (5.86)$$

where $|\lambda_j|$ is the absolute value of λ_j. The following relations can be easily verified

$$\lambda_j = \lambda_j^+ + \lambda_j^- , \quad |\lambda_j| = \lambda_j^+ - \lambda_j^- . \qquad (5.87)$$

Then, the CIR scheme (5.21) applied to each PDE in (5.85) for the characteristic variables reads

$$(v_j)_i^{n+1} = (v_j)_i^n \quad - \quad \frac{\Delta t}{\Delta x}\lambda_j^+ \left[(v_j)_i^n - (v_j)_{i-1}^n\right]$$

$$- \quad \frac{\Delta t}{\Delta x}\lambda_j^- \left[(v_j)_{i+1}^n - (v_j)_i^n\right] \; . \tag{5.88}$$

This is a straight generalisation of the first–order upwind scheme (5.21) to the decoupled linear hyperbolic system (5.84).

Based on definitions (5.86) we can form the positive Λ^+ and negative Λ^- components of the diagonal matrix Λ, namely

$$\Lambda^\pm \equiv \begin{pmatrix} \lambda_1^\pm & & 0 \\ & \ddots & \\ 0 & & \lambda_m^\pm \end{pmatrix} . \tag{5.89}$$

Property (5.81) allows us to introduce the positive and negative components of the coefficient matrix $\mathbf{A}$ as

$$\mathbf{A}^- = \mathbf{K}\Lambda^-\mathbf{K}^{-1} , \quad \mathbf{A}^+ = \mathbf{K}\Lambda^+\mathbf{K}^{-1} . \tag{5.90}$$

Based on properties (5.87), the matrices (5.89), (5.90) can be shown to satisfy the following

$$\left.\begin{array}{l} \Lambda = \Lambda^+ + \Lambda^- , \quad |\,\Lambda\,| = \Lambda^+ - \Lambda^- , \\[2mm] \mathbf{A} = \mathbf{A}^+ + \mathbf{A}^- , \quad |\,\mathbf{A}\,| = \mathbf{A}^+ - \mathbf{A}^- . \end{array}\right\} \tag{5.91}$$

The CIR scheme (5.88) can be written as

$$\left.\begin{array}{l} \mathbf{V}_i^{n+1} = \mathbf{V}_i^n \quad - \quad \frac{\Delta t}{\Delta x}\Lambda^+(\mathbf{V}_i^n - \mathbf{V}_{i-1}^n) \\[2mm] - \quad \frac{\Delta t}{\Delta x}\Lambda^-(\mathbf{V}_{i+1}^n - \mathbf{V}_i^n) \; . \end{array}\right\} \tag{5.92}$$

In terms of the original variables $\mathbf{U} = \mathbf{K}\mathbf{V}$ this scheme may be expressed as

$$\left.\begin{array}{l} \mathbf{U}_i^{n+1} = \mathbf{U}_i^n \quad - \quad \frac{\Delta t}{\Delta x}\mathbf{A}^+ \left[\mathbf{U}_i^n - \mathbf{U}_{i-1}^n\right] \\[2mm] - \quad \frac{\Delta t}{\Delta x}\mathbf{A}^- \left[\mathbf{U}_{i+1}^n - \mathbf{U}_i^n\right] \; . \end{array}\right\} \tag{5.93}$$

This is the generalisation of scheme (5.21) to linear hyperbolic systems (5.80) with constant coefficients. It is left to the reader to verify that the above result can be obtained by multiplying (5.92) from the left by the matrix $\mathbf{K}$ and using (5.90). Note that by splitting the coefficient matrix $\mathbf{A}$ into a positive and a negative component we have been able to retain the basic principle of performing the one–sided spatial differencing according to the sign of the characteristic speeds. The differencing $\mathbf{U}_i^n - \mathbf{U}_{i-1}^n$ is upwind for the coefficient matrix $\mathbf{A}^+$ and $\mathbf{U}_{i+1}^n - \mathbf{U}_i^n$ is upwind for the coefficient matrix $\mathbf{A}^-$.

5.4.2 Godunov's Method

Consider the constant coefficient, linear hyperbolic system (5.80) written in conservation–law form

$$\mathbf{U}_t + \mathbf{F(U)}_x = 0 , \quad \mathbf{F(U)} \equiv \mathbf{AU} . \tag{5.94}$$

The Godunov first–order upwind method utilises the conservative formula (5.76) and requires the solution $\mathbf{U}_{i+\frac{1}{2}}(x/t)$ of the local Riemann problem $RP(\mathbf{U}_i^n, \mathbf{U}_{i+1}^n)$ for (5.94) to compute the intercell numerical flux

$$\mathbf{F}_{i+\frac{1}{2}} = \mathbf{F}(\mathbf{U}_{i+\frac{1}{2}}(0)) . \tag{5.95}$$

See Sect. 5.3.2. Here $\mathbf{U}_{i+\frac{1}{2}}(0)$ is the value of the solution $\mathbf{U}_{i+\frac{1}{2}}(x/t)$ at $x/t = 0$ along the intercell boundary. As seen in Sect. 2.3.3 of Chap. 2, the solution

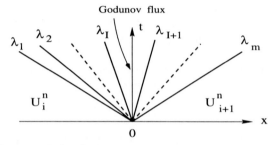

Godunov flux

Fig. 5.8. Evaluation of the Godunov intercell flux for linear hyperbolic systems with constant coefficients

$\mathbf{U}_{i+\frac{1}{2}}(x/t)$ can be easily found by first expanding the initial data $\mathbf{U}_i^n, \mathbf{U}_{i+1}^n$ in terms of the right eigenvectors as

$$\mathbf{U}_i^n = \sum_{j=1}^{m} \alpha_j \mathbf{K}^{(j)} , \quad \mathbf{U}_{i+1}^n = \sum_{j=1}^{m} \beta_j \mathbf{K}^{(j)} . \tag{5.96}$$

The general solution at any point (x, t) is given by

$$\mathbf{U}_{i+\frac{1}{2}}(x/t) = \sum_{j=1}^{I} \beta_j \mathbf{K}^{(j)} + \sum_{j=I+1}^{m} \alpha_j \mathbf{K}^{(j)} , \tag{5.97}$$

where I is the largest integer with $1 \leq I \leq m$ such that $x/t \geq \lambda_I$. The Godunov flux (5.95) requires the solution at $x/t = 0$ in (5.97). See Fig. 5.8. For $x/t = 0$ I is such that $\lambda_I \leq 0$ and $\lambda_{I+1} \geq 0$, then $\mathbf{U}_{i+\frac{1}{2}}(0)$ is obtained by manipulating (5.97), namely

$$\mathbf{U}_{i+\frac{1}{2}}(0) = \mathbf{U}_i^n + \sum_{j=1}^{I} (\beta_j - \alpha_j) \mathbf{K}^{(j)} \tag{5.98}$$

or

$$\mathbf{U}_{i+\frac{1}{2}}(0) = \mathbf{U}_{i+1}^n - \sum_{j=I+1}^m (\beta_j - \alpha_j)\mathbf{K}^{(j)} \ . \tag{5.99}$$

Recall that the jump across wave j with eigenvalue λ_j and eigenvector $\mathbf{K}^{(j)}$ is given by $(\beta_j - \alpha_j)\mathbf{K}^{(j)}$. Note that the solution of the Riemann problem, at $x/t = 0$, as given by (5.98), can be interpreted as being the left data state $\mathbf{U}_i^n$ plus all wave jumps across waves of negative or zero speed. Similarly, the form (5.99) gives the solution as the right data state $\mathbf{U}_{i+1}^n$ minus the wave jumps across all waves of positive or zero speeds. By combining (5.98) and (5.99) we obtain

$$\mathbf{U}_{i+\frac{1}{2}}(0) = \frac{1}{2}(\mathbf{U}_i^n + \mathbf{U}_{i+1}^n) - \frac{1}{2}\sum_{j=1}^m sign(\lambda_j)(\beta_j - \alpha_j)\mathbf{K}^{(j)} \ . \tag{5.100}$$

The Godunov intercell numerical flux (5.95) can now be obtained by evaluating $\mathbf{F}(\mathbf{U})$ at any of the expressions (5.98)–(5.100) for the solution of the Riemann problem. Use of (5.98) gives

$$\mathbf{F}_{i+\frac{1}{2}} = \mathbf{F}_i^n + \sum_{j=1}^I \mathbf{A}(\beta_j - \alpha_j)\mathbf{K}^{(j)} \ , \tag{5.101}$$

and since $\mathbf{A}\mathbf{K}^{(j)} = \lambda_j\mathbf{K}^{(j)}$,

$$\mathbf{F}_{i+\frac{1}{2}} = \mathbf{F}_i^n + \sum_{j=1}^I (\beta_j - \alpha_j)\lambda_j\mathbf{K}^{(j)} \ . \tag{5.102}$$

Similarly, (5.99) gives

$$\mathbf{F}_{i+\frac{1}{2}} = \mathbf{F}_{i+1}^n - \sum_{j=I+1}^m (\beta_j - \alpha_j)\lambda_j\mathbf{K}^{(j)} \ , \tag{5.103}$$

or combining (5.102) and (5.103) we obtain

$$\mathbf{F}_{i+\frac{1}{2}} = \frac{1}{2}\left(\mathbf{F}_i^n + \mathbf{F}_{i+1}^n\right) - \frac{1}{2}\sum_{j=1}^m (\beta_j - \alpha_j)\mid \lambda_j \mid \mathbf{K}^{(j)} \ . \tag{5.104}$$

Next we show that the Godunov flux can also be expressed in two more alternative forms.

Proposition 5.4.1. *The Godunov flux (5.95) to solve (5.94) via (5.76) can be written as*

$$\mathbf{F}_{i+\frac{1}{2}} = \frac{1}{2}(\mathbf{F}_i^n + \mathbf{F}_{i+1}^n) - \frac{1}{2}\mid \mathbf{A} \mid (\mathbf{U}_{i+1}^n - \mathbf{U}_i^n) \ . \tag{5.105}$$

Proof. Starting from (5.104) and using the properties (5.87) and (5.91) one writes

$$
\begin{aligned}
\mathbf{F}_{i+\frac{1}{2}} &= \frac{1}{2}\left(\mathbf{F}_i^n + \mathbf{F}_{i+1}^n\right) - \frac{1}{2}\sum_{j=1}^{m}(\beta_j - \alpha_j)(\lambda_j^+ - \lambda_j^-)\mathbf{K}^{(j)} \\
&= \frac{1}{2}\left(\mathbf{F}_i^n + \mathbf{F}_{i+1}^n\right) - \frac{1}{2}\sum_{j=1}^{m}(\beta_j - \alpha_j)\left[\lambda_j^+\mathbf{K}^{(j)} - \lambda_j^-\mathbf{K}^{(j)}\right] \\
&= \frac{1}{2}\left(\mathbf{F}_i^n + \mathbf{F}_{i+1}^n\right) - \frac{1}{2}\sum_{j=1}^{m}(\beta_j - \alpha_j)\left[\mathbf{A}^+\mathbf{K}^{(j)} - \mathbf{A}^-\mathbf{K}^{(j)}\right] \\
&= \frac{1}{2}\left(\mathbf{F}_i^n + \mathbf{F}_{i+1}^n\right) - \frac{1}{2}\sum_{j=1}^{m}(\beta_j - \alpha_j)\left[\mathbf{A}^+ - \mathbf{A}^-\right]\mathbf{K}^{(j)} \\
&= \frac{1}{2}\left(\mathbf{F}_i^n + \mathbf{F}_{i+1}^n\right) - \frac{1}{2}\mid\mathbf{A}\mid\sum_{j=1}^{m}(\beta_j - \alpha_j)\mathbf{K}^{(j)} .
\end{aligned}
$$

Hence
$$
\mathbf{F}_{i+\frac{1}{2}} = \frac{1}{2}\left(\mathbf{F}_i^n + \mathbf{F}_{i+1}^n\right) - \frac{1}{2}\mid\mathbf{A}\mid(\mathbf{U}_{i+1}^n - \mathbf{U}_i^n)
$$
and the proposition is proved.

Proposition 5.4.2. *The Godunov flux (5.95) for (5.94) can be written in flux–split form as*
$$
\mathbf{F}_{i+\frac{1}{2}} = \mathbf{A}^+\mathbf{U}_i^n + \mathbf{A}^-\mathbf{U}_{i+1}^n . \tag{5.106}
$$

Proof. The result follows directly from manipulating (5.105) and using appropriate definitions. Alternatively we have

$$
\begin{aligned}
\mathbf{F}_{i+\frac{1}{2}} &= \mathbf{A}\mathbf{U}_{i+\frac{1}{2}}(0) \\
&= \sum_{j=1}^{I}\beta_j\mathbf{A}\mathbf{K}^{(j)} + \sum_{j=I+1}^{m}\alpha_j\mathbf{A}\mathbf{K}^{(j)} \\
&= \sum_{j=1}^{I}\beta_j\lambda_j\mathbf{K}^{(j)} + \sum_{j=I+1}^{m}\alpha_j\lambda_j\mathbf{K}^{(j)} \\
&= \sum_{j=1}^{m}\beta_j\lambda_j^-\mathbf{K}^{(j)} + \sum_{j=1}^{m}\alpha_j\lambda_j^+\mathbf{K}^{(j)} \\
&= \sum_{j=1}^{m}\beta_j\mathbf{A}^-\mathbf{K}^{(j)} + \sum_{j=1}^{m}\alpha_j\mathbf{A}^+\mathbf{K}^{(j)} \\
&= \mathbf{A}^+\sum_{j=1}^{m}\alpha_j\mathbf{K}^{(j)} + \mathbf{A}^-\sum_{j=1}^{m}\beta_i\mathbf{K}^{(j)} \\
&= \mathbf{A}^+\mathbf{U}_i^n + \mathbf{A}^-\mathbf{U}_{i+1}^n ,
\end{aligned}
$$

and the proposition is proved.

Remark 5.4.1. The intercell flux has been *split* as

$$\mathbf{F}_{i+\frac{1}{2}} = \mathbf{F}^+_{i+\frac{1}{2}} + \mathbf{F}^-_{i+\frac{1}{2}} , \tag{5.107}$$

where the positive $\mathbf{F}^+_{i+\frac{1}{2}}$ and negative $\mathbf{F}^-_{i+\frac{1}{2}}$ flux components are

$$\mathbf{F}^+_{i+\frac{1}{2}} = \mathbf{A}^+\mathbf{U}^n_i , \quad \mathbf{F}^-_{i+\frac{1}{2}} = \mathbf{A}^-\mathbf{U}^n_i . \tag{5.108}$$

Note that, trivially, the respective Jacobian matrices have eigenvalues that are all positive (or zero) and all negative (or zero).

Exercise 5.4.1. Consider the linearised equations of Gas Dynamics

$$\mathbf{U}_t + \mathbf{A}\mathbf{U}_x = 0 ,$$

with

$$\mathbf{U} = \begin{bmatrix} u_1 \\ u_2 \end{bmatrix} \equiv \begin{bmatrix} \rho \\ u \end{bmatrix} , \quad \mathbf{A} = \begin{bmatrix} 0 & \rho_0 \\ a^2/\rho_0 & 0 \end{bmatrix} .$$

Using the results of Sect. 2.3.4 of Chap. 2

- Find the matrices Λ^-, Λ^+, $\mathbf{A}^-$, $\mathbf{A}^+$.
- Write the scheme (5.93) in full, that is for the two components of the vector of unknowns.
- Compute the Godunov intercell flux directly by using the explicit solution of the Riemann problem in the *Star Region*

$$\left.\begin{aligned} \rho_* &= \tfrac{1}{2}(\rho_L + \rho_R) - \tfrac{1}{2}(u_R - u_L)\rho_0/a , \\[4pt] u_* &= \tfrac{1}{2}(u_L + u_R) - \tfrac{1}{2}(\rho_R - \rho_L)a/\rho_0 . \end{aligned}\right\}$$

How many possible wave patterns do you need to consider here ?
- Write a computer program to solve the linearised equations of Gas Dynamics using the method of Godunov.

Solution 5.4.1. (Left to the reader).

5.5 Sample Numerical Results

To complete this chapter, we present some numerical results obtained by some of the most well known schemes as applied to two model PDEs.

5.5.1 Linear Advection

We apply four schemes to solve

$$u_t + f(u)_x = 0 , \quad f(u) = au , \quad a = \text{constant} \tag{5.109}$$

with two types of initial conditions.

Test 1 for linear advection (smooth data). Here the initial condition is the smooth profile

$$u(x, 0) = \alpha e^{-\beta x^2} . \tag{5.110}$$

In the computations we take $a = 1.0$, $\alpha = 1.0$, $\beta = 8.0$ and a CFL coefficient $C_{cfl} = 0.8$; the initial profile $u(x, 0)$ is evaluated in the interval $-1 \leq x \leq 1$. Computed results are shown in Figs. 5.9 to 5.11; these correspond respectively to the output times $t = 1.0$ unit (125 time steps), $t = 10.0$ units (1250 time steps), $t = 100.0$ units (12499 time steps). In each figure we compare the exact solution (shown by full lines) with the numerical solution (symbols) for the Godunov method, the Lax–Friedrichs method, the Lax–Wendroff method and the Warming–Beam method.

The results of Fig. 5.9 are in many ways representative of the quality of each scheme. Collectively these results are also representative of most of the current successes and limitations of numerical methods for PDEs governing wave propagation. The first–order method of Godunov (CIR scheme) has modified equation of the form (5.24), where α_{cir} is a numerical viscosity coefficient. This is responsible for the *clipping* of the peak values. As seen earlier $\alpha_{cir} < \alpha_{lf}$, which explains the fact that the Lax–Friedrichs scheme gives even more diffused results. For the computational parameters used $\alpha_{cir} = 0.1\Delta x$ and $\alpha_{lf} = 0.225\Delta x$.

The results from the Lax–Wendroff method and the Warming–Beam method, both second–order accurate, are much more accurate than those of the first–order schemes. There are however, slight signs of error in the position of the wave. For the Lax–Wendroff scheme the computed wave *lags behind* the true wave (lagging phase error), while for the Warming–Beam method the computed wave *is ahead* of the true wave (leading phase error). The phase errors of second–order accurate schemes are explained by the dispersive term of the modified equation (5.38).

The limitations of the schemes are more clearly exposed if the solution is evolved for longer times. Fig. 5.10 shows results at the output time $t = 10.0$ units (1250 time steps). Compare with Fig. 5.9. The numerical diffusion inherent in first–order methods has ruined the solution of the Godunov and Lax–Friedrichs schemes. Computed peak values are only of the order of 30 to 40% of the true peak values. The second–order methods are still giving more satisfactory results than their first–order counterparts, but now the numerical dispersion errors are clearly visible. Numerical diffusion is beginning to show its effects too.

Fig. 5.11 shows results at the output time $t = 100.0$ units (12499 time steps). Compare with Figs. 5.9. and 5.10. These results are truly disappointing and clearly expose the limitations of numerical methods for computing solutions to problems involving long time evolution of wave phenomena. In acoustics one may require the computation of (i) very weak signals (ii) over long distances. The combination of these two requirements rules out automatically a wide range of otherwise acceptable numerical methods for PDEs. See

Tam and Webb [341]. The numerical diffusion of the first–order schemes has virtually flattened the wave, while the numerical dispersion of the second–order methods has resulted in unacceptable position errors, in addition to clipping by numerical diffusion.

Test 2 for linear advection (discontinuous data). Now the initial data for (5.109) consists of a *square wave*, namely

$$u(x,0) = \begin{cases} 0 & \text{if} \quad x \leq 0.3 \,, \\ 1 & \text{if} \quad 0.3 \leq x \leq 0.7 \,, \\ 0 & \text{if} \quad x \geq 0.7 \,. \end{cases} \qquad (5.111)$$

The computed results for the three output times are shown in Figs. 5.12 to 5.14. As for Test 1 the effects of numerical diffusion in the first–order methods and the effects of dispersion in the second–order methods lead to visible errors in the numerical solution (symbols), as compared with the exact solution (full line). First–order methods *smear* discontinuities over many computing cells; as expected this error is more apparent in the Lax–Friedrichs scheme. Note also the *pairing* of neighbouring values in the Lax–Friedrichs scheme. Second–order methods reduce the smearing of discontinuities, but at the cost of *overshoots* and *undershoots* in the vicinity of the discontinuities. These *spurious oscillations* are highly undesirable features of second and higher–order methods. We shall return to this theme in Chaps. 13 and 14, where improved methods for dealing with discontinuities will be presented.

Fig. 5.13 shows results for Test 2 at time $t = 10.0$ units (1250 time steps). The errors observed in Fig. 5.12 are now exaggerated. Fig. 5.14 shows results at time $t = 100.0$ units (12499 time steps). Once again first–order methods have *lost* the solution while second–order methods exhibit unacceptable position errors, in addition to spurious oscillations produced near discontinuities.

5.5.2 The Inviscid Burgers Equation

Our Test 3 consists of the inviscid Burgers equation

$$u_t + f(u)_x = 0 \,, \quad f(u) = \frac{1}{2}u^2 \qquad (5.112)$$

in the domain $[0, \frac{3}{2}]$ with initial conditions

$$u(x,0) = \begin{cases} -\frac{1}{2} & \text{if} \quad x \leq \frac{1}{2} \,, \\ 1 & \text{if} \quad \frac{1}{2} \leq x \leq 1 \,, \\ 0 & \text{if} \quad x \geq 1 \,. \end{cases} \qquad (5.113)$$

We solve this problem numerically on a domain of length $L = 1.5$ discretised by $M = 75$ equally spaced cells of width $\Delta x = 0.02$; the CFL coefficient used is 0.8. Fig. 5.15 shows computed results (symbol) along with the exact (line) solution, for the Godunov and Lax–Friedrichs schemes at time $t = 0.5$

units (32 time steps). Two new features are now present in solving non–linear PDEs. First the discontinuity on the right is a shock wave. This satisfies the entropy condition, see Sect. 2.4.2 of Chap. 2, and characteristics on either side of the discontinuity converge into the discontinuity. This compression mechanism helps the more accurate resolution of shock waves. Compare with Fig. 5.12. The Godunov method resolves the shock much more sharply (3 cells) than the Lax–Friedrichs scheme (10 cells). The second new feature to note in this non–linear example is the *entropy glitch* at $x = \frac{1}{2}$. This corresponds to a sonic point, see Sect. 2.4.2 of Chap. 2. The entropy glitch affects the Godunov method and not the Lax–Friedrichs method. A question of crucial importance is the construction of entropy satisfying schemes [252].

More advanced concepts on numerical methods are presented in Chap. 13 for scalar problems. Chaps. 14, 15 and 16 deal with numerical methods for non–linear systems.

5.6 FORTRAN Program for Godunov's Method

A listing of a FORTRAN program to compute the numerical solution to the inviscid Burgers equation is included.

```
*
*-------------------------------------------------------------*
*                                                             *
C      Godunov's method for the inviscid Burgers's            *
C                    equation                                 *
*                                                             *
C       Name of program: HL-B1GOD                             *
*                                                             *
C      Purpose: to solve the inviscid Burgers equation        *
C               using the Godunov first order upwind          *
C               scheme in conjunction with the exact          *
C               Riemann solver                                *
*                                                             *
C      Input  file: b1god.ini                                 *
C      Output file: numer.out (numerical)                     *
*                                                             *
C      Programer: E. F. Toro                                  *
*                                                             *
C      Last revision: February 7th 1999                       *
*                                                             *
C      Theory is found in Section 5.3.3, Chapter 5 of         *
C      Reference 1.                                           *
*                                                             *
C      1. Toro, E. F., "Riemann Solvers and Numerical         *
C                       Methods for Fluid Dynamics"           *
C                       Springer-Verlag,                      *
C                       Second Edition, 1999                  *
*                                                             *
C      This program is part of                                *
*                                                             *
C      NUMERICA                                               *
C      A Library of Source Codes for Teaching,                *
C      Research and Applications,                             *
C      by E. F. Toro                                          *
C      Published by NUMERITEK LTD,                            *
C      Website: www.numeritek.com                             *
*                                                             *
*-------------------------------------------------------------*
*
C      Driver program
*
       IMPLICIT NONE
*
C      Declaration of variables:
```

```
*
        INTEGER ITEST, CELLS, N, NFREQ, NTMAXI
*
        REAL    CFLCOE, DOMLEN, DT, TIME, TIMEOUT, TIMETO
*
        COMMON /DATAIN/ CFLCOE, DOMLEN, ITEST, CELLS,
     &                  NFREQ, NTMAXI, TIMEOUT
        COMMON /DELTAT/ DT
*
        DATA TIMETO /1.0E-07/
*
C       Parameters of problem are read in from
C       file "b1god.ini"
*
        CALL READER
*
C       Initial conditions are set up
*
        CALL INITIA(DOMLEN, ITEST, CELLS)
*
        WRITE(6,*)'-------------------------------------------'
        WRITE(6,*)'   Time step N          TIME          TIMEOUT'
        WRITE(6,*)'-------------------------------------------'
*
C       Time marching procedure
*
        TIME = 0.0
*
        DO 10 N = 1, NTMAXI
*
C          Boundary conditions are set
*
           CALL BCONDI(CELLS)
*
C          Courant-Friedrichs-Lewy (CFL) condition imposed
*
           CALL CFLCON(CFLCOE, CELLS, TIME, TIMEOUT)
*
           TIME = TIME + DT
*
C          Intercell numerical fluxes are computed
*
           CALL FLUXES(CELLS)
*
```

```
C         Solution is updated according to
C         conservative formula
*
          CALL UPDATE(CELLS)
*
          IF(MOD(N,NFREQ).EQ.0)WRITE(6,20)N, TIME
*
C         Check output time
*
          IF(ABS(TIME - TIMEOUT).LE.TIMETO)THEN
*
C             Solution is written to "numer.out' at
C             specified time TIMEOUT
*
              CALL OUTPUT(CELLS)
*
              WRITE(6,*)'----------------------------------'
              WRITE(6,*)'   Number of time steps = ',N
*
              STOP
          ENDIF
*
 10       CONTINUE
*
 20       FORMAT(I12,6X, F12.7)
*
          END
*
*-------------------------------------------------------------*
*
          SUBROUTINE READER
*
C         Purpose: to read initial parameters of the problem
*
          IMPLICIT NONE
*
C         Declaration of variables
*
          INTEGER   ITEST, CELLS, NFREQ, NTMAXI
*
          REAL      CFLCOE, DOMLEN, TIMEOUT
*
          COMMON /DATAIN/ CFLCOE, DOMLEN, ITEST, CELLS, NFREQ,
         &                NTMAXI, TIMEOUT
```

```
*
        OPEN(UNIT = 1,FILE = 'b1god.ini',STATUS = 'UNKNOWN')
*
        READ(1,*)CFLCOE    ! Courant number coefficient
        READ(1,*)DOMLEN    ! Domain length
        READ(1,*)ITEST     ! Test problem
        READ(1,*)CELLS     ! Number of cells in domain
        READ(1,*)NFREQ     ! Output frequency to screen
        READ(1,*)NTMAXI    ! Maximum number of time steps
        READ(1,*)TIMEOUT   ! Output time
*
        CLOSE(1)
*
        WRITE(6,*)'--------------------------------'
        WRITE(6,*)'Data read in is echoed to screen'
        WRITE(6,*)'--------------------------------'
        WRITE(6,*)'CFLCOE  = ',CFLCOE
        WRITE(6,*)'DOMLEN  = ',DOMLEN
        WRITE(6,*)'ITEST   = ',ITEST
        WRITE(6,*)'CELLS   = ',CELLS
        WRITE(6,*)'NFREQ   = ',NFREQ
        WRITE(6,*)'NTMAXI  = ',NTMAXI
        WRITE(6,*)'TIMEOUT = ',TIMEOUT
        WRITE(6,*)'--------------------------------'
*
        RETURN
        END
*
*------------------------------------------------------------*
*
        SUBROUTINE INITIA(DOMLEN, ITEST, CELLS)
*
C       Purpose: to set initial conditions for solution U
C                and initialise other variables. There are
C                two choices of initial conditions,
C                determined by ITEST
*
C       Local variables:
*
C       Name            Description
*       ====            ===========
*
C       DX              Spatial mesh  size
C       I               Variable in do loop
```

```
C       ITEST          Defines test problem
C       FLUX           Array for intercell fluxes
C       U              Array for numerical solution
C       XPOS           Position along x-axis
C       XRIGHT         Left diaphragm
C       XMIDDL         Middle diaphragm
C       XRIGHT         Right diaphragm
*
        IMPLICIT NONE
*
C       Declaration of variables
*
        INTEGER I, ITEST, CELLS, IDIM
*
        REAL    DOMLEN, DX, FLUX, U, XLEFT, XPOS, XMIDDL,
     &          XRIGHT
*
        PARAMETER (IDIM = 1000)
*
        DIMENSION FLUX(0:IDIM + 1), U(0:IDIM + 1)
*
        COMMON /DELTAX/ DX
        COMMON /FLUXFS/ FLUX
        COMMON /SOLUTI/ U
*
C       Calculate mesh size DX
*
        DX = DOMLEN/REAL(CELLS)
*
C       Initialise arrays
*
        DO 10 I = 0, IDIM + 1
           FLUX(I) = 0.0
           U(I)    = 0.0
 10     CONTINUE
*
        IF(ITEST.EQ.1)THEN
*
C          Test 1: smooth profile
*
           XPOS    = -1.0
*
           DO 20 I = 1,  CELLS
              XPOS = XPOS + 2.0/REAL(CELLS)
```

```
                 U(I) = EXP(-8.0*XPOS*XPOS)
 20         CONTINUE
*
       ELSE
*
C          Test 2: square waves
*
            XLEFT  = 0.1*DOMLEN
            XMIDDL = 0.5*DOMLEN
            XRIGHT = 0.9*DOMLEN
*
            DO 30 I = 1, CELLS
*
               XPOS = (REAL(I)-1.0)*DX
*
               IF(XPOS.LT.XLEFT)THEN
                  U(I) = -1.0
               ENDIF
*
               IF(XPOS.GE.XLEFT.AND.XPOS.LE.XMIDDL)THEN
                  U(I) = 1.0
               ENDIF
*
               IF(XPOS.GT.XMIDDL.AND.XPOS.LE.XRIGHT)THEN
                  U(I) = 0.0
               ENDIF
*
               IF(XPOS.GT.XRIGHT)THEN
                  U(I) = -1.0
               ENDIF
*
 30         CONTINUE
*
       ENDIF
*
       RETURN
       END
*
*-----------------------------------------------------------*
*
       SUBROUTINE BCONDI(CELLS)
*
C      Purpose: to apply boundary conditions
*
```

```
      IMPLICIT NONE
*
C     Declaration of variables
*
      INTEGER CELLS, IDIM
*
      REAL    U
*
      PARAMETER (IDIM = 1000)
*
      DIMENSION U(0:IDIM + 1)
*
      COMMON /SOLUTI/ U
*
C     Left boundary, periodic boundary condition
*
      U(0)  = U(CELLS)
*
C     Right boundary, periodic boundary condition
*
      U(CELLS + 1) =  U(1)
*
      RETURN
      END
*
*------------------------------------------------------------*
*
      SUBROUTINE CFLCON(CFLCOE, CELLS, TIME, TIMEOUT)
*
C     Purpose: to apply the CFL condition to compute a
C              stable time step DT
*
      IMPLICIT NONE
*
C     Declaration of variables
*
      INTEGER  I, CELLS, IDIM
*
      REAL     CFLCOE, DT, DX, SMAX, TIME, TIMEOUT, U
*
      PARAMETER (IDIM = 1000)
*
      DIMENSION U(0:IDIM + 1)
*
```

```
      COMMON /SOLUTI/ U
      COMMON /DELTAT/ DT
      COMMON /DELTAX/ DX
*
      SMAX = -1.0E+06
*
C     Find maximum characteristic speed
*
      DO 10 I = 0, CELLS + 1
         IF(ABS(U(I)).GT.SMAX)SMAX = ABS(U(I))
 10   CONTINUE
*
      DT = CFLCOE*DX/SMAX
*
C     Check size of DT to avoid exceeding output time
*
      IF((TIME + DT).GT.TIMEOUT)THEN
*
C        Recompute DT
*
         DT = TIMEOUT - TIME
      ENDIF
*
      RETURN
      END
*
*-------------------------------------------------------------*
*
      SUBROUTINE UPDATE(CELLS)
*
C     Purpose: to update the solution to a new time level
C              using the explicit conservative formula
*
      IMPLICIT NONE
*
C     Declaration of variables
*
      INTEGER I, CELLS, IDIM
*
      REAL    DT, DX, DTODX, FLUX, U
*
      PARAMETER (IDIM = 1000)
*
      DIMENSION U(0:IDIM + 1), FLUX(0:IDIM + 1)
```

```
*
      COMMON /DELTAT/ DT
      COMMON /DELTAX/ DX
      COMMON /FLUXFS/ FLUX
      COMMON /SOLUTI/ U
*
      DTODX = DT/DX
*
      DO 10 I = 1, CELLS
         U(I) = U(I) + DTODX*(FLUX(I-1) - FLUX(I))
 10   CONTINUE
*
      RETURN
      END
*
*-------------------------------------------------------------*
*
      SUBROUTINE OUTPUT(CELLS)
*
C     Purpose: to output the solution at a specified time
C              TIMEOUT
*
      IMPLICIT NONE
*
C     Declaration of variables
*
      INTEGER I, CELLS, IDIM
*
      REAL    DX, U, XPOS
*
      PARAMETER (IDIM = 1000)
*
      DIMENSION U(0:IDIM + 1)
*
      COMMON /DELTAX/ DX
      COMMON /SOLUTI/ U
*
      OPEN(UNIT = 1,FILE = 'numer.out',STATUS = 'UNKNOWN')
*
      DO 10 I = 1, CELLS
         XPOS = REAL(I)*DX
         WRITE(1,20)XPOS, U(I)
 10   CONTINUE
*
```

```
      CLOSE(1)
*
 20   FORMAT(2(4X, F10.5))
*
      RETURN
      END
*
*-----------------------------------------------------------*
*
      SUBROUTINE FLUXES(CELLS)
*
C     Purpose: to compute intercell fluxes according to
C              the Godunov first-order upwind method,
C              in conjunction with the exact Riemann
C              solver
*
      IMPLICIT NONE
*
C     Declaration of variables
*
      INTEGER I, CELLS, IDIM
*
      REAL    FLUX, U, UL, UR, USTAR
*
      PARAMETER (IDIM = 1000)
*
      DIMENSION FLUX(0:IDIM + 1), U(0:IDIM + 1)
*
      COMMON /FLUXFS/ FLUX
      COMMON /SOLUTI/ U
*
C     Compute intercell flux FLUX(I), I = 0, CELLS
C     Solution of Riemann problem RP(I, I+1) is stored
C     in FLUX(I)
*
      DO 10 I = 0, CELLS
*
C        Define states UL (Left) and UR (Right) for local
C        Riemann problem  RP(UL, UR)
*
         UL = U(I)
         UR = U(I+1)
*
C        Solve the Riemann problem RP(UL, UR) exactly
```

```
*
          CALL RIEMANN(UL, UR, USTAR)
*
C         Compute Godunov intercell flux
*
          FLUX(I) = 0.5*USTAR*USTAR
*
 10    CONTINUE
*
       RETURN
       END
*
*-------------------------------------------------------------*
*
       SUBROUTINE RIEMANN(UL, UR, USTAR)
*
C      Purpose: to solve the Riemann problem for the inviscid
C               Burgers equation exactly.
*
C      Local variables:
*
C      Name         Description
*      ====         ===========
*
C      UL           Left data state
C      UR           Right data state
C      S            Shock speed
C      USTAR        Sampled state
*
       IMPLICIT NONE
*
       REAL   S, UL, UR, USTAR
*
       IF(UL.GT.UR)THEN
*
C         Solution is a shock wave
C         Compute shock speed S
*
          S = 0.5*(UL + UR)
*
C         Sample the state along the t-axis
*
          IF(S.GE.0.0)THEN
             USTAR = UL
```

```
            ELSE
               USTAR = UR
            ENDIF
*
        ELSE
*
C           Solution is a rarefaction wave.
C           There are 3 cases:
*
            IF(UL.GE.0.0)THEN
*
C               Right supersonic rarefaction
*
               USTAR = UL
            ENDIF
*
            IF(UR.LE.0.0)THEN
*
C               Left supersonic rarefaction
*
               USTAR = UR
            ENDIF
*
            IF(UL.LE.0.0.AND.UR.GE.0.0)THEN
*
C               Transonic rarefaction
*
               USTAR = 0.0
            ENDIF
*
        ENDIF
*
        RETURN
        END
*
*------------------------------------------------------------*
*
```

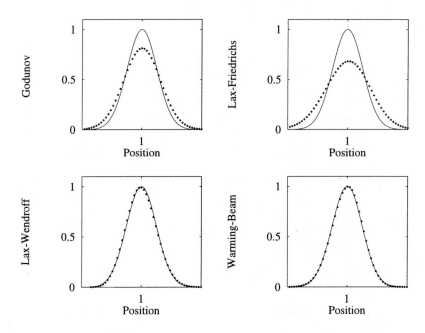

Fig. 5.9. Test 1: Comparison of numerical results for four numerical schemes (symbols) with the exact solution (line) at output time of 1 unit (125 time steps)

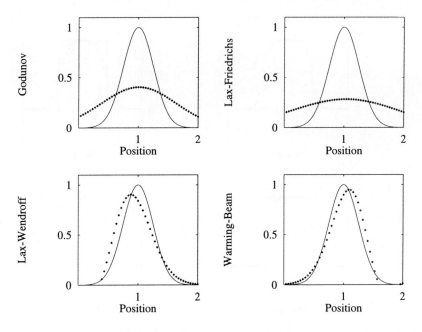

Fig. 5.10. Test 1: Comparison of numerical results for four numerical schemes (symbols) with the exact solution (line) at time 10 units (1250 time steps)

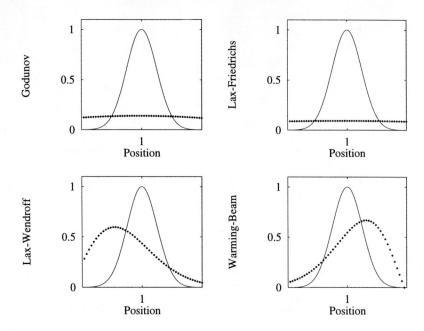

Fig. 5.11. Test 1: Comparison of numerical results for four numerical schemes (symbols) with the exact solution (line) at time 100 units (12499 time steps)

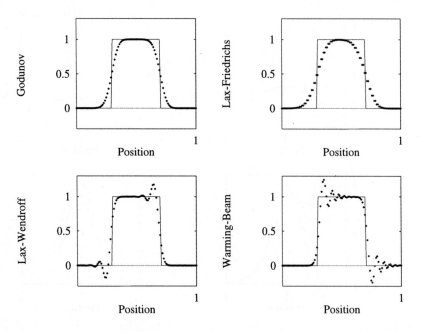

Fig. 5.12. Test 2: Comparison of numerical results for four numerical schemes (symbols) with the exact solution (line) at output time of 1 unit (125 time steps)

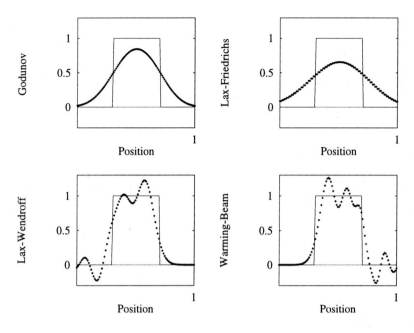

Fig. 5.13. Test 2: Comparison of numerical results for four numerical schemes (symbols) with the exact solution (line) at time 10 units (1250 time steps)

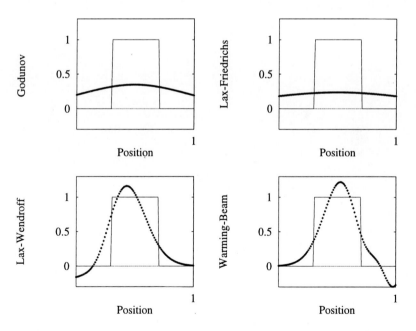

Fig. 5.14. Test 2: Comparison of numerical results for four numerical schemes (symbols) with the exact solution (line) at time 100 units (12499 time steps)

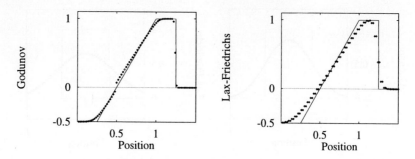

Fig. 5.15. Test 3: Inviscid Burgers's equation. Comparison of numerical results for two schemes (symbols) with the exact solution (line) at time 0.5 units (32 time steps)

6. The Method of Godunov for Non–linear Systems

It was almost 40 years ago when Godunov [145] produced a conservative extension of the first–order upwind scheme of Courant, Isaacson and Rees [98] to non–linear systems of hyperbolic conservation laws. In Chap. 5 we advanced a description of Godunov's method in terms of scalar equations and linear systems with constant coefficients. In this chapter, we describe the scheme for general non–linear hyperbolic systems; in particular, we give a detailed description of the technique as applied to the time–dependent, one dimensional Euler equations. The essential ingredient of Godunov's method is the solution of the Riemann problem, which may be the exact solution or some suitable approximation to it. Here, we present the scheme in terms of the exact solution. In Chaps. 9 to 12 we shall present versions of Godunov's scheme that utilise approximate Riemann solvers; these, if used cautiously, will provide an improvement to the efficiency of the scheme. As seen in Chap. 5 the method is only first–order accurate, which makes it unsuitable for application to practical problems; well–resolved solutions will require the use of very fine meshes, with the associated computing expense. Second and third order extensions of the basic Godunov method will be studied in Chap. 13 for scalar conservation laws; some of these high–order methods are extended to non–linear systems in Chaps. 14 and 16.

Relevant background for studying the Godunov's method is found in all preceding chapters, but detailed review of Chaps. 4 and 5 might be found particularly helpful.

6.1 Bases of Godunov's Method

Consider the general Initial–Boundary Value Problem (IBVP) for non–linear systems of hyperbolic conservation laws

$$\left.\begin{array}{lll} \text{PDEs} & : & \mathbf{U}_t + \mathbf{F}(\mathbf{U})_x = \mathbf{0} \ , \\ \text{ICs} & : & \mathbf{U}(x,0) = \mathbf{U}^{(0)}(x) \ , \\ \text{BCs} & : & \mathbf{U}(0,t) = \mathbf{U}_{\mathrm{l}}(t) \ , \ \mathbf{U}(L,t) = \mathbf{U}_{\mathrm{r}}(t) \ . \end{array}\right\} \tag{6.1}$$

Here, $\mathbf{U}(x,t)$ is the vector of conserved variables; $\mathbf{F}(\mathbf{U})$ is the vector of fluxes; $\mathbf{U}^{(0)}(x)$ is the initial data at time $t = 0$; $[0, L]$ is the spatial domain and boundary conditions are, for the moment, assumed to be represented by the boundary functions $\mathbf{U}_l(t)$ and $\mathbf{U}_r(t)$. We make the assumption that the solution of IVBP (6.1) does exist.

In order to admit discontinuous solutions we must use one of the integral forms of the conservation laws in (6.1). Here we adopt

$$
\left.
\begin{aligned}
\int_{x_1}^{x_2} \mathbf{U}(x, t_2)\, dx \;=\; & \int_{x_1}^{x_2} \mathbf{U}(x, t_1)\, dx + \int_{t_1}^{t_2} \mathbf{F}(\mathbf{U}(x_1, t))\, dt \\
& - \int_{t_1}^{t_2} \mathbf{F}(\mathbf{U}(x_2, t))\, dt \;,
\end{aligned}
\right\}
\tag{6.2}
$$

for any control volume $[x_1, x_2] \times [t_1, t_2]$ in the domain of interest; see Sect. 2.4.1 of Chap. 2.

We discretise the spatial domain $[0, L]$ into M computing cells or finite volumes $I_i = [x_{i-\frac{1}{2}}, x_{i+\frac{1}{2}}]$ of regular size $\Delta x = x_{i+\frac{1}{2}} - x_{i-\frac{1}{2}} = L/M$, with $i = 1, \ldots, M$. For a given cell I_i the location of the cell centre x_i and the cell boundaries $x_{i-\frac{1}{2}}$, $x_{i+\frac{1}{2}}$ are given by

$$
x_{i-\frac{1}{2}} = (i-1)\Delta x \;,\; x_i = (i - \tfrac{1}{2})\Delta x \;,\; x_{i+\frac{1}{2}} = i\Delta x \;.
\tag{6.3}
$$

See Fig. 5.4 of Chap. 5. We denote the temporal domain by $[0, T]$, where T is some output time, not a boundary. The discretisation of the time interval $[0, T]$ is generally done in time steps Δt of variable size; recall that for non–linear systems wave speeds vary in space and time, and thus the choice of Δt is carried out as marching in time proceeds. Given general initial data

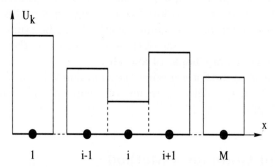

Fig. 6.1. Piece–wise constant distribution of data at time level n, for a single component of the vector **U**

$\widetilde{\mathbf{U}}(x, t^n)$ for (6.1) at time $t = t^n$ say, in order to evolve the solution to a time $t^{n+1} = t^n + \Delta t$, the Godunov method first assumes a *piece–wise constant* distribution of the data. Formally, this is realised by defining *cell averages*

$$
\mathbf{U}_i^n = \frac{1}{\Delta x} \int_{x_{i-\frac{1}{2}}}^{x_{i+\frac{1}{2}}} \widetilde{\mathbf{U}}(x, t^n)\, dx \;,
\tag{6.4}
$$

which produces the desired piecewise constant distribution $\mathbf{U}(x, t^n)$, with

$$\mathbf{U}(x, t^n) = \mathbf{U}_i^n, \text{ for } x \text{ in each cell } I_i = [x_{i-\frac{1}{2}}, x_{i+\frac{1}{2}}], \tag{6.5}$$

as illustrated in Fig. 6.1 for a single component $\mathbf{U}_k$ of the vector of conserved variables. Data now consists of a set $\{\mathbf{U}_i^n\}$ of constant states. Naturally these are in terms of conserved variables, but other variables may be derived to proceed with the implementation of numerical methods. In particular, for the Godunov method we use the solution of the Riemann problem in terms of primitive variables, which for the Euler equations are $\mathbf{W} = (\rho, u, p)^T$; ρ is density, u is velocity and p is pressure.

Once the piece–wise constant distribution of data has been established the next step in the Godunov method is to solve the Initial Value Problem (IVP) for the original conservation laws but with the modified initial data (6.5). Effectively, this generates local Riemann problems $RP(\mathbf{U}_i^n, \mathbf{U}_{i+1}^n)$ with data $\mathbf{U}_i$ (left side) and $\mathbf{U}_{i+1}^n$ (right side), centred at the intercell boundary positions $x_{i+\frac{1}{2}}$. As seen for the Euler equations in Chap. 4, the solution of $RP(\mathbf{U}_i^n, \mathbf{U}_{i+1}^n)$ is a similarity solution and depends on the ratio $\bar{x}/\bar{t}$, see (6.7), and the data states $\mathbf{U}_i^n$, $\mathbf{U}_{i+1}^n$; the solution is denoted by $\mathbf{U}_{i+\frac{1}{2}}(\bar{x}/\bar{t})$, where $(\bar{x}, \bar{t})$ are the local coordinates for the local Riemann problem. Fig. 6.2 shows typical wave patterns emerging from intercell boundaries $x_{i-\frac{1}{2}}$ and $x_{i+\frac{1}{2}}$ when solving the two Riemann problems $RP(\mathbf{U}_{i-1}^n, \mathbf{U}_i^n)$ and $RP(\mathbf{U}_i^n, \mathbf{U}_{i+1}^n)$. For

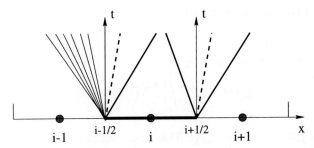

Fig. 6.2. Typical wave patterns emerging from solutions of local Riemann problems at intercell boundaries $i - \frac{1}{2}$ and $i + \frac{1}{2}$

a time step Δt that is sufficiently small, to avoid wave interaction, one can define a global solution $\widetilde{\mathbf{U}}(x, t)$ in the strip $0 \leq x \leq L$, $t^n \leq t \leq t^{n+1}$ in terms of the local solutions as follows

$$\widetilde{\mathbf{U}}(x, t) = \mathbf{U}_{i+\frac{1}{2}}(\bar{x}/\bar{t}), \ x \in [x_i, x_{i+1}], \tag{6.6}$$

where the correspondence between the global (x, t) and local $(\bar{x}, \bar{t})$ coordinates is given by

$$\left. \begin{array}{ll} \bar{x} = x - x_{i+\frac{1}{2}} & , \quad \bar{t} = t - t^n, \\ x \in [x_i, x_{i+1}] & , \quad t \in [t^n, t^{n+1}], \\ \bar{x} \in [-\frac{\Delta x}{2}, \frac{\Delta x}{2}] & , \quad \bar{t} \in [0, \Delta t], \end{array} \right\} \tag{6.7}$$

and is illustrated in Fig. 6.3. Having found a solution $\widetilde{\mathbf{U}}(x,t)$ in terms of solu-

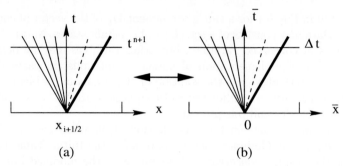

Fig. 6.3. Correspondence between the global (a) and local (b) frames of reference for the solution of the Riemann problem

tions $\mathbf{U}_{i+\frac{1}{2}}(\bar{x}/\bar{t})$ to local Riemann problems, the Godunov method advances the solution to a time $t^{n+1} = t^n + \Delta t$ by defining a new set of average values $\{\mathbf{U}_i^{n+1}\}$, in a way to be described. We shall often use (x,t) to mean the local frame of reference $(\bar{x}, \bar{t})$.

6.2 The Godunov Scheme

The first version of Godunov's method defines new average values $\mathbf{U}_i^{n+1}$ at time $t^{n+1} = t^n + \Delta t$ via the integrals

$$\mathbf{U}_i^{n+1} = \frac{1}{\Delta x} \int_{x_{i-\frac{1}{2}}}^{x_{i+\frac{1}{2}}} \widetilde{\mathbf{U}}(x, t^{n+1}) \, \mathrm{d}x \qquad (6.8)$$

within each cell $I_i = [x_{i-\frac{1}{2}}, x_{i+\frac{1}{2}}]$. This averaging process is illustrated in Fig. 6.4.

Note first that in order to perform the averaging, we need to make the assumption that no wave interaction takes place within cell I_i, in the chosen time Δt. This is satisfied by imposing the following restriction on the size of Δt, namely

$$\Delta t \leq \frac{\frac{1}{2} \Delta x}{S_{\max}^n}, \qquad (6.9)$$

where $S_{\max}^n$ denotes the maximum *wave velocity* present throughout the domain at time t^n. A consequence of this restriction is that only two Riemann problem solutions affect cell I_i, namely the right travelling waves of $\mathbf{U}_{i-\frac{1}{2}}(x/t)$ and the left travelling waves of $\mathbf{U}_{i+\frac{1}{2}}(x/t)$, see Fig. 6.4. Thus $\mathbf{U}_i^{n+1}$, given by (6.8), can be expressed as

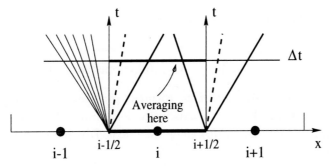

Fig. 6.4. Godunov averaging of local solutions to Riemann problems within cell I_i at a fixed time Δt

$$\mathbf{U}_i^{n+1} = \frac{1}{\Delta x} \int_0^{\frac{1}{2}\Delta x} \mathbf{U}_{i-\frac{1}{2}} \left(\frac{x}{\Delta t}\right) \, dx + \frac{1}{\Delta x} \int_{-\frac{1}{2}\Delta x}^0 \mathbf{U}_{i+\frac{1}{2}} \left(\frac{x}{\Delta t}\right) \, dx , \quad (6.10)$$

after using (6.6) and (6.8). This version of Godunov's method can obviously be implemented as a practical computational scheme. We note however that it has two main drawbacks. First, the CFL–like condition (6.9) is computationally somewhat restrictive on Δt. Second, the evaluation of the integrals in (6.10), although possible, could be involved. Rarefaction waves are bound to add to the complexity of the scheme. The *second version* of Godunov's method is more attractive and is given by the following statement.

Proposition 6.2.1. *The Godunov method can be written in conservative form*

$$\mathbf{U}_i^{n+1} = \mathbf{U}_i^n + \frac{\Delta t}{\Delta x}[\mathbf{F}_{i-\frac{1}{2}} - \mathbf{F}_{i+\frac{1}{2}}] , \qquad (6.11)$$

with intercell numerical flux given by

$$\mathbf{F}_{i+\frac{1}{2}} = \mathbf{F}(\mathbf{U}_{i+\frac{1}{2}}(0)) , \qquad (6.12)$$

if the time step Δt satisfies the condition

$$\Delta t \leq \frac{\Delta x}{S_{\max}^n} . \qquad (6.13)$$

Proof. The integrand $\widetilde{\mathbf{U}}(x,t)$ in (6.8) is an exact solution of the conservation laws, see equation (6.6). We can therefore apply the integral form (6.2) of the conservation laws to any control volume $[x_1, x_2] \times [t_1, t_2]$. In particular, we can apply it to the case in which $x_1 = x_{i-\frac{1}{2}}$, $x_2 = x_{i+\frac{1}{2}}$, $t_1 = t^n$, $t_2 = t^{n+1}$. From (6.4) we then have

$$\left.\begin{array}{rcl} \int_{x_{i-\frac{1}{2}}}^{x_{i+\frac{1}{2}}} \widetilde{\mathbf{U}}(x, t^{n+1}) \, dx & = & \int_{x_{i-\frac{1}{2}}}^{x_{i+\frac{1}{2}}} \widetilde{\mathbf{U}}(x, t^n) \, dx \\[2mm] & & + \int_0^{\Delta t} \mathbf{F}[\widetilde{\mathbf{U}}(x_{i-\frac{1}{2}}, t)] \, dt - \int_0^{\Delta t} \mathbf{F}[\widetilde{\mathbf{U}}(x_{i+\frac{1}{2}}, t)] \, dt . \end{array}\right\}$$
$$(6.14)$$

In terms of local solutions, as in (6.6), and assuming condition (6.13) we have

$$\begin{aligned}
\widetilde{\mathbf{U}}(x_{i-\frac{1}{2}}, t) &= \mathbf{U}_{i-\frac{1}{2}}(0) = \text{constant} , \\
\widetilde{\mathbf{U}}(x_{i+\frac{1}{2}}, t) &= \mathbf{U}_{i+\frac{1}{2}}(0) = \text{constant} ,
\end{aligned} \right\} \tag{6.15}$$

where $\mathbf{U}_{i+\frac{1}{2}}(0)$ is the solution of the Riemann problem $RP(\mathbf{U}_i^n, \mathbf{U}_{i+1}^n)$ along the ray $x/t = 0$, which is the t–axis in the local frame. Similarly, $\mathbf{U}_{i-\frac{1}{2}}(0)$ is the solution of $RP(\mathbf{U}_{i-1}^n, \mathbf{U}_i^n)$ along the t–axis. Division of (6.14) through by Δx gives

$$\frac{1}{\Delta x} \int_{x_{i-\frac{1}{2}}}^{x_{i+\frac{1}{2}}} \widetilde{\mathbf{U}}(x, t^{n+1}) \, dx = \frac{1}{\Delta x} \int_{x_{i-\frac{1}{2}}}^{x_{i+\frac{1}{2}}} \widetilde{\mathbf{U}}(x, t^n) \, dx$$

$$\left. + \frac{\Delta t}{\Delta x} [\mathbf{F}(\mathbf{U}_{i-\frac{1}{2}}(0)) - \mathbf{F}(\mathbf{U}_{i+\frac{1}{2}}(0))] , \right\} \tag{6.16}$$

which by virtue of (6.4) and (6.15) leads to the desired result (6.11)–(6.12), and thus the proposition has been proved.

The following remarks are in order:

- The CFL condition (6.13) for the second version (6.11)–(6.12) of the Godunov method is more generous than (6.9), thus allowing a larger time step. This in turn results in a more efficient time–marching scheme. Here a wave is allowed to travel, at most, a complete cell length Δx in a time Δt.
- Condition (6.13) remains valid even if wave interaction takes place in time Δt within cell I_i, under the assumption that no wave acceleration takes place as a consequence of wave interaction; this is a kind of linearity assumption. Condition (6.13) is necessary in (6.16) when computing the fluxes along the left and right intercell boundaries.
- The second version (6.11)–(6.12) of the Godunov method is the one that is used for practical computations.

6.3 Godunov's Method for the Euler Equations

Here we describe Godunov's method for the specific case of the time–dependent, one–dimensional Euler equations. As data $\{\mathbf{U}_i^n\}$ at time level n is assumed, in order to march the solution to time level $n + 1$ via the conservative formula (6.11) we need to compute the intercell fluxes $\mathbf{F}_{i-\frac{1}{2}}$ and $\mathbf{F}_{i+\frac{1}{2}}$.

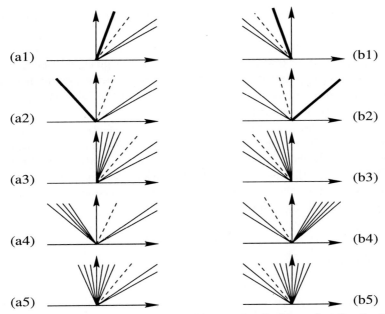

Fig. 6.5. Possible wave patterns in evaluating the Godunov flux for the Euler equations:(a) positive particle speed in the Star Region (b) negative particle speed in the Star Region

6.3.1 Evaluation of the Intercell Fluxes

For a generic cell interface at $x_{i+\frac{1}{2}}$ we compute the Godunov flux $\mathbf{F}_{i+\frac{1}{2}}$ according to (6.12). We therefore require the solution $\mathbf{U}_{i+\frac{1}{2}}(x/t)$ of the Riemann problem $RP(\mathbf{U}_i^n, \mathbf{U}_{i+1}^n)$ evaluated at the point $S = x/t = 0$.

In Chap. 4 we presented the complete exact solution to a general Riemann problem $RP(\mathbf{U}_i^n, \mathbf{U}_{i+1}^n)$ for the Euler equations. In practice we use the solution in terms of the primitive variables, which we denote by $\mathbf{W}_{i+\frac{1}{2}}, (x/t)$. Having found $\mathbf{W}_{i+\frac{1}{2}}(x/t)$ its evaluation at any point $S = x/t$ is carried out by the subroutine SAMPLE in the FORTRAN program given in Sect. 4.9 of Chap. 4. Sampling requires the identification of ten possible wave patterns; these are illustrated in Fig. 6.5. The flow chart of Fig. 4.14 in Chap. 4 relates to the five sub–cases arising from the case in which the sampling point S lies to the left of the contact discontinuity. There is an analogous flow chart for the five sub–cases arising from the case in which the sampling point S lies to the right of the contact discontinuity. For the Godunov method the sampling is performed for the special value $S = x/t = 0$. Unfortunately, this does not simplify the sampling procedure and all ten possible wave patterns must be taken into account; these are shown in Fig. 6.5. Recall that in our convention a shock is a single thick ray, a contact is a dashed line and a rarefaction wave is obviously a fan. A wave of unknown character is represented by a pair of rays emanating from the origin. There are two situations, each of which has five cases, namely (a) $0 \leq u_*$ (positive particle speed in the *Star Region*) and (b) $0 \geq u_*$ (negative particle speed in the *Star Region*). The sampled value $\mathbf{W}_{i+\frac{1}{2}}(0)$ needed for evaluating the Godunov flux is given in Table 6.1 for all ten possible wave patterns; see Fig. 6.5. Consider for example the situation in which u_* is positive. In order to compute correctly the value $\mathbf{W}_{i+\frac{1}{2}}(0)$ along the t–axis (left of contact) we must identify the character of the left wave. This can be a shock, cases (a1) and (a2), or a rarefaction wave, cases (a3), (a4) and (a5). If the left wave is a shock wave we compute the state $\mathbf{W}_{*L}$ between the left shock and the contact using shock relations, see Sect. 4.5.1 of Chap. 4. Then the speed S_L of the left shock is computed. This then allows us to test whether the shock speed is positive (supersonic flow) or negative (subsonic flow). If $S_L \geq 0$ then

$$\mathbf{W}_{i+\frac{1}{2}}(0) = \mathbf{W}_L .$$

If $S_L \leq 0$ then

$$\mathbf{W}_{i+\frac{1}{2}}(0) = \mathbf{W}_{*L} .$$

The analysis for the remaining cases (a3) to (a5) is analogous, as is for the set of cases (b1) to (b5). Details are omitted.

Sub–case	Case (a): positive speed u_*	case (b): negative speed u_*
1	$\mathbf{W}_L$	$\mathbf{W}_R$
2	$\mathbf{W}_{*L}$	$\mathbf{W}_{*R}$
3	$\mathbf{W}_L$	$\mathbf{W}_R$
4	$\mathbf{W}_{*L}$	$\mathbf{W}_{*R}$
5	$\mathbf{W}_{Lfan}$	$\mathbf{W}_{Rfan}$

Table 6.1. Value of $\mathbf{W}_{i+\frac{1}{2}}(0)$ required for evaluating the Godunov flux, for all ten possible wave patterns in the solution of the Riemann problem

Having identified the desired value $\mathbf{W}_{i+\frac{1}{2}}(0)$ the intercell (6.12) becomes

$$\mathbf{F}_{i+\frac{1}{2}} = \mathbf{F}(\mathbf{W}_{i+\frac{1}{2}}(0)) .$$

Exercise 6.3.1. Construct a flow chart for computing the Godunov flux for the time–dependent, one–dimensional Euler equations.

Solution 6.3.1. (Left to the reader).

Exercise 6.3.2. Draw all possible wave patterns required for evaluating the Godunov flux for the isentropic equations of Gas Dynamics; see Sect. 2.4.4 of Chap. 4. Construct a flow chart for computing the Godunov flux.

Solution 6.3.2. (Left to the reader).

6.3.2 Time Step Size

So far we know how to compute the intercell flux (6.12) to be used in the conservative formula (6.11). The spatial discretisation length Δx is chosen on desired accuracy or available computing resources. What remains to be determined in (6.11) is the size of the time step Δt. This is based on the condition (6.13). The time step is then given by

$$\Delta t = \frac{C_{\text{cfl}} \Delta x}{S^n_{\text{max}}} . \tag{6.17}$$

Here C_{cfl} is a Courant or CFL coefficient satisfying

$$0 < C_{\text{cfl}} \le 1 . \tag{6.18}$$

The closer the coefficient C_{cfl} is to 1, the more efficient the time marching scheme is. S^n_{max} is the largest wave speed present throughout the domain at time level n. This means that no wave present in the solution of all Riemann problems travels more than a distance Δx in time Δt. As discussed in Chap. 5 in the context of simple problems, there are various ways of estimating S^n_{max}. For the time–dependent, one dimensional Euler equations a reliable choice is

$$S^n_{\text{max}} = \max_i \left\{ | S^L_{i+\frac{1}{2}} |, | S^R_{i+\frac{1}{2}} | \right\} , \tag{6.19}$$

for $i = 0, \ldots, M$, where $S^L_{i+\frac{1}{2}}$, $S^R_{i+\frac{1}{2}}$ are the wave speeds of the left and right non–linear waves present in the solution of the Riemann problem $RP(\mathbf{U}^n_i, \mathbf{U}^n_{i+1})$. Recall that this Riemann problem generates three waves; the fastest are the non–linear waves, which can be shocks or rarefactions. For rarefaction waves one selects the speed of the head. For shock waves one selects the shock speed, naturally. Note that in sampling the wave speeds one must include the boundaries, as these might generate large wave speeds. Using (6.19) to find S^n_{max} and thus Δt according to (6.17), is a simple and very reliable procedure. As the local solutions of Riemann problems are available for flux evaluation, it is just a question of using this information to find Δt as well. For multi–dimensional problems however, this scheme for estimating the maximum wave speed is unsuitable; see Sect. 16.3.2 of Chap. 16.

A popular alternative for estimating S^n_{max}, which extends to multi-dimensional problems, is

$$S^n_{\text{max}} = \max_i \left\{ | u^n_i | + a^n_i \right\} . \tag{6.20}$$

Only data values for the particle velocity u^n_i and sound speed a^n_i are used here. It is not difficult to see however that (6.20) can lead to an underestimate of S^n_{max}. For instance, assume shock–tube data in which the flow is stationary at time $t = 0$. Then $u^n_i = 0$ and the sound speed is the only contribution to S^n_{max}. Underestimating S^n_{max} results in a choice of Δt that is too large and instabilities may be developed from the beginning of the computations.

A possible way of remedying this, is by choosing the CFL coefficient C_{cfl} in (6.17) cautiously. If S_{max}^n is known reliably then the choice $C_{\text{cfl}} = 1$ is probably adequate, although this implies that waves pass through each other without acceleration, which is a kind of linearity assumption. A practical choice is $C_{\text{cfl}} = 0.9$. If there are uncertainties in the estimate for S_{max}^n, such as when (6.20) is used, a more conservative choice for C_{cfl} is advised. In spite of the alluded disadvantages of choice (6.20), it provides a practical approach when computing solutions to multi–dimensional problems. See Chap. 16.

6.3.3 Boundary Conditions

For a domain $[0, L]$ discretised into M computing cells of length Δx we need boundary conditions at the boundaries $x = 0$ and $x = L$ as illustrated in Fig. 6.6. Numerically, such boundary conditions are expected to provide numerical fluxes $\mathbf{F}_{\frac{1}{2}}$, and $\mathbf{F}_{M+\frac{1}{2}}$. These are required in order to apply the conservative formula (6.11) to update the extreme cells I_1 and I_M to the next time level $n + 1$. The boundary conditions may result in direct prescription of $\mathbf{F}_{\frac{1}{2}}$ and $\mathbf{F}_{M+\frac{1}{2}}$. Alternatively, we may prescribe *fictitious data values* in the fictitious cells I_0 and I_{M+1}, adjacent to I_1 and I_M respectively; see Fig. 6.6. In this way, boundary Riemann problems $RP(\mathbf{U}_0^n, \mathbf{U}_1^n)$ and $RP(\mathbf{U}_M^n, \mathbf{U}_{M+1}^n)$ are solved and the corresponding Godunov fluxes $\mathbf{F}_{\frac{1}{2}}$ and $\mathbf{F}_{M+\frac{1}{2}}$ are computed, as done for the interior cells. The imposition of boundary conditions is, fundamentally,

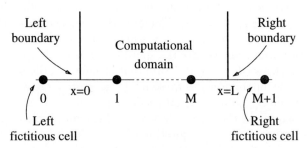

Fig. 6.6. Boundary conditions. Fictitious cells outside the computational domain are created

a physical problem. Great care is required in their numerical implementation. For the Godunov method this task tends to be facilitated by the fact that local Riemann problem solutions are used. Here we consider only two types of boundaries: *reflective* and *transparent* or *transmissive*.

Reflective Boundaries. Consider the boundary at $x = L$ and suppose it physically consists of a fixed, reflective impermeable wall. Then the physical situation is correctly modelled by creating a fictitious state $\mathbf{W}_{M+1}^n$ on the right hand side of the boundary and defining the boundary Riemann problem

$RP(\mathbf{W}_M^n, \mathbf{W}_{M+1}^n)$. The fictitious state $\mathbf{W}_{M+1}^n$ is defined from the known state $\mathbf{W}_M^n$ inside the computational domain, namely

$$\rho_{M+1}^n = \rho_M^n \ , \quad u_{M+1}^n = -u_M^n \ , \quad p_{M+1}^n = p_M^n \ . \tag{6.21}$$

The exact solution of this boundary Riemann problem consists of either (i) two shock waves if $u_M^n > 0$ or (ii) two rarefaction waves if $u_M^n \le 0$. In both cases $u_* = 0$ along the boundary; this is the desired condition at the solid, fixed impermeable boundary. Consequently, the only non–zero contribution to the flux vector at the boundary is in the momentum component and is due to the pressure p_* corresponding to $u_* = 0$. In both cases the solution can be obtained in closed form, no iteration is required. As a matter of fact, a closed–form solution exists for the more general case in which the fluid under consideration obeys the covolume equation of state and the impermeable wall moves with a prescribed speed u_{wall} [351]. The boundary conditions are

$$\rho_{M+1}^n = \rho_M^n \ , \quad u_{M+1}^n = -u_M^n + 2u_{\text{wall}} \ , \quad p_{M+1}^n = p_M^n \ . \tag{6.22}$$

The exact solution of the Riemann problem $RP(\mathbf{W}_M^n, \mathbf{W}_{M+1}^n)$ containing a moving boundary is symmetric about the path of the moving wall and consists of either (a) two shocks or (b) two rarefactions, with the contact wave coinciding with the moving wall, as desired. See Fig. 6.7.

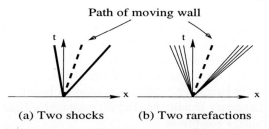

Path of moving wall

(a) Two shocks   (b) Two rarefactions

Fig. 6.7. Boundary Riemann problem for moving wall. Contact surface coincides with moving solid boundary: (a) solution consists of two shocks and the contact (b) solution consists of two rarefactions and the contact

We now find the exact solution for p_* and u_* in the moving–wall Riemann problem. From the analysis of the exact function for pressure, see Sect. 4.3 of Chap 4, it is seen that if $\Delta u = -2(u_M - u_{\text{wall}}) = 0$, that is $u_M = u_{\text{wall}}$, then the solution p_* for pressure at the boundary is $p_* = p_M = p_{M+1}$. For $\Delta u > 0$ we have $p_* < p_M = p_{M+1}$, that is, the solution consists of two rarefaction waves, see Fig. 6.7b. For $\Delta u < 0$ we have $p_* > p_M = p_{M+1}$ and the solution consists of two shocks, see Fig. 6.7a. For the case of two rarefaction waves, $u_M \le u_{\text{wall}}$, direct utilisation of the data in the pressure function $f(p) = 0$, see Sect. 4.3 of Chap 4, gives

$$p_* = p_M \left[1 + \frac{1}{2}(\gamma - 1)\left(\frac{C_M}{a_M}\right) \right]^{\frac{2\gamma}{\gamma - 1}} \ . \tag{6.23}$$

For the case of two shocks, $u_M > u_{\text{wall}}$, we have

$$p_* = p_M + \frac{C_M}{2A_M} \left\{ C_M + \left[C_M^2 + 4A_M (B_M + p_M) \right]^{\frac{1}{2}} \right\} , \qquad (6.24)$$

where

$$A_M = \frac{2}{(\gamma + 1)\rho_M} , \quad B_M = \frac{(\gamma - 1)}{(\gamma + 1)}p_M , \quad C_M = u_M - u_{\text{wall}} . \qquad (6.25)$$

As anticipated, the solution for the velocity u_* in both cases is found to be

$$u_* = u_{\text{wall}} . \qquad (6.26)$$

These closed–form solutions for the pressure and velocity at the boundary (fixed or moving) can also be utilised in the Godunov method even when this is used in conjunction with approximate Riemann solvers, particularly if these are thought to be inaccurate for boundary data Riemann problems. A useful discussion on solid–body boundary conditions for the Euler equations in multi–dimensional domains is given by Rizzi [278]. A recommended paper on boundary conditions for hyperbolic problems is that of Thompson [349].

Transmissive Boundaries. Transmissive, or transparent boundaries arise from the need to define finite, or sufficiently small, computational domains. The corresponding boundary conditions are a numerical attempt to produce boundaries that allow the passage of waves without any effect on them. For one–dimensional problems the objective is reasonably well attained. For multi–dimensional problems this is a substantial area of current research, usually referred to as *open–end* boundary conditions, *transparent* boundary conditions, *far–field* boundary conditions, *radiation* boundary conditions or *non–reflecting* boundary conditions. For a transmissive right boundary we suggest the boundary conditions

$$\rho_{M+1}^n = \rho_M^n , \quad u_{M+1}^n = u_M^n , \quad p_{M+1}^n = p_M^n . \qquad (6.27)$$

This data produces a trivial Riemann problem. No wave of finite strength is produced at the boundary that may affect the flow inside the domain. Useful publications dealing with transparent boundary conditions are those of Giles [134], Bayliss and Turkel [20], Roe [287] and Karni [190].

For an assumed mesh of size Δx, we have defined all details for the practical implementation of the Godunov method, see (6.11)–(6.13). These are

– intercell fluxes
– the maximum wave speed S_{max}^n to compute the time step size Δt, and
– boundary conditions.

Remark 6.3.1. The wave speeds generated at the boundaries, after applying boundary conditions, must be taken into account when selecting the time step Δt.

Exercise 6.3.3. Write a flow chart to implement the Godunov method to solve the one–dimensional, time dependent Euler equations in a tube of constant cross sectional area. Assume the left wall is impermeable and fixed and the right wall is transparent.

Solution 6.3.3. (Left to the reader).

6.4 Numerical Results and Discussion

Here we illustrate the performance of Godunov's first–order upwind method for the Euler equations on test problems with exact solution. For comparison we also show numerical results obtained by the Lax–Friedrichs and Richtmyer (or two–step Lax–Wendroff) methods, discussed in Chap. 5. In all chosen tests, data consists of two constant states $\mathbf{W}_L = (\rho_L, u_L, p_L)^T$ and $\mathbf{W}_R = (\rho_R, u_R, p_R)^T$, separated by a discontinuity at a position $x - x_0$. The states $\mathbf{W}_L$ and $\mathbf{W}_R$ are given in Table 6.2. The ratio of specific heats is chosen to be $\gamma = 1.4$. For all test problems the spatial domain is the interval $[0, 1]$ which is discretised with $M = 100$ computing cells. The Courant number coefficient is $C_{\text{cfl}} = 0.9$; boundary conditions are transmissive and $S_{\max}^n$ is found using the simplified formula (6.20).

Remark 6.4.1. Given that formula (6.20) is not reliable, see discussion in Sect. 6.3.2, in all computations presented here we take, for the the first 5 time steps, a Courant number coefficient C_{cfl} reduced by a factor of 0.2. This allows for waves to begin to form, after which formula (6.20) becomes more reliable.

The exact solutions were found by running the code HE-E1RPEXACT of the library *NUMERICA* [369] and the numerical solutions were obtained by running the code HE-E1GODSTATE of *NUMERICA*.

Test	ρ_L	u_L	p_L	ρ_R	u_R	p_R
1	1.0	0.75	1.0	0.125	0.0	0.1
2	1.0	-2.0	0.4	1.0	2.0	0.4
3	1.0	0.0	1000.0	1.0	0.0	0.01
4	5.99924	19.5975	460.894	5.99242	-6.19633	46.0950
5	1.0	-19.59745	1000.0	1.0	-19.59745	0.01

Table 6.2. Data for five test problems with exact solution. Test 5 is like Test 3 with negative uniform background speed

Test 1 is a *modified version* of the popular Sod's test [318]; the solution consists of a right shock wave, a right travelling contact wave and a left *sonic* rarefaction wave; this test is very useful in assessing the *entropy satisfaction* property of numerical methods. Test 2 has solution consisting of two symmetric rarefaction waves and a trivial contact wave of zero speed; the *Star Region* between the non–linear waves is close to vacuum, which makes this problem a suitable test for assessing the performance of numerical methods for low–density flows; this is the so called *123 problem* introduced in chapter Chap. 4. Test 3 is designed to assess the robustness and accuracy of numerical methods; its solution consists of a strong shock wave, a contact surface and a left rarefaction wave. Test 4 is also designed to test robustness of numerical methods; the solution consists of three strong discontinuities travelling to the right. See Sect. 4.3.3 of Chap. 4 for more details on the exact solution of these test problems. Test 5 is also designed to test the robustness of numerical methods but the main reason for devising this test is to assess the ability of the numerical methods to resolve *slowly– moving contact discontinuities*. The exact solution of Test 5 consists of a left rarefaction wave, a right–travelling shock wave and a *stationary* contact discontinuity. For each test we select a convenient position x_0 of the initial discontinuity and an output time. These are stated in the legend of each figure displaying computational results.

Figs. 6.8 to 6.12 show comparisons between exact solutions (line) and numerical solutions (symbol) at a given output time obtained by the Godunov method, for all five test problems. The quantities shown are density ρ, particle speed u, pressure p and specific internal energy e. For comparison, we also solved these test problems using the Lax–Friedrichs method, see Figs. 6.13 to 6.17, and the Richtmyer (or two–step Lax–Wendroff) method, which failed to produce solutions to Tests 2 to 5. For Test 1 the solution of the Richtmyer scheme is shown in Fig. 6.18.

6.4.1 Numerical Results for Godunov's Method

The results for Test 1, shown in Fig. 6.8, are typical of the Godunov's first–order accurate method described in this chapter.

The numerical approximation of the shock wave, of zero–width transition in the exact solution, has a transition region of width approximately $4\Delta x$; that is, the shock has been *smeared* over 4 computing cells. This spreading of shock waves may seem unsatisfactory, but it is quite typical of numerical solutions; in fact most first–order methods will spread a shock wave even more. A satisfactory feature of the numerical shock wave of Fig. 6.8 is that it is monotone, there are no spurious oscillations in the vicinity of the shock, at least for this example. Monotonicity of shock waves computed by the Godunov method depends on the speed of the shock and it holds in most cases except when the shock speed is very close to zero. The contact discontinuity, seen in the density and internal energy plots, is smeared over 20 cells; generally contact waves are much more difficult to resolve accurately than shock waves. This is due to the linear character of contacts; characteristics on either side of the wave run parallel to the wave. In shock waves, characteristics on either side of the wave run into the shock, a compression mechanism that helps the numerical resolution of shock waves. As for the shock case, the solution for the contact is perfectly monotone.

Another positive feature of the numerical approximation of the discontinuities is that their speed of propagation is correct and thus their average positions are correct. This is a consequence of the conservative character of Godunov's method. The rarefaction wave is a smooth flow feature and is reasonably well approximated by the method except near the head and the tail, where a discontinuity in derivative exists. Another visible error in the rarefaction is the small discontinuous jump within the rarefaction. This is sometimes referred to as the *entropy glitch* and arises only in the presence of *sonic* rarefaction waves, as in the present case. Godunov's method is theoretically entropy satisfying [164] and we therefore expect the size of the jump in the entropy glitch to tend to zero as the mesh size Δx tends to zero. Fig. 6.19 shows the result obtained by refining the mesh by a factor of 5. It appears as if the numerical solution does converge to the exact solution.

The performance of Godunov's method on Test 2, see Fig. 6.9, is generally quite satisfactory as regards the physical variables p, u and ρ but not so much for the specific internal energy, which is computed from ρ and p as $e = p/((\gamma - 1)\rho)$. In this low density example both pressure and density are close to zero and thus small errors will be exaggerated by their ratio. In any case, it is generally accepted that plots of the internal energy e can be quite revealing of the quality of the numerical solution. On the other hand pressure is probably the easiest quantity to get right. The main point of Test 2 is to make the reader aware that this class of low density flows can easily cause numerical methods to fail; even the robust Godunov method fails if used in conjunction with certain approximate Riemann solvers [118]. The Richtmyer (or two–step Lax–Wendroff) method fails to give a solution to this problem.

Test 3 is a very severe problem and is designed to test the robustness of the Godunov method, the results of which are shown in Fig. 6.10. The emerging right travelling shock wave has pressure ratio $p_*/p_R = 46000$ and a corresponding shock Mach number of 198. For flows involving such strong shock waves as this, one would seriously question the validity of the ideal gas equations of state. However, from the point of view of assessing the robustness of numerical schemes, the validity of the test problem as a mathematical/numerical problem still holds. As for Test 1, the resolution of discontinuities is worst for the contact wave; as a consequence of this, post shock values are not attained, as is clearly seen in the density plot. The velocity plot shows a kind of overshoot near the tail of the rarefaction. The Richtmyer (or two–step Lax–Wendroff) scheme failed for this test.

As seen in Fig. 6.11 the solution of Test 4 consists of three discontinuities: two shock waves and a contact. They all travel to the right; the left shock has a small positive speed. The complete wave system has resulted from the interaction of two strong shock waves propagating in opposite directions. The right shock is the fastest wave and is smeared over 5 cells, as seen in the pressure plot; the left, slowly moving shock is sharply resolved (two cells) but is not monotone; there are some *low frequency* spurious oscillations in its vicinity, as seen in the internal energy plot. The contact discontinuity is heavily smeared. Slowly moving shocks are sharply resolved by Godunov's method. In fact, *isolated* shocks and contacts of zero speed are perfectly resolved, if non–defective Riemann solvers are used, see Chap. 10. The phenomenon of spurious oscillations in slowly moving shocks has been studied by Roberts [280] and is so far, to the author's knowledge, an unresolved difficulty. Billett and Toro [40] investigated some possible cures of the problem for the Euler equations. See also the recent papers by Arora and Roe [13] and by Karni and Čanić [193].

Test 5 is like Test 3 but with a uniform, negative background speed so as to obtain a virtually stationary contact discontinuity. In addition to testing the robustness of numerical methods, Test 5 is mainly designed to test the ability of numerical methods to resolve slowly–moving or stationary contact discontinuities. Fig. 6.12 shows the result obtained from the Godunov method as compared with the exact solution. For this test problem the contact discontinuity is *virtually stationary*; the Godunov method resolves this discontinuity very sharply indeed. This result should be compared with that obtained by the Lax–Friedrichs method, Fig. 6.17. Test 5 is a very challenging test problem, as we shall illustrate in subsequent chapters dealing with other numerical methods.

6.4.2 Numerical Results from Other Methods

First we apply the Lax–Friedrichs method, see Sect. 5.3.4 of Chap. 5, to Tests 1 to 5. The numerical results are shown in Figs. 6.13 to 6.17. The results for Test 1 are shown in Fig. 6.13 and are to be compared with those of Godunov's method, Fig. 6.8. The Lax–Friedrichs scheme has the peculiar property of *pairing* cell values, which enhances smearing. The shock wave is resolved with about 8 cells and looks acceptable. The resolution of the rarefaction wave and the contact discontinuity is very poor. The solution for Test 2 is shown in Fig. 6.14. Note how inaccurate the solution for internal energy is; compare with Fig. 6.9 and with the exact solution. A large class of methods are known to have difficulties with this kind of *symmetric* Riemann problems [364]. Fig. 6.15 shows results for Test 3; the scheme is unable to attain the post–shock density values; compare with Fig. 6.10. The results for Test 4 are shown in Fig. 6.16, which are to be compared with those of Fig. 6.11. The Lax–Friedrichs method, although simpler and cheaper, is significantly less accurate than the Godunov method. Fig. 6.17 shows the result from the Lax–Friedrichs scheme for Test 5; this result and that of Fig. 6.12 show the crucial difference between two major classes of numerical methods, namely *centred methods* and *Godunov–type methods*. Test 5 highlights the fact that resolving *linear waves* is perhaps one of the most challenging tasks for numerical methods today.

The second–order Richtmyer (or two–step Lax–Wendroff) method was applied to Test 1 and the results are shown in Fig. 6.18; compare with Figs. 6.8 and 6.13. The solution is generally more accurate in the smooth regions of the flow, as is to be expected from a second–order accurate method; discontinuities are also more sharply resolved but *spurious oscillations* near discontinuities appear; see Chap. 5. This method failed to give a solution at all, for Tests 2 to 5.

In this chapter we have presented the Godunov method as used in conjunction with the exact Riemann solver. Godunov's method can also be used with approximate Riemann solvers. In Chaps. 9 to 12 we present several approximate Riemann solvers for the Euler equations. Second and third order extensions of Godunov's method are presented in Chap. 13 for scalar problems. In Chaps. 14 and 16 we present second–order TVD schemes for non–linear systems.

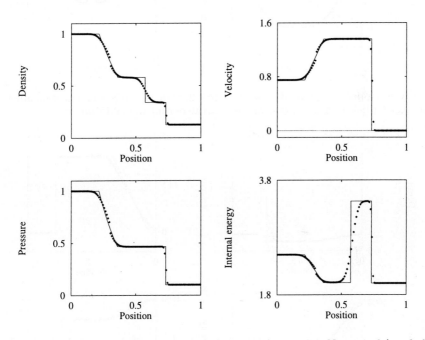

Fig. 6.8. Godunov's method applied to Test 1, with $x_0 = 0.3$. Numerical (symbol) and exact (line) solutions are compared at the output time 0.2 units

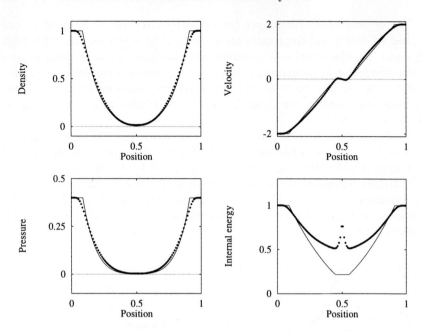

Fig. 6.9. Godunov's method applied to Test 2, with $x_0 = 0.5$. Numerical (symbol) and exact (line) solutions are compared at the output time 0.15 units

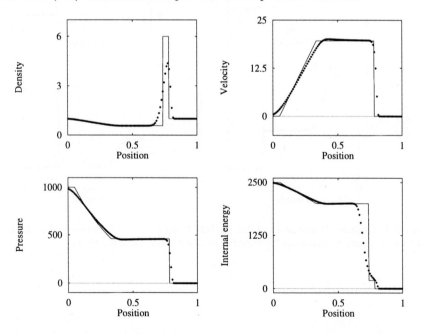

Fig. 6.10. Godunov's method applied to Test 3, with $x_0 = 0.5$. Numerical (symbol) and exact (line) solutions are compared at the output time 0.012 units

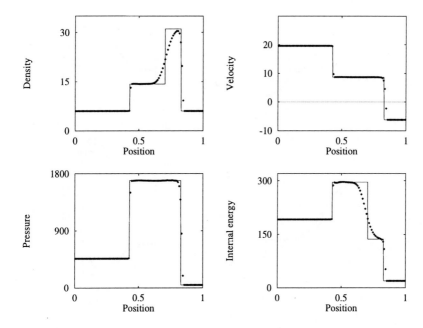

Fig. 6.11. Godunov's method applied to Test 4, with $x_0 = 0.4$. Numerical (symbol) and exact (line) solutions are compared at the output time 0.035 units

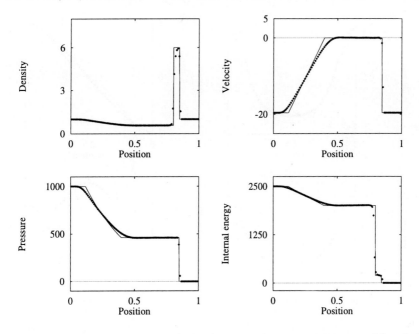

Fig. 6.12. Godunov's method applied to Test 5, with $x_0 = 0.8$. Numerical (symbol) and exact (line) solutions are compared at the output time 0.012

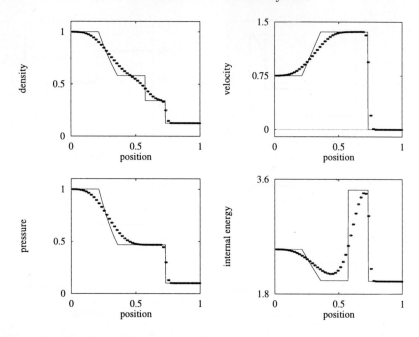

Fig. 6.13. The Lax-Friedrichs method applied to Test 1, with $x_0 = 0.3$. Numerical (symbol) and exact (line) solutions are compared at the output time 0.2 units

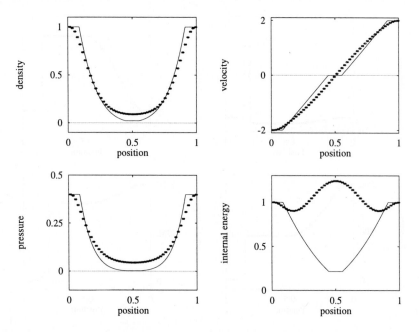

Fig. 6.14. The Lax-Friedrichs method applied to Test 2, with $x_0 = 0.5$. Numerical (symbol) and exact (line) solutions are compared at the output time 0.15 units

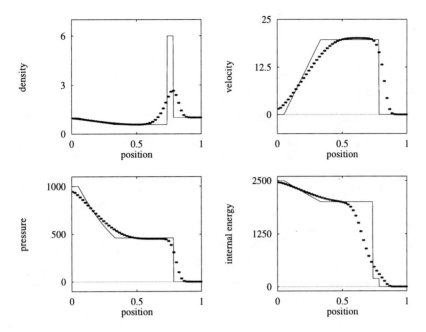

Fig. 6.15. The Lax-Friedrichs method applied to Test 3, with $x_0 = 0.5$. Numerical (symbol) and exact (line) solutions are compared at the output time 0.012 units

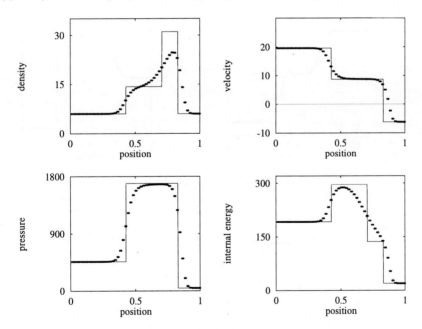

Fig. 6.16. The Lax-Friedrichs method applied to Test 4, with $x_0 = 0.4$. Numerical (symbol) and exact (line) solutions are compared at the output time 0.035 units

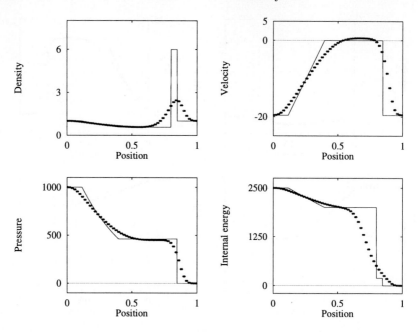

Fig. 6.17. The Lax–Friedrichs method applied to Test 5, with $x_0 = 0.8$. Numerical (symbol) and exact (line) solutions are compared at the output time 0.012

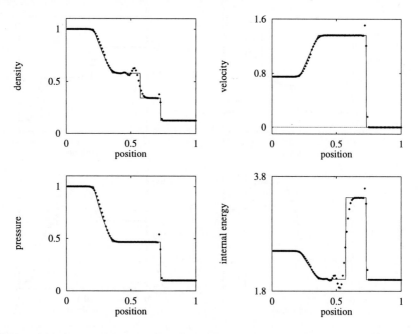

Fig. 6.18. The Richtmyer method applied to Test 1, with $x_0 = 0.3$. Numerical (symbol) and exact (line) solutions are compared at the output time 0.2 units

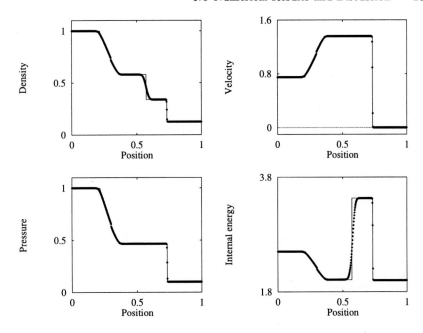

Fig. 6.19. Godunov's method applied to Test 1 with fine mesh, $M = 500$. Numerical (symbol) and exact (line) solutions are compared at time 0.2 units

7. Random Choice and Related Methods

7.1 Introduction

In 1965, Glimm [141] introduced the Random Choice Method (RCM) as part of a constructive proof of existence of solutions to a class of non–linear systems of hyperbolic conservation laws. In 1976, Chorin [73] successfully implemented a modified version of the method, as a computational technique, to solve the Euler equations of Gas Dynamics. In essence, to implement the RCM one requires (i) exact solutions of local Riemann problems and (ii) a random sampling procedure to pick up states to be assigned to the next time level. As we shall see, there is a great deal of commonality between the RCM and the Godunov method presented in Chap. 6. Both schemes use the exact solution of the Riemann problem, although Godunov's method can also be implemented using approximate Riemann solvers, as we shall see in Chaps. 9 to 12. The two methods differ in the way the local Riemann problem solutions are utilised to march to the next time level: the Godunov method takes an integral *average* of local solutions of Riemann problems, while the RCM picks a single state, contained in the local solutions, at *random*. The random sampling procedure is carried out by employing a sequence of random numbers. The statistical properties of these random numbers have a significant effect on the accuracy of the Random Choice Method.

Since the introduction of the RCM as a computational scheme by Chorin, there have been many contributions to the development of the method. Chorin himself [74] extended the RCM to combustion problems; Sod [317] applied the RCM to the one–dimensional Euler equations for cylindrically and spherically symmetric flows, thereby introducing a way of dealing with algebraic source terms. Concus [92] applied the RCM to a non–linear scalar equation governing the two–phase flow of petroleum in underground reservoirs. Major contributions to the method were presented by Colella [85], [86]; these include a better understanding of the method, its strengths and limitations, and improved random sampling techniques. Marshall and Mendez [230] applied the RCM to the one–dimensional shallow water equations. Li and Holt [213] applied the RCM to the study of underwater explosions. Marshall and Plohr [231] applied the RCM to solve the steady supersonic Euler equations, see also Shi and Gottlieb [308], and to the study of shock wave diffraction phenomena. Gottlieb [149] compared the implementation of the

RCM on staggered and non–staggered grids and introduced an effective way of using irregular meshes. Toro [351] applied the RCM to covolume gases with moving boundaries. Applications of the RCM to the study of reactive flows were performed by Saito and Glass [293], Takano [337], Singh and Clarke [314] and Dawes [104]. Olivier and Grönig [248] applied the RCM to solve the two–dimensional time dependent Euler equations to study shock focussing and diffraction phenomena in water and air.

Essentially, the RCM is applicable to scalar problems in any number of dimensions and to non–linear systems in two independent variables. Examples of these systems are the one–dimensional, time dependent Euler equations, the two–dimensional, steady supersonic Euler equations and the one–dimensional shallow water equations. By using splitting schemes, see Chap. 15, one can also solve extensions of these systems to include algebraic source terms or even terms to model viscous diffusion; see Sod [320] and Honma and Glass [174]. A fundamental limitation of the RCM is its inability to solve multi–dimensional non–linear systems via splitting schemes, which usually work well when extending other schemes to multi–dimensional problems; see Chap. 16. An attraction of the RCM is its ability to handle complex wave interaction involving discontinuities such as shock waves and material interfaces; these are resolved as true discontinuities. Most other methods will smear discontinuities over several computing cells, a problem that is particularly serious for contact surfaces. Although computed discontinuities in the RCM have infinite resolution, the position of these waves at any given time has an error, which is random in character. The randomness of the RCM also shows in resolving smooth waves, such as rarefactions. Such randomness is tolerable when solving homogeneous systems, i.e. no source terms. In the presence of source terms however, the randomness tends to be enhanced.

This chapter is primarily devoted to the conventional Random Choice Method, but we also present what appears to be a new random choice method [365] that is analogous to the Lax–Friedrichs (deterministic) scheme. In addition we present a, deterministic, first–order centred (FORCE) scheme based on a reinterpretation of the conventional RCM on a staggered grid. The presentation of the schemes is given in terms of the time–dependent, one dimensional Euler equations. The reader is advised to review Chap. 4 before proceeding with the study of the present chapter.

7.2 RCM on a Non–Staggered Grid

We consider the general Initial Boundary Value Problem (IBVP) for non–linear systems of hyperbolic conservation laws, namely

$$\left.\begin{array}{lll} \text{PDEs} & : & \mathbf{U}_t + \mathbf{F}(\mathbf{U})_x = \mathbf{0} , \\ \text{ICs} & : & \mathbf{U}(x,0) = \mathbf{U}^{(0)}(x) , \\ \text{BCs} & : & \mathbf{U}(0,t) = \mathbf{U}_\mathrm{l}(t) , \ \mathbf{U}(L,t) = \mathbf{U}_\mathrm{r}(t) . \end{array}\right\} \tag{7.1}$$

We assume a solution to this IBVP exists. Here $\mathbf{U}(x,t)$ is the vector of conserved variables, $\mathbf{F}(\mathbf{U})$ is the vector of fluxes, $\mathbf{U}^{(0)}(x)$ is the initial data at time $t = 0$, $[0, L]$ is the spatial domain and boundary conditions are, for the moment, assumed to be represented by the boundary functions $\mathbf{U}_l(t)$ and $\mathbf{U}_r(t)$.

In the RCM the only step in which one is required to work with the vector of conserved variables is at the level of the Riemann problem, when enforcing the Rankine–Hugoniot Conditions at shocks. All other steps of the method are more conveniently performed in terms of the vector of primitive variables, which for the Euler equations are $\mathbf{W} = (\rho, u, p)^T$; ρ is density, u is velocity and p is pressure.

7.2.1 The Scheme for Non–Linear Systems

As for the Godunov method studied in Chap. 6, we discretise the spatial domain $[0, L]$ into M computing cells $I_i = [x_{i-\frac{1}{2}}, x_{i+\frac{1}{2}}]$ of size $\Delta x = x_{i+\frac{1}{2}} - x_{i-\frac{1}{2}} = L/M$, with $i = 1, \dots, M$. For a given cell I_i, the location of the cell centre x_i and the cell boundaries $x_{i-\frac{1}{2}}$, $x_{i+\frac{1}{2}}$ are given by

$$x_{i-\frac{1}{2}} = (i - 1)\Delta x \,, \quad x_i = (i - \frac{1}{2})\Delta x \,, \quad x_{i+\frac{1}{2}} = i\Delta x \,. \tag{7.2}$$

For convenience we choose cells of regular size Δx, but this is not a necessary

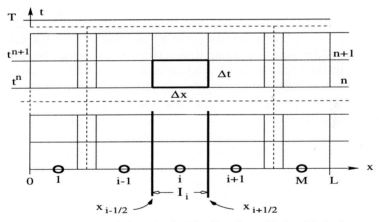

Fig. 7.1. Discretisation of domain for the Random Choice Method on a non–staggered grid

requirement for implementing the RCM. Fig. 7.1 illustrates the non–staggered grid arrangement for this version of the Random Choice Method. The solution is updated at the cell centres x_i, every time step. Obviously, the time step is, in general, of different size for every time step. Given general initial

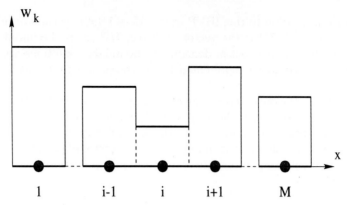

Fig. 7.2. Piece–wise constant distribution of data at time level n. Pairs of neighbouring states define data for local Riemann problems

data $\mathbf{W}(x, t^n)$ at time $t = t^n$ say, in order to evolve the solution to a time $t^{n+1} = t^n + \Delta t$, the Random Choice Method first assumes a *piecewise constant* distribution of the data. Formally, this may be realised by defining *cell averages* as in the Godunov method, see Sect. 6.1 of Chap. 6. For the RCM this is not necessary and we assume that the given data at the cell centres x_i is constant throughout the respective cell I_i. We then have $\mathbf{W}(x, t^n) = \mathbf{W}_i^n$ in each cell I_i. Fig. 7.2 shows the distribution of a typical variable w_k at a given time level n.

The pairs of neighbouring, constant, states $\mathbf{W}_i^n$, $\mathbf{W}_{i+1}^n$ define local Riemann problems $RP(\mathbf{W}_i^n, \mathbf{W}_{i+1}^n)$, which have similarity solutions $\mathbf{W}_{i+\frac{1}{2}}(x/t)$. In Chap. 4 we provided the complete exact solution to the Riemann problem for the Euler equations along with a deterministic sampling procedure contained in the FORTRAN 77 program of Sect. 4.9. Given a time t^*, the sampling routine SAMPLE evaluates $\mathbf{W}_{i+\frac{1}{2}}(x/t^*)$ at any point x in an interval $[x_l, x_r]$ with $x_l \leq 0 \leq x_r$. A detailed understanding of the complete exact solution of the Riemann problem is essential for understanding and implementing the RCM. Fig. 7.3 illustrates the structure of a typical Riemann problem solution and a typical sampling range of the solution at a given time t^*, across the wave structure. In the Random Choice Method the particular point $x = x^*$ is picked up at random within the sampling range $[x_l, x_r]$. The sampling routine SAMPLE evaluates $\mathbf{W}_{i+\frac{1}{2}}(x^*/t^*)$ automatically. See Sect. 4.9 of Chap. 4.

The Random Choice Method updates the solution from the data value $\mathbf{W}_i^n$ in cell I_i at time level n, to the value $\mathbf{W}_i^{n+1}$ at time level $n + 1$, in two steps as follows:

Step I: Solve the Riemann problems $RP(\mathbf{W}_{i-1}^n, \mathbf{W}_i^n)$ and $RP(\mathbf{W}_i^n, \mathbf{W}_{i+1}^n)$ to find their respective solutions $\mathbf{W}_{i-\frac{1}{2}}(x/t)$ and $\mathbf{W}_{i+\frac{1}{2}}(x/t)$. Fig.

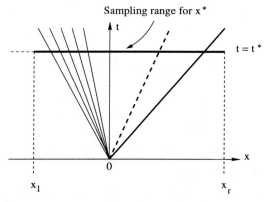

Fig. 7.3. Spacial sampling range in the solution of the Riemann problem at a given time $t = t^*$

7.4 shows typical wave patterns emerging from the intercell boundaries $x_{i-\frac{1}{2}}$ and $x_{i+\frac{1}{2}}$.

Step II: *Random sample* these solutions at time Δt within cell I_i to pick up a state and assign it to cell I_i. The random sampling range is shown in Fig. 7.4 by a thick horizontal line. The picked up state depends on a random, or quasi–random, number θ^n in the interval $[0, 1]$. The updated solution is then

$$
\mathbf{W}_i^{n+1} = \begin{cases} \mathbf{W}_{i-\frac{1}{2}}(\theta^n \Delta x/\Delta t) \text{, if } 0 \le \theta^n \le \frac{1}{2} \text{,} \\[2mm] \mathbf{W}_{i+\frac{1}{2}}((\theta^n - 1)\Delta x/\Delta t) \text{, if } \frac{1}{2} < \theta^n \le 1 \text{.} \end{cases} \tag{7.3}
$$

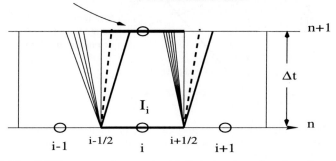

Fig. 7.4. The RCM on non–staggered grid. Solution is updated to time level n by random sampling solutions of Riemann problems within cell I_i at time Δt

The case in which $0 \le \theta^n \le \frac{1}{2}$ is illustrated in Fig. 7.5. Here the updated solution depends on the random sampling procedure applied to the right side of the left Riemann problem solution $\mathbf{W}_{i-\frac{1}{2}}(x/t)$. The particular randomly chosen state is returned by the sampling routine SAMPLE, called with the argument

$$S \equiv x/t = \frac{\theta^n \, \Delta x}{\Delta t} \, .$$

The resulting state is then assigned to the grid point i, which is regarded as the solution in cell I_i for the next time level. A similar procedure is applied if $\frac{1}{2} \le \theta^n \le 1$. In this case one samples the left side of the right Riemann problem solution $\mathbf{W}_{i+\frac{1}{2}}(x/t)$. When programming the non–staggered version of the Random Choice Method there are many ways of organising the tasks of (i) solving of Riemann problems and (ii) random sampling their solutions.

Concerning the use of random numbers in the scheme, Chorin [73] established that one only requires a single random number θ^n for a complete time level n. In Glimm's proof [141] one may take one random number per time step per cell. In Sect. 7.5 we discuss generation of random numbers and their properties.

Finally, we note the relationship between the RCM scheme to obtain the updated value $\mathbf{W}_i^{n+1}$ and the Godunov method, see Chap. 6. The Godunov scheme, instead of random sampling the solution of the relevant Riemann problems, will take the integral average of these local Riemann problem solutions

$$\mathbf{U}_i^{n+1} = \frac{1}{\Delta x} \int_0^{\frac{1}{2}\Delta x} \mathbf{U}_{i-\frac{1}{2}} \left(\frac{x}{\Delta t} \right) \, dx + \frac{1}{\Delta x} \int_{-\frac{1}{2}\Delta x}^0 \mathbf{U}_{i+\frac{1}{2}} \left(\frac{x}{\Delta t} \right) \, dx \, . \quad (7.4)$$

In order to preserve the conservative character of the Godunov method, the

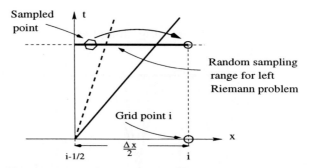

Fig. 7.5. RCM sampling of the right–hand side of the left Riemann problem solution, when $0 \le \theta^n \le \frac{1}{2}$. Sampled state is assigned to the centre of cell I_i

averaging is performed in terms of the conserved variables. The RCM, on the other hand, is not strictly conservative, although one may regard the scheme as being conservative in a statistical sense.

7.2.2 Boundary Conditions and the Time Step Size

The solution updating procedure just described is completely defined for all cells I_i, except for those next to the boundaries, namely I_1 and I_M. In order to update these two cells we apply boundary conditions. This is carried out in exactly the same way as for the Godunov method, see Sects. 6.3.2 and 6.3.3 of Chap. 6. Fictitious states $\mathbf{W}_0$ and $\mathbf{W}_{M+1}$ adjacent to states $\mathbf{W}_1$ and $\mathbf{W}_M$ are defined. The otherwise missing Riemann problem solutions $\mathbf{W}_{\frac{1}{2}}(x/t)$ and $\mathbf{W}_{M+\frac{1}{2}}(x/t)$ at the boundaries are now defined and the random sampling procedure can now be extended to the full computational domain. We consider two types of boundary conditions, as for the one–dimensional time dependent Euler equations.

(I) *Transmissive Boundary Conditions.* Here the fictitious states are given as

$$\left. \begin{array}{c} \rho_0^n = \rho_1^n \ , \ u_0^n = u_1^n \ , \ p_0^n = p_1^n \ , \\[2mm] \rho_{M+1}^n = \rho_M^n \ , \ u_{M+1}^n = u_M^n \ , \ p_{M+1}^n = p_M^n \ . \end{array} \right\} \tag{7.5}$$

(II) *Reflective Boundary Conditions.* Here we state the boundary conditions that apply to reflective left and right boundaries moving with respective speeds u_{wl} and u_{wr}. The fictitious states are given by

$$\left. \begin{array}{c} \rho_0^n = \rho_1^n \ , \ u_0^n = -u_1^n + 2u_{\mathrm{wl}} \ , \ p_0^n = p_1^n \ , \\[2mm] \rho_{M+1}^n = \rho_M^n \ , \ u_{M+1}^n = -u_M^n + 2u_{\mathrm{wr}} \ , \ p_{M+1}^n = p_M^n \ . \end{array} \right\} \tag{7.6}$$

For a more complete discussion on boundary conditions see Sect. 6.3.3 of Chap. 6.

The choice of the time step Δt is determined by a CFL condition. Note first that in order to perform the random sampling described previously, we need to make the assumption that no wave interaction takes place within cell I_i, in the chosen time Δt, see Fig. 7.4. This is satisfied by imposing the following restriction on the size of Δt, namely

$$\Delta t \le \frac{\frac{1}{2}\Delta x}{S_{\mathrm{max}}^n} \ , \tag{7.7}$$

where S_{max}^n denotes the maximum *wave velocity* present throughout the domain at time t^n. A consequence of this restriction is that only two Riemann problem solutions affect cell I_i, namely the right travelling waves of $\mathbf{W}_{i-\frac{1}{2}}(x/t)$ and the left travelling waves of $\mathbf{W}_{i+\frac{1}{2}}(x/t)$. Condition (7.7) may be expressed in the standard form

$$\Delta t = \frac{C_{\mathrm{cfl}}\Delta x}{S_{\mathrm{max}}^n} \ , \tag{7.8}$$

where the CFL coefficient C_{cfl} satisfies

$$0 < C_{\text{cfl}} \leq \frac{1}{2} \, . \tag{7.9}$$

Hence the Random Choice Method has a stability limit of $\frac{1}{2}$, which is only half that of Godunov's method.

Concerning the choice of the maximum wave speed S^n_{max} the reader is referred to Sect. 6.3.2 of Chap. 6. Virtually all relevant remarks made there in the context of the Godunov method apply. For the RCM we recommend the use of the true waves arising from the solutions of local Riemann problems. The collective experience in applying the RCM is that the scheme is not too sensitive to underestimating S^n_{max}. If S^n_{max} is underestimated then the chosen time step Δt will be too large and the likely consequence will not be signs of instabilities, as one would expect, but computed waves will propagate at *obviously* the wrong speed.

We have presented the Random Choice Method on a non–staggered grid as applied to any time–dependent one dimensional non–linear systems of hyperbolic conservations laws. Details of boundary conditions for the time–dependent one dimensional Euler equations have been given, for which numerical results are presented in Sect. 7.6.

7.3 A Random Choice Scheme of the Lax–Friedrichs Type

Here we present a Random Choice Method that arises from interpreting the Lax–Friedrichs scheme as an integral average of solutions of Riemann problems; see [365], [367]. If these averages are transformed by use of the integral form of the conservation laws one recovers the usual Lax–Friedrichs scheme for non–linear systems, thus eliminating the role of the Riemann problem. If the role of the Riemann problem solution is preserved and the integral averages are interpreted in a stochastic sense one obtains a Random Choice Method of the Lax–Friedrichs type.

7.3.1 Review of the Lax–Friedrichs Scheme

As seen in Sect. 5.3.4 of Chap. 5, the Lax–Friedrichs scheme as applied to the linear advection equation

$$u_t + f(u)_x = 0 \, , \; f(u) = au \tag{7.10}$$

reads

$$u_i^{n+1} = \frac{(1+c)}{2} u_{i-1}^n + \frac{(1-c)}{2} u_{i+1}^n \, . \tag{7.11}$$

This is obviously identical to the integral average

$$u_i^{n+1} = \frac{1}{\Delta x} \int_{x_{i-\frac{1}{2}}}^{x_{i+\frac{1}{2}}} \hat{u}_i(x, \tfrac{1}{2}\Delta t)\, dx \qquad (7.12)$$

within cell i, in which $\hat{u}_i(x, t)$ is the solution of the Riemann problem $RP(u_{i-1}^n, u_{i+1}^n)$, that is

$$\hat{u}_i(x/t) = \begin{cases} u_{i-1}^n & \text{if } x/t < a, \\ u_{i+1}^n & \text{if } x/t > a. \end{cases} \qquad (7.13)$$

See Fig. 7.6. The Lax–Friedrichs solution in cell i at time $t^{n+1} = t^n + \Delta t$ is a *weighted average* of the solution of the Riemann problem with the left u_{i-1}^n and right u_{i+1}^n neighbour states as data, at time $t = \frac{1}{2}\Delta t$. Note the two peculiarities of the scheme, (i) the data states do not include u_i^n and (ii) the time for the averaging is half the full time step Δt.

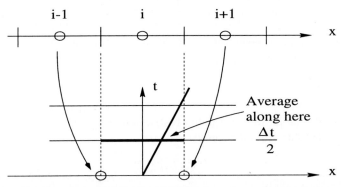

Fig. 7.6. Reinterpretation of the Lax–Friedrichs scheme for the linear advection equation.

7.3.2 The Scheme

We first generalise interpretation (7.12) of the Lax–Friedrichs scheme to non-linear systems of conservation laws

$$\mathbf{U}_t + \mathbf{F}(\mathbf{U})_x = \mathbf{0}. \qquad (7.14)$$

The scheme reads

$$\mathbf{U}_i^{n+1} = \frac{1}{\Delta x} \int_{x_{i-\frac{1}{2}}}^{x_{i+\frac{1}{2}}} \hat{\mathbf{U}}_i(x, \tfrac{1}{2}\Delta t)\, dx, \qquad (7.15)$$

where $\hat{\mathbf{U}}_i(x, t)$ is the solution of the Riemann problem $RP(\mathbf{U}_{i-1}^n, \mathbf{U}_{i+1}^n)$. There are now three possible routes to follow. These are

(i)   Solve Riemann problems and find the updated solution by evaluating (7.15) directly. Numerical results of this scheme are indistinguishable from those obtained from the conventional, much simpler, Lax–Friedrichs scheme. We therefore discard this as a useful scheme.

(ii) Apply the integral form of the conservation laws to (7.15) to reproduce the conventional Lax–Friedrichs scheme for non–linear systems, and thus eliminate the role of the Riemann problem, see Sect. 5.3.4 of Chap. 5.

(iii) Keep the role of the Riemann problem and reinterpret (7.15) in a stochastic sense. We obtain

$$\mathbf{U}_i^{n+1} = \frac{1}{\Delta x} \int_{-\frac{1}{2}\Delta x}^{\frac{1}{2}\Delta x} \hat{\mathbf{U}}_i(\theta^n \Delta x, \frac{1}{2}\Delta t) \, dx \, , \qquad (7.16)$$

where θ^n is a random number satisfying

$$-\frac{1}{2} \leq \theta^n \leq \frac{1}{2} \, . \qquad (7.17)$$

A Random Choice Scheme for updating the solution to the new time level is thus obtained, namely

$$\mathbf{U}_i^{n+1} = \hat{\mathbf{U}}_i(\theta^n \Delta x, \frac{1}{2}\Delta t) \, . \qquad (7.18)$$

The conventional RCM has stability restriction (7.9), while the random choice scheme (7.18) has stability condition

$$0 < C_{cfl} \leq 1 \, , \qquad (7.19)$$

which represents an improvement by a factor of 2.

Fig. 7.7 illustrates the Random Choice Method of the Lax–Friedrichs type as applied to non–linear systems. The programming of the scheme is straightforward. Numerical results will be presented in Sect. 7.6.

7.4 The RCM on a Staggered Grid

As stated earlier the RCM can be implemented on a non–staggered grid and on a staggered grid. The former version was described in Sect. 7.2. Here we describe the RCM on a staggered grid and derive an associated deterministic scheme that is conservative, first–order accurate and centred.

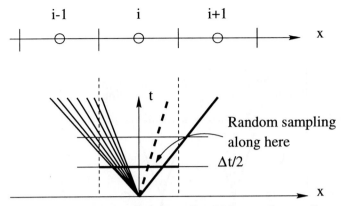

Fig. 7.7. Random choice scheme of the Lax–Friedrichs type for non–linear systems. Updated solution in cell I_i at time Δt is obtained from random sampling solution of Riemann problem $RP(\mathbf{U}_{i-1}^n, \mathbf{U}_{i+1}^n)$ at time $\frac{1}{2}\Delta t$

7.4.1 The Scheme for Non–Linear Systems

The staggered grid version of the RCM to solve (7.14) updates $\mathbf{U}_i^n$ to a new value $\mathbf{U}_i^{n+1}$ in two steps, as illustrated in Fig. 7.8. The steps are:

Step (I) Solve the Riemann problems $RP(\mathbf{U}_{i-1}^n, \mathbf{U}_i^n)$ and $RP(\mathbf{U}_i^n, \mathbf{U}_{i+1}^n)$ to find respective solutions

$$\hat{\mathbf{U}}_{i-\frac{1}{2}}^{n+\frac{1}{2}}(x,t) \ , \quad \hat{\mathbf{U}}_{i+\frac{1}{2}}^{n+\frac{1}{2}}(x,t) \ . \tag{7.20}$$

Random sample these solutions at a stable time $\Delta t^{n+\frac{1}{2}}$, that is

$$\mathbf{U}_{i-\frac{1}{2}}^{n+\frac{1}{2}} = \hat{\mathbf{U}}_{i-\frac{1}{2}}^{n+\frac{1}{2}}(\theta^n \Delta x, \Delta t^{n+\frac{1}{2}}) \ , \quad \mathbf{U}_{i+\frac{1}{2}}^{n+\frac{1}{2}} = \hat{\mathbf{U}}_{i+\frac{1}{2}}^{n+\frac{1}{2}}(\theta^n \Delta x, \Delta t^{n+\frac{1}{2}}) \ . \tag{7.21}$$

Step (II) Solve $RP(\mathbf{U}_{i-\frac{1}{2}}^{n+\frac{1}{2}}, \mathbf{U}_{i+\frac{1}{2}}^{n+\frac{1}{2}})$ to find solution $\hat{\mathbf{U}}_i^{n+1}(x,t)$ and random sample it, at a stable time Δt^{n+1}, to obtain $\mathbf{U}_i^{n+1}$, that is

$$\mathbf{U}_i^{n+1} = \hat{\mathbf{U}}_i^{n+1}(\theta^{n+1} \Delta x, \Delta t^{n+1}) \ . \tag{7.22}$$

The time steps $\Delta t^{n+\frac{1}{2}}$ and Δt^{n+1} need not be the same but must be chosen according to the usual stability restriction (7.8)–(7.9) for the RCM. As for the case of the non–staggered RCM, one may use the primitive variables to describe the staggered grid RCM. However, for the theme of the next section we assume the vector $\mathbf{U}$ in (7.20)–(7.22) to be the vector of conserved variables.

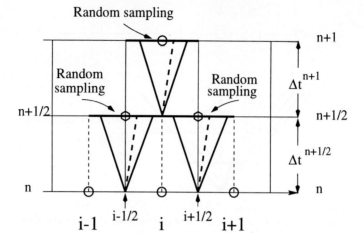

Fig. 7.8. Illustration of the Random Choice Method, on a staggered grid

7.4.2 A Deterministic First–Order Centred Scheme (FORCE)

Here we present a First–Order Centred deterministic scheme (FORCE) [365], [367], that is obtained by replacing the stochastic steps (7.21)–(7.22) by deterministic versions, via integral averages of Riemann problem solutions. We preserve the previous notation and assume

$$\Delta t^{n+\frac{1}{2}} = \Delta t^{n+1} = \frac{1}{2}\Delta t \ .$$

The stochastic integrals (7.21) are replaced by the deterministic integrals

$$\mathbf{U}_{i-\frac{1}{2}}^{n+\frac{1}{2}} = \frac{1}{\Delta x} \int_{-\frac{1}{2}\Delta x}^{\frac{1}{2}\Delta x} \hat{\mathbf{U}}_{i-\frac{1}{2}}^{n+\frac{1}{2}}(x, \frac{\Delta t}{2}) \, dx \tag{7.23}$$

and

$$\mathbf{U}_{i+\frac{1}{2}}^{n+\frac{1}{2}} = \frac{1}{\Delta x} \int_{-\frac{1}{2}\Delta x}^{\frac{1}{2}\Delta x} \hat{\mathbf{U}}_{i+\frac{1}{2}}(x, \frac{\Delta t}{2}) \, dx \ . \tag{7.24}$$

Then we apply the integral form of the conservation laws, see Sect. 2.4.1 of Chap. 2, to expressions (7.23) and (7.24). The result is

$$\mathbf{U}_{i-\frac{1}{2}}^{n+\frac{1}{2}} = \frac{1}{2}\left(\mathbf{U}_{i-1}^{n} + \mathbf{U}_{i}^{n}\right) + \frac{\Delta t}{2\Delta x}\left(\mathbf{F}_{i-1}^{n} - \mathbf{F}_{i}^{n}\right) \ , \tag{7.25}$$

$$\mathbf{U}_{i+\frac{1}{2}}^{n+\frac{1}{2}} = \frac{1}{2}(\mathbf{U}_{i}^{n} + \mathbf{U}_{i+1}^{n}) + \frac{\Delta t}{2\Delta x}\left(\mathbf{F}_{i}^{n} - \mathbf{F}_{i+1}^{n}\right) \ . \tag{7.26}$$

We denote by $\hat{\mathbf{U}}_i(x,t)$ the solution of the Riemann problem $RP(\mathbf{U}_{i-\frac{1}{2}}^{n+\frac{1}{2}}, \mathbf{U}_{i+\frac{1}{2}}^{n+\frac{1}{2}})$ and define an average $\mathbf{U}_i^{n+1}$ at the complete time step Δt in terms of an integral average of $\hat{\mathbf{U}}_i(x,t)$ at the (local) time $t = \frac{1}{2}\Delta t$, namely

$$U_i^{n+1} = \frac{1}{\Delta x} \int_{-\frac{1}{2}\Delta x}^{\frac{1}{2}\Delta x} \hat{U}_i(x, \frac{1}{2}\Delta t) \, dx \; . \tag{7.27}$$

This is the deterministic version of (7.22). Applying the integral form of the conservation laws to the right–hand side of (7.27) gives

$$U_i^{n+1} = \frac{1}{2}\left[U_{i-\frac{1}{2}}^{n+\frac{1}{2}} + U_{i+\frac{1}{2}}^{n+\frac{1}{2}}\right] + \frac{\Delta t}{2\Delta x}\left[F_{i-\frac{1}{2}}^{n+\frac{1}{2}} - F_{i+\frac{1}{2}}^{n+\frac{1}{2}}\right] \; , \tag{7.28}$$

where

$$F_{i+\frac{1}{2}}^{n+\frac{1}{2}} = F(U_{i+\frac{1}{2}}^{n+\frac{1}{2}}) \equiv F_{i+\frac{1}{2}}^{\text{RI}} \tag{7.29}$$

and $F_{i+\frac{1}{2}}^{\text{RI}}$ is the intercell numerical flux for the Richtmyer scheme; see Sect. 5.3.4 of Chap. 5.

Thus the deterministic version of the staggered–grid RCM scheme (7.21)–(7.22) becomes (7.28). The scheme is obviously conservative and when written in conservation form we have

$$U_i^{n+1} = U_i^n + \frac{\Delta t}{\Delta x}\left(F_{i-\frac{1}{2}} - F_{i+\frac{1}{2}}\right) \tag{7.30}$$

with intercell numerical flux

$$F_{i+\frac{1}{2}}^{\text{force}} = \frac{1}{2}\left[F_{i+\frac{1}{2}}^{n+\frac{1}{2}} + \frac{1}{2}\left(F_i^n + F_{i+1}^n\right)\right] + \frac{1}{4}\frac{\Delta x}{\Delta t}\left(U_i^n - U_{i+1}^n\right) \; . \tag{7.31}$$

A surprising outcome is that the intercell flux (7.31) is in fact the arithmetic mean of the fluxes for the Richtmyer and Lax–Friedrichs schemes, namely

$$F_{i+\frac{1}{2}}^{\text{force}} = \frac{1}{2}\left(F_{i+\frac{1}{2}}^{\text{RI}} + F_{i+\frac{1}{2}}^{\text{LF}}\right) \; . \tag{7.32}$$

7.4.3 Analysis of the FORCE Scheme

For the linear advection equation (7.10) the conservative scheme (7.30), (7.32) yields

$$u_i^{n+1} = b_{-1}u_{i-1}^n + b_0 u_i^n + b_1 u_{i+1}^n \; , \tag{7.33}$$

with coefficients given as

$$b_{-1} = \frac{1}{4}(1+c)^2 \; , \; b_0 = \frac{1}{2}(1-c^2) \; , \; b_1 = \frac{1}{4}(1-c)^2 \; . \tag{7.34}$$

Proposition 7.4.1. *The scheme (7.33)–(7.34) is*

− stable, with stability condition

$$0 \leq |c| \leq 1 \; , \tag{7.35}$$

where

$$c = \frac{\Delta t a}{\Delta x} \; : \; \textit{Courant Number} \tag{7.36}$$

- *monotone, and*
- *has modified equation*

$$q_t + a q_x = \alpha_{\mathrm{fo}} q_{xx} \; , \quad \alpha_{\mathrm{fo}} = \frac{1}{4} a \varDelta x \left(\frac{1 - c^2}{c} \right) = \frac{1}{2} \alpha_{\mathrm{lf}} \; , \tag{7.37}$$

where α_{lf} is the coefficient of artificial viscosity for the Lax–Friedrichs scheme.

Proof. The scheme can be shown to be stable with stability condition (7.35) by using standard von Neumann analysis, or more directly by utilising Billett's result, which says [38] that a scheme of the form (7.33)–(7.34) is stable if and only if

1. $b_0 (b_{-1} + b_1) \geq 0$
2. $b_0 (b_{-1} + b_1) + 4 b_{-1} b_1 \equiv B \geq 0.$

The first condition leads to (7.35) directly, while the second condition produces

$$B = \frac{1}{4} \left[\left(1 - c^2 \right) \left(1 + c^2 \right) + \left(1 + c \right)^2 \left(1 - c \right)^2 \right] \; .$$

A sufficient condition for $B \geq 0$ is again $b_0 \geq 0$, which confirms the sought stability restriction (7.35). Concerning monotonicity, by inspection, the scheme is monotone for Courant numbers satisfying the stability condition, i.e. all coefficients are non–negative, see Sect. 5.2.1 of Chap. 5. The result (7.37) is obtained by using standard analysis.

Numerical results of the FORCE scheme for the Euler equations are presented in Sect. 7.6. High–order extensions are presented in Chaps. 13 and 14; these high–order schemes also extend to multi–dimensional problems following the splitting techniques presented in Chap. 16.

7.5 Random Numbers

The quality of the computed RCM solution depends crucially on the random numbers $\{\theta^n\}$. Research in this area has produced some very effective guidelines. For example, it has been established that the more random the generation of $\{\theta^n\}$, the worse the computed RCM results. Colella [85], [86] introduced the use of pseudo–random numbers of the van der Corput type.

7.5.1 Van der Corput Pseudo–Random Numbers

A general van der Corput sequence $\{\theta^n\}$ depends on two parameters k_1, k_2 with $k_1 > k_2 > 0$ both integer and relatively prime. The (k_1, k_2) van der Corput sequence $\{\theta^n\}$ is formally defined [155] as

$$\theta^n = \sum_{i=0}^{m} A_i k_1^{-(i+1)} , \tag{7.38}$$

$$A_i = k_2 a_i (\text{mod } k_1) , \tag{7.39}$$

$$n = \sum_{i=0}^{m} a_i k_1^i . \tag{7.40}$$

To explain the definition of the pseudo–random number θ^n in (7.38) we start from equation (7.40), which gives the non–negative integer n *in scale of notation with radix* k_1. If $k_1 = 2$, (7.40) gives the binary expansion of n. For example, the binary expansion of the integer 3 is

$$3 = 1 \times 2^0 + 1 \times 2^1$$

and $m = 1$.

The next stage is to find the coefficients A_i in (7.38) according to equation (7.39); this says that A_i *is the remainder when dividing the product* $k_2 a_i$ *by* k_1 ($A_i < k_1$). The simplest case is given by $k_2 = 1$, for which $A_i \equiv a_i$, $\forall i$. Having found the number m and the modified coefficients A_i, the random number θ^n corresponding to the integer n is completely determined by the summation (7.38).

Exercise 7.5.1 (Van der Corput sequences). Find m, the coefficients a_i and A_i and the corresponding first 10 random numbers θ^n of the $(2, 1)$ and $(5, 3)$ van der Corput sequences ($n = 1, \ldots, 10$).

Solution 7.5.1. Results are shown in Table 7.1 for the $(2, 1)$ (binary) van der Corput sequence and in Table 7.2 for the $(5, 3)$ van der Corput sequence.

n	m	a_0	a_1	a_2	a_3	A_0	A_1	A_2	A_3	θ^n
1	0	1				1				0.5000
2	1	0	1			0	1			0.2500
3	1	1	1			1	1			0.7500
4	2	0	0	1		0	0	1		0.1250
5	2	1	0	1		1	0	1		0.6250
6	2	0	1	1		0	1	1		0.3750
7	2	1	1	1		1	1	1		0.8750
8	3	0	0	0	1	0	0	0	1	0.0625
9	3	1	0	0	1	1	0	0	1	0.5625
10	3	0	1	0	1	0	1	0	1	0.3125

Table 7.1: Number m, coefficients a_i, A_i and random numbers θ^n for the $(2, 1)$ van der Corput sequence, for $n = 1, \ldots, 10$

7.5.2 Statistical Properties

A desirable statistical property of the sequence of numbers $\{\theta^n\}$ is that $\{\theta^n\}$ be *uniformly distributed* over $[0,1]$. Following Olivier and Grönig [248] we study three statistical quantities that help to characterise the sequence of random numbers $\{\theta^n\}$. These are: the arithmetic mean, the standard deviation and the so called chi–square statistics.

n	m	a_0	a_1	A_0	A_1	θ^n
1	0	1		3		0.6000
2	0	2		1		0.2000
3	0	3		4		0.8000
4	0	4		2		0.4000
5	1	0	1	0	3	0.1200
6	1	1	1	3	3	0.7200
7	1	2	1	1	3	0.3200
8	1	3	1	4	3	0.9200
9	1	4	1	2	3	0.5200
10	1	0	2	0	1	0.0400

Table 7.2: Number m, coefficients a_i, A_i and random numbers θ^n for the $(5,3)$ van der Corput sequence, for $n = 1, \ldots, 10$.

The *arithmetic mean* x_{ar} of the set $\{\theta^n\}_{n=1}^N$ is defined as

$$x_{\mathrm{ar}} = \frac{1}{N} \sum_{n=1}^{N} \theta^n . \tag{7.41}$$

For an optimally equidistributed sequence $\{\theta^n\}$ in $[0,1]$ we expect x_{ar} to be close to $\frac{1}{2}$.

The *standard deviation* is

$$x_{\mathrm{sd}} = \left[\frac{1}{(N-1)} \sum_{n=1}^{N} (\theta^n - x_{\mathrm{ar}})^2 \right]^{\frac{1}{2}} . \tag{7.42}$$

The *chi–square statistic* x_{sq} is computed as follows: the interval $[0,1]$ is subdivided into D equally spaced subintervals R_i. Then we consider a total of N random numbers θ^n and count the number $c(R_i)$ of random numbers θ^n that fall inside R_i, for $i = 1, \ldots, D$. We denote by $p_i = p_i(\theta^n)$ the probability that the number θ^n falls inside the subinterval R_i. The expected number of random numbers in the interval R_i is thus Np_i. As an illustrative example we subdivide $[0,1]$ into $D = 2$ subintervals and choose $N = 4$ random numbers $\{\theta^1, \theta^2, \theta^3, \theta^4\}$. For a uniformly distributed sequence one would expect $p_1(\theta^n) = \frac{1}{2}$ and so the expected number of random numbers θ^n falling into subinterval $R_1 \equiv [0, \frac{1}{2})$ is $Np_1 = 4 \times \frac{1}{2} = 2$. The same holds for subinterval $R_2 \equiv [\frac{1}{2}, 1]$. Then the *chi–square statistic* is defined as

$$x_{sq}^2 = \sum_{i=1}^{N} \frac{(c(R_i) - Np_i)^2}{Np_i} . \tag{7.43}$$

For a uniformly distributed sequence $\{\theta^n\}$ we expect x_{sq} to be small. Better RCM numerical results are obtained with pseudo–random sequences $\{\theta^n\}$ for which x_{sq}^2 is small.

As in [248], we analyse the quantities x_{ar}, x_{sd} and x_{sq} for 6 van der Corput sequences. Table 7.3 shows the statistical properties of $N = 200$ random numbers. For the *chi–square test* the interval $[0, 1]$ was subdivided into $D = 20$ equally spaced subintervals.

(k_1, k_2)	(2,1)	(3,2)	(3,1)	(5,1)	(7,3)	(5,3)
x_{ar}	0.49459	0.49673	0.49457	0.49373	0.49965	0.49598
x_{sd}	0.28885	0.28780	0.28799	0.28914	0.28842	0.28907
x_{sq}^2	1.00000	1.00000	1.00000	1.40000	1.00000	0.20000

Table 7.3: Statistical properties of 6 van de Corput sequences (k_1, k_2)

Our results agree well with those of Olivier and Grönig for the arithmetic mean x_{ar}; for x_{sq}^2 our results are similar but not identical. Our values for the standard deviation x_{sd} are much smaller that those quoted in [248]. Olivier and Grönig also analysed the quantities x_{ar}, x_{sd} and x_{sq} for other sequences of random numbers, including the modified random numbers suggested by Chorin [73]. They observed that in all cases x_{ar} was close to $\frac{1}{2}$; they also observed that all sequences tested had similar values for the standard deviation x_{sd}. The quantity that was different was x_{sq}, which led them to conclude that this was the statistical property of significance. Van der Corput sequences have a small value for x_{sq}, as compared with other sequences, and are known to produce very good computational results when used in the Random Choice Method. Anderson and Gottlieb [6] suggested a sequence of pseudo–random numbers of similar qualities to the van der Corput sequences. See [6] for details. Our tests for the Anderson and Gottlieb numbers give $x_{ar} = 0.4981$, $x_{sd} = 0.2891$ and $x_{sq}^2 = 0.6$. Compare with results of Table 7.3.

7.5.3 Propagation of a Single Shock

Here we consider a single shock wave of positive speed S connecting two constant states $\mathbf{U}_L$ and $\mathbf{U}_R$. We solve the time–dependent Euler equations with initial data

$$\mathbf{U}(x, 0) = \begin{cases} \mathbf{U}_L & \text{if } x < 0 , \\ \mathbf{U}_R & \text{if } x > 0 . \end{cases} \tag{7.44}$$

The exact solution is

$$
\mathbf{U}(x,t) =
\begin{cases}
\mathbf{U}_\mathrm{L} \text{ if } x/t < S , \\[2mm]
\mathbf{U}_\mathrm{R} \text{ if } x/t > S .
\end{cases}
\tag{7.45}
$$

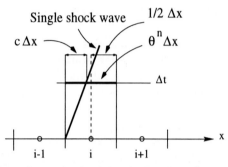

Fig. 7.9. Propagation of a single shock by the Random Choice Method, on a non–staggered grid. Shock propagates by comparing θ^n with c

On applying the Random Choice Method on a non–staggered grid with Courant number $c \le \frac{1}{2}$, based on the shock speed S, the shock wave propagates to the right as a true discontinuity. The situation at any time level is illustrated in Fig. 7.9, where the shock wave is assumed located at the interface $x_{i-\frac{1}{2}}$. The shock wave crosses the line $t = \Delta t$ at $c\Delta x$ and by virtue of the CFL condition this point lies to the left of the middle of cell I_i, namely $\frac{1}{2}\Delta x$. A random position inside cell I_i is given by $\theta^n \Delta x$, with θ^n a random number in the interval $[0,1]$. If $\theta^n \Delta x \le c\Delta x$ then the randomly selected state is the post shock state $\mathbf{U}_\mathrm{L}$, which is then assigned to the whole of the cell I_i for the next time level. This means that the shock moves to the right by a complete distance Δx. If $\theta^n \Delta x > c\Delta x$ then the shock does not move at all. The position X_s of the shock after n time steps is given by

$$
X_s = \sum_{i=1}^{n} \Delta x P_i ,
\tag{7.46}
$$

where

$$
P_i =
\begin{cases}
1 , \text{ if } \theta^i \le c , \\[2mm]
0 , \text{ if } \theta^i > c .
\end{cases}
\tag{7.47}
$$

Table 7.4 shows calculations by hand of the shock position error, normalised by the mesh size Δx, of the RCM solution for 10 time steps. We use two Courant numbers for each of the van der Corput sequences $(2,1)$ and $(5,3)$. For instance, if the RCM is used with the $(2,1)$ van der Corput sequence and a CFL number $c = \frac{1}{4}$, then at the time step $n = 5$ the error in the shock position is $\frac{3}{4}\Delta x$. In the first 10 time steps the largest error takes place at $n = 8$ and is equal to Δx. For the $(2,1)$ van der Corput sequence and a CFL number $c = \frac{1}{2}$ the maximum position error is also Δx. The results for the $(5,3)$ van der Corput sequence are more accurate than those for the $(2,1)$ sequence. For $c = \frac{1}{4}$ the maximum error observed is also $\frac{3}{4}\Delta x$ but there are two time levels at which the solution is exact. For the case $c = \frac{1}{2}$ the maximum position error is $\frac{1}{2}\Delta x$ and the solution is exact every other time step.

n	$(2,1), c=\frac{1}{4}$	$(2,1), c=\frac{1}{2}$	$(5,3), c=\frac{1}{4}$	$(5,3), c=\frac{1}{2}$
1	$\frac{1}{4}$	$\frac{1}{2}$	$\frac{1}{4}$	$\frac{1}{2}$
2	$\frac{1}{2}$	1	$\frac{1}{2}$	0
3	$\frac{1}{4}$	$\frac{1}{2}$	$\frac{1}{4}$	$\frac{1}{2}$
4	1	1	0	0
5	$\frac{3}{4}$	$\frac{1}{2}$	$\frac{3}{4}$	$\frac{1}{2}$
6	$\frac{1}{2}$	1	$\frac{1}{2}$	0
7	$\frac{1}{4}$	$\frac{1}{2}$	$\frac{1}{4}$	$\frac{1}{2}$
8	1	1	0	0
9	$\frac{3}{4}$	$\frac{1}{2}$	$\frac{1}{4}$	$\frac{1}{2}$
10	$\frac{1}{2}$	1	$\frac{1}{2}$	0

Table 7. 4: Position error of shock computed by the RCM. Two van der Corput sequences and two values of the Courant number c are used

Exercise 7.5.2. Verify the results of Table 4 and find the shock position error for the van der Corput sequences $(3,2)$ and $(7,3)$ for Courant numbers $\frac{1}{4}$ and $\frac{1}{3}$.

Solution 7.5.2. (Left to the reader).

7.6 Numerical Results

In this section we show some numerical results for three methods, namely the conventional Random Choice Method on a non–staggered grid associated with the Godunov Method (denoted by RCMG), the Lax–Friedrichs type Random Choice Method (7.18) (denoted by RCMLF) and the First–Order Centred (FORCE) scheme (7.30), (7.32). We solve five test problems with exact solution for the one–dimensional time dependent Euler equations. The initial data consists of two constant states $\mathbf{W}_L = [\rho_L, u_L, p_L]^T$ and $\mathbf{W}_R = [\rho_R, u_R, p_R]^T$. These are given in Table 7.5. For a discussion on the exact solution of these test problems see Sect. 6.4, Chapt. 6. Codes of the library *NUMERICA* [369] were used to obtain the displayed results. The exact solutions were found by running the code *HE-E1RPEXACT*, the numerical solutions using RCMG were obtained by running the code *HE-E1RCM* and the numerical solutions using FORCE were obtained by running the code *HE-E1FOCENT*.

Test	ρ_L	u_L	p_L	ρ_R	u_R	p_R
1	1.0	0.75	1.0	0.125	0.0	0.1
2	1.0	-2.0	0.4	1.0	2.0	0.4
3	1.0	0.0	1000.0	1.0	0.0	0.01
4	5.99924	19.5975	460.894	5.99242	-6.19633	46.0950
5	1.0	-19.59745	1000.0	1.0	-19.59745	0.01

Table 7.5. Data for five test problems with exact solution.

We discretise the domain $[0, 1]$ into $M = 100$ computing cells. For the random choice methods we use a Courant number coefficient $C_{cfl} = 0.45$ and for the FORCE method we use $C_{cfl} = 0.9$.

Figs. 7.10 to 7.13 show results for RCMG applied to Tests 1 to 4. The corresponding results for the RCMLF are shown in Figs. 7.14 to 7.17 and those for the FORCE scheme are shown in Figs. 7.18 to 7.22. All these results are to be compared with those of Chap. 6.

The RCMG results of Fig. 7.10 are, by any standards, very accurate. Jump discontinuities such as shock waves and contacts are resolved as true discontinuities. Also, discontinuities in derivative, such as those along the head and tail of rarefaction waves, are also very well resolved. The complexity of the RCM is comparable to that of the Godunov method, and thus it is fair to compare Fig. 7.10 with Fig. 6.8 of Chap. 6. The entropy glitch inside the sonic rarefaction for the RCM result is smaller than that for the Godunov method. The RCM results for Test 2 are shown in Fig. 7.11; this problem does not contain jump discontinuities and exposes a weakness of the RCM, namely the random noise in smooth parts of the flow. The results of Figs. 7.12 and 7.13 exhibit the true merits of the RCM for computing solutions containing multiple jump discontinuities; compare with the Godunov results of Figs. 6.10 and 6.11.

The computational results for the Lax–Friedrichs type random choice method (RCMLF) are shown in Figs. 7.14 to 7.17. These are somewhat inferior to those of the conventional RCM. However RCMLF has the advantage of having twice the stability range of the conventional RCM. Notice the pairing of neighbouring states, which is also a feature of the conventional Lax–Friedrichs scheme; compare Fig. 7.16 with Fig. 6.12.

Figs. 7.18 to 7.22 show the results for the First–Order Centred scheme FORCE as applied to Tests 1 to 5. Compare with the corresponding results for the Godunov and Lax–Friedrichs methods in Chap. 6. In general, the results for the FORCE scheme look inferior to those of the Godunov method, particularly for Test 5; the former scheme is however significantly simpler and more efficient than the latter. The numerical results of the FORCE scheme are superior to those of the Lax–Friedrichs scheme (see results of Chap. 6), with both schemes being comparable in complexity and efficiency.

7.7 Concluding Remarks

We have presented random choice and related methods to solve time–dependent one dimensional hyperbolic conservation laws. Details have been given for the Euler equations, for which numerical results have been presented. It is well known that the conventional RCM, on staggered and non–staggered grids, is only directly applicable to hyperbolic systems in *two independent variables*. For the time–dependent one dimensional Euler equations these are time t and space x. For the steady supersonic Euler equations the two independent variables are the flow direction x, which is a time–like variable, and space y.

The great merit of the RCM is its ability to resolve discontinuities with infinite resolution, as true discontinuities. The main disadvantage of the RCM is the randomness in the smooth parts of the flow. One way of eliminating this unwanted randomness is by resorting to hybrid approaches, whereby the RCM is used at discontinuities only; elsewhere in the flow one may use some other scheme, see Toro and Roe [378], [379] for instance. A crucial question is this: *can the RCM be extended to in–homogeneous (sources) systems or to systems with more that two independent variables ?*

By applying splitting schemes, see Chap. 15, random choice methods can be extended to solve in–homogeneous systems

$$\mathbf{U}_t + \mathbf{F}(\mathbf{U})_x = \mathbf{S}(\mathbf{U}) \ . \tag{7.48}$$

The *source term vector* $\mathbf{S}(\mathbf{U})$ may be an algebraic function of the flow variables, such as in cylindrically or spherically symmetric flow [318]. $\mathbf{S}(\mathbf{U})$ may also involve higher order spatial derivatives such as in viscous terms; this means that the RCM can be applied to parabolic equations, e.g. the time–dependent, one dimensional Navier–Stokes equations [174]. The inclusion of algebraic source terms retains the infinite resolution of discontinuities but may enhance the randomness in smooth parts of the flow, present in the homogeneous part of the problem. In some special cases, see Glimm et. al. [143], the source term vector $\mathbf{S}(\mathbf{U})$ may be incorporated into the solution of the Riemann problem. This significantly alleviates the problem of enhanced randomness.

Efforts to extend the RCM, retaining its distinctive feature, to solve problems in more that two independent variables, such as the time–dependent, two dimensional Euler equations and the three dimensional steady supersonic Euler equations, have so far proved unsuccessful. For details on how the use of splitting techniques to carry out the extensions fail, were reported by Colella [85]. If resolving shocks with infinite resolution is abandoned then splitting techniques applied to the RCM work reasonably well. Colella [85] introduced artificial viscosity, as in conventional finite different methods. Toro [353] proposed a hybrid approach to extend the RCM to solve the time–dependent, two dimensional Euler equations (three independent variables); here the RCM is used at contacts and shear waves and a high–resolution shock capturing method is used elsewhere. This approach gives up the infinite resolution of shocks but retains infinite resolution of shear waves and material interfaces, features that are exceedingly difficult to resolve by most methods. Recently, Loh et.al. [224] have reported their work to extend the RCM to solve the three–dimensional steady supersonic Euler equations (three independent variables). They also give up infinite resolution of shocks but retain infinite resolution of slip surfaces by virtue of a Lagrangian approach.

At the present time the RCM offers an accurate numerical method for solving non–linear systems of the form (7.48) in conjunction with operator splitting techniques, as presented in Chap. 15.

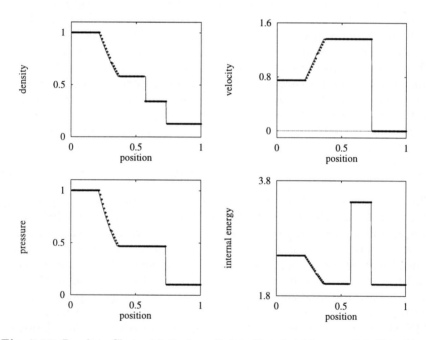

Fig. 7.10. Random Choice Method applied to Test 1, with $x_0 = 0.3$. Numerical (symbol) and exact (line) solutions are compared at the output time 0.2 units

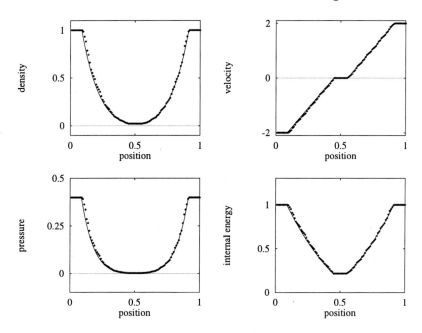

Fig. 7.11. Random Choice Method applied to Test 2, with $x_0 = 0.5$. Numerical (symbol) and exact (line) solutions are compared at the output time 0.15 units

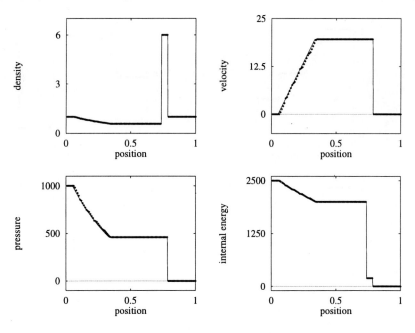

Fig. 7.12. Random Choice Method applied to Test 3, with $x_0 = 0.5$. Numerical (symbol) and exact (line) solutions are compared at the output time 0.012 units

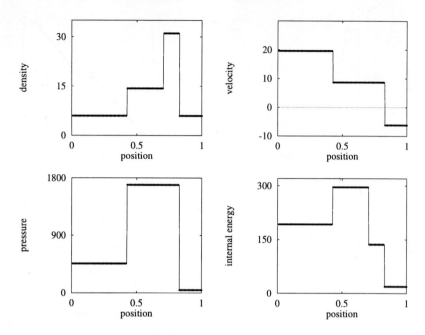

Fig. 7.13. Random Choice Method applied to Test 4, with $x_0 = 0.4$. Numerical (symbol) and exact (line) solutions are compared at the output time 0.035 units

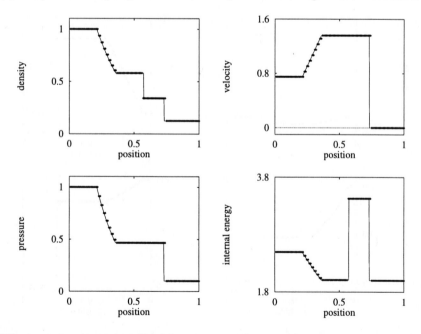

Fig. 7.14. Lax-Friedrichs Random Choice scheme applied to Test 1, with $x_0 = 0.3$. Numerical (symbol) and exact (line) solutions are compared at time 0.2 units

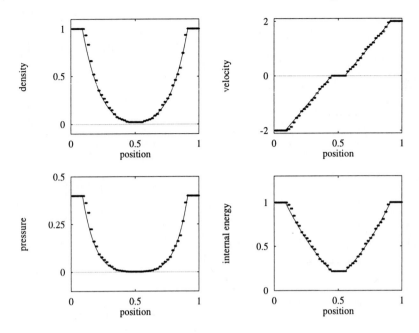

Fig. 7.15. Lax-Friedrichs Random Choice scheme applied to Test 2, with $x_0 = 0.5$. Numerical (symbol) and exact (line) solutions are compared at time 0.15 units

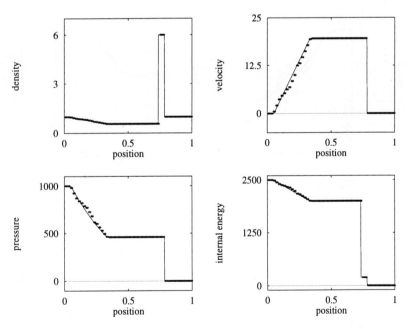

Fig. 7.16. Lax-Friedrichs Random Choice scheme applied to Test 3, with $x_0 = 0.5$. Numerical (symbol) and exact (line) solutions are compared at time 0.012 units

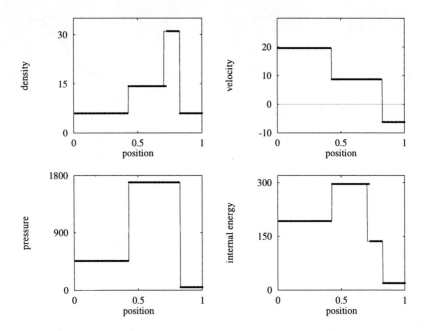

Fig. 7.17. Lax-Friedrichs Random Choice scheme applied to Test 4, with $x_0 = 0.4$. Numerical (symbol) and exact (line) solutions are compared at time 0.035 units

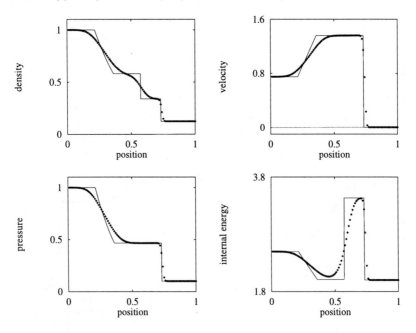

Fig. 7.18. FORCE scheme applied to Test 1, with $x_0 = 0.3$. Numerical (symbol) and exact (line) solutions are compared at the output time 0.2 units

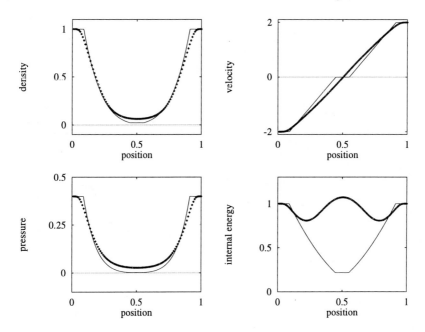

Fig. 7.19. FORCE scheme applied to Test 2, with $x_0 = 0.5$. Numerical (symbol) and exact (line) solutions are compared at the output time 0.15 units

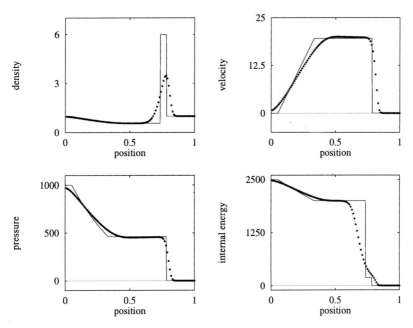

Fig. 7.20. FORCE scheme applied to Test 3, with $x_0 = 0.5$. Numerical (symbol) and exact (line) solutions are compared at the output time 0.012 units

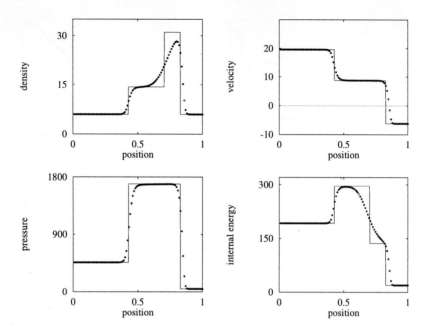

Fig. 7.21. FORCE scheme applied to Test 4, with $x_0 = 0.4$. Numerical (symbol) and exact (line) solutions are compared at the output time 0.035 units

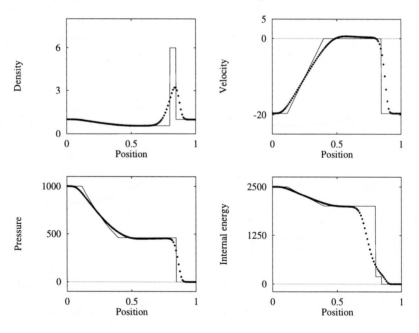

Fig. 7.22. FORCE scheme applied to Test 5, with $x_0 = 0.8$. Numerical (symbol) and exact (line) solutions are compared at the output time 0.012 units

8. Flux Vector Splitting Methods

8.1 Introduction

A distinguishing feature of upwind numerical methods is this: *the discretisation of the equations on a mesh is performed according to the direction of propagation of information on that mesh*. In this way, salient features of the physical phenomena modelled by the equations are incorporated into the discretisation schemes. There are essentially two approaches for identifying upwind directions, namely the *Godunov approach* [145] studied in Chap. 6, and the *Flux Vector Splitting* (FVS) approach [295], [326], [393], [394] to be studied in this chapter. These two approaches are often referred to as the *Riemann approach* and the *Boltzmann approach* [164]. The respective numerical methods derived from these two approaches are often referred to as *Flux Difference Splitting Methods* and *Flux Vector Splitting Methods* . For a review on both of these approaches the paper by Harten, Lax and van Leer [164] is highly recommended. Closely related schemes to FVS, not studied here, are the KFVS or kinetic schemes, see for example Pullin [266], Perthame [261], [262], Mandal and Desphande [227], Xu and Prendergast [416], Xu et. al. [415], Xu [414] and Yang *et. al* [420].

The identification of upwind directions in Flux Vector Splitting Methods is achieved with less effort than in Godunov–type methods, leading to simpler and somewhat more efficient schemes. These two features are very attractive and have made FVS schemes very popular within a large community of practitioners. The Flux Vector Splitting approach is particularly well suited for *implicit methods*; these are popular in Aerodynamics, where the computation of steady solutions is of great practical value. The reduced sophistication of FVS schemes however, as compared with Godunov–type schemes, results in poorer resolution of discontinuities, particularly stationary contact and shear waves. In applications to the Navier–Stokes equations, it is reported by van Leer, Thomas and Roe [397] that their FVS scheme is considerably less accurate than Godunov's method with Roe's approximate Riemann solver [281]. A key feature of the FVS approach is its reliance on a special property of the equations, namely the *homogeneity property*. As seen in Sect. 3.1.1 of Chap. 3, the Euler equations satisfy this property but there are important examples, such as the shallow water equations, that do not. The homogeneity property may however be circumvented so as to be able to apply the FVS approach, see Vázquez–Cendón [400].

The pioneering works of Sanders and Prendergast [295], Steger and Warming [326] and van Leer [393], [394] has been followed by numerous applications as well as by increased research efforts to improve further the technique. See for example the papers [10], [11], [112], [221], [408] and [265], amongst many others.

The purpose of this chapter is to give an elementary introduction to Flux Vector Splitting methods. Sects. 8.2 and 8.3 are devoted to a simple introduction to the FVS approach. In Sect. 8.4 we derive FVS methods for the time–dependent Euler equations following the methodologies of Steger and Warming [326], that of van Leer [393], [394] and the recently proposed approach of Liou and Steffen [221]. Numerical results are presented in Sect. 8.5. Techniques to construct high–order schemes based on FVS are found in Chaps. 13 and 14. In Chap. 15 we show how to solve systems with source terms and in Chap. 16 we deal with approaches to construct multidimensional schemes. Essential background material for reading this chapter is found in Chaps. 2, 3 and 5.

8.2 The Flux Vector Splitting Approach

In this section we introduce the flux vector splitting approach in the simple setting of model hyperbolic systems, namely the small perturbation steady supersonic equations and the isothermal equations of Gas Dynamics; see Sect. 1.6.2 of Chap. 1 and Sects. 2.1 and 2.4.1 of Chap. 2 for details on these systems.

8.2.1 Upwind Differencing

Consider the small perturbation steady supersonic equations

$$u_x - a^2 v_y = 0 \ , \ v_x - u_y = 0 \ , \tag{8.1}$$

where $u = u(x,y)$, $v = v(x,y)$,

$$a = \sqrt{\frac{1}{M_\infty^2 - 1}} \tag{8.2}$$

is the sound speed and M_∞ is the free–stream Mach number, assumed to be greater than unity. Equations (8.1) may be rewritten as

$$\mathbf{U}_x + \mathbf{A}\mathbf{U}_y = \mathbf{0} \ , \tag{8.3}$$

with

$$\mathbf{U} = \begin{bmatrix} u \\ v \end{bmatrix} , \quad \mathbf{A} = \begin{bmatrix} 0 & -a^2 \\ -1 & 0 \end{bmatrix} . \tag{8.4}$$

The eigenvalues of the coefficient matrix $\mathbf{A}$ are

$$\lambda_1 = -a , \quad \lambda_2 = +a , \tag{8.5}$$

with corresponding right eigenvectors

$$\mathbf{K}^{(1)} = \begin{bmatrix} a \\ 1 \end{bmatrix} , \quad \mathbf{K}^{(2)} = \begin{bmatrix} a \\ -1 \end{bmatrix} . \tag{8.6}$$

Given the mixed character of the eigenvalues ($\lambda_1 = -a$ is negative and $\lambda_2 = +a$ is positive), a finite difference discretisation of (8.3) has limited choices for the spatial derivative, if upwind bias is to be applied. Consider, for instance, the *one–sided difference schemes*

$$\mathbf{U}_i^{n+1} = \mathbf{U}_i^n - \frac{\Delta x}{\Delta y} \mathbf{A}[\mathbf{U}_i^n - \mathbf{U}_{i-1}^n] , \tag{8.7}$$

$$\mathbf{U}_i^{n+1} = \mathbf{U}_i^n - \frac{\Delta x}{\Delta y} \mathbf{A}[\mathbf{U}_{i+1}^n - \mathbf{U}_i^n] . \tag{8.8}$$

Clearly scheme (8.7) is upwind relative to the eigenvalue $\lambda_2 = a > 0$ but is downwind, and thus unstable, relative to the eigenvalue $\lambda_1 = -a < 0$. A similar observation applies to scheme (8.8). For the case in which all eigenvalues have the same sign the difficulty of choosing the upwind direction does not arise.

As seen in Sect. 5.4 of Chap. 5, general linear hyperbolic systems with constant coefficients may be solved by the CIR first–order upwind method by decomposing the coefficient matrix $\mathbf{A}$ into a positive component $\mathbf{A}^+$ and a negative component $\mathbf{A}^-$, such that

$$\mathbf{A} = \mathbf{A}^+ + \mathbf{A}^- , \tag{8.9}$$

where $\mathbf{A}^+$ has positive or zero eigenvalues and $\mathbf{A}^-$ has negative or zero eigenvalues. One then has the upwind scheme

$$\mathbf{U}_i^{n+1} = \mathbf{U}_i^n - \frac{\Delta x}{\Delta y} \mathbf{A}^+[\mathbf{U}_i^n - \mathbf{U}_{i-1}^n] - \frac{\Delta x}{\Delta y} \mathbf{A}^-[\mathbf{U}_{i+1}^n - \mathbf{U}_i^n] . \tag{8.10}$$

The *Split–Coefficient Matrix Scheme* of Chakravarthy et. al. [62], [170] is an extension of this procedure to non–linear systems, in non–conservative form.

The CIR upwind scheme, when applied to general linear hyperbolic systems with constant coefficients, may be written in conservative form by defining the flux vector

$$\mathbf{F} = \mathbf{A}\mathbf{U} . \tag{8.11}$$

Then the splitting (8.9) of the coefficient matrix $\mathbf{A}$ results in a natural splitting of the flux vector $\mathbf{F}$, namely

$$\mathbf{F} = \mathbf{F}^+ + \mathbf{F}^- . \tag{8.12}$$

In this way the CIR upwind scheme can be written in conservative form

$$\mathbf{U}_i^{n+1} = \mathbf{U}_i^n - \frac{\Delta t}{\Delta x}[\mathbf{F}_{i+\frac{1}{2}} - \mathbf{F}_{i-\frac{1}{2}}] , \tag{8.13}$$

where the intercell numerical flux

$$\mathbf{F}_{i+\frac{1}{2}} = \mathbf{F}_i^+(\mathbf{U}_i^n) + \mathbf{F}_i^-(\mathbf{U}_{i+1}^n) \tag{8.14}$$

is identical to the Godunov intercell flux. See Sect. 5.4 of Chap. 5 for details. The Flux Vector Splitting Method is a generalisation of this to non–linear systems in conservation form.

8.2.2 The FVS Approach

Here we consider a general system of m non–linear hyperbolic conservation laws

$$\mathbf{U}_t + \mathbf{F}(\mathbf{U})_x = \mathbf{0} . \tag{8.15}$$

From the assumption of hyperbolicity the Jacobian matrix

$$\mathbf{A}(\mathbf{U}) = \frac{\partial \mathbf{F}}{\partial \mathbf{U}} \tag{8.16}$$

may be expressed as

$$\mathbf{A} = \mathbf{K}\boldsymbol{\Lambda}\mathbf{K}^{-1} , \tag{8.17}$$

where $\boldsymbol{\Lambda}$ is the diagonal matrix formed by the eigenvalues of $\mathbf{A}$, namely

$$\boldsymbol{\Lambda} = \begin{bmatrix} \lambda_1 & & 0 \\ & \ddots & \\ 0 & & \lambda_m \end{bmatrix} . \tag{8.18}$$

The matrix $\mathbf{K}$ is

$$\mathbf{K} = [\mathbf{K}^{(1)}, \mathbf{K}^{(2)}, \dots, \mathbf{K}^{(m)}] , \tag{8.19}$$

where the column $\mathbf{K}^{(i)}$ is the right eigenvector of $\mathbf{A}$ corresponding to λ_i and $\mathbf{K}^{-1}$ is the inverse of $\mathbf{K}$. Recall our usual convention of ordering the eigenvalues in *increasing* order.

As anticipated in the previous section, the Flux Vector Splitting method aims at generalising (8.14) to non–linear systems (8.15). That is, FVS requires a splitting of the flux vector $\mathbf{F}$ into two component $\mathbf{F}^+$ and $\mathbf{F}^-$ such that

$$\mathbf{F}(\mathbf{U}) = \mathbf{F}^+(\mathbf{U}) + \mathbf{F}^-(\mathbf{U}) , \tag{8.20}$$

under the restriction that the eigenvalues $\hat{\lambda}_i^+$ and $\hat{\lambda}_i^-$ of the Jacobian matrices

$$\hat{\mathbf{A}}^+ = \frac{\partial \mathbf{F}^+}{\partial \mathbf{U}} , \quad \hat{\mathbf{A}}^- = \frac{\partial \mathbf{F}^-}{\partial \mathbf{U}} \tag{8.21}$$

satisfy the condition

$$\hat{\lambda}_i^+ \geq 0 , \quad \hat{\lambda}_i^- \leq 0 . \tag{8.22}$$

The splitting is also required to reproduce *regular upwinding* when all eigenvalues λ_i of the coefficient matrix $\mathbf{A}$ are one–sided, that is, all positive or zero, or all negative or zero. That is to say

$$\left.\begin{array}{l} \mathbf{F}^+ = \mathbf{F} , \quad \mathbf{F}^- = \mathbf{0} \quad \text{if } \lambda_i \geq 0 \quad \text{for } i = 1,\dots,m , \\[2mm] \mathbf{F}^+ = \mathbf{0} , \quad \mathbf{F}^- = \mathbf{F} \quad \text{if } \lambda_i \leq 0 \quad \text{for } i = 1,\dots,m . \end{array}\right\} \tag{8.23}$$

If in addition to hyperbolicity, the system (8.15) satisfies the *homogeneity property*

$$\mathbf{F}(\mathbf{U}) = \mathbf{A}(\mathbf{U})\mathbf{U} , \tag{8.24}$$

just as in the linear constant coefficient case, then the sought splitting is easily accomplished by identifying a suitable splitting of the Jacobian matrix $\mathbf{A}$. As seen in Sect. 3.1.1 of Chap. 3, the time–dependent Euler equations satisfy the *homogeneity property*.

From the diagonalisation of $\mathbf{A}$ given by (8.17), a splitting of $\mathbf{A}$ may be accomplished by an appropriate splitting of the diagonal matrix $\mathbf{\Lambda}$. This in turn, may be split by identifying a splitting of the eigenvalues $\lambda_i,\, i = 1,\dots,m$ of $\mathbf{A}$. Suppose we may split the eigenvalues λ_i as

$$\lambda_i = \lambda_i^+ + \lambda_i^- , \tag{8.25}$$

such that $\lambda_i^+ \geq 0$ and $\lambda_i^- \leq 0$. Then $\mathbf{\Lambda}$ may be split as

$$\mathbf{\Lambda} = \mathbf{\Lambda}^+ + \mathbf{\Lambda}^- , \tag{8.26}$$

where

$$\mathbf{\Lambda}^+ = \begin{bmatrix} \lambda_1^+ & & 0 \\ & \ddots & \\ 0 & & \lambda_m^+ \end{bmatrix} , \quad \mathbf{\Lambda}^- = \begin{bmatrix} \lambda_1^- & & 0 \\ & \ddots & \\ 0 & & \lambda_m^- \end{bmatrix} . \tag{8.27}$$

A natural splitting of $\mathbf{A}$ results, namely

$$\mathbf{A} = \mathbf{A}^+ + \mathbf{A}^- , \tag{8.28}$$

with

$$\mathbf{A}^+ = \mathbf{K}\mathbf{\Lambda}^+\mathbf{K}^{-1} , \quad \mathbf{A}^- = \mathbf{K}\mathbf{\Lambda}^-\mathbf{K}^{-1} . \tag{8.29}$$

Then, if (8.24) is satisfied, we can split $\mathbf{F}(\mathbf{U})$ as

$$\mathbf{F} = \mathbf{F}^+ + \mathbf{F}^- , \tag{8.30}$$

where

$$\mathbf{F}^+ = \mathbf{A}^+\mathbf{U} , \quad \mathbf{F}^- = \mathbf{A}^-\mathbf{U} . \tag{8.31}$$

Steger and Warming [326] proposed a splitting of the eigenvalues λ_i as in (8.25) with definitions

$$\lambda_i^+ = \frac{1}{2}(\lambda_i + |\lambda_i|) , \quad \lambda_i^- = \frac{1}{2}(\lambda_i - |\lambda_i|) , \qquad (8.32)$$

where $|\lambda_i|$ is the absolute value of λ_i namely,

$$|\lambda_i| = \begin{cases} \lambda_i & if \quad \lambda_i \geq 0 , \\ -\lambda_i & if \quad \lambda_i \leq 0 . \end{cases} \qquad (8.33)$$

Clearly

$$\lambda_i^+ \geq 0 , \quad \lambda_i^- \leq 0 , \quad for \ i = 1, \ldots, m . \qquad (8.34)$$

Exercise 8.2.1. Verify that the following properties are satisfied

$$\left. \begin{array}{llll} \lambda_i &=& \lambda_i^+ + \lambda_i^- & ; & |\lambda_i| &=& \lambda_i^+ - \lambda_i^- , \\ \mathbf{\Lambda} &=& \mathbf{\Lambda}^+ + \mathbf{\Lambda}^- & ; & |\mathbf{\Lambda}| &=& \mathbf{\Lambda}^+ - \mathbf{\Lambda}^- , \\ \mathbf{A} &=& \mathbf{A}^+ + \mathbf{A}^- & ; & |\mathbf{A}| &=& \mathbf{A}^+ - \mathbf{A}^- . \end{array} \right\} \qquad (8.35)$$

Solution 8.2.1. (Left to the reader).

8.3 FVS for the Isothermal Equations

In order to illustrate the FVS approach we consider the isothermal equations of Gas Dynamics

$$\mathbf{U}_t + \mathbf{F}(\mathbf{U})_x = \mathbf{0} , \qquad (8.36)$$

$$\mathbf{U} = \begin{bmatrix} \rho \\ \rho u \end{bmatrix} , \quad \mathbf{F}(\mathbf{U}) = \begin{bmatrix} \rho u \\ \rho u^2 + \rho a^2 \end{bmatrix} , \qquad (8.37)$$

where the sound speed a is a positive constant. For details on the eigenstructure of this system see Sect. 2.4 of Chap. 2. The Jacobian matrix is

$$\mathbf{A} = \frac{\partial \mathbf{F}}{\partial \mathbf{U}} = \begin{bmatrix} 0 & 1 \\ a^2 - u^2 & 2u \end{bmatrix} . \qquad (8.38)$$

The eigenvalues of $\mathbf{A}$ are

$$\lambda_1 = u - a , \quad \lambda_2 = u + a \qquad (8.39)$$

and the matrix $\mathbf{K}$ of corresponding right eigenvectors is

$$\mathbf{K} = \begin{bmatrix} 1 & 1 \\ u - a & u + a \end{bmatrix} . \qquad (8.40)$$

Exercise 8.3.1. Verify that system (8.36)–(8.37) satisfy the *homogeneity property* (8.24).

Solution 8.3.1. (Left to the reader).

8.3.1 Split Fluxes

Given any splitting (8.25) with

$$\boldsymbol{\Lambda}^+ = \begin{bmatrix} \lambda_1^+ & 0 \\ 0 & \lambda_2^+ \end{bmatrix} , \quad \boldsymbol{\Lambda}^- = \begin{bmatrix} \lambda_1^- & 0 \\ 0 & \lambda_2^- \end{bmatrix} , \tag{8.41}$$

we require the computation of the matrices $\mathbf{A}^+$ and $\mathbf{A}^-$ as given by (8.29). One then requires the determination of the inverse $\mathbf{K}^{-1}$ of the matrix $\mathbf{K}$, the products of three matrices as in (8.29) and finally the products (8.31) to find the flux components. For large systems this may be a rather tedious algebraic task. For the isothermal equations we have

$$\mathbf{K}^{-1} = \frac{1}{2a} \begin{bmatrix} u + a & -1 \\ a - u & 1 \end{bmatrix} . \tag{8.42}$$

Now, given any of the two components (8.27) of $\boldsymbol{\Lambda}$, $\mathbf{A}^\alpha$, say, we compute

$$\mathbf{A}^\alpha = \mathbf{K}\boldsymbol{\Lambda}^\alpha\mathbf{K}^{-1} .$$

The result is

$$\mathbf{A}^\alpha = \frac{1}{2a} \begin{bmatrix} \lambda_1^\alpha(u+a) - \lambda_2^\alpha(u-a) & \lambda_2^\alpha - \lambda_1^\alpha \\ (u^2 - a^2)(\lambda_1^\alpha - \lambda_2^\alpha) & \lambda_2^\alpha(u+a) - \lambda_1^\alpha(u-a) \end{bmatrix} . \tag{8.43}$$

Application of (8.31) gives the flux vector component

$$\mathbf{F}^\alpha = \mathbf{A}^\alpha\mathbf{U} ,$$

that is

$$\mathbf{F}^\alpha = \frac{\rho}{2} \begin{bmatrix} \lambda_1^\alpha + \lambda_2^\alpha \\ \lambda_1^\alpha(u-a) + \lambda_2^\alpha(u+a) \end{bmatrix} . \tag{8.44}$$

Note that the expression for the component $\mathbf{F}^\alpha$ given by (8.44) is general. For $\alpha = +$ and $\alpha = -$ the flux components $\mathbf{F}^+$ and $\mathbf{F}^-$ are

$$\mathbf{F}^+ = \frac{\rho}{2} \begin{bmatrix} \lambda_1^+ + \lambda_2^+ \\ \lambda_1^+(u-a) + \lambda_2^+(u+a) \end{bmatrix} , \tag{8.45}$$

and

$$\mathbf{F}^- = \frac{\rho}{2} \begin{bmatrix} \lambda_1^- + \lambda_2^- \\ \lambda_1^-(u-a) + \lambda_2^-(u+a) \end{bmatrix} . \tag{8.46}$$

Exercise 8.3.2. For the split fluxes (8.45)–(8.46), for the case of subsonic flow,

- (i) Find the Jacobian matrices

$$\hat{\mathbf{A}}^+ = \frac{\partial \mathbf{F}^+}{\partial \mathbf{U}} , \quad \hat{\mathbf{A}}^- = \frac{\partial \mathbf{F}^-}{\partial \mathbf{U}} .$$

- (ii) Find the eigenvalues $\hat{\lambda}_i^+$ and $\hat{\lambda}_i^-$.

Solution 8.3.2. For the positive flux component $\mathbf{F}^+$ the Jacobian matrix is

$$\hat{\mathbf{A}}^+ = \frac{\partial \mathbf{F}^+}{\partial \mathbf{U}} = \begin{bmatrix} \frac{1}{2}a & \frac{1}{2} \\ \frac{1}{2}(a^2 - u^2) & u + a \end{bmatrix} .$$

The eigenvalues are the roots of the characteristic polynomial

$$\lambda^2 - (\frac{3}{2}a + u)\lambda + \frac{1}{4}(u + a)^2 = 0 ,$$

namely,

$$\hat{\lambda}_1^+ = \frac{1}{4}a \left[2M + 3 - \sqrt{4M + 5}\right] , \quad \hat{\lambda}_2^+ = \frac{1}{4}a \left[2M + 3 + \sqrt{4M + 5}\right] .$$

Remark 8.3.1. Note that

$$\hat{\mathbf{A}}^+ \neq \mathbf{A}^+$$

and that

$$\hat{\lambda}_i^+ \neq \lambda_i^+ .$$

Note also that $\hat{\lambda}_i^+ > 0$, that is, none of the eigenvalues vanish. Numerically, this particular property is not desirable, and which unfortunately also carries over to the Euler equations. As we shall see in the next section, there are other splitting schemes that remove this difficulty.

8.3.2 FVS Numerical Schemes

The FVS approach can be used to solve (8.36) using the *explicit* conservative scheme

$$\mathbf{U}_i^{n+1} = \mathbf{U}_i^n - \frac{\Delta t}{\Delta x}[\mathbf{F}_{i+\frac{1}{2}} - \mathbf{F}_{i-\frac{1}{2}}] , \tag{8.47}$$

where the FVS numerical flux is given by

$$\mathbf{F}_{i+\frac{1}{2}} = \mathbf{F}_i^+(\mathbf{U}_i^n) + \mathbf{F}_{i+1}^-(\mathbf{U}_{i+1}^n) . \tag{8.48}$$

Fig. 8.1 provides a physical interpretation of (8.48). The intercell numerical flux $\mathbf{F}_{i+\frac{1}{2}}$ is made out from two contributions; one comes from the *forward* component $\mathbf{F}_i^+$ in the *left cell* I_i and the other comes from the *backward* component $\mathbf{F}_{i+1}^-$ in the *right cell* I_{i+1}.

The Steger and Warming [326] splitting (8.32) in a computational set up is as follows: we consider a computing cell I_i at time level n, where $\mathbf{U}_i^n$ is the vector of conserved variables and $\mathbf{F}_i^n \equiv \mathbf{F}(\mathbf{U}_i^n)$ is the vector of fluxes. The three cases to consider are illustrated in Fig. 8.2 and are

Case (a) Left supersonic flow: $\lambda_2 = u_i^n + a_i^n \leq 0$. Fig. 8.2a illustrates the situation in a cell I_i at time level n. Clearly

$$\left. \begin{array}{ll} \lambda_1^+ = 0 \quad, & \lambda_1^- = \lambda_1 = u_i^n - a_i^n , \\ \lambda_2^+ = 0 \quad, & \lambda_2^- = \lambda_2 = u_i^n + a_i^n , \\ \mathbf{F}_i^+ = \mathbf{0} \quad, & \mathbf{F}_i^- = \mathbf{F}_i^n . \end{array} \right\} \tag{8.49}$$

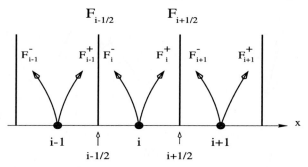

$F_{i-1/2}$ $F_{i+1/2}$

F_{i-1}^{-} F_{i-1}^{+} F_{i}^{-} F_{i}^{+} F_{i+1}^{-} F_{i+1}^{+}

x

i-1   i   i+1

i-1/2   i+1/2

Fig. 8.1. Splitting of the flux function *within* each computing cell I_i at time level n

–   Case (b) Right supersonic flow: $\lambda_1 = u_i^n - a_i^n \geq 0$. See Fig. 8.2b. Obviously

$$
\left.
\begin{array}{lll}
\lambda_1^+ = \lambda_1 = u_i^n - a_i^n & , & \lambda_1^- = 0 , \\
\lambda_2^+ = \lambda_2 = u_i^n + a_i^n & , & \lambda_2^- = 0 , \\
\mathbf{F}_i^+ = \mathbf{F}_i^n & , & \mathbf{F}_i^- = \mathbf{0} .
\end{array}
\right\}
\tag{8.50}
$$

–   Case (c) Subsonic flow: $\lambda_1 = u_i^n - a_i^n \leq 0 \leq \lambda_2 = u_i^n + a_i^n$. See Fig. 8.2c. Evidently

$$
\left.
\begin{array}{lll}
\lambda_1^+ = 0 & , & \lambda_1^- = \lambda_1 = u_i^n - a_i^n , \\
\lambda_2^+ = \lambda_2 = u_i^n + a_i^n & , & \lambda_2^- = 0 .
\end{array}
\right\}
\tag{8.51}
$$

According to (8.45)–(8.46) the fluxes $\mathbf{F}_i^+$ and $\mathbf{F}_i^-$ for the subsonic case are given by

$$
\mathbf{F}_i^+ = \frac{\rho_i^n}{2} \left[\begin{array}{c} u_i^n + a_i^n \\ (u_i^n + a_i^n)^2 \end{array} \right] \;, \quad \mathbf{F}_i^- = \frac{\rho_i^n}{2} \left[\begin{array}{c} u_i^n - a_i^n \\ (u_i^n - a_i^n)^2 \end{array} \right] .
\tag{8.52}
$$

8.4 FVS Applied to the Euler Equations

Here we present three Flux Vector Splitting schemes applied to the time dependent Euler equations.

8.4.1 Recalling the Equations

The one–dimensional Euler Equations in conservation–law form are given by

$$
\mathbf{U}_t + \mathbf{F}(\mathbf{U})_x = \mathbf{0} ,
\tag{8.53}
$$

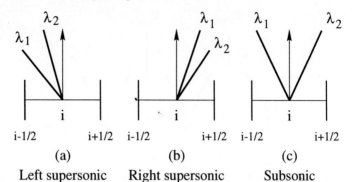

Fig. 8.2. Possible flow patterns in cell I_i at time n: (a) supersonic flow to the left (b) supersonic flow to the right (c) subsonic flow

$$\mathbf{U} = \begin{bmatrix} \rho \\ \rho u \\ E \end{bmatrix}, \quad \mathbf{F(U)} = \begin{bmatrix} \rho u \\ \rho u^2 + p \\ u(E + p) \end{bmatrix}. \tag{8.54}$$

As seen in Sect. 3.1.1 of Chap. 3, the Jacobian matrix $\mathbf{A}$ is given by

$$\mathbf{A} = \begin{bmatrix} 0 & 1 & 0 \\ \frac{1}{2}(\gamma - 3)u^2 & (3 - \gamma)u & \gamma - 1 \\ \frac{1}{2}(\gamma - 2)u^3 - \frac{a^2 u}{\gamma - 1} & \frac{3 - 2\gamma}{2}u^2 + \frac{a^2}{\gamma - 1} & \gamma u \end{bmatrix} \tag{8.55}$$

and the system is hyperbolic with real eigenvalues

$$\lambda_1 = u - a , \quad \lambda_2 = u , \quad \lambda_3 = u + a . \tag{8.56}$$

The matrix $\mathbf{K}$ of corresponding right eigenvectors is

$$\mathbf{K} = \begin{bmatrix} 0 & 1 & 0 \\ u - a & u & u + a \\ H - ua & \frac{1}{2}u^2 & H + ua \end{bmatrix}. \tag{8.57}$$

Here H is the enthalpy

$$H = (E + p)/\rho = \frac{1}{2}u^2 + \frac{a^2}{(\gamma - 1)} . \tag{8.58}$$

As explained in Sect. 3.2.4 of Chap. 3, the three–dimensional Euler equations may be dealt with by only considering the flux component *normal* to the computing cell interface, see also Sect. 16.2 of Chap. 16. In constructing numerical methods for Cartesian geometries it is sufficient to consider the flux in any of the coordinate directions. For general geometries this is modified by use of rotation matrices; see Sect. 3.2 of Chap. 3. We thus state the schemes for the x–split three dimensional Euler equations

$$\mathbf{U}_t + \mathbf{F(U)}_x = \mathbf{0} , \tag{8.59}$$

$$\mathbf{U} = \begin{bmatrix} \rho \\ \rho u \\ \rho v \\ \rho w \\ E \end{bmatrix} , \quad \mathbf{F}(\mathbf{U}) = \begin{bmatrix} \rho u \\ \rho u^2 + p \\ \rho u v \\ \rho u w \\ u(E + p) \end{bmatrix} . \tag{8.60}$$

The Jacobian matrix $\mathbf{A}$, see Sect. 3.2.2 of Chap. 3, is given by

$$\mathbf{A} = \begin{bmatrix} 0 & 1 & 0 & 0 & 0 \\ \hat{\gamma}H - u^2 - a^2 & (3 - \gamma)u & -\hat{\gamma}v & -\hat{\gamma}w & \hat{\gamma} \\ -uv & v & u & 0 & 0 \\ -uw & w & 0 & u & 0 \\ \frac{1}{2}u[(\gamma - 3)H - a^2] & H - \hat{\gamma}u^2 & -\hat{\gamma}uv & -\hat{\gamma}uw & \gamma u \end{bmatrix} , \tag{8.61}$$

where

$$H = (E + p)/\rho = \frac{1}{2}\mathbf{V}^2 + \frac{a^2}{(\gamma - 1)} , \quad \mathbf{V}^2 = u^2 + v^2 + w^2 , \quad \hat{\gamma} = \gamma - 1 . \tag{8.62}$$

This system is hyperbolic with real eigenvalues

$$\lambda_1 = u - a , \quad \lambda_2 = \lambda_3 = \lambda_4 = u , \quad \lambda_5 = u + a . \tag{8.63}$$

The matrix of corresponding right eigenvectors is

$$\mathbf{K} = \begin{bmatrix} 1 & 1 & 0 & 0 & 1 \\ u - a & u & 0 & 0 & u + a \\ v & v & 1 & 0 & v \\ w & w & 0 & 1 & w \\ H - ua & \frac{1}{2}\mathbf{V}^2 & v & w & H + ua \end{bmatrix} \tag{8.64}$$

As seen in Chap. 3 the one–dimensional Euler equations satisfy the *homogeneity property*

$$\mathbf{F}(\mathbf{U}) = \mathbf{A}(\mathbf{U})\mathbf{U} . \tag{8.65}$$

Exercise 8.4.1. Verify that the split three–dimensional Euler equations (8.59)–(8.60) also satisfy the homogeneity property (8.65).

Solution 8.4.1. (Left to the reader).

8.4.2 The Steger–Warming Splitting

For a splitting (8.25)–(8.27) we require an expression for the inverse $\mathbf{K}^{-1}$ of the matrix $\mathbf{K}$, in order to find the split Jacobians (8.29).

The One–Dimensional Case. For the one–dimensional Euler equations we have

$$\mathbf{K}^{-1} = \frac{(\gamma-1)}{2a^2} \begin{bmatrix} \frac{1}{2}u^2 + \frac{ua}{\gamma-1} & -u - \frac{a}{\gamma-1} & 1 \\ \frac{2a^2}{\gamma-1} - u^2 & 2u & -2 \\ \frac{1}{2}u^2 - \frac{ua}{\gamma-1} & \frac{a}{\gamma-1} - u & 1 \end{bmatrix} . \tag{8.66}$$

Then, for any component Λ^α of the two components of Λ in (8.26) the corresponding Jacobian component is

$$\mathbf{A}^\alpha = \mathbf{K}\Lambda^\alpha\mathbf{K}^{-1} .$$

The associated split flux component $\mathbf{F}^\alpha = \mathbf{A}^\alpha\mathbf{U}$ is

$$\mathbf{F}^\alpha = \frac{\rho}{2\gamma} \begin{bmatrix} \lambda_1^\alpha + 2(\gamma-1)\lambda_2^\alpha + \lambda_3^\alpha \\ (u-a)\lambda_1^\alpha + 2(\gamma-1)u\lambda_2^\alpha + (u+a)\lambda_3^\alpha \\ (H-ua)\lambda_1^\alpha + (\gamma-1)u^2\lambda_2^\alpha + (H+ua)\lambda_3^\alpha \end{bmatrix} , \tag{8.67}$$

where the eigenvalues λ_k^α are given by (8.32), for $\alpha = +, -$.

The Three–Dimensional Case. For the three–dimensional case we have

$$\mathbf{K}^{-1} = \frac{(\gamma-1)}{2a^2} \begin{bmatrix} H + \frac{a}{\gamma}(u-a) & -(u+\frac{a}{\gamma}) & -v & -w & 1 \\ -2H + \frac{4}{\gamma}a^2 & 2u & 2v & 2w & -2 \\ -\frac{2va^2}{\gamma} & 0 & \frac{2a^2}{\gamma} & 0 & 0 \\ -\frac{2wa^2}{\gamma} & 0 & 0 & \frac{2a^2}{\gamma} & 0 \\ H - \frac{a}{\gamma}(u+a) & -u+\frac{a}{\gamma} & -v & -w & 1 \end{bmatrix} \tag{8.68}$$

and the resulting split flux component $\mathbf{F}^\alpha = \mathbf{A}^\alpha\mathbf{U}$ is found to be

$$\mathbf{F}^\alpha = \frac{\rho}{2\gamma} \begin{bmatrix} \lambda_1^\alpha + 2(\gamma-1)\lambda_2^\alpha + \lambda_5^\alpha \\ (u-a)\lambda_1^\alpha + 2(\gamma-1)u\lambda_2^\alpha + (u+a)\lambda_5^\alpha \\ v\lambda_1^\alpha + 2(\gamma-1)v\lambda_2^\alpha + v\lambda_5^\alpha \\ w\lambda_1^\alpha + 2(\gamma-1)w\lambda_2^\alpha + w\lambda_5^\alpha \\ (H-ua)\lambda_1^\alpha + (\gamma-1)\mathbf{V}^2\lambda_2^\alpha + (H+ua)\lambda_5^\alpha \end{bmatrix} . \tag{8.69}$$

Exercise 8.4.2. Verify expressions (8.68) and (8.69) above.

Solution 8.4.2. (Left to the reader).

8.4.3 The van Leer Splitting

Van Leer [393], [394] constructed a splitting for the Euler equations that has some extra desirable properties, namely

-    (I) The split Jacobian matrices

$$\hat{\mathbf{A}}^+ = \frac{\partial\mathbf{F}^+}{\partial\mathbf{U}} , \quad \hat{\mathbf{A}}^- = \frac{\partial\mathbf{F}^-}{\partial\mathbf{U}}$$

are required to be continuous.

- (II) The split fluxes are degenerate *for subsonic flow*, that is $\hat{\mathbf{A}}^+$, $\hat{\mathbf{A}}^-$ have a zero eigenvalue.

Van Leer expresses the flux vector $\mathbf{F}$ as a function of density, sound speed and Mach number $M = \frac{u}{a}$, that is

$$\mathbf{F} = \mathbf{F}(\rho, a, M) = \begin{bmatrix} \rho a M \\ \rho a^2 (M^2 + \frac{1}{\gamma}) \\ \rho a^3 M(\frac{1}{2}M^2 + \frac{1}{\gamma-1}) \end{bmatrix} \equiv \begin{bmatrix} f_{\text{mas}} \\ f_{\text{mom}} \\ f_{\text{ene}} \end{bmatrix} . \qquad (8.70)$$

For the mass flux

$$f_{\text{mas}} = \rho a M$$

one requires quadratics in M and the split mass fluxes are

$$f^+_{\text{mas}} = \frac{1}{4}\rho a(1 + M)^2 , \quad f^-_{\text{mas}} = -\frac{1}{4}\rho a(1 - M)^2 . \qquad (8.71)$$

The momentum split fluxes are

$$f^+_{\text{mom}} = f^+_{\text{mas}} \frac{2a}{\gamma}[\frac{(\gamma - 1)}{2}M + 1] , \quad f^-_{\text{mom}} = f^-_{\text{mas}} \frac{2a}{\gamma}[\frac{(\gamma - 1)}{2}M - 1] \qquad (8.72)$$

and the energy split fluxes are

$$f^+_{\text{ene}} = \frac{\gamma^2}{2(\gamma^2 - 1)} \frac{[f^+_{\text{mom}}]^2}{f^+_{\text{mas}}} , \quad f^-_{\text{ene}} = \frac{\gamma^2}{2(\gamma^2 - 1)} \frac{[f^-_{\text{mom}}]^2}{f^-_{\text{mas}}} . \qquad (8.73)$$

In vector form we have

$$\mathbf{F}^+ = \frac{1}{4}\rho a(1 + M)^2 \begin{bmatrix} 1 \\ \frac{2a}{\gamma}(\frac{\gamma-1}{2}M + 1) \\ \frac{2a^2}{\gamma^2-1}(\frac{\gamma-1}{2}M + 1)^2 \end{bmatrix} , \qquad (8.74)$$

$$\mathbf{F}^- = -\frac{1}{4}\rho a(1 - M)^2 \begin{bmatrix} 1 \\ \frac{2a}{\gamma}(\frac{\gamma-1}{2}M - 1) \\ \frac{2a^2}{\gamma^2-1}(\frac{\gamma-1}{2}M - 1)^2 \end{bmatrix} . \qquad (8.75)$$

For the x–split three dimensional Euler equations the split flux formulae are

$$\mathbf{F}^+ = \frac{1}{4}\rho a(1 + M)^2 \begin{bmatrix} 1 \\ \frac{2a}{\gamma}(\frac{\gamma-1}{2}M + 1) \\ v \\ w \\ \frac{2a^2}{\gamma^2-1}(\frac{\gamma-1}{2}M + 1)^2 + \frac{1}{2}(v^2 + w^2) \end{bmatrix} , \qquad (8.76)$$

and

$$\mathbf{F}^- = -\frac{1}{4}\rho a(1-M)^2 \begin{bmatrix} 1 \\ \frac{2a}{\gamma}(\frac{\gamma-1}{2}M-1) \\ v \\ w \\ \frac{2a^2}{\gamma^2-1}(\frac{\gamma-1}{2}M-1)^2 + \frac{1}{2}(v^2+w^2) \end{bmatrix} , \qquad (8.77)$$

where the Mach number is still $M = \frac{u}{a}$.

Concerning stability, van Leer [393] gives the following practical stability condition

$$C_{\text{cfl}} \equiv \frac{\Delta t}{\Delta x}(\mid u \mid +a) \le \frac{2\gamma+ \mid M \mid (3-\gamma)}{\gamma+3} . \qquad (8.78)$$

Note that $C_{\text{cfl}} = C_{\text{cfl}}(M)$ and that when $\gamma = 1.4$ we have

$$C_{\text{cfl}}^{\max} = 1 \text{ for } \mid M \mid = 1 , \quad C_{\text{cfl}}^{\min} = \frac{2\gamma}{\gamma+3} \approx 0.636\ldots , \text{ for } \mid M \mid = 0 . \qquad (8.79)$$

Remark 8.4.1. The CFL condition for the explicit FVS scheme is more restrictive than that for the Godunov method, for which C_{cfl} is close to unity. See Sect. 6.3.2 of Chap. 6 for a discussion on the CFL condition.

8.4.4 The Liou–Steffen Scheme

A recent scheme that attempts to combine features from the Flux Vector Splitting and Godunov approaches is due to Liou and Steffen [221]. The scheme has been formulated in terms of the time–dependent Euler equations and relies on splitting the flux vector $\mathbf{F}$ into a *convective* component $\mathbf{F}^{(c)}$ and a *pressure* component $\mathbf{F}^{(p)}$. For the x–split three dimensional flux we have

$$\mathbf{F(U)} = \begin{bmatrix} \rho u \\ \rho u^2 + p \\ \rho u v \\ \rho u w \\ u(E+p) \end{bmatrix} = \begin{bmatrix} \rho u \\ \rho u^2 \\ \rho u v \\ \rho u w \\ \rho u H \end{bmatrix} + \begin{bmatrix} 0 \\ p \\ 0 \\ 0 \\ 0 \end{bmatrix} \equiv \mathbf{F}^{(c)} + \mathbf{F}^{(p)} , \qquad (8.80)$$

with the obvious definitions for the convective component $\mathbf{F}^{(c)}$ and the pressure component $\mathbf{F}^{(p)}$. By introducing the Mach number and enthalpy

$$M = \frac{u}{a} , \quad H = \frac{E+p}{\rho}$$

we write

$$\mathbf{F}^{(c)} = M \begin{bmatrix} \rho a \\ \rho a u \\ \rho a v \\ \rho a w \\ \rho a H \end{bmatrix} \equiv M \hat{\mathbf{F}}^{(c)} , \qquad (8.81)$$

with the obvious notation for the vector $\hat{\mathbf{F}}^{(c)}$. In defining the intercell numerical flux $\mathbf{F}_{i+\frac{1}{2}}$, Liou and Steffen take

$$\mathbf{F}_{i+\frac{1}{2}} = \mathbf{F}^{(c)}_{i+\frac{1}{2}} + \mathbf{F}^{(p)}_{i+\frac{1}{2}} \;, \tag{8.82}$$

where the *convective flux component* is given by

$$\mathbf{F}^{(c)}_{i+\frac{1}{2}} = M_{i+\frac{1}{2}} \left[\hat{\mathbf{F}}^{(c)} \right]_{i+\frac{1}{2}} \tag{8.83}$$

with definition

$$[\bullet]_{i+\frac{1}{2}} = \left\{ \begin{array}{ll} [\bullet]_i & \text{if} \quad M_{i+\frac{1}{2}} \geq 0 \;, \\ [\bullet]_{i+1} & \text{if} \quad M_{i+\frac{1}{2}} \leq 0 \;. \end{array} \right. \tag{8.84}$$

Note that the flux vector in (8.83) is upwinded according to the sign of the convection, or advection, speed *implied* in the intercell Mach number $M_{i+\frac{1}{2}}$. For this reason Liou and Steffen call their scheme AUSM, which stands for Advection Upstream Splitting Method.

The cell–interface Mach number is given by the splitting

$$M_{i+\frac{1}{2}} = M_i^+ + M_{i+1}^- \tag{8.85}$$

with the positive and negative components yet to be defined. The splitting of the pressure flux component depends on the splitting of the pressure itself, namely

$$p_{i+\frac{1}{2}} = p_i^+ + p_{i+1}^- \;. \tag{8.86}$$

For the splitting of the Mach number Liou and Steffen follow van Leer and set

$$M^{\pm} = \left\{ \begin{array}{ll} \pm\frac{1}{4}(M\pm 1)^2 & \text{if} \quad |M| \leq 1 \;, \\ \frac{1}{2}(M\pm |M|) & \text{if} \quad |M| > 1 \;. \end{array} \right. \tag{8.87}$$

For splitting the pressure they suggest two choices, namely

$$p^{\pm} = \left\{ \begin{array}{ll} \frac{1}{2}p(1\pm M) & \text{if} \quad |M| \leq 1 \\ \frac{1}{2}p\frac{(M\pm|M|)}{M} & \text{if} \quad |M| > 1 \end{array} \right. \tag{8.88}$$

and

$$p^{\pm} = \left\{ \begin{array}{ll} \frac{1}{4}p(M\pm 1)^2(2\mp M) & \text{if} \quad |M| \leq 1 \;, \\ \frac{1}{2}p\frac{(M\pm|M|)}{M} & \text{if} \quad |M| > 1 \;. \end{array} \right. \tag{8.89}$$

For more details see the original paper by Liou and Steffen [221] and the more recent publication of Liou [220].

8.5 Numerical Results

Here we illustrate the performance of three FVS–type schemes on the one-dimensional, time dependent Euler equations for ideal gases with $\gamma = 1.4$, namely the Steger–Warming FVS scheme, the van Leer FVS scheme and the AUSM scheme of Liou and Steffen. Numerical results are compared with the exact solution. The respective results are obtained from running two codes of *NUMERICA* [369], namely HE–E1FVS (FVS schemes) and HE–E1RPEXACT (exact Riemann solver).

8.5.1 Tests

We use five test problems with exact solution. Data consists of two constant states $\mathbf{W}_L = [\rho_L, u_L, p_L]^T$ and $\mathbf{W}_R = [\rho_R, u_R, p_R]^T$, separated by a discontinuity at a position $x = x_0$, and are given in Table 8.1. The exact and numerical solutions are found in the spatial domain $0 \le x \le 1$. The numerical solution is computed with $M = 100$ cells. The Courant number coefficient is taken as $C_{cfl} = 0.9$, except for the van Leer scheme, for which we took $C_{cfl} = 0.6$. In implementing the CFL condition we use the simple formula given by equation 6.20 of Chap. 6 to estimate the maximum wave speed. Therefore, for all methods, we reduce the CFL number further to 0.2 of that given by the CFL condition, for the first 5 time steps. Boundary conditions are transmissive. For each test problem we select a convenient position x_0 of the initial discontinuity and the output time; these are stated in the legend of each figure displaying computational results. All numerical results should be compared with those from Godunov's method, Figs. 6.8 to 6.12 of Chap. 6. For more details on the exact solutions of the test problems see Sect. 4.3.3 of Chap. 4.

Test	ρ_L	u_L	p_L	ρ_R	u_R	p_R
1	1.0	0.75	1.0	0.125	0.0	0.1
2	1.0	-2.0	0.4	1.0	2.0	0.4
3	1.0	0.0	1000.0	1.0	0.0	0.01
4	5.99924	19.5975	460.894	5.99242	-6.19633	46.0950
5	1.0	-19.59745	1000.0	1.0	-19.59745	0.01

Table 8.1. Data for five test problems with exact solution. Test 5 is like Test 3 with negative uniform background speed

8.5.2 Results for Test 1

Test 1 is a *modified* version of the popular Sod's test [318]; the solution consists of a right shock wave, a right travelling contact wave and a left *sonic* rarefaction wave; this test is very useful in assessing the *entropy satisfaction* property of numerical methods. Figs. 8.3 to 8.5 show the results for the three FVS schemes.

In the results from the Steger–Warming scheme, shown in Fig. 8.3, the resolution of the shock and the right travelling contact is comparable with that of Godunov's method, Fig. 6.8 of Chap. 6. The resolution of the left rarefaction is less satisfactory; the head and tail are visibly smeared and the sonic point, as expected, is not handled correctly. The results from the van Leer scheme, shown in Fig. 8.4, are virtually identical to those of Godunov's method of Fig. 6.8 for the rarefaction and contact, but the shock is broader. The performance at the sonic point is comparable with that of Godunov's method and better than that of the Steger–Warming scheme. The results from the Liou and Steffen scheme are shown in Fig. 8.5. In comparison with Godunov's method, the shock wave is more sharply resolved and the contact wave is similar but the resolution of the rarefaction is not as good, particularly near the sonic point.

8.5.3 Results for Test 2

The exact solution of Test 2 consists of two symmetric rarefaction waves and a trivial contact wave of zero speed; the *Star Region* between the non-linear waves is close to vacuum, which makes this problem a suitable test for assessing the performance of numerical methods for low–density flows [118]; this is the so called *123 problem* introduced in Sect. 4.3.3 of Chap. 4. Figs. 8.6 to 8.8 show the results for the three FVS schemes.

The results from the Steger–Warming scheme, shown in Fig. 8.6, are virtually identical to those of the Godunov method, Fig. 6.9 of Chap. 6. The results from the van Leer scheme, shown in Fig. 8.7, are also comparable with those from the Godunov method. The heads of the rarefactions are slightly more diffused. The Liou and Steffen scheme, Fig. 8.8, gives results that are comparable with those of Godunov's method and slightly more accurate than those from van Leer's scheme; in the vicinity of the trivial contact, where both density and pressure are close to zero, the results are somewhat erratic, see velocity and internal energy plots.

In view of the fact that Godunov–type methods with linearised Riemann solvers will fail for this test problem [118], it is quite remarkable to note that all three FVS–type schemes described in this chapter actually run and give, overall, good results.

8.5.4 Results for Test 3

Test 3 is designed to assess the robustness and accuracy of numerical methods; its solution consists of a strong right travelling shock wave of shock Mach number 198, a contact surface and a left rarefaction wave. Figs. 8.9 and 8.10 show the results for two FVS schemes.

The Steger–Warming result, shown in Fig. 8.9, is seen to be overall less accurate than the corresponding result from the Godunov method, shown in Fig. 6.10 of Chap. 6; the numerical solution has an unphysical dip behind the shock wave, which is more clearly seen in the velocity and pressure plots. The results from the van Leer scheme, shown in Fig. 8.10, are also less accurate than those from the Godunov method, but they are more accurate than the results from the Steger–Warming scheme. The Liou and Steffen scheme, as coded by the author, failed to give a solution at all for this very severe test problem, even when reducing the CFL number to a value as low as 0.1.

8.5.5 Results for Test 4

Test 4, as Test 3, is also designed to assess the robustness of numerical methods; data originates from two very strong shock waves travelling towards each other and the solution consists of three strong discontinuities travelling to the right; the left shock wave moves to the right very slowly, which adds another difficulty [280] to numerical methods. Figs. 8.11 to 8.13 show the results for the three FVS schemes.

The results from the Steger and Warming scheme, shown in Fig. 8.11, are overall comparable with those of Godunov's method shown in Fig. 6.11 of Chapter 6. The only visible difference is seen near the left slowly moving shock, and as expected, this is more diffused in the Steger–Warming result; however, it appears as if the low frequency oscillations seen in the Godunov results are significantly reduced in the Steger–Warming scheme. The results from the van Leer scheme, shown in Fig. 8.12, are comparable with those of Godunov's method and are more accurate than those from the Steger–Warming scheme. The slowly–moving shock is resolved with two interior cells, instead of one in the Godunov's method, but low–frequency spurious oscillations are just about visible. The results from the Liou and Steffen scheme, shown in Fig. 8.13, are comparable with the Godunov and van Leer results for this test; the fast right shock is more sharply resolved than with the other methods, but at the cost of an overshoot; the slowly moving left shock is slightly more smeared than in the van Leer result.

8.5.6 Results for Test 5

Test 5 is effectively Test 3, with a negative uniform background speed so as to obtain a stationary contact discontinuity. Test 5 is also designed to test the robustness of numerical methods but the main reason for devising this test is for assessing the ability of numerical methods to resolve *slowly– moving contact discontinuities*. The exact solution of Test 5 consists of a left rarefaction wave, a right–travelling shock wave (slow) and a *stationary* contact discontinuity. Figs. 8.14 to 8.16 show the results for the three FVS schemes and Fig. 8.17 shows the respective results obtained from the Godunov method used in conjunction with the exact Riemann solver, code HE–E1GODSTATE of *NUMERICA* [369]. We note that at the chosen output time, the right travelling shock wave has propagated only 5 cells, in 81 time steps. For this test problem the results from the Steger/Warming and van Leer FVS schemes are similar, in that the contact discontinuity is heavily smeared, even for a relatively short evolution time. The Liou and Steffen FVS scheme, Fig. 8.16, differs from the other two FVS schemes in that it resolves the contact discontinuity more sharply; note however the unphysical oscillations in the vicinity of the shock wave, the contact discontinuity and even near the tail of the rarefaction. For comparison, the results from the Godunov method used in conjuction with the exact Riemann solver are displayed in Fig. 8.17. These are obviously superior to any of the FVS schemes, for this test problem.

The numerical experiments presented suggest that Flux Vector Splitting Schemes give, generally, results of similar quality to those obtained by the Godunov method. The difference between these two upwind approaches is evident when *slowly or stationary* contact waves are present. For multidimensional problems this has important implications for the accurate resolution of shear layers, material interfaces and vortical flows. The Liou and Steffen FVS–type scheme is an exception, as it does resolve contacts more accurately than the Warming–Beam and van Leer schemes, although there are questions marks about its robustness. For Test 3 the Liou and Steffen scheme *crashed* and for Test 5 produced large unphysical oscillations. It is worth remarking that the Godunov method was used in conjunction with the exact Riemann solver, to obtain the numerical results of Fig. 8.17. If the Godunov scheme is used with linearised Riemann solvers, then it would fail for low–density flows, such as Test 2 for example, whereas the FVS–type schemes appear to be much less sensitive; they all produced acceptable results for Test 2. In addition, if the Godunov method is used in conjunction with *incomplete* Riemann solvers, such as those that ignore the presence of linear waves in the structure of the solution of the Riemann problem, then the resolution of contacts will be as poor as that of FVS–type schemes, such as the Warming–Beam and van Leer schemes. The selection of the Riemann solver is crucial to the performance of the Godunov method. See Chaps. 9 to 12.

For details on how to extend FVS–type schemes to higher order of accuracy for one–dimensional homogeneous problems the reader is referred to Chaps. 13 and 14. Methods for treating source terms are given in Chapt. 15 and techniques to extend the methods to solve multidimensional problems are given in Chapt. 16. For multidimensional, steady state, applications of Flux Vector Splitting methods, readers are encouraged to consult, amongst many others, the following references: [10], [11], [429], [112], [408].

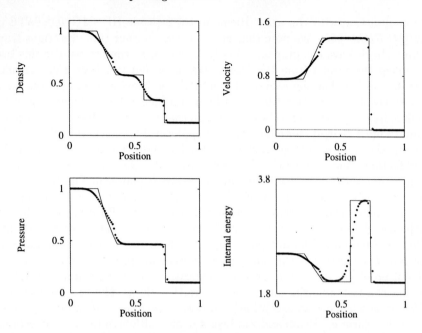

Fig. 8.3. Steger and Warming FVS scheme applied to Test 1, with $x_0 = 0.3$. Numerical (symbol) and exact (line) solutions are compared at time 0.2 units

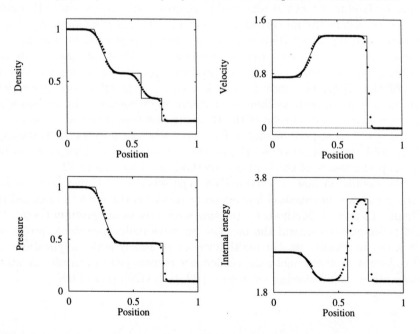

Fig. 8.4. Van Leer FVS scheme applied to Test 1, with $x_0 = 0.3$. Numerical (symbol) and exact (line) solutions are compared at time 0.2 units

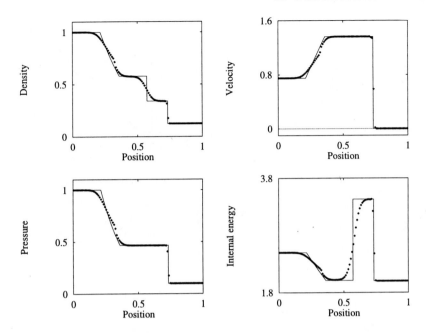

Fig. 8.5. Liou and Steffen scheme applied to Test 1, with $x_0 = 0.3$. Numerical (symbol) and exact (line) solutions are compared at time 0.2 units

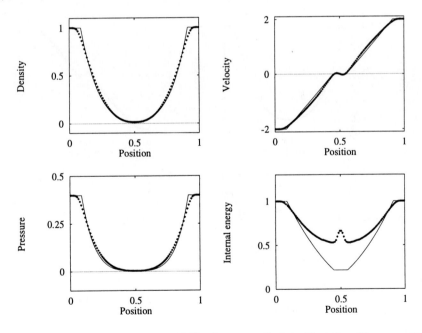

Fig. 8.6. Steger and Warming FVS scheme applied to Test 2, with $x_0 = 0.5$. Numerical (symbol) and exact (line) solutions are compared at time 0.15 units

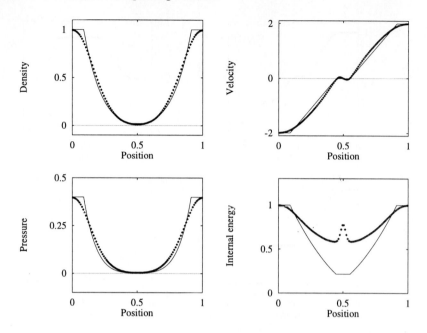

Fig. 8.7. Van Leer FVS scheme applied to Test 2, with $x_0 = 0.5$. Numerical (symbol) and exact (line) solutions are compared at time 0.15 units

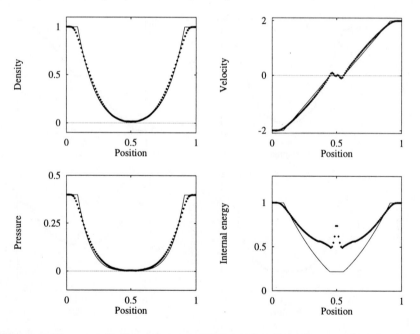

Fig. 8.8. Liou and Steffen scheme applied to Test 2, with $x_0 = 0.5$. Numerical (symbol) and exact (line) solutions are compared at time 0.15 units

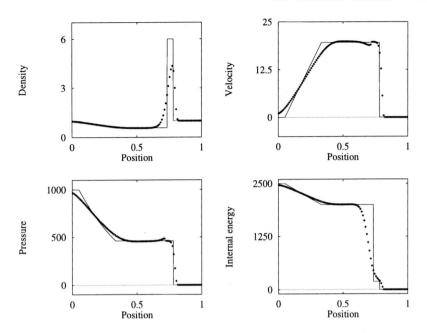

Fig. 8.9. Steger and Warming FVS scheme applied applied to Test 3, with $x_0 = 0.5$. Numerical (symbol) and exact (line) solutions are compared at time 0.012 units

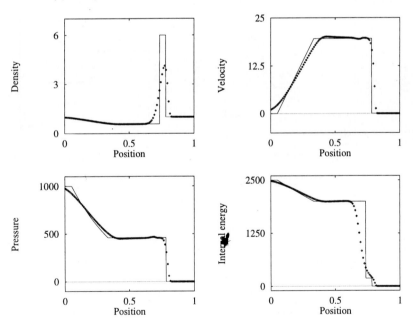

Fig. 8.10. Van Leer FVS scheme applied applied to Test 3, with $x_0 = 0.5$. Numerical (symbol) and exact (line) solutions are compared at time 0.012 units

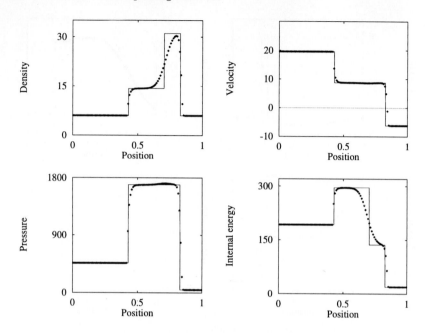

Fig. 8.11. Steger and Warming FVS scheme applied to Test 4, with $x_0 = 0.4$. Numerical (symbol) and exact (line) solutions are compared at time 0.035 units

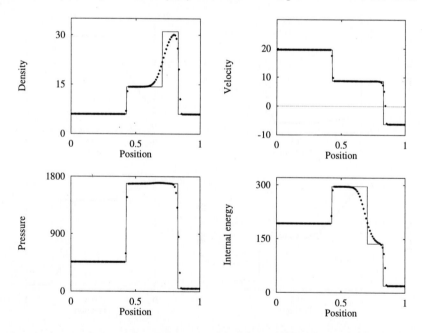

Fig. 8.12. Van Leer FVS scheme applied to Test 4, with $x_0 = 0.4$. Numerical (symbol) and exact (line) solutions are compared at time 0.035 units

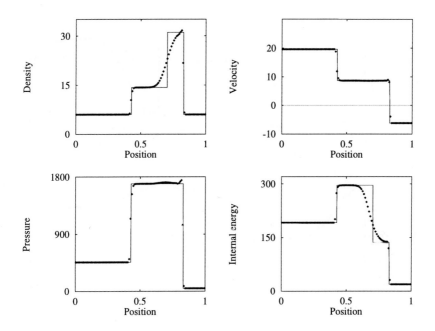

Fig. 8.13. Liou and Steffen scheme applied to Test 4, with $x_0 = 0.4$. Numerical (symbol) and exact (line) solutions are compared at time 0.035 units

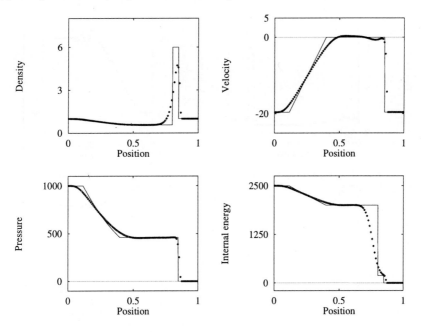

Fig. 8.14. Steger and Warming FVS scheme applied to Test 5, with $x_0 = 0.8$. Numerical (symbol) and exact (line) solutions are compared at time 0.012 units

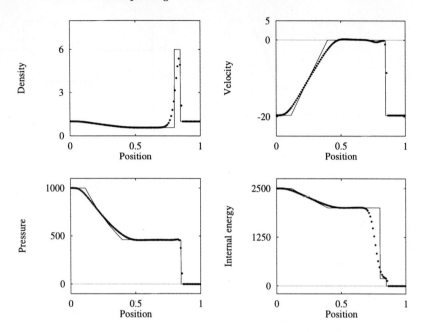

Fig. 8.15. Van Leer FVS scheme applied to Test 5, with $x_0 = 0.8$. Numerical (symbol) and exact (line) solutions are compared at time 0.012 units

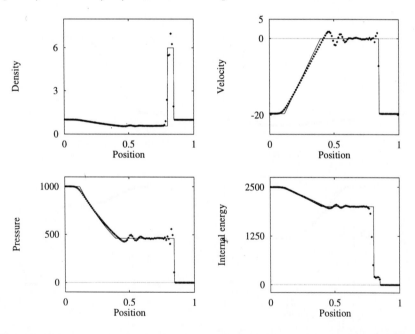

Fig. 8.16. Liou and Steffen scheme applied to Test 5, with $x_0 = 0.8$. Numerical (symbol) and exact (line) solutions are compared at time 0.012 units

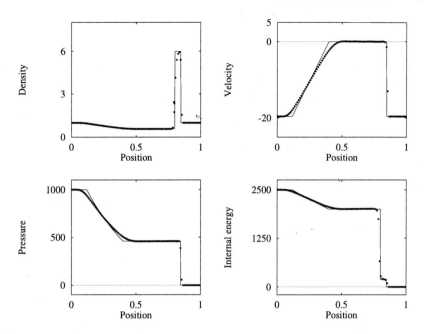

Fig. 8.17. Godunov scheme applied to Test 5, with $x_0 = 0.8$. Numerical (symbol) and exact (line) solutions are compared at time 0.012 units

9. Approximate–State Riemann Solvers

9.1 Introduction

The method of Godunov [145] and its high–order extensions require the solution of the Riemann problem. In a practical computation this is solved billions of times, making the Riemann problem solution process the single most demanding task in the numerical method. In Chap. 4 we provided exact Riemann solvers for the Euler equations for ideal and covolume gases. An iterative procedure is always involved and the associated computational effort may not always be justified. This effort may increase dramatically by equations of state of complicated algebraic form or by the complexity of the particular system of equations being solved, or both. Approximate, non–iterative solutions have the potential to provide the necessary items of information for numerical purposes. There are essentially two ways of extracting approximate information from the solution of the Riemann problem to be used in Godunov–type methods: one approach is to find an *approximation to the numerical flux* employed in the numerical method, directly, see Chaps. 10, 11 and 12; the other approach is to find an *approximation to a state* and then evaluate the physical flux function at this state. It is the latter route the one we follow in this chapter.

We present, approximate, Riemann solvers that do not need an iteration process. We provide an approximate *solution for the state* required to evaluate the Godunov flux. The approximations can be used directly in the first–oder Godunov method and its high–order extensions. Some of the approximations presented are exceedingly simple but not accurate enough to produce *robust numerical methods*. This difficulty is resolved by designing *hybrid schemes* that combine various approximate solvers in and adaptive fashion. There are other uses of the explicit approximate solutions presented here. For instance, the simplest solutions can be used in the *characteristic limiting* of high–order Godunov type methods based on the MUSCL approach; see Sect. 13.4 of Chap. 13. They also provide valuable information of use in other well known approximate Riemann solvers. For instance, Roe's approximate Riemann solver, [281] to be studied in Chap. 11, requires an entropy fix; the results of this chapter may be used to provide the state values in the Harten–Hyman entropy fix [163]. The approximate Riemann solver of Osher [255], to be studied in Chap. 12, requires intersection points for the integration paths; the approximations of this chapter can be used directly. The HLL approach of Harten, Lax and van Leer [164] for deriving approximate solutions to the Riemann problem, to be studied in Chap. 10, requires estimates for the smallest and largest signal velocities in the Riemann problem; again, the pressure–velocity approximation of this chapter can directly lead to estimates for wave speeds. The approximate solutions presented in this chapter may also be of use in other computational approaches, such as in *front tracking schemes* [331], [277]. The techniques discussed here can easily be extended to other systems, such as the shallow water equations, the steady supersonic Euler equations, the artificial compressibility equations (see Sect. 1.6.3 of Chap. 1) and the Euler equations with general equation of state.

Useful background for studying this chapter is found in Chaps. 2, 3, 4, and 6. The rest of this chapter is organised as follows: in Sect. 9.2 we recall the Godunov flux and the Riemann problem solution, in Sect. 9.3 we present very simple Riemann solvers based on primitive variable formulations of the Euler equations. In Sect. 9.4 we study approximations based on the exact function for pressure, namely the two–rarefaction approximation and the two–shock approximation. Hybrid schemes are dealt with in Sect. 9.5 and numerical results are presented in Sect. 9.6.

9.2 The Riemann Problem and the Godunov Flux

We want to solve numerically the general Initial Boundary Value Problem (IBVP)

$$\left.\begin{array}{lll} \text{PDEs} & : & \mathbf{U}_t + \mathbf{F}(\mathbf{U})_x = 0\,, \\ \text{ICs} & : & \mathbf{U}(x,0) = \mathbf{U}^{(0)}(x)\,, \\ \text{BCs} & : & \mathbf{U}(0,t) = \mathbf{U}_l(t)\,,\ \mathbf{U}(L,t) = \mathbf{U}_r(t)\,, \end{array}\right\} \tag{9.1}$$

utilising the explicit conservative formula

$$\mathbf{U}_i^{n+1} = \mathbf{U}_i^n + \frac{\Delta t}{\Delta x}[\mathbf{F}_{i-\frac{1}{2}} - \mathbf{F}_{i+\frac{1}{2}}]\,, \tag{9.2}$$

along with the Godunov intercell numerical flux

$$\mathbf{F}_{i+\frac{1}{2}} = \mathbf{F}(\mathbf{U}_{i+\frac{1}{2}}(0))\,. \tag{9.3}$$

We assume that the solution of IBVP (9.1) exists. Here $\mathbf{U}_{i+\frac{1}{2}}(0)$ is the similarity solution $\mathbf{U}_{i+\frac{1}{2}}(x/t)$ of the Riemann problem

$$\left.\begin{array}{l} \mathbf{U}_t + \mathbf{F}(\mathbf{U})_x = 0\,, \\ \mathbf{U}(x,0) = \left\{\begin{array}{ll} \mathbf{U}_L & \text{if } x < 0\,, \\ \mathbf{U}_R & \text{if } x > 0\,, \end{array}\right. \end{array}\right\} \tag{9.4}$$

evaluated at $x/t = 0$. Fig. 9.1 shows the structure of the exact solution of the Riemann problem for the x–split three–dimensional Euler equations, for which the vectors of conserved variables and fluxes are

$$U = \begin{bmatrix} \rho \\ \rho u \\ \rho v \\ \rho w \\ E \end{bmatrix}, \quad F = \begin{bmatrix} \rho u \\ \rho u^2 + p \\ \rho u v \\ \rho u w \\ u(E + p) \end{bmatrix}. \tag{9.5}$$

The value $x/t = 0$ for the Godunov flux corresponds to the t–axis. See Chap. 6 for details. The piece–wise constant initial data, in terms of primitive variables, is

$$W_L = \begin{bmatrix} \rho_L \\ u_L \\ v_L \\ w_L \\ p_L \end{bmatrix}, \quad W_R = \begin{bmatrix} \rho_R \\ u_R \\ v_R \\ w_R \\ p_R \end{bmatrix}. \tag{9.6}$$

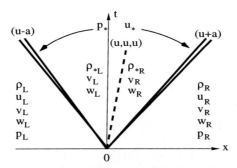

Fig. 9.1. Structure of the solution of the Riemann problem for the x–split, three dimensional Euler equations. Data and solution are given in terms of primitive variables

The purpose of this chapter is to find *approximate* solutions to the Riemann problem in order to evaluate the Godunov flux. As seen in Chap. 6, the evaluation of the flux requires the identification of the appropriate wave pattern in the Riemann problem solution; as depicted in Fig. 9.2, there are ten possibilities to be considered.

In our solution procedure we split the task of solving the complete Riemann problem into three subproblems, namely

(I)   The star values

$$p_* , u_* , \rho_{*L} , \rho_{*R} \tag{9.7}$$

in the *Star Region* between the non–linear waves, see Fig. 9.1.

(II) The solution for the tangential velocity components v and w throughout the wave structure, and

(III) The solution for ρ, u and p inside sonic rarefactions.

Cases (II) and (III) are dealt with in the rest of this section, while case (I) is the subject of the rest of the chapter.

9.2.1 Tangential Velocity Components

Recall that in the exact solution, the values of the tangential velocity components v and w do not change across the non–linear waves but do change, discontinuously, across the *middle* wave. Thus, given an approximate solution u_* for the normal velocity component in the *Star Region*, the solution for the tangential velocity components v and w is

$$v(x,t)\ ,w(x,t) = \begin{cases} v_L\ ,w_L \text{ if } \frac{x}{t} \leq u_* , \\[2mm] v_R\ ,w_R \text{ if } \frac{x}{t} > u_* . \end{cases} \tag{9.8}$$

In this way, the solution for the tangential velocity components is, in a sense, *exact*; the only approximation being that for u_*. As a matter of fact, any *passive* scalar quantity $q(x, y, z, t)$ advected with the fluid will have this property. In the study of multi–component flow, this quantity could be a species concentration; in practical applications there can be many of such quantities. Hence, it is very important that the approximate solution of the Riemann problem preserves the correct behaviour, as in (9.8).

9.2.2 Sonic Rarefactions

Assuming the solution for the star values (9.7) is available, we then need to identify the correct values along the t–axis, in order to evaluate the Godunov flux. The cases (a1) to (a4) and (b1) to (b4) of Fig. 9.2 can be dealt with once solutions for (9.7) and (9.8) are available. The *sonic flow* cases (a5) and (b5) must be treated separately. For these two cases we recommend the use of the exact solution, which, as seen in Sect. 4.4 of Chap. 4 for ideal gases, is non–iterative.

The solution along the t–axis inside a *left sonic rarefaction* is obtained by setting $x/t = 0$ in

$$\mathbf{W}_{\text{Lfan}} = \begin{cases} \rho = \rho_L \left[\frac{2}{(\gamma+1)} + \frac{(\gamma-1)}{(\gamma+1)a_L} \left(u_L - \frac{x}{t} \right) \right]^{\frac{2}{\gamma-1}} , \\[3mm] u = \frac{2}{(\gamma+1)} \left[a_L + \frac{(\gamma-1)}{2} u_L + \frac{x}{t} \right] , \\[3mm] p = p_L \left[\frac{2}{(\gamma+1)} + \frac{(\gamma-1)}{(\gamma+1)a_L} \left(u_L - \frac{x}{t} \right) \right]^{\frac{2\gamma}{\gamma-1}} . \end{cases} \tag{9.9}$$

The solution along the t–axis inside a *right sonic rarefaction* is obtained by setting $x/t = 0$ in

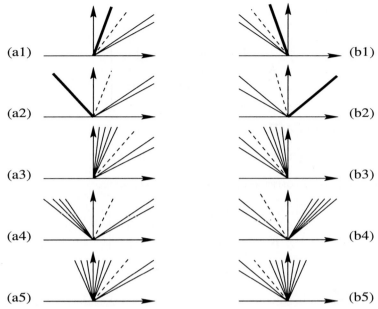

Fig. 9.2. Possible wave patterns in evaluating the Godunov flux for the Euler equations:(a) positive particle speed in the Star Region (b) negative particle speed in the Star Region

$$
\mathbf{W}_{\mathrm{Rfan}} = \begin{cases} \rho = \rho_{\mathrm{R}} \left[\frac{2}{(\gamma+1)} - \frac{(\gamma-1)}{(\gamma+1)a_{\mathrm{R}}} \left(u_{\mathrm{R}} - \frac{x}{t} \right) \right]^{\frac{2}{\gamma-1}} , \\[3mm] u = \frac{2}{(\gamma+1)} \left[-a_{\mathrm{R}} + \frac{(\gamma-1)}{2} u_{\mathrm{R}} + \frac{x}{t} \right] , \\[3mm] p = p_{\mathrm{R}} \left[\frac{2}{(\gamma+1)} - \frac{(\gamma-1)}{(\gamma+1)a_{\mathrm{R}}} \left(u_{\mathrm{R}} - \frac{x}{t} \right) \right]^{\frac{2\gamma}{\gamma-1}} . \end{cases} \qquad (9.10)
$$

The rest of this chapter is devoted to providing approximate solutions for the *star values* (9.7). We study four approaches as well as two adaptive schemes that combine various approximations.

9.3 Primitive Variable Riemann Solvers (PVRS)

A very simple linearised solution to the Riemann problem [355] for the x–split, three dimensional time dependent Euler equations (9.4)–(9.5) can be obtained in terms of the primitive variables ρ, u, v, w, p. The corresponding governing equations, see Sect. 3.2.3 of Chap. 3, are

$$
\mathbf{W}_t + \mathbf{A}(\mathbf{W})\mathbf{W}_x = \mathbf{0} , \qquad (9.11)
$$

where the coefficient matrix $\mathbf{A}(\mathbf{W})$ is given by

$$\mathbf{A} = \begin{bmatrix} u & \rho & 0 & 0 & 0 \\ 0 & u & 0 & 0 & 1/\rho \\ 0 & 0 & u & 0 & 0 \\ 0 & 0 & 0 & u & 0 \\ 0 & \rho a^2 & 0 & 0 & u \end{bmatrix} . \tag{9.12}$$

The eigenvalues of $\mathbf{A}(\mathbf{W})$ are

$$\lambda_1 = u - a , \ \lambda_2 = \lambda_3 = \lambda_4 = u , \ \lambda_5 = u + a \tag{9.13}$$

and the matrix of corresponding right eigenvectors is

$$\mathbf{K} = \begin{bmatrix} \rho & 1 & \rho & \rho & \rho \\ -a & 0 & 0 & 0 & a \\ 0 & v & 1 & v & 0 \\ 0 & w & w & 1 & 0 \\ \rho a^2 & 0 & 0 & 0 & \rho a^2 \end{bmatrix} . \tag{9.14}$$

The difficulty in solving (9.11) is due to the fact that the coefficient matrix $\mathbf{A}(\mathbf{W})$ depends on the solution vector $\mathbf{W}$ itself. If $\mathbf{A}(\mathbf{W})$ were to be *constant*, then we could apply, directly, the various techniques studied in Sect. 2.3.3 of Chap. 2 for solving linear hyperbolic systems with constant coefficients.

Assume that the initial data $\mathbf{W}_L$, $\mathbf{W}_R$ and the solution $\mathbf{W}(x/t)$ are *close* to a constant state $\bar{\mathbf{W}}$. Then, by setting

$$\bar{\mathbf{A}} \equiv \mathbf{A}(\bar{\mathbf{W}}) \tag{9.15}$$

we approximate the Riemann problem for (9.11) by the Riemann problem for the linear hyperbolic systems with constant coefficients

$$\mathbf{W}_t + \bar{\mathbf{A}}\mathbf{W}_x = 0 . \tag{9.16}$$

We now solve this *approximate problem*, with initial data (9.6), *exactly*. In Sect. 2.3.3 of Chap. 2 we studied various techniques that are directly applicable to this problem. One possibility is to apply Rankine–Hugoniot Conditions across each wave of speed $\bar{\lambda}_i$. Thus we treat (9.16) as the system in *conservative form*

$$\mathbf{W}_t + \mathbf{F}(\mathbf{W})_x = 0 , \quad \mathbf{F}(\mathbf{W}) \equiv \bar{\mathbf{A}}\mathbf{W} . \tag{9.17}$$

Then, across a wave of speed $\bar{\lambda}_i$ we have

$$\Delta \mathbf{F} \equiv \bar{\mathbf{A}}\Delta \mathbf{W} = \bar{\lambda}_i \Delta \mathbf{W} . \tag{9.18}$$

Direct application of (9.18) to the $\bar{\lambda}_1$ and $\bar{\lambda}_5$ waves gives four useful relations, namely

$$
\left.\begin{aligned}
(u_* - u_L)\bar{\rho} \;+\; \bar{a}(\rho_{*L} - \rho_L) &= 0 , \\
(p_* - p_L)/\bar{\rho} \;+\; \bar{a}(u_* - u_L) &= 0 , \\
(u_R - u_*)\bar{\rho} \;-\; \bar{a}(\rho_R - \rho_{*R}) &= 0 , \\
(p_R - p_*)/\bar{\rho} \;-\; \bar{a}(u_R - u_*) &= 0 .
\end{aligned}\right\}
\tag{9.19}
$$

The complete solution for the unknowns (9.7) is then given by

$$
\left.\begin{aligned}
p_* &= \tfrac{1}{2}(p_L + p_R) \;+\; \tfrac{1}{2}(u_L - u_R)(\bar{\rho}\bar{a}) , \\[1mm]
u_* &= \tfrac{1}{2}(u_L + u_R) \;+\; \tfrac{1}{2}(p_L - p_R)/(\bar{\rho}\bar{a}) , \\[1mm]
\rho_{*L} &= \rho_L \;+\; (u_L - u_*)(\bar{\rho}/\bar{a}) , \\[1mm]
\rho_{*R} &= \rho_R \;+\; (u_* - u_R)(\bar{\rho}/\bar{a}) .
\end{aligned}\right\}
\tag{9.20}
$$

Notice that in this linearised solution we only need to specify constant values for $\bar{\rho}$ and $\bar{a}$. There is some freedom in making the choice. Selecting some average of the data values ρ_L, ρ_R, a_L, a_R appears sensible. The choice may be constrained to satisfy some desirable properties of the Riemann problem solution, such as exact recognition of particular flow features. Here we suggest to select the simple *arithmetic means*

$$
\bar{\rho} = \frac{1}{2}(\rho_L + \rho_R) , \quad \bar{a} = \frac{1}{2}(a_L + a_R).
\tag{9.21}
$$

Note that if the data states $\mathbf{W}_L$ and $\mathbf{W}_R$ are connected by a *single isolated* contact discontinuity or shear wave, then the solution is actually *exact*, regardless of the particular choice for the averages $\bar{\rho}$ and $\bar{a}$. This is in fact a very important property; contacts and shear waves turn out to be some of the most challenging flow features to resolve correctly by any numerical method.

Another way of obtaining approximate solutions for the star values is to use the *characteristic equations*, see Sect. 3.1.2 of Chap. 3,

$$
dp - \rho a\, du = 0 \text{ along } dx/dt = u - a ,
\tag{9.22}
$$

$$
dp - a^2\, d\rho = 0 \text{ along } dx/dt = u ,
\tag{9.23}
$$

$$
dp + \rho a\, du = 0 \text{ along } dx/dt = u + a .
\tag{9.24}
$$

These differential relations hold true *along characteristic* directions. First we set

$$
C = \rho a.
\tag{9.25}
$$

Then, in order to find the star values we *connect* the state $\mathbf{W}_{*L}$ to the data state $\mathbf{W}_L$, see Fig. 9.1, by integrating (9.24) along the characteristic of speed $u + a$, where C is evaluated at the foot of the characteristic. See Fig. 9.3 The results is

$$
p_* + C_L u_* = p_L + C_L u_L .
\tag{9.26}
$$

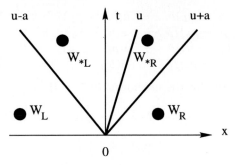

Fig. 9.3. Solution for star values using characteristic equations

Similarly, we connect $\mathbf{W}_{*R}$ to the data state $\mathbf{W}_R$ by integrating (9.22) along the characteristic of speed $u - a$, with C is evaluated at the foot of the characteristic. We obtain

$$p_* - C_R u_* = p_R - C_R u_R \; . \tag{9.27}$$

The values ρ_{*L} and ρ_{*R} are obtained by connecting $\mathbf{W}_{*L}$ to $\mathbf{W}_L$ and $\mathbf{W}_{*R}$ to $\mathbf{W}_R$ via (9.23). The complete solution is

$$\left.\begin{aligned}
p_* &= \tfrac{1}{C_L+C_R}[C_R p_L + C_L p_R + C_L C_R (u_L - u_R)] \; , \\[2mm]
u_* &= \tfrac{1}{C_L+C_R}[C_L u_L + C_R u_R + (p_L - p_R)] \; , \\[2mm]
\rho_{*L} &= \rho_L + (p_* - p_L)/a_L^2 \; , \\[2mm]
\rho_{*R} &= \rho_R + (p_* - p_R)/a_R^2 \; .
\end{aligned}\right\} \tag{9.28}$$

In this approximation we do not need to make a choice for the averages $\bar{\rho}$ and $\bar{a}$; their values are replaced by data values at the foot of the corresponding characteristics. If $C_L = C_R = \bar{\rho}\bar{a}$, then the two approximations (9.20) and (9.28) are identical.

The two linearised approximations (9.20) and (9.28) for the star values are exceedingly simple and may be useful in a variety of ways. The approaches might prove very useful in solving the Riemann problem for complicated sets of equations.

We have now given the complete approximate solution to the sub–problems (9.7)–(9.10). In order to evaluate the Godunov flux (9.3) we need to *sample* the solution to find the value $\mathbf{W}_{i+\frac{1}{2}}(0)$ along the t–axis. This sampling procedure is virtually identical, although simpler, to the sampling procedure for the exact Riemann problem solution presented in Chap. 4. The reader is made aware that the numerical schemes associated with the simple linearised solutions just derived may not be robust enough to be used with absolute confidence under *all flow conditions*. In Sect. 9.5 we study hybrid Riemann solvers, which combine simple and sophisticated solvers to provide schemes that have effectively the computational cost of the simplest Riemann solvers and the robustness of the sophisticated Riemann solvers.

9.4 Approximations Based on the Exact Solver

In Chap. 4 we presented an exact Riemann solver based on the pressure equation

$$f(p) \equiv f_L(p, \mathbf{W}_L) + f_R(p, \mathbf{W}_R) + \Delta u = 0 , \quad \Delta u = u_R - u_L , \quad (9.29)$$

with

$$f_K(p) = \begin{cases} (p - p_K) \left[\dfrac{A_K}{p + B_K} \right]^{\frac{1}{2}} & \text{if } p > p_K \text{ (shock)} , \\[3mm] \dfrac{2a_K}{(\gamma - 1)} \left[\left(\dfrac{p}{p_K} \right)^z - 1 \right] & \text{if } p \le p_K \text{ (rarefaction)} , \end{cases} \quad (9.30)$$

$$z = \frac{\gamma - 1}{2\gamma} , \quad A_K = \frac{2}{(\gamma + 1)\rho_K} , \quad B_K = \left(\frac{\gamma - 1}{\gamma + 1} \right) p_K , \quad K = L, R . \quad (9.31)$$

Various approximations based on $f(p) = 0$ can be obtained, including curve–fitting procedures [361]. Here we give approximations based on the rarefaction and shock branches (9.30) of $f(p)$.

9.4.1 A Two–Rarefaction Riemann Solver (TRRS)

Recall that the non–linear waves in the Riemann problem solution are either shock or rarefaction waves and finding their type is part of the solution procedure. If one assumes *a–priori that both non–linear waves are rarefactions* then (9.29), with the appropriate choice of f_L and f_R in (9.30), becomes

$$\frac{2a_L}{(\gamma - 1)} \left[\left(\frac{p}{p_L} \right)^z - 1 \right] + \frac{2a_R}{(\gamma - 1)} \left[\left(\frac{p}{p_R} \right)^z - 1 \right] + u_R - u_L = 0 .$$

Solving this equation for pressure p_* gives the approximation

$$p_* = \left[\frac{a_L + a_R - \frac{\gamma - 1}{2}(u_R - u_L)}{a_L/p_L^z + a_R/p_R^z} \right]^{\frac{1}{z}} . \quad (9.32)$$

Having found p_* one can obtain the particle velocity u_* from any of the rarefaction wave relations

$$u_* = u_L - \frac{2a_L}{(\gamma - 1)} \left[\left(\frac{p_*}{p_L} \right)^z - 1 \right] \quad (9.33)$$

or

$$u_* = u_R + \frac{2a_R}{(\gamma - 1)} \left[\left(\frac{p_*}{p_R} \right)^z - 1 \right] . \qquad (9.34)$$

Alternatively, one can eliminate p_* from (9.33) and (9.34) to obtain a closed–form solution for the particle velocity

$$u_* = \frac{P_{LR} u_L / a_L + u_R / a_R + 2(P_{LR} - 1)/(\gamma - 1)}{P_{LR}/a_L + 1/a_R} , \quad P_{LR} = \left(\frac{p_L}{p_R} \right)^z . \qquad (9.35)$$

Computing p_* from (9.32) requires the evaluation of 3 fractional powers. A more efficient approach is to calculate u_* from (9.35), which only requires one fractional power, and then evaluate p_* from (9.33) or (9.34), or from a mean value as

$$p_* = \frac{1}{2} \left\{ p_L \left[1 + \frac{(\gamma - 1)}{2a_L} (u_L - u_*) \right]^{\frac{1}{z}} + p_R \left[1 + \frac{(\gamma - 1)}{2a_R} (u_* - u_R) \right]^{\frac{1}{z}} \right\} . \qquad (9.36)$$

Being consistent with the assumption that the two nonlinear waves are rarefaction waves, the computation of the densities ρ_{*L} and ρ_{*R} on either side of the contact discontinuity is obtained from the isentropic law, see Sect. 3.1.2 of Chap. 3. The result is

$$\rho_{*L} = \rho_L \left(\frac{p_*}{p_L} \right)^{\frac{1}{\gamma}} , \quad \rho_{*R} = \rho_R \left(\frac{p_*}{p_R} \right)^{\frac{1}{\gamma}} . \qquad (9.37)$$

An improved version of the two–rarefaction solution is obtained by using exact relations, for given p_* or u_*. For instance, suppose p_* is given by (9.32) say, then u_* can be found from

$$u_* = \frac{1}{2}(u_L + u_R) + \frac{1}{2} [f_R(p_*) - f_L(p_*)] , \qquad (9.38)$$

where the functions f_L and f_R are evaluated according to the exact relations (9.30) by comparing p_* with p_L and p_R. The densities ρ_{*L} and ρ_{*R} can be found from the isentropic law if the K wave is a rarefaction ($p_* \le p_K$) or from the shock relation if the K wave is a shock wave ($p_* > p_K$).

The two–rarefaction approximation is generally quite robust; it is more accurate, although more expensive, than the simple PVRS solutions (9.20) or (9.28) of the previous section. The TRRS is in fact exact when both non–linear waves are actually rarefaction waves. This can be detected a–priori by the condition

$$f(p_{min}) \ge 0 \text{ with } p_{min} = \min(p_L, p_R) . \qquad (9.39)$$

See Sect. 4.3 of Chap. 4 for details on the behaviour of the pressure function.

We have now given another approximate solution to the problem (9.7). The solution for (9.9)–(9.10) follows. The evaluation of the Godunov flux (9.3) requires sampling the solution to find the value $\mathbf{W}_{i+\frac{1}{2}}(0)$ along the t–axis, in the usual way. See Sect. 4.5 of Chap. 4.

9.4.2 A Two–Shock Riemann Solver (TSRS)

By assuming that *both non–linear waves are shock waves* in (9.29)–(9.30) one can derive the two–shock approximation

$$f(p) = (p - p_L)g_L(p) + (p - p_R)g_R(p) + u_R - u_L = 0 , \qquad (9.40)$$

with

$$g_K(p) = \left[\frac{A_K}{p + B_K} \right]^{\frac{1}{2}} \qquad (9.41)$$

and A_K, B_K given by (9.31). Unfortunately, this approximation does not lead to a closed–form solution. Further approximations must be constructed [114], [263], [361]. Obvious approximations to the two–shock approximation involve quadratic equations. These do not generally lead to robust schemes. One difficulty is the non–uniqueness of solutions and making the correct choice; the exact solution, as seen in Chap. 4, is unique. The other problem is the case of complex roots (non–existence) even for data for which the exact problem has a solution; in our experience these can occur very often and is therefore the most serious difficulty of the two–shock approach.

An alternative approach [361] is as follows. First we assume an estimate p_0 for the solution for pressure. Then we insert this estimate in the functions (9.41), which in turn are substituted into equation (9.40). We obtain

$$(p - p_L)g_L(p_0) + (p - p_R)g_R(p_0) + u_R - u_L = 0 .$$

The solution of this equation is immediate:

$$p_* = \frac{g_L(p_0)p_L + g_R(p_0)p_R - (u_R - u_L)}{g_L(p_0) + g_R(p_0)} . \qquad (9.42)$$

Being consistent with the two–shock assumption we derive a solution for the velocity u_* as

$$u_* = \frac{1}{2}(u_L + u_R) + \frac{1}{2}\left[(p_* - p_R)g_R(p_0) - (p_* - p_L)g_L(p_0)\right] . \qquad (9.43)$$

Solution values for $\rho_{*L} and \rho_{*R}$ obtained from shock relations, see Sect. 3.1.3 of Chap. 3, namely

$$\rho_{*L} = \rho_L \left[\frac{\frac{p_*}{p_L} + \frac{(\gamma-1)}{(\gamma+1)}}{\frac{(\gamma-1)}{(\gamma+1)}\frac{p_*}{p_L} + 1} \right] , \quad \rho_{*R} = \rho_R \left[\frac{\frac{p_*}{p_R} + \frac{(\gamma-1)}{(\gamma+1)}}{\frac{(\gamma-1)}{(\gamma+1)}\frac{p_*}{p_R} + 1} \right] . \qquad (9.44)$$

As to the choice for the pressure estimate p_0 we propose

$$p_0 = max(0, p_{pvrs}) , \qquad (9.45)$$

where p_{pvrs} is the solution (9.20) for pressure given by the PVRS solver of Sect. 9.3.

We have just presented another approximate solution to the problem (9.7). As before, the solution for (9.8)–(9.10) follows. The evaluation of the Godunov flux (9.3) requires sampling the solution to find the value $\mathbf{W}_{i+\frac{1}{2}}(0)$ along the $t-$ axis, in the usual way. See Sect. 4.5 of Chap. 4.

The approximation (9.42)–(9.44) to the star values (9.7) is more efficient than the TRRS and only slightly more expensive than the PVRS approximations. Also TSRS is more accurate than TRRS and PVRS for a wider range of flow conditions, except for *near vacuum conditions*, where TRRS is very accurate or indeed exact. As for the case of the TRRS approximation, we can improve the TSRS by using the true wave relations whenever possible. For instance, for given p_* as computed from (9.42), one can obtain u_*, ρ_{*L} and ρ_{*R} from exact wave relations. This is bound to improve the accuracy of the derived quantities.

9.5 Adaptive Riemann Solvers

In a typical flow field the overwhelming majority of local Riemann problems are a representation for smooth flow. Large gradients occur only near shock waves, contact surfaces, shear waves or some other sharp flow features. Large gradients generate Riemann problems with *widely different data states* $\mathbf{W}_L, \mathbf{W}_R$. Generally, it is in this kind of situations where approximate Riemann solvers can be fatally inaccurate, leading to failure of the numerical method being used. The rationale behind the use of hybrid schemes is *the use of simple Riemann solvers in regions of smooth flow and near isolated contacts and shear waves, and more sophisticated Riemann solvers elsewhere, in an adaptive fashion.*

Successful implementations of adaptive schemes involving the PVRS and the exact Riemann solvers were presented in [355] for the two–dimensional, time dependent Euler equations. Toro and Chou [377] extended the idea to the case of the steady supersonic Euler equations. Quirk [275] implemented this Riemann–solver adaptation approach in a MUSCL–type scheme, used in conjunction with adaptive mesh refinement techniques.

Here we present two hybrid schemes to compute the star values (9.7). Problems (9.8)–(9.10) are solved as before and the sampling is handled as described in Sect. 4.5 of Chap. 4.

9.5.1 An Adaptive Iterative Riemann Solver (AIRS)

This adaptive scheme makes use of two Riemann solvers: any of the primitive–variable Riemann solvers PVRS of Sect. 9.3 and the exact Riemann solver of Chap. 4. The PVRS scheme is used if the following two conditions are met:

$$Q = p_{\max}/p_{\min} < Q_{\text{user}} \qquad (9.46)$$

and

$$p_{\min} < p_* < p_{\max} \,, \tag{9.47}$$

where

$$p_{\min} \equiv \min(p_L, p_R) \,, \quad p_{\max} \equiv \max(p_L, p_R) \,, \quad p_* \equiv p_{pvrs} \,. \tag{9.48}$$

Otherwise, the exact Riemann solver is used.

Some remarks on the switching conditions (9.46)–(9.47) are in order. Condition (9.46) ensures that the pressure data values p_L, p_R are not widely different. Condition (9.47) imposes an extra restriction on the use of PVRS. The pressure restriction (9.46) is not sufficient; in fact for $Q \approx 1, (p_L \approx p_R)$ and $\Delta u = u_R - u_L$ negative and large in absolute value, strong shock waves are present in the solution of the Riemann problem, that is $p_* > p_{\max}$. For Δu large and positive $p_* < p_{\min}$ and strong rarefactions are present in the exact solution of the Riemann problem. Condition (9.47) is effectively a condition on Δu and excludes the two–rarefaction and the two–shock cases; both of these cases occur naturally at reflected boundaries, where it would be unwise to employ unreliable approximations. Also, these two cases are inconsistent with condition (9.46) on pressure ratios.

A choice of the switching parameter Q_{user} is to be made. Extensive testing suggests that the value $Q_{\mathrm{user}} = 2$ is perfectly adequate to give both very robust and efficient schemes. Even much larger values of Q_{user} can give accurate solutions, but the gains are not significant and thus the cautious choice of $Q_{\mathrm{user}} = 2$ is recommended. For typical flow conditions and meshes, over 90% of all Riemann problems are handled by the cheap linearised Riemann solver. Effectively, the resulting schemes have the efficiency of the cheapest Riemann solvers and the robustness of the exact Riemann solver. A disadvantage of this hybrid PVRS–EXACT scheme is the iterative character of the robust component of the scheme, namely the exact Riemann solver. This may be inconvenient for some computer architectures. One possibility here is to fix the number of iterations in the exact Riemann solver. In our experience, *one iteration* leads to very accurate values for pressure and subsequent quantities derived. This is due in part to the provision of a sophisticated starting value for the iteration procedure.

9.5.2 An Adaptive Noniterative Riemann Solver (ANRS)

Here we propose to combine a PVRS scheme, as the cheap component, together with the non–iterative TRRS and TSRS solvers of Sects. 9.4.1 and 9.4.2 to provide the robust component of the adaptive scheme. The use of PVRS is again restricted by conditions (9.46)–(9.47) of the previous scheme. If any of conditions (9.46) or (9.47) are not met we use either TRRS or TSRS. The switching between TRRS and TSRS is motivated by the behaviour of the exact function for pressure, see Sect. 4.3.1 of Chap. 4, and is as follows. If

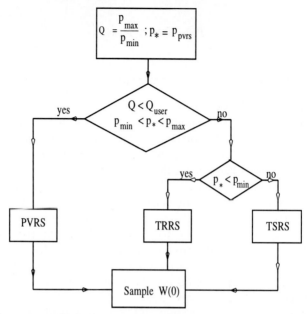

Fig. 9.4. Flow chart for Adaptive Noniterative Riemann Solver (ANRS) involving PVRS, TRRS and TSRS

$$p_{pvrs} \leq p_{min} , \tag{9.49}$$

then we use TRRS, otherwise we use TSRS. The flow chart of Fig. 9.4 illustrates the implementation of this adaptive scheme. The problems of computing the tangential velocity components, handling sonic flow and the sampling procedure to find the Godunov flux are dealt with as described in the previous sections. This adaptive noniterative Riemann solver is recommended for practical applications.

9.6 Numerical Results

Here we assess the performance of Godunov's first–order upwind method used in conjunction with the approximate Riemann solvers presented in this chapter. We select five test problems for the one–dimensional, time dependent Euler equations for ideal gases with $\gamma = 1.4$; these have exact solutions, which are evaluated by running the code HE–E1RPEXACT of *NUMERICA* [369].

In all chosen tests, data consists of two constant states $\mathbf{W}_L = [\rho_L, u_L, p_L]^T$ and $\mathbf{W}_R = [\rho_R, u_R, p_R]^T$, separated by a discontinuity at a position $x = x_0$. The states $\mathbf{W}_L$ and $\mathbf{W}_R$ are given in Table 9.1. The ratio of specific heats is chosen to be $\gamma = 1.4$. For all test problems the spatial domain is the interval $[0, 1]$ which is discretised with $M = 100$ computing cells. The Courant number coefficient is $C_{cfl} = 0.9$; boundary conditions are transmissive and S_{max}^n is found using the simplified formula (6.20) of Chapt. 6. But given that this formula is somewhat unreliable, see discussion of Sect. 6.3.2 of Chapter 6, in all computations presented here we take, for the the first 5 time steps, a Courant number coefficient C_{cfl} reduced by a factor of 0.2.

Test	ρ_L	u_L	p_L	ρ_R	u_R	p_R
1	1.0	0.75	1.0	0.125	0.0	0.1
2	1.0	-2.0	0.4	1.0	2.0	0.4
3	1.0	0.0	1000.0	1.0	0.0	0.01
4	5.99924	19.5975	460.894	5.99242	-6.19633	46.0950
5	1.0	-19.59745	1000.0	1.0	-19.59745	0.01

Table 9.1. Data for five test problems with exact solution. Test 5 is like Test 3 with negative uniform background speed

Test 1 is a *modified version* of the popular Sod's test [318]; the solution consists of a right shock wave, a right travelling contact wave and a left *sonic* rarefaction wave; this test is very useful in assessing the *entropy satisfaction* property of numerical methods. Test 2 has solution consisting of two symmetric rarefaction waves and a trivial contact wave of zero speed; the *Star Region* between the non–linear waves is close to vacuum, which makes this problem a suitable test for assessing the performance of numerical methods for low–density flows; this is the so called *123 problem* introduced in chapter Chap. 4. Test 3 is designed to assess the robustness and accuracy of numerical methods; its solution consists of a strong shock wave, a contact surface and a left rarefaction wave. Test 4 is also designed to test robustness of numerical methods; the solution consists of three strong discontinuities travelling to the right. See Sect. 4.3.3 of Chap. 4 for more details on the exact solution of these test problems. Test 5 is also designed to test the robustness of numerical methods but the main reason for devising this test is to assess the ability of the numerical methods to resolve *slowly– moving contact discontinuities*. The exact solution of Test 5 consists of a left rarefaction wave, a right–travelling shock wave and a *stationary* contact discontinuity. For each test we select a convenient position x_0 of the initial discontinuity and an output time. These are stated in the legend of each figure displaying computational results.

We present numerical results for two of the approximate Riemann solvers presented in this chapter, namely the Two–Shock Riemann solver (TSRS) and the Adaptive Noniterative Riemann Solver (ANRS). The numerical solutions are obtained by running the code HE–E1GODSTATE of *NUMERICA* [369]. The results from TSRS are shown in Figs. 9.5 to 9.9 and those of ANRS are shown in Figs. 9.10 to 9.14. All of these results are to be compared with those obtained from the Godunov scheme used in conjunction with the exact Riemann solver, see Figs. 6.8 to 6.12, Chapt. 6; to plotting accuracy, there is no difference in the computed results. The two approximate Riemann solvers TSRS and ANRS are recommended for practical applications.

The Godunov–type methods based on the approximate–state Riemann solvers of this chapter are extended to second–order of accuracy using the techniques of Chaps. 13 and 14, for one–dimensional problems. Approaches for including source terms are given in Chapt. 15 and for solving multidimensional problems in Chap. 16.

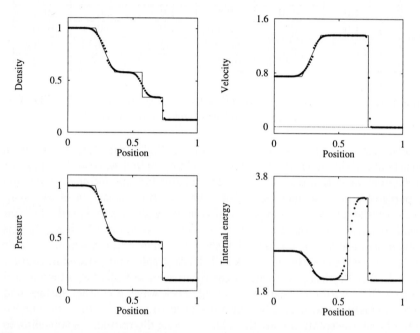

Fig. 9.5. Two–Shock Riemann Solver applied to Test 1, with $x_0 = 0.3$. Numerical (symbol) and exact (line) solutions are compared at time 0.2 units

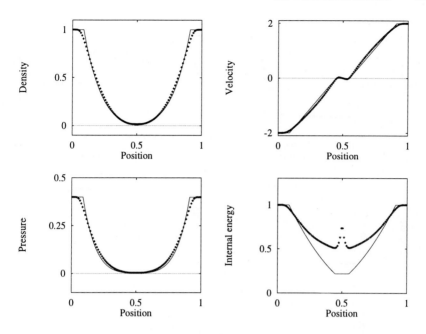

Fig. 9.6. Two–Shock Riemann Solver applied to Test 2, with $x_0 = 0.5$. Numerical (symbol) and exact (line) solutions are compared at time 0.15 units

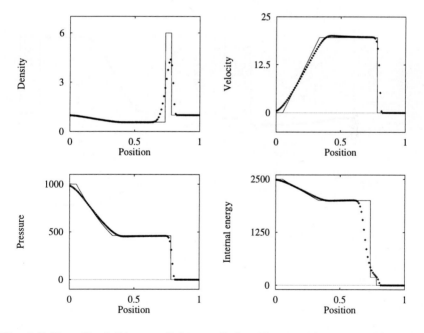

Fig. 9.7. Two–Shock Riemann Solver applied to Test 3, with $x_0 = 0.5$. Numerical (symbol) and exact (line) solutions are compared at time 0.012 units

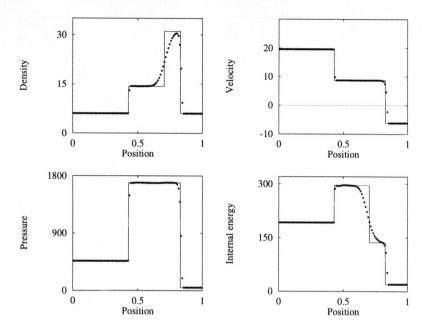

Fig. 9.8. Two–Shock Riemann Solver applied to Test 4, with $x_0 = 0.4$. Numerical (symbol) and exact (line) solutions are compared at time 0.035 units

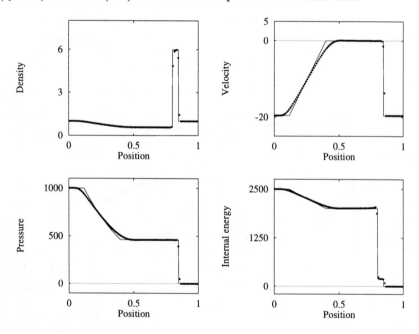

Fig. 9.9. Two–Shock Riemann Solver applied to Test 5, with $x_0 = 0.8$. Numerical (symbol) and exact (line) solutions are compared at time 0.012 units

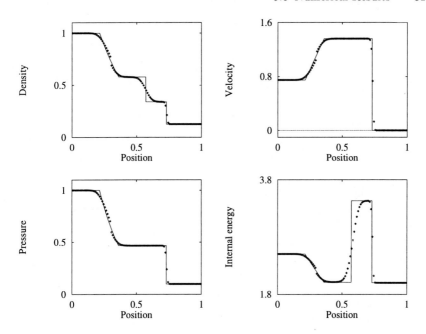

Fig. 9.10. Adaptive Noniterative Riemann Solver applied to Test 1, with $x_0 = 0.3$. Numerical (symbol) and exact (line) solutions are compared at time 0.2 units

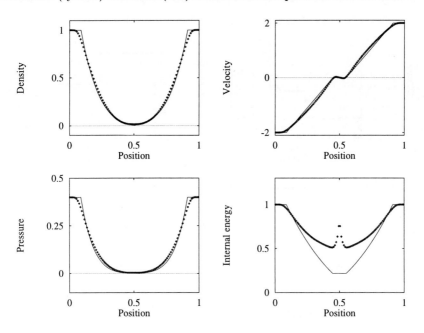

Fig. 9.11. Adaptive Noniterative Riemann Solver applied to Test 2, with $x_0 = 0.5$. Numerical (symbol) and exact (line) solutions are compared at time 0.15 units

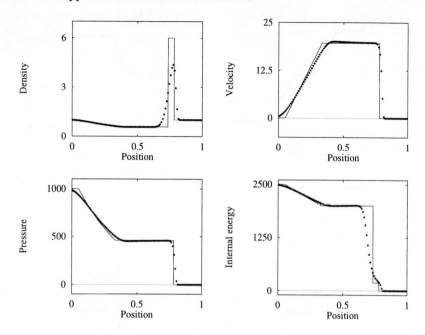

Fig. 9.12. Adaptive Noniterative Riemann Solver applied to Test 3, with $x_0 = 0.5$. Numerical (symbol) and exact (line) solutions are compared at time 0.012 units

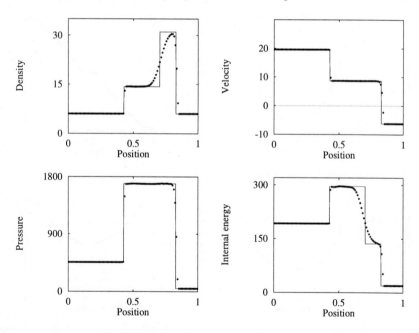

Fig. 9.13. Adaptive Noniterative Riemann Solver applied to Test 4, with $x_0 = 0.4$. Numerical (symbol) and exact (line) solutions are compared at time 0.035 units

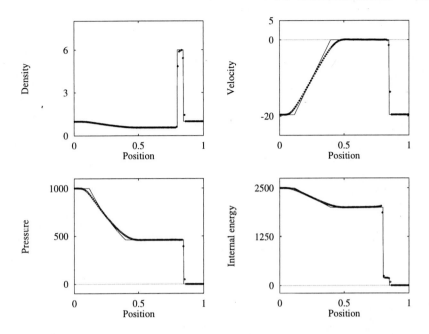

Fig. 9.14. Adaptive Noniterative Riemann Solver applied to Test 5, with $x_0 = 0.8$. Numerical (symbol) and exact (line) solutions are compared at time 0.012 units

10. The HLL and HLLC Riemann Solvers

For the purpose of computing a Godunov flux, Harten, Lax and van Leer [164] presented a novel approach for solving the Riemann problem approximately. The resulting Riemann solvers have become known as HLL Riemann solvers. In this approach an *approximation for the intercell numerical flux is obtained directly*, unlike the Riemann solvers presented previously in Chaps. 4 and 9. The central idea is to assume a wave configuration for the solution that consists of two waves separating three constant states. Assuming that the wave speeds are given by some algorithm, application of the integral form of the conservation laws gives a closed–form, approximate expression for the flux. The approach produced practical schemes after the contributions of Davis [103] and Einfeldt [117], who independently proposed various ways of computing the wave speeds required to completely determine the intercell flux. The resulting HLL Riemann solvers form the bases of very efficient and robust approximate Godunov–type methods. One difficulty with these schemes, however, is the assumption of a two–wave configuration. This is correct only for hyperbolic systems of two equations, such as the one–dimensional shallow water equations. For larger systems, such as the Euler equations or the split two–dimensional shallow water equations for example, the two–wave assumption is incorrect. As a consequence the resolution of physical features such as contact surfaces, shear waves and material interfaces, can be very inaccurate. For the limiting case in which these features are stationary relative to the mesh, the resulting numerical smearing is unacceptable. In view of these shortcomings of the HLL approach, a modification called the HLLC Riemann solver (C stands for Contact) was put forward by Toro, Spruce and Speares [380]. In spite of the limited experience available in using the HLLC scheme, the evidence is that this appears to offer a useful approximate Riemann solver for practical applications. Batten, Leschziner and Goldberg [19] have recently proposed implicit versions of the HLLC Riemann solver, and have applied the scheme to turbulent flows.

In this Chapter we present the HLL and HLLC Riemann solvers as applied to the three–dimensional, time dependent Euler equations. The principles can easily be extended to solve other hyperbolic systems. Useful background reading is found in Chaps. 3, 4, 6 and 9. The rest of this chapter is organised as follows: Sect. 10.1 recalls the Euler equations, the Godunov method and the structure of the exact solution of the Riemann problem for the split three–dimensional equations. In Sect. 10.2 we recall the solution of the Riemann problem and some useful integral relations. In Sect. 10.3 we present the original approach of Harten, Lax and van Leer. In Sect. 10.4 we present the HLLC Riemann solver and in Sect. 10.5 we give various algorithms for computing the required wave speeds. A summary of the HLL and HLLC schemes is presented in Sect. 10.6. In Sect. 10.7 we analyse the behaviour of the approximate Riemann solvers in the presence of contacts and passive scalars. Numerical results are shown in Sect. 10.8.

10.1 The Riemann Problem and the Godunov Flux

We are concerned with solving numerically the general Initial Boundary Value Problem (IBVP)

$$
\left.
\begin{array}{lll}
\text{PDEs} & : & \mathbf{U}_t + \mathbf{F}(\mathbf{U})_x = \mathbf{0} \, , \\
\text{ICs} & : & \mathbf{U}(x,0) = \mathbf{U}^{(0)}(x) \, , \\
\text{BCs} & : & \mathbf{U}(0,t) = \mathbf{U}_\mathrm{l}(t) \, , \ \mathbf{U}(L,t) = \mathbf{U}_\mathrm{r}(t) \, ,
\end{array}
\right\}
\tag{10.1}
$$

in a domain $x_l \leq x \leq x_r$, utilising the explicit conservative formula

$$
\mathbf{U}_i^{n+1} = \mathbf{U}_i^n + \frac{\Delta t}{\Delta x} [\mathbf{F}_{i-\frac{1}{2}} - \mathbf{F}_{i+\frac{1}{2}}] \, .
\tag{10.2}
$$

In Chap. 6 we defined the Godunov intercell numerical flux

$$
\mathbf{F}_{i+\frac{1}{2}} = \mathbf{F}(\mathbf{U}_{i+\frac{1}{2}}(0)) \, ,
\tag{10.3}
$$

in which $\mathbf{U}_{i+\frac{1}{2}}(0)$ is the exact similarity solution $\mathbf{U}_{i+\frac{1}{2}}(x/t)$ of the Riemann problem

$$
\left.
\begin{array}{l}
\mathbf{U}_t + \mathbf{F}(\mathbf{U})_x = \mathbf{0} \, , \\
\mathbf{U}(x,0) = \left\{
\begin{array}{lll}
\mathbf{U}_\mathrm{L} & if & x < 0 \, , \\
\mathbf{U}_\mathrm{R} & if & x > 0 \, ,
\end{array}
\right.
\end{array}
\right\}
\tag{10.4}
$$

evaluated at $x/t = 0$. Fig. 10.1 shows the structure of the exact solution of the Riemann problem for the x–split, three dimensional Euler equations, for which the vectors of conserved variables and fluxes are

$$
\mathbf{U} =
\begin{bmatrix}
\rho \\
\rho u \\
\rho v \\
\rho w \\
E
\end{bmatrix}
, \quad
\mathbf{F} =
\begin{bmatrix}
\rho u \\
\rho u^2 + p \\
\rho u v \\
\rho u w \\
u(E + p)
\end{bmatrix}
.
\tag{10.5}
$$

The value $x/t = 0$ for the Godunov flux corresponds to the t–axis. See Chaps. 4 and 6 for details. The piece–wise constant initial data, in terms of primitive variables, is

$$\mathbf{W}_L = \begin{bmatrix} \rho_L \\ u_L \\ v_L \\ w_L \\ p_L \end{bmatrix} , \quad \mathbf{W}_R = \begin{bmatrix} \rho_R \\ u_R \\ v_R \\ w_R \\ p_R \end{bmatrix} . \tag{10.6}$$

In Chap. 9 we provided approximations to the state $\mathbf{U}_{i+\frac{1}{2}}(x/t)$ and obtained

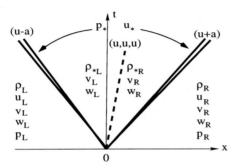

Fig. 10.1. Structure of the exact solution of the Riemann problem for the x–split three dimensional Euler equations. There are five wave families *associated* with the eigenvalues $u - a$, u (of multiplicity 3) and $u + a$

a corresponding approximate Godunov method by evaluating the physical flux function $\mathbf{F}$ at this approximate state. The purpose of this chapter is to find *direct approximations to the flux function* $\mathbf{F}_{i+\frac{1}{2}}$ following the novel approach proposed by Harten, Lax and van Leer [158].

10.2 The Riemann Problem and Integral Relations

Consider Fig. 10.2, in which the whole of the wave structure arising from the exact solution of the Riemann problem is contained in the control volume $[x_L, x_R] \times [0, T]$, that is

$$x_L \leq TS_L , \quad x_R \geq TS_R , \tag{10.7}$$

where S_L and S_R are the *fastest signal velocities* perturbing the initial data states $\mathbf{U}_L$ and $\mathbf{U}_R$ respectively, and T is a chosen time. The integral form of the conservation laws in (10.4), in the control volume $[x_L, x_R] \times [0, T]$ reads

$$\int_{x_L}^{x_R} \mathbf{U}(x, T)dx = \int_{x_L}^{x_R} \mathbf{U}(x, 0)dx + \int_0^T \mathbf{F}(\mathbf{U}(x_L, t))dt - \int_0^T \mathbf{F}(\mathbf{U}(x_R, t))dt .$$
$$\tag{10.8}$$

See Sect. 2.4.1 of Chap. 2 for details on integral forms of conservation laws. Evaluation of the right–hand side of this expression gives

$$\int_{x_L}^{x_R} \mathbf{U}(x, T)dx = x_R\mathbf{U}_R - x_L\mathbf{U}_L + T(\mathbf{F}_L - \mathbf{F}_R) , \qquad (10.9)$$

where $\mathbf{F}_L = \mathbf{F}(\mathbf{U}_L)$ and $\mathbf{F}_R = \mathbf{F}(\mathbf{U}_R)$. We call the integral relation (10.9) the *Consistency Condition*. Now we split the integral on the left–hand side of (10.8) into three integrals, namely

$$\int_{x_L}^{x_R} \mathbf{U}(x, T)dx = \int_{x_L}^{TS_L} \mathbf{U}(x, T)dx + \int_{TS_L}^{TS_R} \mathbf{U}(x, T)dx + \int_{TS_R}^{x_R} \mathbf{U}(x, T)dx$$

and evaluate the first and third terms on the right–hand side. We obtain

$$\int_{x_L}^{x_R} \mathbf{U}(x, T)dx = \int_{TS_L}^{TS_R} \mathbf{U}(x, T)dx + (TS_L - x_L)\mathbf{U}_L + (x_R - TS_R)\mathbf{U}_R .$$
$$(10.10)$$

Comparing (10.10) with (10.9) gives

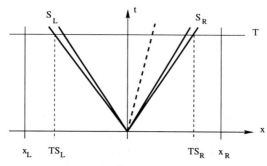

Fig. 10.2. Control volume $[x_L, x_R] \times [0, T]$ on x–t plane. S_L and S_R are the fastest signal velocities arising from the solution of the Riemann problem

$$\int_{TS_L}^{TS_R} \mathbf{U}(x, T)dx = T(S_R\mathbf{U}_R - S_L\mathbf{U}_L + \mathbf{F}_L - \mathbf{F}_R) . \qquad (10.11)$$

On division through by the length $T(S_R - S_L)$, which is the width of the wave system of the solution of the Riemann problem between the slowest and fastest signals at time T, we have

$$\frac{1}{T(S_R - S_L)} \int_{TS_L}^{TS_R} \mathbf{U}(x, T)dx = \frac{S_R\mathbf{U}_R - S_L\mathbf{U}_L + \mathbf{F}_L - \mathbf{F}_R}{S_R - S_L} . \qquad (10.12)$$

Thus, the integral average of the exact solution of the Riemann problem between the slowest and fastest signals at time T is a known constant, provided that the signal speeds S_L and S_R are known; such constant is the right–hand side of (10.12) and we denote it by

$$\mathbf{U}^{hll} = \frac{S_R \mathbf{U}_R - S_L \mathbf{U}_L + \mathbf{F}_L - \mathbf{F}_R}{S_R - S_L} \,. \tag{10.13}$$

We now apply the integral form of the conservation laws to the left portion of Fig. 10.2, that is the control volume $[x_L, 0] \times [0, T]$. We obtain

$$\int_{TS_L}^{0} \mathbf{U}(x, T) dx = -TS_L \mathbf{U}_L + T(\mathbf{F}_L - \mathbf{F}_{0L}) \,, \tag{10.14}$$

where $\mathbf{F}_{0L}$ is the flux $\mathbf{F}(\mathbf{U})$ along the t–axis. Solving for $\mathbf{F}_{0L}$ we find

$$\mathbf{F}_{0L} = \mathbf{F}_L - S_L \mathbf{U}_L - \frac{1}{T} \int_{TS_L}^{0} \mathbf{U}(x, T) dx \,. \tag{10.15}$$

Evaluation of the integral form of the conservation laws on the control volume $[0, x_R] \times [0, T]$ yields

$$\mathbf{F}_{0R} = \mathbf{F}_R - S_R \mathbf{U}_R + \frac{1}{T} \int_{0}^{TS_R} \mathbf{U}(x, T) dx \,. \tag{10.16}$$

The reader can easily verify that the equality

$$\mathbf{F}_{0L} = \mathbf{F}_{0R}$$

results in the *Consistency Condition* (10.9). All relations so far are exact, as we are assuming the exact solution of the Riemann problem.

10.3 The HLL Approximate Riemann Solver

Harten, Lax and van Leer [164] put forward the following approximate Riemann solver

$$\tilde{\mathbf{U}}(x, t) = \begin{cases} \mathbf{U}_L & if & \frac{x}{t} \leq S_L \,, \\ \mathbf{U}^{hll} & if & S_L \leq \frac{x}{t} \leq S_R \,, \\ \mathbf{U}_R & if & \frac{x}{t} \geq S_R \,, \end{cases} \tag{10.17}$$

where $\mathbf{U}^{hll}$ is the constant state vector given by (10.13) and the speeds S_L and S_R are assumed to be known. Fig. 10.3 shows the structure of this approximate solution of the Riemann problem, called the HLL Riemann solver. Note that this approximation consists of just three constant states separated by two waves. The *Star Region* consists of a *single* constant state; all intermediate states separated by intermediate waves are *lumped* into the single state $\mathbf{U}^{hll}$. The corresponding flux $\mathbf{F}^{hll}$ along the t–axis is found from the relations (10.15) or (10.16), with the exact integrand replaced by the approximate solution (10.17). Note that we *do not take* $\mathbf{F}^{hll} = \mathbf{F}(\mathbf{U}^{hll})$. The non–trivial case of interest is the subsonic case $S_L \leq 0 \leq S_R$. Substitution of the integrand in (10.15) or (10.16) by $\mathbf{U}^{hll}$ in (10.13) gives

$$\mathbf{F}^{hll} = \mathbf{F}_L + S_L(\mathbf{U}^{hll} - \mathbf{U}_L) \,, \tag{10.18}$$

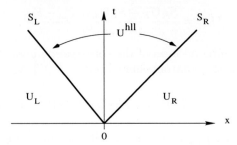

Fig. 10.3. Approximate HLL Riemann solver. Solution in the *Star Region* consists of a single state $\mathbf{U}^{hll}$ separated from data states by two waves of speeds S_L and S_R

or

$$\mathbf{F}^{hll} = \mathbf{F}_R + S_R(\mathbf{U}^{hll} - \mathbf{U}_R) . \tag{10.19}$$

Note that relations (10.18) and (10.19) are also obtained from applying Rankine–Hugoniot Conditions across the left and right waves respectively; see Sect. 2.4.2 of Chap. 2 and Sect. 3.1.3 of Chap. 3 for details on the Rankine–Hugoniot Conditions. Use of (10.13) in (10.18) or (10.19) gives the HLL flux

$$\mathbf{F}^{hll} = \frac{S_R\mathbf{F}_L - S_L\mathbf{F}_R + S_L S_R(\mathbf{U}_R - \mathbf{U}_L)}{S_R - S_L} . \tag{10.20}$$

The corresponding intercell flux for the approximate Godunov method is then given by

$$\mathbf{F}_{i+\frac{1}{2}}^{hll} = \begin{cases} \mathbf{F}_L & if & 0 \le S_L , \\ \frac{S_R\mathbf{F}_L - S_L\mathbf{F}_R + S_L S_R(\mathbf{U}_R - \mathbf{U}_L)}{S_R - S_L} , & if & S_L \le 0 \le S_R , \\ \mathbf{F}_R & if & 0 \ge S_R . \end{cases} \tag{10.21}$$

Given an algorithm to compute the speeds S_L and S_R we have an approximate intercell flux (10.21) to be used in the conservative formula (10.2) to produce an approximate Godunov method. Procedures to estimate the wave speeds S_L and S_R are given in Sect. 10.5. Harten, Lax and van Leer [164] showed that the Godunov scheme (10.2), (10.21), if convergent, converges to the weak solution of the conservation laws. In fact they proved that the converged solution is also the physical, entropy satisfying, solution of the conservation laws. Their results actually apply to a larger class of approximate Riemann solvers. One of the requirements is *consistency with the integral form of the conservation laws*. That is, an approximate solution $\tilde{\mathbf{U}}(x,t)$ is consistent with the integral form of the conservation laws if, when substituted for the exact solution $\mathbf{U}(x,t)$ in the Consistency Condition (10.9), the right–hand side remains unaltered.

A shortcoming of the HLL scheme is exposed by contact discontinuities, shear waves and material interfaces. These waves are associated with the

multiple eigenvalue $\lambda_2 = \lambda_3 = \lambda_4 = u$. See Fig. 10.1. Note that in the integral
(10.12), all that matters is the average across the wave structure, without
regard for the spatial variations of the solution of the Riemann problem in the
Star Region. As pointed out by Harten, Lax and van Leer themselves [164],
this defect of the HLL scheme may be corrected by restoring the missing
waves. Accordingly, Toro, Spruce and Speares [380] proposed the so called
HLLC scheme, where C stands for *Contact*. In this scheme the missing middle
waves are put back into the structure of the approximate Riemann solver.

10.4 The HLLC Approximate Riemann Solver

The HLLC scheme is a modification of the HLL scheme described in the
previous section, whereby the missing contact and shear waves are restored.
The scheme was first presented in terms of the time–dependent, two dimensional Euler equations [380]; later, it was applied to the steady supersonic
two–dimensional Euler equations [376] and to the time–dependent, two dimensional shallow water equations [126], [127].

Consider Fig. 10.2, in which the complete structure of the solution of the
Riemann problem is contained in a sufficiently large control volume $[x_L, x_R] \times$
$[0, T]$. Now, in addition to the slowest and fastest signal speeds S_L and S_R we
include a middle wave of speed S_*, corresponding to the multiple eigenvalue
$\lambda_2 = \lambda_3 = \lambda_4 = u$. See Fig. 10.4. Evaluation of the integral form of the

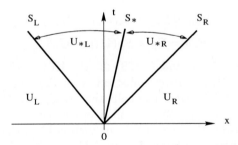

Fig. 10.4. HLLC approximate Riemann solver. Solution in the *Star Region* consists
of two constant states separated from each other by a middle wave of speed S_*

conservation laws in the control volume reproduces the result of equation
(10.12), even if variations of the integrand across the wave of speed S_* are
allowed. Note that the consistency condition (10.9) effectively becomes the
condition (10.12). By splitting the left–hand side of integral (10.12) into two
terms we obtain

$$\left. \frac{1}{T(S_R-S_L)} \int_{TS_L}^{TS_R} \mathbf{U}(x,T)dx \; = \; \frac{1}{T(S_R-S_L)} \int_{TS_L}^{TS_*} \mathbf{U}(x,T)dx \right.$$
$$\left. + \frac{1}{T(S_R-S_L)} \int_{TS_*}^{TS_R} \mathbf{U}(x,T)dx \; . \right\} \quad (10.22)$$

We define the integral averages

$$
\left.
\begin{array}{l}
\mathbf{U}_{*L} = \frac{1}{T(S_R - S_L)} \int_{TS_L}^{TS_*} \mathbf{U}(x, T) dx \ , \\[2mm]
\mathbf{U}_{*R} = \frac{1}{T(S_R - S_L)} \int_{TS_*}^{TS_R} \mathbf{U}(x, T) dx \ .
\end{array}
\right\}
\tag{10.23}
$$

By substitution of (10.23) into (10.22) and use of (10.12), the Consistency Condition (10.9) becomes

$$
\left(\frac{S_* - S_L}{S_R - S_L} \right) \mathbf{U}_{*L} + \left(\frac{S_R - S_*}{S_R - S_L} \right) \mathbf{U}_{*R} = \mathbf{U}^{hll} \ ,
\tag{10.24}
$$

where $\mathbf{U}^{hll}$ is given by (10.13). The HLLC approximate Riemann solver is given as follows

$$
\tilde{\mathbf{U}}(x, t) =
\left\{
\begin{array}{lll}
\mathbf{U}_L & if & \frac{x}{t} \leq S_L \ , \\
\mathbf{U}_{*L} & if & S_L \leq \frac{x}{t} \leq S_* \ , \\
\mathbf{U}_{*R} & if & S_* \leq \frac{x}{t} \leq S_R \ , \\
\mathbf{U}_R & if & \frac{x}{t} \geq S_R \ .
\end{array}
\right.
\tag{10.25}
$$

Fig. 10.4 shows the structure of the HLLC approximate Riemann solver. By integrating over appropriate control volumes, or more directly, by applying Rankine–Hugoniot Conditions across each of the waves of speeds S_L, S_*, S_R, we obtain

$$
\mathbf{F}_{*L} = \mathbf{F}_L + S_L (\mathbf{U}_{*L} - \mathbf{U}_L) \ ,
\tag{10.26}
$$

$$
\mathbf{F}_{*R} = \mathbf{F}_{*L} + S_* (\mathbf{U}_{*R} - \mathbf{U}_{*L}) \ ,
\tag{10.27}
$$

$$
\mathbf{F}_{*R} = \mathbf{F}_R + S_R (\mathbf{U}_{*R} - \mathbf{U}_R) \ .
\tag{10.28}
$$

Compare relations (10.26) and (10.28) for the HLLC scheme with (10.18) and (10.19) for the HLL scheme. Substitution of $\mathbf{F}_{*L}$ from (10.26) and $\mathbf{F}_{*R}$ from (10.28) into (10.27) gives identically the Consistency Condition (10.24). Hence conditions (10.26)–(10.28) are sufficient for ensuring consistency. These are three equations for the four unknowns vectors $\mathbf{U}_{*L}$, $\mathbf{F}_{*L}$, $\mathbf{U}_{*R}$, $\mathbf{F}_{*R}$.

The aim is to find the vectors $\mathbf{U}_{*L}$ and $\mathbf{U}_{*R}$ so that the fluxes $\mathbf{F}_{*L}$ and $\mathbf{F}_{*R}$ can be determined from (10.26) and (10.28) respectively. We impose the following conditions on the approximate Riemann solver

$$
\left.
\begin{array}{l}
u_{*L} = u_{*R} = u_* \ , \\
p_{*L} = p_{*R} = p_* \ , \\
v_{*L} = v_L \ , v_{*R} = v_R \ , \\
w_{*L} = w_L \ , w_{*R} = w_R \ .
\end{array}
\right\}
\tag{10.29}
$$

As seen in Chap. 4, these conditions are satisfied by the exact solution. In addition, it is entirely justified, and convenient, to set

$$
S_* = u_* \ .
\tag{10.30}
$$

Now equations (10.26) and (10.28) can be re–arranged as

$$S_L \mathbf{U}_{*L} - \mathbf{F}_{*L} = \mathbf{Q}_L \, , \qquad (10.31)$$

$$S_R \mathbf{U}_{*R} - \mathbf{F}_{*R} = \mathbf{Q}_R \, , \qquad (10.32)$$

where $\mathbf{Q}_L$ and $\mathbf{Q}_R$ are known constant vectors. Utilisation of conditions (10.29)–(10.30) into (10.31) and (10.32) gives the solution vectors

$$\mathbf{U}_{*K} = \rho_K \left(\frac{S_K - u_K}{S_K - S_*} \right) \begin{bmatrix} 1 \\ S_* \\ v_K \\ w_K \\ \frac{E_K}{\rho_K} + (S_* - u_K)\left[S_* + \frac{p_K}{\rho_K(S_K - u_K)} \right] \end{bmatrix} , \qquad (10.33)$$

for $K = L$ and $K = R$. Therefore the fluxes $\mathbf{F}_{*L}$ and $\mathbf{F}_{*R}$ in (10.26) and (10.28) are completely determined.

In view of (10.25) the HLLC flux for the approximate Godunov method can be written as

$$\mathbf{F}_{i+\frac{1}{2}}^{hllc} = \begin{cases} \mathbf{F}_L & if \quad 0 \leq S_L \, , \\ \mathbf{F}_{*L} = \mathbf{F}_L + S_L(\mathbf{U}_{*L} - \mathbf{U}_L) & if \quad S_L \leq 0 \leq S_* \, , \\ \mathbf{F}_{*R} = \mathbf{F}_R + S_R(\mathbf{U}_{*R} - \mathbf{U}_R) & if \quad S_* \leq 0 \leq S_R \, , \\ \mathbf{F}_R & if \quad 0 \geq S_R \, . \end{cases} \qquad (10.34)$$

where $\mathbf{U}_{*L}$ and $\mathbf{U}_{*R}$ are given by (10.33). We note that for any passive scalar q advected with the fluid speed, the x–split equations will include

$$(\rho q)_t + (\rho q u)_x = 0 \, . \qquad (10.35)$$

The corresponding HLLC state is given by

$$(\rho q)_{*K} = \rho_K \left(\frac{S_K - u_K}{S_K - S_*} \right) q_K \, , \qquad (10.36)$$

for $K = L$ and $K = R$. The tangential velocity components v and w are special cases of passive scalars; see (10.33); compare (10.36) with (10.33) for $q = v$ and $q = w$.

10.5 Wave–Speed Estimates

In order to determine completely the numerical fluxes in both the HLL and HLLC Riemann solvers we need to provide an algorithm for computing the wave speeds. For the HLL solver one requires S_L and S_R. For the HLLC scheme one requires in addition an estimate for the speed of the middle wave S_*. It turns out that one possible scheme for HLLC is to derive S_* from S_L and S_R. There essentially two ways of estimating S_L, S_* and S_R. Historically, the most popular approach has been to estimate the speeds directly. A more recent approach relies on pressure–velocity estimates for the *Star Region*; these are then utilised to obtain S_L, S_* and S_R using exact wave relations.

10.5.1 Direct Wave Speed Estimates

The most well known approach for estimating bounds for the minimum and maximum signal velocities present in the solution of the Riemann problem is to provide, directly, wave speeds S_L and S_R. Davis [103] suggested the simple estimates

$$S_L = u_L - a_L , \quad S_R = u_R + a_R \tag{10.37}$$

and

$$S_L = \min\{u_L - a_L, u_R - a_R\} , \quad S_R = \max\{u_L + a_L, u_R + a_R\} . \tag{10.38}$$

These estimates make use of data values only. Davis [103] and Einfeldt [117], also proposed to use the Roe [281] average eigenvalues for the left and right non–linear waves, that is

$$S_L = \tilde{u} - \tilde{a} , \quad S_R = \tilde{u} + \tilde{a} , \tag{10.39}$$

where $\tilde{u}$ and $\tilde{a}$ are the Roe–average particle and sound speeds respectively, given as follows

$$\tilde{u} = \frac{\sqrt{\rho_L}u_L + \sqrt{\rho_R}u_R}{\sqrt{\rho_L} + \sqrt{\rho_R}} , \quad \tilde{a} = \left[(\gamma - 1)(\tilde{H} - \frac{1}{2}\tilde{u}^2)\right]^{\frac{1}{2}} , \tag{10.40}$$

with the enthalpy $H = (E + p)/\rho$ approximated as

$$\tilde{H} = \frac{\sqrt{\rho_L}H_L + \sqrt{\rho_R}H_R}{\sqrt{\rho_L} + \sqrt{\rho_R}} . \tag{10.41}$$

Complete details of the Roe Riemann solver are given in Chap. 11. Davis made some observations regarding the relationship between the chosen wave speeds and some well–known numerical methods. Suppose that for a given Riemann problem we can identify a positive speed S^+. Then by choosing $S_L = -S^+$ and $S_R = S^+$ in the HLL flux (10.21) one obtains a Rusanov flux [291]

$$\mathbf{F}_{i+1/2} = \frac{1}{2}(\mathbf{F}_L + \mathbf{F}_R) - \frac{1}{2}S^+(\mathbf{U}_R - \mathbf{U}_L) . \tag{10.42}$$

As to the choice of the speed S^+, Davis [103] considered

$$S^+ = \max\{|u_L - a_L|, |u_R - a_R|, |u_L + a_L| , |u_R + a_R|\} .$$

Actually, the above speed is bounded by

$$S^+ = \max\{|u_L| + a_L, |u_R| + a_R\} . \tag{10.43}$$

This choice is likely to produce a more robust scheme and is also simpler than Davis's choice.

Another possible choice is $S^+ = S^n_{max}$, the maximum wave speed present at the appropriate time found by imposing the Courant stability condition;

see Sect. 6.3.2 of Chap. 6. This speed is related to the time step Δt and the grid spacing Δx via

$$S^n_{max} = \frac{C_{cfl}\Delta x}{\Delta t} \,, \tag{10.44}$$

where C_{cfl} is the Courant number coefficient, usually chosen (empirically) to be $C_{cfl} \approx 0.9$. For $C_{cfl} = 1$ one has $S^+ = \frac{\Delta x}{\Delta t}$, which results in the Lax–Friedrichs numerical flux

$$F_{i+1/2} = \frac{1}{2}(F_L + F_R) - \frac{1}{2}\frac{\Delta x}{\Delta t}(U_R - U_L) \,. \tag{10.45}$$

See Sect. 5.3.4 of Chap. 5 and Sect. 7.3.1 of Chap. 7. Motivated by the Roe eigenvalues Einfeldt [117] proposed the estimates

$$S_L = \bar{u} - \bar{d} \,, \quad S_R = \bar{u} + \bar{d} \,, \tag{10.46}$$

where

$$\bar{d}^2 = \frac{\sqrt{\rho_L}a_L^2 + \sqrt{\rho_R}a_R^2}{\sqrt{\rho_L} + \sqrt{\rho_R}} + \eta_2(u_R - u_L)^2 \tag{10.47}$$

and

$$\eta_2 = \frac{1}{2}\frac{\sqrt{\rho_L}\sqrt{\rho_R}}{(\sqrt{\rho_L} + \sqrt{\rho_R})^2} \,. \tag{10.48}$$

These are reported to lead to effective and robust schemes. In the next section we propose a different way of finding wave–speed estimates.

10.5.2 Pressure–Velocity Based Wave Speed Estimates

A different approach for finding wave speed estimates was proposed by Toro et. al. [380], whereby one first finds an estimate for the pressure p_* in the *Star Region* and then one derives estimates for S_L and S_R. This is a simple task and several reliable choices are available. For the HLLC scheme of the previous section one also requires an estimate for S_*; we present two ways of doing this. The first consists of finding an estimate for the particle velocity u_*; this is easily achieved, as approximations for p_* and u_* are closely related. The second approach derives a wave speed estimate S_* from the estimates S_L and S_R using conditions (10.29) in equations (10.31)–(10.32), see Batten et. al. [18].

Suppose we have estimates p_* and u_* for the pressure and particle velocity in the *Star Region*. Then we choose the following wave speeds

$$S_L = u_L - a_L q_L \,, \quad S_* = u_* \,, \quad S_R = u_R + a_R q_R \,, \tag{10.49}$$

where

$$q_K = \begin{cases} 1 & \text{if} \quad p_* \le p_K \\ \left[1 + \frac{\gamma+1}{2\gamma}(p_*/p_K - 1)\right]^{\frac{1}{2}} & \text{if} \quad p_* > p_K \,. \end{cases} \tag{10.50}$$

This choice of wave speeds discriminates between shock and rarefaction waves. If the K wave ($K = L$ or $K = R$) is a rarefaction then the speed S_K corresponds to the characteristic speed of the head of the rarefaction, which carries the fastest signal. If the wave is a shock wave then the speed corresponds to an approximation of the true shock speed; the wave relations used are exact but the pressure ratio across the shock is approximated, because the solution for p_* is an approximation. We propose to use the state approximations of Chap. 9 to find p_* and u_*.

The PVRS approximate Riemann solver [355] presented in Sect. 9.3 of Chap. 9 gives

$$p_{pv} = \frac{1}{2}(p_L + p_R) - \frac{1}{2}(u_R - u_L)\bar{\rho}\bar{a} , \quad u_{pv} = \frac{1}{2}(u_L + u_R) - \frac{1}{2}\frac{(p_R - p_L)}{\bar{\rho}\bar{a}} , \tag{10.51}$$

where

$$\bar{\rho} = \frac{1}{2}(\rho_L + \rho_R) , \quad \bar{a} = \frac{1}{2}(a_L + a_R) . \tag{10.52}$$

These approximations for pressure and velocity can be used directly into (10.49)–(10.50) to obtain wave speed estimates for the HLL and HLLC schemes. See also Eq. (9.28) of Chapt. 9 for alternative estimates for p_* and u_*.

Another choice is furnished by the Two–Rarefaction Riemann solver TRRS of Sect. 9.4.1 of Chap. 9, namely

$$\left.\begin{array}{l} p_{tr} = \left[\dfrac{a_L + a_R - \frac{\gamma-1}{2}(u_R - u_L)}{a_L/p_L^z + a_R/p_R^z}\right]^{\frac{1}{z}} \\[4mm] u_{tr} = \dfrac{P_{LR}u_L/a_L + u_R/a_R + \frac{2(P_{LR}-1)}{(\gamma-1)}}{P_{LR}/a_L + 1/a_R} \end{array}\right\} , \tag{10.53}$$

where

$$P_{LR} = \left(\frac{p_L}{p_R}\right)^z ; \quad z = \frac{\gamma - 1}{2\gamma} . \tag{10.54}$$

The Two–Shock Riemann solver TSRS of Sect. 9.4.2 of Chap. 9 gives

$$\left.\begin{array}{l} p_{ts} = \dfrac{g_L(p_0)p_L + g_R(p_0)p_R - \Delta u}{g_L(p_0) + g_R(p_0)} \\[4mm] u_{ts} = \frac{1}{2}(u_L + u_R) + \frac{1}{2}\left[(p_{ts} - p_R)g_R(p_0) - (p_{ts} - p_L)g_L(p_0)\right] \end{array}\right\} , \tag{10.55}$$

where

$$g_K(p) = \left[\frac{A_K}{p + B_K}\right]^{\frac{1}{2}} , \quad p_0 = max(0, p_{pv}) , \tag{10.56}$$

for $K = L$ and $K = R$.

In computational practice we use the hybrid scheme of Sect. 9.5.2 of Chap. 9 to determine p_* and u_*. See Chap. 9 for full details. The HLL approximate Riemann solver with the hybrid pressure–based wave speed estimates has recently been implemented in the NAG routine D03PXF [214] for Godunov–type methods to solve the time–dependent, one dimensional Euler equations for ideal gases.

As indicated earlier, there is an alternative way of computing the middle wave speed S_* in the HLLC Riemann solver. Given the wave speeds S_L and S_R, by assuming (10.30) in equations (10.31)–(10.32) one obtains the following solutions for pressure in the *Star Region*

$$p_{*L} = p_L + \rho_L(S_L - u_L)(S_* - u_L) , \quad p_{*R} = p_R + \rho_R(S_R - u_R)(S_* - u_R) .$$
(10.57)

From (10.29) $p_{*L} = p_{*R}$, which leads to an expression for the speed S_* purely in terms of the assumed speeds S_L and S_R, namely

$$S_* = \frac{p_R - p_L + \rho_L u_L(S_L - u_L) - \rho_R u_R(S_R - u_R)}{\rho_L(S_L - u_L) - \rho_R(S_R - u_R)} .$$
(10.58)

See Batten et. al. [18], who derived this expression from different considerations, for a discussion of special properties of this particular choice for the middle wave speed.

10.6 Algorithms for Intercell Fluxes

10.6.1 The HLL and HLLC Fluxes

In order to implement the *HLL Riemann solver* in the Godunov method one performs the following steps:

Step I: Compute the wave speeds S_L and S_R according to any of the algorithms of Sect. 10.5.

Step II: Compute the HLL flux according to (10.21) and use it in the conservative formula (10.2).

To implement the *HLLC Riemann solver* in the Godunov method one performs the following steps:

Step I: Compute the wave speeds S_L, S_* and S_R according to any of the algorithms of Sect. 10.5.

Step II: Compute the appropriate states according to (10.33),

Step III: Compute the HLLC flux according to (10.34) and use it in the conservative formula (10.2).

10.6.2 The Rusanov Flux

In order to compute the Rusanov's intercell flux one performs the following steps:

Step I: At each interface compute the wave speeds S_L and S_R according to any of the algorithms of Sect. 10.5.

Step II: Compute a single interface speed as

$$S^+ = \max\{S_L, S_R\}$$

Step III: Compute the Rusanov intercell flux as in (10.42) and use it in the conservative formula (10.2).

10.7 Contact Waves and Passive Scalars

Here we study the special case of a passive scalar $q(x,t)$ transported with the fluid speed $u(x,t)$. The time–dependent, one dimensional Euler equations are augmented by the extra conservation law

$$(\rho q)_t + (\rho q u)_x = 0 . \tag{10.59}$$

Consider the special IVP in which $p = constant$, $\rho = constant$, $u = constant$ and

$$q(x,0) = q_0(x) = \begin{cases} q_L & if \quad x \le 0 , \\ q_R & if \quad x > 0 . \end{cases} \tag{10.60}$$

Clearly, the non–trivial part of the exact solution is

$$q(x,t) = q_0(x - ut) . \tag{10.61}$$

Application of the HLL Riemann solver to this problem gives the following expression for the numerical flux

$$f_{i+\frac{1}{2}}^{hll} = \frac{1}{2}\left(1 + \frac{1}{M}\right) f_i + \frac{1}{2}\left(1 - \frac{1}{M}\right) f_{i+1} , \tag{10.62}$$

where $M = \frac{u}{a}$ is the Mach number and the wave speeds have been taken to be

$$S_L = u - a , \quad S_R = u + a .$$

Obviously, this flux applies only in the subsonic regime $u - a \le 0 \le u + a$. For sonic flow, the flux (10.62) reduces identically to the Godunov flux computed from the exact Riemann solver. For subsonic flow $1/M > 1$ and the resulting scheme is *more diffusive than the Godunov method* when used in conjunction with the exact Riemann solver. For the special case

$$M = \frac{u \Delta t}{\Delta x}$$

the HLL scheme reproduces the Lax–Friedrichs method, which is exceedingly diffusive, see Chaps. 5 and 6. The limiting case of a stationary passive scalar is the worst. Note that the analysis includes the important cases $q = v$ and $q = w$, the tangential velocity components in three–dimensional flow.

The analysis for an isolated contact can be carried out in a similar manner; by using an appropriate choice of the wave speeds the resulting HLL flux is identical to (10.62), and thus the same observations as for a passive scalar apply. The HLLC solver, on the other hand, behaves as the exact Riemann solver; for the limiting case in which the wave is stationary, the HLLC numerical scheme gives infinite resolution; the reader can verify this algebraically. In the next section on numerical results we compare the HLL and HLLC schemes for this type of problems; see Fig. 10.9. The relevance of these observations is that the HLL scheme, unlike the HLLC scheme, will add excessive numerical dissipation to the resolution of special but important flow features such as material interfaces, shear waves and vortices.

10.8 Numerical Results

Here we assess the performance of Godunov's first–order method used in conjunction with the approximate Riemann solvers presented in this chapter. We select seven test problems for the one–dimensional, time dependent Euler equations for ideal gases with $\gamma = 1.4$; these have exact solutions. In all chosen tests, data consists of two constant states $\mathbf{W}_L = [\rho_L, u_L, p_L]^T$ and $\mathbf{W}_R = [\rho_R, u_R, p_R]^T$, separated by a discontinuity at a position $x = x_0$. The states $\mathbf{W}_L$ and $\mathbf{W}_R$ are given in Table 10.1. The exact and numerical solutions are found in the spatial domain $0 \leq x \leq 1$. The numerical solution is computed with $M = 100$ cells and the CFL condition is as for all previous computations, see Chap. 6; the chosen Courant number coefficient is $C_{cfl} = 0.9$; boundary conditions are transmissive.

The exact solutions were found by running the code HE-E1RPEXACT of the library *NUMERICA* [369] and the numerical solutions were obtained by running the code HE-E1GODFLUX of *NUMERICA*.

Test	ρ_L	u_L	p_L	ρ_R	u_R	p_R
1	1.0	0.75	1.0	0.125	0.0	0.1
2	1.0	-2.0	0.4	1.0	2.0	0.4
3	1.0	0.0	1000.0	1.0	0.0	0.01
4	5.99924	19.5975	460.894	5.99242	-6.19633	46.0950
5	1.0	-19.59745	1000.0	1.0	-19.59745	0.01
6	1.4	0.0	1.0	1.0	0.0	1.0
7	1.4	0.1	1.0	1.0	0.1	1.0

Table 10.1. Data for seven test problems with exact solution

Test 1 is a *modified* version of Sod's problem [318]; the solution has a right shock wave, a right travelling contact wave and a left *sonic* rarefaction wave; this test is useful in assessing the *entropy satisfaction* property of numerical methods. The solution of Test 2 consists of two symmetric rarefaction waves and a trivial contact wave; the *Star Region* between the non–linear waves is close to vacuum, which makes this problem a suitable test for assessing the performance of numerical methods for low–density flows. Test 3 is designed to assess the robustness and accuracy of numerical methods; its solution consists of a strong shock wave of shock Mach number 198, a contact surface and a left rarefaction wave. Test 4 is also a very severe test, its solution consists of three strong discontinuities travelling to the right. A detailed discussion on the exact solution of Tests 1 to 4 is found in Sect. 4.3.3 of Chap. 4. Test 5 is also designed to test the robustness of numerical methods but the main reason for devising this test is to assess the ability of the numerical methods to resolve *slowly–moving contact discontinuities*. The exact solution of Test 5 consists of a left rarefaction wave, a right–travelling shock wave and a *stationary* contact discontinuity. Test 6 corresponds to an isolated stationary contact wave and Test 7 corresponds to an isolated contact moving slowly to the right. The purpose of Tests 6 and 7 is to illustrate the likely performance of HLL and HLLC for contacts, shear waves and vortices. For each test problem we select a convenient position x_0 of the initial discontinuity and the output time. These are stated in the legend of each figure displaying computational results.

We compare computed results with the exact solution for three first–order methods, namely the Godunov method used in conjunction with the HLL and HLLC approximate Riemann solvers, and the Rusanov scheme. In all three schemes we compute wave speed estimates by using the adaptive noniterative scheme of Sect. 9.5.2 of Chapt. 9. Figs. 10.5 to 10.9 show results for Godunov's method with the HLLC Riemann solver. Figs. 10.10 to 10.14 show results for the Godunov method with the HLL Riemann solver and Figs. 10.15 to 10.19 show results for Rusanov's method. Fig. 10.20 shows results aimed at comparing the performance of HLL and HLLC for isolated, stationary and slowly moving contact discontinuities.

The numerical results obtained from the Godunov method in conjunction with the HLL and HLLC approximate Riemann solvers are broadly similar to those obtained from Godunov's method in conjunction with the exact Riemann solver. See results of Chapt. 6. Some points to note are the following: the *sonic rarefaction* of Test 1 is better resolved by the HLL and HLLC approximate Riemann solvers than by the exact Riemann solver. The resolution of the stationary contact (non–isolated) of Test 5 by the HLLC Riemann solver is comparable with that of the exact Riemann solver. The HLL Riemann solver however, as anticipated by the analysis of Sect. 10.7, diffuses the contact wave to similar levels seen in the Flux Vector Splitting methods of Steger–Warming and van Leer, see results of Chap. 8. The advantage of HLLC over HLL is the resolution of slowly–moving contact discontinuities; this point is further emphasised by the results of Tests 6 and 7 for an *isolated* contact wave. The HLLC Riemann solver preserves the excellent entropy–satisfaction property of the HLL Riemann solver. The Rusanov scheme is broadly similar to the HLL Riemann solver in that it also diffuses slowly moving contacts. For Test 1 containing a sonic rarefaction however, the Rusanov scheme is clearly inferior to the HLL scheme, compare Fig. 10.15 with Fig. 10.10.

The results of Tests 6 and 7 using both the HLL and the HLLC schemes are shown in Fig. 10.20. As anticipated by the analysis of Sect. 10.7, the HLL scheme will give unacceptably smeared results for stationary and slowly moving contact waves. The HLLC behaves like the exact Riemann solver for this type of problem; it has much less numerical dissipation for slowly moving contacts and it gives infinite resolution for stationary contact waves. The same observations apply to augmented systems of equations containing species equations, and to shear waves and vortices in multi–dimensions.

10.9 Closing Remarks and Extensions

The approximate Riemann solvers of this chapter may be applied in conjunction with the Godunov first–order upwind method presented in Chap. 6. Second–order Total Variation Diminishing (TVD) extensions of the schemes are presented in Chap. 13 for scalar problems and in Chap. 14 for non–linear one dimensional systems. In chap. 15 we present techniques that allow the extension of these schemes to solve problems with source terms. In Chap. 16 we study techniques to extend the methods of this chapter to three–dimensional problems. Implicit versions of the HLL and HLLC Riemann solvers have been developed by Batten, Leschziner and Goldberg [19], who have also applied the schemes to turbulent flows.

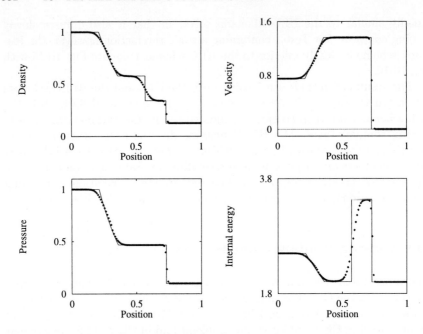

Fig. 10.5. Godunov's method with HLLC Riemann solver applied to Test 1, with $x_0 = 0.3$. Numerical (symbol) and exact (line) solutions are compared at time 0.2

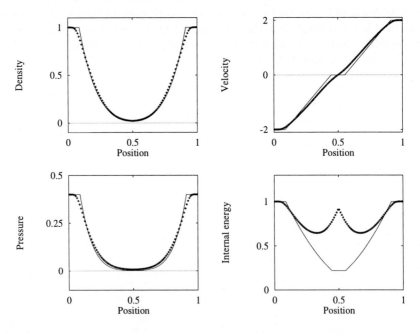

Fig. 10.6. Godunov's method with HLLC Riemann solver applied to Test 2, with $x_0 = 0.5$. Numerical (symbol) and exact (line) solutions are compared at time 0.15

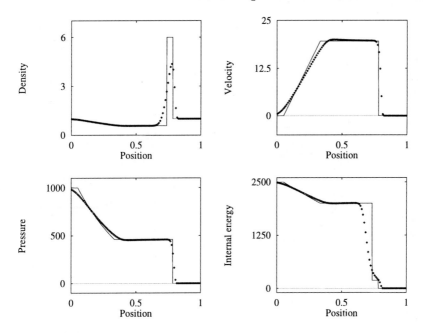

Fig. 10.7. Godunov's method with HLLC Riemann solver applied to Test 3, with $x_0 = 0.5$. Numerical (symbol) and exact (line) solutions are compared at time 0.012

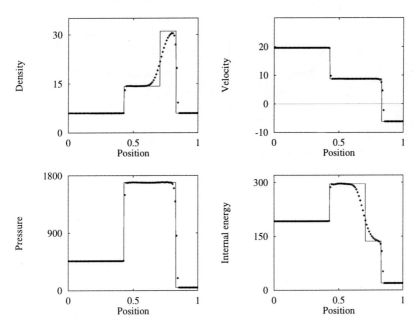

Fig. 10.8. Godunov's method with HLLC Riemann solver applied to Test 4, with $x_0 = 0.4$. Numerical (symbol) and exact (line) solutions are compared at time 0.035

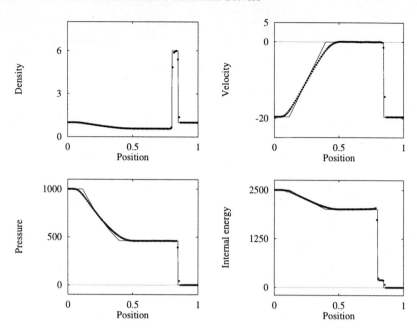

Fig. 10.9. Godunov's method with HLLC Riemann solver applied to Test 5, with $x_0 = 0.8$. Numerical (symbol) and exact (line) solutions are compared at time 0.012

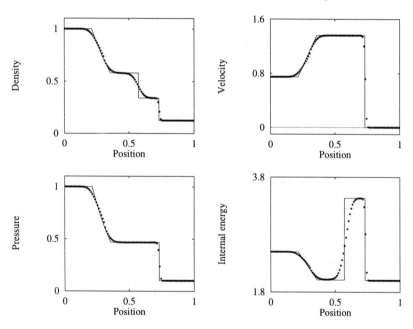

Fig. 10.10. Godunov's method with HLL Riemann solver applied to Test 1, with $x_0 = 0.3$. Numerical (symbol) and exact (line) solutions are compared at time 0.2

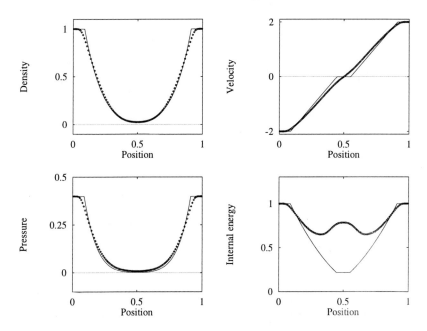

Fig. 10.11. Godunov's method with HLL Riemann solver applied to Test 2, with $x_0 = 0.5$. Numerical (symbol) and exact (line) solutions are compared at time 0.15

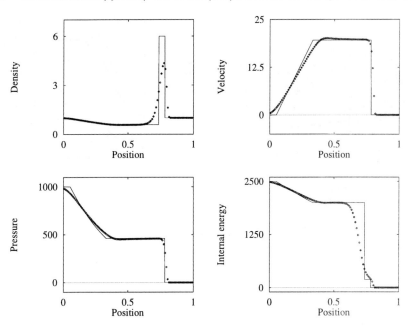

Fig. 10.12. Godunov's method with HLL Riemann solver applied to Test 3, with $x_0 = 0.5$. Numerical (symbol) and exact (line) solutions are compared at time 0.012

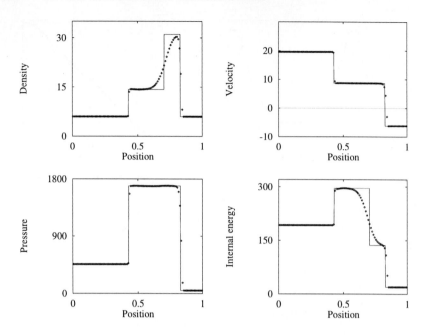

Fig. 10.13. Godunov's method with HLL Riemann solver applied to Test 4, with $x_0 = 0.4$. Numerical (symbol) and exact (line) solutions are compared at time 0.035

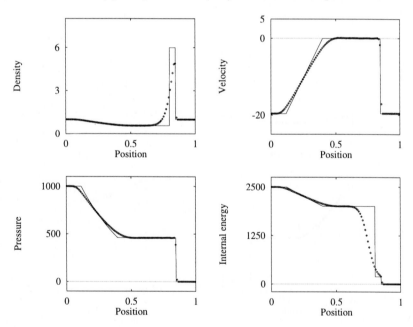

Fig. 10.14. Godunov's method with HLL Riemann solver applied to Test 5, with $x_0 = 0.8$. Numerical (symbol) and exact (line) solutions are compared at time 0.012

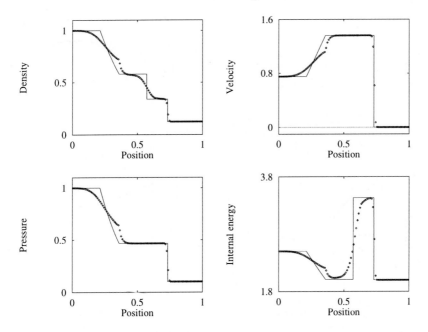

Fig. 10.15. Rusanov's method applied to Test 1, with $x_0 = 0.3$. Numerical (symbol) and exact (line) solutions are compared at time 0.2

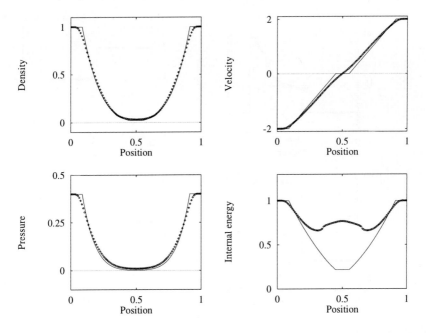

Fig. 10.16. Rusanov's method applied to Test 2, with $x_0 = 0.5$. Numerical (symbol) and exact (line) solutions are compared at time 0.15

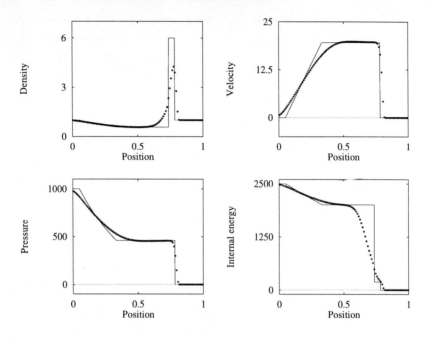

Fig. 10.17. Rusanov's method applied to Test 3, with $x_0 = 0.5$. Numerical (symbol) and exact (line) solutions are compared at time 0.012

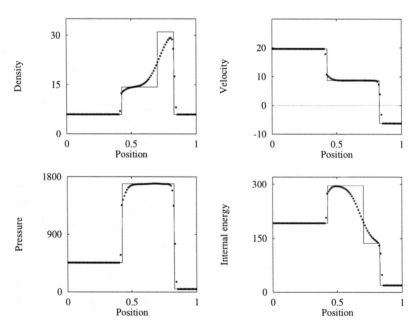

Fig. 10.18. Rusanov's method applied to Test 4, with $x_0 = 0.4$. Numerical (symbol) and exact (line) solutions are compared at time 0.035

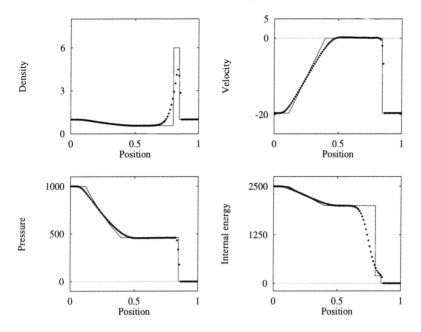

Fig. 10.19. Rusanov's method applied to Test 5, with $x_0 = 0.8$. Numerical (symbol) and exact (line) solutions are compared at time 0.012

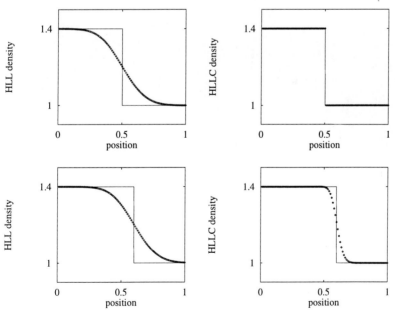

Fig. 10.20. Godunov's method with HLL (left) and HLLC (right) Riemann solvers applied to Tests 6 and 7. Numerical (symbol) and exact (line) solutions are compared at time 2.0

Fig. 10.19. [text too faded to read]

Fig. 10.20. [text too faded to read]

11. The Riemann Solver of Roe

Perhaps, the most well–known of all approximate Riemann solvers today, is the one due to Roe, which was first presented in the open literature in 1981 [281]. Since then, the method has not only been refined, but it has also been applied to a very large variety of physical problems. Refinements to the Roe approach were introduced by Roe and Pike [290], whereby the computation of the necessary items of information does not explicitly require the Roe averaged Jacobian matrix. This second methodology appears to be simpler and is thus useful in solving the Riemann problem for new, complicated sets of hyperbolic conservations laws, or for conventional systems but for complex media. Glaister exploited the Roe–Pike approach to extend Roe's method to the time–dependent Euler equations with a general equation of state [137], [138]. The large body of experience accumulated by many workers over a considerable period of time has led to various improvements of the scheme. As originally presented the Roe scheme computes rarefaction shocks, thus violating the entropy condition. Harten and Hyman [163], Roe and Pike [290], Roe [288], Dubois and Mehlman [113] and others, have produced appropriate modifications to the scheme. Einfeldt et. al. [118] produced corrections to the basic Roe scheme to avoid the so–called *vacuum problem* near low–density flows; they also showed that in fact this anomaly afflicts *all linearised Riemann solvers*.

Ambitious applications of the Roe scheme were presented by Brio and Wu [54], who utilised Roe's method to solve the Magneto–Hydrodynamic equations (MHD). Clarke et. al. [80] applied the method in conjunction with adaptive gridding to the computation of two–dimensional unsteady detonation waves in solid materials. Giraud and Manzini [135] produced parallel implementations of the Roe scheme for two–dimensional Gas Dynamics. LeVeque and Shyue [211] have applied the Roe scheme in the context of *front tracking* in two space dimensions. Marx has applied the Roe scheme to solve the incompressible Navier–Stokes equations [236], [237] and the compressible Navier–Stokes equations [235] using *implicit* versions of the scheme; see also McNeil [239]. The method has also been applied to multiphase flows; Toro [354] solved reactive multi–phase problems in the context of propulsion systems via a phase–splitting procedure; recently, Sainsaulieu [292] has extended the Roe scheme to a class of multiphase flow problems without phase splitting.

The purpose of this chapter is to present the approximate Riemann solver of Roe as applied to the three–dimensional time dependent Euler equations. For the numerical methods considered here, we only need to derive the Riemann solver for the split three–dimensional equations. After a general introduction to the method, we present both the methodology of Roe and that of Roe and Pike. Both methodologies are suitably illustrated via the simpler isothermal equations. Useful background reading is found in Chaps. 2 to 6.

11.1 Bases of the Roe Approach

In this section we describe the Roe approach for a general system of m hyperbolic conservation laws. Detailed application of the scheme to the isothermal and Euler equations are given in subsequent sections.

11.1.1 The Exact Riemann Problem and the Godunov Flux

We are concerned with solving numerically the general Initial Boundary Value Problem (IBVP)

$$\left. \begin{array}{lll} \text{PDEs} & : & \mathbf{U}_t + \mathbf{F}(\mathbf{U})_x = \mathbf{0} \,, \\ \text{ICs} & : & \mathbf{U}(x,0) = \mathbf{U}^{(0)}(x) \,, \\ \text{BCs} & : & \mathbf{U}(0,t) = \mathbf{U}_1(t) \,, \ \mathbf{U}(L,t) = \mathbf{U}_r(t) \,, \end{array} \right\} \tag{11.1}$$

in a domain $x_l \leq x \leq x_r$, utilising the explicit conservative formula

$$\mathbf{U}_i^{n+1} = \mathbf{U}_i^n + \frac{\Delta t}{\Delta x}[\mathbf{F}_{i-\frac{1}{2}} - \mathbf{F}_{i+\frac{1}{2}}] \,. \tag{11.2}$$

We assume the solution of IBVP (11.1) exists. In Chap. 6 we defined the Godunov intercell numerical flux

$$\mathbf{F}_{i+\frac{1}{2}} = \mathbf{F}(\mathbf{U}_{i+\frac{1}{2}}(0)) \,, \tag{11.3}$$

in which $\mathbf{U}_{i+\frac{1}{2}}(0)$ is the exact similarity solution $\mathbf{U}_{i+\frac{1}{2}}(x/t)$ of the Riemann problem

$$\begin{array}{l} \mathbf{U}_t + \mathbf{F}(\mathbf{U})_x = \mathbf{0} \,, \\ \mathbf{U}(x,0) = \left\{ \begin{array}{lll} \mathbf{U}_{\mathrm{L}} & if & x < 0 \,, \\ \mathbf{U}_{\mathrm{R}} & if & x > 0 \end{array} \right. \end{array} \right\} \tag{11.4}$$

evaluated at $x/t = 0$. Fig. 11.1 shows the structure of the exact solution of the Riemann problem for the x–split three dimensional Euler equations, for which the vectors of conserved variables and fluxes are

$$
\mathbf{U} = \begin{bmatrix} \rho \\ \rho u \\ \rho v \\ \rho w \\ E \end{bmatrix} , \quad \mathbf{F} = \begin{bmatrix} \rho u \\ \rho u^2 + p \\ \rho u v \\ \rho u w \\ u(E + p) \end{bmatrix} . \tag{11.5}
$$

The *Star Region* between the left and right waves contains the unknowns of the problem. The particular value at $x/t = 0$ corresponds to the t–axis and is the value required by the Godunov flux. See Chaps. 4 and 6 for details. The piece–wise constant initial data, in terms of primitive variables, is

$$
\mathbf{W}_L = \begin{bmatrix} \rho_L \\ u_L \\ v_L \\ w_L \\ p_L \end{bmatrix} , \quad \mathbf{W}_R = \begin{bmatrix} \rho_R \\ u_R \\ v_R \\ w_R \\ p_R \end{bmatrix} . \tag{11.6}
$$

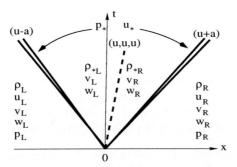

Fig. 11.1. Structure of the solution of the Riemann problem for the x–split three dimensional Euler equations

In Chap. 4 we provided an algorithm to compute the exact solution $\mathbf{U}_{i+\frac{1}{2}}(x/t)$ and in Chap. 6 we utilised this solution in the Godunov method. In Chap. 9 we provided approximations to the state $\mathbf{U}_{i+\frac{1}{2}}(x/t)$ and obtained a corresponding approximate Godunov method by evaluating the physical flux function $\mathbf{F}$ at this approximate state. The purpose of this chapter is to find *direct approximations to the flux function* $\mathbf{F}_{i+\frac{1}{2}}$ following the approach proposed by Roe [281] and Roe and Pike [290].

11.1.2 Approximate Conservation Laws

Roe [281] solved the Riemann problem (11.4) approximately. By introducing the Jacobian matrix

$$
\mathbf{A}(\mathbf{U}) = \frac{\partial \mathbf{F}}{\partial \mathbf{U}} \tag{11.7}
$$

and using the chain rule the conservation laws

$$\mathbf{U}_t + \mathbf{F}(\mathbf{U})_x = \mathbf{0}$$

in (11.4) may be written as

$$\mathbf{U}_t + \mathbf{A}(\mathbf{U})\mathbf{U}_x = \mathbf{0} . \tag{11.8}$$

Roe's approach replaces the Jacobian matrix $\mathbf{A}(\mathbf{U})$ in (11.8) by a *constant* Jacobian matrix

$$\tilde{\mathbf{A}} = \tilde{\mathbf{A}}(\mathbf{U}_L, \mathbf{U}_R) , \tag{11.9}$$

which is a function of the data states $\mathbf{U}_L$, $\mathbf{U}_R$. In this way the original PDEs in (11.4) are replaced by

$$\mathbf{U}_t + \tilde{\mathbf{A}}\mathbf{U}_x = \mathbf{0} . \tag{11.10}$$

This is a linear system with constant coefficients. The original, Riemann problem (11.4) is then replaced by the *approximate Riemann problem*

$$\left. \begin{array}{l} \mathbf{U}_t + \tilde{\mathbf{A}}\mathbf{U}_x = \mathbf{0} \\[4pt] \mathbf{U}(x,0) = \left\{ \begin{array}{ll} \mathbf{U}_L , & x < 0 \\ \mathbf{U}_R , & x > 0 \end{array} \right. \end{array} \right\} , \tag{11.11}$$

which is then *solved exactly*. The approximate problem results from replacing the original non–linear conservation laws by a linearised system with constant coefficients but the initial data of the exact problem is retained.

For a general hyperbolic system of m conservation laws, the Roe Jacobian matrix $\tilde{\mathbf{A}}$ is required to satisfy the following properties:

Property (A): Hyperbolicity of the system. $\tilde{\mathbf{A}}$ is required to have real eigenvalues $\tilde{\lambda}_i = \tilde{\lambda}_i(\mathbf{U}_L, \mathbf{U}_R)$, which we choose to order as

$$\tilde{\lambda}_1 \le \tilde{\lambda}_2 \le \cdots \le \tilde{\lambda}_m , \tag{11.12}$$

and a complete set of linearly independent right eigenvectors

$$\tilde{\mathbf{K}}^{(1)} , \quad \tilde{\mathbf{K}}^{(2)} , \cdots , \quad \tilde{\mathbf{K}}^{(m)} . \tag{11.13}$$

Property (B): Consistency with the exact Jacobian

$$\tilde{\mathbf{A}}(\mathbf{U}, \mathbf{U}) = \mathbf{A}(\mathbf{U}) . \tag{11.14}$$

Property (C): Conservation across discontinuities

$$\mathbf{F}(\mathbf{U}_R) - \mathbf{F}(\mathbf{U}_L) = \tilde{\mathbf{A}} (\mathbf{U}_R - \mathbf{U}_L) . \tag{11.15}$$

Property (A) on hyperbolicity is an obvious requirement; the approximate problem should at the very least preserve the mathematical character of the original non–linear system. Property (B) ensures consistency with the conservation laws. Property (C) ensures conservation. It also ensures exact recognition of isolated discontinuities; that is, if the data $\mathbf{U}_L$, $\mathbf{U}_R$ are connected by a *single, isolated* discontinuity, then the approximate Riemann solver recognises this wave exactly. Note however that this does not mean that the corresponding, approximate, Godunov method with the Roe approximate numerical flux will in general give exact solutions for isolated discontinuities.

The construction of matrices satisfying properties (A)–(C) for *general hyperbolic systems* can be very complicated and thus computationally unattractive. For the *specific* case of the Euler equations of Gas Dynamics Roe [281] proposed a relatively simple way of constructing a matrix $\tilde{\mathbf{A}}$. Later, Roe and Pike [290] proposed a simpler approach, where the explicit construction of $\tilde{\mathbf{A}}$ is actually avoided.

11.1.3 The Approximate Riemann Problem and the Intercell Flux

Once the matrix $\tilde{\mathbf{A}}(\mathbf{U}_L, \mathbf{U}_R)$, its eigenvalues $\tilde{\lambda}_i(\mathbf{U}_L, \mathbf{U}_R)$ and right eigenvectors $\tilde{\mathbf{K}}^{(i)}(\mathbf{U}_L, \mathbf{U}_R)$ are available, one solves the Riemann problem (11.11) by direct application of methods discussed in Sect. 2.3 of Chap. 2 and Sect. 5.4 of Chap. 5, for linear hyperbolic systems with constant coefficients. By projecting the data difference

$$\Delta \mathbf{U} = \mathbf{U}_R - \mathbf{U}_L$$

onto the right eigenvectors we write

$$\Delta \mathbf{U} = \mathbf{U}_R - \mathbf{U}_L = \sum_{i=1}^{m} \tilde{\alpha}_i \tilde{\mathbf{K}}^{(i)} \,, \tag{11.16}$$

from which one finds the *wave strengths* $\tilde{\alpha}_i = \tilde{\alpha}_i(\mathbf{U}_L, \mathbf{U}_R)$. The solution $\mathbf{U}_{i+\frac{1}{2}}(x/t)$ evaluated along the t–axis, $x/t = 0$, is given by

$$\mathbf{U}_{i+\frac{1}{2}}(0) = \mathbf{U}_L + \sum_{\tilde{\lambda}_i \leq 0} \tilde{\alpha}_i \tilde{\mathbf{K}}^{(i)} \,, \tag{11.17}$$

or

$$\mathbf{U}_{i+\frac{1}{2}}(0) = \mathbf{U}_R - \sum_{\tilde{\lambda}_i \geq 0} \tilde{\alpha}_i \tilde{\mathbf{K}}^{(i)} \,. \tag{11.18}$$

We now find the corresponding numerical flux. Recall that we have replaced the original set of conservation laws in (11.4) by the constant coefficient linear system (11.10); this can be viewed as a modified system of conservation laws

$$\overline{\mathbf{U}}_t + \overline{\mathbf{F}}(\overline{\mathbf{U}})_x = \mathbf{0} \,, \tag{11.19}$$

with flux function

$$\mathbf{F}(\overline{\mathbf{U}}) = \tilde{\mathbf{A}}\overline{\mathbf{U}} . \tag{11.20}$$

The corresponding numerical flux, see (11.3), *is not* the obvious choice

$$\mathbf{F}_{i+\frac{1}{2}} = \tilde{\mathbf{A}}\,\overline{\mathbf{U}}_{i+\frac{1}{2}}(0) ,$$

where $\overline{\mathbf{U}}_{i+\frac{1}{2}}(0)$ is given by any of (11.17)–(11.18). That this would be incorrect becomes obvious when, for instance, assuming right supersonic flow in (11.17) one would compute an intercell flux $\mathbf{F}_{i+\frac{1}{2}} \neq \mathbf{F}_L$. Instead, the correct expression for the corresponding numerical flux is obtained from any of the integral relations

$$\mathbf{F}_{0L} = \mathbf{F}_L - S_L \mathbf{U}_L - \frac{1}{T}\int_{TS_L}^{0}\mathbf{U}(x,T)dx , \tag{11.21}$$

$$\mathbf{F}_{0R} = \mathbf{F}_R - S_R \mathbf{U}_R + \frac{1}{T}\int_{0}^{TS_R}\mathbf{U}(x,T)dx , \tag{11.22}$$

derived in Sect. 10.2 of Chap. 10. Here S_L, S_R are the smallest and largest signal speeds in the exact solution of the Riemann problem with data $\mathbf{U}_L, \mathbf{U}_R$ and T is a positive time. If the integrand $\mathbf{U}(x,t)$ in (11.21) or (11.22) is replaced by some approximate solution, then equality of the fluxes $\mathbf{F}_{0L}$ and $\mathbf{F}_{0R}$ requires the approximate solution to satisfy a *Consistency Condition*, see Sect. 10.2 of Chap. 10.

If $\overline{\mathbf{U}}_{i+\frac{1}{2}}(x,t)$ is the solution of the Riemann problem for the modified conservation laws (11.19) with data $\mathbf{U}_L, \mathbf{U}_R$, then the integrals in (11.21) and (11.22) respectively, are

$$\int_{TS_L}^{0}\overline{\mathbf{U}}_{i+\frac{1}{2}}(x,T)dx = T[\overline{\mathbf{F}}(\mathbf{U}_L) - \overline{\mathbf{F}}(\overline{\mathbf{U}}_{i+\frac{1}{2}}(0))] - TS_L\mathbf{U}_L \tag{11.23}$$

and

$$\int_{0}^{TS_R}\overline{\mathbf{U}}_{i+\frac{1}{2}}(x,T)dx = T[\overline{\mathbf{F}}(\overline{\mathbf{U}}_{i+\frac{1}{2}}(0)) - \overline{\mathbf{F}}(\mathbf{U}_R)] + TS_R\mathbf{U}_R . \tag{11.24}$$

Substitution of (11.23) and (11.24) into (11.21) and (11.22) gives

$$\mathbf{F}_{0L} = \overline{\mathbf{F}}(\overline{\mathbf{U}}_{i+\frac{1}{2}}(0)) + \mathbf{F}(\mathbf{U}_L) - \overline{\mathbf{F}}(\mathbf{U}_L) \tag{11.25}$$

and

$$\mathbf{F}_{0R} = \overline{\mathbf{F}}(\overline{\mathbf{U}}_{i+\frac{1}{2}}(0)) + \mathbf{F}(\mathbf{U}_R) - \overline{\mathbf{F}}(\mathbf{U}_R) . \tag{11.26}$$

Finally, by using $\overline{\mathbf{U}}_{i+\frac{1}{2}}(0)$ as given by (11.17) or (11.18) and the definition of the flux $\overline{\mathbf{F}} = \tilde{\mathbf{A}}\,\overline{\mathbf{U}}$ we obtain the numerical flux as

$$\mathbf{F}_{i+\frac{1}{2}} = \mathbf{F}_L + \sum_{\tilde{\lambda}_i \leq 0}\tilde{\alpha}_i\tilde{\lambda}_i\tilde{\mathbf{K}}^{(i)} , \tag{11.27}$$

or

$$\mathbf{F}_{i+\frac{1}{2}} = \mathbf{F}_R - \sum_{\tilde{\lambda}_i \geq 0} \tilde{\alpha}_i \tilde{\lambda}_i \tilde{\mathbf{K}}^{(i)} . \tag{11.28}$$

Alternatively, we may also write

$$\mathbf{F}_{i+\frac{1}{2}} = \frac{1}{2}(\mathbf{F}_L + \mathbf{F}_R) - \frac{1}{2}\sum_{i=1}^{m} \tilde{\alpha}_i |\tilde{\lambda}_i| \tilde{\mathbf{K}}^{(i)} . \tag{11.29}$$

We remark that all previous relations (11.19)–(11.29) are valid for any hyperbolic system and any linearisation of it. In order to compute Roe's numerical flux for a particular system of hyperbolic conservation laws, one requires expressions for the wave strengths $\tilde{\alpha}_i$, the eigenvalues $\tilde{\lambda}_i$ and the right eigenvectors $\tilde{\mathbf{K}}^{(i)}$ in any of the flux expressions (11.27)–(11.29). Note that the Jacobian matrix $\tilde{\mathbf{A}}(\mathbf{U}_L, \mathbf{U}_R)$ is not explicitly required by the numerical flux. In the next two sections we give details on methodologies to find $\tilde{\alpha}_i$, $\tilde{\lambda}_i$ and $\tilde{\mathbf{K}}^{(i)}$. There are two approaches, namely the original approach presented by Roe in 1981 [281] and the Roe–Pike approach [290].

11.2 The Original Roe Method

In order for the approximate Godunov method based on (11.2) with the Roe-type numerical flux (11.27)–(11.29) to be completely determined, we need to find the average eigenvalues $\tilde{\lambda}_i$, the corresponding averaged right eigenvectors $\tilde{\mathbf{K}}^{(i)}$ and averaged wave strengths $\tilde{\alpha}_i$. In his original paper [281] Roe finds an averaged Jacobian matrix $\tilde{\mathbf{A}}$, the Roe matrix, from which $\tilde{\lambda}_i$, $\tilde{\mathbf{K}}^{(i)}$ and $\tilde{\alpha}_i$ follow directly. In constructing a matrix $\tilde{\mathbf{A}}$ the properties (A)–(C), equations (11.12)–(11.15), are enforced. It is not difficult to think of candidates $\tilde{\mathbf{A}}$ that satisfy the first two properties. Property C is crucial and is the one that narrows down the choices. Roe showed that the existence of a matrix $\tilde{\mathbf{A}}$ satisfying Property C is assured by the mean value theorem. An early line of attack in constructing a matrix $\tilde{\mathbf{A}}$ satisfying all desirable properties is reported by Sells [305]. Roe identifies some disadvantages of this approach; it is argued, for instance, that the construction is far from unique and that the resulting schemes are too complicated.

A breakthrough in constructing $\tilde{\mathbf{A}}$ resulted from Roe's ingenious idea of introducing a *parameter vector* $\mathbf{Q}$, such that both the vector of conserved variables $\mathbf{U}$ and the flux vector $\mathbf{F}(\mathbf{U})$ could be expressed in terms of $\mathbf{Q}$. That is

$$\mathbf{U} = \mathbf{U}(\mathbf{Q}) , \quad \mathbf{F} = \mathbf{F}(\mathbf{Q}) . \tag{11.30}$$

Two important steps then follow. First, the changes

$$\Delta\mathbf{U} = \mathbf{U}_R - \mathbf{U}_L , \quad \Delta\mathbf{F} = \mathbf{F}(\mathbf{U}_R) - \mathbf{F}(\mathbf{U}_L) \tag{11.31}$$

can be expressed in terms of the change $\Delta\mathbf{Q} = \mathbf{Q}_R - \mathbf{Q}_L$. Then, averages are obtained in terms of *simple arithmetic means* of $\mathbf{Q}$. Next, we illustrate the technique as applied to a simple set of conservation laws.

11.2.1 The Isothermal Equations

Consider the isothermal equations

$$\mathbf{U}_t + \mathbf{F}(\mathbf{U})_x = \mathbf{0} \,,$$

$$\left. \mathbf{U} \equiv \begin{bmatrix} u_1 \\ u_2 \end{bmatrix} \equiv \begin{bmatrix} \rho \\ \rho u \end{bmatrix} ; \quad \mathbf{F} \equiv \begin{bmatrix} f_1 \\ f_2 \end{bmatrix} \equiv \begin{bmatrix} \rho u \\ \rho u^2 + a^2 \rho \end{bmatrix} , \right\} \tag{11.32}$$

where a is a constant sound speed. See Sect. 1.6.2 of Chap. 1. See also Sect. 2.4.1 of Chap. 2, where the eigenstructure of the equations is given. The exact Jacobian, eigenvalues and corresponding right eigenvectors are

$$\left. \begin{aligned} \mathbf{A}(\mathbf{U}) &= \begin{bmatrix} 0 & 1 \\ a^2 - u^2 & 2u \end{bmatrix} , \\[2mm] \lambda_1 &= u - a , \quad \lambda_2 = u + a , \\[2mm] \mathbf{K}^{(1)} &= \begin{bmatrix} 1 \\ u - a \end{bmatrix} , \quad \mathbf{K}^{(2)} = \begin{bmatrix} 1 \\ u + a \end{bmatrix} . \end{aligned} \right\} \tag{11.33}$$

Choose the *parameter vector*

$$\mathbf{Q} \equiv \begin{bmatrix} q_1 \\ q_2 \end{bmatrix} \equiv \frac{\mathbf{U}}{\sqrt{\rho}} = \begin{bmatrix} \sqrt{\rho} \\ \sqrt{\rho} u \end{bmatrix} . \tag{11.34}$$

Then $\mathbf{U}$ and $\mathbf{F}$ can be expressed in terms of the components q_1, q_2 of $\mathbf{Q}$, namely

$$\mathbf{U} \equiv \begin{bmatrix} u_1 \\ u_2 \end{bmatrix} \equiv q_1 \mathbf{Q} = \begin{bmatrix} q_1^2 \\ q_1 q_2 \end{bmatrix} \tag{11.35}$$

and

$$\mathbf{F} \equiv \begin{bmatrix} f_1 \\ f_2 \end{bmatrix} \equiv \begin{bmatrix} q_1 q_2 \\ q_2^2 + a^2 q_1^2 \end{bmatrix} . \tag{11.36}$$

One now looks for an averaged vector $\tilde{\mathbf{Q}} = (\tilde{q}_1, \tilde{q}_2)^T$. This is found by simple *arithmetic averaging*

$$\tilde{\mathbf{Q}} = \begin{bmatrix} \tilde{q}_1 \\ \tilde{q}_2 \end{bmatrix} = \frac{1}{2}(\mathbf{Q}_L + \mathbf{Q}_R) = \frac{1}{2} \begin{bmatrix} \sqrt{\rho_L} + \sqrt{\rho_R} \\ \sqrt{\rho_L} u_L + \sqrt{\rho_R} u_R \end{bmatrix} . \tag{11.37}$$

Then two matrices $\tilde{\mathbf{B}} = \tilde{\mathbf{B}}(\tilde{\mathbf{Q}})$ and $\tilde{\mathbf{C}} = \tilde{\mathbf{C}}(\tilde{\mathbf{Q}})$ are found, such that the jumps $\Delta\mathbf{U}$ and $\Delta\mathbf{F}$ in (11.31) can be expressed in terms of the jump $\Delta\mathbf{Q}$, namely

$$\Delta\mathbf{U} = \tilde{\mathbf{B}}\Delta\mathbf{Q} ; \quad \Delta\mathbf{F} = \tilde{\mathbf{C}}\Delta\mathbf{Q} . \tag{11.38}$$

Use of these two expressions produces

$$\Delta \mathbf{F} = (\tilde{\mathbf{C}}\tilde{\mathbf{B}}^{-1})\Delta \mathbf{U} ,\qquad (11.39)$$

which if compared with condition (C), equation (11.15), produces the Roe averaged matrix

$$\tilde{\mathbf{A}} = \tilde{\mathbf{C}}\tilde{\mathbf{B}}^{-1} .\qquad (11.40)$$

Matrices $\tilde{\mathbf{B}}$ and $\tilde{\mathbf{C}}$ satisfying (11.38) are

$$\tilde{\mathbf{B}} = \begin{bmatrix} 2\tilde{q}_1 & 0 \\ \tilde{q}_2 & \tilde{q}_1 \end{bmatrix} ; \quad \tilde{\mathbf{C}} = \begin{bmatrix} \tilde{q}_2 & \tilde{q}_1 \\ 2a^2\tilde{q}_1 & 2\tilde{q}_2^2 \end{bmatrix} ,\qquad (11.41)$$

which the reader can easily verify. The sought Roe matrix is then

$$\tilde{\mathbf{A}} = \begin{bmatrix} 0 & 1 \\ a^2 - \tilde{u}^2 & 2\tilde{u} \end{bmatrix} ,\qquad (11.42)$$

where $\tilde{u}$ is the *Roe averaged velocity* and is given by

$$\tilde{u} = \frac{\sqrt{\rho_L}u_L + \sqrt{\rho_R}u_R}{\sqrt{\rho_L} + \sqrt{\rho_R}} .\qquad (11.43)$$

Compare (11.42) with the matrix in (11.33). As the sound speed a is constant, no averaged $\tilde{\rho}$ is required.

Having found $\tilde{\mathbf{A}}$ one computes the averaged eigenvalues, eigenvectors and wave strengths. The eigenvalues of $\tilde{\mathbf{A}}$ are

$$\tilde{\lambda}_1 = \tilde{u} - a ; \quad \tilde{\lambda}_2 = \tilde{u} + a \qquad (11.44)$$

and are all real. The corresponding averaged right eigenvectors are

$$\tilde{\mathbf{K}}^{(1)} = \begin{bmatrix} 1 \\ \tilde{u} - a \end{bmatrix} ; \quad \tilde{\mathbf{K}}^{(2)} = \begin{bmatrix} 1 \\ \tilde{u} + a \end{bmatrix} \qquad (11.45)$$

and are easily seen to be linearly independent. Thus condition (A) is satisfied. To find the wave strengths $\tilde{\alpha}_i$ we solve the 2×2 linear system, see (11.16),

$$\Delta \mathbf{U} \equiv \begin{bmatrix} \Delta u_1 \\ \Delta u_2 \end{bmatrix} = \sum_{i=1}^{2} \tilde{\alpha}_i \tilde{\mathbf{K}}^{(i)} .$$

The solution is easily verified to be

$$\left.\begin{aligned} \tilde{\alpha}_1 &= \frac{\Delta u_1(\tilde{u} + a) - \Delta u_2}{2a} , \\ \tilde{\alpha}_2 &= \frac{-\Delta u_1(\tilde{u} - a) + \Delta u_2}{2a} , \end{aligned}\right\} \qquad (11.46)$$

with the obvious definitions $\Delta u_1 \equiv \rho_R - \rho_L$, $\Delta u_2 \equiv \rho_R u_R - \rho_L u_L$. The corresponding Roe numerical flux $\mathbf{F}_{i+\frac{1}{2}}$ now follows from using (11.43)–(11.46) into any of the expressions (11.27)–(11.29).

11.2.2 The Euler Equations

Here we present the Roe Riemann solver as applied to the Riemann problem
(11.4)–(11.5) for the x–split three dimensional time dependent Euler equations for ideal gases. Details of the Euler equations are found in Sect. 1.1
and Sect. 1.2 of Chap. 1; mathematical properties of the Euler equations are
studied in Chap. 3.

The exact, x–direction Jacobian matrix $\mathbf{A}(\mathbf{U})$ is

$$
\mathbf{A} =
\begin{bmatrix}
0 & 1 & 0 & 0 & 0 \\
\hat{\gamma}H - u^2 - a^2 & (3-\gamma)u & -\hat{\gamma}v & -\hat{\gamma}w & \hat{\gamma} \\
-uv & v & u & 0 & 0 \\
-uw & w & 0 & u & 0 \\
\frac{1}{2}u[(\gamma-3)H - a^2] & H - \hat{\gamma}u^2 & -\hat{\gamma}uv & -\hat{\gamma}uw & \gamma u
\end{bmatrix} , \qquad (11.47)
$$

where $\hat{\gamma} = \gamma - 1$. The eigenvalues are

$$
\lambda_1 = u - a , \quad \lambda_2 = \lambda_3 = \lambda_4 = u , \quad \lambda_5 = u + a , \qquad (11.48)
$$

where $a = \sqrt{\gamma p/\rho}$ is the sound speed. The corresponding right eigenvectors
are

$$
\mathbf{K}^{(1)} =
\begin{bmatrix} 1 \\ u-a \\ v \\ w \\ H-ua \end{bmatrix} ; \quad
\mathbf{K}^{(2)} =
\begin{bmatrix} 1 \\ u \\ v \\ w \\ \frac{1}{2}V^2 \end{bmatrix} ; \quad
\mathbf{K}^{(3)} =
\begin{bmatrix} 0 \\ 0 \\ 1 \\ 0 \\ v \end{bmatrix}
$$

$$
\mathbf{K}^{(4)} =
\begin{bmatrix} 0 \\ 0 \\ 0 \\ 1 \\ w \end{bmatrix} ; \quad
\mathbf{K}^{(5)} =
\begin{bmatrix} 1 \\ u+a \\ v \\ w \\ H+ua \end{bmatrix} .
$$

$$(11.49)$$

Here H is the total enthalpy

$$
H = \frac{E+p}{\rho} \qquad (11.50)
$$

and E is the total energy per unit volume

$$
E = \frac{1}{2}\rho \mathbf{V}^2 + \rho e , \qquad (11.51)
$$

with

$$
\mathbf{V}^2 = u^2 + v^2 + w^2 \qquad (11.52)
$$

and e denoting the specific internal energy, which for ideal gases, see Sect.
1.2 of Chap. 1, is

$$e = \frac{p}{(\gamma - 1)\rho} \, .$$ (11.53)

Roe chooses the *parameter vector*

$$\mathbf{Q} \equiv \begin{bmatrix} q_1 \\ q_2 \\ q_3 \\ q_4 \\ q_5 \end{bmatrix} \equiv \sqrt{\rho} \begin{bmatrix} 1 \\ u \\ v \\ w \\ H \end{bmatrix} ,$$ (11.54)

which has the property that every component u_i of $\mathbf{U}$ and every component f_i of $\mathbf{F}(\mathbf{U})$ in (11.4)–(11.5) is a quadratic in the components q_i of $\mathbf{Q}$. For instance $u_1 = q_1^2$ and $f_1 = q_1 q_2$, etc. Actually, the property is also valid for the components of the $\mathbf{G}$ and $\mathbf{H}$ fluxes for the full three–dimensional Euler equations.

As done for the isothermal equations, see equations (11.38), one can express the jumps $\Delta\mathbf{U}$ and $\Delta\mathbf{F}$ in terms of the jump $\Delta\mathbf{Q}$ via two matrices $\tilde{\mathbf{B}}$ and $\tilde{\mathbf{C}}$. Roe [281] gives the following expressions

$$\tilde{\mathbf{B}} = \begin{pmatrix} 2\tilde{q}_1 & 0 & 0 & 0 & 0 \\ \tilde{q}_2 & \tilde{q}_1 & 0 & 0 & 0 \\ \tilde{q}_3 & 0 & \tilde{q}_1 & 0 & 0 \\ \tilde{q}_4 & 0 & 0 & \tilde{q}_1 & 0 \\ \dfrac{\tilde{q}_5}{\gamma} & \dfrac{\gamma-1}{\gamma}\tilde{q}_2 & \dfrac{\gamma-1}{\gamma}\tilde{q}_3 & \dfrac{\gamma-1}{\gamma}\tilde{q}_4 & \dfrac{\tilde{q}_1}{\gamma} \end{pmatrix}$$ (11.55)

and

$$\tilde{\mathbf{C}} = \begin{pmatrix} \tilde{q}_2 & \tilde{q}_1 & 0 & 0 & 0 \\ \dfrac{\gamma-1}{\gamma}\tilde{q}_5 & \dfrac{\gamma+1}{\gamma}\tilde{q}_2 & -\dfrac{\gamma-1}{\gamma}\tilde{q}_3 & -\dfrac{\gamma-1}{\gamma}\tilde{q}_4 & \dfrac{\gamma-1}{\gamma}\tilde{q}_1 \\ 0 & \tilde{q}_3 & \tilde{q}_2 & 0 & 0 \\ 0 & \tilde{q}_4 & 0 & \tilde{q}_2 & 0 \\ 0 & \tilde{q}_5 & 0 & 0 & \tilde{q}_2 \end{pmatrix} .$$ (11.56)

The sought Roe matrix is then given by

$$\tilde{\mathbf{A}} = \tilde{\mathbf{B}}\tilde{\mathbf{C}}^{-1} \, .$$ (11.57)

The eigenvalues of $\tilde{\mathbf{A}}$ are

$$\tilde{\lambda}_1 = \tilde{u} - \tilde{a} , \quad \tilde{\lambda}_2 = \tilde{\lambda}_3 = \tilde{\lambda}_4 = \tilde{u} , \quad \tilde{\lambda}_5 = \tilde{u} + \tilde{a}$$ (11.58)

and the corresponding right eigenvectors are

$$
\tilde{\mathbf{K}}^{(1)} = \begin{bmatrix} 1 \\ \tilde{u} - \tilde{a} \\ \tilde{v} \\ \tilde{w} \\ \tilde{H} - \tilde{u}\tilde{a} \end{bmatrix} \; ; \quad
\tilde{\mathbf{K}}^{(2)} = \begin{bmatrix} 1 \\ \tilde{u} \\ \tilde{v} \\ \tilde{w} \\ \frac{1}{2}\tilde{V}^2 \end{bmatrix} \; ; \quad
\tilde{\mathbf{K}}^{(3)} = \begin{bmatrix} 0 \\ 0 \\ 1 \\ 0 \\ \tilde{v} \end{bmatrix}
$$

$$
\tilde{\mathbf{K}}^{(4)} = \begin{bmatrix} 0 \\ 0 \\ 0 \\ 1 \\ \tilde{w} \end{bmatrix} \; ; \quad
\tilde{\mathbf{K}}^{(5)} = \begin{bmatrix} 1 \\ \tilde{u} + \tilde{a} \\ \tilde{v} \\ \tilde{w} \\ \tilde{H} + \tilde{u}\tilde{a} \end{bmatrix} \; .
$$

$$(11.59)$$

The symbol $\tilde{r}$ in (11.58), (11.59) denotes a Roe average for a variable r. The relevant averages are given as follows

$$
\begin{aligned}
\tilde{u} &= \frac{\sqrt{\rho_L}u_L + \sqrt{\rho_R}u_R}{\sqrt{\rho_L} + \sqrt{\rho_R}} \; , \\[4pt]
\tilde{v} &= \frac{\sqrt{\rho_L}v_L + \sqrt{\rho_R}v_R}{\sqrt{\rho_L} + \sqrt{\rho_R}} \; , \\[4pt]
\tilde{w} &= \frac{\sqrt{\rho_L}w_L + \sqrt{\rho_R}w_R}{\sqrt{\rho_L} + \sqrt{\rho_R}} \; , \\[4pt]
\tilde{H} &= \frac{\sqrt{\rho_L}H_L + \sqrt{\rho_R}H_R}{\sqrt{\rho_L} + \sqrt{\rho_R}} \; , \\[4pt]
\tilde{a} &= (\gamma - 1)[\tilde{H} - \tfrac{1}{2}\tilde{V}^2]^{\frac{1}{2}} \; ,
\end{aligned}
$$

$$(11.60)$$

where $\tilde{V}^2 = \tilde{u}^2 + \tilde{v}^2 + \tilde{w}^2$.

In order to determine completely the Roe numerical flux $\mathbf{F}_{i+\frac{1}{2}}$ we need, in addition, the wave strengths $\tilde{\alpha}_i$. These are obtained by projecting the jump $\Delta\mathbf{U}$ onto the right, averaged eigenvectors (11.59), namely

$$
\Delta\mathbf{U} = \sum_{i=1}^{5} \tilde{\alpha}_i \tilde{\mathbf{K}}^{(i)} \; .
$$

$$(11.61)$$

When written in full these equations read

$$
\tilde{\alpha}_1 + \tilde{\alpha}_2 + \tilde{\alpha}_5 = \Delta u_1 \; ,
$$

$$(11.62)$$

$$
\tilde{\alpha}_1(\tilde{u} - \tilde{a}) + \tilde{\alpha}_2\tilde{u} + \tilde{\alpha}_5(\tilde{u} + \tilde{a}) = \Delta u_2 \; ,
$$

$$(11.63)$$

$$
\tilde{\alpha}_1\tilde{v} + \tilde{\alpha}_2\tilde{v} + \tilde{\alpha}_3 + \tilde{\alpha}_5\tilde{v} = \Delta u_3 \; ,
$$

$$(11.64)$$

$$
\tilde{\alpha}_1\tilde{w} + \tilde{\alpha}_2\tilde{w} + \tilde{\alpha}_4 + \tilde{\alpha}_5\tilde{w} = \Delta u_4 \; ,
$$

$$(11.65)$$

$$\tilde{\alpha}_1(\tilde{H} - \tilde{u}\tilde{a}) + \frac{1}{2}\tilde{\mathbf{V}}^2\tilde{\alpha}_2 + \tilde{\alpha}_3\tilde{v} + \tilde{\alpha}_4\tilde{w} + \tilde{\alpha}_5(\tilde{H} + \tilde{u}\tilde{a}) = \Delta u_5 \ . \qquad (11.66)$$

Here the right–hand side terms of equations (11.62)–(11.66) are known: they are jumps Δu_i in the conserved quantity u_i, namely

$$\Delta u_i = (u_i)_R - (u_i)_L \ .$$

Before solving these equations we note that in the purely one–dimensional case

$$\tilde{v} = \tilde{w} = 0 \ , \quad \tilde{\alpha}_3 = \tilde{\alpha}_4 = 0 \ , \quad \tilde{\mathbf{K}}^{(3)} = \tilde{\mathbf{K}}^{(4)} = \mathbf{0} \qquad (11.67)$$

and the problem reduces to solving (11.62), (11.63) and (11.66) for $\tilde{\alpha}_1, \tilde{\alpha}_2$ and $\tilde{\alpha}_5$, with terms involving $\tilde{\alpha}_3$ and $\tilde{\alpha}_4$ being absent.

For the x–split three dimensional problem the system (11.62)–(11.66) may be viewed in exactly the same manner as for the one–dimensional case. Use of equation (11.62) into (11.64) and (11.65) gives directly

$$\tilde{\alpha}_3 = \Delta u_3 - \tilde{v}\Delta u_1 \ ; \quad \tilde{\alpha}_4 = \Delta u_4 - \tilde{w}\Delta u_1 \ . \qquad (11.68)$$

Then one solves (11.62), (11.63) and (11.66) for $\tilde{\alpha}_1, \tilde{\alpha}_2, \tilde{\alpha}_5$. Computationally, it is convenient to arrange the solution as follows

$$\left. \begin{aligned} \tilde{\alpha}_2 &= \frac{\gamma - 1}{\tilde{a}^2}\left[\Delta u_1(\tilde{H} - \tilde{u}^2) + \tilde{u}\Delta u_2 - \overline{\Delta u_5}\right] \ , \\ \tilde{\alpha}_1 &= \frac{1}{2\tilde{a}}\left[\Delta u_1(\tilde{u} + \tilde{a}) - \Delta u_2 - \tilde{a}\tilde{\alpha}_2\right] \ , \\ \tilde{\alpha}_5 &= \Delta u_1 - (\tilde{\alpha}_1 + \tilde{\alpha}_2) \ , \end{aligned} \right\} \qquad (11.69)$$

where

$$\overline{\Delta u_5} = \Delta u_5 - (\Delta u_3 - \tilde{v}\Delta u_1)\tilde{v} - (\Delta u_4 - \tilde{w}\Delta u_1)\tilde{w} \ . \qquad (11.70)$$

An Algorithm. To compute the Roe numerical flux $\mathbf{F}_{i+\frac{1}{2}}$ according to any of the formulae (11.27)–(11.29) we do the following:

(1) Compute the Roe average values for $\tilde{u}, \tilde{v}, \tilde{w}, \tilde{H}$ and $\tilde{a}$ according to (11.60).
(2) Compute the averaged eigenvalues $\tilde{\lambda}_i$ according to (11.58).
(3) Compute the averaged right eigenvectors $\tilde{\mathbf{K}}^{(i)}$ according to (11.59).
(4) Compute the wave strengths $\tilde{\alpha}_i$ according to (11.68)–(11.70).
(5) Use all of the above quantities to compute $\mathbf{F}_{i+\frac{1}{2}}$, according to any of the formulae (11.27)–(11.29).

For the pure one–dimensional case, virtually all the required information for the application of the above algorithm is contained in this Chapter. An entropy fix is given in Sect. 11.4. The remaining items such as choosing the time step size and boundary conditions are found in Chap. 6. For two and three dimensional applications the reader requires the additional information provided in Chap. 16.

11.3 The Roe–Pike Method

Recall that solving the Riemann problem (11.4) approximately using Roe's method means finding *averaged* eigenvalues $\tilde{\lambda}_i$, right eigenvectors $\tilde{\mathbf{K}}^{(i)}$ and wave strengths $\tilde{\alpha}_i$, so that the Roe numerical flux may be evaluated by any of the formulae (11.27)–(11.29). In the previous section this task was carried out by following the original Roe approach, where the averaged Jacobian matrix $\tilde{\mathbf{A}}$ is first sought. In this section we present a different approach, due to Roe and Pike [290], whereby the construction of $\tilde{\mathbf{A}}$ is avoided; instead, one seeks directly averages of a set of scalar quantities that can then be used to evaluate the eigenvalues, right eigenvectors and wave strengths needed in formulae (11.27)–(11.29).

11.3.1 The Approach

The approach assumes, of course, that the appropriate original system is hyperbolic and that analytical expressions for the eigenvalues λ_i and the set of linearly independent right eigenvectors $\mathbf{K}^{(i)}$ are available. Analytical expressions $\hat{\alpha}_i$ for the wave strengths require extra work via an *extra linearisation*. One then selects a suitable vector of scalar quantities, typically the vector $\mathbf{W}$ of primitive variables in (11.6) or variations of it, for which an average $\tilde{\mathbf{W}}$ is to be found. The values of $\tilde{\lambda}_i$, $\tilde{\mathbf{K}}^{(i)}$ and $\tilde{\alpha}_i$ are then found by direct evaluation of the analytical expressions for λ_i, $\mathbf{K}^{(i)}$ and $\hat{\alpha}_i$ at the state $\tilde{\mathbf{W}}$. There are two distinct steps in the Roe–Pike approach.

Linearisation about a Reference State. To find analytical expressions for the wave strengths α_i Roe and Pike assume a linearised form of the governing equations based on the assumption that the data states $\mathbf{U}_L$ and $\mathbf{U}_R$ are close to a reference state $\hat{\mathbf{U}}$, to order $O(\Delta^2)$. Linearisation of the conservation laws in (11.4) about this state $\hat{\mathbf{U}}$ gives

$$\mathbf{U}_t + \mathbf{F}(\mathbf{U})_x \equiv \mathbf{U}_t + \left(\frac{\partial \mathbf{F}}{\partial \mathbf{U}}\right)\mathbf{U}_x \approx \mathbf{U}_t + \hat{\mathbf{A}}\mathbf{U}_x ,$$

where

$$\mathbf{U}_t + \hat{\mathbf{A}}\mathbf{U}_x = 0 \tag{11.71}$$

is an approximation to the original conservation laws. Here $\hat{\mathbf{A}}$ is the Jacobian matrix, assumed available, computed at the reference state $\hat{\mathbf{U}}$. Eigenvalues and right eigenvectors follow. Analytical expressions for the wave strengths $\hat{\alpha}_i$ in the solution of the linear Riemann problem

$$\left. \begin{array}{l} \mathbf{U}_t + \hat{\mathbf{A}}\mathbf{U}_x = 0 , \\ \mathbf{U}(x, 0) = \left\{ \begin{array}{ll} \mathbf{U}_L & if \quad x < 0 , \\ \mathbf{U}_R & if \quad x > 0 , \end{array} \right. \end{array} \right\} \tag{11.72}$$

are found by decomposing the data jump $\Delta \mathbf{U}$ onto the right eigenvectors, in the usual way; see Sect. 2.3 of Chap. 2 and Sect. 5.4 of Chap. 5. That is we solve

$$\Delta \mathbf{U} = \mathbf{U}_R - \mathbf{U}_L = \sum_{k=1}^{m} \hat{\alpha}_k \hat{\mathbf{K}}^{(k)} \,. \tag{11.73}$$

Before proceeding, we note that this linearisation *is not* the Roe linearisation resulting from the Roe matrix $\tilde{\mathbf{A}}$; it is merely a step to find some sufficiently simple analytical expressions for the wave strengths, which can then be evaluated at the unknown *Roe–Pike average state* $\tilde{\mathbf{W}}$, yet to be found.

The Algebraic Problem for the Average State. The sought Roe–Pike average vector $\tilde{\mathbf{W}}$ is then found by first setting

$$\tilde{\alpha}_i = \hat{\alpha}_i(\tilde{\mathbf{W}}) \,, \quad \tilde{\lambda}_i = \lambda_i(\tilde{\mathbf{W}}) \,, \quad \tilde{\mathbf{K}}^{(i)} = \mathbf{K}^{(i)}(\tilde{\mathbf{W}}) \,; \tag{11.74}$$

the analytical expressions for λ_i, $\mathbf{K}^{(i)}$ and $\hat{\alpha}_i$ are evaluated at the unknown average state $\tilde{\mathbf{W}}$. Then $\tilde{\mathbf{W}}$ is found by solving the algebraic problem posed by the following two sets of equations

$$\Delta \mathbf{U} = \mathbf{U}_R - \mathbf{U}_L = \sum_{k=1}^{m} \tilde{\alpha}_k \tilde{\mathbf{K}}^{(k)} \tag{11.75}$$

and

$$\Delta \mathbf{F} = \mathbf{F}_R - \mathbf{F}_L = \sum_{k=1}^{m} \tilde{\alpha}_k \tilde{\lambda}_k \tilde{\mathbf{K}}^{(k)} \,. \tag{11.76}$$

In the following section we illustrate the Roe–Pike approach in terms of a simple system of conservation laws.

11.3.2 The Isothermal Equations

We solve the Riemann problem

$$\mathbf{U}_t + \mathbf{F}(\mathbf{U})_x = \mathbf{0} \,, \\ \mathbf{U}(x,0) = \left\{ \begin{array}{ll} \mathbf{U}_L & if \quad x < 0 \,, \\ \mathbf{U}_R & if \quad x > 0 \,, \end{array} \right\} \tag{11.77}$$

for the isothermal equations using the Roe–Pike approach; the vectors $\mathbf{U}$ and $\mathbf{F}$ are given in (11.32). The exact Jacobian matrix, eigenvalues and right eigenvectors are

$$\mathbf{A}(\mathbf{U}) = \left[\begin{array}{cc} 0 & 1 \\ a^2 - u^2 & 2u \end{array} \right] \,, \\ \lambda_1 = u - a \,, \quad \lambda_2 = u + a \,, \\ \mathbf{K}^{(1)} = \left[\begin{array}{c} 1 \\ u - a \end{array} \right] \,, \quad \mathbf{K}^{(2)} = \left[\begin{array}{c} 1 \\ u + a \end{array} \right] \,. \tag{11.78}$$

Linearisation about a Reference State. Assume that the data states $\mathbf{U}_L$ and $\mathbf{U}_R$ are close to a state $\hat{\mathbf{U}}$ to order $O(\Delta^2)$. Linearisation of the conservation laws in (11.77) about this state $\hat{\mathbf{U}}$ gives linear Riemann problem

$$\left.\begin{array}{l} \mathbf{U}_t + \hat{\mathbf{A}}\mathbf{U}_x = \mathbf{0} \,, \\ \mathbf{U}(x,0) = \left\{ \begin{array}{ll} \mathbf{U}_L & if \quad x < 0 \,, \\ \mathbf{U}_R & if \quad x > 0 \,. \end{array} \right. \end{array}\right\} \tag{11.79}$$

Here $\hat{\mathbf{A}}$ is the Jacobian $\mathbf{A}$ evaluated at the reference state $\hat{\mathbf{U}}$, which in terms of primitive variables is denoted by $\hat{\mathbf{W}} = (\hat{\rho}, \hat{u})^T$. The complete eigenstructure is

$$\left.\begin{array}{c} \mathbf{A}(\mathbf{U}) = \left[\begin{array}{cc} 0 & 1 \\ a^2 - \hat{u}^2 & 2\hat{u} \end{array} \right] \,, \\[2mm] \hat{\lambda}_1 = \hat{u} - a \,, \quad \hat{\lambda}_2 = \hat{u} + a \,, \\[2mm] \hat{\mathbf{K}}^{(1)} = \left[\begin{array}{c} 1 \\ \hat{u} - a \end{array} \right] \,, \quad \hat{\mathbf{K}}^{(2)} = \left[\begin{array}{c} 1 \\ \hat{u} + a \end{array} \right] \,. \end{array}\right\} \tag{11.80}$$

Recall that the sound speed a is constant. We look for solutions of (11.79). The system is linear with constant coefficients. One can therefore deploy appropriate techniques studied in Sect. 2.3 of Chap. 2 and Sect. 5.4 of Chap. 5. We decompose the data jump $\Delta\mathbf{U}$ onto the right eigenvectors as follows

$$\Delta\mathbf{U} = \mathbf{U}_R - \mathbf{U}_L = \sum_{k=1}^{2} \hat{\alpha}_k \hat{\mathbf{K}}^{(k)} = \hat{\alpha}_1 \hat{\mathbf{K}}^{(1)} + \hat{\alpha}_2 \hat{\mathbf{K}}^{(2)} \,, \tag{11.81}$$

where analytical expressions for the coefficients $\hat{\alpha}_1$, $\hat{\alpha}_2$ are to be found. Writing (11.81) in full gives

$$\Delta\rho = \rho_R - \rho_L = \hat{\alpha}_1 + \hat{\alpha}_2 \,, \tag{11.82}$$

$$\Delta(\rho u) = (\rho u)_R - (\rho u)_L = \hat{\alpha}_1(\hat{u} - a) + \hat{\alpha}_2(\hat{u} + a) \,. \tag{11.83}$$

It can easily be shown that

$$\Delta(\rho u) = \hat{\rho}\Delta u + \hat{u}\Delta\rho + O(\Delta^2) \,, \tag{11.84}$$

where the leading term in $O(\Delta^2)$ is

$$(\rho_R - \hat{\rho})(u_R - \hat{u}) - (\rho_L - \hat{\rho})(u_L - \hat{u}) \,.$$

By neglecting $O(\Delta^2)$, (11.83) becomes

$$\hat{\rho}\Delta u + \hat{u}\Delta\rho = \hat{\alpha}_1(\hat{u} - a) + \hat{\alpha}_2(\hat{u} + a) \,. \tag{11.85}$$

Solving equations (11.82) and (11.85) gives the sought analytical expressions for $\hat{\alpha}_1$ and $\hat{\alpha}_2$, namely

$$\hat{\alpha}_1 = \frac{1}{2}\left[\varDelta\rho - \hat{\rho}\frac{\varDelta u}{a}\right] , \quad \hat{\alpha}_2 = \frac{1}{2}\left[\varDelta\rho + \hat{\rho}\frac{\varDelta u}{a}\right] . \tag{11.86}$$

Compare these with expressions (11.46). The reader may easily verify that, to within $O(\varDelta^2)$, the following two sets of equations are identically satisfied

$$\varDelta\mathbf{U} = \mathbf{U}_R - \mathbf{U}_L = \sum_{k=1}^{2} \hat{\alpha}_k \hat{\mathbf{K}}^{(k)} , \quad \varDelta\mathbf{F} = \mathbf{F}_R - \mathbf{F}_L = \sum_{k=1}^{2} \hat{\alpha}_k \hat{\lambda}_k \hat{\mathbf{K}}^{(k)} . \tag{11.87}$$

Here we give details for the second set. In full, these equations read

$$\varDelta(\rho u) = \hat{\alpha}_1 \hat{\lambda}_1 + \hat{\alpha}_2 \hat{\lambda}_2 , \tag{11.88}$$

$$\varDelta(\rho u^2 + \rho a^2) = \hat{\alpha}_1 \hat{\lambda}_1 (\hat{u} - a) + \hat{\alpha}_2 \hat{\lambda}_2 (\hat{u} + a) . \tag{11.89}$$

Equation (11.88) may be written as

$$\hat{\rho}\varDelta u + \hat{u}\varDelta\rho = \hat{u}(\hat{\alpha}_1 + \hat{\alpha}_2) + a(\hat{\alpha}_2 - \hat{\alpha}_1) ,$$

which after using (11.86) becomes an identity. To prove (11.89) we first expand its left–hand side

$$\varDelta(\rho u^2 + \rho a^2) = 2\hat{\rho}\hat{u}\varDelta u + \hat{u}^2\varDelta\rho + a^2\varDelta\rho .$$

The right–hand side of (11.89) can be expressed as

$$(\hat{\alpha}_1 + \hat{\alpha}_2)(\hat{u}^2 + a^2) + 2\hat{u}a(\hat{\alpha}_2 - \hat{\alpha}_1) .$$

Therefore, after use of (11.86), equation (11.89) becomes an identity and thus the second set of equations in (11.87), to order $O(\varDelta^2)$, is identically satisfied.

The Algebraic Problem for the Average State. For the general case in which the data states $\mathbf{U}_L$ and $\mathbf{U}_R$ are not necessarily close, the Roe–Pike approach proposes the algebraic problem of finding the Roe–Pike averages $\tilde{\rho}$ and $\tilde{u}$ such that the two conditions (11.75) and (11.76) are valid, namely

$$\varDelta\mathbf{U} = \sum_{k=1}^{2} \tilde{\alpha}_k \tilde{\mathbf{K}}^{(k)} , \quad \varDelta\mathbf{F} = \sum_{k=1}^{2} \tilde{\alpha}_k \tilde{\lambda}_k \tilde{\mathbf{K}}^{(k)} . \tag{11.90}$$

Here, according to (11.74), $\tilde{\alpha}_k, \tilde{\lambda}_k$ and $\tilde{\mathbf{K}}^{(k)}$ are obtained by evaluating the available analytical expressions at the sought averages $\tilde{\rho}, \tilde{u}$. For the wave strengths these are given by (11.86). For the eigenvalues and right eigenvectors they are given by (11.78). We then set

$$\tilde{\alpha}_1 = \frac{1}{2}\left[\varDelta\rho - \tilde{\rho}\frac{\varDelta u}{a}\right] , \quad \tilde{\alpha}_2 = \frac{1}{2}\left[\varDelta\rho + \tilde{\rho}\frac{\varDelta u}{a}\right] , \tag{11.91}$$

$$\tilde{\lambda}_1 = \tilde{u} - a , \quad \tilde{\lambda}_2 = \tilde{u} + a , \tag{11.92}$$

$$\tilde{\mathbf{K}}^{(1)} = \left[\begin{array}{c} 1 \\ \tilde{u} - a \end{array} \right] , \quad \tilde{\mathbf{K}}^{(2)} = \left[\begin{array}{c} 1 \\ \tilde{u} + a \end{array} \right] . \tag{11.93}$$

Writing conditions (11.90) in full produces

$$\Delta \rho = \tilde{\alpha}_1 + \tilde{\alpha}_2 , \tag{11.94}$$

$$\Delta(\rho u) = \tilde{\alpha}_1 (\tilde{u} - a) + \tilde{\alpha}_2 (\tilde{u} + a) , \tag{11.95}$$

$$\Delta(\rho u) = \tilde{\lambda}_1 \tilde{\alpha}_1 + \tilde{\lambda}_2 \tilde{\alpha}_2 , \tag{11.96}$$

$$\Delta(\rho u^2 + a^2 \rho) = \tilde{\lambda}_1 \tilde{\alpha}_1 (\tilde{u} - a) + \tilde{\lambda}_2 \tilde{\alpha}_2 (\tilde{u} + a) . \tag{11.97}$$

These are a set of four non–linear algebraic equations for the two unknowns $\tilde{\rho}$ and $\tilde{u}$. Note however that, by virtue of (11.91), (11.94) is an identity, for any average value $\tilde{\rho}$. Also, (11.95) is identical to (11.96) and thus we work with (11.96) and (11.97) only. From equation (11.96) one obtains

$$\Delta(\rho u) = \tilde{u}(\tilde{\alpha}_1 + \tilde{\alpha}_2) + a(\tilde{\alpha}_2 - \tilde{\alpha}_1).$$

Use of (11.91) here leads to

$$\Delta(\rho u) = \tilde{\rho} \Delta u + \tilde{u} \Delta \rho . \tag{11.98}$$

From (11.97) we write

$$\Delta(\rho u^2 + \rho a^2) = (\tilde{\alpha}_1 + \tilde{\alpha}_2)(\tilde{u}^2 + a^2) + 2a\tilde{u}(\tilde{\alpha}_2 - \tilde{\alpha}_1) ,$$

which after using (11.91) and the exact relation

$$\Delta(\rho u^2 + \rho a^2) = \Delta(\rho u^2) + a^2 \Delta \rho$$

leads to the result

$$\Delta(\rho u^2) = 2\tilde{u} \tilde{\rho} \Delta u + \tilde{u}^2 \Delta \rho . \tag{11.99}$$

Elimination of $\tilde{\rho}$ from (11.98) and (11.99) leads to a quadratic equation for $\tilde{u}$, namely

$$\Delta \rho \tilde{u}^2 - 2\Delta(\rho u)\tilde{u} + \Delta(\rho u^2) = 0 . \tag{11.100}$$

This equation has two solutions, namely

$$\tilde{u} = \frac{\Delta(\rho u) \pm \sqrt{[\Delta(\rho u)]^2 - \Delta \rho \Delta(\rho u^2)}}{\Delta \rho} . \tag{11.101}$$

After using the definition $\Delta r = r_R - r_L$ the discriminant is found to be

$$\rho_L \rho_R (\Delta u)^2 ,$$

which simplifies (11.101) to

$$\tilde{u} = \frac{\Delta(\rho u) \pm \Delta u \sqrt{\rho_L \rho_R}}{\Delta \rho} . \tag{11.102}$$

The root obtained by taking the *negative* sign in (11.102) produces the Roe–averaged velocity

$$\tilde{u} = \frac{\sqrt{\rho_L} u_L + \sqrt{\rho_R} u_R}{\sqrt{\rho_L} + \sqrt{\rho_R}} \ . \tag{11.103}$$

Compare (11.103) with (11.43). From (11.98) we obtain

$$\tilde{\rho} = \sqrt{\rho_L \rho_R} \ . \tag{11.104}$$

We have thus found algebraic expressions for the sought Roe–Pike averages $\tilde{\rho}$ and $\tilde{u}$. We observe that the second root obtained by taking the *positive* sign in (11.102) leads to the *spurious* solution

$$\tilde{u} = \frac{\sqrt{\rho_R} u_R - \sqrt{\rho_L} u_L}{\sqrt{\rho_R} - \sqrt{\rho_L}} \ . \tag{11.105}$$

There is a very good reason for rejecting this as a *useful solution*; in the trivial case $\rho_L = \rho_R$, $u_L \neq u_R$ the solution $\tilde{u}$ is not even defined.

Having found the Roe–Pike averages $\tilde{\rho}$ and $\tilde{u}$ we can then compute the wave strengths $\tilde{\alpha}_k$, the eigenvalues $\tilde{\lambda}_k$ and the right eigenvectors $\tilde{\mathbf{K}}^{(k)}$ according to expressions (11.91)–(11.93). The Roe numerical flux $\mathbf{F}_{i+\frac{1}{2}}$ to be used in the conservative formula (11.2) can now be obtained from any of the relations (11.27)–(11.29).

11.3.3 The Euler Equations

We solve the Riemann problem (11.4) for the x–split, three dimensional Euler equations using the Roe–Pike method. Assuming the analytical expressions (11.48)–(11.49) for the eigenvalues and eigenvectors, one then linearises the equations about a state $\hat{\mathbf{U}}$ to find analytical expressions for the wave strengths; this is done under the assumption that both data states $\mathbf{U}_L, \mathbf{U}_R$ are close to $\hat{\mathbf{U}}$ to $O(\Delta^2)$. This leads to the linear system

$$\left. \begin{array}{l} \mathbf{U}_t + \hat{\mathbf{A}} \mathbf{U}_x = \mathbf{0} \ , \\[2mm] \mathbf{U}(x,t) = \left\{ \begin{array}{ll} \mathbf{U}_L \ , & x < 0 \ , \\ \mathbf{U}_R \ , & x > 0 \ . \end{array} \right. \end{array} \right\} \tag{11.106}$$

The Jacobian matrix $\hat{\mathbf{A}}$ is obtained by evaluating the exact Jacobian matrix (11.47) at the state $\hat{\mathbf{U}}$; the eigenvalues $\hat{\lambda}_i$ are

$$\hat{\lambda}_1 = \hat{u} - \hat{a} \ , \quad \hat{\lambda}_2 = \hat{\lambda}_3 = \hat{\lambda}_4 = \hat{u} \ , \quad \hat{\lambda}_5 = \hat{u} + \hat{a} \tag{11.107}$$

and the right eigenvectors $\hat{\mathbf{K}}^{(i)}$ are

$$\hat{\mathbf{K}}^{(1)} = \begin{bmatrix} 1 \\ \hat{u} - \hat{a} \\ \hat{v} \\ \hat{w} \\ \hat{H} - \hat{u}\hat{a} \end{bmatrix} \quad ; \quad \hat{\mathbf{K}}^{(2)} = \begin{bmatrix} 1 \\ \hat{u} \\ \hat{v} \\ \hat{w} \\ \frac{1}{2}\hat{V}^2 \end{bmatrix} \quad ; \quad \hat{\mathbf{K}}^{(3)} = \begin{bmatrix} 0 \\ 0 \\ 1 \\ 0 \\ \hat{v} \end{bmatrix}$$

$$\hat{\mathbf{K}}^{(4)} = \begin{bmatrix} 0 \\ 0 \\ 0 \\ 1 \\ \hat{w} \end{bmatrix} \quad ; \quad \hat{\mathbf{K}}^{(5)} = \begin{bmatrix} 1 \\ \hat{u} + \hat{a} \\ \hat{v} \\ \hat{w} \\ \hat{H} + \hat{u}\hat{a} \end{bmatrix} .$$

$$(11.108)$$

By expanding the data jump $\Delta\mathbf{U}$ onto the right eigenvectors we write

$$\Delta\mathbf{U} = \sum_{i=1}^{5} \hat{\alpha}_i \hat{\mathbf{K}}^{(i)} \qquad (11.109)$$

The solution of this 5×5 linear system will provide analytical expressions for the wave strengths $\hat{\alpha}_i$. As a matter of fact we can use the solution for the wave strengths obtained in the Roe original method, (11.68)–(11.70), and reinterpret the solution appropriately. These are

$$\begin{aligned}
\hat{\alpha}_3 &= \Delta u_3 - \hat{v}\Delta u_1 , \\
\hat{\alpha}_4 &= \Delta u_4 - \hat{w}\Delta u_1 , \\
\hat{\alpha}_2 &= \tfrac{\gamma-1}{\hat{a}^2}[\Delta u_1(\hat{H} - \hat{u}^2) + \hat{u}\Delta u_2 - \overline{\Delta u_5}] , \\
\hat{\alpha}_1 &= \tfrac{1}{2\hat{a}}[\Delta u_1(\hat{u} + \hat{a}) - \Delta u_2 - \hat{a}\hat{\alpha}_2] , \\
\hat{\alpha}_5 &= \Delta u_1 - (\hat{\alpha}_1 + \hat{\alpha}_2) ,
\end{aligned} \right\} \qquad (11.110)$$

where

$$\overline{\Delta u_5} = \Delta u_5 - (\Delta u_3 - \hat{v}\Delta u_1)\hat{v} - (\Delta u_4 - \hat{w}\Delta u_1)\hat{w} . \qquad (11.111)$$

By applying the operator

$$\Delta(rs) = \hat{r}\Delta s + \hat{s}\Delta r + O(\Delta^2) \qquad (11.112)$$

and neglecting $O(\Delta^2)$ we arrive at the following solution:

$$\begin{aligned}
\hat{\alpha}_1 &= \frac{1}{2\hat{a}^2}[\Delta p - \hat{\rho}\hat{a}\Delta u] , \\
\hat{\alpha}_2 &= \Delta\rho - \Delta p/\hat{a}^2 , \\
\hat{\alpha}_3 &= \hat{\rho}\Delta v , \\
\hat{\alpha}_4 &= \hat{\rho}\Delta w , \\
\hat{\alpha}_5 &= \frac{1}{2\hat{a}^2}[\Delta p + \hat{\rho}\hat{a}\Delta u]
\end{aligned} \right\} \qquad (11.113)$$

The second step in the Roe–Pike method is to find an average state

$$\tilde{\mathbf{W}} = (\tilde{\rho}, \tilde{u}, \tilde{v}, \tilde{w}, \tilde{a})^T ,\tag{11.114}$$

such that the algebraic problem posed by the following two sets of equations

$$\Delta \mathbf{U} = \sum_{i=1}^{5} \tilde{\alpha}_i \tilde{\mathbf{K}}^{(i)} ,\tag{11.115}$$

$$\Delta \mathbf{F} = \sum_{i=1}^{5} \tilde{\alpha}_i \tilde{\lambda}_i \tilde{\mathbf{K}}^{(i)} ,\tag{11.116}$$

is satisfied, where

$$\tilde{\alpha}_i = \hat{\alpha}_i(\tilde{\mathbf{W}}) , \quad \tilde{\lambda}_i = \lambda_i(\tilde{\mathbf{W}}) , \quad \tilde{\mathbf{K}}^{(i)} = \mathbf{K}^{(i)}(\tilde{\mathbf{W}}) ,\tag{11.117}$$

with λ_i and $\mathbf{K}^{(i)}$ given by (11.48)–(11.49) and $\hat{\alpha}_i$ given by (11.113). Details of the algebra for the one–dimensional case are given by Roe and Pike [290]. For the x–split three dimensional case the solution for the average vector $\tilde{\mathbf{W}}$ is

$$\left. \begin{array}{c} \tilde{\rho} = \sqrt{\rho_L \rho_R} , \\[2mm] \tilde{u} = \dfrac{\sqrt{\rho_L} u_L + \sqrt{\rho_R} u_R}{\sqrt{\rho_L} + \sqrt{\rho_R}} , \\[4mm] \tilde{v} = \dfrac{\sqrt{\rho_L} v_L + \sqrt{\rho_R} v_R}{\sqrt{\rho_L} + \sqrt{\rho_R}} , \\[4mm] \tilde{w} = \dfrac{\sqrt{\rho_L} w_L + \sqrt{\rho_R} w_R}{\sqrt{\rho_L} + \sqrt{\rho_R}} , \\[4mm] \tilde{H} = \dfrac{\sqrt{\rho_L} H_L + \sqrt{\rho_R} H_R}{\sqrt{\rho_L} + \sqrt{\rho_R}} , \\[4mm] \tilde{a} = [(\gamma - 1)(\tilde{H} - \tfrac{1}{2}\tilde{\mathbf{V}}^2)]^{\frac{1}{2}} , \end{array} \right\} \tag{11.118}$$

where $\tilde{\mathbf{V}}^2 = \tilde{u}^2 + \tilde{v}^2 + \tilde{w}^2$. These are identical to the Roe averages obtained by the original Roe method, see (11.60). Now $\tilde{\alpha}_i, \tilde{\lambda}_i$ and $\tilde{\mathbf{K}}^{(i)}$ are computed according to (11.117) and then the Roe intercell flux $\mathbf{F}_{i+\frac{1}{2}}$ follows from any of the formulae (11.27)–(11.29).

An Algorithm. To compute the Roe numerical flux $\mathbf{F}_{i+\frac{1}{2}}$ according to any of the formulae (11.27)–(11.29) we do the following:

(1) Compute the Roe average values according to (11.118).
(2) Compute the eigenvalues $\tilde{\lambda}_i$ using the analytical expressions (11.107) evaluated on the averages (11.118).

(3) Compute the right eigenvectors using the analytical expressions (11.108) evaluated on the averages (11.118).
(4) Compute the wave strengths using the analytical expressions (11.113) evaluated on the averages (11.118).
(5) Use all of the above quantities to compute $\mathbf{F}_{i+\frac{1}{2}}$, according to any of the formulae (11.27)–(11.29).

Before applying the scheme as described to practical problems, a modification to handle *sonic flow* correctly is required. This is the subject of the next section.

11.4 An Entropy Fix

Linearised Riemann problem solutions consist of discontinuous jumps only. See Sect. 2.3 of Chap. 2. This can be a good approximation for contacts and shocks, in that the discontinuous character of the wave is correct, although the size of the jump may not be correctly approximated by the linearised solution. Rarefaction waves, on the other hand, carry a continuous change in flow variables, and as time increases, they tend to spread; that is spatial gradients tend to decay. Quite clearly then, the linearised approximation via discontinuous jumps is grossly incorrect. In a practical computational set up however, it is only in the case in which the rarefaction wave is *transonic*, or *sonic*, where linearised approximations encounter difficulties; these show up in the form of unphysical, entropy violating discontinuous waves, sometimes called *rarefaction shocks*.

11.4.1 The Entropy Problem

Consider the Riemann problem whose initial data is that of Test 1 in Table 11.1. The structure of the exact solution of this problem, depicted in Fig. 11.2, contains a left *sonic rarefaction*, a contact discontinuity of speed u_* and a right shock wave. As the left rarefaction is *sonic* the eigenvalue $\lambda_1 = u - a$ changes from negative to positive, as the wave is crossed from left to right. There is a point at which $\lambda_1 = u - a = 0$, giving the sonic flow condition $u = a$.

$$\lambda_1(\mathbf{U}_L) = S_{HL} = u_L - a_L < 0$$

is the speed of the *head* of the rarefaction and

$$\lambda_1(\mathbf{U}_{*L}) = S_{TL} = u_* - a_{*L} > 0$$

is the speed of the *tail*. Fig. 11.4 shows the numerical (symbols) and exact (line) solutions of this problem, where the numerical solution is obtained by Roe's method as described so far. The numerical solution within the rarefaction exhibits a discontinuity within the wave; this discontinuity is unphysical, it violates the entropy condition. See Sect. 2.4.2 of Chap. 2. Recall that a physically admissible discontinuity of speed S requires $S_b \geq S \geq S_a$ where S_b and S_a are characteristic speeds behind and ahead of the wave respectively. That is, characteristics move into the discontinuity; the limiting case of parallel characteristic speeds is that of a contact discontinuity. For the example above, the opposite happens. See Sect. 2.4.2 of Chap. 2, for a discussion on entropy–violating solutions.

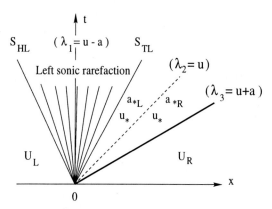

Fig. 11.2. Left transonic rarefaction wave. Left eigenvalue $\lambda_1 = u - a$ changes sign as the wave is crossed from left to right

Roe's solver can be modified so as to avoid entropy violating solutions. This is usually referred to as an *entropy fix*. Harten and Hyman [163] suggested an entropy fix for Roe's method, which has widespread use. Other ways of correcting the scheme have been discussed by Roe and Pike [290], Roe [288], Sweby [332] and Dubois and Mehlman [113], amongst others. Here we present the details of the Harten–Hyman approach.

11.4.2 The Harten–Hyman Entropy Fix

The general approach is presented in the original paper of Harten and Hyman of 1983 [163]. A description can also be found in [207]. The presentation here is tailored specifically to the time–dependent Euler equations, for which we only need to consider the left and right non–linear waves associated with the eigenvalues $\lambda_1 = u - a$ and $\lambda_5 = u + a$ respectively. Our version of the Harten–Hyman entropy fix relies on estimates for particle velocity u_* and sound speeds a_{*L}, a_{*R} in the *Star Region*; see Figs. 11.1 and 11.2. Various ways of finding these are given in Sect. 11.4.3.

Left Transonic Rarefaction. Consider the situation depicted in Fig. 11.2. Assuming u_* and a_{*L} are available, we compute the speeds

$$\lambda_1^L = u_L - a_L \; ; \quad \lambda_1^R = u_* - a_{*L} \; . \tag{11.119}$$

If

$$\lambda_1^L < 0 < \lambda_1^R ,\qquad(11.120)$$

then the left wave is a *transonic, or sonic, rarefaction wave*. In these circumstances the entropy fix is required and is enforced as follows. The single jump

$$\mathbf{U}_{*L} - \mathbf{U}_L = \tilde{\alpha}_1 \tilde{\mathbf{K}}^{(1)}\qquad(11.121)$$

travelling with speed $\tilde{\lambda}_1$ is split into two smaller jumps $\mathbf{U}_{SL} - \mathbf{U}_L$ and $\mathbf{U}_{*L} - \mathbf{U}_{SL}$ travelling respectively at speeds λ_1^L and λ_1^R, where $\mathbf{U}_{SL}$ is a *transonic state* yet to be found; see Fig. 11.3. Application of the integral form of the conservation laws, see Chaps. 3 and 10, gives

$$\lambda_1^R(\mathbf{U}_{SL} - \mathbf{U}_{*L}) + \lambda_1^L(\mathbf{U}_L - \mathbf{U}_{SL}) = \tilde{\lambda}_1(\mathbf{U}_L - \mathbf{U}_{*L}) ,\qquad(11.122)$$

from which we obtain

$$\mathbf{U}_{SL} = \frac{(\tilde{\lambda}_1 - \lambda_1^L)\mathbf{U}_L + (\lambda_1^R - \tilde{\lambda}_1)\mathbf{U}_{*L}}{\lambda_1^R - \lambda_1^L} .\qquad(11.123)$$

To compute the Roe intercell flux we adopt the one–sided formulae (11.27),

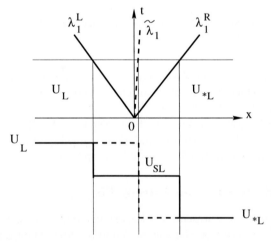

Fig. 11.3. Entropy fix for left transonic rarefaction wave. Single jump $\mathbf{U}_{*L} - \mathbf{U}_L$ travelling with speed $\tilde{\lambda}_1$ is split into the two jumps $\mathbf{U}_{SL} - \mathbf{U}_L$ and $\mathbf{U}_{*L} - \mathbf{U}_{SL}$ travelling with speeds λ_1^L and λ_1^R. Profile shown is a representation a single variable

namely

$$\mathbf{F}_{i+\frac{1}{2}} = \mathbf{F}_L + \sum_{\tilde{\lambda}_k \leq 0} \tilde{\lambda}_k \tilde{\alpha}_k \tilde{\mathbf{K}}^{(k)} ,\qquad(11.124)$$

where in the present case the summation applies to the single jump $\mathbf{U}_{SL} - \mathbf{U}_L$ travelling with speed $\lambda_1^L < 0$; in view of (11.122) the jump is

$$\mathbf{U}_{SL} - \mathbf{U}_L = \frac{(\lambda_1^R - \tilde{\lambda}_1)}{(\lambda_1^R - \lambda_1^L)}(\mathbf{U}_{*L} - \mathbf{U}_L) \ . \tag{11.125}$$

But the Roe approximation gives

$$\mathbf{U}_{*L} - \mathbf{U}_L = \tilde{\alpha}_1 \tilde{\mathbf{K}}^{(1)} \tag{11.126}$$

and thus the flux jump $(\varDelta \mathbf{F})_1^L$ across the wave of speed λ_1^L is

$$(\varDelta \mathbf{F})_1^L = \lambda_1^L \left(\frac{\lambda_1^R - \tilde{\lambda}_1}{\lambda_1^R - \lambda_1^L} \right) \tilde{\alpha}_1 \tilde{\mathbf{K}}^{(1)} \ . \tag{11.127}$$

By defining the new wave speed

$$\bar{\lambda}_1 = \lambda_1^L \left(\frac{\lambda_1^R - \tilde{\lambda}_1}{\lambda_1^R - \lambda_1^L} \right) \ , \tag{11.128}$$

the intercell flux (11.27) becomes

$$\mathbf{F}_{i+\frac{1}{2}} = \mathbf{F}_L + \bar{\lambda}_1 \tilde{\alpha}_1 \tilde{\mathbf{K}}^{(1)} \ . \tag{11.129}$$

Right Transonic Rarefaction. For a right transonic rarefaction, the entropy fix procedure is entirely analogous to the left rarefaction case. Assuming the speeds u_* and a_{*R} are available, we compute the two wave speeds

$$\lambda_5^L = u_* + a_{*R} \ , \quad \lambda_5^R = u_R + a_R \ . \tag{11.130}$$

If

$$\lambda_5^L < 0 < \lambda_5^R \tag{11.131}$$

then the right wave is a *transonic rarefaction wave*. The transonic state $\mathbf{U}_{SR}$ is defined between the waves of speeds λ_5^L and λ_5^R and is given by

$$\mathbf{U}_{SR} = \frac{(\lambda_5^R - \tilde{\lambda}_5)\mathbf{U}_R + (\tilde{\lambda}_5 - \lambda_5^L)\mathbf{U}_{*R}}{\lambda_5^R - \lambda_5^L} \ . \tag{11.132}$$

Next we define the new wave speed

$$\bar{\lambda}_5 = \lambda_5^R \left(\frac{\tilde{\lambda}_5 - \lambda_5^L}{\lambda_5^R - \lambda_5^L} \right) \tag{11.133}$$

and then use the one–sided flux formula (11.28) to compute the numerical flux. The resulting Roe numerical flux is

$$\mathbf{F}_{i+\frac{1}{2}} = \mathbf{F}_R - \bar{\lambda}_5 \tilde{\alpha}_5 \tilde{\mathbf{K}}^{(5)} \ . \tag{11.134}$$

In the present version of the Harten–Hyman entropy fix we have used the one–sided flux formulae (11.27) and (11.28). The procedure can be easily adapted for use in conjunction with the centred formulae (11.29), if desired.

Next we discuss ways of finding the speeds u_*, a_{*L} and a_{*R} needed to implement the entropy fix.

11.4.3 The Speeds u_*, a_{*L}, a_{*R}

The *star* states $\mathbf{U}_{*L}$, $\mathbf{U}_{*R}$ are required in order to obtain the speeds u_*, a_{*L}, a_{*R} and thus the characteristic speeds in (11.119) and (11.130). We present various possible alternatives.

The Roe–Averaged States. Given the Roe–averaged $\tilde{\alpha}_i$ and $\tilde{\mathbf{K}}^{(i)}$ one can find the state $\mathbf{U}_{*L}$ as

$$\mathbf{U}_{*L} = \mathbf{U}_L + \tilde{\alpha}_1 \tilde{\mathbf{K}}^{(1)} , \tag{11.135}$$

which leads to

$$\left.\begin{array}{l} \rho_{*L} = \rho_L + \tilde{\alpha}_1 , \quad u_* = \dfrac{\rho_L u_L + \tilde{\alpha}_1(\tilde{u} - \tilde{a})}{\rho_L + \tilde{\alpha}_1} , \\[3mm] p_* = (\gamma - 1)\left[E_L + \tilde{\alpha}_1(\tilde{H} - \tilde{u}\tilde{a}) - \tfrac{1}{2}\rho_{*L}u_*^2\right] . \end{array}\right\} \tag{11.136}$$

Then we compute the sound speed

$$a_{*L} = \sqrt{\frac{\gamma p_*}{\rho_{*L}}} \tag{11.137}$$

and thus the speeds λ_1^L and λ_1^R in Eq. (11.119) follow. For the right wave one has

$$\mathbf{U}_{*R} = \mathbf{U}_R - \tilde{\alpha}_5 \tilde{\mathbf{K}}^{(5)} , \tag{11.138}$$

which produces

$$\left.\begin{array}{l} \rho_{*R} = \rho_R - \tilde{\alpha}_5 , \quad u_* = \dfrac{\rho_R u_R - \tilde{\alpha}_5(\tilde{u} + \tilde{a})}{\rho_R - \tilde{\alpha}_5} , \\[3mm] p_* = (\gamma - 1)\left[E_R - \tilde{\alpha}_5(\tilde{H} + \tilde{u}\tilde{a}) - \tfrac{1}{2}\rho_{*R}u_*^2\right] . \end{array}\right\} \tag{11.139}$$

The sound speed follows as $a_{*R} = \sqrt{\frac{\gamma p_*}{\rho_{*R}}}$ and thus the wave speeds λ_5^L and λ_5^R in Eq. (11.130) are determined.

The PVRS Approximation. Another way of estimating the required wave speeds is by using the Primitive–Variable Riemann Solver (PVRS) of Toro [355] presented in Sect. 9.3 of Chap. 9. The relevant solution values are

$$\left.\begin{array}{lll} p_* & = & \tfrac{1}{2}(p_L + p_R) & + & \tfrac{1}{2}(u_L - u_R)\bar{\rho}\bar{a} , \\[1mm] u_* & = & \tfrac{1}{2}(u_L + u_R) & + & \tfrac{1}{2}(p_L - p_R)/(\bar{\rho}\bar{a}) , \\[1mm] \rho_{*L} & = & \rho_L & + & (u_L - u_*)\bar{\rho}/\bar{a} , \\[1mm] \rho_{*R} & = & \rho_R & + & (u_* - u_R)\bar{\rho}/\bar{a} , \end{array}\right\} \tag{11.140}$$

with

$$\bar{\rho} = \frac{1}{2}(\rho_L + \rho_R) , \quad \bar{a} = \frac{1}{2}(a_L + a_R). \tag{11.141}$$

In order to avoid negative pressures we recommend replacing the linearised solution p_* by $\max\{0, p_*\}$. The sound speeds a_{*L}, a_{*R} are then computed in the usual way.

TRRS Approximation. Another possibility is to use the Two–Rarefaction Riemann Solver (TRRS) discussed in Chap. 9, Sect. 9.4.1. The pressure p_* is given by

$$p_* = \left[\frac{a_L + a_R - \frac{\gamma-1}{2}(u_R - u_L)}{a_L/p_L^z + a_R/p_R^z} \right]^{\frac{1}{z}}, \qquad (11.142)$$

with $z = \frac{\gamma-1}{2\gamma}$. For the left non–linear wave the sound speed and particle velocity follow directly as

$$a_{*L} = a_L(p_*/p_L)^z, \quad u_* = u_L + \frac{2}{(\gamma-1)}(a_L - a_{*L}). \qquad (11.143)$$

For the right non–linear wave we have

$$a_{*R} = a_R(p_*/p_R)^z, \quad u_* = u_R + \frac{2}{(\gamma-1)}(a_{*R} - a_R). \qquad (11.144)$$

Hence speeds (11.119) and (11.130) are determined.

Other Alternatives. Both the PVRS and the Roe linearised solutions for the speeds u_*, a_{*L}, a_{*R} may fail in the vicinity of low density flow [118]. The TRRS approximation presented above would not suffer from such difficulties; in fact, in the case in which both non–linear waves are rarefactions such an approximation would be exact. But as seen in equations (11.142)–(11.144) there are four fractional powers to be computed in each case, which makes this approximation rather expensive to use. A robust and yet more efficient scheme is the Two–Shock Riemann Solver (TSRS) [361] of Sect. 9.4.2, Chap. 9. Even better is the adaptive Riemann solver scheme of Sect. 9.5.2, Chap. 9.

11.5 Numerical Results and Discussion

Here we illustrate the performance of Godunov's first–order upwind method used in conjunction with the Roe approximate Riemann solver, discuss the results and point directions for extending the method.

11.5.1 The Tests

We select five test problems for the one–dimensional, time dependent Euler equations for ideal gases with $\gamma = 1.4$; these have exact solutions. In all chosen tests, data consists of two constant states $\mathbf{W}_L = [\rho_L, u_L, p_L]^T$ and $\mathbf{W}_R = [\rho_R, u_R, p_R]^T$, separated by a discontinuity at a position $x = x_0$. The states $\mathbf{W}_L$ and $\mathbf{W}_R$ are given in Table 11.1. The exact and numerical solutions are found in the spatial domain $0 \le x \le 1$. The numerical solution is computed with $M = 100$ cells and the CFL condition is as for all previous computations, see Chap. 6; the chosen Courant number coefficient is $C_{cfl} = 0.9$; boundary conditions are transmissive.

The exact solutions were found by running the code HE-E1RPEXACT of the library *NUMERICA* [369] and the numerical solutions were obtained by running the code HE-E1GODFLUX of *NUMERICA*.

Test	ρ_L	u_L	p_L	ρ_R	u_R	p_R
1	1.0	0.75	1.0	0.125	0.0	0.1
2	1.0	-2.0	0.4	1.0	2.0	0.4
3	1.0	0.0	1000.0	1.0	0.0	0.01
4	5.99924	19.5975	460.894	5.99242	-6.19633	46.0950
5	1.0	-19.59745	1000.0	1.0	-19.59745	0.01

Table 11.1. Data for five test problems with exact solution, for the time–dependent, one dimensional Euler equations

Test 1 is a *modified* version of Sod's problem [318]; the solution has a right shock wave, a right travelling contact wave and a left *sonic* rarefaction wave; this test is useful in assessing the *entropy satisfaction* property of numerical methods. The solution of Test 2 consists of two symmetric rarefaction waves and a trivial contact wave; the *Star Region* between the non–linear waves is close to vacuum, which makes this problem a suitable test for assessing the performance of numerical methods for low–density flows. Test 3 is designed to assess the robustness and accuracy of numerical methods; its solution consists of a strong shock wave of shock Mach number 198, a contact surface and a left rarefaction wave. Test 4 is also a very severe test, its solution consists of three strong discontinuities travelling to the right. A detailed discussion on the exact solution of the test problems is found in Sect. 4.3.3 of Chap. 4. Test 5 is also designed to test the robustness of numerical methods but the main reason for devising this test is to assess the ability of numerical methods to resolve *slowly–moving contact discontinuities*. The exact solution of Test 5 consists of a left rarefaction wave, a right–travelling shock wave and a *stationary* contact discontinuity. For each test problem we select a convenient position x_0 of the initial discontinuity and the output time. These are stated in the legend of each figure displaying computational results.

11.5.2 The Results

The computed results for Tests 1 to 5 using the Godunov first–order method in conjunction with the Roe approximate Riemann solver are shown in Figs. 11.4–11.8, where the numerical solution is shown by the symbols and the full line denotes the exact solution. As discussed earlier, Fig. 11.4 shows the results obtained from the Roe Riemann solver without the entropy fix and, as expected, the computed solution is obviously incorrect. Fig. 11.5 shows the corresponding results from the modified scheme using the Harten–Hyman entropy fix presented in the previous section. These results are, to plotting accuracy, almost indistinguishable from those obtained by the Godunov method in conjunction with the exact Riemann solver; see Fig. 6.8, Chap. 6. As a matter of fact, near the sonic point, the modified Roe solution looks slightly better; it also looks better than the Flux Vector Splitting solution, with the van Leer splitting, see Fig. 8.4 of Chap. 8. The HLL and HLLC solutions of Chap. 10, still seem to be the most accurate near sonic points. Compare also with the Osher results of Chap. 12. As anticipated, the Roe solver will fail near low–density flows; Test 2 contains two strong rarefactions with a low density and low pressure region in the middle and the Roe method, as described, does actually fail on this test. To compute successfully this kind of flows one must modify the Roe solver following the methodology of Einfeldt et. al. [118]. The results for Tests 3 and 4 are virtually identical to those of Godunov's method with the exact Riemann solver, as the reader can verify by comparing Figs. 11.6 and 11.7 with Figs. 6.10 and 6.11 of Chap. 6. The results for Test 5 are also very similar to those obtained from the Godunov method with the exact Riemann solver; note however that the (non–isolated) stationary contact is not as sharply resolved as with the approximate HLLC Riemann solver of Chapt. 10, see Fig. 10.9. As expected of course, the resolution of the stationary contact is better than that of the Flux Vector Splitting Method with the Steger–Warming splitting and that with the van Leer splitting, see Figs. 8.14 and 8.15 of Chap. 8.

11.6 Extensions

The Roe approximate Riemann solver, following the original method of Roe and that of Roe and Pike, has been presented and illustrated via the isothermal equations of gas dynamics and the split three–dimensional, time dependent Euler equations. Details of the Roe solver for the three–dimensional steady supersonic Euler equations are found in the original paper of Roe [281]. For one–dimensional applications all the required information is contained in this chapter and Chap. 6. Second–order Total Variation Diminishing (TVD) extensions of the schemes are presented in. Chap. 13 for scalar problems and in Chap. 14 for non–linear one dimensional systems. In chap. 15 we present techniques that allow the extension of these schemes to solve problems with source terms. In Chap. 16 we study techniques to extend the methods of this chapter to three–dimensional problems.

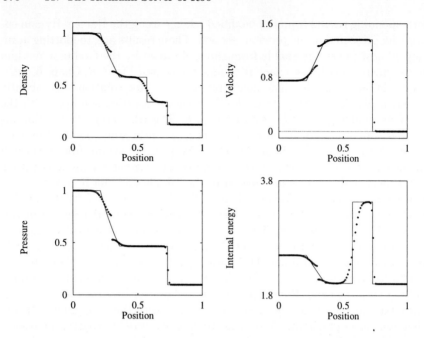

Fig. 11.4. Godunov's method with Roe's Riemann solver (no entropy fix) for Test 1, $x_0 = 0.3$. Numerical (symbol) and exact (line) solutions compared at time 0.2

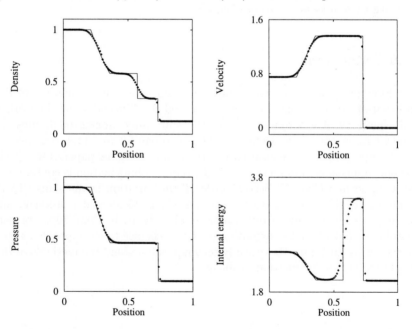

Fig. 11.5. Godunov's method with Roe's Riemann solver applied to Test 1, with $x_0 = 0.3$. Numerical (symbol) and exact (line) solutions are compared at time 0.2

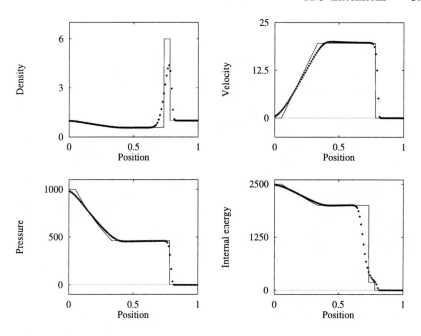

Fig. 11.6. Godunov's method with Roe's Riemann solver applied to Test 3, with $x_0 = 0.5$. Numerical (symbol) and exact (line) solutions are compared at time 0.012

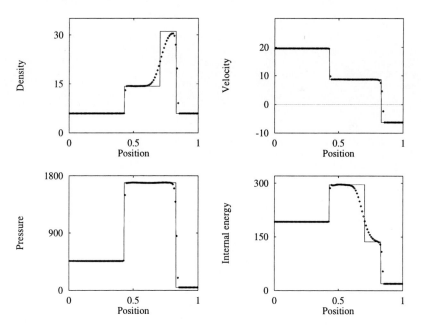

Fig. 11.7. Godunov's method with Roe's Riemann solver applied to Test 4, with $x_0 = 0.4$. Numerical (symbol) and exact (line) solutions are compared at time 0.035

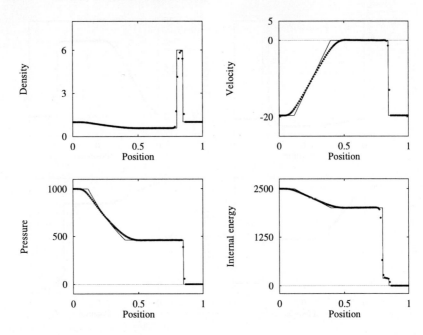

Fig. 11.8. Godunov's method with Roe's Riemann solver applied to Test 5, with $x_0 = 0.8$. Numerical (symbol) and exact (line) solutions are compared at time 0.012

12. The Riemann Solver of Osher

Osher's approximate Riemann solver is one of the earliest in the literature. The bases of the approach were communicated in the papers by Engquist and Osher in 1981 [120] and Osher and Solomon the following year [255]. Applications to the Euler equations were published later in a paper by Osher and Chakravarthy [253]. Since then the scheme has gained increasing popularity, particularly within the CFD community concerned with Steady Aerodynamics; see for example the works of Spekreijse [323], [324], Hemker and Spekreijse [167], Koren and Spekreijse [195], Qin et. al. [270], [271], [272], [273], [267], [268], [269]. One of the attractions of Osher's scheme is the smoothness of the numerical flux; the scheme has also been proved to be entropy satisfying and in practical computations it is seen to handle sonic flow well. A distinguishing feature of the Osher scheme is its performance near slowly–moving shock waves; see Roberts [280], Billett and Toro [40] and Arora and Roe [13]. The scheme is closely related to the Flux Vector Splitting approach described in Chap. 8 and, as Godunov's method of Chap. 6, it is a generalisation of the CIR scheme described in Chap. 5 for linear hyperbolic systems with constant coefficients. For a scalar conservation law, van Leer [395] studied in detail the relationship between the Osher scheme and some other Riemann solvers available at the time. Useful background material for reading this chapter is found in the previous Chaps. 2, 3, 5, 6, 8 and 9.

The derivation of the Osher intercell numerical flux depends on integration in phase space. Such operation involves the choice of *integration paths*, *intersection points* and *sonic points*. The integration paths are taken to be *integral curves* associated with the set of right eigenvectors and to date there are essentially two ways of *ordering* these integration paths. The most recent approach orders the integration paths such that these correspond to physically meaningful relations across wave families in *physical space*. In the current literature this is called *physical ordering* or *P–ordering*. Osher's original scheme utilises the ordering of paths that is precisely the inverse of the P–ordering; this is usually called *O–ordering*. Intersection and sonic points are computed via *Generalised Riemann Invariants*.

The purpose of this chapter is to present the Osher scheme in a way that can be directly applied to practical problems involving hyperbolic conservation laws. We first describe the principles behind Osher's method as applied to any non–linear system of hyperbolic conservation laws. We then give detailed applications of the approach to a single scalar equation, to the isentropic equations of Gas Dynamics and to the split three–dimensional time dependent Euler equations for ideal gases. For one–dimensional applications, all the required information is found in this chapter and in Chap. 6.

12.1 Osher's Scheme for a General System

Here we give, in a self–contained manner, some of the basic aspects of the Osher scheme for a general non–linear system of hyperbolic conservation laws.

12.1.1 Mathematical Bases

Osher's approach to upwind differencing provides an approximation to the Godunov numerical flux of Chap. 6 and results from evaluating the physical flux $\mathbf{F}(\mathbf{U})$ at various states $\mathbf{U}_k$; these include the data states $\mathbf{U}_L$, $\mathbf{U}_R$, intersection points and sonic points. Consider a system of m hyperbolic conservation laws

$$\mathbf{U}_t + \mathbf{F}(\mathbf{U})_x = 0 \tag{12.1}$$

and the conservative scheme

$$\mathbf{U}_i^{n+1} = \mathbf{U}_i^n + \frac{\Delta t}{\Delta x} \left[\mathbf{F}_{i-\frac{1}{2}} - \mathbf{F}_{i+\frac{1}{2}} \right] \tag{12.2}$$

to solve it numerically. The objective of this chapter is to provide an expression for the numerical flux $\mathbf{F}_{i+\frac{1}{2}}$ following Osher's approach.

We assume (12.1) to be *strictly hyperbolic* with eigenvalues

$$\lambda_1(\mathbf{U}) < \lambda_2(\mathbf{U}) < \cdots < \lambda_m(\mathbf{U}) \tag{12.3}$$

and corresponding right eigenvectors

$$\mathbf{K}^{(1)}(\mathbf{U}), \ \mathbf{K}^{(2)}(\mathbf{U}), \ \cdots, \ \mathbf{K}^{(m)}(\mathbf{U}).$$

From hyperbolicity, the Jacobian matrix

$$\mathbf{A}(\mathbf{U}) = \partial \mathbf{F} / \partial \mathbf{U} \tag{12.4}$$

is diagonalisable, that is

$$\mathbf{A}(\mathbf{U}) = \mathbf{K}(\mathbf{U}) \Lambda(\mathbf{U}) \mathbf{K}^{-1}(\mathbf{U}), \tag{12.5}$$

where $\mathbf{K}(\mathbf{U})$ is the non–singular matrix whose columns are the right eigenvectors of $\mathbf{A}(\mathbf{U})$, that is

$$\mathbf{K}(\mathbf{U}) = \left[\mathbf{K}^{(1)}(\mathbf{U}); \mathbf{K}^{(2)}(\mathbf{U}); \cdots; \mathbf{K}^{(m)}(\mathbf{U}) \right] \tag{12.6}$$

and $\boldsymbol{\Lambda}(\mathbf{U})$ is the diagonal matrix formed by the eigenvalues $\lambda_i(\mathbf{U})$

$$\boldsymbol{\Lambda}(\mathbf{U}) = \begin{pmatrix} \lambda_1(\mathbf{U}) & \cdots & 0 \\ \vdots & & \vdots \\ 0 & \cdots & \lambda_m(\mathbf{U}) \end{pmatrix}. \tag{12.7}$$

See Sect. 2.3 of Chap. 2 and Sect. 3.2 of Chap. 3. As done in Sect. 2.3 of Chap. 2 for linear systems with constant coefficients, we introduce the following notation

$$\lambda_i^+(\mathbf{U}) = \max(\lambda_i(\mathbf{U}), 0) \ ; \quad \lambda_i^-(\mathbf{U}) = \min(\lambda_i(\mathbf{U}), 0) \tag{12.8}$$

to define diagonal matrices

$$\boldsymbol{\Lambda}^+(\mathbf{U}) = \begin{pmatrix} \lambda_1^+(\mathbf{U}) & \cdots & 0 \\ \vdots & & \vdots \\ 0 & \cdots & \lambda_m^+(\mathbf{U}) \end{pmatrix}, \tag{12.9}$$

$$\boldsymbol{\Lambda}^-(\mathbf{U}) = \begin{pmatrix} \lambda_1^-(\mathbf{U}) & \cdots & 0 \\ \vdots & & \vdots \\ 0 & \cdots & \lambda_m^-(\mathbf{U}) \end{pmatrix}, \tag{12.10}$$

and

$$|\boldsymbol{\Lambda}(\mathbf{U})| = \begin{pmatrix} |\lambda_1(\mathbf{U})| & \cdots & 0 \\ \vdots & & \vdots \\ 0 & \cdots & |\lambda_m(\mathbf{U})| \end{pmatrix}. \tag{12.11}$$

We also introduce

$$|\mathbf{A}(\mathbf{U})| = \mathbf{K}(\mathbf{U})|\boldsymbol{\Lambda}(\mathbf{U})|\mathbf{K}^{-1}(\mathbf{U}) . \tag{12.12}$$

But

$$|\lambda_i(\mathbf{U})| = \lambda_i^+(\mathbf{U}) - \lambda_i^-(\mathbf{U})$$

and hence

$$|\boldsymbol{\Lambda}(\mathbf{U})| = \boldsymbol{\Lambda}^+(\mathbf{U}) - \boldsymbol{\Lambda}^-(\mathbf{U}) ,$$

which if substituted in (12.12) gives

$$|\mathbf{A}(\mathbf{U})| = \mathbf{A}^+(\mathbf{U}) - \mathbf{A}^-(\mathbf{U}) , \tag{12.13}$$

with

$$\mathbf{A}^+(\mathbf{U}) = \mathbf{K}(\mathbf{U})\boldsymbol{\Lambda}^+(\mathbf{U})\mathbf{K}^{-1}(\mathbf{U}) \tag{12.14}$$

and

$$\mathbf{A}^-(\mathbf{U}) = \mathbf{K}(\mathbf{U})\boldsymbol{\Lambda}^-(\mathbf{U})\mathbf{K}^{-1}(\mathbf{U}) \ . \tag{12.15}$$

These matrices produce a splitting of the Jacobian matrix as

$$\mathbf{A}(\mathbf{U}) = \mathbf{A}^+(\mathbf{U}) + \mathbf{A}^-(\mathbf{U}) \ , \tag{12.16}$$

where $\mathbf{A}^+(\mathbf{U})$ has positive or zero eigenvalues and $\mathbf{A}^-(\mathbf{U})$ has negative or zero eigenvalues. Note that this is a direct generalisation to non–linear systems of the Jacobian splitting for linear systems with constant coefficients performed in Sect. 5.4 of Chap. 5. See also Sect. 8.2.2 of Chap. 8 on the Flux Vector Splitting approach.

12.1.2 Osher's Numerical Flux

Osher's approach assumes that there exist vector–valued functions $\mathbf{F}^+(\mathbf{U})$ and $\mathbf{F}^-(\mathbf{U})$ that satisfy

$$\mathbf{F}(\mathbf{U}) = \mathbf{F}^+(\mathbf{U}) + \mathbf{F}^-(\mathbf{U}) \tag{12.17}$$

and

$$\frac{\partial \mathbf{F}^+}{\partial \mathbf{U}} = \mathbf{A}^+(\mathbf{U}) \ ; \quad \frac{\partial \mathbf{F}^-}{\partial \mathbf{U}} = \mathbf{A}^-(\mathbf{U}) \ . \tag{12.18}$$

If the initial data $\mathbf{U}_L$, $\mathbf{U}_R$ of the Riemann problem for the conservation laws (12.1) is denoted by

$$\mathbf{U}_0 \equiv \mathbf{U}_L = \mathbf{U}_i^n \ ; \quad \mathbf{U}_1 \equiv \mathbf{U}_R = \mathbf{U}_{i+1}^n \ , \tag{12.19}$$

then the corresponding numerical flux to be used in (12.2) is

$$\mathbf{F}_{i+\frac{1}{2}} = \mathbf{F}^+(\mathbf{U}_0) + \mathbf{F}^-(\mathbf{U}_1) \ . \tag{12.20}$$

Sect. 8.2.2 of Chap. 8. Using the integral relations

$$\int_{\mathbf{U}_0}^{\mathbf{U}_1} \mathbf{A}^-(\mathbf{U})d\mathbf{U} = \mathbf{F}^-(\mathbf{U}_1) - \mathbf{F}^-(\mathbf{U}_0)$$

and

$$\int_{\mathbf{U}_0}^{\mathbf{U}_1} \mathbf{A}^+(\mathbf{U})d\mathbf{U} = \mathbf{F}^+(\mathbf{U}_1) - \mathbf{F}^+(\mathbf{U}_0) \ ,$$

we can express (12.20) in three different forms, namely

$$\mathbf{F}_{i+\frac{1}{2}} = \mathbf{F}(\mathbf{U}_0) + \int_{\mathbf{U}_0}^{\mathbf{U}_1} \mathbf{A}^-(\mathbf{U})d\mathbf{U} \ , \tag{12.21}$$

$$\mathbf{F}_{i+\frac{1}{2}} = \mathbf{F}(\mathbf{U}_1) - \int_{\mathbf{U}_0}^{\mathbf{U}_1} \mathbf{A}^+(\mathbf{U})d\mathbf{U} \tag{12.22}$$

and

$$\mathbf{F}_{i+\frac{1}{2}} = \frac{1}{2}[\mathbf{F}(\mathbf{U}_0) + \mathbf{F}(\mathbf{U}_1)] - \frac{1}{2}\int_{\mathbf{U}_0}^{\mathbf{U}_1} |\mathbf{A}(\mathbf{U})| d\mathbf{U} . \qquad (12.23)$$

Compare these flux formulae with those of Sect. 5.4.2 of Chap. 5 for linear systems with constant coefficients and with those for the Roe flux in Sect. 11.1.3 of Chap. 11.

The integration with respect to $\mathbf{U}$ in (12.21)–(12.23) is carried out in phase space $\mathbf{R}^m$. Elements of this vector space are vectors

$$\mathbf{U} = [u_1, u_2, \cdots, u_m]^T , \qquad (12.24)$$

whose components u_i are real numbers. In general, the integrals (12.21)–(12.23) *depend on the integration path chosen*. Osher's approach is to select particular integration paths *so as to make the actual integration tractable*.

12.1.3 Osher's Flux for the Single–Wave Case

The solution of the Riemann problem for (12.1) with data $\mathbf{U}_0, \mathbf{U}_1$ has m waves, in general. Osher's scheme utilises partial information on the solution to provide integration paths to evaluate the integrals (12.21)–(12.23), which in turn produce an expression for the numerical flux. Consider first the simplest case in which all waves in the solution of the Riemann problem, except for that associated with the eigenvalue $\lambda_j(\mathbf{U})$ and eigenvector $\mathbf{K}^{(j)}(\mathbf{U})$, are trivial. That is, the states $\mathbf{U}_0$ and $\mathbf{U}_1$ are connected by a single j–wave. Associated with any vector field $\mathbf{K}^{(k)}(\mathbf{U})$, there are *integral curves*. These have the property that their tangent lies in the direction of the eigenvector $\mathbf{K}^{(k)}(\mathbf{U})$, at any point $\mathbf{U}$ in phase space. For background on integral curves see [207] and [425]. An integration path $I_k(\mathbf{U})$ is now taken to be an integral curve of $\mathbf{K}^{(k)}(\mathbf{U})$ connecting $\mathbf{U}_0$ and $\mathbf{U}_1$. It is important to note here that the eigenvector $\mathbf{K}^{(k)}(\mathbf{U})$ and the eigenvalue $\lambda_k(\mathbf{U})$ associated with the relevant integration path *are not necessarily* those corresponding to the non–trivial j–wave family in question.

Suppose $I_k(\mathbf{U})$ is parameterised by $\mathbf{U}(\xi)$, $0 \leq \xi \leq \xi_1$, and

$$\mathbf{U}_0 = \mathbf{U}(0) , \quad \mathbf{U}_1 = \mathbf{U}(\xi_1) ; \qquad (12.25)$$

then

$$\frac{d\mathbf{U}(\xi)}{d\xi} = \mathbf{K}^{(k)}(\mathbf{U}(\xi)) . \qquad (12.26)$$

By performing a change of variables, utilising (12.25)–(12.26) and the fact that $\lambda_k^-(\mathbf{U})$ is an eigenvalue of $\mathbf{A}^-(\mathbf{U})$ with eigenvector $\mathbf{K}^{(k)}(\mathbf{U})$, we have

$$\int_{\mathbf{U}_0}^{\mathbf{U}_1} \mathbf{A}^-(\mathbf{U}) d\mathbf{U} = \int_0^{\xi_1} \lambda_k^-[\mathbf{U}(\xi)]\mathbf{K}^{(k)}[\mathbf{U}(\xi)]d\xi . \qquad (12.27)$$

We first consider the case in which the k–th field is *linearly degenerate*, that is $\lambda_k(\mathbf{U})$ is constant along $I_k(\mathbf{U})$; see Sect. 2.4.3 of Chap. 2 and Sect. 3.1.3

of Chap. 3. If $\lambda_k(\mathbf{U}) \geq 0$ $\forall \mathbf{U}$ along $I_k(\mathbf{U})$, then $\lambda_k^-(\mathbf{U}) = 0$, see (12.8), and thus from (12.27) we have

$$\int_{\mathbf{U}_0}^{\mathbf{U}_1} \mathbf{A}^-(\mathbf{U}) d\mathbf{U} = \mathbf{0} \ . \tag{12.28}$$

If $\lambda_k(\mathbf{U}) \leq 0$ $\forall \mathbf{U}$ along $I_k(\mathbf{U})$, then manipulations of (12.27) give

$$\int_{\mathbf{U}_0}^{\mathbf{U}_1} \mathbf{A}^-(\mathbf{U}) d\mathbf{U} = \int_0^{\xi_1} \frac{d\mathbf{F}}{d\mathbf{U}}[\mathbf{U}(\xi)] \frac{d\mathbf{U}}{d\xi} d\xi$$

and thus

$$\int_{\mathbf{U}_0}^{\mathbf{U}_1} \mathbf{A}^-(\mathbf{U}) d\mathbf{U} = \mathbf{F}(\mathbf{U}_1) - \mathbf{F}(\mathbf{U}_0) \ . \tag{12.29}$$

Hence, if the k–field is linearly degenerate, use of (12.28) and (12.29) in the one–sided flux formula (12.21) gives the Osher's intercell flux as

$$\mathbf{F}_{i+\frac{1}{2}} = \begin{cases} \mathbf{F}(\mathbf{U}_0) & \text{if } \lambda_k \geq 0 \\ \\ \mathbf{F}(\mathbf{U}_1) & \text{if } \lambda_k < 0 \ . \end{cases} \tag{12.30}$$

An entirely equivalent derivation of the Osher flux results from using the flux formulae (12.22) or (12.23). Note that in the case of a single wave associated with a linearly degenerate field, the Osher flux is identical to the Godunov flux if $j = k$.

We next consider the case in which the k–th field is *genuinely non–linear*, that is the eigenvalue $\lambda_k(\mathbf{U})$ is *monotone* along $I_k(\mathbf{U})$. This means that $\lambda_k(\mathbf{U})$ changes sign *at most once*, along $I_k(\mathbf{U})$. If $\lambda_k(\mathbf{U})$ does not change sign along $I_k(\mathbf{U})$ then this is simply like the linearly degenerate case above and the flux is given by (12.30). If $\lambda_k(\mathbf{U})$ changes sign at $\xi = \xi_S$, then there are two cases to consider. First assume

$$\lambda_k[\mathbf{U}(\xi)] \geq 0 \ , \quad \forall \xi \in [0, \xi_S] \ ; \quad \lambda_k[\mathbf{U}(\xi)] \leq 0 \ , \quad \forall \xi \in [\xi_S, \xi_1] \ .$$

Then (12.27) can be split into an integral between 0 and ξ_S and an integral between ξ_S and ξ_1. The first integral is zero, as $\lambda_k^-(\mathbf{U}) = \min(\lambda_k(\mathbf{U}), 0)$ and $\lambda_k(\mathbf{U}) \geq 0$, see (12.8). The second integral gives

$$\int_{\xi_S}^{\xi_1} \frac{d\mathbf{F}}{d\mathbf{U}}[\mathbf{U}(\xi)] \frac{d\mathbf{U}}{d\xi} d\xi = \mathbf{F}(\mathbf{U}_1) - \mathbf{F}(\mathbf{U}_S) \ ,$$

where $\mathbf{U}_S = \mathbf{U}(\xi_S)$ and is called a *sonic point*. The case

$$\lambda_k[\mathbf{U}(\xi)] \leq 0 \ , \quad \forall \xi \in [0, \xi_S] \ ; \quad \lambda_k[\mathbf{U}(\xi)] \geq 0 \ , \quad \forall \xi \in [\xi_S, \xi_1]$$

can be treated in a similar way to give

$$\int_{\mathbf{U}_0}^{\mathbf{U}_1} \mathbf{A}^-(\mathbf{U})d\mathbf{U} = \mathbf{F}(\mathbf{U}_S) - \mathbf{F}(\mathbf{U}_0) .$$

Collecting results, for a single wave, we have

$$\int_{\mathbf{U}_0}^{\mathbf{U}_1} \mathbf{A}^-(\mathbf{U})d\mathbf{U} = \begin{cases} \mathbf{0} & \text{if } \lambda_k(\mathbf{U}) \geq 0 , \\ \mathbf{F}(\mathbf{U}_1) - \mathbf{F}(\mathbf{U}_0) & \text{if } \lambda_k(\mathbf{U}) \leq 0 , \\ \mathbf{F}(\mathbf{U}_1) - \mathbf{F}(\mathbf{U}_S) & \text{if } \lambda_k(\mathbf{U}_0) \geq 0 , \lambda_k(\mathbf{U}_1) \leq 0 , \\ & \qquad \lambda_k(\mathbf{U}_S) = 0 , \\ \mathbf{F}(\mathbf{U}_S) - \mathbf{F}(\mathbf{U}_0) & \text{if } \lambda_k(\mathbf{U}_0) \leq 0 , \lambda_k(\mathbf{U}_1) \geq 0 , \\ & \qquad \lambda_k(\mathbf{U}_S) = 0 . \end{cases}$$

By substituting these expressions into the one–sided flux formula (12.21) we obtain the Osher intercell flux, for the case in which the states $\mathbf{U}_0$ and $\mathbf{U}_1$ are connected by the single j–wave.

$$\mathbf{F}_{i+\frac{1}{2}} = \begin{cases} \mathbf{F}(\mathbf{U}_0) , & \lambda_k(\mathbf{U}) \geq 0 , \\ \mathbf{F}(\mathbf{U}_1) , & \lambda_k(\mathbf{U}) \leq 0 , \\ \mathbf{F}(\mathbf{U}_0) + \mathbf{F}(\mathbf{U}_1) - \mathbf{F}(\mathbf{U}_S) , & \lambda_k(\mathbf{U}_0) \geq 0 , \lambda_k(\mathbf{U}_1) \leq 0 , \\ & \qquad \lambda_k(\mathbf{U}_S) = 0 , \\ \mathbf{F}(\mathbf{U}_S) , & \lambda_k(\mathbf{U}_0) \leq 0 , \lambda_k(\mathbf{U}_1) \geq 0 , \\ & \qquad \lambda_k(\mathbf{U}_S) = 0 . \end{cases}$$

$$(12.31)$$

These results can be applied directly to any scalar, non–linear conservation law

$$u_t + f(u)_x = 0 . \qquad (12.32)$$

12.1.4 Osher's Flux for the Inviscid Burgers Equation

Consider (12.32) with flux function

$$f(u) = \frac{1}{2}u^2 . \qquad (12.33)$$

This gives the inviscid Burgers equation; see Sect. 2.4.2 of Chap. 2. Recall that the exact solution of the Riemann problem for (12.32)–(12.33) with data $u_L \equiv u_0$, $u_R \equiv u_1$ is

$$u(x/t) = \begin{cases} u_0 & \text{if } x/t \leq S = \frac{1}{2}(u_0 + u_1) \\ u_1 & \text{if } x/t > S \end{cases} \qquad (12.34)$$

when $u_0 > u_1$ (shock case), and

$$u(x/t) = \begin{cases} u_0 & \text{if } x/t \leq u_0 , \\ x/t & \text{if } u_0 \leq x/t \leq u_1 , \\ u_1 & \text{if } x/t \geq u_1 , \end{cases} \qquad (12.35)$$

when $u_0 \leq u_1$ (rarefaction case). As, trivially, the only eigenvalue is the characteristic speed $\lambda = u$, direct application of (12.31) gives

$$\mathbf{F}_{i+\frac{1}{2}} = \begin{cases} f(u_0) & \text{if } u_0, u_1 > 0 , \\ f(u_1) & \text{if } u_0, u_1 < 0 , \\ f(u_0) + f(u_1) & \text{if } u_0 \geq 0 \geq u_1 , \\ 0 & \text{if } u_0 \leq 0 \leq u_1 . \end{cases} \qquad (12.36)$$

At the sonic point $u_S = 0$ and thus $f(u_S) = 0$. Note that the Godunov flux $f_{i+\frac{1}{2}}^{god}$, obtained from the exact solution to the Riemann problem, is identical to Osher's flux (12.36), except in the case of a *transonic shock* ($u_0 \geq 0 \geq u_1$), where

$$f_{i+\frac{1}{2}}^{god} = \begin{cases} f(u_0) & \text{if } S \geq 0 , \\ f(u_1) & \text{otherwise} . \end{cases}$$

For a full discussion on the relationship between the Godunov scheme, with the exact Riemann solver, and approximate Riemann solvers, including Osher's scheme, see the paper by van Leer [395].

12.1.5 Osher's Flux for the General Case

In the previous section we analysed Osher's numerical flux for the case of two states $\mathbf{U}_0, \mathbf{U}_1$ connected by a single wave, where an integration path was chosen to be tangential to a right eigenvector. For the general case of m wave families Osher chooses a set $\{I_k(\mathbf{U})\}$, $k = 1, \cdots, m$, of *partial* integration paths such that $I_k(\mathbf{U})$ is tangential to $\mathbf{K}^{(k)}(\mathbf{U})$ and two successive paths $I_k(\mathbf{U})$ and $I_{k+1}(\mathbf{U})$ intersect at a *single* point

$$\mathbf{U}_{k/m} = I_k(\mathbf{U}) \cap I_{k+1}(\mathbf{U}) \qquad (12.37)$$

in phase space, called an *intersection point*. The data points $\mathbf{U}_0$ and $\mathbf{U}_1$ are now to be interpreted as

$$\mathbf{U}_0 \equiv \mathbf{U}_{(k-1)/m} , \quad \mathbf{U}_1 \equiv \mathbf{U}_{k/m} .$$

The total integration path is the union of all partial paths, namely

$$I(\mathbf{U}) = \cup I_k(\mathbf{U}) . \qquad (12.38)$$

Fig. 12.1 illustrates a choice of integration paths for the case of a 3×3 hyperbolic system such as the one–dimensional, time dependent Euler equations.

The vectors $\mathbf{U}_0, \mathbf{U}_{1/3}, \mathbf{U}_{2/3}$ and $\mathbf{U}_1$ are vectors in phase space $\mathbf{R}^3$ and can be thought of as being the four constant states arising in the exact solution to the corresponding Riemann problem represented in physical space x–t in Fig. 12.1. See Chap. 4 for details of the exact solution of the Riemann problem for the Euler equations. The points $\mathbf{U}_{\frac{1}{3}}$ and $\mathbf{U}_{\frac{2}{3}}$ are the *intersection points* in phase space, and the points $\mathbf{U}_{S0}$ and $\mathbf{U}_{S1}$ are a representation of the potential *sonic points* that may arise from the non–linear fields associated with $\lambda_1(\mathbf{U})$ and $\lambda_3(\mathbf{U})$.

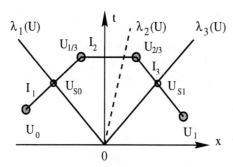

Fig. 12.1. Possible configuration of integration paths $I_k(\mathbf{U})$, intersection points $\mathbf{U}_{\frac{1}{3}}$, $\mathbf{U}_{\frac{2}{3}}$ and sonic points $\mathbf{U}_{S0}$, $\mathbf{U}_{S1}$ in physical space x–t for a 3×3 system

For the general case, and assuming for the moment that the intersection points (12.37) and sonic points are known, the integration along $I(\mathbf{U})$ to evaluate (12.21), say, can now be performed by integrating along each partial integration path $I_k(\mathbf{U})$. But since these have been chosen to be tangential to the corresponding eigenvector $\mathbf{K}^{(k)}(\mathbf{U})$, the results of the previous section, see (12.31), can now be applied directly. The determination of the intersection points $\mathbf{U}_{k/m}$ and sonic points requires extra information about the solution of the Riemann problem. Traditionally, these have been determined by the use of Generalised Riemann Invariants, in at least two different ways, as we shall see.

12.2 Osher's Flux for the Isothermal Equations

The isothermal equations

$$\mathbf{U}_t + \mathbf{F}(\mathbf{U})_x = 0 \,, \tag{12.39}$$

$$\mathbf{U} = \begin{bmatrix} \rho \\ \rho u \end{bmatrix}, \quad \mathbf{F}(\mathbf{U}) = \begin{bmatrix} \rho u \\ \rho u^2 + a^2 \rho \end{bmatrix}, \tag{12.40}$$

have Jacobian matrix

$$\mathbf{A}(\mathbf{U}) = \begin{bmatrix} 0 & 1 \\ a^2 - u^2 & 2u \end{bmatrix} \tag{12.41}$$

with eigenvalues

$$\lambda_1(\mathbf{U}) = u - a \,, \quad \lambda_2(\mathbf{U}) = u + a \,, \tag{12.42}$$

and corresponding right eigenvectors

$$\mathbf{K}^{(1)}(\mathbf{U}) = \begin{bmatrix} 1 \\ u - a \end{bmatrix} \,, \quad \mathbf{K}^{(2)}(\mathbf{U}) = \begin{bmatrix} 1 \\ u + a \end{bmatrix} \,. \tag{12.43}$$

Recall that the sound speed a is constant here; for details on the isothermal equations see Sect. 1.6.2 of Chap. 1 and Sect. 2.4.1 of Chap. 2. The structure of the exact solution to the Riemann problem for (12.39)–(12.40) with initial data $\mathbf{U}_0, \mathbf{U}_1$ is depicted in Fig. 12.2 in the x–t plane, where the intersection point $\mathbf{U}_{\frac{1}{2}}$ is identified with the solution of the Riemann problem between the non–linear waves; potential sonic points are also shown.

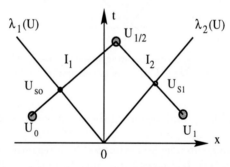

Fig. 12.2. Structure of the solution of the Riemann problem for the isothermal equations in physical space x–t. Integration paths are I_1 and I_2, intersection point is $\mathbf{U}_{\frac{1}{2}}$, potential sonic points are $\mathbf{U}_{SO}$ and $\mathbf{U}_{S1}$; P–ordering

12.2.1 Osher's Flux with P–Ordering

There are two ways of ordering the integration paths in Osher's scheme, namely the *original* Osher ordering, or *O–ordering*, and the *physical* ordering following valid relations across waves in physical space, or *P–ordering*. First we apply *P–ordering*. It is easy to check, see Sect. 2.4.3 of Chap. 2, that the Riemann Invariants along the path $I_1(\mathbf{U})$, across the left wave in physical space, give

$$u_{\frac{1}{2}} + a \, ln(\rho_{\frac{1}{2}}) = u_0 + a \, ln(\rho_0) \,. \tag{12.44}$$

Similarly, across the right wave

$$u_{\frac{1}{2}} - a \, ln(\rho_{\frac{1}{2}}) = u_0 - a \, ln(\rho_1) \,. \tag{12.45}$$

The simultaneous solution for $\rho_{\frac{1}{2}}$ and $u_{\frac{1}{2}}$ is

$$u_{\frac{1}{2}} = \frac{1}{2}(u_0 + u_1) + \frac{1}{2}a \, ln(\rho_0/\rho_1) \,, \tag{12.46}$$

$$\rho_{\frac{1}{2}} = \sqrt{\rho_0 \rho_1} \, \exp\left[-\frac{u_1 - u_0}{2a}\right] \tag{12.47}$$

and thus the intersection point

$$\mathbf{U}_{\frac{1}{2}} = \left[\begin{array}{c} \rho_{\frac{1}{2}} \\ \rho_{\frac{1}{2}} u_{\frac{1}{2}} \end{array} \right] \tag{12.48}$$

is determined.

	Behaviour of $\lambda_1(\mathbf{U}) = u - a$	$\int_{\mathbf{U}_0}^{\mathbf{U}_{\frac{1}{2}}} \mathbf{A}^-(\mathbf{U})d\mathbf{U}$
1	$u_0 - a \geq 0 \,, \; u_{\frac{1}{2}} - a \geq 0$	$\mathbf{0}$
2	$u_0 - a \leq 0 \,, \; u_{\frac{1}{2}} - a \leq 0$	$\mathbf{F}_{\frac{1}{2}} - \mathbf{F}_0$
3	$u_0 - a \geq 0 \,, \; u_{\frac{1}{2}} - a \leq 0$	$\mathbf{F}_{\frac{1}{2}} - \mathbf{F}_{S0}$
4	$u_0 - a \leq 0 \,, \; u_{\frac{1}{2}} - a \geq 0$	$\mathbf{F}_{S0} - \mathbf{F}_0$

Table 12.1. Evaluation of integral along integration path $I_1(\mathbf{U})$ with P–ordering, $\mathbf{F}_k \equiv \mathbf{F}(\mathbf{U}_k)$

The left and right sonic points $\mathbf{U}_{S0}$ and $\mathbf{U}_{S1}$ can easily be found by using the sonic conditions $u_{S0} = a$ from $\lambda_1 = u - a = 0$, $u_{S1} = -a$ from $\lambda_2 = u + a = 0$ and the Riemann Invariants. The result is

$$\rho_{S0} = \rho_0 \exp\left[\frac{u_0 - a}{a}\right] \,, \quad u_{S0} = a \,, \tag{12.49}$$

$$\rho_{S1} = \rho_1 \exp\left[-\frac{u_1 + a}{a}\right] \,, \quad u_{S1} = -a \,. \tag{12.50}$$

It is worth remarking at this stage, that the solution (12.46)–(12.47) for the intersection point is exact when the left and right waves are both rarefaction waves; in the general case it is an approximation. It is in fact the Two–Rarefaction approximation TRRS presented in Sect. 9.4.1 of Chap. 9 for the Euler equations.

	Behaviour of $\lambda_2(\mathbf{U}) = u + a$	$\int_{\mathbf{U}_{\frac{1}{2}}}^{\mathbf{U}_1} \mathbf{A}^-(\mathbf{U})d\mathbf{U}$
1	$u_{\frac{1}{2}} + a \geq 0$, $u_1 + a \geq 0$	$\mathbf{0}$
2	$u_{\frac{1}{2}} + a \leq 0$, $u_1 + a \leq 0$	$\mathbf{F}_1 - \mathbf{F}_{\frac{1}{2}}$
3	$u_{\frac{1}{2}} + a \geq 0$, $u_1 + a \leq 0$	$\mathbf{F}_1 - \mathbf{F}_{S1}$
4	$u_{\frac{1}{2}} + a \leq 0$, $u_1 + a \geq 0$	$\mathbf{F}_{S1} - \mathbf{F}_{\frac{1}{2}}$

Table 12.2. Evaluation of integral along integration path $I_2(\mathbf{U})$ with P–ordering, $\mathbf{F}_k \equiv \mathbf{F}(\mathbf{U}_k)$

Integration Along Partial Paths. In order to compute the Osher flux we use the one–sided flux formula

$$\mathbf{F}_{i+\frac{1}{2}} = \mathbf{F}_0 + \int_{\mathbf{U}_0}^{\mathbf{U}_1} \mathbf{A}^-(\mathbf{U})d\mathbf{U} \, , \tag{12.51}$$

where the integral is evaluated along each of the partial integration paths $I_1(\mathbf{U})$ and $I_2(\mathbf{U})$ shown in Fig. 12.2. For each case the integration is performed according to the local characteristic configuration. Tables 12.1 and 12.2 show the results for $I_1(\mathbf{U})$ and $I_2(\mathbf{U})$ respectively. As

$$\int_{\mathbf{U}_0}^{\mathbf{U}_1} \mathbf{A}^-(\mathbf{U})d\mathbf{U} = \int_{\mathbf{U}_0}^{\mathbf{U}_{\frac{1}{2}}} \mathbf{A}^-(\mathbf{U})d\mathbf{U} + \int_{\mathbf{U}_{\frac{1}{2}}}^{\mathbf{U}_1} \mathbf{A}^-(\mathbf{U})d\mathbf{U} \, , \tag{12.52}$$

strictly speaking, one should consider all 16 possible characteristic configurations (i, j) that result from Tables 12.1 and 12.2. Closer examination of all cases reveals that 4 possibilities are unrealisable. These are (1,2), (1,4), (4,2) and (4,4). For instance case (1,2) contains the requirements

$$u_{\frac{1}{2}} - a \geq 0 \, ; \quad u_{\frac{1}{2}} + a \leq 0 \, . \tag{12.53}$$

But $a > 0$ and thus these two conditions are contradictory. The remaining 12 cases can be tabulated as in the 3×4 Table 12.3, which gives the Osher intercell numerical flux $\mathbf{F}_{i+\frac{1}{2}}$. We separate conditions on the intersection point $\mathbf{U}_{\frac{1}{2}}$ (first column) from conditions on the data points $\mathbf{U}_0$ and $\mathbf{U}_1$ (top row). Note that in general $\mathbf{F}_{i+\frac{1}{2}}$ is a combination of physical flux values at several points in phase space. In contrast, the Godunov flux obtained from the exact Riemann solver of Chap. 4 and from the approximate Riemann solvers of Chaps. 9, 10 and 11, consists of a *single value*. This has a bearing on the simplicity and computational efficiency of the schemes.

12.2.2 Osher's Flux with O–Ordering

In the original Osher scheme the ordering of the partial integration paths $I_k(\mathbf{U})$ is inverted. Fig. 12.3 illustrates the corresponding path configuration for the isothermal equations. The integration path connecting $\mathbf{U}_0$ to $\mathbf{U}_{\frac{1}{2}}$ is tangential to the eigenvector $\mathbf{K}^{(2)}(\mathbf{U})$ and that connecting $\mathbf{U}_{\frac{1}{2}}$ to $\mathbf{U}_1$ is tangential to $\mathbf{K}^{(1)}(\mathbf{U})$. Compare with Fig. 12.2. The O–ordering of the Osher scheme is used both for the integration paths as well as for the determination of the intersection point $\mathbf{U}_{\frac{1}{2}}$ and the sonic points $\mathbf{U}_{S0}, \mathbf{U}_{S1}$. The data state

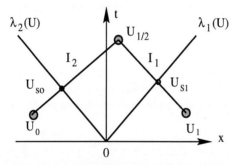

Fig. 12.3. Configuration of integration paths $I_k(\mathbf{U})$, intersection point $\mathbf{U}_{\frac{1}{2}}$ and potential sonic points $\mathbf{U}_{S0}, \mathbf{U}_{S1}$ in physical space x–t when using the O–ordering

$\mathbf{U}_0$ is connected to $\mathbf{U}_{\frac{1}{2}}$ via the *right* (relation valid across the right wave) Riemann Invariant to give

$$u_{\frac{1}{2}} - a \, ln(\rho_{\frac{1}{2}}) = u_0 - a \, ln(\rho_0) \ .$$

Compare with (12.44). The data point $\mathbf{U}_1$ is connected to $\mathbf{U}_{\frac{1}{2}}$ via the *left* (relation valid across the left wave) Riemann Invariant to produce

$$u_{\frac{1}{2}} + a \, ln(\rho_{\frac{1}{2}}) = u_1 + a \, ln(\rho_1) \ .$$

Compare with (12.45). The resulting solution is

$$u_{\frac{1}{2}} = \frac{1}{2}(u_0 + u_1) + \frac{1}{2}a \, ln(\rho_1/\rho_0) \ , \tag{12.54}$$

$$\rho_{\frac{1}{2}} = \sqrt{\rho_0 \rho_1} \, \exp\left[\frac{u_1 - u_0}{2a}\right] \tag{12.55}$$

	$u_0 - a \geq 0$ $u_1 + a \geq 0$	$u_0 - a \geq 0$ $u_1 + a \leq 0$	$u_0 - a \leq 0$ $u_1 + a \leq 0$	$u_0 - a \leq 0$ $u_1 + a \geq 0$
$u_{\frac{1}{2}} - a \geq 0$	$\mathbf{F}_0$	$\mathbf{F}_0 + \mathbf{F}_1$ $-\mathbf{F}_{S1}$	$\mathbf{F}_{S0} + \mathbf{F}_1$ $-\mathbf{F}_{S1}$	$\mathbf{F}_{S0}$
$u_{\frac{1}{2}} - a \leq 0$ $u_{\frac{1}{2}} + a \geq 0$	$\mathbf{F}_0 + \mathbf{F}_{\frac{1}{2}}$ $-\mathbf{F}_{S0}$	$\mathbf{F}_0 + \mathbf{F}_{\frac{1}{2}}$ $-\mathbf{F}_{S0} + \mathbf{F}_1$ $-\mathbf{F}_{S1}$	$\mathbf{F}_{\frac{1}{2}} + \mathbf{F}_1$ $-\mathbf{F}_{S1}$	$\mathbf{F}_{\frac{1}{2}}$
$u_{\frac{1}{2}} + a \leq 0$	$\mathbf{F}_0 + \mathbf{F}_{S1}$ $-\mathbf{F}_{S0}$	$\mathbf{F}_0 + \mathbf{F}_1$ $-\mathbf{F}_{S0}$	$\mathbf{F}_1$	$\mathbf{F}_{S1}$

Table 12.3. Osher's intercell flux for the isothermal equations using P–ordering, $\mathbf{F}_k \equiv \mathbf{F}(\mathbf{U}_k)$

and thus the intersection point $\mathbf{U}_{\frac{1}{2}}$ in (12.48) is determined. The left and right sonic points are evaluated using the sonic conditions $\lambda_2 = u + a = 0$ and $\lambda_1 = u - a = 0$ and the right and left Riemann Invariants, respectively. The result is

$$u_{S0} = -a , \quad \rho_{S0} = \rho_0 \exp\left[-\frac{u_0 + a}{a}\right] , \tag{12.56}$$

$$u_{S1} = a , \quad \rho_{S1} = \rho_1 \exp\left[\frac{u_1 - a}{a}\right] . \tag{12.57}$$

Compare solutions (12.54)–(12.57) with (12.46)–(12.50) obtained with the P–ordering of the Osher scheme. Note that the solution for the intersection point given by the O–ordering has no physical meaning as a solution for the *Star Region* between the left and right wave families. If the exact solution of the Riemann problem consists of two rarefactions, then solution (12.47) is exact, in which case the density is expected to decrease; solution (12.55) gives an increase in density in the *Star Region*, which is obviously incorrect. We expand on this point when dealing with the Euler equations later in Sect. 12.3.3.

Integration along the path $I_2(\mathbf{U})$ connecting $\mathbf{U}_0$ to $\mathbf{U}_{\frac{1}{2}}$ produces the results of Table 12.4. The results of integrating along the path $I_1(\mathbf{U})$ connecting $\mathbf{U}_{\frac{1}{2}}$ to $\mathbf{U}_1$ are given in Table 12.5.

Out of the 16 possible combinations 12 cases are realisable. Table 12.6 gives the Osher numerical flux $\mathbf{F}_{i+\frac{1}{2}}$ for all 12 cases. Compare with Table 12.3 in which the flux is computed using the P–ordering.

	Behaviour of $\lambda_2(\mathbf{U}) = u + a$	$\int_{\mathbf{U}_0}^{\mathbf{U}_{\frac{1}{2}}} \mathbf{A}^-(\mathbf{U})d\mathbf{U}$
1	$u_0 + a \geq 0$, $u_{\frac{1}{2}} + a \geq 0$	$\mathbf{0}$
2	$u_0 + a \leq 0$, $u_{\frac{1}{2}} + a \leq 0$	$\mathbf{F}_{\frac{1}{2}} - \mathbf{F}_0$
3	$u_0 + a \geq 0$, $u_{\frac{1}{2}} + a \leq 0$	$\mathbf{F}_{\frac{1}{2}} - \mathbf{F}_{S0}$
4	$u_0 + a \leq 0$, $u_{\frac{1}{2}} + a \geq 0$	$\mathbf{F}_{S0} - \mathbf{F}_0$

Table 12.4. Integration along path $I_2(\mathbf{U})$ connecting $\mathbf{U}_0$ to $\mathbf{U}_{\frac{1}{2}}$ (O–ordering), $\mathbf{F}_k \equiv \mathbf{F}(\mathbf{U}_k)$

	Behaviour of $\lambda_1(\mathbf{U}) = u - a$	$\int_{\mathbf{U}_{\frac{1}{2}}}^{\mathbf{U}_0} \mathbf{A}^-(\mathbf{U})d\mathbf{U}$
1	$u_1 - a \geq 0$, $u_{\frac{1}{2}} - a \geq 0$	$\mathbf{0}$
2	$u_1 - a \leq 0$, $u_{\frac{1}{2}} - a \leq 0$	$\mathbf{F}_1 - \mathbf{F}_{\frac{1}{2}}$
3	$u_1 - a \leq 0$, $u_{\frac{1}{2}} - a \geq 0$	$\mathbf{F}_1 - \mathbf{F}_{S1}$
4	$u_1 - a \geq 0$, $u_{\frac{1}{2}} - a \leq 0$	$\mathbf{F}_{S1} - \mathbf{F}_{\frac{1}{2}}$

Table 12.5. Integration along path $I_1(\mathbf{U})$ connecting $\mathbf{U}_{\frac{1}{2}}$ to $\mathbf{U}_1$ (O–ordering), $\mathbf{F}_k \equiv \mathbf{F}(\mathbf{U}_k)$

	$\begin{array}{l} u_0 + a \geq 0 \\ u_1 - a \geq 0 \end{array}$	$\begin{array}{l} u_0 + a \geq 0 \\ u_1 - a \leq 0 \end{array}$	$\begin{array}{l} u_0 + a \leq 0 \\ u_1 - a \leq 0 \end{array}$	$\begin{array}{l} u_0 + a \leq 0 \\ u_1 - a \geq 0 \end{array}$
$u_{\frac{1}{2}} - a \geq 0$	$\mathbf{F}_0$	$\begin{array}{l} \mathbf{F}_0 + \mathbf{F}_1 \\ -\mathbf{F}_{S1} \end{array}$	$\begin{array}{l} \mathbf{F}_{S0} + \mathbf{F}_1 \\ -\mathbf{F}_{S1} \end{array}$	$\mathbf{F}_{S0}$
$\begin{array}{l} u_{\frac{1}{2}} - a \leq 0 \\ u_{\frac{1}{2}} + a \geq 0 \end{array}$	$\begin{array}{l} \mathbf{F}_0 + \mathbf{F}_{S1} \\ -\mathbf{F}_{\frac{1}{2}} \end{array}$	$\begin{array}{l} \mathbf{F}_0 + \mathbf{F}_1 \\ -\mathbf{F}_{\frac{1}{2}} \end{array}$	$\begin{array}{l} \mathbf{F}_{S0} + \mathbf{F}_1 \\ -\mathbf{F}_{\frac{1}{2}} \end{array}$	$\begin{array}{l} \mathbf{F}_{S0} + \mathbf{F}_{S1} \\ -\mathbf{F}_{\frac{1}{2}} \end{array}$
$u_{\frac{1}{2}} + a \leq 0$	$\begin{array}{l} \mathbf{F}_0 + \mathbf{F}_{S1} \\ -\mathbf{F}_{S0} \end{array}$	$\begin{array}{l} \mathbf{F}_0 + \mathbf{F}_1 \\ -\mathbf{F}_{S0} \end{array}$	$\mathbf{F}_1$	$\mathbf{F}_{S1}$

Table 12.6. Osher's intercell flux for the isothermal equations using O–ordering of integration paths, $\mathbf{F}_k = \mathbf{F}(\mathbf{U}_k)$

12.3 Osher's Scheme for the Euler Equations

Here we develop in detail the Osher scheme, with both P and O orderings, for the time–dependent Euler equations. We first consider the one–dimensional case

$$\mathbf{U}_t + \mathbf{F}(\mathbf{U})_x = \mathbf{0} , \qquad (12.58)$$

$$\mathbf{U} = \begin{bmatrix} \rho \\ \rho u \\ E \end{bmatrix} , \quad \mathbf{F}(\mathbf{U}) = \begin{bmatrix} \rho u \\ \rho u^2 + p \\ u(E + p) \end{bmatrix} . \qquad (12.59)$$

Details of the Euler equations are found in Sect. 1.1 of Chap. 1 and Chap. 3. We require an expression for the intercell flux $\mathbf{F}_{i+\frac{1}{2}}$ in the explicit conservative formula

$$\mathbf{U}_i^{n+1} = \mathbf{U}_i^n + \frac{\Delta t}{\Delta x}[\mathbf{F}_{i-\frac{1}{2}} - \mathbf{F}_{i+\frac{1}{2}}] . \qquad (12.60)$$

Recall that the Jacobian matrix $\mathbf{A}(\mathbf{U})$, see Sect. 3.1.2 of Chap. 3, has eigenvalues

$$\lambda_1 = u - a , \quad \lambda_2 = u, \ , \quad \lambda_3 = u + a \qquad (12.61)$$

and right eigenvectors

$$\mathbf{K}^{(1)} = \begin{bmatrix} 1 \\ u - a \\ H - ua \end{bmatrix} , \quad \mathbf{K}^{(2)} = \begin{bmatrix} 1 \\ u \\ \frac{1}{2}u^2 \end{bmatrix} , \quad \mathbf{K}^{(3)} = \begin{bmatrix} 1 \\ u + a \\ H + ua \end{bmatrix} . \qquad (12.62)$$

It is instructive to relate Osher's scheme to the solution of the Riemann problem in the classical sense, see Chap. 4 for details. Fig. 12.1 shows the structure of the solution of the Riemann problem with data $\mathbf{U}_0$, $\mathbf{U}_1$ in the x–t plane. Also shown there are the partial integration paths $I_1(\mathbf{U})$, $I_2(\mathbf{U})$, $I_3(\mathbf{U})$, the intersection points $\mathbf{U}_{\frac{1}{3}}$, $\mathbf{U}_{\frac{2}{3}}$ and the sonic points $\mathbf{U}_{S0}$, $\mathbf{U}_{S1}$ in phase space; the illustrated paths follow the *P–ordering*. There are essentially two steps in obtaining the Osher flux formulae. First the intersection points $\mathbf{U}_{\frac{1}{3}}$ and $\mathbf{U}_{\frac{2}{3}}$ are obtained. We identify these points with the states $\mathbf{U}_{*L}$ and $\mathbf{U}_{*R}$ in the solution of the Riemann problem, see Chaps. 4 and 9. The second step consists of evaluating the integral in (12.21), for instance, along the integration paths $I_k(\mathbf{U})$ to obtain the intercell flux.

12.3.1 Osher's Flux with P–Ordering

The physical or *P–ordering* of integration paths for the Euler equations is illustrated in Fig. 12.1. States $\mathbf{U}_0$ and $\mathbf{U}_{\frac{1}{3}}$ are connected by the partial integration path $I_1(\mathbf{U})$, which is taken to be tangential to the right eigenvector $\mathbf{K}^{(1)}(\mathbf{U})$ in (12.62). Similarly, $I_2(\mathbf{U})$ connects $\mathbf{U}_{\frac{1}{3}}$ to $\mathbf{U}_{\frac{2}{3}}$ and $I_3(\mathbf{U})$ connects $\mathbf{U}_{\frac{2}{3}}$ to $\mathbf{U}_1$. In the P–ordering the intersection points $\mathbf{U}_{\frac{1}{3}}$, $\mathbf{U}_{\frac{2}{3}}$ and the sonic points $\mathbf{U}_{S0}$, $\mathbf{U}_{S1}$ are obtained by using the physically correct Generalised Riemann Invariants; see Sect. 3.1.3 of Chap. 3 for details.

Intersection Points and Sonic Points. Effectively, the intersection points $\mathbf{U}_{\frac{1}{3}}$, $\mathbf{U}_{\frac{2}{3}}$ can be taken to be the solution of the Riemann problem with data $\mathbf{U}_0$, $\mathbf{U}_1$ in the conventional sense. In the spirit of Osher's scheme we obtain $\mathbf{U}_{\frac{1}{3}}$ and $\mathbf{U}_{\frac{2}{3}}$ utilising the Generalised Riemann Invariants

$$I_L \equiv u + \frac{2a}{\gamma - 1} = \text{constant} \tag{12.63}$$

and

$$I_R \equiv u - \frac{2a}{\gamma - 1} = \text{constant} \tag{12.64}$$

to relate $\mathbf{U}_0$ to $\mathbf{U}_{\frac{1}{3}}$ and $\mathbf{U}_{\frac{2}{3}}$ to $\mathbf{U}_1$ respectively. Recall that if the left and right non–linear waves are rarefaction waves then relations (12.63)–(12.64) are exact. These waves can be either shock or rarefactions and thus the derived intersection points $\mathbf{U}_{\frac{1}{3}}$ and $\mathbf{U}_{\frac{2}{3}}$ are, in general, approximations. The underlying assumption is that in the solution of the Riemann problem with data $\mathbf{U}_0$, $\mathbf{U}_1$, both non–linear waves are rarefaction waves. This corresponds to the Two–Rarefaction approximation TRRS presented in Sect. 9.4.1 of Chap. 9. Using (12.63) across the left wave gives

$$u_* + \frac{2a_{\frac{1}{3}}}{\gamma - 1} = u_0 + \frac{2a_0}{\gamma - 1} \tag{12.65}$$

and use of (12.64) across the right wave gives

$$u_* - \frac{2a_{\frac{2}{3}}}{\gamma - 1} = u_1 - \frac{2a_1}{\gamma - 1} \ . \tag{12.66}$$

Here u_* is the common particle velocity for $\mathbf{U}_{\frac{1}{3}}$ and $\mathbf{U}_{\frac{2}{3}}$. Recall that, see Chaps. 3 and 4, the pressure p_* is also common, that is

$$u_{\frac{1}{3}} = u_{\frac{2}{3}} = u_* = constant \,, \quad p_{\frac{1}{3}} = p_{\frac{2}{3}} = p_* = constant \,. \tag{12.67}$$

In addition to (12.65)–(12.66) the isentropic law applied to the left and right waves gives

$$a_{\frac{1}{3}} = a_0(p_*/p_0)^z \,, \quad a_{\frac{2}{3}} = a_1(p_*/p_1)^z \,, \tag{12.68}$$

with

$$z = \frac{\gamma - 1}{2\gamma} \,. \tag{12.69}$$

From (12.65) and (12.68) we obtain

$$u_* = u_0 - \frac{2a_0}{\gamma - 1} \left[\left(\frac{p_*}{p_0} \right)^z - 1 \right] \,. \tag{12.70}$$

Similarly, use of (12.66) and (12.68) gives

$$u_* = u_1 + \frac{2a_1}{\gamma - 1} \left[\left(\frac{p_*}{p_1} \right)^z - 1 \right] \,. \tag{12.71}$$

Solving for p_* and u_* gives

$$p_* = \left[\frac{a_0 + a_1 - (u_1 - u_0)(\gamma - 1)/2}{a_0/p_0^z + a_1/p_1^z} \right]^{\frac{1}{z}} \,, \tag{12.72}$$

$$u_* = \frac{Hu_0/a_0 + u_1/a_1 + 2(H - 1)/(\gamma - 1)}{H/a_0 + 1/a_1} \,, \tag{12.73}$$

with

$$H = (p_0/p_1)^z \,.$$

The density values $\rho_{\frac{1}{3}}, \rho_{\frac{2}{3}}$ could be obtained from (12.68)–(12.69) or more directly as

$$\rho_{\frac{1}{3}} = \rho_0 \left(\frac{p_*}{p_0} \right)^{\frac{1}{\gamma}} \,, \quad \rho_{\frac{2}{3}} = \rho_1 \left(\frac{p_*}{p_1} \right)^{\frac{1}{\gamma}} \,. \tag{12.74}$$

The complete solution for $\mathbf{U}_{\frac{1}{3}}, \mathbf{U}_{\frac{2}{3}}$ is given by (12.72)–(12.74). This is identical to the two–rarefaction approximation of Sect. 9.4.1 of Chap. 9. The computation of the sonic points $\mathbf{U}_{S0}$ (left wave) and $\mathbf{U}_{S1}$ (right wave) is performed by first enforcing the sonic conditions $\lambda_1 = u - a = 0$ and $\lambda_3 = u + a = 0$, respectively and then applying the *corresponding* Generalised Riemann Invariants. The solution for the left sonic point is

$$\left. \begin{array}{cc} u_{S0} = \frac{\gamma - 1}{\gamma + 1} u_0 + \frac{2a_0}{\gamma + 1} \,, & a_{S0} = u_{S0} \,, \\[2ex] \rho_{S0} = \rho_0 \left(\frac{a_{S0}}{a_0} \right)^{\frac{2}{\gamma - 1}} \,, & p_{S0} = p_0 \left(\frac{\rho_{S0}}{\rho_0} \right)^\gamma \,. \end{array} \right\} \tag{12.75}$$

For the right sonic point the solution is

$$u_{S1} = \frac{\gamma-1}{\gamma+1}u_1 - \frac{2a_1}{\gamma+1} , \qquad a_{S1} = -u_{S1} ,$$

$$\rho_{S1} = \rho_1 \left(\frac{a_{S1}}{a_1}\right)^{\frac{2}{\gamma-1}} , \qquad p_{S1} = p_1 \left(\frac{\rho_{S1}}{\rho_1}\right)^{\gamma} . \tag{12.76}$$

Integration Along Partial Paths. We adopt expression (12.21) for the Osher intercell flux

$$\mathbf{F}_{i+\frac{1}{2}} = \mathbf{F}_0 + \int_{\mathbf{U}_0}^{\mathbf{U}_1} \mathbf{A}^-(\mathbf{U})d\mathbf{U} ,$$

where the integral in phase space along the path

$$I(\mathbf{U}) = I_1(\mathbf{U}) \cup I_2(\mathbf{U}) \cup I_3(\mathbf{U})$$

gives

$$\mathbf{F}_{i+\frac{1}{2}} = \mathbf{F}_0 + \int_{\mathbf{U}_0}^{\mathbf{U}_{\frac{1}{3}}} \mathbf{A}^-(\mathbf{U})d\mathbf{U} + \int_{\mathbf{U}_{\frac{1}{3}}}^{\mathbf{U}_{\frac{2}{3}}} \mathbf{A}^-(\mathbf{U})d\mathbf{U} + \int_{\mathbf{U}_{\frac{2}{3}}}^{\mathbf{U}_1} \mathbf{A}^-(\mathbf{U})d\mathbf{U} . \tag{12.77}$$

The integration along each partial path $I_k(\mathbf{U})$ is performed individually and the results are added to produce $\mathbf{F}_{i+\frac{1}{2}}$. The partial integrations are easily performed following the methodology presented in previous sections. For any given path there is only one wave involved. The left and right waves define genuinely non–linear fields, see Sect. 3.1.3 of Chap. 3, and the corresponding eigenvalue changes sign at most once, generating the sonic–point values. The second field (middle wave) is linearly degenerate, see Sect. 3.1.3 of Chap. 3, and thus the eigenvalue $\lambda_2(\mathbf{U})$ is constant along the path $I_2(\mathbf{U})$.

The integration results for each path are given in Table 12.7 and are labelled A, B and C respectively. To obtain the intercell flux $\mathbf{F}_{i+\frac{1}{2}}$ the integral terms in (12.77) must be selected according to the behaviour of the eigenvalue $\lambda_k(\mathbf{U})$ along the path $I_k(\mathbf{U})$. As seen in Table 12.7 there are 32 possible combinations i, j, l; of these only 16 are realisable, which are tabulated as in Table 12.8. Here, in order to identify the correct expression for the resulting intercell flux $\mathbf{F}_{i+\frac{1}{2}}$, we split conditions on the intersection points, first column, from conditions on the data, top row.

12.3.2 Osher's Flux with O–Ordering

As originally presented, Osher's scheme uses the O–ordering of integration paths; this is precisely the opposite of the P–ordering described previously. The approach is used consistently to determine the intersection points, the sonic points and for performing the integration in (12.77). Fig. 12.4 illustrates the O–ordering of Osher's scheme as applied to a 3 × 3 system, such as the

$A:$	Behaviour of $\lambda_1(\mathbf{U}) = u - a$	$\int_{\mathbf{U}_0}^{\mathbf{U}_{\frac{1}{3}}} \mathbf{A}^-(\mathbf{U})d\mathbf{U}$
1	$u_0 - a_0 \geq 0 , \quad u_* - a_{\frac{1}{3}} \geq 0$	$\mathbf{0}$
2	$u_0 - a_0 \leq 0 , \quad u_* - a_{\frac{1}{3}} \leq 0$	$\mathbf{F}_{\frac{1}{3}} - \mathbf{F}_0$
3	$u_0 - a_0 \geq 0 , \quad u_* - a_{\frac{1}{3}} \leq 0$	$\mathbf{F}_{\frac{1}{3}} - \mathbf{F}_{S0}$
4	$u_0 - a_0 \leq 0 , \quad u_* - a_{\frac{1}{3}} \geq 0$	$\mathbf{F}_{S0} - \mathbf{F}_0$
$B:$	Behaviour of $\lambda_2(\mathbf{U}) = u$	$\int_{\mathbf{U}_{\frac{1}{3}}}^{\mathbf{U}_{\frac{2}{3}}} \mathbf{A}^-(\mathbf{U})d\mathbf{U}$
1	$u_* \geq 0$	$\mathbf{0}$
2	$u_* < 0$	$\mathbf{F}_{\frac{2}{3}} - \mathbf{F}_{\frac{1}{3}}$
$C:$	Behaviour of $\lambda_3(\mathbf{U}) = u + a$	$\int_{\mathbf{U}_{\frac{2}{3}}}^{\mathbf{U}_1} \mathbf{A}^-(\mathbf{U})d\mathbf{U}$
1	$u_1 + a_1 \geq 0 , \quad u_* + a_{\frac{2}{3}} \geq 0$	$\mathbf{0}$
2	$u_1 + a_1 \leq 0 , \quad u_* + a_{\frac{2}{3}} \leq 0$	$\mathbf{F}_1 - \mathbf{F}_{\frac{2}{3}}$
3	$u_* + a_{\frac{2}{3}} \geq 0 , \quad u_1 + a_1 \leq 0$	$\mathbf{F}_1 - \mathbf{F}_{S1}$
4	$u_* + a_{\frac{2}{3}} \leq 0 , \quad u_1 + a_1 \geq 0$	$\mathbf{F}_{S1} - \mathbf{F}_{\frac{2}{3}}$

Table 12.7. Integration along partial paths for the Euler equations following the P–ordering, $\mathbf{F}_k \equiv \mathbf{F}(\mathbf{U}_k)$

	$u_0 - a_0 \geq 0$ $u_1 + a_1 \geq 0$	$u_0 - a_0 \geq 0$ $u_1 + a_1 \leq 0$	$u_0 - a_0 \leq 0$ $u_1 + a_1 \geq 0$	$u_0 - a_0 \leq 0$ $u_1 + a_1 \leq 0$
$u^* \geq 0$ $u^* - a_{\frac{1}{3}} \geq 0$	$\mathbf{F}_0$	$\mathbf{F}_0 + \mathbf{F}_1$ $-\mathbf{F}_{S1}$	$\mathbf{F}_{S0}$	$\mathbf{F}_{S0} - \mathbf{F}_{S1}$ $+\mathbf{F}_1$
$u^* \geq 0$ $u^* - a_{\frac{1}{3}} \leq 0$	$\mathbf{F}_0 - \mathbf{F}_{S0}$ $+\mathbf{F}_{\frac{1}{3}}$	$\mathbf{F}_0 - \mathbf{F}_{S0}$ $+\mathbf{F}_{\frac{1}{3}} - \mathbf{F}_{S1}$ $+\mathbf{F}_1$	$\mathbf{F}_{\frac{1}{3}}$	$\mathbf{F}_1 + \mathbf{F}_{\frac{1}{3}}$ $-\mathbf{F}_{S1}$
$u^* \leq 0$ $u^* + a_{\frac{2}{3}} \geq 0$	$\mathbf{F}_0 - \mathbf{F}_{S0}$ $+\mathbf{F}_{\frac{2}{3}}$	$\mathbf{F}_0 - \mathbf{F}_{S0}$ $+\mathbf{F}_{\frac{2}{3}} - \mathbf{F}_{S1}$ $+\mathbf{F}_1$	$\mathbf{F}_{\frac{2}{3}}$	$\mathbf{F}_{\frac{2}{3}} - \mathbf{F}_{S1}$ $+\mathbf{F}_1$
$u^* \leq 0$ $u^* + a_{\frac{2}{3}} \leq 0$	$\mathbf{F}_0 - \mathbf{F}_{S0}$ $+\mathbf{F}_{S1}$	$\mathbf{F}_0 - \mathbf{F}_{S0}$ $+\mathbf{F}_1$	$\mathbf{F}_{S1}$	$\mathbf{F}_1$

Table 12.8. Osher's flux formulae for the Euler equations using P–ordering of integration paths, $\mathbf{F}_k \equiv \mathbf{F}(\mathbf{U}_k)$

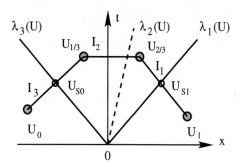

Fig. 12.4. Configuration of integration paths $I_k(\mathbf{U})$, intersection points $\mathbf{U}_{\frac{1}{3}}$, $\mathbf{U}_{\frac{2}{3}}$ and sonic points $\mathbf{U}_{S0}$, $\mathbf{U}_{S1}$ in physical space x–t for a 3×3 system, following the O–ordering

one–dimensional Euler equations. We combine the configuration for the integration paths, the intersection points and the sonic points in phase space with the Riemann problem solution with data $\mathbf{U}_0, \mathbf{U}_1$ in physical space x–t. The O–ordering can now be interpreted as assigning the eigenvalue $\lambda_3(\mathbf{U})$ and eigenvector $\mathbf{K}^{(3)}(\mathbf{U})$ to the wave family with eigenvalue $\lambda_1(\mathbf{U})$ and eigenvector $\mathbf{K}^{(1)}(\mathbf{U})$ (left wave), and vice–versa.

Intersection Points and Sonic Points. First, the intersection points $\mathbf{U}_{\frac{1}{3}}$ and $\mathbf{U}_{\frac{2}{3}}$ are determined by using Generalised Riemann Invariants, as done in Sect. 12.1.3 with P–ordering. The difference is that $\mathbf{U}_0$ and $\mathbf{U}_{\frac{1}{3}}$ are connected using the *right* Riemann Invariant and $\mathbf{U}_{\frac{2}{3}}$ is connected to $\mathbf{U}_1$ using the *left* Riemann Invariant. Thus we write

$$I_R \equiv u_* - \frac{2}{\gamma - 1} a_{\frac{1}{3}} = u_0 - \frac{2}{\gamma - 1} a_0 , \tag{12.78}$$

$$I_L \equiv u_* + \frac{2}{\gamma - 1} a_{\frac{2}{3}} = u_1 + \frac{2}{\gamma - 1} a_1 . \tag{12.79}$$

See (12.63)–(12.66). Again, we assume u_* and p_* are the common particle velocity and pressure at points $\mathbf{U}_{\frac{1}{3}}$ and $\mathbf{U}_{\frac{2}{3}}$. Use of (12.68) into (12.78) gives

$$u_* = u_0 + \frac{2}{(\gamma - 1)} a_0 \left[\left(\frac{p_*}{p_0} \right)^z - 1 \right] , \tag{12.80}$$

and use of (12.68) into (12.79) produces

$$u_* = u_1 - \frac{2}{(\gamma - 1)} a_1 \left[\left(\frac{p_*}{p_1} \right)^z - 1 \right] . \tag{12.81}$$

Solving for p_* gives

$$p_* = \left[\frac{a_0 + a_1 + (u_1 - u_0)(\gamma - 1)/2}{a_0/p_0^z + a_1/p_1^z} \right]^{\frac{1}{z}} . \tag{12.82}$$

Compare with solution (12.72) using P–ordering. Equations (12.80) and (12.81) can be rearranged as

$$p_* = p_0 \left[\frac{\gamma - 1}{2a_0} (u_* - u_0) + 1 \right]^{\frac{1}{z}} , \tag{12.83}$$

$$p_* = p_1 \left[\frac{\gamma - 1}{2a_1} (u_1 - u_*) + 1 \right]^{\frac{1}{z}} , \tag{12.84}$$

whose solution for u_* is

$$u_* = \frac{H u_0/a_0 + u_1/a_1 - 2(H - 1)/(\gamma - 1)}{H/a_0 + 1/a_1} , \tag{12.85}$$

with

$$H = (p_0/p_1)^z , \quad z = \frac{\gamma - 1}{2\gamma} .$$

Compare with solution (12.73) using the P–ordering. The solution for $\rho_{\frac{1}{3}}$ and $\rho_{\frac{2}{3}}$ is, using the isentropic law,

$$\rho_{\frac{1}{3}} = \rho_0 (p_*/p_0)^{\frac{1}{\gamma}} , \quad \rho_{\frac{2}{3}} = \rho_1 (p_*/p_1)^{\frac{1}{\gamma}} . \tag{12.86}$$

To find the sonic points $\mathbf{U}_{S0}$ and $\mathbf{U}_{S1}$ we first connect $\mathbf{U}_0$ to $\mathbf{U}_{\frac{1}{3}}$ via the *right* Riemann Invariant to obtain

$$u_{S0} = \frac{2}{\gamma - 1} a_{S0} + u_0 - \frac{2}{\gamma - 1} a_0 .$$

Then by enforcing the sonic condition

$$\lambda_3 (\mathbf{U}) = u_{S0} + a_{S0} = 0$$

along $I_3(\mathbf{U})$ and applying the isentropic law one obtains the solution

$$\left. \begin{array}{ll} u_{S0} = \frac{\gamma-1}{\gamma+1} u_0 - \frac{2a_0}{\gamma+1} , & a_{S0} = -u_{S0} , \\[2mm] \rho_{S0} = \rho_0 \left(\frac{a_{S0}}{a_0} \right)^{\frac{2}{\gamma-1}} , & p_{S0} = p_0 \left(\frac{\rho_{S0}}{\rho_0} \right)^{\gamma} . \end{array} \right\} \tag{12.87}$$

The solution for the right sonic point $\mathbf{U}_{S1}$ is

$$\left. \begin{array}{ll} u_{S1} = \frac{\gamma-1}{\gamma+1} u_1 + \frac{2}{\gamma+1} a_1 , & a_{S1} = u_{S1} , \\[2mm] \rho_{S1} = \rho_1 \left(\frac{a_{S1}}{a_1} \right)^{\frac{2}{\gamma-1}} , & p_{S1} = p_1 \left(\frac{\rho_{S1}}{\rho_1} \right)^{\gamma} . \end{array} \right\} \tag{12.88}$$

Integration Along Partial Paths. To compute Osher's intercell flux

$$\mathbf{F}_{i+\frac{1}{2}} = \mathbf{F}_0 + \int_{\mathbf{U}_0}^{\mathbf{U}_1} \mathbf{A}^-(\mathbf{U}) d\mathbf{U} , \tag{12.89}$$

we need to find the integral

$$\int_{\mathbf{U}_0}^{\mathbf{U}_1} \mathbf{A}^-(\mathbf{U}) d\mathbf{U} = \int_{\mathbf{U}_0}^{\mathbf{U}_{\frac{1}{3}}} \mathbf{A}^-(\mathbf{U}) d\mathbf{U} + \int_{\mathbf{U}_{\frac{1}{3}}}^{\mathbf{U}_{\frac{2}{3}}} \mathbf{A}^-(\mathbf{U}) d\mathbf{U} + \int_{\mathbf{U}_{\frac{2}{3}}}^{\mathbf{U}_1} \mathbf{A}^-(\mathbf{U}) d\mathbf{U} . \tag{12.90}$$

Using the O–ordering one has

$$\int_{\mathbf{U}_0}^{\mathbf{U}_{\frac{1}{3}}} \mathbf{A}^-(\mathbf{U}) d\mathbf{U} = \int_{I_3(\mathbf{U})} \mathbf{A}^-(\mathbf{U}) d\mathbf{U} , \tag{12.91}$$

$$\int_{\mathbf{U}_{\frac{1}{3}}}^{\mathbf{U}_{\frac{2}{3}}} \mathbf{A}^-(\mathbf{U}) d\mathbf{U} = \int_{I_2(\mathbf{U})} \mathbf{A}^-(\mathbf{U}) d\mathbf{U} , \tag{12.92}$$

$$\int_{U_\frac{2}{3}}^{U_1} \mathbf{A}^-(\mathbf{U})d\mathbf{U} = \int_{I_1(\mathbf{U})} \mathbf{A}^-(\mathbf{U})d\mathbf{U} .$$ (12.93)

The evaluation of the three terms in (12.90) according to (12.91)–(12.93) is given in Table 12.9, by cases labelled A, B and C respectively. The first column contains the sub–case numbers. The second column contains the behaviour of the eigenvalue along the corresponding partial path. For instance in case A one considers the sign of the eigenvalue $\lambda_3(\mathbf{U})$ along the path $I_3(\mathbf{U})$ joining the points $\mathbf{U}_0$ and $\mathbf{U}_\frac{1}{3}$. The third column contains the resulting integral in (12.91)–(12.93). There are 32 combinations, of which only 16 are realisable.

Just to illustrate the method of analysis, consider first the combination (A_1, B_1, C_1), that is

$$A_1: \quad u_0 + a_0 \geq 0, \quad u_* + a_\frac{1}{3} \geq 0,$$

$$B_1: \quad\quad\quad u_* \geq 0,$$

$$C_1: \quad u_1 - a_1 \geq , \quad u_* - a_\frac{2}{3} \geq 0.$$

As all eigenvalues are positive throughout, the integrals are all zero. The combination (A_1, B_2, C_1) gives

$$A_1: \quad u_0 + a_0 \geq 0, \quad u_* + a_\frac{1}{3} \geq 0,$$

$$B_2: \quad\quad\quad u_* \leq 0,$$

$$C_1: \quad u_1 - a_1 \geq 0, \quad u_* - a_\frac{2}{3} \geq 0.$$

Clearly these conditions cannot be satisfied simultaneously, $u_* \leq 0$ contradicts $u_* - a_\frac{2}{3} \geq 0$, as $a_\frac{2}{3} > 0$.

The Osher intercell flux with O–ordering is given in table 12.10. The logic involved can be organised so as to test conditions on the data points (top row) and conditions on the intersection points (first column). The Osher's flux formulae with P and O ordering given by Tables 12.8 and 12.10 respectively, are valid for any 3×3 non–linear hyperbolic system. The specific properties of a particular system enter in the determination of the sonic and intersection points only. The formulae can be directly applied, for instance, to the split two–dimensional shallow water equations.

12.3.3 Remarks on Path Orderings

It is useful to compare the pressure solutions (12.72) and (12.82) when computing the intersection points using the P and O orderings for the integration paths. Let us redefine the respective solutions as

$$p_*^{(P)} = \left[\frac{a_0 + a_1 - \Delta u(\gamma - 1)/2}{a_0/p_0^z + a_1/p_1^z}\right]^{1/z}$$ (12.94)

$A:$	Behaviour of $\lambda_3(\mathbf{U}) = u - a$	$\int_{\mathbf{U}_0}^{\mathbf{U}_{\frac{1}{3}}} \mathbf{A}^-(\mathbf{U})d\mathbf{U}$
1	$u_0 + a_0 \geq 0 \,,\;\; u_* + a_{\frac{1}{3}} \geq 0$	$\mathbf{0}$
2	$u_0 + a_0 \leq 0 \,,\;\; u_* + a_{\frac{1}{3}} \leq 0$	$\mathbf{F}_{\frac{1}{3}} - \mathbf{F}_0$
3	$u_0 + a_0 \geq 0 \,,\;\; u_* + a_{\frac{1}{3}} \leq 0$	$\mathbf{F}_{\frac{1}{3}} - \mathbf{F}_{S0}$
4	$u_0 + a_0 \leq 0 \,,\;\; u_* + a_{\frac{1}{3}} \geq 0$	$\mathbf{F}_{S0} - \mathbf{F}_0$
$B:$	Behaviour of $\lambda_2(\mathbf{U}) = u$	
1	$u_* \geq 0$	$\mathbf{0}$
2	$u_* \leq 0$	$\mathbf{F}_{\frac{2}{3}} - \mathbf{F}_{\frac{1}{3}}$
$C:$	Behaviour of $\lambda_1(\mathbf{U}) = u - a$	
1	$u_1 - a_1 \geq 0 \,,\;\; u_* - a_{\frac{2}{3}} \geq 0$	$\mathbf{0}$
2	$u_1 - a_1 \leq 0 \,,\;\; u_* - a_{\frac{2}{3}} \leq 0$	$\mathbf{F}_1 - \mathbf{F}_{\frac{2}{3}}$
3	$u_1 - a_1 \geq 0 \,,\;\; u_* - a_{\frac{2}{3}} \leq 0$	$\mathbf{F}_{S1} - \mathbf{F}_{\frac{2}{3}}$
4	$u_1 - a_1 \leq 0 \,,\;\; u_* - a_{\frac{2}{3}} \geq 0$	$\mathbf{F}_1 - \mathbf{F}_{S1}$

Table 12.9. Osher's flux for the Euler equations. Evaluation of integral following O–ordering, $\mathbf{F}_k \equiv \mathbf{F}(\mathbf{U}_k)$

	$u_0 + a_0 \geq 0$ $u_1 - a_1 \geq 0$	$u_0 + a_0 \geq 0$ $u_1 - a_1 \leq 0$	$u_0 + a_0 \leq 0$ $u_1 - a_1 \geq 0$	$u_0 + a_0 \leq 0$ $u_1 - a_1 \leq 0$
$u_* + a_{\frac{1}{3}} \leq 0$	$\mathbf{F}_0 - \mathbf{F}_{S0}$ $+\mathbf{F}_{S1}$	$\mathbf{F}_0 - \mathbf{F}_{S0}$ $+\mathbf{F}_1$	$\mathbf{F}_{S1}$	$\mathbf{F}_1$
$u_* + a_{\frac{1}{3}} \geq 0$ $u_* \leq 0$	$\mathbf{F}_0 - \mathbf{F}_{\frac{1}{3}}$ $+\mathbf{F}_{S1}$	$\mathbf{F}_0 - \mathbf{F}_{\frac{1}{3}}$ $+\mathbf{F}_1$	$\mathbf{F}_{S0} - \mathbf{F}_{\frac{1}{3}}$ $+\mathbf{F}_{S1}$	$\mathbf{F}_{S0} - \mathbf{F}_{\frac{1}{3}}$ $+\mathbf{F}_1$
$u_* - a_{\frac{2}{3}} \geq 0$	$\mathbf{F}_0$	$\mathbf{F}_0 - \mathbf{F}_{S1}$ $+\mathbf{F}_1$	$\mathbf{F}_{S0}$	$\mathbf{F}_{S0} - \mathbf{F}_{S1}$ $+\mathbf{F}_1$
$u_* - a_{\frac{2}{3}} \leq 0$ $u_* \geq 0$	$\mathbf{F}_0 - \mathbf{F}_{\frac{2}{3}}$ $+\mathbf{F}_{S1}$	$\mathbf{F}_0 - \mathbf{F}_{\frac{2}{3}}$ $+\mathbf{F}_1$	$\mathbf{F}_{S0} - \mathbf{F}_{\frac{2}{3}}$ $+\mathbf{F}_{S1}$	$\mathbf{F}_{S0} - \mathbf{F}_{\frac{2}{3}}$ $+\mathbf{F}_1$

Table 12.10. Osher's flux formulae for the Euler equations using O–ordering of integration paths, $\mathbf{F}_k = \mathbf{F}(\mathbf{U}_k)$

and

$$p_*^{(O)} = \left[\frac{a_0 + a_1 + \Delta u (\gamma - 1)/2}{a_0/p_0^z + a_1/p_1^z} \right]^{1/z}, \qquad (12.95)$$

where $\Delta u = u_1 - u_0$ is the velocity difference in the data. The reader is encouraged to review Sect. 4.3.1 of Chap. 4, in which a detailed discussion is given on the influence of Δu on the solution for pressure in the exact solution. First we note that $p_*^{(P)} = p_*^{(O)}$ when $\Delta u = 0$ and that for $\Delta u \neq 0$ the two solutions are not only different but more importantly, they have *very different behaviour*. This does not seem to matter too much in practical computations, except for special but important situations. There are at least two cases that deserve special attention.

The case $\Delta u \gg 0$, see Sect. 4.3.1 of Chap. 4, is associated with strong rarefaction waves. In fact there is a limit for which the pressure p_* becomes negative and is associate with the *pressure positivity condition* stated in Sect. 4.3.1 of Chap. 4. In the *incipient cavitation case* the pressure is 0. The solution $p_*^{(P)}$ will correctly reflect this physical situation of low–density flow, including the detection of *vacuum*. The author is not aware of this having been exploited in the context of Osher's Riemann solver with P–ordering. On the other hand, the solution $p_*^{(O)}$ for $\Delta u \gg 0$ obtained with the O–ordering, will give unrealistically large values for the pressure at the intersection points, which is more consistent with the presence of strong shock waves, rather than strong rarefaction waves. Such large pressure values will lead the Osher scheme to be very inaccurate or simply to fail for low–density flows, just as linearised Riemann solvers do; see Einfeldt et. al. [118] for a discussion on numerical difficulties for low–density flows. We illustrate this point through Test 2 of Table 12.11, for which the scheme actually fails.

The case $\Delta u \ll 0$, see Sect. 4.3.1 of Chap. 4, is associated with strong shock waves. Again the P–ordering solution (12.94) will correctly reflect this. However the O–ordering solution (12.95) will not. More importantly, there will be a limiting strong shock situation for which the O–ordering pressure is *undefined* and the scheme will again fail; see Test 5 in Table 12.11. The failure condition in this case is analytic, namely

$$a_0 + a_1 + \Delta u(\gamma - 1)/2 \leq 0. \tag{12.96}$$

It is paradoxical that the scheme fails in the presence of strong shocks, which is consistent with large pressure, through a pressure solution $p_*^{(O)}$ that is undefined for being so close to vacuum conditions. One could devise some sort of fix to remedy this situation. One possibility is to abandon the O–ordering altogether or switch to the P–ordering locally, in a kind of adaptive ordering.

12.3.4 The Split Three–Dimensional Case

The extension of the Osher scheme to two and three–dimensional problems is straightforward. All methods considered here require expressions for the split fluxes. In the x–split, three dimensional case, for instance, we require in addition the y and z momentum flux components ρuv and ρuw. In turn, this requires the extra components ρv and ρw in the vector of conserved variables $\mathbf{U}$. These new components will be needed in the intersection points and sonic points. But, as seen in Sect. 4.8 of Chap. 4, the exact solution for v and w in the Riemann problem is given by

$$q(x,t) = \begin{cases} q_L & \text{if } x/t < u_* \\ \\ q_R & \text{if } x/t > u_* \end{cases}, \tag{12.97}$$

where $q = v$ and $q = w$. The approximate solution for the tangential components of velocity obtained from using Generalised Riemann Invariants in the Osher scheme preserves the form of the exact solution (12.97), the only approximation being that of the normal velocity component u_*. Therefore, for computing $\mathbf{U}_{S0}$ and $\mathbf{U}_{\frac{1}{3}}$ we take v_L and w_L; for computing $\mathbf{U}_{S1}$ and $\mathbf{U}_{\frac{2}{3}}$ we take v_R and w_R. Solution (12.97) ensures that contact waves and shear waves (and shear layers in Navier–Stokes applications) are well resolved by Osher's scheme, a property that is common to the exact Riemann solver of Chap. 6, the approximate–state Riemann solvers of Chap. 9, the HLLC Riemann solver of Chap. 10 and Roe's solver of Chap. 11.

12.4 Numerical Results and Discussion

Here we illustrate the performance of Godunov's first–order method used in conjunction with Osher–type approximate Riemann solvers. We select six test problems for the one–dimensional time dependent Euler equations for ideal gases with $\gamma = 1.4$; these have exact solutions. In all chosen tests, data consists of two constant states $\mathbf{W}_L = [\rho_L, u_L, p_L]^T$ and $\mathbf{W}_R = [\rho_R, u_R, p_R]^T$, separated by a discontinuity at a position $x = x_0$. The states $\mathbf{W}_L$ and $\mathbf{W}_R$ are given in Table 12.11. The exact and numerical solutions are found in the spatial domain $0 \leq x \leq 1$. The numerical solution is computed with $M = 100$ cells and the CFL condition is as for all previous computations, see Chap. 6; the chosen Courant number coefficient is $C_{cfl} = 0.9$; boundary conditions are transmissive. The exact solutions were found by running the code *HE-E1RPEXACT* of the library *NUMERICA* [369] and the numerical solutions were obtained by running the code *HE-E1GODOSHER* of *NUMERICA*.

Test	ρ_L	u_L	p_L	ρ_R	u_R	p_R
1	1.0	0.75	1.0	0.125	0.0	0.1
2	1.0	-2.0	0.4	1.0	2.0	0.4
3	1.0	0.0	1000.0	1.0	0.0	0.01
4	5.99924	19.5975	460.894	5.99242	-6.19633	46.0950
5	1.0	-19.59745	1000.0	1.0	-19.59745	0.01
6	1.0	2.0	0.1	1.0	-2.0	0.1

Table 12.11. Data for six test problems with exact solution, for the time–dependent one dimensional Euler equations

Test 1 is a *modified* version of Sod's problem [318]; the solution has a right shock wave, a right travelling contact wave and a left *sonic* rarefaction wave; this test is useful in assessing the *entropy satisfaction* property of numerical methods. The solution of Test 2 consists of two symmetric rarefaction waves and a trivial contact wave; the *Star Region* between the non–linear waves is close to vacuum, which makes this problem a suitable test for assessing the performance of numerical methods for low–density flows. Test 3 is designed to assess the robustness and accuracy of numerical methods; its solution consists of a strong right shock wave of shock Mach number 198, a contact surface and a left rarefaction wave. Test 4 is also a very severe test, its solution consists of three strong discontinuities travelling to the right. A detailed discussion on the exact solution of the test problems is found in Chap. 4. Test 5 is also designed to test the robustness of numerical methods but the main reason for devising this test is to assess the ability of numerical methods to resolve *slowly–moving contact discontinuities*. The exact solution of Test 5 consists of a left rarefaction wave, a right–travelling shock wave and a *stationary* contact discontinuity. Test 6 simulates the collision of two uniform streams; the exact solution consists of two strong symmetric shock waves and a trivial contact discontinuity. The purpose of Test 6 is to illustrate the fact that Osher's scheme with O–ordering will fail for a range of problems of this kind. For each test problem we select a convenient position x_0 of the initial discontinuity and the output time. These are stated in the legend of each figure displaying computational results.

The computed results for Tests 1, 3, 4 and 5 using the Godunov first–order method in conjunction with the Osher approximate Riemann solver with O–ordering are shown in Figs. 12.5–12.8. The scheme failed for Tests 2 and 6. The result of Fig. 12.5 is virtually identical to that of the Godunov method with an exact Riemann solver, Fig. 6.8 of Chap. 6; note also that both schemes give comparable performance in the vicinity of the sonic point. The results for Tests 3 and 4, shown in Figs. 12.6 and 12.7, obtained with O–ordering compare well with those of the exact Riemann solver, except for the fact that the slowly moving shock in Fig. 12.7 does not have the spurious oscillations that other Riemann solvers produce [280], [40], [13], [193]; this is a distinguishing property of the Osher approach.

Results for Tests 1 to 6 from the Osher scheme with P–ordering are shown in Figs. 12.9–12.14. The results for Tests 1 to 4 are, overall, as accurate as those obtained from an exact Riemann solver; the P–ordering scheme actually works for Tests 2 and 6, whereas O–ordering scheme does not; note however the spurious oscillations in Test 6 near the shocks, a feature of P–ordering schemes already noted by Osher. For Test 5 the P–ordering scheme gives the incorrect solution; compare Fig. 12.13 with Fig. 12.8 and note the scales. First, there are very large unphysical overshoots in density, velocity and pressure. More intriguing is the fact that the expected right–travelling shock does not propagate at all; no signal propagates to the right of the initial discontinuity at $x = 0.8$. As a matter of fact, it looks as if the shock wave does not actually form and the state behind the contact discontinuity has the incorrect values, see pressure and velocity plots. My numerical results have been confirmed by Dr N. Qin (private communication), who used an independently written code to solve Test 5.

Earlier we made some remarks concerning a modified Osher scheme based on an adaptive ordering. In view of the numerical results it seems as if O–ordering could be used in all cases except for the two situations in which the solution of the Riemann problem contains either two shocks or two rarefactions. Such situations could be identified reliably and cheaply by using any of the approximate state Riemann solvers of Chap. 9.

12.5 Extensions

The approximate Riemann solvers of this chapter may be applied in conjunction with the Godunov first–order upwind method presented in Chap. 6. Second–order Total Variation Diminishing (TVD) extensions of the schemes are presented in Chap. 13 for scalar problems and in Chap. 14 for non–linear one dimensional systems. In chap. 15 we present techniques that allow the extension of these schemes to solve problems with source terms. In Chap. 16 we study techniques to extend the methods of this chapter to three–dimensional problems.

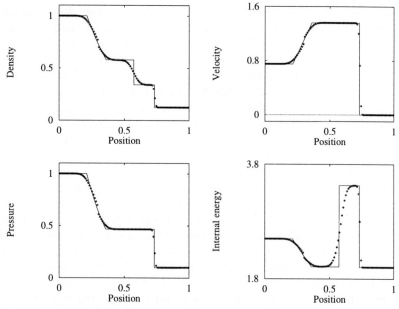

Fig. 12.5. Osher scheme with O–ordering applied to Test 1, with $x_0 = 0.3$. Numerical (symbol) and exact (line) solutions are compared at time 0.2 units

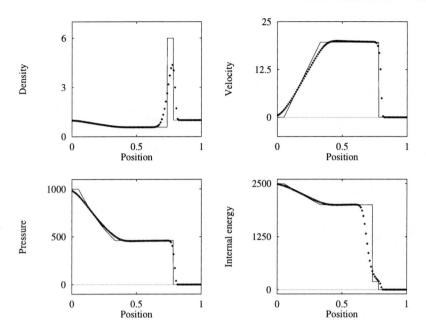

Fig. 12.6. Osher scheme with O–ordering applied to Test 3, with $x_0 = 0.5$. Numerical (symbol) and exact (line) solutions are compared at time 0.012 units

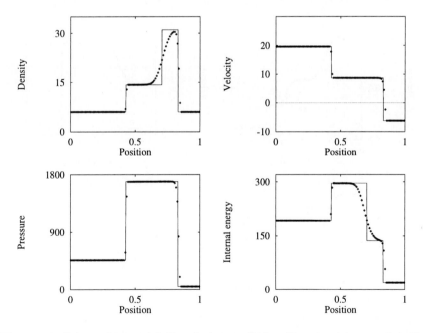

Fig. 12.7. Osher scheme with O–ordering applied to Test 4, with $x_0 = 0.4$. Numerical (symbol) and exact (line) solutions are compared at time 0.035 units

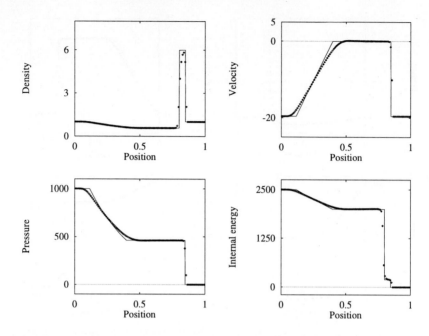

Fig. 12.8. Osher scheme with O–ordering applied to Test 5, with $x_0 = 0.8$. Numerical (symbol) and exact (line) solutions are compared at time 0.012 units

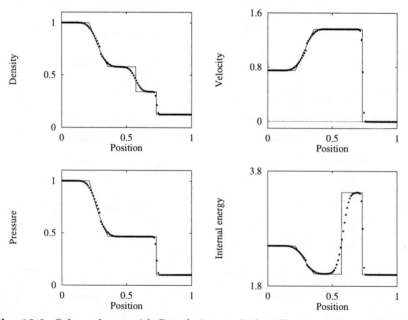

Fig. 12.9. Osher scheme with P–ordering applied to Test 1, with $x_0 = 0.3$. Numerical (symbol) and exact (line) solutions are compared at time 0.2 units

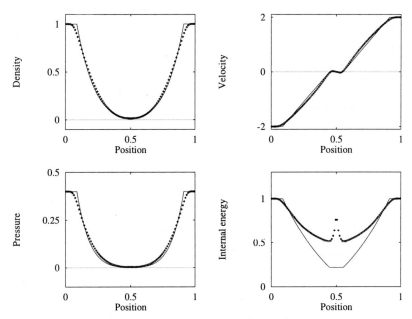

Fig. 12.10. Osher scheme with P–ordering applied to Test 2, with $x_0 = 0.3$. Numerical (symbol) and exact (line) solutions are compared at time 0.2 units

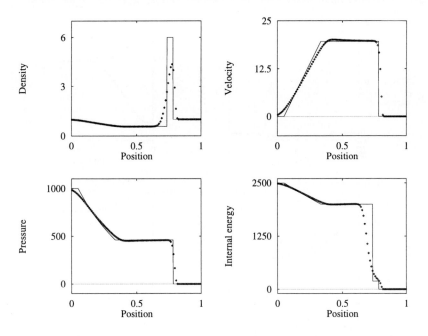

Fig. 12.11. Osher scheme with P–ordering applied to Test 3, with $x_0 = 0.5$. Numerical (symbol) and exact (line) solutions are compared at time 0.012 units

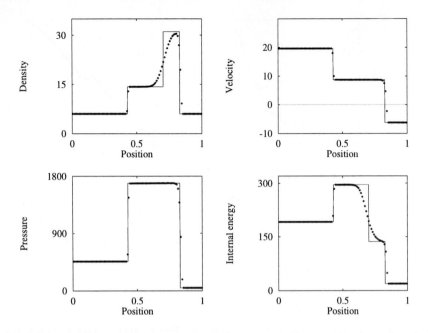

Fig. 12.12. Osher scheme with P–ordering applied to Test 4, with $x_0 = 0.4$. Numerical (symbol) and exact (line) solutions are compared at time 0.035 units

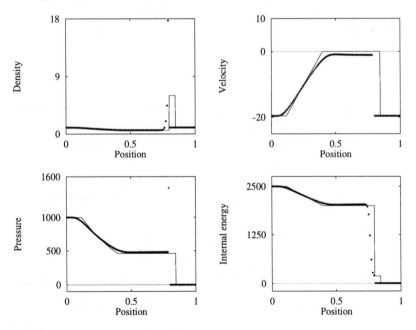

Fig. 12.13. Osher scheme with P–ordering applied to Test 5, with $x_0 = 0.8$. Numerical (symbol) and exact (line) solutions are compared at time 0.012 units

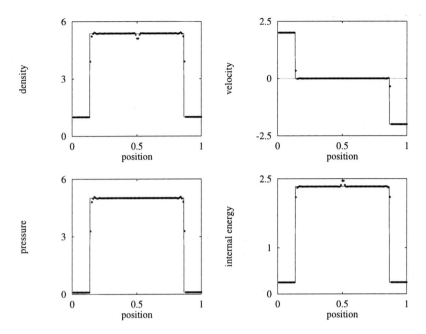

Fig. 12.14. Osher scheme with P–ordering applied to Test 6, with $x_0 = 0.5$. Numerical (symbol) and exact (line) solutions are compared at time 0.8 units

13. High–Order and TVD Methods for Scalar Equations

Central to this chapter is the resolution of two contradictory requirements on numerical methods, namely high–order of accuracy and absence of *spurious* (unphysical) oscillations in the vicinity of large gradients. It is well–known that high–order linear (constant coefficients) schemes produce unphysical oscillations in the vicinity of large gradients. This was illustrated by some numerical results shown in Chap. 5. On the other hand, the class of *monotone methods*, defined in Sect. 5.2 of Chap. 5, do not produce unphysical oscillations. However, *monotone methods are at most first order accurate* and are therefore of limited use. These difficulties are embodied in the statement of Godunov's theorem [145] to be studied in Sect. 13.5.3. One way of resolving the contradiction between linear schemes of high–order of accuracy and absence of spurious oscillations is by constructing non–linear methods. *Total Variation Diminishing Methods*, or TVD Methods for short, are a prominent class of non–linear methods, which we shall study in detail in this chapter.

13.1 Introduction

The purpose of this chapter is to introduce and develop the concept of TVD methods in a simple setting, so as to aid the reader to extend the techniques to more general problems. TVD methods are one of the most significant achievements in the development of numerical methods for partial differential equations (PDEs) in the last 20 years or so, although the preliminary ideas can be traced as far back as 1959, with the pioneering work of Godunov and later by van Leer [387], the Flux Corrected Transport (FCT) approach of Boris and Book [49], [48], [50] and many others. The theoretical bases of TVD methods are sound for scalar one–dimensional problems only. In practical, non–linear, multidimensional problems, the accumulated experience of numerous applications has demonstrated that the one–dimensional scalar theory serves well as a guideline for extending the ideas, on a more or less empirical basis. TVD schemes have their roots on the fundamental question of *convergence* of schemes for *non–linear* problems. Total–Variation Stable schemes have been proved to be convergent. TVD methods are a class of Total–Variation Stable schemes and they are based on the requirement that the total variation of the numerical solution be non–increasing in time. As a matter of fact the *exact* solution of a scalar conservation law does possess this property and therefore TVD numerical methods just mimic a property of the exact solution. Apart from being a class of methods proved to be convergent, for scalar problems, the TVD concept is also extremely useful in *designing schemes*, as we shall see in this chapter.

TVD schemes are intimately linked to traditional *Artifical Viscosity Methods* [276]. Both TVD and Artifical Viscosity Methods attempt to circumvent Godunov's theorem by constructing schemes of accuracy $p > 1$, such that *spurious oscillations near high gradients* are eliminated or controlled. Both classes of schemes resort effectively to the same mechanism, namely the addition of *artificial viscosity*. Artifical Viscosity Methods do this *explicitly*, by adding extra terms to the PDEs. In TVD methods, on the other hand, the artificial viscosity is inherent in the scheme itself, but the way this is activated is rather sophisticated.

The TVD approach has managed to circumvent Godunov's theorem by resorting to *non–linear schemes*, even when applied to linear problems. The inherent artificial viscosity of the schemes is controlled in a very precise manner so that spurious oscillations are eliminated and high–order accuracy in smooth parts of the solution is retained. Historically, TVD methods have almost exclusively been associated with high–order *upwind* methods. Over the last few years, however, the TVD concept has been extended to *centred* schemes, as distinct from *upwind* schemes. This development has effectively provided Artificial Viscosity Methods with a more rational basis and has tended to unify the developments on numerical methods for conservation laws.

Relevant contributions to the general area of TVD and related methods can be found in the papers by Godunov [145]; van Leer [387]; Boris and Book [49]; Book, Boris and Hain [48]; Boris and Book [50]; Majda and Osher [226]; Zalesak [426]; Crandall and Majda [99]; Roe [282], [283], [284]; van Albada, van Leer and Roberts [386]; Harten [158], [159]; Sweby [333], [334]; Baines and Sweby [16]; Davis [102]; Osher[252]; Osher and Chakravarthy [254]; Jameson and Lax [181]; Yee, Warming and Harten [424]; Shu [311]; Yee [421], [422]; Tadmor [336]; Leonard [206]; Gaskell and Lau [130]; Swanson and Turkel [330]; Jorgenson and Turkel [187]; Cockburn and Shu [83]; Liu and Lax [223]; Billett and Toro [45]; Toro and Billett [374]; and many others.

In this chapter, after a brief review of some basic properties of schemes in Sect. 13.2, we present several approaches for constructing high–order methods. The emphasis is on approaches that extend to non–linear systems and can be suitably modified according to some TVD criteria; see Sects. 13.3 and 13.4. General theoretical aspects relevant to TVD methods are discussed in Sect. 13.5. In Sect. 13.6 we introduce TVD methods. We then present two approaches for constructing TVD methods, namely the *Flux Limiter Approach* of Sect. 13.7 and the *Slope Limiter Approach* of Sect. 13.8. For each of these two classes of TVD methods we present *upwind*–based schemes and *centred* schemes. Numerical results on model problems are included.

The reader is encouraged to review Chaps. 2 and 5 before proceeding with this chapter. Application of the TVD concepts to non–linear systems is found in Chaps. 14 and 16.

13.2 Basic Properties of Selected Schemes

The purpose of this section is to review a selected number of schemes and discuss some of their main properties. In the main, the discussion is centred on the linear advection equation

$$u_t + f(u)_x = 0 , \quad f(u) = au , \tag{13.1}$$

where $u = u(x,t)$ and a is a constant wave propagation speed.

13.2.1 Selected Schemes

In Chap. 5 we introduced a number of well known schemes to solve (13.1) numerically. Prominent examples are the Godunov first–order upwind method (the CIR scheme for linear systems)

$$\left. \begin{array}{l} u_i^{n+1} = u_i^n - c(u_i^n - u_{i-1}^n) , \quad a > 0 , \\ u_i^{n+1} = u_i^n - c(u_{i+1}^n - u_i^n) , \quad a < 0 , \end{array} \right\} \tag{13.2}$$

and the Lax–Wendroff method

$$u_i^{n+1} = \frac{1}{2}c(1+c)u_{i-1}^n + (1-c^2)u_i^n - \frac{1}{2}c(1-c)u_{i+1}^n . \tag{13.3}$$

Here c is the CFL or Courant number

$$c = \frac{a\Delta t}{\Delta x} . \tag{13.4}$$

For background on these well–known schemes see Sect. 5.2 of Chap. 5. A general form of these schemes is

$$u_i^{n+1} = \sum_{k=-k_L}^{k_R} b_k u_{i+k}^n , \tag{13.5}$$

where k_L and k_R are two non–negative integers. We assume that these schemes (13.5) are *linear*, that is all the coefficients b_k are constant. Any scheme (13.5) is entirely determined by the coefficients b_k. Table 13.1 lists the coefficients for several schemes. The first on the list is the Lax–Friedrichs method (LF); the second scheme is the First–Order Centred scheme (FORCE) introduced in Sect. 7.4.2 of Chap. 7; the third and fourth schemes are the two versions of the Godunov first–order upwind method (GODu). The fifth scheme is the Godunov first–order centred method (GODc) [146]. LW is the Lax–Wendroff method; WB is the Warming–Beam scheme and FR is the Fromm scheme.

Properties of the various schemes are stated in terms of the coefficients b_k. An important feature of a given scheme is its *support*; this is defined as the number $k_L + k_R + 1$ of data mesh point values u_i^n involved in the summation (13.5). One speaks of a *compact scheme* when its support is small. For example the support of the GODu scheme has only two points, whereas the support of the Fromm scheme has four points.

	b_{-2}	b_{-1}	b_0	b_1	b_2
LF	0	$\frac{1}{2}(1+c)$	0	$\frac{1}{2}(1-c)$	0
FORCE	0	$\frac{1}{4}(1+c)^2$	$\frac{1}{2}(1-c^2)$	$\frac{1}{4}(1-c)^2$	0
GODu $a>0$	0	c	$1-c$	0	0
GODu $a<0$	0	0	$1+c$	$-c$	0
GODc	0	$\frac{1}{2}c(1+2c)$	$1-2c^2$	$-\frac{1}{2}c(1-2c)$	0
LW	0	$\frac{1}{2}c(1+c)$	$1-c^2$	$-\frac{1}{2}c(1-c)$	0
WB, $a>0$	$\frac{1}{2}c(c-1)$	$(2-c)c$	$\frac{1}{2}(c-2)(c-1)$	0	0
WB, $a<0$	0	0	$\frac{1}{2}(c+2)(c+1)$	$-(c+2)c$	$\frac{1}{2}c(c+1)$
FR, $a>0$	$-\frac{1}{4}c(1-c)$	$\frac{1}{4}c(5-c)$	$\frac{1}{4}(1-c)(4+c)$	$-\frac{1}{4}c(1-c)$	0
FR, $a<0$	0	$\frac{1}{4}c(1+c)$	$\frac{1}{4}(1+c)(4-c)$	$-\frac{1}{4}c(5+c)$	$\frac{1}{4}c(1+c)$

Table 13.1. Coefficients b_k for several first and second order linear schemes

Another feature of a scheme is the *bias* of its support. Notice for instance that the support of the GODu scheme is biased towards the left of the centre point u_i^n or to the right, depending on the sign of the wave speed a in the conservation law (13.1). For $a > 0$ the support is biased to the left; if $a < 0$ the support is biased to the right. For this reason we say that the GODu scheme is *upwind*. By contrast the Lax–Friedrichs scheme has a *centred support*; there is always one point to the left and one point to the right of the centre of the stencil, regardless of the sign of a. Note however that the upwind coefficient of the Lax–Friedrichs scheme is always larger than the downwind coefficient; schemes whose support configuration does not depend on the sign of characteristic speeds are called *centred schemes*. Examples of centred schemes are Lax–Friedrichs, FORCE, Godunov's first–order centred method (GODc) and the Lax–Wendroff method.

Given a numerical method, there are four fundamental properties associated with it, namely consistency, stability, convergence and accuracy. A most useful reference in this regard is Hirsch, [170]. See also Smith [315], Mitchell and Griffiths [241] and Hoffmann [172].

13.2.2 Accuracy

The first five schemes listed in Table 13.1 are first–order accurate. The remaining schemes are second–order accurate. Here we quantify the accuracy of a scheme by the form of the leading term in its *local truncation error*. Appropriate definitions and techniques to analyse the accuracy of a scheme are found in any textbook on numerical methods for PDEs. See for example the references listed above.

Next we state a theorem due to Roe [282] that facilitates the verification of the accuracy of any scheme of the form (13.5).

Theorem 13.2.1 (Roe [282]). *A scheme of the form (13.5) is p–th (p $\geq$ 0) order accurate in space and time if and only if*

$$\sum_{k=-k_L}^{k_R} k^q b_k = (-c)^q , \quad 0 \leq q \leq p . \tag{13.6}$$

Proof. (Exercise)

Remark 13.2.1. For each integer value of q, with $0 \leq q \leq p$, we verify that the sum of terms $k^q b_k$, for all integers k with $-k_L \leq k \leq k_R$, reproduces identically the power $(-c)^q$, where c is the Courant number (13.4). The condition for $q = 0$ is simply the *Consistency Condition* of a scheme.

Example 13.2.1 (Godunov's upwind method). The GODu scheme for $a \geq 0$ has coefficients $b_{-1} = c$, $b_0 = 1 - c$. For $q = 0$ we have the summation

$$\sum_{k=-1}^{0} k^0 b_k = (-1)^0 \times c + 0^0 \times (1 - c) = 1 = (-c)^0 .$$

For $q = 1$

$$\sum_{k=-1}^{0} k^1 b_k = (-1)^1 \times c + 0^1 \times (1 - c) = -c = (-c)^1 .$$

Thus all conditions for first order accuracy are satisfied. Note that the GODu scheme is *not* second–order accurate, as for $q = 2$

$$\sum_{k=-1}^{0} k^2 b_k = (-1)^2 \times c + 0^2 \times (1 - c) = c \neq (-c)^2 .$$

Example 13.2.2 (The Lax–Wendroff method). Application of (13.6) with $q = 0, 1, 2$ gives

$$\sum_{k=-1}^{1} k^0 b_k = 1 = (-c)^0 \ , \quad \sum_{k=-1}^{1} k^1 b_k = -c = (-c)^1 \ , \quad \sum_{k=-1}^{1} k^2 b_k = (-c)^2 \ .$$

Therefore the Lax–Wendroff method is second–order accurate in space and time. This scheme is *not* third–order accurate, as

$$\sum_{k=-1}^{1} k^3 b_k = -c \neq (-c)^3 \ .$$

Exercise 13.2.1. Apply (13.6) to prove that the Lax–Friedrichs, FORCE and Godunov's first order centred (GODc) methods are first order accurate in space and time. Show also that the Warming–Beam and Fromm schemes are second order accurate in space and time.

Solution 13.2.1. (Left to the reader).

Exercise 13.2.2 (The Lax–Wendroff Scheme). Use (13.6) to show that any centred three–point support scheme of the form (13.5) to solve (13.1) that is second–order accurate in space and time must be identical to the Lax–Wendroff method.

Solution 13.2.2. (Left to the reader).

13.2.3 Stability

Background reading on stability analysis of numerical methods is found in any text book on numerical methods for differential equations. We particularly recommend Hirsch [170] Chaps. 7–10, Smith [315] Chaps. 3 and 4, Mitchell and Griffiths [241] Chap. 4 and Hoffmann [172] Chap. 4.

   Naturally, the only useful numerical methods are those that are *convergent*. Unfortunately, it is difficult or impossible to prove, theoretically, the convergence of a particular numerical method. For linear problems a most useful result is the *Lax Equivalence Theorem*. This states that *the only convergent schemes are those that are both consistent and stable*. There are various techniques available for studying stability. However, even for simple problems and schemes, the algebra involved may be cumbersome and simply not lead to analytical conditions on the parameters of the scheme to assume stability or otherwise. For non–linear problems there is a severe lack of theorems, and in practice one relies heavily on linear analysis and on numerical experimentation as guidance. A useful practice when dealing with non–linear problems is to compute the numerical solution on a sequence of meshes of decreasing mesh size and observe the behaviour of the numerical solutions as the mesh is refined. Exact solutions, even for simplified initial conditions, can also be very helpful here in assessing the performance of numerical methods.

For any centred three–point scheme of the form (13.5) we quote a useful result on stability derived for the model advection–diffusion equation [38]. Here we state the special case applicable to the model equation (13.1).

Proposition 13.2.1. *Any centred, three–point scheme*

$$u_i^{n+1} = b_{-1}u_{i-1}^n + b_0 u_i^n + b_1 u_{i+1}^n$$

for the model equation (13.1) is stable if the following two conditions on the coefficients b_k are met, namely

$$\left.\begin{array}{ll} (i) & b_0(b_{-1} + b_1) \geq 0 , \\[2mm] (ii) & b_0(b_{-1} + b_1) + 4b_{-1}b_1 \geq 0 . \end{array}\right\} \qquad (13.7)$$

Proof. (Exercise).

Example 13.2.3 (The Godunov first–order upwind method GODu). For $a > 0$, $b_{-1} = c$, $b_0 = 1 - c$, $b_1 = 0$. Thus (i) and (ii) become identical; they both give $c(1 - c) \geq 0$. But $a > 0$ and so $c > 0$. Thus $1 - c > 0$ and therefore $0 \leq c \leq 1$ is the stability condition for the GODu scheme when $a \geq 0$. For $a < 0$ we obtain the condition $-1 \leq c \leq 0$ and thus for both positive and negative speed a the stability condition for the GODu scheme becomes

$$0 \leq |c| \leq 1 . \qquad (13.8)$$

Exercise 13.2.3 (Stability Conditions). Apply (13.7) to prove that the stability condition of the Lax–Friedrichs and FORCE schemes is as given by (13.8) and that the stability condition for the Godunov first order centred method GODc is

$$0 \leq |c| \leq \frac{1}{2}\sqrt{2} . \qquad (13.9)$$

Solution 13.2.3. Left to the reader.

In the next two sections we study various ways of constructing high order numerical methods. The emphasis is on explicit, fully discrete schemes that are based on the solution of the Riemann problem and that can be extended to non–linear systems of conservation laws. We shall consider two approaches. One is the Weighted Average Flux approach and the other is the MUSCL or Variable Extrapolation approach.

13.3 WAF–Type High Order Schemes

The Weighted Average Flux (or WAF for short) approach is a generalisation of the Lax–Wendroff and the Godunov first–order upwind schemes to non–linear systems of conservation laws. It is also a generalisation of the Warming–Beam method. The WAF approach has its origins in the random flux scheme [350], which is second–order accurate in space and time in a statistical sense [378]. The WAF approach is *deterministic* and leads to fully discrete, explicit second order accurate schemes. See Refs. [352], [358], [357] [373], [43].

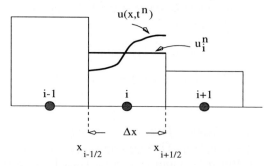

Fig. 13.1. Integral averages produce piece–wise constant data u_i^n at time level n

In this section we present the approach as applied to a general scalar conservation law

$$u_t + f(u)_x = 0 \,, \tag{13.10}$$

where $u = u(x,t)$ and $f(u)$ is a convex flux function, see Sect. 2.4 of Chap. 2. The scheme is based on the explicit conservative formula

$$u_i^{n+1} = u_i^n + \frac{\Delta t}{\Delta x} \left[f_{i-\frac{1}{2}} - f_{i+\frac{1}{2}} \right] \,. \tag{13.11}$$

Full details for the linear advection case (13.1) will be developed. Useful background reading is found in Sect. 5.3 of Chap. 5.

13.3.1 The Basic WAF Scheme

The WAF method assumes piece–wise constant data $\{u_i^n\}$, as in the Godunov first–order upwind method; that is, u_i^n is an integral average of the solution $u(x,t^n)$ within a cell $I_i = [x_{i-\frac{1}{2}}, x_{i+\frac{1}{2}}]$ at time $t = t^n$, namely

$$u_i^n = \frac{1}{\Delta x} \int_{x_{i-\frac{1}{2}}}^{x_{i+\frac{1}{2}}} u(x,t^n) dx \,. \tag{13.12}$$

Fig. 13.1 illustrates the cell average u_i^n in cell I_i of length $x_{i+\frac{1}{2}} - x_{i-\frac{1}{2}} = \Delta x$. In practice, the formality of defining u_i^n is not necessary. The initial data

$\{u_i^0\}$ at time $t = 0$ is regarded as cell averages and for all subsequent time levels these averages are produced automatically by formula (13.11). Thus, all we require is a definition of the numerical flux. In the original presentation

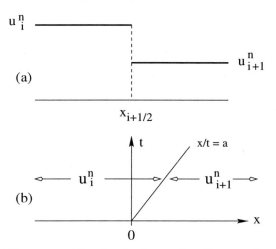

Fig. 13.2. The Riemann problem for the linear advection equation: (a) piece–wise constant initial data, (b) exact solution on x–t plane for $a > 0$

of the WAF method, the intercell flux was defined as an integral average of the flux function $f(u)$, namely

$$f_{i+\frac{1}{2}}^{waf} = \frac{1}{\Delta x} \int_{-\frac{1}{2}\Delta x}^{\frac{1}{2}\Delta x} f[u_{i+\frac{1}{2}}(x, \frac{1}{2}\Delta t)]dx \ . \tag{13.13}$$

Here, the integration range goes from the middle of cell I_i to the middle of cell I_{i+1}. The integrand is the physical flux function $f(u)$ in the conservation law (13.10) (linear or non–linear) evaluated at $u_{i+\frac{1}{2}}(x, \frac{\Delta t}{2})$, where $u_{i+\frac{1}{2}}(x, t)$ is the solution of the Riemann problem with *piece–wise constant data* (u_i^n, u_{i+1}^n). In defining (13.13) we have assumed a regular mesh of size Δx, but this definition can easily be altered to include the case of irregular meshes.

Now we develop, in full detail, the WAF scheme as applied to the linear advection equation (13.1). As seen in Sect. 2.2 of Chap. 2, the solution $u_{i+\frac{1}{2}}(x, t)$ of the Riemann problem for (13.1) is trivial and consists of the two data states u_i^n, u_{i+1}^n separated by a wave of speed a. Fig. 13.2(a) shows the initial data and Fig. 13.2(b) shows the solution $u_{i+\frac{1}{2}}(x, t)$ for $a > 0$. Algebraically

$$u_{i+\frac{1}{2}}(x, t) = \begin{cases} u_i^n \ , & \dfrac{x}{t} < a \ , \\ u_{i+1}^n \ , & \dfrac{x}{t} > a \ . \end{cases} \tag{13.14}$$

Obviously the two states present in the solution $u_{i+\frac{1}{2}}(x,t)$ are constant and thus the exact evaluation of the integral (13.13) is trivial, see Fig. 13.3. The integration domain is the line AC, which is subdivided into the segments AB and BC. Note that the lengths $|AB|$ and $|BC|$, when normalised by Δx, are respectively given by

$$\beta_1 = \frac{1}{2}(1+c) , \quad \beta_2 = \frac{1}{2}(1-c) . \tag{13.15}$$

The integral (13.13) becomes the summation

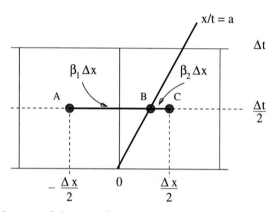

Fig. 13.3. Evaluation of the WAF flux for the linear advection equation, for $a > 0$. Weights normalised by distance Δx are β_1, β_2

$$f_{i+\frac{1}{2}}^{waf} = \frac{1}{\Delta x}\left[\frac{1}{2}(1+c)\Delta x(au_i^n)\right] + \frac{1}{\Delta x}\left[\frac{1}{2}(1-c)\Delta x(au_{i+1}^n)\right] ,$$

or

$$f_{i+\frac{1}{2}}^{waf} = \frac{1}{2}(1+c)(au_i^n) + \frac{1}{2}(1-c)(au_{i+1}^n) . \tag{13.16}$$

This numerical flux is identical to that of the Lax–Wendroff method in conservation form. For non–linear conservation laws (13.10) $f_{i+\frac{1}{2}}^{waf}$, as given by (13.13), becomes a Riemann–problem based generalisation of the Lax–Wendroff method, see Sect 5.3.4 of Chap. 5.

Note that for $a > 0$ the WAF flux as given by (13.16) is a *weighted average* of the upwind flux $f_i = au_i^n$, with weight $\beta_1 = \frac{1}{2}(1+c)$ and the downwind flux $f_{i+1} = au_{i+1}^n$ with weight $\beta_2 = \frac{1}{2}(1-c)$. For $a < 0$ the derivation is analogous and the result is in fact identical to (13.16), but now $\beta_1 = \frac{1}{2}(1+c)$ is the downwind weight and $\beta_2 = \frac{1}{2}(1-c)$ is the upwind weight. The upwind weight is always larger than the downwind weight and thus the WAF method is, in this sense, upwind biased. The WAF flux contains the Godunov flux in one of its terms, and it may thus be also regarded as an extension of the Godunov first order upwind method.

Exercise 13.3.1. Evaluate the WAF numerical flux (13.13) for the linear advection equation (13.1) for the case $a < 0$.

Solution 13.3.1. Left to the reader.

It is convenient to express intercell numerical fluxes for the linear advection equation (13.1) as a linear combination of data flux values $f_i = au_i^n$, namely

$$f_{i+\frac{1}{2}} = \sum_{k=-1}^{2} \beta_k (au_{i+k}^n) . \qquad (13.17)$$

Table 13.2 lists the coefficients β_k for a range of schemes.

	β_{-1}	β_0	β_1	β_2
LF	0	$\frac{1}{2c}(1+c)$	$-\frac{1}{2c}(1-c)$	0
FORCE	0	$\frac{1}{4c}(1+c)^2$	$-\frac{1}{4c}(1-c)^2$	0
GODu	0	$\frac{1}{2}(1+\text{sign}(c))$	$\frac{1}{2}(1-\text{sign}(c))$	0
GODc	0	$\frac{1}{2}(1+2c)$	$\frac{1}{2}(1-2c)$	0
LW	0	$\frac{1}{2}(1+c)$	$\frac{1}{2}(1-c)$	0
WB, $a > 0$	$-\frac{1}{2}(1-c)$	$\frac{1}{2}(3-c)$	0	0
WB, $a < 0$	0	0	$\frac{1}{2}(3+c)$	$-\frac{1}{2}(1+c)$
FR, $a > 0$	$-\frac{1}{4}(1-c)$	1	$\frac{1}{4}(1-c)$	0
FR, $a < 0$	0	$\frac{1}{4}(1+c)$	1	$-\frac{1}{4}(1+c)$
SLIC	$-\frac{1}{16}(1-c)$ $(1+c)^2$	$\frac{1}{4c}(1+c)^2$ $+\frac{1}{16c}(1+c)$ $(1-c)^2$	$\frac{1}{16c}(1-c)$ $(1+c)^2$ $+\frac{1}{4c}(1-c)^2$	$-\frac{1}{16c}(1+c)$ $(1-c)^2$

Table 13.2. Coefficients β_k in the expression for the intercell flux for various schemes. See also Table 13.1

Exercise 13.3.2. Evaluate the WAF flux, according to (13.13), for the inviscid Burgers equation, $f(u) = \frac{1}{2}u^2$ in (13.10). Recall that the Riemann problem at the interface $i + \frac{1}{2}$, see Sect. 2.4.2 of Chap. 2, has two cases, namely the shock case $u_i^n > u_{i+1}^n$ and the rarefaction case $u_i^n \leq u_{i+1}^n$.

Solution 13.3.2. Left to the reader.

13.3.2 Generalisations of the WAF Scheme

Generalisations of the basic WAF flux (13.13) may be carried out in a number of ways. The definition of the flux to include irregular meshes of size $(\Delta x)_i$ is obvious,

$$f_{i+\frac{1}{2}}^{waf} = \frac{1}{\frac{1}{2}[(\Delta x)_i + (\Delta x)_{i+1}]} \int_{-\frac{1}{2}(\Delta x)_i}^{\frac{1}{2}(\Delta x)_{i+1}} f[u_{i+\frac{1}{2}}(x, \frac{\Delta t}{2})]dx \qquad (13.18)$$

and definitions (13.13) and (13.18) are actually special cases of the more general formula

$$f_{i+\frac{1}{2}}^{waf} = \frac{1}{t_2 - t_1} \frac{1}{x_2 - x_1} \int_{t_1}^{t_2} \int_{x_1}^{x_2} f[\hat{u}_{i+\frac{1}{2}}(x, t)]dx dt . \qquad (13.19)$$

Here we integrate $f(\hat{u}_{i+\frac{1}{2}})$ in a box $[x_1, x_2] \times [t_1, t_2]$ in the x–t plane, where $\hat{u}_{i+\frac{1}{2}}(x, t)$ is the solution of some relevant initial value problem.

Let us consider the following conditions on the integral (13.19) as applied to the linear advection equation (13.1):

(i) $x_1 = -\frac{1}{2}\Delta x$, $x_2 = \frac{1}{2}\Delta x$,
(ii) $t_1 = 0$, $t_2 = \Delta t$,
(iii) the time integral in (13.19) is approximated by the mid–point rule in time,
(iv) $\hat{u}_{i+\frac{1}{2}} \equiv u_{i+\frac{1}{2}}(x, t)$ is the solution of the Riemann problem with data u_i^n, u_{i+1}^n,
(v) Δt is such that $|c| = \frac{\Delta t}{\Delta x}|a| < 1$.

If all these conditions are fulfilled, then (13.19) reduces identically to (13.13). If in condition (iii) above we replace the approximate integration by exact integration, then two second–order accurate methods result; for $c \leq \frac{1}{2}$ one obtains the Lax–Wendroff method; for $\frac{1}{2} \leq c \leq 1$ one obtains a second–order method which appears to be new.

Consider the case in which conditions (i)–(iv) are imposed but condition (v) is weakened so that the CFL number $|c|$ may be larger than unity. A consequence of relaxing (v) in this way is that $\hat{u}_{i+\frac{1}{2}}$ may become the solution of more than one piece–wise constant data Riemann problem. For the linear advection equation this results in an explicit method with arbitrarily large time step Δt, see Casulli and Toro [61]. Extensions of this large–time step scheme to non–linear problems is, however, very difficult. The particular case in which $|c| \leq 2$ is of interest and we call the resulting scheme WAF CFL2 [373]; this scheme is effectively a combined, Riemann–problem based extension of the Lax–Wendroff scheme and the Warming–Beam scheme [404], namely

$$f_{i+\frac{1}{2}}^{wafcfl2} = \begin{cases} f_{i+\frac{1}{2}}^{lw} = \frac{1}{2}(1+c)(au_i^n) + \frac{1}{2}(1-c)(au_{i+1}^n) , & |c| < 1 , \\ f_{i+\frac{1}{2}}^{wb} , & 1 \leq |c| \leq 2 , \end{cases} \qquad (13.20)$$

where $f_{i+\frac{1}{2}}^{wb}$ is the Warming–Beam numerical flux

$$f_{i+\frac{1}{2}}^{wb} = \begin{cases} -\frac{1}{2}(1-c)(au_{i-1}^n) + \frac{1}{2}(3-c)(au_i^n) , & \text{if } a > 0 , \\ \frac{1}{2}(3+c)(au_{i+1}^n) - \frac{1}{2}(1+c)(au_{i+2}) , & \text{if } a < 0 . \end{cases} \qquad (13.21)$$

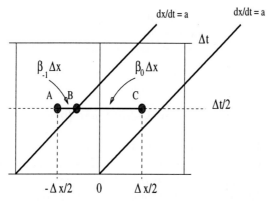

Fig. 13.4. Evaluation of the WAF flux for the linear advection equation, for $a > 0$ and $1 \le c \le 2$

Fig. 13.4 illustrates the flux evaluation for $a > 0$ and $1 \le c \le 2$. Now the normalised lengths $|AB|$ and $|BC|$ for $a > 0$ are $\beta_{-1} = -\frac{1}{2}(1 - c)$, $\beta_0 = \frac{1}{2}(3 - c)$. For $a < 0$ the weights are $\beta_1 = \frac{1}{2}(3 + c)$, $\beta_2 = -\frac{1}{2}(1 + c)$. See (13.17) and Table 13.2.

The WAF CFL2 scheme of Toro and Billett has been extended to non–linear systems, with a suitable TVD condition in [373], where it is applied to the time–dependent two dimensional Euler equations. A final remark on the WAF CFL2 scheme is this. If the integral (13.19), under the conditions following it, is evaluated by the mid–point rule in space and exact integration in time, then a first–order upwind method of *Courant number two* stability restriction is produced. This has numerical flux

$$
f^{tb}_{i+\frac{1}{2}} = \begin{cases} f^{GODu}_{i+\frac{1}{2}} = \frac{1}{2}(1 + \text{sign}(c))(au^n_i) \\ \qquad + \frac{1}{2}(1 - \text{sign}(c))(au^n_{i+1}) \, , & \text{if } |c| \le 1 \, , \\ \frac{(c-1)}{c}(au^n_{i-1}) + \frac{1}{c}(au^n_i) \, , & \text{if } a > 0 \, , 1 \le c \le 2 \, , \\ -\frac{1}{c}(au^n_{i+1}) + \frac{(c+1)}{c}(au^n_{i+2}) \, , & \text{if } a < 0 \, , 1 \le |c| \le 2 \, . \end{cases}
$$

$$(13.22)$$

Exercise 13.3.3. Substitute the above numerical fluxes (13.22) into the conservative formula (13.11) to derive the scheme in the form (13.5). Check the accuracy of the scheme using formula (13.6) and verify that the scheme produces the exact solution for $|c| = 1$ and $|c| = 2$.

Solution 13.3.3. Left to the reader.

A Total Variation Diminishing (TVD) modification of the WAF method is constructed in Sect. 13.7.1. In Chap. 14 we extend the WAF method to non–linear systems of conservations laws; in Chap. 15 we extend the scheme further to deal with source terms and in Chap. 16 we extend the scheme to deal with multi–dimensional problems.

In the following section we present another approach for deriving high–order numerical methods.

13.4 MUSCL–Type High–Order Methods

Van Leer, see [390], [391], [392], introduced the idea of *modifying* the piece–wise constant data (13.12) in the first–order Godunov method, as a first step to achieve higher order of accuracy. This approach has become known as the MUSCL or Variable Extrapolation approach, where MUSCL stands for Monotone Upstream–centred Scheme for Conservation Laws. The MUSCL approach is routinely used in practice today; it allows the construction of very high order methods, fully discrete, semi–discrete and also implicit methods. As we shall see later in this chapter, high–order (linear) schemes produce spurious oscillations. See numerical results of Chap. 5. The MUSCL approach implies (i) high–order of accuracy obtained by data reconstruction and (ii) the reconstruction is constrained so as to avoid spurious oscillations, and thus the justification of the word *monotone* in the name of the scheme. In this section we shall only be concerned with the first aspect of MUSCL, that is as a method for increasing the accuracy of first–order schemes.

13.4.1 Data Reconstruction

The simplest way of modifying the piecewise constant data $\{u_i^n\}$ is to replace the constant states u_i^n by piecewise linear functions $u_i(x)$. As for the first–order Godunov method, one assumes that u_i^n represents an integral average in cell $I_i = [x_{i-\frac{1}{2}}, x_{i+\frac{1}{2}}]$ as given by (13.12).

A piece–wise linear, *local* reconstruction of u_i^n is

$$u_i(x) = u_i^n + \frac{(x - x_i)}{\Delta x} \Delta_i , \quad x \in [0, \Delta x] , \tag{13.23}$$

where $\frac{\Delta_i}{\Delta x}$ is a suitably chosen *slope* of $u_i(x)$ in cell I_i. In what follows we call Δ_i a slope. Note that the linear function $u_i(x)$ is defined *locally* in cell I_i, that is $x \in [0, \Delta x]$. Fig. 13.5 illustrates the situation. The centre of the cell x_i in local co–ordinates is $x = \frac{1}{2}\Delta x$ and $u_i(x_i) = u_i^n$. The values of $u_i(x)$ at the extreme points play a fundamental role; they are given by

$$u_i^L = u_i(0) = u_i^n - \frac{1}{2}\Delta_i ; \quad u_i^R = u_i(\Delta x) = u_i^n + \frac{1}{2}\Delta_i \tag{13.24}$$

and are usually called *boundary extrapolated values*; hence the alternative name of *Variable Extrapolation Method*. Note that the integral of $u_i(x)$ in cell I_i is identical to that of u_i^n and thus the reconstruction process retains

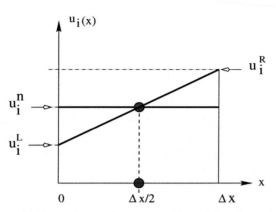

Fig. 13.5. Piece–wise linear MUSCL reconstruction of data in a single computing cell I_i; boundary extrapolated values are u_i^L, u_i^R

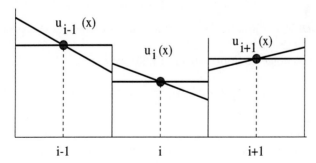

Fig. 13.6. Piece–wise linear MUSCL reconstruction for three successive computing cells I_{i-1}, I_i , I_{i+1}

conservation. Fig. 13.6 illustrates the piece–wise linear reconstruction process applied to three successive cells. As a consequence of having modified the data, at each interface $i + \frac{1}{2}$ one now may consider the so called *Generalised Riemann Problem* (or GRP)

$$
\left.
\begin{array}{l}
u_t + f(u)_x = 0 \, , \\[2mm]
u(x,0) = \left\{
\begin{array}{ll}
u_i(x) \, , & x < 0 \, , \\[2mm]
u_{i+1}(x) \, , & x > 0 \, ,
\end{array}
\right.
\end{array}
\right\}
\tag{13.25}
$$

to compute an intercell Godunov–type flux $f_{i+\frac{1}{2}}$. Fig. 13.7 illustrates the initial data and the structure of the solution of the Generalised Riemann problem. The solution no longer contains uniform regions as in the conventional Riemann problem in which the data is piece–wise constant; wave paths are now curved in x–t space.

Ben–Artzi and Falcovitz [23] were the first to develop a method based on the GRP. They applied their scheme to the Euler equations. Naturally, for non–linear systems the exact solution of the generalised Riemann problem is exceedingly complicated, but for the purpose of flux evaluation, approximate information may be obtained. Most approaches do in fact give up the solution of the generalised Riemann problem and rely instead on judicious use of the boundary extrapolated values u_i^L, u_i^R in (13.24), for each function $u_i(x)$ (13.23). In this way, one may instead consider the *piece–wise constant data Riemann problem*

$$
\left.
\begin{array}{l}
u_t + f(u)_x = 0 \, , \\[2mm]
u(x,0) = \left\{
\begin{array}{ll}
u_i^R \, , & x < 0 \, , \\[2mm]
u_{i+1}^L \, , & x > 0 \, ,
\end{array}
\right.
\end{array}
\right\}
\tag{13.26}
$$

which if used in conjunction with some other procedures can produce useful high–order methods.

As to the choice of slopes Δ_i in (13.23) we define

$$
\Delta_i = \frac{1}{2}(1 + \omega)\Delta u_{i-\frac{1}{2}} + \frac{1}{2}(1 - \omega)\Delta u_{i+\frac{1}{2}} \, ,
\tag{13.27}
$$

where as usual

$$
\Delta u_{i-\frac{1}{2}} \equiv u_i^n - u_{i-1}^n \, , \quad \Delta u_{i+\frac{1}{2}} \equiv u_{i+1}^n - u_i^n
\tag{13.28}
$$

and ω is a free parameter in the real interval $[-1, 1]$. For $\omega = 0$, Δ_i is a central–difference approximation, multiplied by Δx, to the first spatial derivative of the numerical solution at time level n.

MUSCL data reconstructions that are more accurate than the piece–wise linear reconstruction (13.23) are possible. A piece–wise quadratic reconstruction [170] is

$$u_i(x) = u_i^n + \frac{(x-x_i)}{\Delta x}\Delta_i^{(1)} + \frac{3\kappa}{2(\Delta x)^2}\left[(x-x_i)^2 - \frac{(\Delta x)^2}{12}\right]\Delta_i^{(2)}, \quad (13.29)$$

where $\Delta_i^{(1)} = \Delta_i$ as in (13.27), $\Delta_i^{(2)}$ is associated with an estimate for the second space derivative of $u_i(x)$ in cell I_i and κ is a parameter. For $\kappa = \frac{1}{3}$ this piece–wise quadratic reconstruction leads to third–order accurate *space* discretisations. The Piece–wise Parabolic Method (PPM) of Colella and Woodward [91], [413] is based on a piece–wise parabolic reconstruction of the data.

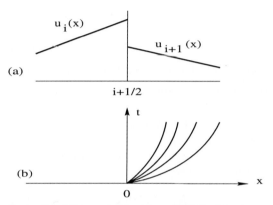

Fig. 13.7. The Generalised Riemann Problem (GRP): (a) piece–wise linear data (b) structure of the solution in x–t plane

The next sections present various methodologies based on MUSCL data reconstruction to derive high–order methods. Three of the approaches use, in addition, solutions of associated Riemann problems. We also present a high–order scheme that uses the MUSCL reconstruction step, but instead of using the Riemann problem solution it uses the First–Order Centred Scheme FORCE introduced in Sect. 7.4.2 Chap. 7, giving rise to a centred high–order method.

13.4.2 The MUSCL–Hancock Method (MHM)

Van Leer [395] attributes this method to S. Hancock, and we therefore adopt the name MUSCL–Hancock Method, or MHM, for this approach. MHM has three distinct steps to construct fully discrete second–order accurate schemes based on (13.11) to solve (13.10). These are:

Step (I)   Data reconstruction as in (13.23) with boundary extrapolated values as in (13.24), namely

$$u_i^L = u_i^n - \frac{1}{2}\Delta_i \; ; \quad u_i^R = u_i^n + \frac{1}{2}\Delta_i \;.$$

Step (II)   Evolution of u_i^L, u_i^R by a time $\frac{1}{2}\Delta t$ according to

$$\left. \begin{array}{l} \overline{u}_i^L = u_i^L + \dfrac{1}{2}\dfrac{\Delta t}{\Delta x}[f(u_i^L) - f(u_i^R)]\,, \\[2mm] \overline{u}_i^R = u_i^R + \dfrac{1}{2}\dfrac{\Delta t}{\Delta x}[f(u_i^L) - f(u_i^R)]\,. \end{array} \right\} \qquad (13.30)$$

Step (III) Solution of the piece–wise constant data Riemann problem

$$\left. \begin{array}{l} u_t + f(u)_x = 0\,, \\[3mm] u(x,0) = \left\{ \begin{array}{ll} \overline{u}_i^R\,, & x < 0\,, \\[2mm] \overline{u}_{i+1}^L\,, & x > 0\,, \end{array} \right. \end{array} \right\} \qquad (13.31)$$

to find the *similarity* solution $u_{i+\frac{1}{2}}(x/t)$. Fig. 13.8 illustrates steps (I) and (II) at the intercell boundary position $i+\frac{1}{2}$; the boundary extrapolated values u_i^R, u_{i+1}^L are evolved to $\overline{u}_i^R$, $\overline{u}_{i+1}^L$. These form the *piece–wise constant data* for a conventional Riemann problem at the cell interface $i + \frac{1}{2}$ with solution $u_{i+\frac{1}{2}}(x/t)$. The intercell numerical flux $f_{i+\frac{1}{2}}$ is then obtained in exactly the same way as in the Godunov first–order upwind method, see Sect. 5.3.2 of Chaps. 5 and Chap. 6, namely

$$f_{i+\frac{1}{2}} = f(u_{i+\frac{1}{2}}(0))\,. \qquad (13.32)$$

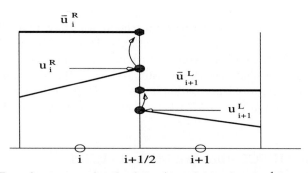

Fig. 13.8. Boundary extrapolated values. At each interface $i+\frac{1}{2}$ boundary extrapolated values u_i^R, u_{i+1}^L are evolved to $\overline{u}_i^R$, $\overline{u}_{i+1}^L$, to form the piece–wise constant data for a conventional Riemann problem at the intercell boundary

Now we develop the details of the scheme for the linear advection equation (13.1). Step (I) gives the boundary extrapolated values (13.24). In step (II) we need the fluxes $f(u_i^L), f(u_i^R)$. As for the linear advection equation $f(u) = au$, then we have

$$f(u_i^L) = a[u_i^n - \frac{1}{2}\Delta_i] \; ; \quad f(u_i^R) = a[u_i^n + \frac{1}{2}\Delta_i] \; .$$

Substitution of these into (13.30) gives the evolved boundary extrapolated values in cell I_i, namely

$$\overline{u}_i^L = u_i^n - \frac{1}{2}(1+c)\Delta_i \; ; \quad \overline{u}_i^R = u_i^n + \frac{1}{2}(1-c)\Delta_i \; . \tag{13.33}$$

In step (III) one solves the conventional Riemann problem at the interface $i + \frac{1}{2}$ with data

$$(\overline{u}_i^R, \overline{u}_{i+1}^L)$$

the solution of which is

$$u_{i+\frac{1}{2}}(x/t) = \begin{cases} \overline{u}_i^R = u_i^n + \frac{1}{2}(1-c)\Delta_i \; , & \text{if } x/t < a \; , \\[2mm] \overline{u}_{i+1}^L = u_{i+1}^n - \frac{1}{2}(1+c)\Delta_{i+1} \; , & \text{if } x/t > a \; . \end{cases} \tag{13.34}$$

The intercell flux is now as in the first–order Godunov method (13.32). The result is

$$f_{i+\frac{1}{2}}^{mhm} = \frac{1}{2}(1 + sign(c))f(\overline{u}_i^R) + \frac{1}{2}(1 - sign(c))f(\overline{u}_{i+1}^L) \; , \tag{13.35}$$

where

$$f(\overline{u}_i^R) = a[u_i^n + \frac{1}{2}(1-c)\Delta_i] \; ; \quad f(\overline{u}_{i+1}^L) = a[u_{i+1}^n - \frac{1}{2}(1+c)\Delta_{i+1}] \; . \tag{13.36}$$

Recall that $sign(x) = 1$ if $x > 0$ and $sign(x) = -1$ if $x < 0$, for any real number x.

When expressed as in (13.17) the MHM flux for the linear advection equation (13.1) has coefficients β_k that depend on the Courant number c and the parameter ω.

Exercise 13.4.1. Verify that the coefficients β_k in (13.35), for $a > 0$, are

$$\beta_{-1} = -\frac{1}{4}(1-c)(1+\omega) \; ; \beta_0 = 1 + \frac{1}{2}(1-c)\omega \; ; \beta_1 = \frac{1}{4}(1-c)(1-\omega) \; ; \beta_2 = 0 \tag{13.37}$$

and that for $a < 0$ they are

$$\beta_{-1} = 0 \; ; \beta_0 = \frac{1}{4}(1+c)(1+\omega) \; ; \beta_1 = 1 - \frac{1}{2}(1+c)\omega \; ; \beta_2 = -\frac{1}{4}(1+c)(1-\omega) \; . \tag{13.38}$$

Solution 13.4.1. Left to the reader.

Exercise 13.4.2. Verify that if the parameter ω in the slopes (13.27) is $\omega = 0$, then MHM reproduces the Fromm scheme, see Tables 13.1 and 13.2. Assume $a > 0$ and verify that for $\omega = 1$ the scheme is identical to the Warming–Beam scheme and that for $\omega = -1$ the scheme is identical to the Lax–Wendroff method; see Tables 13.1 and 13.2.

Solution 13.4.2. Left to the reader.

Exercise 13.4.3. Apply (13.6) to show that for any value of ω in the slopes Δ_i in (13.27) MHM is second–order accurate in space and time, for both $a < 0$ and $a > 0$.

Solution 13.4.3. Left to the reader.

Exercise 13.4.4. Apply (13.6) to show that MHM is third–order accurate in space and time if the parameter ω in (13.27) is chosen to be

$$\omega = \frac{1}{3}(2c - sign(c)) . \tag{13.39}$$

Solution 13.4.4. Left to the reader.

A Total Variation Diminishing (TVD) version of the MHM is constructed in Sect. 13.8. Extensions to non–linear systems in one space dimension are given in Chap. 14. Application of the method to systems involving source terms is discussed in Chap. 15 and extension to multidimensional problems is given in Chap. 16.

13.4.3 The Piece–Wise Linear Method (PLM)

Colella [87] proposed the Piece–wise Linear Method, or PLM, to obtain second–order accurate generalisations of the Godunov first–order upwind method. PLM is also based on the MUSCL data reconstruction step discussed in Sect. 13.4.1. For any conservation law (13.10) as solved by (13.11), PLM defines the numerical flux $f_{i+\frac{1}{2}}$ as

$$f_{i+\frac{1}{2}}^{PLM} = \frac{1}{2}[f(u_{i+\frac{1}{2}}^{grp}(0,0)) + f(u_{i+\frac{1}{2}}^{grp}(0,\Delta t))] , \tag{13.40}$$

where $u_{i+\frac{1}{2}}^{grp}(x,t)$ is the solution of the Generalised Riemann problem (13.25). See Fig. 13.7. PLM does not solve the Generalised Riemann problem directly; instead, it uses a trapezium rule approximation in time to the integral

$$f_{i+\frac{1}{2}} = \frac{1}{\Delta t} \int_0^{\Delta t} f[u_{i+\frac{1}{2}}^{grp}(0,t)]dt , \tag{13.41}$$

where $u_{i+\frac{1}{2}}^{grp}(0,t)$ is the solution of the Generalised Riemann Problem along the cell interface position (t–axis). In order to determine the two terms in the PLM flux (13.40), one proceeds as follows

Step (I)   Data reconstruction in each cell I_i as in (13.23) with slopes as in (13.27) and computation of the boundary extrapolated values u_i^L, u_i^R, for each cell I_i.

Step (II)   The first term is found by solving the conventional Riemann problem

$$u_t + f(u)_x = 0 \, ,$$

$$u(x,0) = \left\{ \begin{array}{ll} u_i^R \, , & x < 0 \, , \\ u_{i+1}^L \, , & x > 0 \, , \end{array} \right\} \tag{13.42}$$

with piece–wise constant data u_i^R, u_{i+1}^L. As $u_{i+\frac{1}{2}}^{grp}(0,0)$ is solely determined by the interaction of the extreme data values u_i^R, u_{i+1}^L in the Generalised Riemann Problem (13.25), the solution $u_{i+\frac{1}{2}}(x/t)$ of (13.42) produces the sought result, namely $u_{i+\frac{1}{2}}^{grp}(0,0) = u_{i+\frac{1}{2}}(0)$.

Step (III)   The second term in (13.40) depends on $u_{i+\frac{1}{2}}^{grp}(0, \Delta t)$, which is found by tracing characteristics from the point $(0, \Delta t)$ back to the *reconstructed piece–wise linear data* (13.23); Fig. 13.9 illustrates this step for the linear advection equation (13.1). Then $u_{i+\frac{1}{2}}^{grp}(0, \Delta t) = u_k(\hat{x})$, where k is either i or $i + 1$ and $\hat{x}$ is the point of intersection of the characteristic with the x–axis.

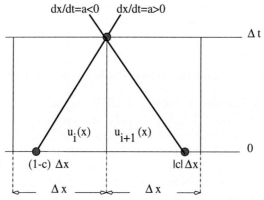

Fig. 13.9. Tracing characteristic $dx/dt = a$ back to the reconstructed data u_i to find the constant value at the interface position $i + \frac{1}{2}$ at time Δt

We now work out the details for the linear advection equation (13.1). Step (I) produces the extrapolated values (13.24). Step (II) gives the solution of (13.42), which, when evaluated along the t–axis gives

$$u_{i+\frac{1}{2}}(0) = \left\{ \begin{array}{llll} u_i^R & = & u_i^n + \frac{1}{2}\Delta_i \, , & \text{if } a > 0 \, , \\ u_{i+1}^L & = & u_{i+1}^n - \frac{1}{2}\Delta_{i+1} \, , & \text{if } a < 0 \, , \end{array} \right.$$

which in turn leads to the first term in (13.40). In step (III), one traces the characteristic $dx/dt = a$ from the point $(0, \Delta t)$ back to cell I_i if $a > 0$ or to cell I_{i+1} if $a < 0$, to find the intersection point $\hat{x}$; this is

$$\hat{x} = \begin{cases} c\Delta x & , \text{ if } a \leq 0 , \\ (1-c)\Delta x & , \text{ if } a \geq 0 . \end{cases}$$

See Fig. 13.9. For $a > 0$, evaluation of $u_i(x)$ at $\hat{x} = (1-c)\Delta x$ in (13.23) gives

$$u_i[(1-c)\Delta x] = u_i^n + (\frac{1}{2} - c)\Delta_i .$$

As this value is constant along the characteristic we have $u_{i+\frac{1}{2}}^{grp}(0, \Delta t) = u_i^n + (\frac{1}{2} - c)\Delta_i$, which gives the second term for the PLM flux (13.40).

Exercise 13.4.5. Verify that for both $a > 0$ and $a < 0$ the PLM flux is

$$f_{i+\frac{1}{2}}^{plm} = \begin{cases} a[u_i^n + \frac{1}{2}(1-c)\Delta_i] , & a > 0 , \\ a[u_{i+1}^n - \frac{1}{2}(1+c)\Delta_{i+1}] , & a < 0 . \end{cases} \tag{13.43}$$

Solution 13.4.5. Left to the reader.

The PLM intercell flux is identical to the MHM flux (13.35)–(13.36), for the linear advection equation and leads to the same numerical fluxes (13.17) with coefficients (13.37)–(13.38), with third–order accuracy for the parameter value given by (13.39). See Colella [87] for details on the extension of PLM to non–linear systems. A simplified version of the PLM to construct conservative and non–conservative methods is found in [368].

13.4.4 The Generalised Riemann Problem (GRP) Method

The Generalised Riemann Problem, or GRP, method was first put forward by Ben–Artzi and Falcovitz [23], [24], [22], who applied it to the Euler equations. Here we present the scheme as applied to the scalar conservation law (13.10), as solved by (13.11). The basic ingredient of the GRP method is the solution of the Generalised Riemann Problem (13.25) to obtain a Godunov–type numerical flux that yields a second–order accurate scheme. The GRP method defines a numerical flux as

$$f_{i+\frac{1}{2}}^{grp} = f(u_{i+\frac{1}{2}}^{grp}(0, \frac{1}{2}\Delta t)) , \tag{13.44}$$

where $u_{i+\frac{1}{2}}^{grp}(x, t)$ is the solution of the Generalised Riemann Problem (13.25) and $u_{i+\frac{1}{2}}^{grp}(0, \frac{1}{2}\Delta t)$ is the mid–point rule approximation in time to the integral (13.41). As a matter of fact, a further approximation to $u_{i+\frac{1}{2}}^{grp}(0, \frac{1}{2}\Delta t)$ is required to obtain the numerical flux (13.44). The GRP scheme then has the following steps

Step (I)   Data reconstruction in each cell I_i as in (13.23) with slopes as in (13.27).

Step (II)   Taylor expansion of $u_{i+\frac{1}{2}}^{grp}(x,t)$ about $(0,0)$ to obtain an approximation to $u_{i+\frac{1}{2}}^{grp}(0,\frac{1}{2}\Delta t)$, namely

$$u_{i+\frac{1}{2}}^{grp}\left(0,\frac{1}{2}\Delta t\right) = u_{i+\frac{1}{2}}^{grp}(0,0) + \frac{1}{2}\Delta t \frac{\partial}{\partial t}u_{i+\frac{1}{2}}^{grp}(0,0) , \qquad (13.45)$$

with error terms $O(\Delta t^2)$.

Step (III)   Determination of the two terms on the right hand side of (13.45).

The first term $u_{i+\frac{1}{2}}^{grp}(0,0)$ in the expansion (13.45) is the value of $u_{i+\frac{1}{2}}^{grp}(x,t)$ immediately after the interaction of the piece–wise linear states $u_i(x), u_{i+1}(x)$ in the GRP (13.25). This value is solely determined by the extrapolated values (13.24). Thus $u_{i+\frac{1}{2}}^{grp}(0,0)$ is the solution of the piece–wise constant data Riemann problem (13.42) evaluated along the t–axis, namely

$$u_{i+\frac{1}{2}}^{grp}(0) = \begin{cases} u_i^R & = & u_i^n + \frac{1}{2}\Delta_i , & \text{if } a > 0 , \\ u_{i+1}^L & = & u_{i+1}^n - \frac{1}{2}\Delta_{i+1} , & \text{if } a < 0 . \end{cases} \qquad (13.46)$$

This is identical to the step leading to the first term in the PLM flux (13.40). To determine the second term in (13.45) there are various possibilities.

Here we suggest a modification of the original GRP method whereby the time derivative in (13.45) is replaced by a space derivative. The required value along the cell interface results from solving an extra Riemann problem for gradients. For the linear advection equation (13.1) the modification is based on the following result

Proposition 13.4.1. *For any p–th order spatial derivative $v = \partial^p u/\partial x^p$, where $u(x,t)$ is a solution of the linear advection equation (13.1), we have*

$$v_t + av_x = 0 .$$

Proof. (Left as an exercise).

The proposition states that any spatial gradient of $u(x,t)$, if defined, obeys the original PDE. Hence one may pose Riemann problems for gradients $\partial^p u/\partial x^p$, which if assumed piece–wise constant, lead to conventional Riemann problems.

For the linear advection equation (13.1) we have $u_t = -au_x$ and thus the time derivative in the second term of (13.45) can be replaced by a space derivative. This in turn, by virtue of the above proposition, can be found by solving the gradient $(p=1)$ Riemann problem

$$v_t + av_x = 0 ,$$

$$\left. v(x,0) \equiv u_x(x,0) = \begin{cases} \dfrac{\Delta_i}{\Delta x} , & x < 0 , \\[2mm] \dfrac{\Delta_{i+1}}{\Delta x} , & x > 0 . \end{cases} \right\} \qquad (13.47)$$

The solution is

$$
v(x,t) \equiv u_x(x,t) =
\begin{cases}
\dfrac{\Delta_i}{\Delta x}, & \dfrac{x}{t} < a, \\[2ex]
\dfrac{\Delta_{i+1}}{\Delta x}, & \dfrac{x}{t} > a.
\end{cases}
\tag{13.48}
$$

Therefore, the GRP flux for the linear advection equation (13.1) is

$$
f_{i+\frac{1}{2}}^{GRP} =
\begin{cases}
a[u_i^n + \frac{1}{2}(1-c)\Delta_i], & a > 0, \\[2ex]
a[u_{i+1}^n - \frac{1}{2}(1+c)\Delta_{i+1}], & a < 0.
\end{cases}
\tag{13.49}
$$

The GRP flux is identical to those of the MUSCL Hancock Method (13.35) and of Colella's PLM scheme (13.43). Therefore all the schemes of the form (13.5) with coefficients (13.37)–(13.38) are reproduced for any value of the slope parameter ω in (13.27). Third order accuracy is achieved by the special value (13.39).

Remark 13.4.1. The modification of the GRP method to include solutions of Riemann problems for spatial derivatives has potential for constructing higher–order GRP–type schemes. Some preliminary results on a third–order scheme were reported by former MSc student Cáceres [58]. The approach also has potential for constructing schemes for convection–diffusion PDEs, where a viscous flux component involves space derivatives. Some preliminary results were reported by former MSc student Cheney [67], who derived schemes with enlarged stability regions.

One extension of the GRP scheme to the Euler equations is found in the original paper of Ben–Artzi and Falcovitz [23]. Hillier [169] has successfully applied the GRP scheme to study shock diffraction problems; see also [24] and [22]. The modified GRP scheme is applied in [368] to construct non–conservative and conservative schemes for non–linear systems.

13.4.5 Slope–Limiter Centred (SLIC) Schemes

Recall that one aspect of the MUSCL approach allows the construction of high–order extensions of the Godunov first–order upwind method via reconstruction of the data. Here we extend the approach for constructing high–order versions of any low–order scheme with numerical flux

$$
f_{i+\frac{1}{2}}^{LO} = f_{i+\frac{1}{2}}^{LO}(u_L, u_R)
$$

that depends on the two states u_L and u_R on the left and right of the interface respectively. For the linear advection equation (13.1), Table 13.2 shows four possible choices for the low–order flux, including the Godunov first–order upwind method. In this section we are interested in low–order schemes that avoid the solution of the Riemann problem, and thus possible choices for the low–order scheme are the Godunov first–order centred, the Lax–Friedrichs and the FORCE schemes, all of which extend to non–linear systems. As we are fundamentally interested in oscillation–free versions of these high–order schemes, we construct Slope–Limiter versions of these centred schemes; we therefore adopt the name of SLIC methods [374]. We describe the SLIC schemes in terms of the scalar conservation law (13.10) as solved by the explicit conservative formula (13.11). The schemes have three steps, the first two being identical to those of the MUSCL–Hancock method of Sect. 13.4.2. The steps are

Step (I)   Data reconstruction as in (13.23) with boundary extrapolated values as in (13.24), namely

$$u_i^L = u_i^n - \frac{1}{2}\Delta_i \; ; \quad u_i^R = u_i^n + \frac{1}{2}\Delta_i \; . \tag{13.50}$$

Step (II)   Evolution of u_i^L, u_i^R by a time $\frac{1}{2}\Delta t$ according to

$$\left. \begin{aligned} \bar{u}_i^L &= u_i^L + \frac{1}{2}\frac{\Delta t}{\Delta x}[f(u_i^L) - f(u_i^R)] \; , \\ \bar{u}_i^R &= u_i^R + \frac{1}{2}\frac{\Delta t}{\Delta x}[f(u_i^L) - f(u_i^R)] \; . \end{aligned} \right\} \tag{13.51}$$

Step (III) Now, instead of solving the Riemann problem with data $(\bar{u}_i^R, \bar{u}_{i+1}^L)$ to find the Godunov first–order upwind flux, we compute a low–order flux with data arguments $(\bar{u}_i^R, \bar{u}_{i+1}^L)$ and thus we have

$$f_{i+\frac{1}{2}}^{slic} = f_{i+\frac{1}{2}}^{LO}(\bar{u}_i^R, \bar{u}_{i+1}^L) \; . \tag{13.52}$$

One possible choice for the low–order flux is the FORCE flux. Recall that for data (u_L, u_R) at the cell interface $i + \frac{1}{2}$ the FORCE flux, see Sect. 7.4.2 of Chap. 7, is

$$f_{i+\frac{1}{2}}^{force}(u_L, u_R) = \frac{1}{2}[f_{i+\frac{1}{2}}^{RI}(u_L, u_R) + f_{i+\frac{1}{2}}^{LF}(u_L, u_R)] \; , \tag{13.53}$$

where $f_{i+\frac{1}{2}}^{RI}$ is the Richtmyer flux and $f_{i+\frac{1}{2}}^{LF}$ is the Lax–Friedrichs flux.

We consider a general low–order flux

$$f_{i+\frac{1}{2}}^{LO}(u_L, u_R) = \beta_0(au_L) + \beta_1(au_R) \tag{13.54}$$

for the linear advection equation in which the flux function is $f(u) = au$. See Table 13.2 for examples of low–order fluxes. Application of (13.54) in (13.52) gives the high–order flux

$$f_{i+\frac{1}{2}}^{slic} = \beta_0 a[u_i^n + \frac{1}{2}(1-c)\Delta_i] + \beta_1 a[u_{i+1}^n - \frac{1}{2}(1+c)\Delta_{i+1}] \; . \tag{13.55}$$

Compare this with the MUSCL–Hancock flux (13.35). Direct substitution of these fluxes into (13.11) gives a five–point support *centred* scheme based on the MUSCL approach, namely

$$u_i^{n+1} = \sum_{k=-2}^{2} b_k u_{i+k}^n \qquad (13.56)$$

with coefficients

$$
\begin{aligned}
b_{-2} &= -\tfrac{1}{4}\beta_0 c(1-c)(1+\omega)\,, \\[4pt]
b_{-1} &= \beta_0 c(1+\tfrac{1}{4}(1-c)(1+3\omega)) + \tfrac{1}{4}\beta_1 c(1+c)(1+\omega)\,, \\[4pt]
b_0 &= 1 - \beta_0 c(1 - \tfrac{1}{4}(1-c)(1-3\omega)) + \beta_1 c(1 - \tfrac{1}{4}(1+c)(1+3\omega))\,, \\[4pt]
b_1 &= -\beta_1 c(1 + \tfrac{1}{4}(1+c)(1-3\omega)) - \tfrac{1}{4}\beta_0 c(1-c)(1-\omega)\,, \\[4pt]
b_2 &= \tfrac{1}{4}\beta_1 c(1+c)(1-\omega)\,.
\end{aligned}
$$

$$(13.57)$$

Theorem 13.4.1. *Schemes (13.56)–(13.57) are second order accurate in space and time for any value ω and for any coefficients β_0, β_1 that give a consistent low–order flux (13.54). Provided β_0 and β_1 also give a first order flux, schemes (13.56)–(13.57) are third order accurate in space and time when*

$$\omega = \frac{3c(\beta_0 - \beta_1) - 2c^2 - 1}{3(\beta_0 - \beta_1 - c)}\,. \qquad (13.58)$$

Proof. Conventional truncation error analysis gives the desired result. A more direct alternative is to apply Roe's theorem (13.2.1), [282], formula (13.16).

Remark 13.4.2. Note that (13.58) always gives a well–defined ω under the conditions of the theorem because the denominator is only zero when

$$\beta_0 = \frac{1}{2}(1+c) \quad \text{and} \quad \beta_1 = \frac{1}{2}(1-c)\,,$$

which are the coefficients for the second order Lax–Wendroff flux.

Exercise 13.4.6. Consider schemes (13.56)–(13.57). Verify that the following statements are true.

1. If the first order scheme is the Godunov upwind scheme, third order accuracy is achieved for

$$\omega = \frac{1}{3}(2c - \text{sign}(c))\,.$$

2. If the first order scheme is the Godunov centred scheme, third order accuracy is achieved for

$$\omega = \frac{4c^2 - 1}{3c}.$$

3. If the first order scheme is the FORCE scheme, third order accuracy is achieved for

$$\omega = \frac{1}{3}c.$$

4. If the first order scheme is the Lax–Friedrichs scheme, third order accuracy is achieved for

$$\omega = \frac{2c}{3}.$$

5. If the coefficients β_0, β_1 are chosen as

$$\beta_0 = \frac{1}{6}\left(\frac{c+1}{c-\omega}\right)(1 + 2c - 3\omega) ; \quad \beta_1 = \frac{1}{6}\left(\frac{c-1}{c-\omega}\right)(1 - 2c + 3\omega) ,$$

then the scheme (13.56)–(13.57) is third order accurate $\forall\omega$.

Solution 13.4.6. (Left to the reader).

A Total Variation Diminishing (TVD) version of these centred schemes will be constructed in Sect. 13.8. In Chap. 14 we extend the schemes to non–linear systems of conservation laws.

13.4.6 Other Approaches

There are two more approaches for constructing numerical methods, which are closely related to the explicit, fully discrete schemes studied previously; these are the semi–discrete methods and implicit methods. Both families of methods admit first–order schemes as special cases. Here we give a brief description.

13.4.7 Semi–Discrete Schemes

Consider the scalar conservation law (13.10). In the semi–discrete approach, or *method of lines*, one separates the space and time discretisation processes. First one assumes some discretisation in space, while leaving the problem continuous in time. This results in an Ordinary Differential Equation (ODE) in time, namely

$$\frac{du}{dt} = \frac{1}{\Delta x}(f_{i-\frac{1}{2}} - f_{i+\frac{1}{2}}) . \tag{13.59}$$

Here

$$f_{i+\frac{1}{2}} = f_{i+\frac{1}{2}}(\{u_i(t)\}, t) \tag{13.60}$$

is an intercell numerical flux, an approximation to the true flux $f(u(x_{i+\frac{1}{2}}, t))$ at the interface $i + \frac{1}{2}$ at time t.

A possible choice for $f_{i+\frac{1}{2}}$ in (13.59) is the Godunov first–order upwind flux. A second–order space discretisation may be obtained by first applying the MUSCL reconstruction step (13.23) followed by the solution of the Riemann problem (13.42), where the initial data are the right and left boundary extrapolated values (13.24). Higher–order reconstructions lead to corresponding high–order space discretisations.

The time discretisation in (13.10) results from solving the ODE (13.59). There are many techniques for solving ODEs numerically; see Refs. [199] and [170]. The simplest method for solving Ordinary Differential Equations is the first–order Euler method, which if applied to (13.59) leads to the explicit conservative formula (13.11). Runge–Kutta methods are particularly popular for solving ODEs, in the context of the semi–discrete approach. See Jameson et. al. [182] and Gottlieb and Shu [151]. Related works based on the method of lines are those of Berzins [36], [37].

The separation of the time and space discretisation processes in the semi–discrete approach allows enormous flexibility and is well suited for deriving very high–order schemes, such as the families of UNO and ENO schemes. See [165], [162], [160], [161].

13.4.8 Implicit Methods

Natural extensions of the second–order fully discrete schemes based on (13.11) to solve (13.10) are conservative schemes

$$u_i^{n+1} = u_i^n + \frac{\Delta t}{\Delta x}(f_{i-\frac{1}{2}} - f_{i+\frac{1}{2}}) , \tag{13.61}$$

where the intercell numerical flux $f_{i+\frac{1}{2}}$ depends on both data values $\{u_i^n\}$ and unknown values $\{u_i^{n+1}\}$, that is

$$f_{i+\frac{1}{2}} = f_{i+\frac{1}{2}}(u_{i-r+1}^n, \cdots, u_{i+r}^n; u_{i-s+1}^{n+1}, \cdots, u_{i+s}^{n+1}) , \tag{13.62}$$

with r and s two non–negative integers.

See Harten [159] for approaches for constructing Total Variation Diminishing (TVD) implicit schemes. A most useful reference on implicit TVD methods is the VKI Lecture Notes by Helen Yee [422]; see also [424].

Implicit methods involve the solution of systems of algebraic equations at each time step, which is an expensive process on both CPU time and storage. The advantage of implicit schemes over explicit schemes is that the choice of the time step Δt is, at least in theory, not restricted by stability considerations. Implicit schemes are particularly well suited for solving steady state problems by marching in time. Steady Aerodynamics is perhaps the most prominent area of application of implicit schemes [182], [424].

Several ways of constructing high–order methods have been presented. While high order of accuracy is a very desirable feature of numerical methods, we shall see that, if unmodified, these schemes produce *spurious oscillations* in the vicinity of high gradients of the solution, see Fig. 13.10. The next three sections are devoted to study ways of eliminating the unwanted spurious oscillations while preserving high order of accuracy in the smooth parts of the solution.

13.5 Monotone Schemes and Accuracy

Here we study the class of monotone methods and their relation to accuracy. We prove the important theorem due to Godunov that asserts that *monotone, linear schemes are at most first–order accurate*.

13.5.1 Monotone Schemes

A useful class of methods for non–linear scalar conservation laws (13.10) are those which are *monotone*. These were defined in Chap. 5 but we recall the definition here.

Definition 13.5.1 (Monotone Schemes). *A scheme*

$$u_i^{n+1} = H(u_{i-k_L+1}^n , \cdots , u_{i+k_R}^n) , \tag{13.63}$$

with k_L and k_R two non–negative integers, is said to be monotone if

$$\frac{\partial H}{\partial u_j^n} \geq 0 \; \forall j . \tag{13.64}$$

That is, H is a non–decreasing function of each of its arguments.

This definition of a monotone scheme is actually equivalent to the following property:

$$\text{if } v_i^n \geq u_i^n \; \forall i \text{ then } v_i^{n+1} \geq u_i^{n+1} . \tag{13.65}$$

This property in turn is the discrete version of the following property of the exact solution of the conservation law (13.10): if two initial data functions $v_0(x)$ and $u_0(x)$ for (13.10) satisfy $v_0(x) \geq u_0(x) \; \forall x$, then their corresponding solutions $v(x,t)$ and $u(x,t)$ satisfy $v(x,t) \geq u(x,t)$, $t > 0$. Hence *monotone schemes mimic a basic property of exact solutions of conservation laws* (13.10). See [158] and [207].

Next we state a useful property of monotone schemes when applied to the scalar non–linear conservation law (13.10).

Theorem 13.5.1. *Given the data set $\{u_i^n\}$, if the solution set $\{u_i^{n+1}\}$ is obtained with a monotone method (13.63) then*

$$\max_i\{u_i^{n+1}\} \leq \max_i\{u_i^n\} ; \quad \min_i\{u_i^{n+1}\} \geq \min_i\{u_i^n\} . \tag{13.66}$$

Proof. Define $v_i^n = \max_j\{u_j^n\} = \text{constant} \quad \forall i$. Then application of (13.63) gives $v_i^{n+1} = v_i^n$. As $v_i^n \geq u_i^n$, application of (13.65) gives $v_i^{n+1} = v_i^n \geq u_i^{n+1}$ and therefore $\max_i\{u_i^{n+1}\} \leq \max_i\{u_i^n\}$. The second inequality in (13.66) follows analogously.

An obvious consequence of (13.66) is

$$\left.\begin{array}{ccccc}\max_i\{u^n\} & \leq & \max_i\{u_i^{n-1}\} & \leq \cdots \leq & \max_i\{u_i^0\}\,, \\ \min_i\{u^n\} & \geq & \min_i\{u_i^{n-1}\} & \geq \cdots \geq & \min_i\{u_i^0\}\,. \end{array}\right\} \qquad (13.67)$$

This result says that *no new extrema are created* and thus spurious *oscillations* do not appear; see Fig. 13.10.

In numerical solutions computed with monotone methods, minima increase and maxima decrease as time evolves. This results in *clipping of extrema*, which is in fact a disadvantage of monotone methods. See Figs. 5.10 and 5.11 of Chap. 5, which show results for two monotone methods, namely the Lax–Friedrichs method and the Godunov first–order upwind method.

Another very useful consequence of (13.66) is that solutions of monotone schemes satisfy

$$\min_j\{u_j^n\} \leq u_i^{n+1} \leq \max_j\{u_j^n\}\,. \qquad (13.68)$$

This follows from noting that

$$\min_j\{u_j^{n+1}\} \leq u_i^{n+1} \leq \max_j\{u_j^{n+1}\}$$

and direct application of (13.66). Condition (13.68) says that the solution at any point i is bounded by the minimum and maximum of the data.

The following result applies to all three–point schemes for non–linear scalar conservation laws (13.10).

Theorem 13.5.2 (Monotonicity and the Flux). *A three point scheme of the form*

$$u_i^{n+1} = u_i^n + \frac{\Delta t}{\Delta x}[f_{i-\frac{1}{2}} - f_{i+\frac{1}{2}}] \qquad (13.69)$$

for the non–linear conservation law (13.10) is monotone if

$$\frac{\partial}{\partial u_i^n} f_{i+\frac{1}{2}}(u_i^n, u_{i+1}^n) \geq 0 \ \text{ and } \ \frac{\partial}{\partial u_{i+1}^n} f_{i+\frac{1}{2}}(u_i^n, u_{i+1}^n) \leq 0\,. \qquad (13.70)$$

That is, the numerical flux $f_{i+\frac{1}{2}}(u_i^n, u_{i+1}^n)$ is an increasing (meaning non–decreasing) function of its first argument and a decreasing (meaning non–increasing) function of its second argument.

Proof. In (13.69) we define

$$H(u_{i-1}^n, u_i^n, u_{i+1}^n) \equiv u_i^n + \frac{\Delta t}{\Delta x}[f_{i-\frac{1}{2}}(u_{i-1}^n, u_i^n) - f_{i+\frac{1}{2}}(u_i^n, u_{i+1}^n)]\,.$$

The requirement $\dfrac{\partial H}{\partial u_{i-1}^n} \geq 0$ implies $\dfrac{\partial f_{i-\frac{1}{2}}}{\partial u_{i-1}^n}(u_{i-1}^n, u_i^n) \geq 0$, while the require-

ment $\dfrac{\partial H}{\partial u_{i+1}^n} \geq 0$ implies $\dfrac{\partial f_{i+\frac{1}{2}}}{\partial u_{i+1}^n}(u_i^n, u_{i+1}^n) \leq 0$, and the result follows.

Example 13.5.1 (The Lax–Friedrichs Scheme). For a general non–linear conservation law (13.10) the Lax–Friedrichs flux is

$$f_{i+\frac{1}{2}}^{LF}(u_i^n, u_{i+1}^n) = \frac{1}{2}[f(u_i^n) + f(u_{i+1}^n)] + \frac{1}{2}\frac{\Delta x}{\Delta t}(u_i^n - u_{i+1}^n) \, .$$

Application of conditions (13.70) to the Lax–Friedrichs flux shows that monotonicity is ensured provided

$$-1 \leq \frac{\Delta t \lambda_{max}(u)}{\Delta x} \leq 1 \, ,$$

where $\lambda(u) = \partial f/\partial u$ is the characteristic speed and $\lambda_{max}(u)$ is the maximum. That is, provided the CFL stability condition is enforced properly,
the Lax–Friedrichs method is monotone, when applied to non–linear scalar
conservation laws (13.10).

Exercise 13.5.1. Derive conditions for the FORCE flux

$$f_{i+\frac{1}{2}}^{force} = \frac{1}{2}[f_{i+\frac{1}{2}}^{LF} + f_{i+\frac{1}{2}}^{RI}] \, ,$$

where $f_{i+\frac{1}{2}}^{RI}$ is the Richtmyer flux, see Sect. 7.4.2 of Chap. 7, to give a monotone scheme, when applied to the inviscid Burgers equation

$$u_t + f(u)_x = 0 \, , \quad f(u) = \frac{1}{2}u^2 \, .$$

Solution 13.5.1. Left to the reader.

Next we state a result for linear schemes (13.5) to solve the linear advection equation (13.1).

Theorem 13.5.3 (Positivity of Coefficients). *A scheme*

$$u_i^{n+1} = \sum_{k=-k_L}^{k_R} b_k u_{i+k}^n \tag{13.71}$$

*for the linear advection equation (13.1) is monotone if and only if all coeffi-
cients b_k are non–negative, that is $b_k \geq 0 \ \forall k$.*

Proof. By defining

$$H \equiv \sum_{k=-k_L}^{k_R} b_k u_{i+k}^n$$

and using (13.70) the result $b_k \geq 0$ follows immediately.

Example 13.5.2. Note that the first–order upwind scheme (see Table 13.1) is always monotone, whereas the Lax–Wendroff method is not monotone. As a matter of fact, none of the second–order accurate schemes, whose coefficients b_k are listed in Table 13.1, is monotone.

Exercise 13.5.2. Show that the Godunov first–order centred method, see Table 13.1, *is non–monotone* for the range of Courant numbers $0 \leq |c| \leq \frac{1}{2}$ and is monotone for the range $\frac{1}{2} \leq |c| \leq \frac{1}{2}\sqrt{2}$.

Solution 13.5.2. Left to the reader.

13.5.2 A Motivating Example

Suppose we want to solve the linear advection equation (13.1) with a conservative scheme (13.11) in conjunction with a WAF–type flux

$$f_{i+\frac{1}{2}}^{\alpha} = \frac{1}{\Delta x} \int_{-\frac{1}{2}\Delta x}^{\frac{1}{2}\Delta x} f[u_{i+\frac{1}{2}}(x, \alpha\Delta t)]dx \;, \tag{13.72}$$

where $\alpha \geq 0$ is a dimensionless parameter and $u_{i+\frac{1}{2}}(x, t)$ is the solution of the Riemann problem with data (u_i^n, u_{i+1}^n). Compare (13.72) with (13.13) and (13.19). In fact $\alpha = \frac{1}{2}$ reproduces the WAF flux (13.16) that leads to the Lax–Wendroff method; see Sect. 13.3.1. Evaluation of (13.72), see Fig. 13.3, for the linear advection equation gives

$$f_{i+\frac{1}{2}}^{\alpha} = \frac{1}{2}(1 + 2\alpha c)(au_i^n) + \frac{1}{2}(1 - 2\alpha c)(au_{i+1}^n) \;. \tag{13.73}$$

Appropriate choices for α give familiar schemes. For instance

$$\alpha = \begin{cases} \dfrac{1}{2} & : & \text{Lax–Wendroff method;} \\[2mm] \dfrac{1}{2|c|} & : & \text{Godunov first–order upwind method;} \\[2mm] 1 & : & \text{Godunov first–order centred method;} \\[2mm] \dfrac{1}{4c^2}(1 + c^2) & : & \text{FORCE scheme;} \\[2mm] \dfrac{1}{2c^2} & : & \text{Lax–Friedrichs method.} \end{cases} \tag{13.74}$$

Substitution of (13.73) into (13.11) gives

$$u_i^{n+1} = b_{-1}u_{i-1}^n + b_0 u_i^n + b_1 u_{i+1}^n \;, \tag{13.75}$$

with coefficients

$$b_{-1} = \frac{1}{2}(1 + 2\alpha c)c \;; \quad b_0 = 1 - 2\alpha c^2 \;; \quad b_1 = -\frac{1}{2}(1 - 2\alpha c)c \;. \tag{13.76}$$

A stability analysis of the scheme (13.75)–(13.76) using (13.7) shows that the scheme is stable for the following values of the parameter α:

$$\frac{1}{2} \le \alpha \le \frac{1}{2c^2} . \tag{13.77}$$

The flux (13.73) is a weighted average of the upwind and downwind fluxes, with α controlling the relative size of their contributions. For $\alpha > \frac{1}{2}$ the upwind flux contribution is increased and the downwind contribution is decreased. The minimum upwind contribution consistent with a stable scheme is $\alpha = \frac{1}{2}$, the Lax–Wendroff method. Also note that the upper limit for stability is $\alpha = \frac{1}{2c^2}$; that is, the maximum upwind contribution results in the Lax–Friedrichs method. Any attempt at increasing the upwind contribution further results in an unstable method.

An accuracy analysis of (13.75)–(13.76) using (13.6) says that the scheme is second–order accurate in space and time for the single value $\alpha = \frac{1}{2}$ (the Lax–Wendroff method), and for any value of α with $\frac{1}{2} < \alpha \le \frac{1}{2c^2}$ the scheme is first–order accurate. As soon as the upwind contribution in the Lax–Wendroff scheme is slightly increased, the accuracy of the scheme drops to first order.

According to Theorem (13.5.3), monotonicity of the scheme (13.75)–(13.76) holds if the coefficients (13.76) are non–negative.

Exercise 13.5.3 (Monotonicity). Verify that scheme (13.75)–(13.76) is monotone for values of α in the range

$$\frac{1}{2|c|} \le \alpha \le \frac{1}{2c^2} . \tag{13.78}$$

Solution 13.5.3. Left to the reader.

An important observation follows from (13.77) and (13.78): for the range of values

$$\frac{1}{2} < \alpha \le \frac{1}{2|c|} \tag{13.79}$$

of the parameter α the scheme (13.75) is stable, first–order accurate, but *not monotone*.

Remark 13.5.1. There are infinitely many first–order accurate schemes that are stable but not monotone. In (13.75) we have one example for each value of α in the range given by (13.79).

Exercise 13.5.4. On the c–α plane, identify the regions of stability, instability and monotonicity of scheme (13.75)–(13.76). Note that the Godunov first–order centred scheme is stable for $0 \le |c| \le \frac{1}{2}\sqrt{2}$ and is monotone only in the range $\frac{1}{2} \le |c| \le \frac{1}{2}\sqrt{2}$.

Solution 13.5.4. Left to the reader.

Fig. 13.10 shows numerical results (symbols) compared with the exact solution (line), for the linear advection equation (13.1) with a square wave as initial condition; see Test 2, Sect. 5.5.1 of Chap. 5. The computational domain is $[0, 10]$ which is discretised with $M = 1000$ cells; the CFL number used is $c = \frac{1}{4}$ and the output time is $t = 1.0$. The schemes used result from the conservative formula (13.11) and the numerical flux (13.73), with six fixed values for the parameter α; see legend to figure. A relevant observation to make is that the second–order Lax–Wendroff method (a) produces high resolution of the discontinuities (narrow transition zone) but also produces spurious oscillations in their vicinity; the result labelled (b) is some first–order accurate method resulting from the choice $\alpha = \frac{3}{4}$ and the result confirms the fact that the scheme is not monotone, as predicted by (13.79). This first–order accurate scheme is not monotone, as remarked earlier. The result labelled (c) corresponds to the Godunov first–order centred scheme, which for the CFL number used is not monotone, in agreement with condition (13.79). The result (d) is oscillation–free and corresponds to a scheme with a larger value of α ($\alpha = 2$) so that the Godunov first–order upwind method is reproduced. Larger values of α produce other schemes; (e) corresponds to the First–Order Centred (FORCE) scheme ($\alpha = 4.25$) and (f) corresponds to the Lax–Friedrichs method ($\alpha = 8$), both giving oscillation–free results.

Remark 13.5.2. Monotonicity and Accuracy. Our empirical observation based on the numerical results of Chap. 5 and those of Fig. 13.10 is that schemes that are second order accurate produce spurious oscillations, and schemes that are monotone are inaccurate. The observation is in fact true in general, for fixed coefficient (linear) schemes, as we shall prove in the next section. A possible way forward is to vary the parameter α so as to maintain accuracy in smooth parts of the solution and monotonicity near high gradients. This would require α to be a function of the data set $\{u_i^n\}$, which would then lead to non–linear schemes; see Sect. 13.7.1.

13.5.3 Monotone Schemes and Godunov's Theorem

Our empirical observations so far reveal that *linear* second–order accurate schemes (13.5) to solve the linear advection equation (13.1) are distinctly better than first–order methods for problems with smooth solutions; see computational results of Figs. 5.9 to 5.11 of Chap. 5. However, for solutions involving high gradients, such as near a discontinuity, these methods produce spurious oscillations; see Figs. 5.12 to 5.14 of Chap. 5 and Fig. 13.10. On the other hand, monotone methods avoid spurious oscillations near high gradients; see results of Figs. 5.12 to 5.14 of Chap. 5. Their disadvantage though, seems to be their limited accuracy, which is clearly exposed in the results of Figs. 5.9 to 5.11 of Chap. 5.

Godunov's theorem establishes, theoretically, that the desirable properties of accuracy and monotonicity are, for linear schemes, contradictory require-

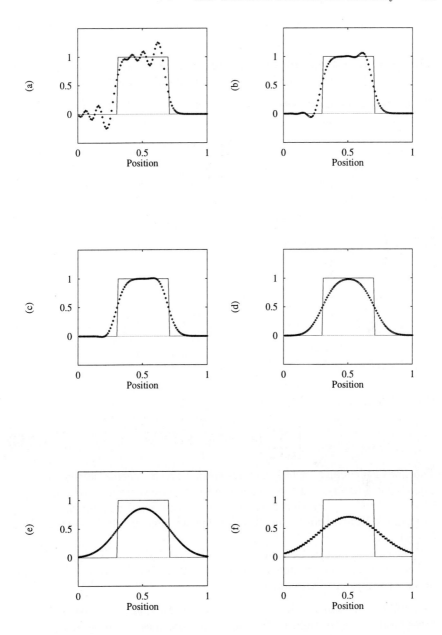

Fig. 13.10. Numerical results at time $t = 1$ for six numerical schemes contained in the motivating example of Sect. 13.5.2; symbols denote numerical solution and lines denote the exact solution; CFL number used is $c = \frac{1}{4}$. (a): $\alpha = \frac{1}{2}$ (Lax–Wendroff); (b): $\alpha = \frac{3}{4}$ (Some first–order non–monotone method); (c): $\alpha = 1$ (Godunov first–order centred, non–monotone); (d): $\alpha = 2$ (Godunov first–order upwind); (e): $\alpha = 4.25$ (First–Order Centred, FORCE); (f): $\alpha = 8$ (Lax–Friedrichs)

ments. The following result applies to the linear advection equation (13.1) and linear schemes (13.5).

Theorem 13.5.4 (Godunov's Theorem). *There are no monotone, linear schemes (13.5) for (13.1) of second or higher order of accuracy.*

Proof. The proof given here is based on the accuracy relation (13.6). Denote by s_q the summation

$$s_q = \sum_{k=-k_L}^{k_R} k^q b_k \,, \tag{13.80}$$

where b_k are the coefficients (constant) of the (linear) scheme (13.5). For second–order accuracy one requires

$$s_0 = 1 \,, \quad s_1 = -c \,, \quad s_2 = c^2 \,. \tag{13.81}$$

From definition (13.80)

$$
\begin{aligned}
s_2 &= \sum_{k=-k_L}^{k_R} k^2 b_k \\
&= \sum_{k=-k_L}^{k_R} (k+c)^2 b_k - 2c \sum_{k=-k_L}^{k_R} k b_k - c^2 \sum_{k=-k_L}^{k_R} b_k \\
&= \left[\sum_{k=-k_L}^{k_R} (k+c)^2 b_k\right] - 2c s_1 - c^2 s_0 \,.
\end{aligned}
\tag{13.82}
$$

Use of (13.81) into (13.82) gives

$$\left[\sum_{k=-k_L}^{k_R} (k+c)^2 b_k\right] + c^2 \geq c^2 \,. \tag{13.83}$$

The above inequality holds, as a monotone linear scheme satisfies $b_k \geq 0$. Equality in (13.83), and thus second order accuracy, is only possible if $b_k = 0 \ \forall k$ or when $c = -k_0$, that is for integer Courant numbers, and $b_k = 0 \ \forall k \neq k_0$. Thus we have proved the theorem for schemes satisfying the condition $0 \leq |c| \leq 1$.

The case of integer Courant numbers larger that unity is only of theoretical interest, as for non–linear systems this is an impossible requirement to impose on numerical methods.

Remark 13.5.3. Consequences of Godunov's Theorem. Another way to express Godunov's theorem is that monotone schemes are at most first–order accurate. First–order methods are too inaccurate to be of practical interest. We therefore must search for other classes of schemes that, ideally, allow for both the oscillation–free property of monotone schemes and the accuracy of high order methods to coexist. In other words we must look for ways of circumventing Godunov's theorem. The key to this lies on the assumption made in the theorem that the schemes have fixed coefficients (linear schemes).

13.5.4 Spurious Oscillations and High Resolution

The problem of spurious oscillations in the vicinity of high gradients is depicted in the sketch of Fig. 13.11, where the full line denotes the exact solution and the dotted line denotes the numerical solution obtained by some linear method of second or higher order of accuracy. Different methods will produce different patterns for the oscillatory profile. The Lax–Wendroff method will produce spurious oscillations *behind* the wave, whereas the Warming–Beam method will produce spurious oscillations *ahead* of the wave; see Fig. 5.12 of Chap. 5. This is related to the form of the leading term in the local truncation error of the method.

The data u_i^n and the exact solution are monotone decreasing functions but the numerical solution is *not*; new local extrema have been introduced, thus violating a fundamental property of solutions of scalar conservation laws. See Sect. 13.6.1 and the statements preceding (13.94). See also conditions (13.94).

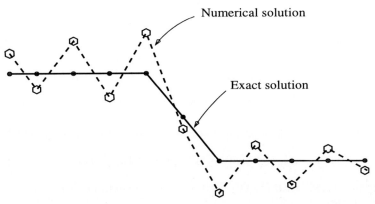

Fig. 13.11. Illustration of the numerical phenomenon of spurious oscillations near high gradients.

The aim of the rest of this chapter is to provide the necessary theoretical background and construct numerical methods that avoid the problem of spurious oscillations, sometimes called the Gibb phenomena, while retaining accuracy. In fact we aim at constructing numerical methods that have the following properties

- (i) The schemes have second or higher order of accuracy in smooth parts of the solution
- (ii) The schemes produce numerical solutions free from spurious oscillations
- (iii) The schemes produce *high–resolution* of discontinuities, that is the number of mesh points in the transition zone containing the numerical wave is narrow in comparison with that of first–order monotone methods.

Schemes satisfying the above properties are called *high–resolution methods*, after Harten [158].

13.5.5 Data Compatibility

One of the first reported attempts at providing a rational approach to circumventing Godunov's theorem is due to Roe [282]. The central idea is to construct adaptive algorithms that would adjust themselves to the local nature of the solution. This results, obviously, in variable coefficient (non–linear) schemes, even when applied to linear PDEs. The coefficients must be functions of the data.

   The following definition relates an algorithm for the linear equation (13.1) with particular classes of data.

Definition 13.5.2 (Data Compatible Algorithm). *A scheme is compatible with a given data set $\{u_i^n\}$ if the solution u_i^{n+1} at each point i, as given by the algorithm, is bounded by the* upwind pair (u_{i-s}^n, u_i^n), *where $s \equiv sign(c) = sign(a)$.*

Remark 13.5.4. For a given point i the upwind pair is (u_{i-1}^n, u_i^n) for $c > 0$ and (u_i^n, u_{i+1}^n) for $c < 0$. The characteristic through $(i, n+1)$ of the PDE (13.1) intersects the real line at a point between the upwind pair.

   We may therefore rephrase the above definition as follows: a scheme is compatible with the data if u_i^{n+1} lies between u_{i-s}^n and u_i^n. The data compatibility requirement of an algorithm is conveniently expressed as follows:

$$\min\{u_{i-s}^n, u_i^n\} \leq u_i^{n+1} \leq \max\{u_{i-s}^n, u_i^n\} \,. \tag{13.84}$$

Compare (13.84) with (13.68) derived from Theorem (13.5.1).

Proposition 13.5.1 (Data Compatibility). *The data compatibility condition (13.84) is equivalent to requiring*

$$0 \leq \frac{u_i^{n+1} - u_i^n}{u_{i-s}^n - u_i^n} \leq 1 \,. \tag{13.85}$$

Proof. First assume $u_{i-s}^n \leq u_i^n$. Imposing (13.84) gives $u_{i-s}^n \leq u_i^{n+1} \leq u_i^n$, that is $u_{i-s}^n - u_i^n \leq u_i^{n+1} - u_i^n \leq 0$. Division through by $u_{i-s}^n - u_i^n < 0$ gives (13.85). The case $u_{i-s}^n \geq u_i^n$ follows similarly.

Example 13.5.3. The flux for the Godunov first–order upwind method to solve (13.1) is

$$f_{i+\frac{1}{2}} = \frac{1}{2}(1 + s)(au_i^n) + \frac{1}{2}(1 - s)(au_{i+1}^n) \,.$$

Substitution of $f_{i+\frac{1}{2}}$ and $f_{i-\frac{1}{2}}$ into (13.11) gives

$$u_i^{n+1} = u_i^n + sc(u_{i-s}^n - u_i^n) \,. \tag{13.86}$$

The data compatibility condition (13.85) for (13.86) is

$$0 \leq \frac{u_i^{n+1} - u_i^n}{u_{i-s}^n - u_i^n} = s|c| \leq 1 \,,$$

which is *always* satisfied. That is, the Godunov first–order upwind method is compatible with all possible sets of data. This is not a property enjoyed by the Lax–Wendroff method, for instance.

Next we consider data compatibility conditions for any three point scheme

$$u_i^{n+1} = b_{-1} u_{i-1}^n + b_0 u_i^n + b_1 u_{i+1}^n \tag{13.87}$$

applied to the linear advection equation (13.1). We first note that (13.87) can be written in *incremental form* as

$$u_i^{n+1} = u_i^n - b_{-1} \Delta u_{i-\frac{1}{2}} + b_1 \Delta u_{i+\frac{1}{2}} \,, \tag{13.88}$$

where as usual $\Delta u_{i-\frac{1}{2}} = u_i^n - u_{i-1}^n$ and $\Delta u_{i+\frac{1}{2}} = u_{i+1}^n - u_i^n$.

It is easily verified that any three–point scheme (13.87) or (13.88) for (13.1) has data compatibility conditions

$$\begin{aligned} 0 &\leq b_{-1} - b_1/r_i \leq 1 \,, && \text{if } c > 0 \,, \\ 0 &\leq b_1 - b_{-1} r_i \leq 1 \,, && \text{if } c < 0 \,, \end{aligned} \right\} \tag{13.89}$$

where

$$r_i = \frac{\Delta u_{i-\frac{1}{2}}}{\Delta u_{i+\frac{1}{2}}} \,. \tag{13.90}$$

Example 13.5.4 (The Lax–Wendroff Method). Manipulations of the previous result (13.89), (13.90) show that the Lax–Wendroff method is *not* compatible with all possible sets of data, it is compatible only with data satisfying

$$\begin{aligned} -\tfrac{1+c}{1-c} &\leq 1/r_i \leq \tfrac{c+2}{c} \,; && \text{if } 0 \leq c \leq 1 \,, \\ -\tfrac{1-c}{1+c} &\leq r_i \leq \tfrac{c-2}{c} \,; && \text{if } -1 \leq c \leq 0 \,. \end{aligned} \right\}$$

Example 13.5.5 (The Warming and Beam Method). Roe [282] showed that the Warming–Beam Method is *not* compatible with all possible sets of data, it is compatible only with data satisfying

$$\begin{aligned} \tfrac{c-2}{c} &\leq r_i \leq \tfrac{c-3}{c-1} \,; && \text{if } 0 \leq c \leq 1 \,, \\ \tfrac{c}{c-2} &\leq 1/r_i \leq \tfrac{c+1}{c-1} \,; && \text{if } 1 \leq c \leq 2 \end{aligned} \right\}$$

and

$$\begin{aligned} \tfrac{c+2}{c} &\leq 1/r_i \leq \tfrac{c+3}{c+1} \,; && \text{if } -1 \leq c \leq 0 \,, \\ \tfrac{c+2}{c} &\leq r_i \leq \tfrac{c-1}{c} \,; && \text{if } -2 \leq c \leq -1 \,. \end{aligned} \right\}$$

Roe [282] also observed that given the set S_{DC} of all possible data states one can select combinations of schemes such that the subsets for which the schemes satisfy the data compatibility requirement do cover the full set S_{DC}; in this way he constructed *adaptive, non–linear schemes that are monotone and second order accurate*, thus circumventing Godunov's Theorem.

More recent applications of the data compatibility concept impose directly the data compatibility requirement (13.85) on schemes that depend on some free parameter to allow for variable coefficients. In Sect. 13.7.1 we illustrate this approach to modify the scheme (13.11) with flux (13.73) so as to avoid spurious oscillations and maintain second–order accuracy in smooth parts of the solution.

13.6 Total Variation Diminishing (TVD) Methods

The real issue concerning numerical methods is convergence. For non–linear systems convergence proofs rely on *non–linear stability*, the theory of which relies on functional analysis concepts, such as *compactness*. Sets of functions whose total variation is bounded lead to compact sets. Total Variation Stable methods are then defined as those whose mesh–dependent approximations lie in compact function sets. It can then be proved that Total Variation Stable methods are convergent. See Harten [158] and LeVeque [207] for details.

A subclass of Total Variation Stable methods are those whose total variation does not increase in time; these are commonly referred to as *Total Variation Diminishing* (TVD) methods, or Total Variation Non–Increasing (TVNI) methods. See Harten [158], [159] and Sweby [333]. Here we are interested in studying and designing TVD methods.

13.6.1 The Total Variation

Given a function $u = u(x)$, the total variation of u is defined as

$$TV(u) = \lim_{\delta \to 0} \sup \frac{1}{\delta} \int_{-\infty}^{\infty} |u(x+\delta) - u(x)| dx \ . \tag{13.91}$$

See Apostol [12] for definitions of supremum (sup) and related concepts of Real Analysis. If $u(x)$ is smooth then (13.91) is identical to

$$TV(u) = \int_{-\infty}^{\infty} |u'(x)| dx \ . \tag{13.92}$$

In fact, even for discontinuous $u(x)$, (13.92) is still correct in the sense of Distribution Theory, see Smoller [316], Chap. 7. If $u = u(x,t)$ then one can generalise definitions (13.91) or (13.92), but in fact for convergence purposes, it suffices to define the total variation of $u(x,t)$ at fixed times $t = t^n$, which we denote by $TV(u(t))$. Moreover, if $u^n = \{u_i^n\}$ is a *mesh function*, then the total variation of u^n is defined as

$$TV(u^n) = \sum_{i=-\infty}^{\infty} |u_{i+1}^n - u_i^n| . \tag{13.93}$$

Obviously, in order for $TV(u^n)$ to be finite, one must assume $u_i^n = 0$ or $u_i^n = $ constant as $i \to \pm\infty$. For a moment, let us just consider the results of Fig. 13.10 again, where computed results at time $t = 1$ are shown. Clearly, for the Lax–Wendroff result (a) the total variation of the numerical solution is larger than that of the exact solution; highly oscillatory solutions have large total variation. Of the numerical solutions shown, that of the Lax–Friedrichs method (f) has the smallest total variation. It is easy to see that as time evolves the total variation of the Lax–Friedrichs solution will decrease; the total variation of the exact solution of the linear equation (13.1) will remain constant.

A fundamental property of the exact solution of the IVP for the non–linear scalar conservation law (13.10), when the initial data $u(x,0)$ has bounded total variation, is

- (i) No new local extrema in x may be created
- (ii) The value of a local minimum increases (it does not decrease) and the value of a local maximum decreases (it does not increase).

From this it follows that the Total Variation $TV(u(t))$ is a decreasing function of time (non–increasing); that is,

$$TV(u(t_2)) \leq TV(u(t_1)) \; \forall t_2 \geq t_1 . \tag{13.94}$$

See Harten [158] and Lax [203]. This property of the exact solution is the one that we want to mimic when designing numerical methods.

13.6.2 TVD and Monotonicity Preserving Schemes

Consider a numerical scheme of the form

$$u_i^{n+1} = H(u_{i-r+1}^n, \cdots, u_{i+s}^n) , \tag{13.95}$$

with r and s two non–negative integers, to solve the scalar conservation law (13.10). Motivated by property (13.94) we introduce the following definition

Definition 13.6.1 (TVD Schemes). *Scheme (13.95) is said to be a Total Variation Diminishing (TVD) scheme, or Total Variation Non–Increasing (TVNI) scheme, if*

$$TV(u^{n+1}) \leq TV(u^n) , \quad \forall n . \tag{13.96}$$

An obvious consequence of the above definition is that

$$TV(u^n) \leq TV(u^{n-1}) \leq \cdots \leq TV(u^0) \,, \tag{13.97}$$

where $\{u_i^0\}$ is data at time $t = 0$. Next we define another class of numerical methods.

Definition 13.6.2. *Schemes of the form (13.95) for the scalar, nonlinear conservation law (13.10) are said to be Monotonicity Preserving Schemes if whenever the data $\{u_i^n\}$ is monotone the solution set $\{u_i^{n+1}\}$ is monotone in the same sense. That is, if $\{u_i^n\}$ is monotone increasing so is $\{u_i^{n+1}\}$ and if $\{u_i^n\}$ is monotone decreasing so is $\{u_i^{n+1}\}$.*

An important result that relates monotone, TVD and monotonicity preserving schemes for non–linear scalar conservation laws (13.10) is given by the following theorem, the proof of which we omit here.

Theorem 13.6.1. *In general, the set S_{mon} of monotone schemes is contained in the set S_{tvd} of TVD schemes and this in turn is contained in the set S_{mpr} of monotonicity preserving schemes, that is*

$$S_{mon} \subseteq S_{tvd} \subseteq S_{mpr} \,. \tag{13.98}$$

Proof. See Harten [158]; see also [207].

For linear schemes (13.5) to solve the linear advection equation (13.1) one can prove that monotone schemes are equivalent to monotonicity preserving schemes, as the following theorem states.

Theorem 13.6.2. *A linear scheme (13.5) as applied to the linear advection equation (13.1) is Monotonicity Preserving if and only if the coefficients b_k are non–negative, i.e. $b_k \geq 0 \; \forall k$.*

Proof. In the first part of the proof, we assume the scheme is monotonicity preserving and show that $b_k \geq 0$. Applying the scheme (13.5) to two consecutive points leads to

$$u_{i+1}^{n+1} - u_i^{n+1} = \sum_{k=-k_L}^{k_R} b_k (u_{i+1+k}^n - u_{i+k}^n) \,. \tag{13.99}$$

In proving this part by contradiction we assume $b_K < 0$ for some K. The objective now is to contradict the hypothesis that the scheme is Monotonicity Preserving. This is achieved by considering the special case

$$u_j^n = \begin{cases} 0 & j < i + 1 + K \,, \\ 1 & j \geq i + 1 + K \,. \end{cases} \tag{13.100}$$

Thus

$$u_{i+1}^{n+1} - u_i^{n+1} = b_K < 0 \,.$$

But if $\{u_i^n\}$ is monotone increasing for instance, then by hypothesis, we must have $u_{i+1}^{n+1} - u_i^{n+1} \geq 0$. Therefore $b_k \geq 0$, $\forall k$.

In the second part of the proof we assume $b_k \geq 0$ in (13.5) and prove that the scheme is Monotonicity Preserving. If $\{u_i^n\}$ is monotone then all terms in the summation (13.99) are of the same sign, as $b_k \geq 0$; consequently the left hand side of (13.99) is of the same sign and thus $\{u_i^{n+1}\}$ is monotone in the same sense as $\{u_i^n\}$.

Remark 13.6.1. Recall that a monotone linear scheme (13.5) to solve (13.1) is monotone if and only if $b_k \geq 0$. Hence, in this case, monotone schemes are identical to monotonicity preserving schemes and by (13.98) they are also equivalent to TVD schemes.

Remark 13.6.2. The most useful class is that of TVD methods, which have some precise mathematical properties that allow proofs of convergence. For an example of convergence proofs see for instance the work of Sweby and Baines [335] and relevant references cited there. The TVD conditions also allow the practical construction of numerical methods having the TVD property.

Harten [158] considered the class of non–linear schemes

$$u_i^{n+1} = u_i^n - C_{i-\frac{1}{2}}\Delta u_{i-\frac{1}{2}} + D_{i+\frac{1}{2}}\Delta u_{i+\frac{1}{2}} , \qquad (13.101)$$

where $\Delta u_{i+\frac{1}{2}} = u_{i+1}^n - u_i^n$ and the coefficients $C_{i-\frac{1}{2}}$, $D_{i+\frac{1}{2}}$ are in general assumed to be functions of the data. Harten [158] proved the following important result

Theorem 13.6.3 (Harten). *For any scheme of the form (13.101) to solve (13.10), a sufficient condition for the scheme to be TVD is that the coefficients satisfy*

$$\left. \begin{array}{l} C_{i+\frac{1}{2}} \geq 0 ; \quad D_{i+\frac{1}{2}} \geq 0 , \\ 0 \leq C_{i+\frac{1}{2}} + D_{i+\frac{1}{2}} \leq 1 . \end{array} \right\} \qquad (13.102)$$

Proof. Apply the scheme (13.101) to two consecutive cells i and $i + 1$. We obtain

$$u_i^{n+1} = u_i^n - C_{i-\frac{1}{2}}(u_i^n - u_{i-1}^n) + D_{i+\frac{1}{2}}(u_{i+1}^n - u_i^n) \qquad (13.103)$$

and

$$u_{i+1}^{n+1} = u_{i+1}^n - C_{i+\frac{1}{2}}(u_{i+1}^n - u_i^n) + D_{i+\frac{3}{2}}(u_{i+2}^n - u_{i+1}^n) . \qquad (13.104)$$

Subtracting (13.103) from (13.104) gives

$$u_{i+1}^{n+1} - u_i^{n+1} = (u_{i+1}^n - u_i^n)(1 - C_{i+\frac{1}{2}} - D_{i+\frac{1}{2}})$$

$$+ C_{i-\frac{1}{2}}(u_i^n - u_{i-1}^n) + D_{i+\frac{3}{2}}(u_{i+2}^n - u_{i+1}^n) .$$

Taking absolute values on both sides of the above inequality leads to

$$|u_{i+1}^{n+1} - u_i^{n+1}| \leq |1 - C_{i+\frac{1}{2}} - D_{i+\frac{1}{2}}||u_{i+1}^n - u_i^n|$$

$$+|C_{i-\frac{1}{2}}||u_i^n - u_{i-1}^n| + |D_{i+\frac{3}{2}}||u_{i+2}^n - u_{i+1}^n| .$$

As $|r| = r$ for any non–negative real number r, application of the sufficient conditions (13.102) gives

$$|u_{i+1}^{n+1} - u_i^{n+1}| \leq |u_{i+1}^n - u_i^n| + C_{i-\frac{1}{2}}|u_i^n - u_{i-1}^n| - C_{i+\frac{1}{2}}|u_{i+1}^n - u_i^n|$$

$$+D_{i+\frac{3}{2}}|u_{i+2}^n - u_{i+1}^n| - D_{i+\frac{1}{2}}|u_{i+1}^n - u_i^n| .$$

Summing over i gives

$$\sum_i |u_{i+1}^{n+1} - u_i^{n+1}| \leq \sum_i |u_{i+1}^n - u_i^n| + \sum_i C_{i-\frac{1}{2}}|u_i^n - u_{i-1}^n|$$

$$- \sum_i C_{i+\frac{1}{2}}|u_{i+1}^n - u_i^n| + \sum_i D_{i+\frac{3}{2}}|u_{i+2}^n - u_{i+1}^n|$$

$$- \sum_i D_{i+\frac{1}{2}}|u_{i+1}^n - u_i^n| .$$

Clearly the second and third summations on the right–hand side of the above inequality cancel and so do the fourth and fifth summations, leading to the sought result

$$TV(u^{n+1}) \leq TV(u^n) .$$

Remark 13.6.3. During the proof of the above theorem we assumed

$$C_{i-\frac{1}{2}} \geq 0 , \quad 1 - C_{i+\frac{1}{2}} - D_{i+\frac{1}{2}} \geq 0 , \quad D_{i+\frac{3}{2}} \geq 0.$$

By appropriate shifting of the position at which expressions (13.103),(13.104) are applied, the above conditions are equivalent to the conditions (13.102) of the theorem.

Remark 13.6.4. The coefficients $C_{i+\frac{1}{2}}$, $D_{i+\frac{1}{2}}$ in Harten's theorem, may in general, be data dependent. The theorem therefore applies to nonlinear schemes. This fact can be then used to circumvent Godunov's theorem (13.5.4), which applies to linear schemes only. Harten's theorem (13.6.3) offers a very useful tool for constructing high resolution schemes.

Remark 13.6.5. Schemes that allow a controlled increase in the total variation have also been constructed. They are usually referred to as Total Variation Bounded Schemes, or TVB schemes. See for instance the work of Shu [311], [310].

13.7 Flux Limiter Methods

In this section we construct Total Variation Diminishing (TVD) methods following the *flux limiter approach*, see Sweby [333]. In Sect. 13.7.1 we first consider a particular method, namely a TVD version of the Weighted Average Flux (WAF) method described in Sect. 13.3. We then present the general flux limiter approach as applied to families of numerical methods; in Sect. 13.7.2 we present *upwind* based flux limiter methods and in Sect. 13.7.4 we present *centred* (non–upwind) flux limiter methods.

13.7.1 TVD Version of the WAF Method

Recall the motivating example of Sect. 13.5.2 in which the conservative scheme (13.11) has numerical flux (13.73), which depends on a free parameter α. Various, fixed values of α reproduce familiar schemes. One may exploit α by regarding it as a function of data u_i^n and constrained by some TVD requirement to construct non–linear oscillation–free schemes of second order accuracy in smooth parts of the solution. This is in fact the approach we follow here to construct a TVD version of the basic WAF scheme presented in Sect. 13.3.1 to solve the model linear advection equation (13.1).

For convenience we introduce a new parameter ϕ and re–write the WAF flux (13.16) as

$$f_{i+\frac{1}{2}} = \frac{1}{2}(1 + \phi)(au_i^n) + \frac{1}{2}(1 - \phi)(au_{i+1}^n) . \tag{13.105}$$

Compare with (13.73). Three particular choices of ϕ are $\phi = |c|$ (which leads to the Lax–Wendroff method), $\phi = s \equiv sign(a) = sign(c)$ (which produces the Godunov first–order upwind method) and $\phi = -s$ (which gives the downwind method, unconditionally unstable).

The purpose is to find appropriate ranges for ϕ as a function of some data–dependent variables that produce a TVD version of the WAF scheme. Based on the extreme cases $\phi = s$ and $\phi = -s$ above we restrict ϕ to satisfy the following requirement

$$-1 \leq \phi \leq 1 . \tag{13.106}$$

Other choices for the upper and lower bounds of ϕ are of course also possible. By introducing appropriate notation for ϕ at the intercell boundaries the relevant intercell fluxes become

$$\left. \begin{array}{l} f_{i+\frac{1}{2}} = \frac{1}{2}(1 + \phi_{i+\frac{1}{2}})(au_i^n) + \frac{1}{2}(1 - \phi_{i+\frac{1}{2}})(au_{i+1}^n) , \\[2mm] f_{i-\frac{1}{2}} = \frac{1}{2}(1 + \phi_{i-\frac{1}{2}})(au_{i-1}^n) + \frac{1}{2}(1 - \phi_{i-\frac{1}{2}})(au_i^n) . \end{array} \right\} \tag{13.107}$$

Thus, for each intercell boundary position $i + \frac{1}{2}$ we seek a value $\phi_{i+\frac{1}{2}}$ that will produce a modified WAF flux leading to a TVD scheme.

Application of Data Compatibility. Instead of using Harten's Theorem we impose the slightly stronger constraint embodied in the data compatibility condition (13.85). First we assume that $a > 0$ in the PDE (13.1). Substitution of the fluxes (13.107) into the conservative formula (13.11) followed by some obvious algebraic manipulations, to reproduce the ratio in (13.85), give

$$\frac{u_i^{n+1} - u_i^n}{u_{i-1}^n - u_i^n} = \frac{1}{2}c\left[\frac{1}{r_{i+\frac{1}{2}}}\left(1 - \phi_{i+\frac{1}{2}}\right) + \phi_{i-\frac{1}{2}} + 1\right] , \qquad (13.108)$$

where

$$r_{i+\frac{1}{2}} = \frac{u_i^n - u_{i-1}^n}{u_{i+1}^n - u_i^n} . \qquad (13.109)$$

Compare the left–hand side of (13.108) with the ratio of consecutive data changes in the data compatibility condition (13.85). Direct application of condition (13.85) to the right–hand side of (13.108) leads to

$$-1 \le \frac{1}{r_{i+\frac{1}{2}}}\left(1 - \phi_{i+\frac{1}{2}}\right) + \phi_{i-\frac{1}{2}} \le \frac{2 - c}{c} . \qquad (13.110)$$

For negative speed a in the PDE (13.1) ($c < 0$) one arrives at the same result (13.110), with the CFL number c replaced by its absolute value and $r_{i+\frac{1}{2}}$ defined as

$$r_{i+\frac{1}{2}} = \frac{u_{i+2}^n - u_{i+1}^n}{u_{i+1}^n - u_i^n} . \qquad (13.111)$$

Thus, for both positive and negative waves speeds in (13.1) we obtain

$$-1 \le \frac{1}{r_{i+\frac{1}{2}}}\left(1 - \phi_{i+\frac{1}{2}}\right) + \phi_{i-\frac{1}{2}} \le \frac{2 - |c|}{|c|} , \qquad (13.112)$$

with

$$r_{i+\frac{1}{2}} = \begin{cases} \frac{u_i^n - u_{i-1}^n}{u_{i+1}^n - u_i^n} & , \quad a > 0 , \\[2mm] \frac{u_{i+2}^n - u_{i+1}^n}{u_{i+1}^n - u_i^n} & , \quad a < 0 . \end{cases} \qquad (13.113)$$

Thus $r_{i+\frac{1}{2}}$ is always the ratio of the *upwind change*, $\Delta u_{i-\frac{1}{2}} \equiv u_i^n - u_{i-1}^n$ for $a > 0$ or $\Delta u_{i+\frac{3}{2}} \equiv u_{i+2}^n - u_{i+1}^n$ for $a < 0$, to the *local change* $\Delta u_{i+\frac{1}{2}} = u_{i+1}^n - u_i^n$. That is

$$r_{i+\frac{1}{2}} = \frac{\Delta_{upw}}{\Delta_{loc}} ,$$

with the obvious definitions for Δ_{upw} and Δ_{loc}. Fig. 13.12. illustrates the identification of local and upwind changes across appropriate waves. The ratio $r_{i+\frac{1}{2}}$ of local changes in the data is now regarded as a *flow parameter* that will cause the value of the sought function ϕ to adjust to local conditions on the data. So far we have identified two quantities that may be regarded as the independent variables of the function ϕ, namely $|c|$ and the flow parameter $r_{i+\frac{1}{2}}$. Thus we define

$$\phi_{i+\frac{1}{2}} = \phi_{i+\frac{1}{2}}(r_{i+\frac{1}{2}}, |c|) . \qquad (13.114)$$

Occasionally, we shall omit the argument $|c|$.

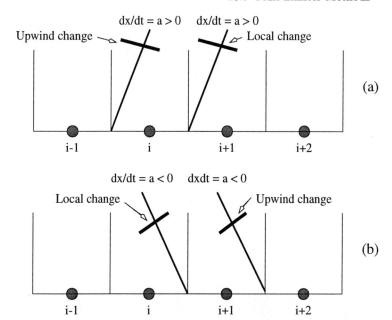

Fig. 13.12. Identification of upwind direction to form the ratio $r_{i+\frac{1}{2}}$ of upwind change to local change: (a) positive wave speed, upwind change lies on the left (b) negative wave speed, upwind change lies on the right

Construction of the TVD Region. The purpose here is to find a suitable range of values for $\phi_{i+\frac{1}{2}}$ as a function of $r_{i+\frac{1}{2}}$ and $|c|$. To this end we select two inequalities, one for $\phi_{i+\frac{1}{2}}$ and one for $\phi_{i-\frac{1}{2}}$ so that both inequalities in (13.112) are automatically satisfied. We take

$$-1 - L \leq \frac{1}{r_{i+\frac{1}{2}}}(1 - \phi_{i+\frac{1}{2}}) \leq \frac{2(1 - |c|)}{|c|} \qquad (13.115)$$

and

$$L \leq \phi_{i-\frac{1}{2}} \leq 1 , \qquad (13.116)$$

with the lower bound L in the interval $[-1, |c|]$. This will allow some freedom in selecting how much *downwinding* is to be allowed. For $L = -1$, see (13.112), full downwinding is allowed; for $L = 0$ no downwinding is allowed. Note that by adding (13.115) and (13.116) we reproduce (13.112). We study (13.115) in detail. Subscripts are omitted, for convenience.

The *left inequality* in (13.115): $-(1 + L) \leq \frac{1}{r}(1 - \phi)$ contains two cases, namely

−   Positive r: if $r > 0$ then $-(1 + L)r \leq 1 - \phi$, which implies

$$\phi \leq 1 + (1 + L)r \equiv \phi_L(r, |c|) . \qquad (13.117)$$

–   Negative r: if $r < 0$ then $-(1 + L)r \geq 1 - \phi$, which implies

$$\phi \geq 1 + (1 + L)r \equiv \phi_L(r, |c|) \ . \tag{13.118}$$

The *right inequality* in (13.115): $\frac{1}{r}(1 - \phi) \leq \frac{2(1 - |c|)}{|c|}$ has two cases, namely

–   Positive r: if $r > 0$ we obtain

$$\phi \geq 1 - \frac{2(1 - |c|)}{|c|} r \equiv \phi_R(r, |c|) \ . \tag{13.119}$$

–   Negative r: if $r < 0$ we obtain

$$\phi \leq 1 - \frac{2(1 - |c|)}{|c|} r \equiv \phi_R(r, |c|) \ . \tag{13.120}$$

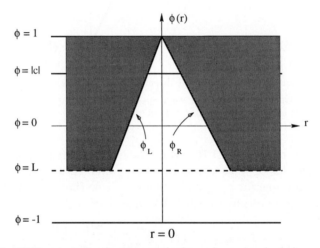

Fig. 13.13. TVD regions for the WAF method are shown by shaded area in the r–ϕ plane, for fixed values of the CFL number c

Fig. 13.13 illustrates the derived TVD regions (dark zone) in the plane $r - \phi$ that satisfies conditions (13.115)–(13.116). There is a TVD region for $r < 0$ and a TVD region for $r > 0$. The choice of the lower bound L determines the left TVD region; for $L = -1$ this region coalesces to the single line $\phi_R(r, |c|) = 1$ (Godunov first–order upwind). Any fixed value of the limiter function $\phi(r, |c|)$ results in a scheme with a numerical viscosity α_ϕ, see Sect. 5.2 of Chap. 5. For $|c| \leq \phi(r, |c|) \leq 1$ this numerical viscosity is positive and results in *spreading of discontinuities* and *clipping of extrema*. For $L \leq \phi(r, |c|) \leq |c|$ the numerical viscosity is negative and results in steepening of discontinuities. Obviously the case $\phi(r, |c|) = |c|$ does not add any extra numerical viscosity and corresponds to the Lax–Wendroff method.

A *limiter function* $\phi = \phi(r, |c|)$ can be constructed within the TVD region and any choice will produce an oscillation–free scheme (13.11) with flux (13.105); the only restriction is

$$\phi(1, |c|) = |c| \ . \tag{13.121}$$

This ensures second–order accuracy for values of r close to 1, that is when the upwind change is comparable to the local change (smooth part of the solution), see (13.113).

Construction of WAF **Limiter Functions.** There is scope for the imagination in constructing functions $\phi(r, |c|)$. Here we give five WAF limiter functions, namely

$$\phi_{ua}(r, |c|) = \begin{cases} 1 & \text{if } r \leq 0 \,, \\ 1 - \frac{2(1-|c|)r}{|c|} & \text{if } 0 \leq r \leq \frac{|c|}{1-|c|} \,, \\ -1 & \text{if } r \geq \frac{|c|}{1-|c|} \,, \end{cases} \tag{13.122}$$

$$\phi_{sa}(r, |c|) = \begin{cases} 1 & \text{if } r \leq 0 \,, \\ 1 - 2(1 - |c|)r & \text{if } 0 \leq r \leq \frac{1}{2} \,, \\ |c| & \text{if } \frac{1}{2} \leq r \leq 1 \,, \\ 1 - (1 - |c|)r & \text{if } 1 \leq r \leq 2 \,, \\ 2|c| - 1 & \text{if } r \geq 2 \end{cases} \tag{13.123}$$

$$\phi_{vl}(r, |c|) = \begin{cases} 1 & \text{if } r \leq 0 \,, \\ 1 - \frac{2(1-|c|)r}{1+r} & \text{if } r \geq 0 \,, \end{cases} \tag{13.124}$$

$$\phi_{va}(r, |c|) = \begin{cases} 1 & \text{if } r \leq 0 \,, \\ 1 - \frac{(1-|c|)r(1+r)}{1+r^2} & \text{if } r \geq 0 \,, \end{cases} \tag{13.125}$$

and

$$\phi_{ma}(r, |c|) = \begin{cases} 1 & \text{if } r \leq 0 \,, \\ 1 - (1 - |c|)r & \text{if } 0 \leq r \leq 1 \,, \\ |c| & \text{if } r \geq 1 \,. \end{cases} \tag{13.126}$$

Each of these WAF limiter functions above corresponds to the well known flux limiters ULTRABEE, SUPERBEE, VANLEER, VANALBADA and MIN-BEE, to be studied in Sect. 13.7.2. We therefore give the WAF limiters analogous names. The function $\phi_{ua}(r, |c|)$ (ULTRAA) corresponds to the flux limiter ULTRABEE of Roe [283]; $\phi_{sa}(r, |c|)$ (SUPERA) corresponds to the flux limiter SUPERBEE of Roe [283] ; $\phi_{vl}(r, |c|)$ (VANLEER) corresponds to the flux limiter of van Leer [387], [388]; $\phi_{va}(r, |c|)$ (VANALBADA) corresponds to the limiter proposed by van Albada [386]; $\phi_{ma}(r, |c|)$ (MINA) corresponds to the flux limiter MINBEE of Roe [283]. Fig. 13.14 illustrates the limiter function SUPERA. Note that $\phi_{ua}(r, |c|)$ does not satisfy the second–order requirement (13.121).

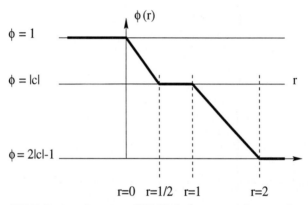

Fig. 13.14. WAF limiter function SUPERA (corresponding to flux limiter SU-PERBEE) is a piece–wise linear function shown by thick line

Numerical Experiments. We consider the same two test problems for the linear advection (13.1) as in Chap. 5. In the computations we take $a = 1.0$ and a CFL coefficient $C_{cfl} = 0.8$. Computed results are shown at the output times $t = 1.0$ unit (125 time steps) and $t = 10.0$ unit (1250 time steps). In each figure we compare the exact solution (shown by full lines) with the numerical solution (symbols).

The initial conditions for Test 1 (smooth data) and Test 2 (discontinuous data, square wave) are respectively given by

$$u(x,0) = \alpha e^{-\beta x^2} \; ; \; u(x,0) = \begin{cases} 0 & \text{if} \quad x \leq 0.3 \,, \\ 1 & \text{if} \quad 0.3 \leq x \leq 0.7 \,, \\ 0 & \text{if} \quad x \geq 0.7 \,. \end{cases} \qquad (13.127)$$

In the computations for Test 1 we take $\alpha = 1.0$, $\beta = 8.0$; the initial profile $u(x,0)$ is evaluated in the interval $-1 \leq x \leq 1$. The mesh used is $\Delta x = 0.02$. For Test 2 the mesh used is $\Delta x = 0.01$. In both test problems the initial profile and that at the output time is resolved by 100 computing cells.

Results for Test 1 are shown in Figs. 13.15 (t=1) and 13.16 (t=10). Results for Test 2 are shown in Figs. 13.17 (t=1) and 13.18 (t=10). In each figure we first show the linear schemes corresponding to $\phi(r, |c|) = |c| = constant$ (Lax–Wendroff) and to $\phi(r, |c|) = 1 = constant$ (Godunov first order upwind). The other solutions correspond to TVD versions of the WAF method for four choices of the limiter function $\phi(r, |c|)$; these are: ULTRAA, SUPERA, VANLEER and MINA, given respectively by (13.122)–(13.124) and (13.126). In analysing the results of Fig. 13.15 one must bear in mind that these are for a test with smooth initial data and that the solution has been evolved for a short time (125 time steps). The results from the Lax–Wendroff (or WAF without the TVD property) are excellent and no apparent improvements are offered by the TVD version of WAF using any of the limiter functions tested. The result of the Godunov first order upwind method shows the effects of the *numerical viscosity* inherent in this method; as a consequence extrema are unduly smoothed out. Of the four TVD results shown, that of the limiter ULTRAA is the worst of the group. As noted earlier this function does not pass through the point of second–order accuracy; it adds *negative numerical viscosity* (locally) and results in the wrong *steepening* and *squaring* of the solution profiles. Clearly this limiter is unsuitable, at least for smooth solutions. All other TVD results are very satisfactory. SUPERA shows a tendency to *squaring* of the solution profile near extrema, similar to ULTRAA. This is in fact a well known feature of SUPERA. The limiters VANLEER and MINA give similar results, at least for this output time.

Compare the results for Fig. 13.15 with those of Fig. 13.16 (t=10). The results of Fig. 13.16 (1250 time steps) begin to show the effect of the dispersive errors in the Lax–Wendroff method, the computed solution lags behind (the speed a is positive) the true solution. For very long evolution times this error becomes unacceptable. The Godunov first–order upwind method now clearly shows the effects of its numerical viscosity, which increases as the number of time steps increases. As for $t = 1$, the TVD results that are acceptable are those produced by SUPERA, VANLEER AND MINA. The result of ULTRAA is completely unacceptable. The tendency of the limiter SUPERA to *square* smooth profiles unduly is now visible. The limiters VANLEER and MINA are beginning to show the effect of their inherent numerical viscosity acting near extrema.

The main conclusion from the results of Figs. 13.15 and 13.16 is that the TVD results of the limiter functions SUPERA and VANLEER are the best, a conclusion that is also valid for non–linear systems (see Chap. 14).

In discussing the results of Fig. 13.17 one must bear in mind that the solution contains two discontinuities and that the output time is short (125 time steps). Now, unlike Test 1, the schemes (linear schemes) Lax–Wendroff and Godunov first–order upwind give the worse results. As noted earlier, the Lax–Wendroff scheme resolves steep fronts more sharply but at the cost of introducing *spurious oscillations*. Godunov's first–order upwind method on the other hand, while avoiding the production of spurious oscillations, *smears* sharp fronts excessively. It is in fact the limitations of these (linear) methods that motivates the introduction of TVD methods. These have the best features of both classes of schemes.

Of the four limiter functions tested in Fig. 13.17, ULTRAA produces the best results; the corresponding results of Figs. 13.15 and 13.16 are the worst of their group. It is worth noting that at even time steps the discontinuities in the ULTRAA profile have one intermediate point and none at odd time steps. The results from the other limiter functions are also satisfactory. Recall that the larger the limiter function is (within the TVD region), the more numerical viscosity is added; thus, of the four TVD results the one given by MINA is the one with the largest numerical viscosity, which explains the smearing of the discontinuities. Compare results with those of Fig. 13.18, which correspond to 1250 time steps; these results clearly expose the limitations of Lax–Wendroff and Godunov first–order upwind method. All TVD results are superior, with ULTRAA giving the best result followed by SUPERA, VANLEER and MINA. In practical computations the solutions will contain smooth parts as well as discontinuities. It is found that SUPERA and VANLEER give the best results. Extension of TVD methods to non–linear systems is carried out in Chap. 14.

13.7.2 The General Flux–Limiter Approach

A well established approach for constructing high–order Total Variation Diminishing (TVD) schemes is the *flux limiter approach* [49], [48], [50], [333], [283]. This requires a high–order flux $f_{i+\frac{1}{2}}^{HI}$ associated with a scheme of accuracy greater than or equal to two and a low–order flux $f_{i+\frac{1}{2}}^{LO}$ associated with a *monotone*, first–order scheme. We present the approach in terms of the model conservation law

$$u_t + f(u)_x = 0 \; ; \quad f(u) = au \tag{13.128}$$

as solved by

$$u_i^{n+1} = u_i^n + \frac{\Delta t}{\Delta x}[f_{i-\frac{1}{2}} - f_{i+\frac{1}{2}}] \; . \tag{13.129}$$

One then defines a high–order TVD flux as

$$f_{i+\frac{1}{2}}^{TVD} = f_{i+\frac{1}{2}}^{LO} + \phi_{i+\frac{1}{2}}[f_{i+\frac{1}{2}}^{HI} - f_{i+\frac{1}{2}}^{LO}] \; , \tag{13.130}$$

where $\phi_{i+\frac{1}{2}}$ is a flux limiter function yet to be determined. To preserve some generality we assume that $f_{i+\frac{1}{2}}^{HI}$ and $f_{i+\frac{1}{2}}^{LO}$ are respectively of the form

$$\left. \begin{array}{l} f_{i+\frac{1}{2}}^{LO} = \alpha_0 \, au_i^n + \alpha_1 \, au_{i+1}^n \; , \\[2mm] f_{i+\frac{1}{2}}^{HI} = \beta_0 \, au_i^n + \beta_1 \, au_{i+1}^n \; . \end{array} \right\} \tag{13.131}$$

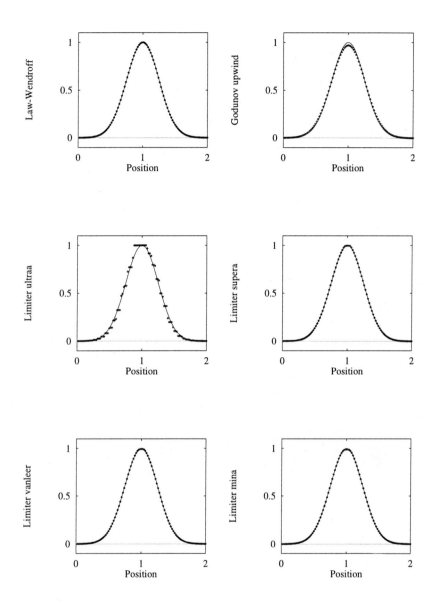

Fig. 13.15. Numerical (symbols) and exact (line) solutions for Test 1 at time $t = 1$. The results of the linear schemes Lax–Wendroff and Godunov first order upwind are compared with results from TVD version of the WAF method using four limiter functions, namely: ULTRAA, SUPERA, VANLEER and MINA. Compare the results with those of Fig. 13.16.

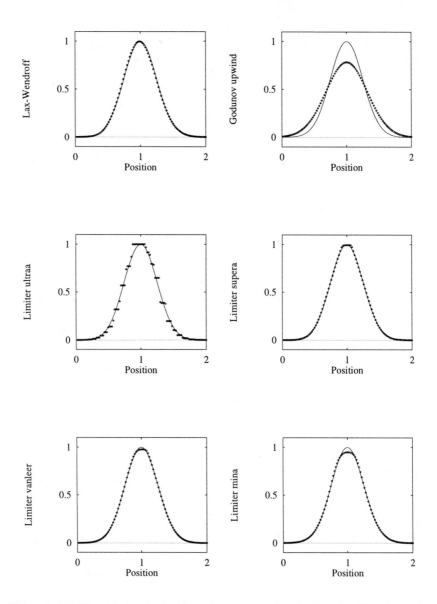

Fig. 13.16. Numerical (symbols) and exact (line) solutions for Test 1 at time $t = 10$. The results of the linear schemes Lax–Wendroff and Godunov first order upwind are compared with results from TVD version of the WAF method using four limiter functions, namely: ULTRAA, SUPERA, VANLEER and MINA. Compare the results with those of Fig. 13.15.

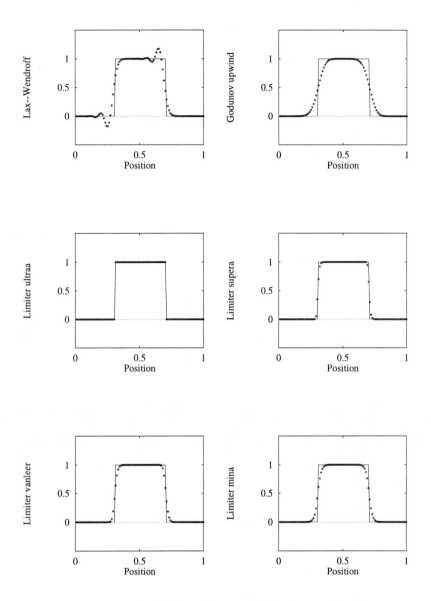

Fig. 13.17. Numerical (symbols) and exact (line) solutions for Test 2 at time $t = 1$. The results of the linear schemes Lax–Wendroff and Godunov first order upwind are compared with results from TVD version of the WAF method using four limiter functions, namely: ULTRAA, SUPERA, VANLEER and MINA. Compare the results with those of Fig. 13.18.

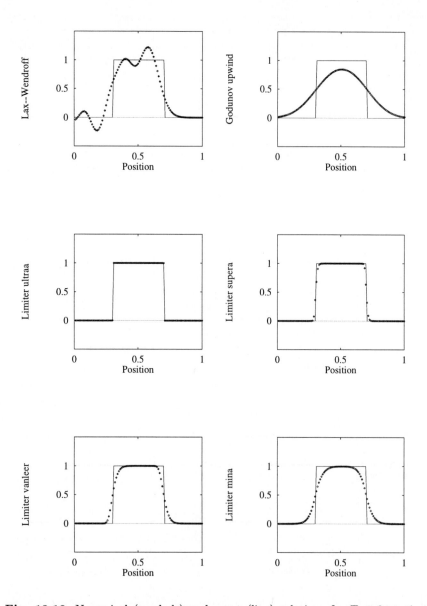

Fig. 13.18. Numerical (symbols) and exact (line) solutions for Test 2 at time $t = 10$. The results of the linear schemes Lax–Wendroff and Godunov first order upwind are compared with results from TVD version of the WAF method using four limiter functions, namely: ULTRAA, SUPERA, VANLEER and MINA. Compare the results with those of Fig. 13.17.

The choice

$$\alpha_0 = \frac{1}{2}(1+s) \; , \quad \alpha_1 = \frac{1}{2}(1-s) \; , \tag{13.132}$$

with $s = \text{sign}(a)$, reduces $f_{i+\frac{1}{2}}^{LO}$ to the Godunov first order upwind flux. For the choice

$$\alpha_0 = \frac{1}{4c}(1+c)^2 \; ; \quad \alpha_1 = -\frac{1}{4c}(1-c)^2 \tag{13.133}$$

$f_{i+\frac{1}{2}}^{LO}$ becomes the FORCE flux and the choice

$$\alpha_0 = \frac{1}{2c}(1+c) \; ; \quad \alpha_1 = -\frac{1}{2c}(1-c) \tag{13.134}$$

reproduces the Lax–Friedrichs flux; see (13.17) and Table 13.2. The two–point support for the flux $f_{i+\frac{1}{2}}^{HI}$ means that this flux must be the Lax–Wendroff flux, for which

$$\beta_0 = \frac{1}{2}(1+c) \; ; \quad \beta_1 = \frac{1}{2}(1-c) \; . \tag{13.135}$$

Therefore the coefficients for β_0 and β_1 for the high–order flux are fixed but the coefficients α_0 and α_1 for the low–order flux are general. Substitution of (13.131) into (13.130) gives

$$f_{i+\frac{1}{2}}^{TVD} = [\alpha_0 + (\beta_0 - \alpha_0)\phi_{i+\frac{1}{2}}](au_i^n) + [\alpha_1 + (\beta_1 - \alpha_1)\phi_{i+\frac{1}{2}}](au_{i+1}^n) \; , \tag{13.136}$$

which if substituted into (13.129) produces

$$\left.\begin{aligned}
u_i^{n+1} &= u_i^n - C\Delta u_{i-\frac{1}{2}} + D\Delta u_{i+\frac{1}{2}} \; , \\
C &= c[\alpha_0 + (\beta_0 - \alpha_0)\phi_{i-\frac{1}{2}}] \; , \\
D &= -c[\alpha_1 + (\beta_1 - \alpha_1)\phi_{i+\frac{1}{2}}] \; , \\
\Delta u_{i-\frac{1}{2}} &= u_i^n - u_{i-1}^n \; ; \quad \Delta u_{i+\frac{1}{2}} = u_{i+1}^n - u_i^n \; .
\end{aligned}\right\} \tag{13.137}$$

In the next section we specialise the flux limiter approach to particular choices of low and high order schemes.

13.7.3 TVD Upwind Flux Limiter Schemes

Here we assume that the low–order flux is that of the Godunov first–order upwind scheme and the high order flux is that of the Lax–Wendroff scheme. The derivation of the TVD scheme relies on identifying *upwind directions*. Flux limiters are now denoted by $\psi_{i+\frac{1}{2}}(r)$.

First assume $a > 0$ in the model conservation law (13.128). Then

$$\left.\begin{aligned}
\alpha_0 &= 1 \; ; \quad \alpha_1 = 0 \; ; \quad c > 0 \; , \\
C &= c[1 + (\beta_0 - 1)\psi_{i-\frac{1}{2}}] \; ; \quad D = -c\beta_1\psi_{i+\frac{1}{2}} \; , \\
\beta_0 &= \tfrac{1}{2}(1+c) \; ; \quad \beta_1 = \tfrac{1}{2}(1-c) \; .
\end{aligned}\right\} \tag{13.138}$$

Now we rewrite (13.137) as

$$u_i^{n+1} = u_i^n - \hat{C}\Delta u_{i-\frac{1}{2}} \ , \left.\begin{array}{c} \\ \\ \end{array}\right\}$$
$$\hat{C} = C - D/r \ ; \quad r = \Delta u_{i-\frac{1}{2}}/\Delta u_{i+\frac{1}{2}} \ .$$
(13.139)

Application of the TVD condition (13.102) of Harten's theorem gives the inequalities

$$0 \le c \left[1 + (\beta_0 - 1)\psi_{i-\frac{1}{2}} + \beta_1 \psi_{i+\frac{1}{2}} \frac{1}{r} \right] \le 1 \ .$$
(13.140)

We now impose a global constraint, *independent* of r, on the sought limiter functions, namely

$$\psi_B \le \psi_{i-\frac{1}{2}} \le \psi_T \ , \quad \forall i \, , \forall r \ .$$
(13.141)

This constraint may be re–written as

$$c[1 + (\beta_0 - 1)\psi_T] \le c[1 + (\beta_0 - 1)\psi_{i-\frac{1}{2}}] \le c[1 + (\beta_0 - 1)\psi_B] \ .$$
(13.142)

Now we consider the following two inequalities

$$- c[1 + (\beta_0 - 1)\psi_T] \le c\beta_1 \psi_{i+\frac{1}{2}} \frac{1}{r} \le 1 - c[1 + (\beta_0 - 1)\psi_B] \ .$$
(13.143)

Note that (13.142) and (13.143) reproduce the TVD condition (13.140) identically. As (13.142) is only a re–statement of (13.141) we only work with inequalities (13.143) to find the limiter function $\psi_{i+\frac{1}{2}}$ at the intercell boundary $i + \frac{1}{2}$.

Analysis of the left inequality in (13.143) leads to

$$\psi_{i+\frac{1}{2}}(r) \begin{cases} \ge \psi_L(r) \ , & \text{if } r > 0 \ , \\ \le \psi_L(r) \ , & \text{if } r < 0 \ , \end{cases}$$
(13.144)

where

$$\psi_L(r) = (\psi_T - 1/\beta_1)r \ .$$
(13.145)

Analysis of the right inequality in (13.143) produces the constraints

$$\psi_{i+\frac{1}{2}}(r) \begin{cases} \le \psi_R(r) \ , & r > 0 \ , \\ \ge \psi_R(r) \ , & r < 0 \ , \end{cases}$$
(13.146)

where

$$\psi_R(r) = \left[\psi_B + \frac{(1-c)}{c\beta_1} \right] r \ .$$
(13.147)

Using constraints (13.141) together with (13.144)–(13.147), one can draw the TVD region for the flux limiter $\psi_{i+\frac{1}{2}}(r)$ for the case $a > 0$.

Before doing that we consider the case $a < 0$. Now the upwind direction is on the right hand side of the relevant intercell boundary. The scheme may now be written as

$$u_i^{n+1} = u_i^n + \hat{D}\Delta u_{i+\frac{1}{2}} \; ; \\ \hat{D} = D - C/r \; ; \quad r = \frac{\Delta u_{i+\frac{1}{2}}}{\Delta u_{i-\frac{1}{2}}} \; . \left.\begin{array}{c} \\ \\ \end{array}\right\} \qquad (13.148)$$

Now we regard the position $i - \frac{1}{2}$ as the local position so that r remains the ratio of upwind to local changes. Harten's TVD conditions (13.102) lead to

$$0 \leq -c\left[1 + (\beta_1 - 1)\psi_{i+\frac{1}{2}} + \beta_0\psi_{i-\frac{1}{2}}\frac{1}{r}\right] \leq 1 \; . \qquad (13.149)$$

Now we impose the global constraint

$$\psi_B \leq \psi_{i+\frac{1}{2}} \leq \psi_T \; . \qquad (13.150)$$

Following the same steps as for the case $a > 0$ we end up with the conditions

$$c[1 + (\beta_1 - 1)\psi_T] \leq -c\beta_0\psi_{i-\frac{1}{2}}\frac{1}{r} \leq 1 + c[1 + (\beta_1 - 1)\psi_B] \qquad (13.151)$$

for the flux limiter function $\psi_{i-\frac{1}{2}}$ at the intercell boundary $i - \frac{1}{2}$. Note that by replacing c with its absolute value $|c|$ conditions (13.151) are identical to conditions (13.143), with the appropriate interpretation for the *local* intercell position.

Thus conditions (13.141) together with (13.144)–(13.147) apply to both $a > 0$ and $a < 0$, provided we replace c by $|c|$ and correctly interpret the ratio r. Hence, the limiter function at a general intercell position $i + \frac{1}{2}$ satisfies

$$\psi_B \leq \psi(r) \leq \psi_T \; , \qquad (13.152)$$

$$\psi(r)\begin{cases} \geq \psi_L(r) \; , & \text{if } r > 0 \; , \\ \leq \psi_L(r) \; , & \text{if } r < 0 \; , \end{cases} \qquad (13.153)$$

$$\psi(r)\begin{cases} \leq \psi_R(r) \; , & \text{if } r > 0 \; , \\ \geq \psi_R(r) \; , & \text{if } r < 0 \; , \end{cases} \qquad (13.154)$$

$$\psi_L(r) = \left(\psi_T - \frac{2}{1 - |c|}\right)r \; , \qquad (13.155)$$

$$\psi_R(r) = \left(\psi_B + \frac{2}{|c|}\right)r \; , \qquad (13.156)$$

$$r = \frac{\Delta_{upw}}{\Delta_{loc}} = \begin{cases} \dfrac{u_i^n - u_{i-1}^n}{u_{i+1}^n - u_i^n} \; , & a > 0 \; , \\[2ex] \dfrac{u_{i+2}^n - u_{i+1}^n}{u_{i+1}^n - u_i^n} \; , & a < 0 \; . \end{cases} \qquad (13.157)$$

The choice of the bottom and top bounds ψ_B and ψ_T in (13.152) determines the functions ψ_L and ψ_R in (13.155)–(13.156) and thus the TVD region. Fig. 13.19 shows the resulting TVD region for

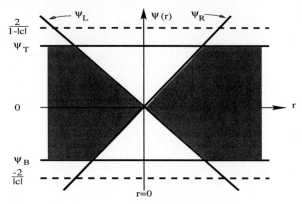

Fig. 13.19. General TVD region (dark zone) for flux limiter schemes based on the Godunov first–order upwind method and the Lax–Wendroff scheme. Compare with Fig. 13.13 showing the TVD region for the WAF method

$$\psi_B \geq -\frac{2}{|c|} \;;\; \psi_T \leq \frac{2}{1-|c|} \;. \tag{13.158}$$

As for the derivation of the TVD regions for the WAF method in Sect. 13.7.1, the choice of the bottom and top boundaries ψ_B and ψ_T depends on what schemes one wants to reproduce as *boundary schemes*. The choice

$$\psi_B = 0 \;;\; \psi_T = \frac{2}{1-|c|} \;. \tag{13.159}$$

allows for all schemes between the Godunov first–order upwind scheme and the downwind scheme. In this case the TVD region for $r < 0$ coalesces to the single line $\psi(r) = 0$ (Godunov first–order upwind for negative r).

The Sweby TVD region [333] is reproduced if we take

$$\psi_T = 2 \leq \frac{2}{1-|c|} \;,\quad \forall c \;;\quad \psi_B = 0 \tag{13.160}$$

and replace $2/|c|$ by 2 in ψ_R. See Fig. 13.20. Five flux limiter functions are the following: ULTRABEE is given by

$$\psi_{ub}(r) = \begin{cases} 0\,, & r \leq 0\,, \\[2mm] \dfrac{2}{|c|}r\,, & 0 \leq r \leq \dfrac{|c|}{1-|c|}\,, \\[3mm] \dfrac{2}{1-|c|}\,, & r \geq \dfrac{|c|}{1-|c|}\,, \end{cases} \tag{13.161}$$

SUPERBEE is given by

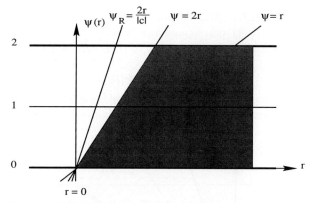

Fig. 13.20. Sweby's TVD Region, a special case of that shown in Fig. 13.19. Here the bottom and top boundaries are 0 and 2 respectively, so that the TVD region and limiters are independent of CFL number

$$
\psi_{sb}(r) = \begin{cases} 0, & r \leq 0, \\ 2r, & 0 \leq r \leq \frac{1}{2}, \\ 1, & \frac{1}{2} \leq r \leq 1, \\ r, & 1 \leq r \leq 2, \\ 2, & r \geq 2, \end{cases} \tag{13.162}
$$

VANLEER is given by

$$
\psi_{vl}(r) = \begin{cases} 0, & r \leq 0, \\ \dfrac{2r}{1+r}, & r \geq 0, \end{cases} \tag{13.163}
$$

VANALBADA is given by

$$
\psi_{va}(r) = \begin{cases} 0, & r \leq 0, \\ \frac{r(1+r)}{1+r^2}, & r \geq 0, \end{cases} \tag{13.164}
$$

and MINBEE (or MINMOD) is given by

$$
\psi_{mb}(r) = \begin{cases} 0, & r \leq 0, \\ r, & 0 \leq r \leq 1, \\ 1, & r \geq 1. \end{cases} \tag{13.165}
$$

Fig. 13.21 illustrates four of these flux limiters constructed from Sweby's TVD region, namely: ULTRABEE, SUPERBEE, VANLEER and MINBEE.

These are related to the WAF limiter functions given by (13.122)–(13.126). There is in fact a direct correspondence between the WAF limiter functions $\phi(r)$ and the flux limiter functions $\psi(r)$ of this section. Such correspondence is given by

$$\phi(r) = 1 - (1 - |c|)\psi(r) , \qquad (13.166)$$

so that for any given conventional flux limiter function $\psi(r)$ there is a corresponding WAF limiter function $\phi(r)$, and vice–versa.

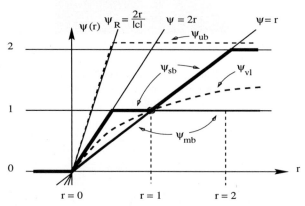

Fig. 13.21. Four flux limiter functions constructed from TVD region of Fig. 13.20: ULTRABEE, SUPERBEE, VANLEER and MINBEE. Compare with WAF limiter of Fig. 13.14

13.7.4 TVD Centred Flux Limiter Schemes

We follow the general flux limiter approach discussed in Sect. 13.7.2. Now, the low–order flux $f_{i+\frac{1}{2}}^{LO}$ is assumed to be a *centred flux*, with general coefficients α_0 and α_1 in (13.131). The high–order flux is still assumed to be that of the Lax–Wendroff method; see equations (13.128) to (13.137).

TVD Criteria for Centred Schemes. Convenient TVD conditions for constructing centred TVD schemes are first presented. Such conditions are a generalisation of the classical data compatibility conditions of Sect. 13.5.5 and the necessary conditions given by Harten's Theorem (13.6.3). We assume scheme (13.129) to be expressed as in (13.137), namely

$$u_i^{n+1} = u_i^n - C\Delta u_{i-\frac{1}{2}} + D\Delta u_{i+\frac{1}{2}} , \qquad (13.167)$$

where the coefficients C and D are, in general, data–dependent. By defining

$$r_i \equiv \frac{\Delta u_{upw}}{\Delta u_{dow}} \equiv \frac{\Delta u_{i-\frac{s}{2}}}{\Delta u_{i+\frac{s}{2}}} = \begin{cases} \dfrac{\Delta u_{i-\frac{1}{2}}}{\Delta u_{i+\frac{1}{2}}} , & \text{if } a > 0 , \\[3mm] \dfrac{\Delta u_{i+\frac{1}{2}}}{\Delta u_{i-\frac{1}{2}}} , & \text{if } a < 0 , \end{cases} \qquad (13.168)$$

scheme (13.167) produces

$$R_i^{(s)} \equiv \frac{u_i^{n+1} - u_i^n}{u_{i-s}^n - u_i^n} = \begin{cases} C - D/r_i \,, & \text{if } a > 0 \,, \\ D - C/r_i \,, & \text{if } a < 0 \,, \end{cases} \tag{13.169}$$

where $s = \text{sign}(a)$.

Next we state a result giving sufficient conditions for scheme (13.167) to be TVD [374].

Theorem 13.7.1. *Scheme (13.167) for (13.128) is TVD if*

$$-\frac{\epsilon_R}{r_i} \leq R_i^{(s)} \leq \epsilon_L \,, \quad \text{if } r_i > 0 \,, \tag{13.170}$$

$$0 \leq R_i^{(s)} \leq 1 - \epsilon_0 - \epsilon_0/r_i \,, \quad \text{if } r_i < 0 \,, \tag{13.171}$$

where ϵ_L , ϵ_0 , ϵ_R are real numbers satisfying

$$0 \leq \epsilon_L \,, \epsilon_0 \,, \epsilon_R \leq 1 \,. \tag{13.172}$$

Proof. We first prove result (13.170), which refers to monotone increasing or monotone decreasing data. Monotonicity is ensured, and thus the TVD property, if conditions (13.68) are enforced, namely

$$\min_k \{u_k^n\} \leq u_i^{n+1} \leq \max_k \{u_k^n\} \,. \tag{13.173}$$

First consider the case of monotone decreasing data

$$u_{i-1}^n > u_i^n \geq u_{i+1}^n$$

and assume

$$(1 - \epsilon_R)u_i^n + \epsilon_R u_{i+1}^n \leq u_i^{n+1} \leq (1 - \epsilon_L)u_i^n + \epsilon_L u_{i-1}^n \,, \tag{13.174}$$

which, in a sense, is a special case of (13.173), as it is more restrictive than (13.173). Assuming $a > 0$, subtracting u_i^n and dividing through by $u_{i-1}^n - u_i^n > 0$ leads to the sought result (13.170). For $a < 0$ the proof is similar. The case of monotone increasing data follows in an analogous manner. Condition (13.171) refers to extrema, local maxima or local minima. First assume a local minimum, i.e.

$$u_i^n \leq u_{i-1}^n \,; \quad u_i^n \leq u_{i+1}^n \,.$$

We impose the following monotonicity condition

$$u_i^n \leq u_i^{n+1} \leq (1 - \epsilon_0)u_{i-1}^n + \epsilon_0 u_{i+1}^n \,,$$

which is a special case of (13.173). Simple manipulations lead to the sought result (13.171). The case of a local maximum leads to the same result and thus the TVD conditions (13.170), (13.171) are proved.

Remark 13.7.1. The TVD condition (13.170) relates to points away from extrema and is more relaxed than the Data Compatibility condition (13.85), namely

$$0 \leq R_i^{(s)} \leq 1 , \tag{13.175}$$

which, as discussed previously, is perfectly adequate for deriving limiter functions for the case in which the underlying first–order scheme is the Godunov first order *upwind* method [145]. However, for the case in which the underlying first order scheme is *centred*, direct application of (13.175) leads to over restrictive TVD regions that may actually exclude the underlying monotone first–order scheme! The TVD condition (13.171) is more restrictive than (13.75) but since this condition relates to extrema, for which the scheme is locally first–order accurate, the TVD conditions (13.170)– (13.171) are overall more relaxed than (13.175).

TVD Regions. We first prove a result that concerns the construction of the TVD region for centred flux limiter schemes.

Theorem 13.7.2. *In order to ensure a TVD scheme, flux limiters $\phi(r)$ must lie in a region (the TVD region) satisfying the following constraints*

$$\phi_B \leq \phi(r) \leq \phi_T , \tag{13.176}$$

$$\phi_L^+(r) \leq \phi(r) \leq \phi_R^+(r) , \quad r > 0 , \tag{13.177}$$

$$\phi_R^- \leq \phi(r) \leq \phi_L^-(r) , \quad r \leq 0 , \tag{13.178}$$

$$\phi_L^+(r) = (S_L + \phi_T)r , \tag{13.179}$$

$$\phi_R^+(r) = \phi_g + (S_R^+ + \phi_B)r , \tag{13.180}$$

$$\phi_L^-(r) = \phi_g + (S_L + \phi_T)r , \tag{13.181}$$

$$\phi_R^-(r) = (S_R^- + \phi_B)r . \tag{13.182}$$

Here ϕ_B, ϕ_T are global lower and upper bounds and definitions for the parameters $S_L, \phi_g, S_R^+, S_R^-$ are given in Table 13.3.

Proof. (For details of proof see Toro and Billett [374]).

A few remarks are in order. The TVD region is determined by inequalities involving six straight lines, namely the horizontal bounds ϕ_B, ϕ_T and the functions defined by (13.179)–(13.182). Fig. 13.22 depicts the general TVD region defined by the theorem. We have chosen $\phi_B = 0$ and ϕ_T consistent with a non–positive slope for $\phi_L^-(r)$ and $\phi_L^+(r)$. In fact it suffices to take these functions to be constant and we may therefore set

$$\phi_T = -\frac{\alpha_0}{\alpha_1 - \beta_1} ; \quad a > 0 ; \quad \phi_T = -\frac{\alpha_1}{\alpha_0 - \beta_0} ; \quad a < 0 . \tag{13.183}$$

and

$$\phi_L^+(r) = 0 ; \quad \phi_L^-(r) \equiv \phi_g = \begin{cases} \frac{\alpha_1}{\alpha_1 - \beta_1} , & a > 0 , \\ \frac{\alpha_0}{\alpha_0 - \beta_0} , & a < 0 . \end{cases} \tag{13.184}$$

Table 13.3 lists expressions for various quantities involved, for three first–order schemes, including the Godunov first order upwind method.

	$a > 0$	$a < 0$	FORCE	Lax–Friedrichs	Godunov
S_L	$\dfrac{\alpha_0}{\alpha_1 - \beta_1}$	$\dfrac{\alpha_1}{\alpha_0 - \beta_0}$	$\dfrac{1 + |c|}{1 - |c|}$	$\dfrac{1}{1 - |c|}$	$\dfrac{2}{1 - |c|}$
ϕ_g	$\dfrac{\alpha_1}{\alpha_1 - \beta_1}$	$\dfrac{\alpha_0}{\alpha_0 - \beta_0}$	$\dfrac{1 - |c|}{1 + |c|}$	$\dfrac{1}{1 + |c|}$	0
S_R^+	$\dfrac{\alpha_0 - 1/|c|}{\alpha_1 - \beta_1}$	$\dfrac{\alpha_1 - 1/|c|}{\alpha_0 - \beta_0}$	$\dfrac{3 + |c|}{1 + |c|}$	$\dfrac{1}{1 + |c|}$	$\dfrac{2}{|c|}$
S_R^-	$\dfrac{\alpha_0 - 1/|c| - \alpha_1}{\alpha_1 - \beta_1}$	$\dfrac{\alpha_1 - 1/|c| - \alpha_0}{\alpha_0 - \beta_0}$	2	0	$\dfrac{2}{|c|}$

Table 13.3. Values of useful quantities for the FORCE scheme, the Lax– Friedrichs scheme and the Godunov first order upwind method.

Remark 13.7.2. In the course of the proof of the above theorem [374] it is found that the convenient values for the parameters $\epsilon_L, \epsilon_0, \epsilon_R$ in (13.170), (13.171) are

$$\epsilon_L = 1 , \quad \epsilon_0 = \epsilon_R = -c\alpha_1 . \tag{13.185}$$

Had we enforced the usual Data Compatibility Conditions (13.175), we would

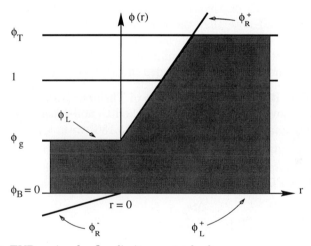

Fig. 13.22. TVD region for flux limiter centred schemes

have had $\epsilon_0 = \epsilon_R = 0$ and thus the TVD region in Fig. 13.22 would have excluded limiter functions $\phi(r)$ with $\phi(r) \leq \phi_g > 0$ except for the special case $|c| = 1$, for which $\phi_g = 0$. Even the basic case $\phi(r) = 0$ that reproduces the low–order monotone scheme in (13.130) would be excluded. For upwind methods, this difficulty does not arise, as the downwind coefficient (α_1 for $a > 0$ and α_0 for $a < 0$) is zero, and thus $\phi_g = 0 \; \forall c$.

Construction of Limiters. The task ahead is to construct flux limiters $\phi(r)$ to be used in the flux limiter scheme (13.130). In this connection the following result applies.

Theorem 13.7.3. *The centred TVD flux*

$$f^{(c)}_{i+\frac{1}{2}} = f^{LO}_{i+\frac{1}{2}} + \phi_{i+\frac{1}{2}}(r)(f^{LW}_{i+\frac{1}{2}} - f^{LO}_{i+\frac{1}{2}}) \,, \qquad (13.186)$$

where $f^{LO}_{i+\frac{1}{2}}$ is the flux for some first–order centred monotone scheme and $f^{LW}_{i+\frac{1}{2}}$ is the Lax–Wendroff flux, reduces to the flux for the Godunov first–order upwind flux

$$f^{(g)}_{i+\frac{1}{2}} = \frac{1}{2}(1 + s)(au^n_i) + \frac{1}{2}(1 - s)(au^n_{i+1}) \,, \qquad (13.187)$$

with $s = sign(a)$, when

$$\phi(r) = \phi_g \,, \qquad (13.188)$$

and upwind flux limiters $\psi(r)$ are related to centred flux limiters $\phi(r)$ by the equation

$$\phi(r) = \phi_g + (1 - \phi_g)\psi(r) \,. \qquad (13.189)$$

Proof. (Left as an exercise).

Remark 13.7.3. Based on the above result, we call ϕ_g the Godunov point, see Table 13.3, and note that the conventional TVD conditions (13.175), while suitable for upwind TVD schemes, do not admit flux limiters below ϕ_g. That is, schemes that have larger numerical viscosity such as FORCE and the Lax–Friedrichs method, are not included in the TVD region.

Remark 13.7.4. Based on the first result of Theorem (13.7.3) one can generalise the Godunov first–order upwind method to non–linear scalar conservation laws by simply defining a flux as in (13.186) with $\phi_{i+\frac{1}{2}}(r) = \phi_g$. Such extension depends on the particular centred low order flux $f^{LO}_{i+\frac{1}{2}}$ used and on a Courant number c; this may be obtained from the characteristic speed (local) or the CFL coefficient (global). A local choice is

$$c_{i+\frac{1}{2}} = \frac{\Delta t s_{i+\frac{1}{2}}}{\Delta x} \,, \quad s_{i+\frac{1}{2}} \equiv |a_{i+\frac{1}{2}}| \,, \qquad (13.190)$$

where $a_{i+\frac{1}{2}}$ is a characteristic speed at the cell interface $i + \frac{1}{2}$. We note that such extension *does not require the solution of a local Riemann problem*, as does the Godunov first–order upwind method.

Before constructing flux limiters $\phi(r)$ we note that the parameters S_L, ϕ_g, S^+_R, S^-_R in (13.179)–(13.182) define the TVD region, and they depend on the Courant number $|c|$; see Table 13.3. By identifying possible minimum and maximum values of $|c|$, i.e.

$$c_{min} = \min |c| \,, \quad c_{max} = \max |c| \,, \qquad (13.191)$$

we construct a TVD region based on the most restrictive conditions. The corresponding values for the relevant parameters are

$$\phi_T = \phi_T(c_{min}) , \quad \phi_g = \phi_g(c_{max}) , \quad S_R^+ = S_R^+(c_{max}) . \qquad (13.192)$$

Recall that we set $\phi_L^+(r) = 0$; $\phi_L^-(r) = \phi_g$ and S_L in (13.179), (13.181) is defined once ϕ_T is chosen. For the model equation (13.128) we interpret (13.191) as meaning $c_{min} = 0$ and $c_{max} = 1$, leading to a Courant–number independent TVD region depicted in Fig. 13.23. It follows that the only

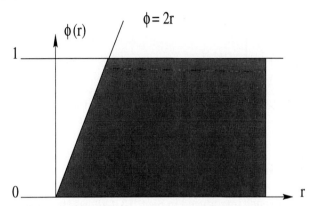

Fig. 13.23. TVD region for flux limiter centred schemes, that is independent of Courant number

possible flux limiters are of the MINBEE type [333], [283], the most prominent example being

$$\phi_{mb}(r) = \begin{cases} 0 , & r \leq 0 , \\ r , & 0 \leq r \leq 1 , \\ 1 , & r \geq 1 . \end{cases} \qquad (13.193)$$

Note that $\phi_{mb}(r)$ is the conventional upwind flux limiter MINBEE (13.165) [333], [283]. This is perfectly adequate for most realistic applications but has the disadvantage of being over diffusive. In general, larger values of $\phi(r)$ add less numerical viscosity than smaller ones. It is therefore desirable to preserve at least part of the TVD region above $\phi = 1$. Using the relation (13.189) between conventional upwind flux limiters $\psi(r)$ and the sought centred flux limiters $\phi(r)$ we construct these as follows

$$\phi(r) = \hat{\phi}_g + (1 - \hat{\phi}_g)\psi(r) , \qquad (13.194)$$

with

$$\hat{\phi}_g = \begin{cases} 0 , & r \leq 1 , \\ \phi_g \equiv \phi_g(c_{max}) , & r \geq 1 . \end{cases} \qquad (13.195)$$

By choosing the upwind flux limiters ULTRABEE, SUPERBEE, VANLEER and VANALBADA we obtain *corresponding* centred flux limiters

$$\phi_{ub}(r) = \begin{cases} 0 \,, & r \leq 0 \,, \\ min \left\{ \phi_T, \phi_g + S_R^+ r \right\} \,, & r > 0 \,, \end{cases} \qquad (13.196)$$

$$\phi_{sb}(r) = \begin{cases} 0 \,, & r \leq 0 \,, \\ 2r \,, & 0 \leq r \leq \frac{1}{2} \,, \\ 1 \,, & \frac{1}{2} \leq r \leq 1 \,, \\ min \left\{ 2, \phi_g + (1 - \phi_g)r \right\} \,, & r > 1 \,, \end{cases} \qquad (13.197)$$

$$\phi_{vl}(r) = \begin{cases} 0 \,, & r \leq 0 \,, \\ \frac{2r}{1+r} \,, & 0 \leq r \leq 1 \,, \\ \phi_g + \frac{2(1-\phi_g)r}{1+r} \,, & r \geq 1 \,, \end{cases} \qquad (13.198)$$

$$\phi_{va}(r) = \begin{cases} 0 \,, & r \leq 0 \,, \\ \frac{r(1+r)}{1+r^2} \,, & 0 \leq r \leq 1 \,, \\ \phi_g + \frac{(1-\phi_g)r(1+r)}{1+r^2} \,, & r \geq 1 \,, \end{cases} \qquad (13.199)$$

We stressed that the corresponding centred limiters (13.196)–(13.199) are only *analogous* to the upwind limiters, they are *not equivalent*.

13.8 Slope Limiter Methods

The MUSCL approach introduced in Sect. 13.4 allows the construction of high order methods. In Sects. 13.4.2 to 13.4.4 we constructed second and third order accurate extensions of the Godunov first–order upwind method using this approach. In Sect. 13.4.5 we constructed second and third order accurate extensions of *centred* first order schemes following the MUSCL–Hancock methodology. All of these high order schemes, as stated, will produce spurious oscillations in the vicinity of high gradients, see Godunov's theorem in Sect. 13.5.3. In this section we construct non–linear versions of these schemes by replacing the slopes Δ_i in the data reconstruction step (13.23) by limited slopes $\overline{\Delta}_i$, according to some TVD constraints.

13.8.1 TVD Conditions

First we note that in constructing fully discrete second–order explicit TVD methods, restricting the boundary extrapolated values u_i^L, u_i^R in (13.24) to satisfy

$$\left. \begin{aligned} min\{u_{i-1}^n, u_i^n\} \leq u_i^L \leq max\{u_{i-1}^n, u_i^n\} \,, \\ min\{u_{i+1}^n, u_i^n\} \leq u_i^R \leq max\{u_i^n, u_{i+1}^n\} \,. \end{aligned} \right\} \qquad (13.200)$$

does not lead to useful results. Instead, we impose a restriction on the *evolved* boundary extrapolated values $\overline{u}_i^L, \overline{u}_i^R$, see (13.30) or (13.33), and prove the following result.

Theorem 13.8.1. *If the evolved boundary extrapolated values* $\{\overline{u}_k^{L,R}\}$ *satisfy*

$$\left.\begin{array}{l} \min\{u_{i-1}^n, u_i^n\} \leq \overline{u}_i^L \leq \max\{u_{i-1}^n, u_i^n\} \;,\; \forall i\;, \\[2mm] \min\{u_i^n, u_{i+1}^n\} \leq \overline{u}_i^R \leq \max\{u_i^n, u_{i+1}^n\} \;,\; \forall i\;, \end{array}\right\} \tag{13.201}$$

then, for any monotone *scheme of the form*

$$v_i^{n+1} = \sum_{k=-1}^{1} b_k v_{i+k}^n \tag{13.202}$$

applied to $\{\overline{u}_k^{L,R}\}$ *one has*

$$\min_k\{u_k^n\}_{k=i-2}^{i+2} \leq u_i^{n+1} \leq \max_k\{u_k^n\}_{k=i-1}^{i+2} \;. \tag{13.203}$$

Hence the corresponding MUSCL scheme is TVD.

Proof. From conditions (13.201) we have

$$\min_k\{u_k^n\}_{k=i-1}^{i+1} \leq \overline{u}_i^{L,R} \leq \max_k\{u_k^n\}_{k=i-1}^{i+1} \;,\; \forall i\;. \tag{13.204}$$

For any monotone scheme (13.202) applied to $\{u_i^n; \overline{u}_k^{L,R}\}_{k=i-1}^{i+1}$ we will have

$$\min_k\{u_i^n; \overline{u}_k^{L,R}\}_{k=i-1}^{i+1} \leq \overline{u}_i^{n+1} \leq \max_k\{u_i^n; \overline{u}_k^{L,R}\}_{k=i-1}^{i+1} \;. \tag{13.205}$$

But from (13.201)

$$\left.\begin{array}{l} \min\{u_k^n\}_{k=i-2}^{i+2} \leq \min\{u_i^n; \overline{u}_k^{L,R}\}_{k=i-1}^{i+1} \;, \\[2mm] \max\{u_i^n; \overline{u}_k^{L,R}\}_{k=i-1}^{i+1} \leq \max\{u_k^n\}_{k=i-2}^{i+2} \;. \end{array}\right\} \tag{13.206}$$

From (13.205) and (13.206) inequalities (13.203) follow and the result is thus proved.

13.8.2 Construction of TVD Slopes

Now we construct *limited* slopes $\overline{\Delta}_i$ to replace Δ_i in (13.23), or (13.33), according to the TVD constraints (13.201) and prove the following result.

Theorem 13.8.2. *If the limited slopes* $\overline{\Delta}_i$ *are chosen according to*

$$\left.\begin{array}{c} \overline{\Delta}_i = \frac{1}{2}[sign(\Delta_{i-\frac{1}{2}}) + sign(\Delta_{i+\frac{1}{2}})] \times \min[\beta_{i-\frac{1}{2}}|\Delta_{i-\frac{1}{2}}|, \beta_{i+\frac{1}{2}}|\Delta_{i+\frac{1}{2}}|] \\[3mm] \beta_{i-\frac{1}{2}} = \dfrac{2}{1+c} \;,\quad \beta_{i+\frac{1}{2}} = \dfrac{2}{1-c} \;, \end{array}\right\} \tag{13.207}$$

then the resulting MUSCL scheme is TVD.

Proof. First assume $u^n_{i-1} \leq u^n_i$. Application of (13.201) leads to

$$u^n_{i-1} \leq u^n_i - \frac{1}{2}(1+c)\overline{\Delta}_i \leq u^n_i .$$

Use of these two inequalities produces

$$0 \leq \overline{\Delta}_i \leq \frac{2}{1+c}\Delta_{i-\frac{1}{2}} ; \quad \Delta_{i-\frac{1}{2}} \geq 0 . \tag{13.208}$$

The case in which $u^n_{i-1} \geq u^n_i$ leads to

$$\frac{2}{1+c}\Delta_{i-\frac{1}{2}} \leq \overline{\Delta}_i \leq 0 ; \quad \Delta_{i-\frac{1}{2}} \leq 0 . \tag{13.209}$$

Similarly, for $u^n_i \leq u^n_{i+1}$ one obtains

$$0 \leq \overline{\Delta}_i \leq \frac{2}{1-c}\Delta_{i+\frac{1}{2}} ; \quad \Delta_{i+\frac{1}{2}} \geq 0 \tag{13.210}$$

and for $u^n_i \geq u^n_{i+1}$ one obtains

$$\frac{2}{1-c}\Delta_{i+\frac{1}{2}} \leq \overline{\Delta}_i \leq 0 ; \quad \Delta_{i+\frac{1}{2}} \leq 0 . \tag{13.211}$$

From inspection of conditions (13.208) and (13.210) when $\Delta_{i-\frac{1}{2}} \geq 0$, $\Delta_{i+\frac{1}{2}} \geq 0$, one requires $\overline{\Delta}_i$ to be bounded by zero and the minimum of $\frac{2}{1+c}\Delta_{i-\frac{1}{2}}$ and $\frac{2}{1-c}\Delta_{i+\frac{1}{2}}$. Conditions (13.209) and (13.211), when $\Delta_{i-\frac{1}{2}} \leq 0$, $\Delta_{i+\frac{1}{2}} \leq 0$, require $\overline{\Delta}_i$ to lie between

$$\text{sign}(\Delta_{i-\frac{1}{2}}) \times \min\{\frac{2}{1+c}|\Delta_{i-\frac{1}{2}}|, \frac{2}{1-c}|\Delta_{i+\frac{1}{2}}|\}$$

and zero. When $\Delta_{i-\frac{1}{2}}$ and $\Delta_{i+\frac{1}{2}}$ are of opposite sign, the only possible choice is $\overline{\Delta}_i = 0$. Hence all the above constraints may then be written as in (13.207).

The reader may easily verify that (13.207) may also be written as

$$\overline{\Delta}_i = \text{sign}(\Delta_{i+\frac{1}{2}}) \times \max\{0, \min[\beta_{i-\frac{1}{2}}\Delta_{i-\frac{1}{2}}\text{sign}(\Delta_{i+\frac{1}{2}}), \beta_{i+\frac{1}{2}}\Delta_{i+\frac{1}{2}}]\} . \tag{13.212}$$

13.8.3 Slope Limiters

The TVD analysis of the previous section is sufficient to construct *upwind* and *centred* slope limiter methods. A re–interpretation of this analysis in terms of a *slope limiter* ξ_i such that

$$\overline{\Delta}_i = \xi_i \Delta_i \,, \tag{13.213}$$

with Δ_i as given by

$$\Delta_i = \frac{1}{2}(1+\omega)\Delta u_{i-\frac{1}{2}} + \frac{1}{2}(1-\omega)\Delta u_{i+\frac{1}{2}} \,, \tag{13.214}$$

is useful in a number of ways. It leads to a TVD region for $\xi(r)$ given as follows

$$\xi(r) = 0 \text{ for } r \leq 0 \,, \ 0 \leq \xi(r) \leq \min\{\xi_L(r), \xi_R(r)\} \text{ for } r > 0 \,, \tag{13.215}$$

where

$$\left.\begin{aligned}
\xi_L(r) &= \frac{2\beta_{i-\frac{1}{2}}r}{1-\omega+(1+\omega)r} \,, \\
\xi_R(r) &= \frac{2\beta_{i+\frac{1}{2}}}{1-\omega+(1+\omega)r} \,, \\
r &= \frac{\Delta_{i-\frac{1}{2}}}{\Delta_{i+\frac{1}{2}}} \,.
\end{aligned}\right\} \tag{13.216}$$

The corresponding TVD region is depicted in Fig. 13.24, see also [16]. A

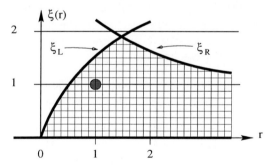

Fig. 13.24. TVD region for slope limiters. For negative r TVD region is single line $\xi = 0$ and for positive r TVD region lies between 0 and $\min\{\xi_L(r), \xi_R(r)\}$.

possible slope limiter function $\xi(r)$ is

$$\xi(r) = \left\{\begin{array}{ll} 0 \,, & r \leq 0 \\ \min\{\xi_L(r), \xi_R(r)\} \,, & r > 0 \end{array}\right. \tag{13.217}$$

This slope limiter is analagous to the flux limiter ULTRABEE, in that one follows the boundaries of the TVD region. It is expected to be over compressive for smooth solutions but to give satisfactory results for discontinuities. A slope limiter that is *analagous* (not equivalent) to the SUPERBEE flux limiter is

$$
\xi(r) = \begin{cases} 0\,, & \text{if } r \le 0\,, \\ 2r\,, & \text{if } 0 \le r \le \frac{1}{2}\,, \\ 1\,, & \text{if } \frac{1}{2} \le r \le 1\,, \\ \min\{r, \xi_R(r), 2\}\,, & \text{if } r \ge 1\,. \end{cases} \tag{13.218}
$$

A van Leer–type slope limiter is

$$
\xi(r) = \begin{cases} 0\,, & r \le 0\,, \\ \min\{\dfrac{2r}{1+r}, \xi_R(r)\}\,, & r \ge 0\,. \end{cases} \tag{13.219}
$$

A van Albada–type slope limiter is

$$
\xi(r) = \begin{cases} 0\,, & r \le 0\,, \\ \min\{\dfrac{r(1+r)}{1+r^2}, \xi_R(r)\}\,, & r \ge 0\,. \end{cases} \tag{13.220}
$$

A MINBEE–type slope limiter is

$$
\xi(r) = \begin{cases} 0\,, & r \le 0\,, \\ r\,, & 0 \le r \le 1\,, \\ \min\{1, \xi_R(r)\}\,, & r \ge 1\,. \end{cases} \tag{13.221}
$$

One issue is the choice of the coefficients $\beta_{i-\frac{1}{2}}$, $\beta_{i+\frac{1}{2}}$ in (13.207) and (13.216). There are two cases to consider, namely the case of positive Courant number c and the case of negative Courant number c. For schemes that are stable in the range $|c| \le 1$ one has

$$
\begin{cases} \beta_{i-\frac{1}{2}} \in [1,2] & \text{and} & \beta_{i+\frac{1}{2}} \in [2,\infty) & \text{when } a > 0\,, \\ \beta_{i-\frac{1}{2}} \in [2,\infty) & \text{and} & \beta_{i+\frac{1}{2}} \in [1,2] & \text{when } a < 0\,. \end{cases} \tag{13.222}
$$

For upwind methods one may exploit the upwind information contained in the Courant number c. For *centred* (non–upwind) methods such information is unavailable in general and we therefore recommend the choice

$$
\beta_{i-\frac{1}{2}} = \beta_{i+\frac{1}{2}} = 1\,, \tag{13.223}
$$

which gives a *smaller* TVD region, valid for all Courant numbers in the range $|c| \le 1$. One may improve this by utilising a limited amount of wave information via the CFL coefficient, which must be available for any explicit scheme to solve any hyperbolic system. See Sect. 6.3.2 of Chap. 6 on the CFL condition.

13.8.4 Limited Slopes Obtained from Flux Limiters

By establishing a relationship between upwind–based flux limiter schemes and MUSCL–type schemes for the model linear advection equation (13.1), one may select the *limited slopes* $\overline{\Delta}_i$ in the reconstruction step so as to reproduce conventional (upwind) flux–limiters $\psi_{i+\frac{1}{2}}$. From (13.130), the intercell flux for the flux limiter method resulting from the Godunov first order upwind and the Lax–Wendroff schemes is

$$f_{i+\frac{1}{2}} = \begin{cases} au_i^n + \psi_{i+\frac{1}{2}}[\frac{1}{2}(1-c)\Delta_{i+\frac{1}{2}}]a \,, & a > 0 \,, \\ au_{i+1}^n - \psi_{i+\frac{1}{2}}[\frac{1}{2}(1+c)\Delta_{i+\frac{1}{2}}]a \,, & a < 0 \,, \end{cases} \tag{13.224}$$

where $\psi_{i+\frac{1}{2}}$ is an upwind flux limiter, see Sect. 13.7.3. An upwind–based slope limiter method has intercell flux

$$f_{i+\frac{1}{2}} = \begin{cases} au_i^n + \frac{1}{2}(1-c)a\overline{\Delta}_i \,, & a > 0 \,, \\ au_{i+1} - \frac{1}{2}(1+c)a\overline{\Delta}_{i+1} \,, & a < 0 \,, \end{cases} \tag{13.225}$$

where $\overline{\Delta}_i$ and $\overline{\Delta}_{i+1}$ are *limited* slopes in cells i and $i+1$ respectively. See (13.35), (13.36), (13.43) and (13.49). By comparing (13.224) and (13.225) one obtains

$$\psi_{i+\frac{1}{2}}\Delta_{i+\frac{1}{2}} = \begin{cases} \overline{\Delta}_i \,, & a > 0 \,, \\ \overline{\Delta}_{i+1} \,, & a < 0 \,. \end{cases} \tag{13.226}$$

As $\psi_{i+\frac{1}{2}} = 1 \,\, \forall i$ reproduces the Lax–Wendroff method in the flux–limiter scheme, the choice of slopes

$$\Delta_i = \begin{cases} \Delta_{i+\frac{1}{2}} \,, & a > 0 \,, \\ \Delta_{i-\frac{1}{2}} \,, & a < 0 \end{cases} \tag{13.227}$$

in the MUSCL schemes also reproduces the Lax–Wendroff method. Note that this would be obtained by an appropriate value of the parameter ω in (13.27). Using this relation one can construct *limited slopes*

$$\overline{\Delta}_i = \overline{\Delta}_i(\Delta_{i-\frac{1}{2}}, \Delta_{i+\frac{1}{2}}) \tag{13.228}$$

so as to reproduce conventional (upwind) flux limiters. The reader can easily verify that two methods for obtaining *limited slopes* result from

$$\overline{\Delta}_i = \begin{cases} max[0, min(\beta\Delta_{i-\frac{1}{2}}, \Delta_{i+\frac{1}{2}}), min(\Delta_{i-\frac{1}{2}}, \beta\Delta_{i+\frac{1}{2}})] \,, & \Delta_{i+\frac{1}{2}} > 0 \,, \\ min[0, max(\beta\Delta_{i-\frac{1}{2}}, \Delta_{i+\frac{1}{2}}), max(\Delta_{i-\frac{1}{2}}, \beta\Delta_{i+\frac{1}{2}})] \,, & \Delta_{i+\frac{1}{2}} < 0 \end{cases} \tag{13.229}$$

for particular values of the parameter β. The value $\beta = 1$ reproduces the MINBEE flux limiter (13.165), which may also be written as

$$\psi_{mi}(r) = max[0, min(1, r)] \ . \tag{13.230}$$

$\beta = 2$ reproduces the SUPERBEE flux limiter (13.162), which may also be written as

$$\psi_{sb}(r) = max[0, min(2r, 1), min(r, 2)] \ . \tag{13.231}$$

The parameter r is defined in (13.157).

Remark 13.8.1. Note that upwind MUSCL type schemes based on the limited slopes (13.229) have the Lax–Wendroff as the base second–order scheme, when applied to the linear advection equation (13.1), and *not* the Fromm scheme; see Sect. 13.4.

13.9 Extensions of TVD Methods

In this chapter we have studied a variety of TVD schemes for the scalar, homogeneous linear partial differential equation (13.1). Crucial questions concern the extension of these methods to scalar inhomogeneous (source terms) equations, convection–diffusion equations and non–linear systems. The extension of the schemes to homogeneous non–linear systems is carried out in Chap. 14. Here we make some remarks concerning model PDEs with source terms and diffusion terms.

13.9.1 TVD Schemes in the Presence of Source Terms

Sweby [334] considered the problem of devising TVD schemes to solve the inhomogeneous PDE

$$u_t + au_x = s(u) \ , \tag{13.232}$$

where $s(u)$ is an algebraic function of the unknown $u = u(x, t)$, usually called a *source term*, or forcing term. In fact an even more basic problem is that of just devising numerical methods to solve (13.232) after some fashion. In Chap. 15 we present splitting methods, whereby the full inhomogeneous problem is *split* into a homogenous problem, for which a TVD scheme can be applied directly, and an Ordinary Differential Equation (ODE) that can be solved by some appropriate ODE solver. This approach carries over to non–linear inhomogeneous systems of PDEs. For special cases in which a change of variables allows us to re–write the PDEs as a, new, homogeneous problem, one applies TVD methods directly to the homogeneous problem and thus to the full original inhomogeneous problem. See Sweby [334] for a discussion on this approach and Watson et. al. [406] for an application to the non–linear shallow water equations with variable bed elevation.

On the question of whether one can construct TVD schemes for general inhomogeneous PDEs of the form (13.232), it should first be realised that the TVD concept itself is inappropriate here. The effect of the source term may cause the Total Variation of the exact solution to increase and it would therefore be absurd to attempt to produce numerical methods to compute approximate solutions with properties not enjoyed by the exact solution.

For general problems one resorts to the splitting schemes presented in Chap. 15.

13.9.2 TVD Schemes in the Presence of Diffusion Terms

Consider the question of devising TVD schemes to solve the convection–diffusion equation

$$u_t + au_x = \alpha_{phy} u_{xx} , \qquad (13.233)$$

where α_{phy} is a physical or natural viscosity coefficient, assumed constant here. A very simple way of solving this PDE is to deploy the splitting schemes studied in Chap. 15. These splitting schemes apply a TVD method to the convection part $u_t + au_x = 0$ and the parabolic PDE $u_t = \alpha_{phy} u_{xx}$ is solved by some other appropriate method. In general these methods are found to be quite successful for solving the Navier–Stokes equations [56], [375].

Recall however that when a TVD method is applied to $u_t + au_x = 0$, one effectively adds a numerical viscosity term to this pure convection PDE, so that one solves

$$u_t + au_x = \alpha_{num} u_{xx} , \qquad (13.234)$$

where α_{num} is a numerical or artificial viscosity coefficient; see Sect. 5.2 of Chap. 5. The value of α_{num} depends on the flux or slope limiter used and its purpose is to avoid spurious oscillations in the vicinity of high gradients in the numerical solution. As the complete PDE (13.233) to be solved contains a physical viscosity term, an obvious question is this: is the physical viscosity by itself sufficient to guarantee oscillation free solutions ? The answer is no, in general. Toro [359] considered this problem and found that for certain TVD discretisation schemes one does not require artificial viscosity only if

$$R_{cell} \leq \frac{2}{1 - |c|} , \qquad (13.235)$$

where R_{cell} is the cell Reynolds number defined as $R_{cell} = \frac{|c|}{d}$; c is the Courant number and $d = \frac{\alpha_{phy} \Delta t}{\Delta x^2}$ is the diffusion number; in this case the physical viscosity is sufficient to guarantee oscillation–free solutions. Otherwise, numerical viscosity is also needed. Viscous flux limiters were constructed in [359], whereby the absolute minimum of *extra numerical viscosity* is added to the physical viscosity to guarantee oscillation–free solutions. These viscous limiters use fully the physical viscosity and minimise the amount of numerical viscosity; in fact this is turned off when condition (13.235) is satisfied. See [359] for details on preliminary results for model problems.

In the next section we present some numerical results for the linear advection equation.

13.10 Numerical Results for Linear Advection

We consider the same two test problems for the linear advection (13.1) as in Sect. 13.7.2. We take $a = 1$ and a CFL coefficient $C_{\text{cfl}} = 0.8$. Computed results are shown at the output time $t = 10$ units (1250 time steps). In each figure we compare the exact solution (shown by full lines) with the numerical solution (symbols).

The initial conditions for Test 1 (smooth data) and Test 2 (discontinuous data, square wave) are respectively given by

$$u(x,0) = \alpha e^{-\beta x^2} \; ; \; u(x,0) = \begin{cases} 0 & \text{if} \quad x \le 0.3 \,, \\ 1 & \text{if} \quad 0.3 \le x \le 0.7 \,, \\ 0 & \text{if} \quad x \ge 0.7 \,. \end{cases} \qquad (13.236)$$

For Test 1 we take $\alpha = 1.0$ and $\beta = 8.0$; the initial profile $u(x,0)$ is evaluated in the interval $-1 \le x \le 1$; the mesh used is $\Delta x = 0.02$. For Test 2 the mesh used is $\Delta x = 0.01$. In both test problems the initial profile and that at the output time is resolved by 100 computing cells.

Results are presented for two classes of methods, namely *flux limiter methods* and *slope limiter methods*. For each of these two classes we consider one *upwind*–based method and one *centred* method. The upwind schemes are high–order, TVD extensions of the Godunov first–order upwind scheme. The centred schemes used are second–order, TVD extensions of the FORCE scheme (with $\omega = 0$), see Sect. 13.2.1. The centred schemes are the **Flux LI**miter **C**entred, or FLIC, scheme (Sect. 13.7.4) and the **Slope LI**miter **C**entred, or SLIC, scheme (Sect. 13.8). In the centred TVD schemes we use the FORCE flux as the low–order flux. All of these schemes are extended to non–linear systems in Chap. 14.

Results from the flux limiter methods are shown in Figs. 13.25 and 13.26 for Tests 1 and 2 respectively. Results from the slope limiter methods are shown in Figs. 13.27 and 13.28 for the same tests. The results of the left column in each of the figures are those of the upwind based scheme and those on the right column are obtained with the centred TVD scheme. Four *types* of limiter functions have been used, namely SUPERBEE, VANLEER, VANALBADA and MINBEE; *note that the limiters of the upwind scheme are not equivalent to those of the centred scheme.*

The main conclusion that may be drawn from the results is that there appears to be no obvious advantages, at least for this scalar problem, in using the upwind based TVD scheme. We note however, as will be seen in Chap. 14, that for non–linear systems, upwind methods give more accurate solutions.

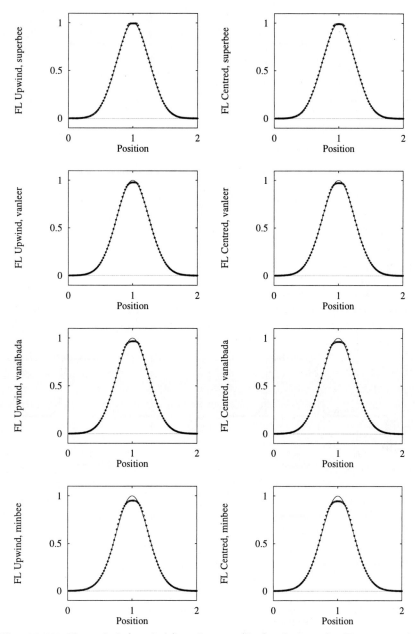

Fig. 13.25. Numerical (symbols) and exact (line) solutions for Tests 1 at time $t = 10$. Results from Flux–Limiter Upwind and Flux–Limiter Centred (FORCE) are compared using various flux limiters.

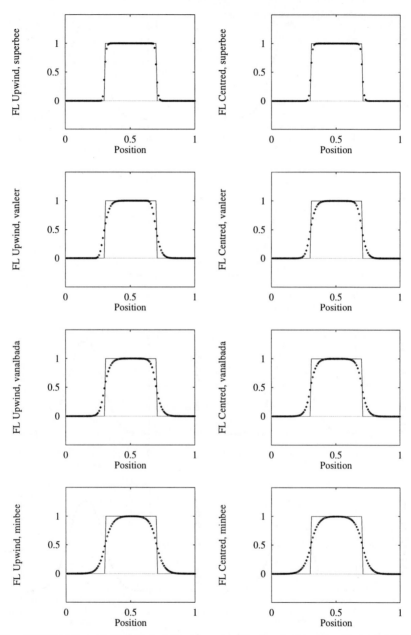

Fig. 13.26. Numerical (symbols) and exact (line) solutions for Tests 2 at time $t = 10$. Results from Flux–Limiter Upwind and Flux–Limiter Centred (FORCE) are compared using various flux limiters.

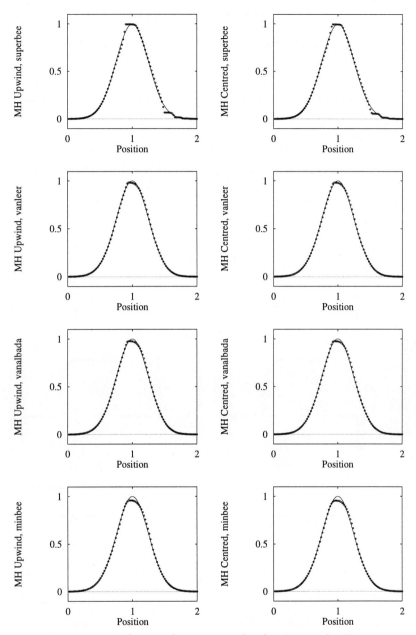

Fig. 13.27. Numerical (symbols) and exact (line) solutions for Tests 2 at time $t = 10$. Results from MUSCL–Hancock Upwind and MUSCL–Hancock Centred (FORCE) are compared using various slope limiters.

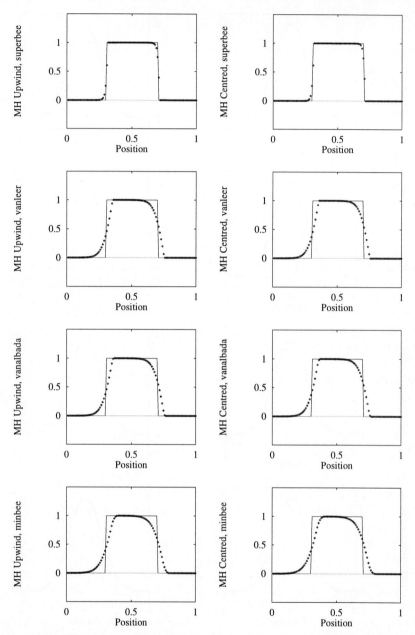

Fig. 13.28. Numerical (symbols) and exact (line) solutions for Tests 2 at time $t = 10$. Results from MUSCL–Hancock Upwind and MUSCL–Hancock Centred (FORCE) are compared using various slope limiters.

14. High–Order and TVD Schemes for Non–Linear Systems

This chapter is concerned with TVD *upwind* and *centred* schemes for non–linear systems of conservation laws that depend on time t, or a time–like variable t, and one space dimension x. The upwind schemes are extensions of the Godunov first order upwind method of Chap. 6 and can be applied with any of the Riemann solvers presented in Chap. 4 (exact) and Chaps. 9 to 12 (approximate); they can also be used with the Flux Vector Splitting flux of Chap. 8. The centred schemes are extensions of the First Order Centred (FORCE) method presented in Chap. 7. All the TVD schemes are in effect the culmination of work carried out in all previous chapters, particularly Chap. 13, where the TVD concept was developed in the context of simple scalar problems. The schemes are presented in terms of the time–dependent one dimensional Euler equations for ideal gases, which are introduced in Chap. 1 and studied in detail in Chap. 3. Applications to other systems may be easily accomplished. Techniques for extending the methods to systems with source terms, as for reactive flows for instance, are given in Chap. 15 and to multidimensional systems in Chap. 16.

14.1 Introduction

We study fully discrete, explicit methods that are suitable for time–dependent problems. *Upwind* TVD schemes are more accurate than their *centred* TVD counterparts; a disadvantage of upwind schemes is their complexity and the computing cost due to the solution of the Riemann problem, see Chaps. 4, 6, 8, 9, 10, 11 and 12. For ideal gases the extra expense of upwind schemes is not significant, if efficient Riemann solvers are used. The TVD *centred* schemes presented are considerably simpler to implement and somewhat more efficient to apply. They are particularly recommended to the reader who does not want to get involved with the details of the Riemann problem. Also, for complicated sets of conservation laws for which the solution of the Riemann problem is not available, or is complex, or is prohibitly expensive to compute, then TVD centred schemes are definitely an option to consider.

The methods are applied to the Initial–Boundary Value Problem

$$
\begin{array}{lll}
PDEs & : & \mathbf{U}_t + \mathbf{F}(\mathbf{U})_x = \mathbf{0} \ , \\
ICs & : & \mathbf{U}(x,0) = \mathbf{U}^{(0)}(x) \ , \\
BCs & : & \mathbf{U}(0,t) = \mathbf{U}_l(t) \ ; \quad \mathbf{U}(L,t) = \mathbf{U}_r(t) \ .
\end{array}
\left.\vphantom{\begin{array}{l}a\\b\\c\end{array}}\right\} \qquad (14.1)
$$

We assume the spatial domain $[0, L]$ to be discretised into M computing cells $I_i = [x_{i-\frac{1}{2}}, x_{i+\frac{1}{2}}]$, and for simplicity we assume the mesh size $\Delta x = x_{i+\frac{1}{2}} - x_{i-\frac{1}{2}}$ to be constant. The time–dependent Euler equations are typical conservation laws in (14.1), for which the vectors $\mathbf{U}$ of conserved variables and $\mathbf{F} = \mathbf{F}(\mathbf{U})$ of fluxes are

$$
\mathbf{U} = \begin{bmatrix} \rho \\ \rho u \\ E \end{bmatrix} \ ; \quad \mathbf{F}(\mathbf{U}) = \begin{bmatrix} \rho u \\ \rho u^2 + p \\ u(E + p) \end{bmatrix} \qquad (14.2)
$$

respectively. Here ρ is density, u is velocity, p is pressure and E is total energy per unit volume. See Chaps. 1 and 3 for background on the Euler equations.

We solve the general IBVP (14.1) using the explicit, conservative fully discrete scheme

$$
\mathbf{U}_i^{n+1} = \mathbf{U}_i^n + \frac{\Delta t}{\Delta x}[\mathbf{F}_{i-\frac{1}{2}} - \mathbf{F}_{i+\frac{1}{2}}] \ , \qquad (14.3)
$$

where $\mathbf{F}_{i+\frac{1}{2}}$ is the numerical flux at the cell interface position $x_{i+\frac{1}{2}}$. For background on the meaning of the conservative formula (14.3) the reader is referred to Sect. 5.3 of Chap. 5 and Chap. 6. The choice of the time step Δt in (14.3) and the implementation of boundary conditions is addressed in Sect. 14.2. The choice of the numerical flux $\mathbf{F}_{i+\frac{1}{2}}$ in (14.3) determines the scheme. Here we study two upwind–based second order TVD schemes and two centred second–order TVD schemes. These are extensions of schemes presented in detail for scalar conservation laws in Chap. 13. In addition, we present two upwind second order TVD schemes based on *primitive variables* and adaptive primitive–conservative variations of these.

The rest of this chapter is organised as follows: Sect. 14.2 discusses the application of the CFL condition to compute the time step Δt and the implementation of boundary conditions. Sects. 14.3 and 14.4 present two upwind TVD schemes, namely the WAF method and the MUSCL–Hancock scheme respectively. Sect. 14.5 presents two centred TVD schemes; the first method, called FLIC, is a flux limiter extension of the First Order Centred (FORCE) scheme; the other, called SLIC, is a slope limiter extension of FORCE. Sect. 14.6 presents primitive variable schemes and an adaptive primitive–conservative scheme. Numerical results are shown in Sect. 14.7.

14.2 CFL and Boundary Conditions

Before describing particular schemes contained in (14.3) by specifying the numerical flux, we briefly address two basic problems, namely the choice of the time step Δt and application of boundary conditions. See Sects. 6.3.2 and 6.3.3 of Chap. 6 for full discussion on these topics, in the context of the Godunov first order upwind method.

All methods considered in this chapter have linearised stability constraint $|c| \leq 1$, where c is the Courant number. For non–linear systems we implement this condition as

$$\Delta t = C_{cfl} \frac{\Delta x}{S_{max}^{(n)}} ,\qquad (14.4)$$

where Δx is the mesh spacing, $S_{max}^{(n)}$ is the maximum wave speed present at time level n and C_{cfl} is the CFL coefficient, with $C_{cfl} \in (0,1]$. See Sect. 6.3.2 of Chap. 6 for details on possible choices for $S_{max}^{(n)}$. A practical choice, which must be used with caution, is

$$S_{max}^{(n)} = \max_i \{|u_i^n| + a_i^n\} ,\qquad (14.5)$$

where the range for i must include data arising from boundary conditions. As remarked in Sect. 6.3.2 of Chap. 6, inappropriate choices for $S_{max}^{(n)}$ in (14.4) can result in the scheme becoming unstable, even for *small* values of C_{cfl}.

A detailed discussion on boundary conditions for the first–order Godunov method is given in Sect. 6.3.3 of Chap. 6, which the reader is encouraged to consult before proceeding. For the second order methods discussed in this chapter the application of boundary conditions is fundamentally the same as for the Godunov method. We assume the computational domain $[0, L]$ to be discretised by M cells I_i, so that cells $i = 1, \cdots, M$ lie within the computational domain. In applying boundary conditions we now require *two* fictitious cells next to each boundary. For the left boundary at $x = 0$ the fictitious cells are denoted by $i = -1$ and $i = 0$ and for the right boundary they are denoted by $i = M + 1$ and $i = M + 2$. We discuss two types of boundary conditions.

Transmissive boundary conditions are given by

$$\left. \begin{array}{llll} \mathbf{W}_0^n & = & \mathbf{W}_1^n ; & \mathbf{W}_{-1}^n & = & \mathbf{W}_2^n , \\ \mathbf{W}_{M+1}^n & = & \mathbf{W}_M^n ; & \mathbf{W}_{M+2}^n & = & \mathbf{W}_{M-1}^n , \end{array} \right\} \qquad (14.6)$$

where $\mathbf{W}$ may be the vector of conserved variables or some other variables, such as the primitive variables.

Reflective boundary conditions are applied at reflective, moving or stationary, boundaries. Suppose the speed of a reflective solid boundary at $x = L$ is u_{wall}, then the reflective boundary conditions for the Euler equations are applied as

$$\left. \begin{array}{lll} \rho_{M+1}^n = \rho_M^n ; & u_{M+1}^n = -u_M^n + 2u_{wall} ; & p_{M+1}^n = p_M^n , \\ \rho_{M+2}^n = \rho_{M-1}^n ; & u_{M+2}^n = -u_{M-1}^n + 2u_{wall} ; & p_{M+2}^n = p_{M-1}^n . \end{array} \right\} \qquad (14.7)$$

For a moving reflective left boundary the situation is entirely analagous to the case just given.

Having decided on the application of boundary conditions and on a strategy to compute the time step Δt, all that remains to be discussed is the selection of the intercell flux $\mathbf{F}_{i+\frac{1}{2}}$, for a scheme (14.3) to be completely determined. Sects. 14.3 to 14.6 present various ways of choosing the intercell flux.

14.3 Weighted Average Flux (WAF) Schemes

The Weighted Average Flux (WAF) approach for obtaining second–order extensions of the Godunov first–order upwind method has its origins in the Random Flux Method [350], [378]. WAF is a deterministic approach and has two versions. In the first the intercell flux is an integral average of the physical flux across the full structure of the solution of a local Riemann problem. The second version computes an averaged state and then $\mathbf{F}_{i+\frac{1}{2}}$ results from evaluating the physical flux $\mathbf{F}$ at this state.

Useful background is found in Sect. 13.3 of Chap. 13, where the WAF approach is presented in detail, when applied to scalar equations.

14.3.1 The Original Version of WAF

The simplest WAF flux is given as

$$\mathbf{F}_{i+\frac{1}{2}} = \frac{1}{\Delta x} \int_{-\frac{1}{2}\Delta x}^{\frac{1}{2}\Delta x} \mathbf{F}(\mathbf{U}_{i+\frac{1}{2}}(x, \frac{\Delta t}{2}))dx , \qquad (14.8)$$

where $\mathbf{U}_{i+\frac{1}{2}}(x, t)$ is the solution of the Riemann problem with *piece–wise constant data* $\mathbf{U}_i^n, \mathbf{U}_{i+1}^n$ at the interface position $i + \frac{1}{2}$. $\mathbf{U}_{i+\frac{1}{2}}(x, t)$ is exactly the same solution as for the Godunov first–order upwind method, see Chap. 6. Chap. 4 gives the detailed, exact solution of the Riemann problem for the Euler equations. The approximate Riemann solvers of Chaps. 9 to 12 may also be used here. Fig. 14.1 depicts the structure of the solution of the Riemann problem for the Euler equations. There are three waves of speeds S_1, S_2, S_3

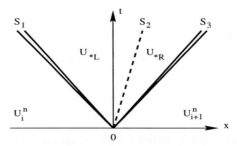

Fig. 14.1. Structure of the solution of the Riemann problem with data $\mathbf{U}_i^n$, $\mathbf{U}_{i+1}^n$

that separate four constant states: $\mathbf{U}_i^n, \mathbf{U}_{*L}, \mathbf{U}_{*R}, \mathbf{U}_{i+1}^n$. The left and right waves may be shock or rarefaction waves. In the latter case there will be additional, non–uniform states present, namely those inside rarefaction fans. Fig 14.2 illustrates the evaluation of the integral average (14.8), for a wave structure assumed to contain no rarefaction waves. By setting

$$\mathbf{U}^{(1)} = \mathbf{U}_i^n \; ; \quad \mathbf{U}^{(2)} = \mathbf{U}_{*L} \; ; \quad \mathbf{U}^{(3)} = \mathbf{U}_{*R} \; ; \quad \mathbf{U}^{(4)} = \mathbf{U}_{i+1}^n \qquad (14.9)$$

the integral (14.8) gives

$$\mathbf{F}_{i+\frac{1}{2}} = \sum_{k=1}^{N+1} \beta_k \mathbf{F}_{i+\frac{1}{2}}^{(k)} \; , \qquad (14.10)$$

where $\mathbf{F}_{i+\frac{1}{2}}^{(k)} = \mathbf{F}(\mathbf{U}^{(k)})$, N is the number of waves in the solution of the Riemann problem and β_k, $k = 1, \cdots, 4$, are the normalised lengths of the segments $A_{k-1}A_k$,

$$\beta_k = \frac{|A_{k-1}A_k|}{\Delta x} \; .$$

It can be seen easily that, in terms of the wave speeds S_k, the weights are

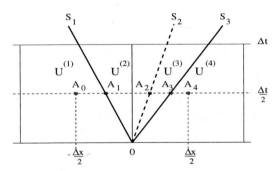

Fig. 14.2. Evaluation of the WAF intercell flux. Integral average becomes summation involving flux terms in regions 1 to 4 with weights obtained by normalising lengths $|A_0A_1|$ to $|A_3A_4|$

$$\left.\begin{array}{l} \beta_k = \frac{1}{2}(c_k - c_{k-1}) \; , \\[2mm] c_k = \frac{\Delta t S_k}{\Delta x} \; , \quad c_0 = -1 \; , \quad c_{N+1} = 1 \; . \end{array}\right\} \qquad (14.11)$$

Here c_k is the Courant number for wave k of speed S_k. Formula (14.10) is a *weighted average flux*, and thus the name of the scheme.

Substitution of β_k in (14.11) into (14.10) gives an alternative form for the WAF flux

$$\mathbf{F}_{i+\frac{1}{2}} = \frac{1}{2}(\mathbf{F}_i + \mathbf{F}_{i+1}) - \frac{1}{2}\sum_{k=1}^{N} c_k \Delta\mathbf{F}_{i+\frac{1}{2}}^{(k)} , \qquad (14.12)$$

where

$$\Delta\mathbf{F}_{i+\frac{1}{2}}^{(k)} = \mathbf{F}_{i+\frac{1}{2}}^{(k+1)} - \mathbf{F}_{i+\frac{1}{2}}^{(k)} \qquad (14.13)$$

is the flux jump across wave k of CFL number c_k.

14.3.2 A Weighted Average State Version

A possible variant of the scheme is obtained by first defining a *weighted average state* as

$$\overline{\mathbf{W}}_{i+\frac{1}{2}} = \frac{1}{\Delta x} \int_{-\frac{1}{2}\Delta x}^{\frac{1}{2}\Delta x} \mathbf{W}_{i+\frac{1}{2}}(x, \tfrac{1}{2}\Delta t)dx , \qquad (14.14)$$

where $\mathbf{W}$ is a suitable vector of variables and $\mathbf{W}_{i+\frac{1}{2}}$ is the solution of the Riemann problem with data $\mathbf{W}_i^n, \mathbf{W}_{i+1}^n$. For the assumed wave structure of Fig. 14.2 this integral becomes

$$\overline{\mathbf{W}}_{i+\frac{1}{2}} = \sum_{k=1}^{N} \beta_k \mathbf{W}_{i+\frac{1}{2}}^{(k)} , \qquad (14.15)$$

where the weights β_k are as given by (14.11) and $\mathbf{W}_{i+\frac{1}{2}}^{(k)}$ is the value of $\mathbf{W}_{i+\frac{1}{2}}$ in region k. Expanding (14.15) gives

$$\overline{\mathbf{W}}_{i+\frac{1}{2}} = \frac{1}{2}(\mathbf{W}_i^n + \mathbf{W}_{i+1}^n) - \frac{1}{2}\sum_{k=1}^{N} c_k[\mathbf{W}_{i+\frac{1}{2}}^{(k+1)} - \mathbf{W}_{i+\frac{1}{2}}^{(k)}] . \qquad (14.16)$$

Compare (14.15), (14.16) with (14.10), (14.12). An intercell flux is now defined as

$$\mathbf{F}_{i+\frac{1}{2}} = \mathbf{F}(\overline{\mathbf{W}}_{i+\frac{1}{2}}) . \qquad (14.17)$$

As to the choice of the vector $\mathbf{W}$ in (14.17) we recommend the set of primitive variables, i.e. $\mathbf{W} = (\rho, u, p)^T$.

Computationally, formulation (14.17), (14.16) is more efficient than formulation (14.12); it involves only one flux evaluation per component of the system.

14.3.3 Rarefactions in State Riemann Solvers

When the Riemann solver provides a state at which the physical flux is to be evaluated, such as in the exact Riemann solver of Chap. 4 and the approximate state Riemann solvers of Chap. 9, one requires a special treatment of rarefaction waves in the WAF method. For other Riemann solvers such as HLL and HLLC of Chap. 10 such special treatment is not needed.

Let us consider the case of the exact Riemann solver. In the presence of rarefaction waves in the solution $\mathbf{U}_{i+\frac{1}{2}}(x,t)$ of the Riemann problem, definitions (14.8) and (14.14) still apply. As the exact solution across rarefactions is available, see Chap. 4, the exact evaluation of (14.8) or (14.14) is possible and the result is again a summation like (14.10) or (14.15), with extra terms to account for regions occupied by rarefactions.

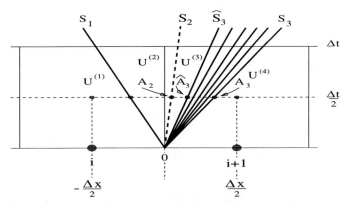

Fig. 14.3. Treatment of right non–sonic rarefaction when evaluating the WAF flux

For the Euler equations, waves 1 (left) and 3 (right) may be shocks or rarefactions; wave 2 (middle) is always a contact discontinuity of speed $S_2 = u_*$. See Sect. 3.1.3 of Chap. 3. Fig. 14.3 depicts the case in which the right wave is a rarefaction, whose bounding characteristics have speeds

$$\hat{S}_3 = S_{TR} = u_* + a_{*R} \quad \text{(Tail)} ; \quad S_3 = S_{HR} = u_{i+1}^n + a_{i+1}^n \quad \text{(Head)} .$$

Evaluation of the integral (14.8) gives one term for the zone across the rarefaction, namely

$$\frac{1}{\Delta x} \int_{\hat{A}_3}^{A_3} \mathbf{F}(\mathbf{U}_{i+\frac{1}{2}}(x, \frac{\Delta t}{2}))dx , \tag{14.18}$$

which can be found exactly, if desired. This results in one extra state, one extra weight β_k and one extra wave speed $\hat{S}_3$.

Computational experience suggests that for the purpose of evaluating the integral (14.8) or (14.14) one may 'lump' the rarefaction state together with the closest *constant* state in the direction of the t–axis. The integral in the resulting enlarged zone, $\mathbf{A}_2$ to $\mathbf{A}_3$ in Fig. 14.3, is then approximated as

$$\frac{1}{\Delta x} \int_{\mathbf{A}_2}^{\mathbf{A}_3} \mathbf{F}(\mathbf{U}_{i+\frac{1}{2}}(x, \frac{\Delta t}{2}))dx \approx \frac{1}{2}(c_3 - c_2)\mathbf{F}_{i+\frac{1}{2}}^{(3)} \ , \tag{14.19}$$

where

$$\left.\begin{array}{l} S_3 = S_{HR} = u_{i+1}^n + a_{i+1}^n \ , \\[2mm] \mathbf{F}_{i+\frac{1}{2}}^{(3)} = \left\{ \begin{array}{ll} \mathbf{F}(\mathbf{U}^{(3)}) & : \ \text{non–sonic rarefaction} \ , \\ \mathbf{F}(\mathbf{U}_{i+\frac{1}{2}}(0)) & : \ \text{sonic rarefaction} \ . \end{array} \right. \end{array}\right\} \tag{14.20}$$

Fig. 14.4 illustrates the case in which the right rarefaction is sonic. The required state lies inside the rarefaction along the t–axis. Recall that for ideal gases the solution inside rarefactions is given in closed form, see Sect. 4.4 of Chap. 4.

For a left rarefaction we have

$$\frac{1}{\Delta x} \int_{\mathbf{A}_1}^{\mathbf{A}_2} \mathbf{F}(\mathbf{U}_{i+\frac{1}{2}}(x, \frac{\Delta t}{2}))dx \approx \frac{1}{2}(c_2 - c_1)\mathbf{F}_{i+\frac{1}{2}}^{(2)} \ , \tag{14.21}$$

where

$$\left.\begin{array}{l} S_1 = S_{HL} = u_i^n - a_i^n \ , \\[2mm] \mathbf{F}_{i+\frac{1}{2}}^{(2)} = \left\{ \begin{array}{ll} \mathbf{F}(\mathbf{U}^{(2)}) & : \ \text{non–sonic rarefaction} \ , \\ \mathbf{F}(\mathbf{U}_{i+\frac{1}{2}}(0)) & : \ \text{sonic rarefaction} \ . \end{array} \right. \end{array}\right\} \tag{14.22}$$

Note that the flux term evaluated at $\mathbf{U}_{i+\frac{1}{2}}(0)$ is the Godunov flux, which is

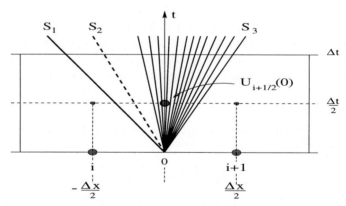

Fig. 14.4. Treatment of right sonic rarefaction when evaluating the WAF flux

always part of the WAF flux.

The special treatment of rarefactions, just described, also applies when using the Approximate–State Riemann Solvers of Chap. 9. When using the HLL and HLLC Riemann solvers of Chap. 10, no need for this special treatment arises; this is also the case if one was to use the Flux Vector Splitting intercell flux of Chap. 8.

14.3.4 TVD Version of WAF Schemes

As seen in Sect. 13.3 of Chap. 13, the WAF schemes are second order accurate in space and time. According to Godunov's Theorem, Sect. 13.5.3 of Chap. 13, spurious oscillations in the vicinity of high gradients are to be expected. Here we extend the Total Variation Diminishing (TVD) version of the WAF method derived in Sect. 13.7.1 to non–linear systems.

Strictly speaking, the extension to non–linear systems is somewhat empirical but is found to work well in practice. In the scalar case, for which the TVD property was enforced rigourously, there is one conservation law and one wave. In the case of the Euler equations in (14.1), (14.2) there are three conservation laws and three wave families. For the purpose of applying a TVD constraint to non–linear systems, a valuable, though empirical, observation is that the solution to the complete system may be characterised by *jumps in a single quantity q*, across each of the three waves. We therefore require the computation of three limiter functions $\phi_{i+\frac{1}{2}}(r)$ per intercell boundary. The reader is encouraged to review carefully Sect 13.7.1 before proceeding with the rest of this section.

The TVD modification of the WAF flux (14.12) is

$$
\mathbf{F}_{i+\frac{1}{2}} = \frac{1}{2}(\mathbf{F}_i + \mathbf{F}_{i+1}) - \frac{1}{2}\sum_{k=1}^{N} \text{sign}(c_k)\phi_{i+\frac{1}{2}}^{(k)} \Delta \mathbf{F}_{i+\frac{1}{2}}^{(k)} , \tag{14.23}
$$

where

$$
\phi_{i+\frac{1}{2}}^{(k)} = \phi_{i+\frac{1}{2}}(r^{(k)}) \tag{14.24}
$$

is a WAF limiter function, as derived in Sect. 13.7.1 of Chap. 13. The flow parameter $r^{(k)}$ refers to wave k in the solution $\mathbf{U}_{i+\frac{1}{2}}(x,t)$ of the Riemann problem and is the ratio

$$
r^{(k)} = \begin{cases} \dfrac{\Delta q_{i-\frac{1}{2}}^{(k)}}{\Delta q_{i+\frac{1}{2}}^{(k)}} , & \text{if } c_k > 0 , \\[3ex] \dfrac{\Delta q_{i+\frac{3}{2}}^{(k)}}{\Delta q_{i+\frac{1}{2}}^{(k)}} , & \text{if } c_k < 0 . \end{cases} \tag{14.25}
$$

As has already been indicated, one selects a single quantity q which is known to change across each wave family in the solution of the Riemann problem. For the Euler equations the choices $q \equiv \rho$ (density) or $q \equiv e$ (specific internal energy) give very satisfactory results. For multidimensional problems, see Chap. 16, jumps in tangential velocity components must also be used. $\Delta q_{i-\frac{1}{2}}^{(k)}$ denotes the jump in q across wave k in the solution $\mathbf{U}_{i-\frac{1}{2}}(x,t)$ of the Riemann problem with data $(\mathbf{U}_{i-1}^n, \mathbf{U}_i^n)$, $\Delta q_{i+\frac{3}{2}}^{(k)}$ is the jump in q across wave k in the solution $\mathbf{U}_{i+\frac{3}{2}}(x,t)$ of the Riemann problem with data $(\mathbf{U}_{i+1}^n, \mathbf{U}_{i+2}^n)$; $\Delta q_{i+\frac{1}{2}}^{(k)}$ is the corresponding jump across wave k in the *local* Riemann problem solution $\mathbf{U}_{i+\frac{1}{2}}(x,t)$ with data $(\mathbf{U}_i^n, \mathbf{U}_{i+1}^n)$.

The TVD version of the weighted average state version of the scheme has

$$\overline{\mathbf{W}}_{i+\frac{1}{2}} = \frac{1}{2}(\mathbf{W}_i^n + \mathbf{W}_{i+1}^n) - \frac{1}{2}\sum_{k=1}^{N} \text{sign}(c_k)\phi_k(\mathbf{W}_{i+\frac{1}{2}}^{(k+1)} - \mathbf{W}_{i+\frac{1}{2}}^{(k)}) \quad (14.26)$$

and the TVD flux is

$$\mathbf{F}_{i+\frac{1}{2}} = \mathbf{F}(\overline{\mathbf{W}}_{i+\frac{1}{2}}) .$$

As to the WAF limiters ϕ one may use any of the following

$$\phi_{sa}(r, |c|) = \begin{cases} 1 & \text{if } r \leq 0 , \\ 1 - 2(1 - |c|)r & \text{if } 0 \leq r \leq \frac{1}{2} , \\ |c| & \text{if } \frac{1}{2} \leq r \leq 1 , \\ 1 - (1 - |c|)r & \text{if } 1 \leq r \leq 2 , \\ 2|c| - 1 & \text{if } r \geq 2 , \end{cases} \quad (14.27)$$

$$\phi_{vl}(r, |c|) = \begin{cases} 1 & \text{if } r \leq 0 , \\ 1 - \frac{(1-|c|)2r}{1+r} & \text{if } r \geq 0 , \end{cases} \quad (14.28)$$

$$\phi_{va}(r, |c|) = \begin{cases} 1 & \text{if } r \leq 0 , \\ 1 - \frac{(1-|c|)r(1+r)}{(1+r^2)} & \text{if } r \geq 0 , \end{cases} \quad (14.29)$$

$$\phi_{ma}(r, |c|) = \begin{cases} 1 & \text{if } r \leq 0 , \\ 1 - (1 - |c|)r & \text{if } 0 \leq r \leq 1 , \\ |c| & \text{if } r \geq 1 . \end{cases} \quad (14.30)$$

For convenience, subscripts and superscripts have been omitted. The WAF limiter functions are related, and are entirely equivalent, to conventional flux limiters $\psi(r)$ via

$$\phi(r) = 1 - (1 - |c|)\psi(r) .$$

ϕ_{sa} is related to SUPERBEE, ϕ_{vl} is related to van Leer's limiter, ϕ_{va} is related to van Albada's limiter and ϕ_{ma} is related to MINBEE. See Sect. 13.7.1 of Chap. 13, for details on the derivation of limiter functions for the WAF method.

14.3.5 Riemann Solvers

The WAF method may be used with the exact Riemann solver or approximate Riemann solvers. An exact Riemann solver is given in Chap. 4 for the Euler equations. Approximate Riemann solvers for the Euler equations are given in Chaps. 9 to 12; one may also use the Flux–Vector splitting intercell flux given in Chap. 8. We particularly recommend the adaptive, approximate–state Riemann solvers of Chap. 9 and the HLLC approximate Riemann solver presented in Chap. 10 for the Euler equations. The WAF approach may be generalised to irregular grids. A Courant–number 2 extension of the WAF method is reported in [373]. Numerical results are presented in Sect. 14.7

14.3.6 Summary of the WAF Method

For a given domain $[0, L]$ and a mesh size Δx, determined by the choice of the number M of computing cells, one first sets the initial conditions $\mathbf{U}^0 \equiv \mathbf{U}(x, 0)$ at time $t = 0$. Then, for each time step n one performs the following operations

Operation (I):   **Boundary conditions.** This is carried out according to (14.6) or (14.7), for example.

Operation (II):   **Computation of time step.** This is carried out according to (14.4) and (14.5). A choice of C_{cfl} must be made at the beginning of the computations. One usually takes $C_{cfl} = 0.9$. Recall however that the choice of $S_{max}^{(n)}$ is crucial and given that (14.5) produces somewhat unreliable estimates for the true speeds we recommend that when solving problems with shock–tube like data, the CFL coefficient C_{cfl} be set to small number, e.g. 0.2, for a few time steps, e.g. 5.

Operation (III): **Solution of Riemann problem.** For each pair of data states $(\mathbf{U}_i^n, \mathbf{U}_{i+1}^n)$, compute the solution of Riemann problem and compute: states $\mathbf{U}^k$, fluxes $\mathbf{F}_{i+\frac{1}{2}}^k$, wave speeds S_k and wave jumps $\Delta q_{i+\frac{1}{2}}$, where q may be chosen to be density, when solving the Euler equations.

Operation (IV):   **TVD intercell flux.** At each cell interface $i + \frac{1}{2}$ compute the ratios r^k according to (14.25), compute Courant numbers c_k according to (14.11), compute limiter functions according to any of (14.27) to (14.30), and, finally compute the intercell flux at position $i + \frac{1}{2}$ according to (14.23); alternatively one computes the intercell flux according to (14.17), (14.26).

Operation (V):   **Updating of solution.** Proceed to update the conserved variables according to the conservative formula (14.3)

Operation (VI):   **Next time level.** Go to (I).

14.4 The MUSCL–Hancock Scheme

Here we present a scheme that achieves second order accuracy by applying the MUSCL approach [387], [388], [389], [390], [391], [392], [395]. MUSCL stands for Monotonic Upstream–Centred Scheme for Conservation Laws. An introduction to this general approach was given in Sect. 13.4 of Chap. 13, in the context of scalar conservation laws. Three MUSCL type schemes were presented in detail, namely the MUSCL–Hancock method [395], [275], the Piece–Wise Linear method (PLM) of Colella [87] and the Generalised Riemann Problem (GRP) approach of Ben–Artzi and Falcovitz [23], [24], [22]. The reader is encouraged to review Sects. 13.4.1 and 13.4.2 of Chap. 13 before proceeding.

14.4.1 The Basic Scheme

In this section we present the extension of the MUSCL–Hancock scheme to any system of non–linear hyperbolic conservation laws

$$\mathbf{U}_t + \mathbf{F}(\mathbf{U})_x = \mathbf{0} \qquad (14.31)$$

and discuss the details when (14.31) are the Euler equations, with the vector of conserved variables $\mathbf{U}$ and fluxes $\mathbf{F}(\mathbf{U})$ as given by (14.2). The objective is to construct a flux $\mathbf{F}_{i+\frac{1}{2}}$ in (14.3) that is a second order extension of the Godunov first–order upwind method, Chap. 6. Given that the bases of the scheme were discussed in great detail in Sects. 13.4.1 and 13.4.2 of Chap. 13, we restrict ourselves here to a succinct, algorithmic presentation of the method.

The MUSCL–Hancock approach achieves a second order extension of the Godunov first order upwind method if the intercell flux $\mathbf{F}_{i+\frac{1}{2}}$ is computed according to the following steps:

Step (I):   **Data Reconstruction.** In the following, we assume a choice of variables $\mathbf{W}$ has been made. One possibility is the conserved variables $\mathbf{W} = \mathbf{U}$ of course; another choice is offered by the *primitive or physical* variables. For the Euler equations (14.1), (14.2) these are $\mathbf{W} = (\rho, u, p)^T$. Here we present the scheme in terms of the conserved variables $\mathbf{U}$. In the data reconstruction step, data cell average values $\mathbf{U}_i^n$ are locally replaced by piece–wise linear functions in each cell $I_i = [x_{i-\frac{1}{2}}, x_{i+\frac{1}{2}}]$, namely

$$\mathbf{U}_i(x) = \mathbf{U}_i^n + \frac{(x - x_i)}{\Delta x} \Delta_i , \quad x \in [0, \Delta x] . \tag{14.32}$$

Δ_i is a suitably chosen *slope* vector (actually a difference) of $\mathbf{U}_i(x)$ in cell I_i. The extreme points $x = 0$ and $x = \Delta x$, in local co–ordinates, correspond to the intercell boundaries $x_{i-\frac{1}{2}}$ and $x_{i+\frac{1}{2}}$, in global co–ordinates, respectively. The values of $\mathbf{U}_i(x)$ at the extreme points are

$$\mathbf{U}_i^L = \mathbf{U}_i^n - \frac{1}{2}\Delta_i ; \quad \mathbf{U}_i^R = \mathbf{U}_i^n + \frac{1}{2}\Delta_i \tag{14.33}$$

and are usually called *boundary extrapolated values*. Note that $\mathbf{U}$ and Δ_i are vectors of three components for the Euler equations and thus there are six scalar extrapolated values in (14.33).

Step (II):   **Evolution.** For each cell I_i, the boundary extrapolated values $\mathbf{U}_i^L, \mathbf{U}_i^R$ in (14.33) are *evolved* by a time $\frac{1}{2}\Delta t$, according to

$$\left.\begin{array}{l} \overline{\mathbf{U}}_i^L = \mathbf{U}_i^L + \dfrac{1}{2}\dfrac{\Delta t}{\Delta x}[\mathbf{F}(\mathbf{U}_i^L) - \mathbf{F}(\mathbf{U}_i^R)] , \\[2mm] \overline{\mathbf{U}}_i^R = \mathbf{U}_i^R + \dfrac{1}{2}\dfrac{\Delta t}{\Delta x}[\mathbf{F}(\mathbf{U}_i^L) - \mathbf{F}(\mathbf{U}_i^R)] . \end{array}\right\} \tag{14.34}$$

Note that this evolution step is entirely contained in each cell I_i, as the *intercell* fluxes are evaluated at the boundary extrapolated values of each cell. At each intercell position $i + \frac{1}{2}$ there are two fluxes, namely $\mathbf{F}(\mathbf{U}_i^R)$ and $\mathbf{F}(\mathbf{U}_{i+1}^L)$, which are in general distinct. This does not really affect the *conservative* character of the overall method, as this step is only an intermediate step; the intercell flux $\mathbf{F}_{i+\frac{1}{2}}$ to be used in (14.3) is yet to be evaluated.

Step (III):   **The Riemann Problem.** To compute the intercell flux $\mathbf{F}_{i+\frac{1}{2}}$ one now solves the conventional Riemann problem with data

$$\mathbf{U}_L \equiv \overline{\mathbf{U}}_i^R ; \quad \mathbf{U}_R \equiv \overline{\mathbf{U}}_{i+1}^L \tag{14.35}$$

to obtain the similarity solution $\mathbf{U}_{i+\frac{1}{2}}(x/t)$. The intercell flux $\mathbf{F}_{i+\frac{1}{2}}$ is now computed in *exactly the same way as in the Godunov first order upwind method*, namely

$$\mathbf{F}_{i+\frac{1}{2}} = \mathbf{F}(\mathbf{U}_{i+\frac{1}{2}}(0)) . \tag{14.36}$$

Here $\mathbf{U}_{i+\frac{1}{2}}(0)$ denotes the value of $\mathbf{U}_{i+\frac{1}{2}}(x/t)$ at $x/t = 0$, i.e. the value of $\mathbf{U}_{i+\frac{1}{2}}(x/t)$ along the t–axis. Fig. 14.5 illustrates only one of the ten possible wave patterns in the solution of the Riemann problem that must be taken into account when computing the Godunov flux. See Chaps. 4 and 6 for details.

A possible choice for the *slope* vector Δ_i in (14.32), (14.33) is

$$\Delta_i = \frac{1}{2}(1+w)\Delta_{i-\frac{1}{2}} + \frac{1}{2}(1-w)\Delta_{i+\frac{1}{2}} , \qquad (14.37)$$

where

$$\Delta_{i-\frac{1}{2}} \equiv \mathbf{U}_i^n - \mathbf{U}_{i-1}^n ; \quad \Delta_{i+\frac{1}{2}} \equiv \mathbf{U}_{i+1}^n - \mathbf{U}_i^n . \qquad (14.38)$$

and $w \in [-1, 1]$.

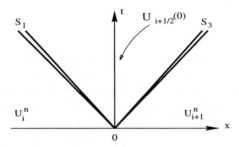

Fig. 14.5. Evaluation of the Godunov flux for a particular wave pattern occuring in the solution of the local Riemann problem

14.4.2 A Variant of the Scheme

It is possible to modify steps (I) and (II) above by formulating the reconstruction process in terms of a vector $\mathbf{W}$ of primitive variables. For the Euler equations we may choose $\mathbf{W} = (\rho, u, p)^T$ and we have

$$\mathbf{W}_t + \mathbf{A}(\mathbf{W})\mathbf{W}_x = 0 . \qquad (14.39)$$

Then steps (I) and (II) can be reduced to a single step. As the evolution step does not need to be performed in terms of the conserved variables, we replace (14.34) by the scheme

$$\overline{\mathbf{W}}_i^{L,R} = \mathbf{W}_i^{L,R} + \frac{1}{2}\frac{\Delta t}{\Delta x}\mathbf{A}(\mathbf{W}_i^n)[\mathbf{W}_i^L - \mathbf{W}_i^R] . \qquad (14.40)$$

Here $\mathbf{A}(\mathbf{W}_i^n)$ denotes the matrix $\mathbf{A}$ in (14.39) evaluated at the data cell average $\mathbf{W}_i^n$. Substitution of $\mathbf{W}_i^L$, $\mathbf{W}_i^R$, as obtained from (14.33), into (14.40) gives

$$\left.\begin{array}{l}
\overline{\mathbf{W}}_i^L = \mathbf{W}_i^n - \dfrac{1}{2}[\mathbf{I} + \dfrac{\Delta t}{\Delta x}\mathbf{A}(\mathbf{W}_i^n)]\Delta_i \; , \\[3mm]
\overline{\mathbf{W}}_i^R = \mathbf{W}_i^n + \dfrac{1}{2}[\mathbf{I} - \dfrac{\Delta t}{\Delta x}\mathbf{A}(\mathbf{W}_i^n)]\Delta_i \; ,
\end{array}\right\} \tag{14.41}$$

where $\mathbf{I}$ is the identity matrix. Thus the complete MUSCL–Hancock scheme reduces to solving the Riemann problem with data

$$\left.\begin{array}{l}
\mathbf{W}_L = \mathbf{W}_i^n + \dfrac{1}{2}[\mathbf{I} - \dfrac{\Delta t}{\Delta x}\mathbf{A}(\mathbf{W}_i^n)]\Delta_i \; , \\[3mm]
\mathbf{W}_R = \mathbf{W}_{i+1}^n - \dfrac{1}{2}[\mathbf{I} + \dfrac{\Delta t}{\Delta x}\mathbf{A}(\mathbf{W}_{i+1}^n)]\Delta_{i+1} \; ,
\end{array}\right\} \tag{14.42}$$

to find $\mathbf{W}_{i+\frac{1}{2}}(x/t)$ and evaluate the intercell flux as

$$\mathbf{F}_{i+\frac{1}{2}} = \mathbf{F}(\mathbf{W}_{i+\frac{1}{2}}(0)) \; ,$$

in which the vector $\mathbf{U}_{i+\frac{1}{2}}(0)$ in (14.36) is replaced by $\mathbf{W}_{i+\frac{1}{2}}(0)$. See also Sect. 14.6.3.

For primitive variables, in the one–dimensional time dependent Euler equations, we have

$$\mathbf{A} = \begin{pmatrix} u & \rho & 0 \\ 0 & u & \dfrac{1}{\rho} \\ 0 & \rho a^2 & u \end{pmatrix} \; , \tag{14.43}$$

where a is the speed of sound. See Chap. 3 for matrices for the Euler equations in three space dimensions. See also Chap. 8.

14.4.3 TVD Version of the Scheme

As seen in Sect. 13.4.2 of Chap. 13, the MUSCL–Hancock scheme is second–order accurate in space and time for any value of the parameter ω in the slopes (14.37), when applied to the linear advection equation $u_t + au_x = 0$. For $\omega = 0$ the scheme reduces to the Fromm method. It follows from Godunov's Theorem, see Sect. 13.5.3 of Chap. 13, that spurious oscillations will be produced in the vicinity of strong gradients. The purpose here is to remedy this problem using some Total Variation Diminishing (TVD) constraint.

We present TVD versions of the scheme. These are extensions of the analysis of Sect. 13.8 of Chap. 13, for the linear advection equation, to non–linear systems of conservation laws (14.31). The extension is somewhat empirical; the rigour of the scalar case can not be preserved for non–linear systems. In spite of this, the TVD versions of the schemes work well and computed solutions are, in the main, oscillation free. The TVD version of the scheme is obtained by replacing the slopes Δ_i in (14.33) by *limited* slopes $\overline{\Delta}_i$. As discussed in Sect. 13.8, there are essentially two ways of achieving this.

Limited Slopes. A well established approach [10], [275] is to select *limited slopes* directly from forcing equivalence of the schemes with conventional flux limiter methods, for the model scalar equation. Limited slopes are thus obtained from

$$
\overline{\Delta}_i = \begin{cases}
max[0, min(\beta\Delta_{i-\frac{1}{2}}, \Delta_{i+\frac{1}{2}}), min(\Delta_{i-\frac{1}{2}}, \beta\Delta_{i+\frac{1}{2}})] , & \Delta_{i+\frac{1}{2}} > 0 , \\
min[0, max(\beta\Delta_{i-\frac{1}{2}}, \Delta_{i+\frac{1}{2}}), max(\Delta_{i-\frac{1}{2}}, \beta\Delta_{i+\frac{1}{2}})] , & \Delta_{i+\frac{1}{2}} < 0
\end{cases}
\tag{14.44}
$$

for particular values of the parameter β. The value $\beta = 1$ reproduces the MINBEE (or MINMOD) flux limiter and $\beta = 2$ reproduces the SUPERBEE flux limiter. Recall that $\overline{\Delta}_i$ is a vector and thus the limiting process is applied to each component of the system being solved.

As noted in Sect. 13.4 of Chap. 13, upwind MUSCL type schemes based on the limited slopes (14.44) have the Lax–Wendroff as the base second–order scheme, when applied to the linear advection equation, and *not* the Fromm scheme.

A refinement of the above approach is the so called *characteristic limiting* or *wave–by–wave limiting*; see for instance [275] and [362]. First note that the slope Δ_i in cell i is a function of the jumps $\Delta_{i-\frac{1}{2}}$, $\Delta_{i+\frac{1}{2}}$ given in (14.38), namely

$$
\overline{\Delta}_i = \overline{\Delta}_i(\Delta_{i-\frac{1}{2}}, \Delta_{i+\frac{1}{2}}) .
\tag{14.45}
$$

In this approach one decomposes the arguments $\Delta_{i-\frac{1}{2}}$, $\Delta_{i+\frac{1}{2}}$ above into jumps across characteristic fields emerging respectively from the Riemann problem solutions with data $(\mathbf{U}^n_{i-1}, \mathbf{U}^n_i)$ and $(\mathbf{U}^n_i, \mathbf{U}^n_{i+1})$, namely

$$
\Delta_{i-\frac{1}{2}} = \sum_{k=1}^{N} \Delta^{(k)}_{i-\frac{1}{2}} ; \quad \Delta_{i+\frac{1}{2}} = \sum_{k=1}^{N} \Delta^{(k)}_{i+\frac{1}{2}}
\tag{14.46}
$$

where $\Delta^{(k)}_{i+\frac{1}{2}}$ is the jump across wave k at interface $i+\frac{1}{2}$ and N is the number of waves. See Fig. 14.6. One then replaces (14.45) by the summation

$$
\overline{\Delta}_i = \sum_{k=1}^{N} \overline{\Delta}^{(k)}_i(\Delta^{(k)}_{i-\frac{1}{2}}, \Delta^{(k)}_{i+\frac{1}{2}}) ,
\tag{14.47}
$$

where

$$
\overline{\Delta}^{(k)}_i = \overline{\Delta}^{(k)}_i(\Delta^{(k)}_{i-\frac{1}{2}}, \Delta^{(k)}_{i+\frac{1}{2}})
\tag{14.48}
$$

is simply a *limited slope* as applied to a scalar problem, obtained from (14.44), for instance.

The application of characteristic limiting requires the solution of the Riemann problem with data $(\mathbf{U}^n_i, \mathbf{U}^n_{i+1})$ at each cell interface $i + \frac{1}{2}$. Note that this means that the MUSCL–Hancock scheme, or some other MUSCL–type schemes, requires the solution of *two* Riemann problems per cell interface per

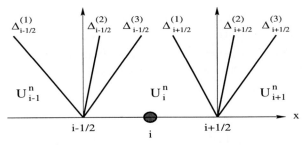

Fig. 14.6. Characteristic limiting. Wave–by–wave decomposition of gradients obtained from local solutions of Riemann problems, with cell averages as initial data

time step. As for the purpose of characteristic limiting, accuracy and robustness of the Riemann solver are not of paramount importance, one may use the simplest Riemann solver available. Quirk [275] reports satisfactory results when using the Primitive–Variable Linearised Riemann solver presented in Sect. 9.3 of Chap. 9; see [355]. The collective experience of using characteristic limiting is generally a positive one and is well worth the extra cost, particularly when high resolution of contact and shear waves is required.

Use of Slope Limiters. Another approach is to find a *slope limiter* ξ_i such that

$$\overline{\Delta}_i = \xi_i \Delta_i \,, \tag{14.49}$$

with Δ_i as given by (14.37), for example. This approach leads to a TVD region for $\xi(r)$ given as follows:

$$\xi(r) = 0 \text{ for } r \leq 0\,, \quad 0 \leq \xi(r) \leq \min\{\xi_L(r), \xi_R(r)\} \text{ for } r > 0\,, \tag{14.50}$$

where

$$\left. \begin{aligned}
\xi_L(r) &= \frac{2\beta_{i-\frac{1}{2}} r}{1 - \omega + (1 + \omega)r}\,, \\
\xi_R(r) &= \frac{2\beta_{i+\frac{1}{2}}}{1 - \omega + (1 + \omega)r}\,, \\
r &= \frac{\Delta_{i-\frac{1}{2}}}{\Delta_{i+\frac{1}{2}}}\,,
\end{aligned} \right\} \tag{14.51}$$

and

$$\beta_{i-\frac{1}{2}} = \frac{2}{1+c}\,, \quad \beta_{i+\frac{1}{2}} = \frac{2}{1-c}\,. \tag{14.52}$$

The coefficients $\beta_{i-\frac{1}{2}}$ and $\beta_{i+\frac{1}{2}}$ above are derived for the scalar case in Sect. 13.8.3 of Chap. 13 and c is the Courant number for the single wave present. By considering the limit values of $\beta_{i-\frac{1}{2}}$ and $\beta_{i+\frac{1}{2}}$ one may eliminate the dependence on c and simply set $\beta_{i-\frac{1}{2}} = \beta_{i+\frac{1}{2}} = 1$ in (14.52). A more refined approach would be to adopt the characteristic limiting approach described earlier in this section. This would require solving extra Riemann problems to find jumps across each wave and their respective Courant numbers, so that one may define coefficients $\beta_{i+\frac{1}{2}}$ as functions of Courant numbers $c_{i+\frac{1}{2}}^{(k)}$.

In Chap. 13 we constructed slope limiters that are analogous to conventional flux limiters, such as SUPERBEE and MINBEE. We stress however that they are only analogous, not equivalent. A slope limiter that is analogous to the SUPERBEE flux limiter is

$$
\xi_{sb}(r) = \begin{cases} 0 \, , & \text{if } r \leq 0 \, , \\ 2r \, , & \text{if } 0 \leq r \leq \frac{1}{2} \, , \\ 1 \, , & \text{if } \frac{1}{2} \leq r \leq 1 \, , \\ \min\{r, \xi_R(r), 2\} \, , & \text{if } r \geq 1 \, . \end{cases} \tag{14.53}
$$

A van Leer–type slope limiter is

$$
\xi_{vl}(r) = \begin{cases} 0 \, , & r \leq 0 \, , \\ \min\{\dfrac{2r}{1+r}, \xi_R(r)\} \, , & r \geq 0 \, . \end{cases} \tag{14.54}
$$

A van Albada–type slope limiter is

$$
\xi_{va}(r) = \begin{cases} 0 \, , & r \leq 0 \, , \\ \min\{\dfrac{r(1+r)}{1+r^2}, \xi_R(r)\} \, , & r \geq 0 \, . \end{cases} \tag{14.55}
$$

A MINBEE–type slope limiter is

$$
\xi_{mb}(r) = \begin{cases} 0 \, , & r \leq 0 \, , \\ r \, , & 0 \leq r \leq 1 \, , \\ \min\{1, \xi_R(r)\} \, , & r \geq 1 \, . \end{cases} \tag{14.56}
$$

As to Riemann solvers to be used the reader is referred to Sect. 14.3.5.

14.4.4 Summary of the MUSCL–Hancock Method

For a given domain $[0, L]$ and a mesh size Δx, determined by the choice of the number M of computing cells, one first sets the initial conditions $\mathbf{U}^0 \equiv \mathbf{U}(x, 0)$ at time $t = 0$. Then, for each time step n one performs the following operations

Operation (I):   **Boundary conditions.** This is carried out according to (14.6) or (14.7), for example.

Operation (II):   **Computation of time step.** This is carried out according to (14.4) and (14.5). A choice of C_{cfl} must be made at the beginning of the computations. One usually takes $C_{cfl} = 0.9$. Recall however that the choice of $S_{max}^{(n)}$ is crucial and given that (14.5) produces somewhat unreliable estimates for the true speeds we recommend that when solving problems with shock–tube like data, the CFL coefficient C_{cfl} be set to small number, e.g. 0.2, for a few time steps, e.g. 5.

Operation (III):   **Boundary extrapolated values**. This involves Steps (I) and (II) of Sect. 14.4.1, where the slopes Δ_i in (14.32), (14.33) are replaced by the limited slopes $\overline{\Delta}_i$, which are to be obtained by applying the methods of Sect. 14.4.3. The operation results in TVD, evolved boundary extrapolated values $\overline{\mathbf{U}}_i^{L,R}$, for each cell i.

Operation (IV):   **Solution of Riemann problem**. At each intercell position $i + \frac{1}{2}$ one finds the solution of Riemann problem with data $(\overline{\mathbf{U}}_i^R, \overline{\mathbf{U}}_{i+1}^L)$ and computes the intercell flux according to (14.36).

Operation (V):   **Updating of solution**. Proceed to update the conserved variables according to conservative formula (14.3)

Operation (VI):   **Next time level**. Go to (I).

In the next section we present *centred* (non upwind) TVD schemes for non–linear systems of conservation laws.

14.5 Centred TVD Schemes

In this section we present two second–order TVD schemes that are extensions of the First–Order Centred (FORCE) scheme [365], [367]. The first scheme is a flux limiter blending of the FORCE scheme and the Richtmyer scheme [276]. The second scheme is of the slope limiter type and results from replacing the Godunov flux by the FORCE flux in the MUSCL–Hancock scheme of Sect. 14.4. Details on the construction and theory of these TVD centred schemes are found in Toro and Billett [372], [374].

A key issue in constructing TVD *centred* schemes is the fact that the dependence of the scheme, via the limiter functions, on the direction of wave propagation is impossible to implement or is simply undesirable. Upwind based TVD schemes have plenty of local information on directionality supplied by the Riemann problem or some other type of characteristic information. This upwind information is usually put to good use in constructing limiter functions with a minimum of artificial viscosity.

14.5.1 Review of the FORCE Flux

We solve systems of hyperbolic conservation laws

$$\mathbf{U}_t + \mathbf{F}(\mathbf{U})_x = \mathbf{0} \tag{14.57}$$

such as the Euler equations, for which $\mathbf{U}$ and $\mathbf{F}(\mathbf{U})$ are given by (14.2). A classical scheme of first order accuracy to solve (14.57) is that of Lax–Friedrichs, whose numerical flux at the interface of two states $\mathbf{U}_L, \mathbf{U}_R$ is

$$\mathbf{F}_{i+\frac{1}{2}}^{LF} = \mathbf{F}_{i+\frac{1}{2}}^{LF}(\mathbf{U}_L, \mathbf{U}_R) = \frac{1}{2}[\mathbf{F}(\mathbf{U}_L) + \mathbf{F}(\mathbf{U}_R)] + \frac{1}{2}\frac{\Delta x}{\Delta t}[\mathbf{U}_L - \mathbf{U}_R] \, . \quad (14.58)$$

A second–order accurate scheme of interest to us here is the Richtmyer scheme [276], which computes a numerical flux by first defining an intermediate state

$$\mathbf{U}_{i+\frac{1}{2}}^{RI} \equiv \mathbf{U}_{i+\frac{1}{2}}^{RI}(\mathbf{U}_L, \mathbf{U}_R) = \frac{1}{2}(\mathbf{U}_L + \mathbf{U}_R) + \frac{1}{2}\frac{\Delta t}{\Delta x}[\mathbf{F}(\mathbf{U}_L) - \mathbf{F}(\mathbf{U}_R)] \quad (14.59)$$

and then setting

$$\mathbf{F}_{i+\frac{1}{2}}^{RI} = \mathbf{F}(\mathbf{U}_{i+\frac{1}{2}}^{RI}) \, . \quad (14.60)$$

The First Order Centred (FORCE) scheme [365] was derived in Chap. 7 for non–linear systems (14.57) and has numerical flux

$$\mathbf{F}_{i+\frac{1}{2}}^{force} = \mathbf{F}_{i+\frac{1}{2}}^{force}(\mathbf{U}_L, \mathbf{U}_R) = \frac{1}{2}[\mathbf{F}_{i+\frac{1}{2}}^{LF}(\mathbf{U}_L, \mathbf{U}_R) + \mathbf{F}_{i+\frac{1}{2}}^{RI}(\mathbf{U}_L, \mathbf{U}_R)] \, . \quad (14.61)$$

Remark 14.5.1. We note that the FORCE flux results from replacing random values in the staggered–grid version of the Random Choice Method by integral averages; an unexpected outcome is that it turns out to be a simple mean value of the Lax–Friedrichs and Richtmyer fluxes. Motivated by the averaging outcome, one may find the *optimal* averaging of the Lax–Friedrichs and Richtmyer fluxes to reproduce the Godunov first order upwind method for model problems; however, this depends on *upwind* information.

In the next two sections we construct second–order TVD extensions of the FORCE scheme for non–linear systems.

14.5.2 A Flux Limiter Centred (FLIC) Scheme

The general flux limiter approach combines a low order monotone flux $\mathbf{F}_{i+\frac{1}{2}}^{LO}$ and a high order flux $\mathbf{F}_{i+\frac{1}{2}}^{HI}$ as

$$\mathbf{F}_{i+\frac{1}{2}} = \mathbf{F}_{i+\frac{1}{2}}^{LO} + \phi_{i+\frac{1}{2}}[\mathbf{F}_{i+\frac{1}{2}}^{HI} - \mathbf{F}_{i+\frac{1}{2}}^{LO}] \, , \quad (14.62)$$

where $\phi_{i+\frac{1}{2}}$ is a flux limiter. For details on the flux limiter approach as applied to scalar problems see Sect. 13.7.2 of Chap. 13.

Here we construct a **Flux LImiter Centred** (FLIC) scheme by taking

$$\mathbf{F}_{i+\frac{1}{2}}^{LO} \equiv \mathbf{F}_{i+\frac{1}{2}}^{force} \, ; \quad \mathbf{F}_{i+\frac{1}{2}}^{HI} \equiv \mathbf{F}_{i+\frac{1}{2}}^{RI} \quad (14.63)$$

where $\mathbf{F}_{i+\frac{1}{2}}^{force}$ and $\mathbf{F}_{i+\frac{1}{2}}^{RI}$ are the fluxes for the FORCE and Richtmyer schemes given respectively by (14.61) and (14.60). The scalar version of this scheme was studied in detail in Sect. 13.7.4, where a Total Variation Diminishing (TVD) version of the scheme was also constructed. Here we extend that analysis to non–linear systems in a somewhat empirical manner.

The analysis for the scalar case contains wave propagation information via the Courant number c. For non–linear systems it will be actually possible to retain, at no cost, some of this information by making use of the CFL coefficient C_{cfl} in (14.4) required to compute a stable time step Δt in (14.3).

In Sect. 13.7.4 of Chap. 13 we established a relationship between conventional upwind flux limiters $\psi(r)$ and centred flux limiters $\phi(r)$, so that these may be found as follows

$$\phi(r) = \hat{\phi}_g + (1 - \hat{\phi}_g)\psi(r) , \qquad (14.64)$$

with

$$\hat{\phi}_g = \begin{cases} 0 , & r \leq 1 , \\ \phi_g \equiv (1 - c_{max})/(1 + c_{max}) , & r \geq 1 . \end{cases} \qquad (14.65)$$

The function ϕ_g above retains some upwinding at no extra cost, as one may set for instance $c_{max} = C_{cfl}$. First order centred schemes other than FORCE may also be used, in which case the expression for ϕ_g is different. See Sect. 13.7.4 of Chap. 13 for details. By choosing the upwind flux limiters SUPERBEE, VANLEER, VANALBADA and MINBEE we obtain corresponding *centred flux limiters*

$$\phi_{sb}(r) = \begin{cases} 0 , & r \leq 0 , \\ 2r , & 0 \leq r \leq \frac{1}{2} , \\ 1 , & \frac{1}{2} \leq r \leq 1 , \\ min\,\{2, \phi_g + (1 - \phi_g)r\} , & r > 1 , \end{cases} \qquad (14.66)$$

$$\phi_{vl}(r) = \begin{cases} 0 , & r \leq 0 , \\ \frac{2r}{1+r} , & 0 \leq r \leq 1 , \\ \phi_g + \frac{2(1-\phi_g)r}{1+r} , & r \geq 1 , \end{cases} \qquad (14.67)$$

$$\phi_{va}(r) = \begin{cases} 0 , & r \leq 0 , \\ \frac{r(1+r)}{1+r^2} , & 0 \leq r \leq 1 , \\ \phi_g + \frac{(1-\phi_g)r(1+r)}{1+r^2} , & r \geq 1 , \end{cases} \qquad (14.68)$$

$$\phi_{mb}(r) = \begin{cases} 0 , & r \leq 0 , \\ r , & 0 \leq r \leq 1 , \\ 1 , & r \geq 1 . \end{cases} \qquad (14.69)$$

For the Euler equations we recommend the following procedure: we first define $q \equiv E$ (total energy) and set

$$r^L_{i+\frac{1}{2}} = \frac{\Delta q_{i-\frac{1}{2}}}{\Delta q_{i+\frac{1}{2}}} ; \quad r^R_{i+\frac{1}{2}} = \frac{\Delta q_{i+\frac{3}{2}}}{\Delta q_{i+\frac{1}{2}}} . \qquad (14.70)$$

Then we compute a single flux limiter

$$\phi^{LR} = \min\{\phi(r^L_{i+\frac{1}{2}}), \phi(r^R_{i+\frac{1}{2}})\} , \tag{14.71}$$

where ϕ is any of the limiter functions (14.66)–(14.69). Then the single limiter ϕ^{LR} is applied to all three flux components in (14.62). For other systems, some experimentation might be necessary before arriving at some satisfactory procedure. A fairly general approach for a system whose vector of conserved variables has components u_k, $k = 1, \ldots, m$, is to first apply (14.70)–(14.71) to every component u_k and obtain a limiter ϕ^{LR}_k. Then, one can select the final limiter as

$$\phi^{LR} = \min_k \left(\phi^{LR}_k\right) , \quad k = 1, \ldots, m .$$

Methods for extending the FLIC scheme to systems with source terms are given in Chap. 15 and techniques for solving multidimensional problems are given in Chap. 16. Numerical results are presented in Sect. 14.7.

In the next section we present another centred TVD scheme.

14.5.3 A Slope Limiter Centred (SLIC) Scheme

Here we construct another second order extension of the FORCE scheme [365]. This results from replacing the Godunov upwind flux in the MUSCL–Hancock scheme [395] by the FORCE flux. The scheme as applied to the scalar case was presented in Sect. 13.8 of Chap 13. The TVD version, derived in Sect. 13.8.2 and 13.8.3, results from limiting the slopes in the data reconstruction step.

The Slope LImiter Centred (SLIC) scheme has three steps. Steps (I) and (II) are exactly the same as in the MUSCL–Hancock scheme, see Sect. 14.3. The first step results in boundary extrapolated values

$$\mathbf{U}^L_i = \mathbf{U}^n_i - \frac{1}{2}\Delta_i ; \quad \mathbf{U}^R_i = \mathbf{U}^n_i + \frac{1}{2}\Delta_i \tag{14.72}$$

in each cell $I_i = [x_{i-\frac{1}{2}}, x_{i+\frac{1}{2}}]$, where Δ_i is a slope vector (a difference). The second step evolves $\mathbf{U}^L_i, \mathbf{U}^R_i$ by a time $\frac{1}{2}\Delta t$ according to

$$\left.\begin{array}{l} \overline{\mathbf{U}}^L_i = \mathbf{U}^L_i + \frac{1}{2}\frac{\Delta t}{\Delta x}[\mathbf{F}(\mathbf{U}^L_i) - \mathbf{F}(\mathbf{U}^R_i)] , \\[2mm] \overline{\mathbf{U}}^R_i = \mathbf{U}^R_i + \frac{1}{2}\frac{\Delta t}{\Delta x}[\mathbf{F}(\mathbf{U}^L_i) - \mathbf{F}(\mathbf{U}^R_i)] . \end{array}\right\} \tag{14.73}$$

Now, instead of solving the Riemann problem with data

$$\mathbf{W}_L \equiv \overline{\mathbf{U}}^R_i ; \quad \mathbf{W}_R \equiv \overline{\mathbf{U}}^L_{i+1} \tag{14.74}$$

to find the Godunov flux at the intercell position $i+\frac{1}{2}$, we evaluate the FORCE flux

$$\mathbf{F}^{force}_{i+\frac{1}{2}} = \mathbf{F}^{force}_{i+\frac{1}{2}}(\overline{\mathbf{U}}^R_i, \overline{\mathbf{U}}^L_{i+1}) , \tag{14.75}$$

as given by (14.61). For the slopes Δ_i in (14.72) we take

$$\Delta_i = \frac{1}{2}(1+\omega)\Delta_{i-\frac{1}{2}} + \frac{1}{2}(1-\omega)\Delta_{i+\frac{1}{2}} , \qquad (14.76)$$

where $\omega \in [-1, 1]$ is a parameter and $\Delta_{i-\frac{1}{2}}, \Delta_{i+\frac{1}{2}}$ denote jumps across interfaces $i - \frac{1}{2}$ and $i + \frac{1}{2}$ respectively, see (14.38).

As seen in section 13.4.5 of Chap. 13, this scheme is second–order accurate in space and time and stable with Courant number c satisfying $|c| \le 1$, for any $\omega \in [-1, 1]$, when applied to the scalar, linear advection equation. According to Godunov's theorem, see Sect. 13.5.3, spurious oscillations will be produced in the vicinity of strong gradients. To avoid this difficulty, a Total Variation Diminishing (TVD) version of the scheme for the scalar case was constructed in Sect. 13.8. This was achieved by replacing the slopes Δ_i by *limited* slopes $\overline{\Delta}_i$. The extension to non–linear systems is somewhat empirical. Any of the slope limiters given by (14.53) to (14.56), with $\beta_{i-\frac{1}{2}} = \beta_{i+\frac{1}{2}} = 1$ in (14.51) will lead to satisfactory results.

Remark 14.5.2. The choice of limited slopes given by (14.44) will not in general give a TVD method when the underlying first order scheme is centred, and therefore (14.44) is not a choice of limiting to be considered here.

In the next section we present upwind TVD schemes based on non–conservative formulations of the equations.

14.6 Primitive–Variable Schemes

Here we construct two second–order TVD *upwind* schemes for hyperbolic systems expressed in *non–conservative form*. The schemes are respectively based on the WAF and MUSCL–Hancock approaches presented in Sects. 14.3 and 14.4 for constructing conservative methods. Modifications of the PLM [87] and the GRP [23] approaches have also been used to construct primitive and conservative schemes [362], [179]. Full details on the construction of primitive, conservative and adaptive schemes are given in [368].

14.6.1 Formulation of the Equations and Primitive Schemes

We shall call *primitive–variable method* or simply primitive method, any scheme that is based on the *non-conservative* form of the equations, namely

$$\mathbf{W}_t + \mathbf{A}(\mathbf{W})\mathbf{W}_x = 0 , \qquad (14.77)$$

where $\mathbf{A}$ is a coefficient matrix that depends on the particular choice of variables $\mathbf{W}$. See Sect. 3.2.2 of Chap. 3, for various possible formulations of the governing equations. Note that (14.77) includes the case in which $\mathbf{W}$ is the vector of conserved variables.

It is well known that *primitive–variable* schemes will compute shock waves with the wrong strength and thus the wrong speed of propagation. At the numerical level there is some recent work by Hou and Le Floch [177] which is also relevant here. However, if the problem is shock free then primitive–variable schemes are perfectly adequate. Conservative methods, on the other hand, have been reported to give erroneous solutions when computing material interfaces by Clarke et. al. [80] and by Karni [191]. As a matter of fact, there are classes of initial–value problems (IVPs) for which conservative methods give erroneous solutions [360], [364], while primitive–variable schemes successfully avoid some of these difficulties. These IVPs include material interfaces, shear waves and *symmetric data* Riemann problems. Other relevant examples in which primitive methods would be adequate are (i) the artificial compressibility equations associated with the incompressible Navier–Stokes equations, [366] (ii) linear constant coefficient hyperbolic systems, see Chap. 5, (iii) systems admitting very weak shock waves such as in Acoustics and (iv) hyperbolic systems with unknown conservative formulations [383]; we note however that if such systems contain shocks, then their primitive form will not give the correct solutions. There is a large body of work concerned with primitive–variable schemes and their combination with shock–fitting techniques; see for instance the works of Moretti and co–workers [244], [242], [243], [107]. Upwind TVD primitive–variable schemes have been constructed by Karni [191], Toro [360], [363], [362] and more recently by Abgrall [1]. Obviously, if shock waves are present then primitive schemes must be ruled out. However, by making use of adaptive schemes, where shock waves are treated by a conservative method, one can preserve the advantages of both approaches. Adaptive primitive–conservative schemes have been put forward by Toro [360], [363], Karni [192] and Abgrall [1]; see also Ivings et. al. [179].

We describe the schemes for the *augmented*, one dimensional time dependent Euler equations, see Chaps. 1 and 3. We express the equations in terms of the primitive variables $\mathbf{W} = (\rho, u, p, q)^T$, where q is any *passive scalar*. The governing equations are of the form (14.77), where the coefficient matrix $\mathbf{A}(\mathbf{W})$ is

$$\mathbf{A}(\mathbf{W}) = \begin{pmatrix} u & \rho & 0 & 0 \\ 0 & u & \frac{1}{\rho} & 0 \\ 0 & \rho a^2 & u & 0 \\ 0 & 0 & 0 & u \end{pmatrix} . \tag{14.78}$$

The variable $q = q(x,t)$ may represent a reaction progress variable when modelling chemically active flows, a variable associated with a material interface, the concentration of some chemical species, or simply the tangential velocity component in two–dimensional gas dynamics.

Now we propose a two–step finite–difference scheme to solve (14.77), namely

$$\mathbf{W}_i^{n+1} = \mathbf{W}_i^n + \frac{\Delta t}{\Delta x} \overline{\mathbf{A}}_i \left[\mathbf{W}_{i-\frac{1}{2}}^{n+\frac{1}{2}} - \mathbf{W}_{i+\frac{1}{2}}^{n+\frac{1}{2}} \right] . \tag{14.79}$$

Fig. 14.7 shows the stencil of the scheme. At the half–time level $n + \frac{1}{2}$, the matrix $\overline{\mathbf{A}}_i$ is defined at the grid point i and the *intermediate* states $\mathbf{W}_{i+\frac{1}{2}}^{n+\frac{1}{2}}$ are defined at the intermediate positions $i + \frac{1}{2}$. The scheme is completely determined once $\overline{\mathbf{A}}_i$ and $\mathbf{W}_{i+\frac{1}{2}}^{n+\frac{1}{2}}$ are defined. We note that the problem of finding $\mathbf{W}_{i+\frac{1}{2}}^{n+\frac{1}{2}}$ is similar to that of finding an intercell flux in conservative methods, see equation (14.3). We shall study two approaches.

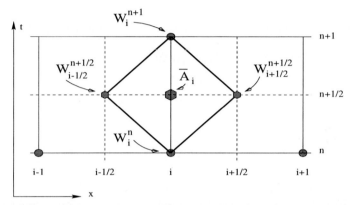

Fig. 14.7. Stencil for primitive–variable schemes on a staggered grid. Point values $\mathbf{W}_i^n$ are updated to $\mathbf{W}_i^{n+1}$, intermediate values are $\mathbf{W}_{i-\frac{1}{2}}^{n+\frac{1}{2}}$ and $\mathbf{W}_{i+\frac{1}{2}}^{n+\frac{1}{2}}$

14.6.2 A WAF–Type Primitive Variable Scheme

Here we follow the WAF approach of Sect. 14.3 for constructing the intermediate vectors $\mathbf{W}_{i+\frac{1}{2}}^{n+\frac{1}{2}}$ and the coefficient matrix $\overline{\mathbf{A}}_i$. First we define

$$\mathbf{W}_{i+\frac{1}{2}}^{n+\frac{1}{2}} = \frac{1}{\Delta x} \int_{-\frac{1}{2}\Delta x}^{\frac{1}{2}\Delta x} \mathbf{W}_{i+\frac{1}{2}}\left(x, \frac{1}{2}\Delta t\right) dx . \tag{14.80}$$

See equation (14.14). So far this is identical to the Weighted Average State version of WAF, see Sect. 14.3.2. Here $\mathbf{W}_{i+\frac{1}{2}}(x, t)$ is the solution of the conventional Riemann problem with data $(\mathbf{W}_i^n, \mathbf{W}_{i+1}^n)$. An approximation to the integral is

$$\mathbf{W}_{i+\frac{1}{2}}^{n+\frac{1}{2}} = \sum_{k=1}^{N} \beta_k \mathbf{W}_{i+\frac{1}{2}}^{(k)} , \tag{14.81}$$

where N is the total number of waves in the solution of the Riemann problem ($N = 4$ in the present case). The weights or coefficients are given by

$$\beta_k = \tfrac{1}{2}(c_k - c_{k-1}) \,, \left.\begin{array}{c} \\ \\ \end{array}\right\} \quad (14.82)$$

$$c_k = \frac{\Delta t S_k}{\Delta x} \,, \quad c_0 = -1 \,, \quad c_{N+1} = 1 \,,$$

where c_k is the Courant number for wave k of speed S_k. Manipulation of (14.81)–(14.82) leads to

$$\mathbf{W}_{i+\frac{1}{2}}^{n+\frac{1}{2}} = \frac{1}{2}(\mathbf{W}_i^n + \mathbf{W}_{i+1}^n) - \frac{1}{2}\sum_{k=1}^{N} c_k[\mathbf{W}_{i+\frac{1}{2}}^{(k+1)} - \mathbf{W}_{i+\frac{1}{2}}^{(k)}] \,. \quad (14.83)$$

To compute the coefficient matrix $\overline{\mathbf{A}}_i$ we first define a state $\overline{\mathbf{W}}_i$ and then set

$$\overline{\mathbf{A}}_i = \mathbf{A}(\overline{\mathbf{W}}_i) \,. \quad (14.84)$$

The simplest choice is obviously $\overline{\mathbf{W}}_i = \mathbf{W}_i^n$, which we do not believe to be sufficiently robust to be used with confidence. A sophisticated choice for $\overline{\mathbf{A}}_i$ is

$$\overline{\mathbf{W}}_i = \frac{1}{\Delta x}\int_0^{\frac{1}{2}\Delta x} \mathbf{W}_{i-\frac{1}{2}}(x, \tfrac{1}{2}\Delta t)dx + \frac{1}{\Delta x}\int_{-\frac{1}{2}\Delta x}^0 \mathbf{W}_{i+\frac{1}{2}}(x, \tfrac{1}{2}\Delta t)dx \,, \quad (14.85)$$

which is an integral average of the solutions $\mathbf{W}_{i-\frac{1}{2}}(x/t)$ and $\mathbf{W}_{i+\frac{1}{2}}(x/t)$ of the two Riemann problems *affecting* mesh point i, at the half–time level. See Fig. 14.8. A suitable approximation to this integral is

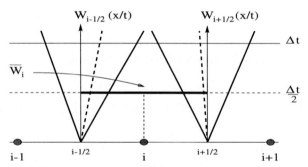

Fig. 14.8. State $\overline{\mathbf{W}}_i$ to compute the coefficient matrix is and integral average of solutions of Riemann problems affecting mesh point i

$$\overline{\mathbf{W}}_i = \sum_{j=j_L}^{N+1} \beta_{i-\frac{1}{2}}^{(j)}\mathbf{W}_{i-\frac{1}{2}}^{(j)} + \sum_{j=1}^{j_R} \beta_{i+\frac{1}{2}}^{(j)}\mathbf{W}_{i+\frac{1}{2}}^{(j)} \,, \quad (14.86)$$

with the coefficients defined as

$$\left.\begin{array}{rcl}
\beta^{(j)}_{i-\frac{1}{2}} &=& \frac{1}{2}(c_j - c_{j-1}) \text{ for } j > j_L \ ; \ \beta^{j_L}_{i-\frac{1}{2}} = \frac{1}{2}c_{j_L} \ , \\[2mm]
\beta^{(j)}_{i+\frac{1}{2}} &=& \frac{1}{2}(c_j - c_{j-1}) \text{ for } j < j_R \ ; \ \beta^{j_R}_{i+\frac{1}{2}} = -\frac{1}{2}c_{j_R} \ .
\end{array}\right\} \qquad (14.87)$$

The superscript j_L refers to the value of k in the solution of the Riemann problem with data $(\mathbf{W}^n_{i-1}, \mathbf{W}^n_i)$ such that the speed S_{k-1} is negative and S_k is positive. A similar interpretation holds for j_R. Another approximation to (14.85) that is reported to work well [179] is

$$\overline{\mathbf{W}}_i = \frac{1}{2}(\mathbf{W}^{n+\frac{1}{2}}_{i-\frac{1}{2}} + \mathbf{W}^{n+\frac{1}{2}}_{i+\frac{1}{2}}) \ , \qquad (14.88)$$

which makes use of information made available by (14.83).

In approximating the integrals (14.80) and (14.85) by the summations (14.81) and (14.86) we have assumed the simplified wave structure of Figs. 14.2 and 14.8. The treatment of rarefactions is entirely analogous to the conservative version of the scheme and is discussed in Sect. 14.3.3.

Exercise 14.6.1. Apply the primitive scheme (14.79) to the model equation

$$u_t + au_x = 0 \ , \quad a = constant. \qquad (14.89)$$

Verify that for $a > 0$ and $a < 0$ the resulting scheme is the Lax–Wendroff method, see Chap. 13, which is second–order accurate in space and time.

Show also that if the intermediate state (14.80) is computed by a mid–point rule, then the scheme reduces to the Godunov first–order upwind scheme, see Chap. 6, which is monotone.

Remark 14.6.1. If the above observation is generalised and we compute the intermediate state vectors $\overline{\mathbf{W}}^{n+\frac{1}{2}}_{i+\frac{1}{2}}$ in (14.80) by a mid–point rule approximation, then one obtains a first–order accurate primitive scheme, that is the non–conservative counterpart of Godunov's conservative method studied in Chap. 6.

Based on the first–order primitive scheme above one can construct a Total Variation Diminishing (TVD) version of the second–order primitive method. This is most easily accomplished by replacing (14.83) by

$$\mathbf{W}^{n+\frac{1}{2}}_{i+\frac{1}{2}} = \frac{1}{2}(\mathbf{W}^n_i + \mathbf{W}^n_{i+1}) - \frac{1}{2}\sum_{k=1}^{N} \text{sign}(c_k)\phi_k[\mathbf{W}^{(k+1)}_{i+\frac{1}{2}} - \mathbf{W}^{(k)}_{i+\frac{1}{2}}] \ , \qquad (14.90)$$

where ϕ_k is any of the WAF limiter functions (14.27)–(14.30). See Chap. 13 for general background on TVD schemes and limiter functions.

We summarise the application of the primitive WAF scheme as applied to (14.77): One first solves the Riemann problems with data $(\mathbf{W}^n_i, \mathbf{W}^n_{i+1})$ to find $\mathbf{W}_{i+\frac{1}{2}}(x/t)$; one then computes the intermediate state vectors $\mathbf{W}^{n+\frac{1}{2}}_{i+\frac{1}{2}}$ according to (14.90) and the coefficient matrix $\overline{\mathbf{A}}_i$ according to (14.84), with

$\overline{\mathbf{W}}$ obtained from (14.85) or (14.88); the primitive scheme (14.79) is thus completely determined. The linearised stability condition of the scheme is $|c| \leq 1$.

Remark 14.6.2. The primitive WAF scheme is most easily converted into the conservative WAF scheme of Sect. 14.3.2 by simply computing an intercell flux as $\mathbf{F}_{i+\frac{1}{2}} = \mathbf{F}(\mathbf{W}_{i+\frac{1}{2}}^{n+\frac{1}{2}})$ and advancing the solution via the conservative formula (14.3). See Sect. 14.3.2.

Concerning Riemann solvers the reader is referred to Sect. 14.3.5.

14.6.3 A MUSCL–Hancock Primitive Scheme

A different primitive scheme results if the computation of $\mathbf{W}_{i+\frac{1}{2}}^{n+\frac{1}{2}}$ and the coefficient matrix $\overline{\mathbf{A}}_i$ in (14.79) is carried out via the MUSCL–Hancock approach, first put forward for constructing conservative methods [395]. See Sect. 14.4 for details on this approach.

First we define a procedure for computing the intermediate values $\mathbf{W}_{i+\frac{1}{2}}^{n+\frac{1}{2}}$. This is done in essentially the same way as for conservative schemes. After the reconstruction step (14.32) in terms of $\mathbf{W}$, there are three distinct steps, namely

Step (I):   **Extrapolation.** Using the reconstructed solution in (14.32) in terms of the vector of primitive variables $\mathbf{W}$ we obtain the boundary extrapolated values

$$\mathbf{W}_i^L = \mathbf{W}_i^n - \frac{1}{2}\Delta_i \; ; \quad \mathbf{W}_i^R = \mathbf{W}_i^n + \frac{1}{2}\Delta_i \; . \tag{14.91}$$

Step (II):   **Evolution.** We evolve the values $\mathbf{W}_i^{L,R}$ by a time $\frac{1}{2}\Delta t$ according to

$$\overline{\mathbf{W}}_i^{L,R} = \mathbf{W}_i^{L,R} + \frac{1}{2}\frac{\Delta t}{\Delta x}\mathbf{A}(\mathbf{W}_i^n)(\mathbf{W}_i^L - \mathbf{W}_i^R) \; . \tag{14.92}$$

Step (III):   **The Riemann problem.** In order to compute $\mathbf{W}_{i+\frac{1}{2}}^{n+\frac{1}{2}}$ we find the solution $\mathbf{W}_{i+\frac{1}{2}}(x/t)$ of the Riemann problem with piece–wise constant data $(\overline{\mathbf{W}}_i^R, \overline{\mathbf{W}}_{i+1}^L)$ and set

$$\mathbf{W}_{i+\frac{1}{2}}^{n+\frac{1}{2}} = \mathbf{W}_{i+\frac{1}{2}}(0) \; , \tag{14.93}$$

which requires a solution sampling procedure entirely analogous to that of the Godunov method described in Chap. 6. See also Chap. 9.

There are several possible choices for the matrix $\overline{\mathbf{A}}_i$ in (14.79). By using (14.84) one only requires the computation of a state $\overline{\mathbf{W}}_i$. A possible choice is

$$\overline{\mathbf{W}}_i = \frac{1}{2}(\overline{\mathbf{W}}_i^L + \overline{\mathbf{W}}_i^R) . \tag{14.94}$$

Some algebra gives the simple expression

$$\overline{\mathbf{W}}_i = \mathbf{W}_i^n - \frac{1}{2}\frac{\Delta t}{\Delta x}\mathbf{A}(\mathbf{W}_i^n)\Delta_i . \tag{14.95}$$

Another choice is given by (14.85), where now $\mathbf{W}_{i+\frac{1}{2}}(x/t)$ is the solution of the Riemann problem in Step (III) above. The integral may now be approximated as in (14.86), (14.87). A trapezium–rule approximation gives another choice,

$$\overline{\mathbf{W}}_i = \frac{1}{2}(\mathbf{W}_{i-\frac{1}{2}}^{n+\frac{1}{2}} + \mathbf{W}_{i+\frac{1}{2}}^{n+\frac{1}{2}}) . \tag{14.96}$$

Exercise 14.6.2. Apply the primitive MUSCL–Hancock scheme described to the model equation (14.89). Verify that for $a > 0$ and $a < 0$ the resulting scheme is the Fromm scheme, see Chap. 13, which is second–order accurate in space and time.

Verify also that if the slopes Δ_i in (14.91) are identically zero, then the scheme reduces to the Godunov first–order upwind scheme (the CIR scheme), see Chap. 13, which is monotone.

Solution 14.6.1. Left to the reader.

Remark 14.6.3. Based on the remark above one can construct a first–order primitive scheme for non–linear systems by setting the slopes to zero. The resulting scheme is the primitive counterpart of Godunov's conservative method studied in Chap. 6 and which is identical to the primitive first order scheme obtained via the WAF approach in Sect. 14.6.2.

A Total Variation Diminishing (TVD) version of the second–order primitive MUSCL–Hancock method is easily constructed. To this end one replaces the slopes Δ_i by *limited slopes* $\overline{\Delta}_i$ in the same way as in the conservative MUSCL–Hancock scheme described in Sect. 14.4.3.

Remark 14.6.4. The primitive MUSCL–Hancock scheme is most easily converted into the conservative MUSCL–Hancock scheme by simply computing an intercell flux as $\mathbf{F}_{i+\frac{1}{2}} = \mathbf{F}(\mathbf{W}_{i+\frac{1}{2}}^{n+\frac{1}{2}})$, where $\mathbf{W}_{i+\frac{1}{2}}^{n+\frac{1}{2}}$ is given by (14.93), and advancing the solution via the conservative formula (14.3). See Sects. 14.4.1 and 14.4.2.

Concerning Riemann solvers the reader is referred to Sect. 14.3.5.

14.6.4 Adaptive Primitive–Conservative Schemes

Here we present an adaptive procedure whereby a conservative scheme is used at shocks only and a primitive scheme is used elsewhere. Details of the approach are found in [360], [363], [368]; see also Ivings et. al. [179]. The experience reported in [378], [379] in constructing adaptive schemes is useful here.

Consider a mesh point i as depicted in Fig. 14.8. This mesh point is updated by a conservative method *only if* any of the two neighbouring Riemann problem solutions $\mathbf{W}_{i-\frac{1}{2}}(x/t)$, $\mathbf{W}_{i+\frac{1}{2}}(x/t)$ contains a shock travelling in the direction of the mesh point i. The local solutions $\mathbf{W}_{i-\frac{1}{2}}(x/t)$, $\mathbf{W}_{i+\frac{1}{2}}(x/t)$ contain all the required information and thus the adaptive scheme is easily implemented. Denote by $p^*_{i-\frac{1}{2}}$ and $p^*_{i+\frac{1}{2}}$ the solutions for pressure in the star regions of $\mathbf{W}_{i-\frac{1}{2}}(x/t)$ and $\mathbf{W}_{i+\frac{1}{2}}(x/t)$ respectively. See Chaps. 4 and 9 for details on the structure of the solution of the Riemann problem. Define

$$S_{str} = 1 + PTOL , \tag{14.97}$$

where $PTOL$ is a small positive quantity yet to be defined. If

$$\left. \begin{array}{l} p^*_{i-\frac{1}{2}}/p^n_i > S_{str} \quad \text{and } S_{i-\frac{1}{2}} > 0 , \\ \text{or if} \\ p^*_{i+\frac{1}{2}}/p^n_i > S_{str} \quad \text{and } S_{i+\frac{1}{2}} < 0 \end{array} \right\} \tag{14.98}$$

then mesh point i is advanced via a conservative method (14.3); otherwise one advances the solution via a primitive scheme (14.79) of Sect. 14.6.1.

In (14.98) $S_{i-\frac{1}{2}}$ denotes the speed of a shock wave contained in the solution $\mathbf{W}_{i-\frac{1}{2}}(x/t)$ of the Riemann problem and $S_{i+\frac{1}{2}}$ denotes the speed of a shock wave contained in the solution $\mathbf{W}_{i+\frac{1}{2}}(x/t)$. Experience reported in [360] and [179] suggests that the choice of $PTOL$ in (14.97) is not too critical and any value in the range $(0, 0.1)$ gives satisfactory results. If $PTOL$ is too large one risks computing shocks of *moderate* strength with the primitive scheme, which will lead to shock waves being in the wrong position. Small values of $PTOL$ simply mean that the conservative method is used almost everywhere.

Note that these adaptive schemes are *essentially primitive*, whereby conservative methods are only used at isolated shocked points. Given that any of the primitive schemes presented can most easily be converted into their conservative counterparts, the adaptive primitive–conservative approach is easily implemented. Alternative hybrid schemes have been presented by Karni [192] and independently by Abgrall [1].

In the next section we present some numerical results for a test problem with exact solution.

14.7 Some Numerical Results

Here we present numerical results for some of the schemes studied in this chapter, as applied to the Euler equations. As a test problem with exact solution we use Test 1 of Sect. 6.4 of Chap. 6, which has initial data consisting of two constant states $\mathbf{W}_L = [\rho_L, u_L, p_L]^T$ and $\mathbf{W}_R = [\rho_R, u_R, p_R]^T$, separated by a discontinuity at a position $x = 0.3$. The data states are $\mathbf{W}_L = [1.0, 0.0, 1.0]^T$ and $\mathbf{W}_R = [0.125, 0.0, 0.1]^T$. The ratio of specific heats is chosen to be $\gamma = 1.4$. The exact and numerical solutions are found in the spatial domain $0 \leq x \leq 1$. The numerical solution is computed with $M = 100$ cells and the Courant number coefficient is $C_{cfl} = 0.9$; boundary conditions are transmissive and $S_{max}^{(n)}$ is found using the simplified formula (14.5). The test problem is a modified version of the popular Sod's test [318]; the solution consists of a right shock wave, a right travelling contact wave and a left *sonic* rarefaction wave, a feature that is very useful for assessing the *entropy satisfaction* property of numerical methods. The exact solution was found by running the code *HE-E1RPEXACT* of the library *NUMERICA* [369]. The numerical solutions were obtained by running the codes *HE-E1WAF, HE-E1MUSHAN* and *HE-E1SLIC* of *NUMERICA*.

We present four sets of results. The first two sets correspond to *upwind* methods and include the WAF method of Sect. 14.3, see Figs. 14.9 and 14.10, and the MUSCL–Hancock method of Sect. 14.4, see Figs. 14.11 to 14.14. All upwind results have been obtained with the HLLC Riemann solver of Sect. 10.4 of Chap. 10. The third and fourth sets of results correspond to *centred* methods and include the Flux Limiter Centred (FLIC) scheme of Sect. 14.5.2, see Figs. 14.15 and 14.16, and the Slope Limiter Centred (SLIC) scheme of Sect. 14.5.3, see Figs. 14.17 and 14.18. For each method we show results for at least two limiter functions.

14.7.1 Upwind TVD Methods

Figs. 14.9 and 14.10 show results for the WAF method using the limiter functions that are *equivalent* to the flux limiters SUPERBEE (14.27) and MINBEE (14.30); the first result gives very sharp resolution of discontinuities but small spurious oscillations are just visible on the plots; the second result using MINBEE has no signs of spurious oscillations but discontinuities are somewhat smeared, especially the contact discontinuity. In practice one tends to use some limiter in between MINBEE and SUPERBEE, such as VANLEER or VANALBADA, given respectively by (14.28) and (14.29).

Figs. 14.11 to 14.14 show results from the MUSCL–Hancock method. The results of Figs. 14.11 and 14.12 were obtained from the *limited slopes* (14.44), the first one corresponds to the flux limiter SUPERBEE and the second one corresponds to the flux limiter MINBEE. Compare with Figs. 14.9 and 14.10 respectively. The result from SUPERBEE gives sharp discontinuities but has visible spurious oscillations, specially in the internal energy plot. The details of the particular Riemann solver being used may also have an effect. The result from MINBEE has virtually eliminated the spurious oscillations but discontinuities are somewhat smeared. The results of Fig. 14.13 are obtained from the slope limiter (14.53), which is *analogous* to SUPERBEE, and the results of Fig. 14.14 are obtained from the slope limiter (14.56), which is analogous to MINBEE. In both slope limiters we have set $c = C_{cfl}$ in (14.51), (14.52). Overall, the WAF method appears to give better results than the MUSCL–Hancock scheme, although the use of *characteristic limiting*, see Sect. 14.4.3, would improve the quality of the MUSCL scheme, but at the cost of solving an extra, though cheaper, Riemann problem per cell interface per time step. All results should be compared with those obtained from the Godunov first order upwind method of Chap. 6; see Fig. 6.8.

14.7.2 Centred TVD Methods

Figs. 14.15 and 14.16 show results from the Flux Limiter *Centred* method, FLIC, for two *centred* flux limiter functions, namely (14.66) and (14.69); these are *analogous* to SUPERBEE and MINBEE respectively. Overall these results are less accurate than those from the *upwind* methods. Note however that the results of Fig. 14.15 are comparable with those of Figs. 14.10, 14.12 and 14.14, obtained from the most diffusive of upwind limiters used. Also, the most compressive limiter for upwind methods tends to overshoot, which does not happen with the most compressive limiter for the centred method.

Figs. 14.17 and 14.18 show results from the Slope Limiter *Centred* method, SLIC, for two slope limiter functions, namely (14.53) and (14.56). In fact these limiters are the same as those used for the upwind MUSCL–Hancock scheme to produce the results of Figs. 14.13 and 14.14. Overall these results are very similar to those obtained from the other centred method, FLIC, and thus the same remarks apply.

In order to appreciate the gains in going from first–order schemes to second–order TVD schemes, the reader should compare all the results of this chapter with those obtained from (i) the Godunov method of Chap. 6 (Fig. 6.8), (ii) the Lax–Friedrichs scheme of Chap. 5 (Fig. 6.13) and (iii) the FORCE scheme of Chap. 7 (Fig. 7.18). They should also be compared with those obtained from a non–TVD second order method, such as the Richtmyer scheme of Chap. 5 (Fig. 6.18).

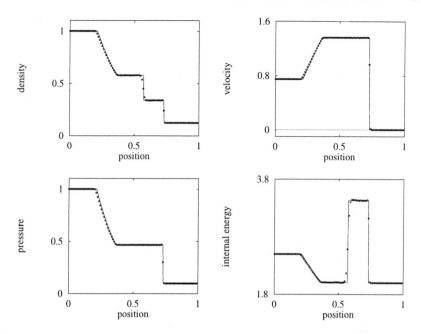

Fig. 14.9. WAF Scheme with HLLC Riemann solver and SUPERBEE applied to Test 1. Numerical (symbol) and exact (line) solutions at time 0.2 units

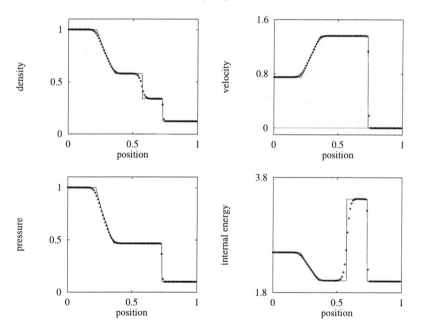

Fig. 14.10. WAF Scheme with HLLC Riemann solver and MINBEE applied to Test 1. Numerical (symbol) and exact (line) solutions at time 0.2 units

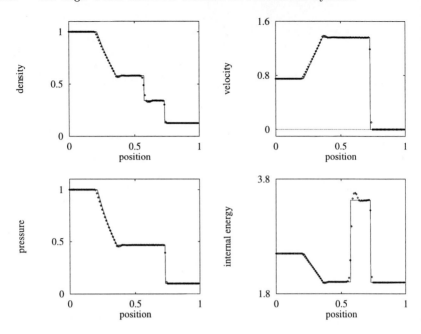

Fig. 14.11. MUSCL–Hancock scheme, HLLC Riemann solver and SUPERBEE for to Test 1. Numerical (symbol) and exact (line) solutions at time 0.2 units

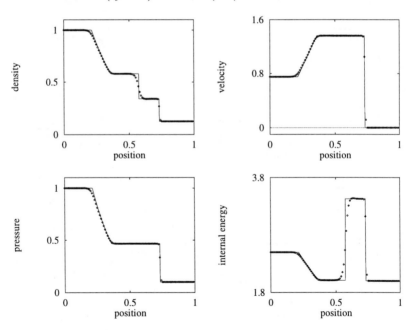

Fig. 14.12. MUSCL–Hancock Scheme with HLLC Riemann solver and MINBEE applied to Test 1. Numerical (symbol) and exact (line) solutions at time 0.2 units

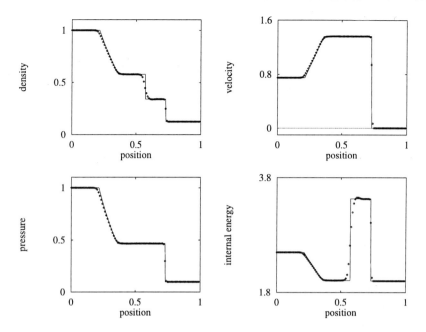

Fig. 14.13. MUSCL–Hancock Scheme, HLLC Riemann solver and SUPERBEEsl for Test 1. Numerical (symbol) and exact (line) solutions at time 0.2 units

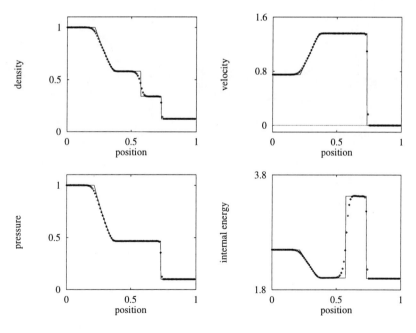

Fig. 14.14. MUSCL–Hancock Scheme with HLLC Riemann solver and MINBEEsl applied to Test 1. Numerical (symbol) and exact (line) solutions at time 0.2 units

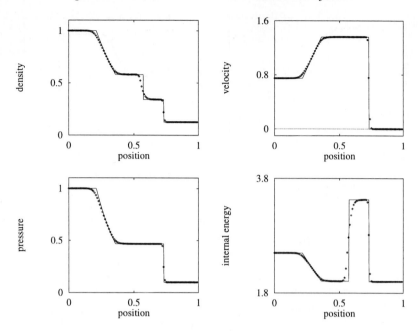

Fig. 14.15. Flux Limiter Centred (FLIC) scheme with Superbee applied to Test 1. Numerical (symbol) and exact (line) solutions at time 0.2 units

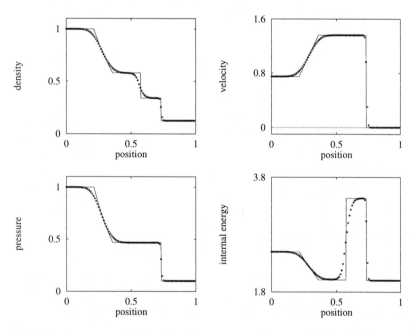

Fig. 14.16. Flux Limiter Centred (FLIC) scheme with Minbee applied to Test 1. Numerical (symbol) and exact (line) solutions at time 0.2 units

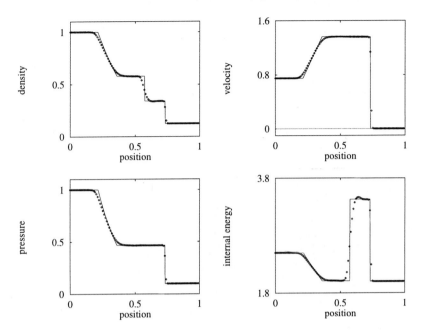

Fig. 14.17. Slope Limiter Centred (SLIC) scheme with Superbee applied to Test 1. Numerical (symbol) and exact (line) solutions at time 0.2 units

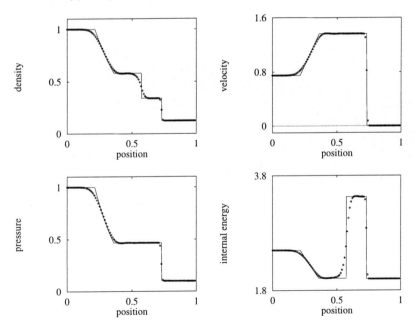

Fig. 14.18. Slope Limiter Centred (SLIC) scheme with Minbee applied to Test 1. Numerical (symbol) and exact (line) solutions at time 0.2 units

Fig. 10.17.

Fig. 10.18.

15. Splitting Schemes for PDEs with Source Terms

15.1 Introduction

This chapter is concerned with numerical methods for solving non–linear systems of hyperbolic conservation laws with source terms

$$\mathbf{U}_t + \mathbf{F}(\mathbf{U})_x = \mathbf{S}(\mathbf{U}) \,. \tag{15.1}$$

$\mathbf{U}$ is the vector of unknowns, $\mathbf{F}(\mathbf{U})$ is the vector of fluxes and $\mathbf{S}(\mathbf{U})$ is a vector of *sources*, which in general is an algebraic function of $\mathbf{U}$ or other physical parameters of the problem at hand. In Chap. 2 we studied some properties for the *pure advection* hyperbolic problem

$$\mathbf{U}_t + \mathbf{F}(\mathbf{U})_x = \mathbf{0} \,. \tag{15.2}$$

This *homogeneous system*, in which $\mathbf{S}(\mathbf{U}) \equiv \mathbf{0}$ (no sources) is a simplified version of (15.1). The time–dependent one dimensional Euler equations of Chap. 3 are one example of a homogeneous system of this kind. Another simplification of (15.1) results from the assumption of no spatial variations, $\mathbf{F}(\mathbf{U})_x = \mathbf{0}$, in which case one obtains

$$\frac{d}{dt}\mathbf{U} = \mathbf{S}(\mathbf{U}) \,, \tag{15.3}$$

which is a system of Ordinary Differential Equations (ODEs).

Inhomogeneous systems of the form (15.1), $\mathbf{S}(\mathbf{U}) \neq \mathbf{0}$, arise naturally in many problems of practical interest. A whole class of inhomogeneous systems are derived when reducing the spatial dimensionality of multidimensional problems. For example, under the assumption of spherically or cylindrically symmetric flow, the three or two dimensional Euler equations become a one–dimensional system of the form (15.1); see Sect. 1.6.2 of Chap. 1. In this case the source terms are geometric in character. Sources of similar type are present in the shallow water equations for flow on non–horizontal channels; see Sect. 1.6.3 of Chap. 1.

Important examples of inhomogeneous systems of the form (15.1) arise in the study of the fluid dynamics of reactive gaseous mixtures, where in addition to the fluid dynamics governed by a system like (15.2), there are chemical reactions between the constituent gases, which in the absence of fluid flow may be modelled by systems of the form (15.3). Examples of problems of this kind arise in the study of hypersonic flows [81], [7] and detonation waves, see for instance [124], [51], [53], [80] and [247].

Chemically active flows contain a range of widely varying time scales, which leads to *stiff* ODEs of the form (15.3), [199]. The problem of *stiffness* in ODEs may be resolved by resorting to *implicit methods*. For chemically active flow models (15.1), stiffness may not be resolved by simply using implicit methods. If the mesh is not sufficiently fine in both space and time, then spurious solutions travelling at unphysical speeds may be computed [90], [212], [32], [53], [152]. See also the recent review paper by Yee and Sweby [423]. There are still a number of unresolved problems in solving systems like (15.2), which are the subject of current research.

There are essentially two ways of constructing methods to solve inhomogeneous systems of the form (15.1). One approach attempts to preserve some coupling between the two processes in (15.1). These two processes might be represented by the systems (15.2) (advection) and (15.3) (reaction–like). LeVeque and Yee [212] report on a predictor–corrector scheme of the MacCormack type with a TVD constraint. The idea of *upwinding the source terms* [286], may be seen as an attempt to couple the two processes involved, although the eigenstructure used in projecting the source terms is oblivious to the influence of these.

Another approach is to *split* (15.1), for a time Δt, into the advection problem (15.2) and the source problem (15.3). At first sight this might appear unreasonable. However, for the case of a model inhomogeneous PDE, splitting is actually exact. For more general problems, the fact that one can construct high–order splitting schemes following this approach is also somewhat reassuring. In addition, computational experience suggests that splitting is a viable approach, *if used with caution*. The main attraction of splitting schemes is in the fact that one can deploy the optimal, existing schemes for each subproblem. For instance, to solve the homogeneous subproblem (15.2) one may use directly any of the schemes presented in Chaps. 6 to 14, or any other appropriate method. To solve the subproblem (15.3) one may use directly any of the ODE solvers available. If the system is known to be stiff, then *stiff* solvers must be used.

This chapter is concerned with splitting schemes to solve (15.1). In Sect. 15.2 we show that for a model equation, splitting is exact; in Sect. 15.3 we present numerical schemes based on the splitting approach; in Sect. 15.4 we briefly review some basic aspects of numerical methods for Ordinary Differential Equations (ODEs). Concluding remarks are given in Sect. 15.5.

15.2 Splitting for a Model Equation

The simplest model hyperbolic equation of the form (15.1) is given by

$$u_t + au_x = \lambda u \, , \tag{15.4}$$

where a is constant wave propagation speed and λ is a constant parameter. This simple model equation will prove very useful in discussing possible strategies for solving (15.1) numerically.

Consider the initial value problem (IVP) for (15.4), namely

$$\left.\begin{array}{lll} \text{PDE} & : & u_t + au_x = \lambda u \, , \\ \text{IC} & : & u(x,0) = u_0(x) \, . \end{array}\right\} \tag{15.5}$$

Here $u = u(x,t)$, $-\infty < x < \infty$, $t > 0$ and $u_0(x)$ is the initial data for the problem at $t = 0$. It is easy to verify that the exact solution of IVP (15.5) is

$$u(x,t) = u_0(x - at)e^{\lambda t} \, . \tag{15.6}$$

Note in particular that if $\lambda = 0$ we recover the exact solution for the homogeneous equation $u_t + au_x = 0$, namely $u_0(x - at)$; see Sect. 2.2 of Chap. 2.

A geometric interpretation of the original IVP (15.5) results if we view (15.5) as an IVP involving an ODE along characteristics, namely

$$\left.\begin{array}{lll} \frac{d}{dt}u = \lambda u & ; & u(0) = u(x_0) \, , \\ \frac{d}{dt}x = a & ; & x(0) = x_0 \, . \end{array}\right\} \tag{15.7}$$

Fig. 15. 1 illustrates the situation. The ODE in (15.7) requires initial data at

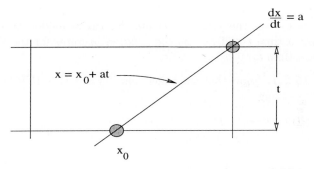

Fig. 15.1. Illustration of operator splitting scheme for model PDE with source term

the foot of the characteristic curve $x = x_0 + at$, namely the point x_0. Actually the initial data is then $u(0) = u(x_0) = u(x - at)$, which is the solution of the homogeneous problem in (15.5), $\lambda = 0$.

Next we show that the exact solution (15.6) can also be obtained by solving exactly a pair of initial value problems in succession.

Theorem 15.2.1 (Splitting of Source Term). *The exact solution (15.6) of the inhomogeneous IVP (15.5) can be found by solving exactly the following pair of IVP's.*

$$\left.\begin{array}{llll} PDE & : & r_t + ar_x = 0 \,, \\ IC & : & r(x,0) = u_0(x) \,, \end{array}\right\} \Longrightarrow r(x,t) \tag{15.8}$$

$$\left.\begin{array}{llll} ODE & : & \frac{d}{dt}s = \lambda s \,, \\ IC & : & s(0) = r(x,t) \,, \end{array}\right\} \Longrightarrow u(x,t) \tag{15.9}$$

Note here that the initial condition of IVP (15.8) is the actual initial condition for the original IVP (15.5) and the initial condition for IVP (15.9) is the solution $r(x,t)$ of IVP (15.8).

Proof. Clearly the solution of IVP (15.8) is $r(x,t) = u_0(x - at)$, while the exact solution of IVP (15.9) is $s(x(t),t) = s(0)e^{\lambda t}$. But $s(0) = r(x,t) = u_0(x - at)$ and thus the resulting solution of IVPs (15.8) and (15.9) is

$$s(x,t) = u_0(x - at)e^{\lambda t} \,,$$

which is the exact solution (15.6) of the original inhomogeneous IVP (15.5), and the theorem is thus proved.

The result on the splitting scheme obtained by solving in succession (15.8), (15.9) can be expressed in the succinct form

$$u(x,t) = S^{(t)}C^{(t)}[u_0(x)] \,. \tag{15.10}$$

We interpret $C^{(t)}$ as the *solution operator* for the advection problem (15.8) applied over a time t and $S^{(t)}$ as the *solution operator* for the ODE (15.9) applied for a time t.

Exercise 15.2.1. Show that the exact solution $u(x,t)$ of IVP (15.5) can be obtained by solving

$$\left.\begin{array}{llll} PDE & : & u_t + f(u)_x = 0 \,; f(u) = au \,, \\ IC & : & u(x,0) = u_0(x) \end{array}\right\} \Longrightarrow \overline{u}^{n+1} \tag{15.11}$$

and

$$\left.\begin{array}{llll} ODE & : & \frac{d}{dx}f(u) = \lambda u \,, \\ IC & : & \overline{u}^{n+1} \end{array}\right\} \Longrightarrow u^{n+1} \tag{15.12}$$

in succession.

This result says that the splitting scheme (15.8), (15.9) modified by replacing the ODE in time by an ODE in space also gives the exact solution. This splitting scheme can be expressed in the succinct form

$$u(x,t) = S^{(x)}C^{(t)}[u_0(x)] \,, \tag{15.13}$$

where $S^{(x)}$ denotes the solution operator for the ODE in x in (15.12).

Solution 15.2.1. Left to the reader.

Glimm, Marshall and Plohr [143] constructed numerical splitting schemes of the form (15.11), (15.12) to solve numerically one–dimensional flows with area variation. In this case the source term involves a spatial derivative and does not depend on time.

In the next section we construct numerical methods based on the splitting approach, or fractional step approach [329], [417].

15.3 Numerical Methods Based on Splitting

We have shown that for the model inhomogeneous PDE (15.4) the splitting approach, as described in the previous section, is *exact*. For non–linear systems (15.1) this result is no longer valid. However, approximate, numerical schemes based on the splitting approach can be constructed. We first consider the scalar case.

15.3.1 Model Equations

The splitting scheme (15.8)–(15.9), represented by (15.10), is exact if the operators C and S are exact. Here we are interested in constructing numerical methods for the scalar IVP

$$\left. \begin{array}{llll} \text{PDE} & : & u_t + f(u)_x = s(u) & : 0 \leq x \leq L, \\ \text{IC} & : & u(x, t^n) = u^n, \end{array} \right\} \qquad (15.14)$$

To this end we replace the exact operators $C^{(t)}$ and $S^{(t)}$ in (15.10) by approximate operators and re–state the problem in a numerical context. Given the IVP (15.14), we want to evolve the solution from its initial value u^n at a time t^n, by one time step of size Δt, to a value u^{n+1} at time $t^{n+1} = t^n + \Delta t$. We assume the spatial domain $[0, L]$ has been discretised into a finite number M of cells i (finite volume approach) or grid points i (finite difference approach). Here u^n is a set of discrete values u_i^n at time t^n. The discrete analogue of the splitting scheme (15.8)–(15.9) is now

$$\left. \begin{array}{lll} \text{PDE} & : & u_t + f(u)_x = 0, \\ \text{IC} & : & u(x, t^n) = u^n, \end{array} \right\} \Longrightarrow \bar{u}^{n+1} \qquad (15.15)$$

$$\left. \begin{array}{lll} \text{ODE} & : & \frac{d}{dt}u = s(u), \\ \text{IC} & : & \bar{u}^{n+1}. \end{array} \right\} \Longrightarrow u^{n+1} \qquad (15.16)$$

The initial condition for the advection problem (15.15) is the initial condition for the complete problem (15.14). The solution of (15.15) after a time Δt is denoted by $\bar{u}^{n+1}$ and is used as the initial condition for the second IVP (15.16). This second IVP accounts for the presence of the source term $s(u)$ and is also solved for a complete time step Δt; this solution is then regarded as an approximation to the solution u^{n+1} of the full problem (15.14) at a time $t^{n+1} = t^n + \Delta t$. If the numerical analogues of $S^{(t)}$ and $C^{(t)}$ in (15.10) are still denoted by S and C, then we can write the splitting of (15.14) into (15.15)–(15.16) as

$$u^{n+1} = S^{(\Delta t)} C^{(\Delta t)}(u^n) . \tag{15.17}$$

Each numerical sub–problem (15.15), (15.16) is dealt with separately, for a time step Δt. One requires a numerical method to solve the homogeneous advection problem (15.15) and another numerical method to solve the ordinary differential equation in (15.16), with the initial data taken from the solution of (15.15). This procedure for solving inhomogeneous systems is exceedingly simple but is only first–order accurate in time, when S and C are at least first–order accurate solution operators. A second–order accurate scheme is

$$u^{n+1} = S^{(\frac{1}{2}\Delta t)} C^{(\Delta t)} S^{(\frac{1}{2}\Delta t)}(u^n) , \tag{15.18}$$

where S and C are at least second–order accurate solution operators.

15.3.2 Schemes for Systems

Here we extend the application of the splitting scheme of the previous section to non–linear systems of the form (15.1). The generalisation of (15.15), (15.16) to solve (15.1) is straightforward. Given the IVP

$$\left.\begin{array}{lll} \text{PDE's} & : & \mathbf{U}_t + \mathbf{F}(\mathbf{U})_x = \mathbf{S}(\mathbf{U}) \; ; 0 \le x \le L \, , \\ \text{IC} & : & \mathbf{U}(x, t^n) = \mathbf{U}^n \, , \end{array}\right\} \tag{15.19}$$

we want to evolve $\mathbf{U}^n$ from time $t = t^n$ to the new value $\mathbf{U}^{n+1}$ at $t = t^{n+1}$ in a time step $\Delta t = t^{n+1} - t^n$. The splitting (15.15), (15.16) becomes

$$\left.\begin{array}{lll} \text{PDE's} & : & \mathbf{U}_t + \mathbf{F}(\mathbf{U})_x = \mathbf{0} \, , \\ \text{IC} & : & \mathbf{U}(x, t^n) = \mathbf{U}^n \end{array}\right\} \Longrightarrow \overline{\mathbf{U}}^{n+1} , \tag{15.20}$$

$$\left.\begin{array}{lll} \text{PDE's} & : & \frac{d}{dt}\mathbf{U} = \mathbf{S}(\mathbf{U}) \, , \\ \text{IC's} & : & \overline{\mathbf{U}}^{n+1} \end{array}\right\} \Longrightarrow \mathbf{U}^{n+1} . \tag{15.21}$$

The analogue of the first–order scheme (15.17) is

$$\mathbf{U}^{n+1} = S^{(\Delta t)} C^{(\Delta t)}(\mathbf{U}^n) . \tag{15.22}$$

A second–order accurate scheme for systems is

$$\mathbf{U}^{n+1} = S^{(\frac{1}{2}\Delta t)} C^{(\Delta t)} S^{(\frac{1}{2}\Delta t)}(\mathbf{U}^n) . \tag{15.23}$$

A splitting scheme based on (15.11), (15.12) is

$$\left.\begin{array}{lll} \text{PDEs} & : & \mathbf{U}_t + \mathbf{F}(\mathbf{U})_x = \mathbf{0} \, , \\ \text{IC} & : & \mathbf{U}(x, t^n) = \mathbf{U}^n \, , \end{array}\right\} \Longrightarrow \overline{\mathbf{U}}^{n+1} \tag{15.24}$$

$$\left. \begin{array}{lll} \text{ODEs} & : & \frac{d}{dx}\mathbf{F}(\mathbf{U}) = \mathbf{S}(\mathbf{U}) \ , \\ \text{IC} & : & \mathbf{U}^{n+1} \ . \end{array} \right\} \Longrightarrow \mathbf{U}^{n+1} \qquad (15.25)$$

There appears to be little experience in using this approach. For source terms that are independent of time or involve spatial derivatives, this approach may be advantageous. See [143].

The attraction of splitting schemes is in the freedom available in choosing the numerical operators S and C. In general, one may choose the best scheme for each type of problems. For solving the advection (homogeneous) IVP (15.20) one can, for instance, use any of the schemes studied in Chaps. 6 to 14, or some other method. For solving the ODEs in (15.21), (15.25) one may choose some appropriate ODE solver, see next section.

15.4 Remarks on ODE Solvers

There is a vast literature on ODEs and on numerical methods for solving ODEs. For theoretical properties of ODEs see for example Brown [55], Ince and Sneddon [178], Sánchez [294] and Coddington and Levinson [84]. Almost any textbook on Numerical Analysis will contain some chapter on schemes for ODEs. See for example Hildebrand [168]; Mathews [238]; Conte and de Boor [93]; Maron and Lopez [228]; Johnson and Riess [186]; Kahaner, Moler and Nash [188]. Advanced textbooks are those of Gear [131], Lambert [199] and Shampine [306].

15.4.1 First–Order Systems of ODEs

Here we recall some very basic facts about first–order systems of ODEs

$$\frac{d}{dt}\mathbf{U}(t) \equiv \mathbf{U}' = \mathbf{S}(t, \mathbf{U}(t)) \ . \qquad (15.26)$$

Here $\mathbf{U} = \mathbf{U}(t)$ and $\mathbf{S}(t, \mathbf{U}(t))$ are vector–valued functions of m components

$$\mathbf{U} = [u_1, u_2, \ldots, u_m]^T \ ; \quad \mathbf{S} = [s_1, s_2, \ldots, s_m]^T \qquad (15.27)$$

and the independent variable t is a time–like variable. The Jacobian $\mathbf{A}(\mathbf{U})$ is defined as the matrix

$$\mathbf{A}(\mathbf{U}) = \partial \mathbf{S}/\partial \mathbf{U} = \begin{bmatrix} \partial s_1/\partial u_1 & \cdots & \partial s_1/\partial u_m \\ \partial s_2/\partial u_1 & \cdots & \partial s_2/\partial u_m \\ \vdots & \vdots & \vdots \\ \partial s_m/\partial u_1 & \cdots & \partial s_m/\partial u_m \end{bmatrix} \ . \qquad (15.28)$$

The entries a_{ij} of $\mathbf{A}(\mathbf{U})$ are partial derivatives of the components s_i of the vector $\mathbf{S}$ with respect to the components u_j of the vector $\mathbf{U}$, that is

$a_{ij} = \partial s_i/\partial u_j$. The eigenvalues λ_i of $\mathbf{A}$ are the solutions of the *characteristic polynomial*

$$|\mathbf{A} - \lambda\mathbf{I}| = \det(\mathbf{A} - \lambda\mathbf{I}) = 0 \,, \tag{15.29}$$

where $\mathbf{I}$ is the identity matrix. Generally, the eigenvalues are complex numbers. Trivially, the eigenvalue of the model ODE

$$u'(t) = \lambda u(t) \tag{15.30}$$

is λ. The behaviour of a system of ODEs is, in the main, determined by the behaviour of its eigenvalues. For instance, the exact solution of (15.30) with initial condition $u(0) = 1$ is $u(t) = e^{\lambda t}$. For t close to 0 the solution varies rapidly if the eigenvalue is negative and large in absolute value. For t away from zero the solution is almost indistinguishable from 0.

An important property of ODEs is that of *stability*. Generally speaking stable solutions are bounded. Note that the solution of the linear ODE (15.30) is bounded only if $\lambda < 0$. Geometrically, a solution $\mathbf{U}(t)$ is stable if any other solution of the ODE whose initial condition is sufficiently close to that of $\mathbf{U}(t)$ remains in a *tube* enclosing $\mathbf{U}(t)$. If the diameter of the tube tends to 0 as t tends to ∞, the solution is said to be *asymptotically stable*. Stability of solutions $\mathbf{U}(t)$ is characterised in terms of the eigenvalues λ_j of the Jacobian matrix. In particular if the real part of every eigenvalue is negative the solution is asymptotically stable.

Another feature of ODEs is that of *stiffness*. Stiff ODEs are usually associated with processes operating on disparate time scales. Chemical kinetics is a classical source of stiff ODEs. The stiffness of a system is generally determined by the behaviour of the eigenvalues of the system. In addition, the time interval over which the solution is sought is also a consideration in determining the stiffness of the system. There will be intervals of *rapid variations* (transient) of the solution and intervals of *slow variation*. The single ODE (15.30) is stiff for $\lambda \ll 0$ and for time t in the vicinity of 0.

Following Lambert [199], a nonlinear system of the form (15.26) is said to be stiff if

- (i) $Re(\lambda_j) < 0 \,, \quad j = 1, 2, \ldots, m$ and
- (ii) $\lambda_{max} \equiv \max_j |Re(\lambda_j)| \gg \lambda_{min} \equiv \min_j |Re(\lambda_j)|$.

Here $Re(\lambda_j)$ denotes the real part of the complex number λ_j. The stiffness ratio is defined as $R_{stif} = \lambda_{max}/\lambda_{min}$. Modest values of R_{stif}, e.g. 20, are sufficient to cause serious numerical difficulties to explicit methods. In real applications R_{stif} may be as large as 10^6.

Before thinking of numerical methods to solve ODEs, a fundamental question is to investigate whether the ODEs are stiff or not; this will determine the appropriate numerical methods to be used for solving the equations. See Kahaner, Moler and Nash [188], Gear [131] and Lambert [199].

15.4.2 Numerical Methods

We are interested in solving the Initial Value Problem (IVP) for (15.26) with initial condition

$$\mathbf{U}(t^0) = \mathbf{U}_0. \tag{15.31}$$

Discretise the domain of integration $[t^0, t^f]$ through the partition $t^0 < t^1 < t^2 \ldots < t^n < t^{n+1} \ldots < t^f$. One way of constructing numerical methods to solve the IVP (15.26), (15.31) is by using Taylor series expansions. Another way is to integrate (15.26) between t^n and t^{n+1} to obtain

$$\mathbf{U}(t^{n+1}) = \mathbf{U}(t^n) + \int_{t^n}^{t^{n+1}} \mathbf{S}(t, \mathbf{U}(t)) dt . \tag{15.32}$$

Various numerical methods are obtained depending on the way the integral is evaluated. The *Euler Method* results from evaluating the integral at the *old* time,

$$\mathbf{U}^{n+1} = \mathbf{U}^n + \Delta t \mathbf{S}(t^n, \mathbf{U}^n) . \tag{15.33}$$

where $\Delta t = t^{n+1} - t^n$ is the time step and $\mathbf{U}^n \approx \mathbf{U}(t^n)$. The Euler method is *explicit* and first–order accurate. The *Backward Euler Method,* also first order accurate but *implicit,* results from evaluating the integral at the *new* time t^{n+1}, namely

$$\mathbf{U}^{n+1} = \mathbf{U}^n + \Delta t \mathbf{S}(t^{n+1}, \mathbf{U}^{n+1}) . \tag{15.34}$$

A second–order implicit method results from a *trapezium rule* approximation to the integral, giving the *Trapezoidal Method*

$$\mathbf{U}^{n+1} = \mathbf{U}^n + \frac{1}{2} \Delta t [\mathbf{S}(t^n, \mathbf{U}^n) + \mathbf{S}(t^{n+1}, \mathbf{U}^{n+1})] . \tag{15.35}$$

A second–order, two stage *Runge–Kutta method* (explicit) is

$$\left. \begin{aligned} \mathbf{K}_1 &= \Delta t \mathbf{S}(t^n, \mathbf{U}^n) , \\ \mathbf{K}_2 &= \Delta t \mathbf{S}(t^n + \Delta t, \mathbf{U}^n + \mathbf{K}_1) , \\ \mathbf{U}^{n+1} &= \mathbf{U}^n + \tfrac{1}{2} [\mathbf{K}_1 + \mathbf{K}_2] . \end{aligned} \right\} \tag{15.36}$$

A fourth–order, four stage Runge–Kutta method (explicit) is

$$\left. \begin{aligned} \mathbf{K}_1 &= \Delta t \mathbf{S}(t^n, \mathbf{U}^n) , \\ \mathbf{K}_2 &= \Delta t \mathbf{S}(t^n + \tfrac{1}{2}\Delta t, \mathbf{U}^n + \tfrac{1}{2}\mathbf{K}_1) , \\ \mathbf{K}_3 &= \Delta t \mathbf{S}(t^n + \tfrac{1}{2}\Delta t, \mathbf{U}^n + \tfrac{1}{2}\mathbf{K}_2) , \\ \mathbf{K}_4 &= \Delta t \mathbf{S}(t^n + \Delta t, \mathbf{U}^n + \mathbf{K}_3) , \\ \mathbf{U}^{n+1} &= \mathbf{U}^n + \tfrac{1}{6} [\mathbf{K}_1 + 2\mathbf{K}_2 + 2\mathbf{K}_3 + \mathbf{K}_4] . \end{aligned} \right\} \tag{15.37}$$

Stability of numerical methods is a most important issue. To illustrate this point consider the model ODE (15.30) as solved by the explicit Euler method (15.33). The scheme reads

$$u^{n+1} = (1 + \Delta t \lambda) u^n . \tag{15.38}$$

Clearly, for stability one requires that the ODE itself be stable, $\lambda < 0$, and that the *amplification factor* satisfy $|1 + \lambda \Delta t| \le 1$. Therefore the time step Δt must satisfy the stability restriction

$$\Delta t \le \frac{2}{|\lambda|} . \tag{15.39}$$

For large $|\lambda|$ (stiff ODE) Δt can be extremely small, which means that the method becomes very inefficient or even useless in practice.

On the other hand, the Trapezoidal method (15.35), which is implicit, gives

$$u^{n+1} = \frac{(1 + \frac{1}{2} \Delta t \lambda)}{(1 - \frac{1}{2} \Delta t \lambda)} u^n . \tag{15.40}$$

This is stable for any Δt, provided $\lambda \le 0$, that is whenever the ODE itself is stable.

Explicit methods are much simpler to use than implicit methods. The latter require the solution of non–linear algebraic equations at each time step and are therefore much more expensive. However, as illustrated, for stiff problems implicit methods are the only methods to use in any practical situation.

15.4.3 Implementation Details for Split Schemes

There are two facts that need to be emphasised when solving ODEs in the context of the splitting schemes described in Sect. 15.3. First, at every time t^n, at each mesh point i one has a system of ODEs to solve; second, the time evolution of the ODEs is generally short and is dictated by the time step in the overall splitting scheme. This second point is relevant when choosing ODE solvers.

Before selecting a method, an analysis of the ODEs must be performed. If the problem is non–stiff then a high–order explicit method is recommended. A stability analysis of the method must be carried out and enforced when selecting the size of the time step Δt. For simplicity, let us assume we want to implement the first–order splitting scheme (15.22). A practical problem is to determine the time step Δt. One first determines the time step Δt_c for the advection problem (15.20). If this problem is solved by some explicit method, e.g. Godunov's first order upwind method (see Chap. 6), then Δt_c is found from some stability constraint, i.e. the Courant condition, see Chap. 6. The solution of the advection problem is found at every mesh point i and this gives $\overline{\mathbf{U}}^{n+1}$, which is then used as the initial data for the ODE step (15.21). If the ODEs are solved by some implicit method, then there will be no stability restriction on the time step, and therefore one can advance the solution via the ODE solver by a time $\Delta t_s = \Delta t_c$, in *one go*. However, if an explicit method is used to solve the ODEs, then a stable time step Δt_s must first be found. If $\Delta t_s \geq \Delta t_c$, then one may again advance the solution via the ODEs by a time Δt_c in *one go*. Hence the final solution at time t^{n+1} has been advanced by a time $\Delta t = \Delta t_c$. If $\Delta t_s < \Delta t_c$, then one possibility is to update via the ODEs in k steps of size $\Delta t_k = \Delta t_c/k$, where k is a positive integer such such Δt_k is a stable time step for the ODE solver. The previous observations apply directly when implementing the second–order splitting scheme (15.23). A useful reference is Chiang and Hoffmann [71].

15.5 Concluding Remarks

We have only presented one approach for treating source terms. There are other approaches, but at the present time there appears to be no clear, and sufficiently general, alternative to splitting. The idea of *upwinding* the source terms proposed by Roe [286] appears to work well for certain problems. See for instance the work of Vázquez [400] and that of Bermúdez and Vázquez [34]. See also the recent paper of Vázquez [401], which addresses the issue of geometric and friction source terms in shallow water models. For steady–state problems computed by time–marching schemes the reader should consult the recent paper by LeVeque [209].

The reader is strongly encouraged to utilise problems with exact solutions, whenever available, to carefully assess the numerical methods before applying them to *the real problem*. The simplest test problem is the IVP (15.5) with exact solution (15.6). More scalar test problems are found in [212], [90], [152]. A test problem with exact solution for a 2×2 non–linear system is the so called Fickett detonation analogue [124], see also example 2.4.3 of Sect. 2.4.2 in Chap. 2. This problem is exploited in [80] for testing numerical methods for detonation waves in high–energy solids. For the Euler equations, a test problem with exact solution is reported in [82]. Details of the solution are given in [77] and applications are also shown in [286] and [334].

For certain types of problems, such as detonation waves, there are serious difficulties in designing numerical methods to properly account for the fluid dynamics and the chemistry. For sufficiently fine meshes such difficulties may be overcome but at a cost that is impossible to meet with current computing resources, if realistic problems in multidimensions are to be solved. Since the early papers by Colella, Majda and Roytburd [90] and that of LeVeque and Yee [212], there has been a noticeable increase in the interest for hyperbolic systems with source terms, both numerically and theoretically. See, amongst others: Griffiths, Stuart and Yee [152]; Berkenbosch, Kaasschieter and Boonkkamp [32]; Bourlioux, Majda and Roytburd [53]; Fey, Jeltsch and Müller [123]; Chalabi [63], [64], [65], [66]; Benkhaldoun and Chalabi [27]; Pember [258], [259]; Schroll and Tveito [301]; Schroll and Winther [303]; Schroll, Tveito and Winther [302]; Corberán and Gascón [94]; Lorenz and Schroll

[225]; Lafon and Yee [197], [198]; and especially the review paper of Yee and Sweby [423].

The splitting approach may also be applied to treat diffusion like terms [56], [375] in exactly the same manner as for algebraic source terms. The splitting approach also offers one way of solving multidimensional problems; this topic is dealt with in Chap. 16.

16. Methods for Multi–Dimensional PDEs

16.1 Introduction

This chapter is concerned with numerical methods for solving non–linear systems of hyperbolic conservation laws in multidimensions. For Cartesian geometries one may write the equations of our interest here as

$$\mathbf{U}_t + \mathbf{F(U)}_x + \mathbf{G(U)}_y + \mathbf{H(U)}_z = \mathbf{0} \,, \tag{16.1}$$

where t denotes time or a time–like variable and x, y, z are Cartesian coordinate directions. $\mathbf{U}$ is the vector of conserved variables and $\mathbf{F(U)}$, $\mathbf{G(U)}$, $\mathbf{H(U)}$ are vectors of fluxes in the x, y, z directions respectively. A prominent example are the time–dependent three dimensional Euler equations studied in Sect. 3.2 of Chap. 3. Other examples are the time–dependent two dimensional shallow water equations and the artificial compressibility equations. See Sect. 1.6.3 of Chap. 1.

We shall present two ways of solving (16.1) numerically. The first approach is *dimensional splitting* or *method of fractional steps* [329], [417]. This method is also known as time–operator splitting. In this approach one applies one–dimensional methods in each coordinate direction. A simple extension of the one–dimensional methods is required to solve the one–dimensional conservation laws *augmented* by the extra components of velocity present in multidimensional problems. Any of the methods studied in Chaps. 6 to 14 can be applied to solve (16.1), in conjunction with dimensional splitting. The other approach studied here is the *finite volume method*, whereby in updating the solution within some control volume one includes all the intercell flux contributions in a single step. Only some of the one–dimensional approaches discussed in Chaps. 6 to 14 can be extended to multidimensional problems following the (unsplit) finite volume approach. There is a close relationship between dimensional splitting and the finite volume method. If the schemes used in the dimensional splitting approach are of the finite–volume type in one space dimension, see Chaps. 5 to 12 and 14, then the resulting splitting schemes may be viewed as being predictor–corrector type finite volume schemes in multidimensions. Both the dimensional splitting and the finite volume approaches may be extended to deal with non–Cartesian geometries.

16.2 Dimensional Splitting

Before introducing splitting schemes for multidimensional systems of conservation laws, we study the splitting approach as applied to a model conservation law in three space dimensions, for which the dimensional splitting method can be shown to be exact.

16.2.1 Splitting for a Model Problem

Consider the Initial–Value Problem (IVP) for the linear advection equation in three space dimensions

$$\begin{array}{ll}
\text{PDE} & : \quad u_t + a_1 u_x + a_2 u_y + a_3 u_z = 0 \,, \\
\text{IC} & : \quad u(x, y, z, t^n) = u_0(x, y, z) \equiv u^n \,.
\end{array} \left. \right\} \qquad (16.2)$$

Here a_1, a_2 and a_3 are the three velocity components of a constant velocity vector $\mathbf{V} = (a_1, a_2, a_3)$. Let us consider three one–dimensional IVPs in the x, y and z directions respectively. The IVP in the x–direction, or x sweep, is

$$\begin{array}{ll}
\text{PDE} & : \quad u_t + a_1 u_x = 0 \\
\text{IC} & : \quad u^n
\end{array} \left. \right\} \overset{\Delta t}{\Longrightarrow} u^{n+\frac{1}{3}} \,. \qquad (16.3)$$

The initial data for this IVP is the initial data for the original full IVP (16.2) and its solution after a time Δt is denoted by $u^{n+\frac{1}{3}}$. The y–direction IVP, or y sweep, is

$$\begin{array}{ll}
\text{PDE} & : \quad u_t + a_2 u_y = 0 \\
\text{IC} & : \quad u^{n+\frac{1}{3}}
\end{array} \left. \right\} \overset{\Delta t}{\Longrightarrow} u^{n+\frac{2}{3}} \,. \qquad (16.4)$$

The initial data for this IVP is the solution $u^{n+\frac{1}{3}}$ of IVP (16.3) and its solution after a time Δt is denoted by $u^{n+\frac{2}{3}}$. The z–direction IVP, or z sweep, is

$$\begin{array}{ll}
\text{PDE} & : \quad u_t + a_3 u_z = 0 \\
\text{IC} & : \quad u^{n+\frac{2}{3}}
\end{array} \left. \right\} \overset{\Delta t}{\Longrightarrow} u^{n+1} \,. \qquad (16.5)$$

The initial data for this IVP is the solution $u^{n+\frac{2}{3}}$ of IVP (16.4) and its solution after a time Δt is denoted by u^{n+1}, which is then regarded as the solution to the full problem (16.2) after a time Δt.

Let us denote by $\mathcal{X}^{(t)}$, $\mathcal{Y}^{(t)}$ and $\mathcal{Z}^{(t)}$ the solution operators for IVPs (16.3), (16.4) and (16.5) respectively, when solved *exactly* for a time t. Then the following result can be proved.

Proposition 16.2.1 (Dimensional Splitting). *The exact solution u^{n+1} of the three–dimensional IVP (16.2) after a time Δt can be obtained by solving exactly the sequence of three one–dimensional IVPs (16.3)–(16.5), each for a time Δt, that is*

$$u^{n+1} = \mathcal{Z}^{(\Delta t)} \mathcal{Y}^{(\Delta t)} \mathcal{X}^{(\Delta t)} (u^n) \,. \qquad (16.6)$$

Proof. (Left as an exercise. See [207]).

16.2.2 Splitting Schemes for Two–Dimensional Systems

For non–linear systems, dimensional splitting is not exact but one may construct approximate splitting schemes. Consider the two–dimensional initial value problem

$$\left.\begin{array}{lll} \text{PDE} & : & \mathbf{U}_t + \mathbf{F}(\mathbf{U})_x + \mathbf{G}(\mathbf{U})_y = \mathbf{0} \ , \\ \text{IC} & : & \mathbf{U}(x, y, t^n) = \mathbf{U}^n \ . \end{array}\right\} \tag{16.7}$$

The initial data at a time t^n is given by the set $\mathbf{U}^n$ of discrete cell average values $\mathbf{U}^n_{i,j}$; index i refers to the x–coordinate direction and index j refers to the y–coordinate direction. Fig. 16.1 illustrates a finite volume discretisation of a two–dimensional domain in the x–y plane.

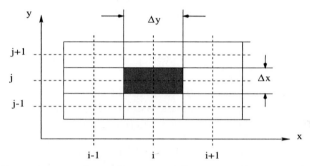

Fig. 16.1. Discretisation of two–dimensional Cartesian domain into finite volumes I_{ij} of area $\Delta x \times \Delta y$

The dimensional splitting approach replaces (16.7) by a pair of one–dimensional IVPs. The simplest version of the approach replaces (16.7) by the sequence of IVPs

$$\left.\begin{array}{lll} \text{PDEs} & : & \mathbf{U}_t + \mathbf{F}(\mathbf{U})_x = \mathbf{0} \\ \text{ICs} & : & \mathbf{U}^n \end{array}\right\} \xRightarrow{\Delta t} \mathbf{U}^{n+\frac{1}{2}} \tag{16.8}$$

and

$$\left.\begin{array}{lll} \text{PDEs} & : & \mathbf{U}_t + \mathbf{G}(\mathbf{U})_y = \mathbf{0} \\ \text{ICs} & : & \mathbf{U}^{n+\frac{1}{2}} \end{array}\right\} \xRightarrow{\Delta t} \mathbf{U}^{n+1} \ . \tag{16.9}$$

In the first IVP (16.8) one solves a one–dimensional problem in the x–direction for a time step Δt. This is called the x *sweep* and its solution is denoted by $\mathbf{U}^{n+\frac{1}{2}}$. Note that *for each strip labelled j*, see Fig. 16.1, one solves the one–dimensional problem (16.8). In the next IVP (16.9) one solves a one–dimensional problem in the y–direction, also for a time step Δt. This is called the y *sweep*. The initial condition for the second IVP (16.9) is the solution $\mathbf{U}^{n+\frac{1}{2}}$ of IVP (16.8). Note here that *for each strip labelled i*, see

Fig. 16.1, one solves the one–dimensional problem (16.9). Both sweeps have a common time step Δt; in Sec. 16.3.2 we discuss ways of determining Δt.

If $\mathcal{X}^{(t)}$ and $\mathcal{Y}^{(t)}$ are approximate solution operators for IVPs (16.8) and (16.9), then the splitting (16.8), (16.9) of the original two–dimensional IVP (16.7) can be written thus

$$\mathbf{U}^{n+1} = \mathcal{Y}^{(\Delta t)} \mathcal{X}^{(\Delta t)} (\mathbf{U}^n) . \tag{16.10}$$

There is no particular reason for applying the operators in the order just described. An equivalent scheme is

$$\mathbf{U}^{n+1} = \mathcal{X}^{(\Delta t)} \mathcal{Y}^{(\Delta t)} (\mathbf{U}^n) . \tag{16.11}$$

It can be shown, see Strang [329], that for general systems both splitting schemes (16.10) and (16.11) are first–order accurate in time if the individual operators $\mathcal{X}$ and $\mathcal{Y}$ are at least first–order accurate accurate in time. A second–order accurate splitting [329] is

$$\mathbf{U}^{n+1} = \frac{1}{2} \left[\mathcal{X}^{(\Delta t)} \mathcal{Y}^{(\Delta t)} + \mathcal{Y}^{(\Delta t)} \mathcal{X}^{(\Delta t)} \right] (\mathbf{U}^n) , \tag{16.12}$$

provided each of the solution operators are at least second–order accurate in time. Note however that this scheme requires double the amount of work of schemes (16.10) and (16.11). More attractive second–order schemes are

$$\mathbf{U}^{n+1} = \mathcal{X}^{(\frac{1}{2}\Delta t)} \mathcal{Y}^{(\Delta t)} \mathcal{X}^{(\frac{1}{2}\Delta t)} (\mathbf{U}^n) \tag{16.13}$$

and

$$\mathbf{U}^{n+1} = \mathcal{Y}^{(\frac{1}{2}\Delta t)} \mathcal{X}^{(\Delta t)} \mathcal{Y}^{(\frac{1}{2}\Delta t)} (\mathbf{U}^n) . \tag{16.14}$$

Schemes (16.13) and (16.14) require about 50% more work than the first order schemes (16.10) and (16.11). Strang [329] suggested a modification to schemes (16.13) and (16.14) so that they become as efficient as (16.10) or (16.11), while still preserving formal second–order accuracy. Suppose (16.13) is applied over m time steps of size Δt

$$\mathbf{U}^{n+m} = \left(\mathcal{X}^{(\frac{1}{2}\Delta t)} \mathcal{Y}^{(\Delta t)} \mathcal{X}^{(\frac{1}{2}\Delta t)} \right)^m (\mathbf{U}^n) . \tag{16.15}$$

For example, two successive applications of (16.13) give

$$\mathcal{X}^{(\frac{1}{2}\Delta t)} \mathcal{Y}^{(\Delta t)} \mathcal{X}^{(\frac{1}{2}\Delta t)} \mathcal{X}^{(\frac{1}{2}\Delta t)} \mathcal{Y}^{(\Delta t)} \mathcal{X}^{(\frac{1}{2}\Delta t)} (\mathbf{U}^n) .$$

The two successive applications of $\mathcal{X}^{(\frac{1}{2}\Delta t)}$ can be combined into a single application of $\mathcal{X}^{(\Delta t)}$. It is then easy to see that (16.15) can be expressed as

$$\mathbf{U}^{n+m} = \mathcal{X}^{(\frac{1}{2}\Delta t)} \left[\mathcal{Y}^{(\Delta t)} \mathcal{X}^{(\Delta t)} \right]^{m-1} \mathcal{Y}^{(\Delta t)} \mathcal{X}^{(\frac{1}{2}\Delta t)} (\mathbf{U}^n) \tag{16.16}$$

or

$$\mathbf{U}^{n+m} = \mathcal{X}^{(\frac{1}{2}\Delta t)} \mathcal{Y}^{(\Delta t)} \left[\mathcal{X}^{(\Delta t)} \mathcal{Y}^{(\Delta t)} \right]^{m-1} \mathcal{X}^{(\frac{1}{2}\Delta t)} (\mathbf{U}^n) . \tag{16.17}$$

Note that in the manipulations above we have assumed that the size of the time step Δt is constant for at least m time steps. For linear systems this can certainly be imposed but not for non–linear systems, as Δt depends on the wave speeds. In practice one still applies the above schemes (16.16), (16.17), where the choice of m depends on the required output times. In implementing these schemes it is reasonable to ensure that the size of the time step Δt in the operators outside the squared brackets is constant. These splitting schemes are almost as efficient as the first–order splittings (16.10), (16.11).

Two more second–order accurate schemes are

$$\mathbf{U}^{n+2} = \mathcal{X}^{(\Delta t)} \mathcal{Y}^{(\Delta t)} \mathcal{Y}^{(\Delta t)} \mathcal{X}^{(\Delta t)} (\mathbf{U}^n) \tag{16.18}$$

and

$$\mathbf{U}^{n+2} = \mathcal{Y}^{(\Delta t)} \mathcal{X}^{(\Delta t)} \mathcal{X}^{(\Delta t)} \mathcal{Y}^{(\Delta t)} (\mathbf{U}^n) \ . \tag{16.19}$$

These can be shown to be second–order accurate every other time step, see Warming and Beam [404], and are as efficient as the first–order schemes (16.10) and (16.11).

The principle of dimensional splitting applies directly to three–dimensional problems, which is the subject of the next section.

16.2.3 Splitting Schemes for Three–Dimensional Systems

Consider the three–dimensional IVP

$$\begin{array}{rl} \text{PDE} & : \quad \mathbf{U}_t + \mathbf{F}(\mathbf{U})_x + \mathbf{G}(\mathbf{U})_y + \mathbf{H}(\mathbf{U})_z = \mathbf{0} \ , \\ \text{IC} & : \quad \mathbf{U}(x,y,z,t^n) = \mathbf{U}^n \ . \end{array} \left.\begin{array}{r} \\ \\ \end{array}\right\} \tag{16.20}$$

The initial data $\mathbf{U}^n$ at a time t^n is a set of discrete cell averages $\mathbf{U}^n_{i,j,k}$ on a three–dimensional Cartesian computational domain. The simplest version of the dimensional splitting approach replaces (16.20) by three one–dimensional IVPs, namely

$$\begin{array}{rl} \text{PDEs} & : \quad \mathbf{U}_t + \mathbf{F}(\mathbf{U})_x = \mathbf{0} \\ \text{ICs} & : \quad \mathbf{U}^n \end{array} \left.\begin{array}{r} \\ \\ \end{array}\right\} \stackrel{\Delta t}{\Longrightarrow} \mathbf{U}^{n+\frac{1}{3}} \ , \tag{16.21}$$

$$\begin{array}{rl} \text{PDEs} & : \quad \mathbf{U}_t + \mathbf{G}(\mathbf{U})_y = \mathbf{0} \\ \text{ICs} & : \quad \mathbf{U}^{n+\frac{1}{3}} \end{array} \left.\begin{array}{r} \\ \\ \end{array}\right\} \stackrel{\Delta t}{\Longrightarrow} \mathbf{U}^{n+\frac{2}{3}} \ , \tag{16.22}$$

and

$$\begin{array}{rl} \text{PDEs} & : \quad \mathbf{U}_t + \mathbf{H}(\mathbf{U})_z = \mathbf{0} \\ \text{ICs} & : \quad \mathbf{U}^{n+\frac{2}{3}} \end{array} \left.\begin{array}{r} \\ \\ \end{array}\right\} \stackrel{\Delta t}{\Longrightarrow} \mathbf{U}^{n+1} \ . \tag{16.23}$$

The first IVP (16.21) solves an *augmented* one–dimensional problem in the x–direction, the x *sweep*, for a time step Δt and the solution is denoted by $\mathbf{U}^{n+\frac{1}{3}}$. The next IVP (16.22) solves a one–dimensional (augmented) problem in the y–direction, the y *sweep*, for a time step Δt; the initial condition for the second IVP (16.22) is the solution of IVP (16.21). The next IVP (16.23) solves an augmented one–dimensional problem in the z–direction, the z *sweep*, for a time step Δt. The initial condition for the third IVP (16.23) is the solution of IVP (16.22). It is important to realise that the size of the time step Δt in each one–dimensional IVP is constant during one time step. $\mathbf{U}^{n+1}$ is regarded as the solution of IVP (16.20) after a time Δt.

If $\mathcal{X}^{(t)}$, $\mathcal{Y}^{(t)}$ and $\mathcal{Z}^{(t)}$ are approximate solution operators for IVPs (16.21), (16.22) and (16.23) respectively, then the simplest, first–order accurate, splitting based on (16.21)–(16.23) can be written thus

$$\mathbf{U}^{n+1} = \mathcal{Z}^{(\Delta t)} \mathcal{Y}^{(\Delta t)} \mathcal{X}^{(\Delta t)} (\mathbf{U}^n) \ . \tag{16.24}$$

Naturally, other orderings of the operators are possible to produce other first–order splitting schemes. A three–dimensional splitting, see Shang [307], that is second–order accurate every other time step is

$$\mathbf{U}^{n+2} = \mathcal{X}^{(\Delta t)} \mathcal{Y}^{(\Delta t)} \mathcal{Z}^{(\Delta t)} \mathcal{Z}^{(\Delta t)} \mathcal{Y}^{(\Delta t)} \mathcal{X}^{(\Delta t)} (\mathbf{U}^n) \ . \tag{16.25}$$

A three–dimensional splitting scheme that is second–order accurate every time step [41] is

$$\mathbf{U}^{n+1} = \mathcal{X}^{(\frac{1}{2}\Delta t)} \mathcal{Y}^{(\frac{1}{2}\Delta t)} \mathcal{Z}^{(\Delta t)} \mathcal{Y}^{(\frac{1}{2}\Delta t)} \mathcal{X}^{(\frac{1}{2}\Delta t)} (\mathbf{U}^n) \ . \tag{16.26}$$

Other combinations of one–dimensional operators will produce other second order accurate splitting schemes. The tools for constructing splitting schemes and analysing their accuracy, see Strang [329] and Shang [307], rely on Taylor series expansions and therefore the derived splitting schemes are strictly valid only for problems with smooth solutions. For discontinuous solutions splitting schemes may produce erroneous results. Ironically, the failure of dimensional splitting procedures is closely related to how successful the one–dimensional schemes used are for computing discontinuous solutions. For instance, the Random Choice Method (RCM), see Chap. 7, has the unique property of computing contact surfaces and shock waves as true discontinuities. However, dimensional splitting schemes based on the RCM have been shown to be a complete failure, see Colella [86]. An exception to these difficulties are contact surfaces; these waves can be correctly preserved by splitting schemes based on the RCM, see Toro [353].

In general, schemes that smear discontinuities in one space dimension work well in dimensional splitting procedures. We particularly recommend the use of the one–dimensional schemes presented in Chaps. 6 to 14, except for the Random Choice method of Chap. 7.

In the next section we illustrate via an example, some of the practical details involved when implementing dimensional splitting schemes to solve three–dimensional time dependent problems.

16.3 Practical Implementation of Splitting Schemes in Three Dimensions

As an example, we consider the time–dependent, three dimensional Euler equations

$$\mathbf{U}_t + \mathbf{F}(\mathbf{U})_x + \mathbf{G}(\mathbf{U})_y + \mathbf{H}(\mathbf{U})_z = \mathbf{0} , \tag{16.27}$$

with

$$
\left.
\begin{array}{c}
\mathbf{U} = \begin{bmatrix} \rho \\ \rho u \\ \rho v \\ \rho w \\ E \end{bmatrix} , \quad
\mathbf{F} = \begin{bmatrix} \rho u \\ \rho u^2 + p \\ \rho u v \\ \rho u w \\ u(E + p) \end{bmatrix} , \\[3em]
\mathbf{G} = \begin{bmatrix} \rho v \\ \rho u v \\ \rho v^2 + p \\ \rho v w \\ v(E + p) \end{bmatrix} , \quad
\mathbf{H} = \begin{bmatrix} \rho w \\ \rho u w \\ \rho v w \\ \rho w^2 + p \\ w(E + p) \end{bmatrix} .
\end{array}
\right\} \tag{16.28}
$$

See Chaps. 1 and 3 for details on the Euler equations.

16.3.1 Handling the Sweeps by a Single Subroutine

For simplicity let us assume we apply the first–order splitting scheme (16.24). For convenience we re–order the equations so that every sweep is handled by *a single one–dimensional subroutine*. In the x sweep we solve

$$
\begin{bmatrix} \rho \\ \rho u \\ E \\ \rho v \\ \rho w \end{bmatrix}_t
+
\begin{bmatrix} \rho u \\ \rho u^2 + p \\ u(E + p) \\ \rho u v \\ \rho u w \end{bmatrix}_x = \mathbf{0} . \tag{16.29}
$$

In the the y sweep one solves

$$
\begin{bmatrix} \rho \\ \rho v \\ E \\ \rho u \\ \rho w \end{bmatrix}_t
+
\begin{bmatrix} \rho v \\ \rho v^2 + p \\ v(E + p) \\ \rho v u \\ \rho v w \end{bmatrix}_y = \mathbf{0} . \tag{16.30}
$$

In the z sweep one solves

$$
\begin{bmatrix} \rho \\ \rho w \\ E \\ \rho u \\ \rho v \end{bmatrix}_t
+
\begin{bmatrix} \rho w \\ \rho w^2 + p \\ w(E + p) \\ \rho w u \\ \rho w v \end{bmatrix}_z = \mathbf{0} . \tag{16.31}
$$

In the x sweep (16.29) u is the *normal* velocity components and the first three equations look identical to the pure one–dimensional problem in x. Note however that the full problem (16.29) differs from the pure one–dimensional problem in two respects, namely (i) there are two extra equations for momentum in the y and z directions and (ii) the total energy E involves contributions from the tangential velocity components v and w via the kinetic energy.

As seen in Sect. 3.1.2 of Chap. 3, use of the continuity equation into the expanded form of the y and z momentum equations gives the two advection equations

$$v_t + uv_x = 0 , \quad w_t + uw_x = 0 . \tag{16.32}$$

These say that the *tangential velocity components* v and w are passively advected with the normal velocity component u. As a matter of fact, the extra equations for momentum in the y and z directions have exactly the same form as species equations present in the study of reactive mixtures.

For the y sweep (16.30) the normal velocity component is v and the tangential velocity components are u and w. For the z sweep (16.31) the normal velocity component is w and the tangential velocity components are u and v.

It is now obvious that a single subroutine can be used to deal with all three sweeps in a dimensional splitting scheme. Suppose we are solving the x sweep. Then an explicit conservative method, see Chaps. 5 to 14, reads

$$\mathbf{U}_{i,j,k}^{n+\frac{1}{3}} = \mathbf{U}_{i,j,k}^{n} + \frac{\Delta t}{\Delta x}[\mathbf{F}_{i-\frac{1}{2},j,k}^{n} - \mathbf{F}_{i+\frac{1}{2},j,k}^{n}] , \forall j , \forall k , \tag{16.33}$$

where $\mathbf{F}_{i+\frac{1}{2},j,k}^{n}$ is the numerical flux at the cell interface position $x_{i+\frac{1}{2}}$. For background on the meaning of the conservative formula (16.33) the reader is referred to Chap. 5, Sect. 5.3 and Chap. 6. For the y sweep (16.30) the updating conservative formula reads

$$\mathbf{U}_{i,j,k}^{n+\frac{2}{3}} = \mathbf{U}_{i,j,k}^{n+\frac{1}{3}} + \frac{\Delta t}{\Delta y}[\mathbf{G}_{i,j-\frac{1}{2},k}^{n+\frac{1}{3}} - \mathbf{G}_{i,j+\frac{1}{2},k}^{n+\frac{1}{3}}] , \forall i , \forall k , \tag{16.34}$$

and for the z sweep (16.31) the updating conservative formula reads

$$\mathbf{U}_{i,j,k}^{n+1} = \mathbf{U}_{i,j,k}^{n+\frac{2}{3}} + \frac{\Delta t}{\Delta z}[\mathbf{H}_{i,j,k-\frac{1}{2}}^{n+\frac{2}{3}} - \mathbf{H}_{i,j,k+\frac{1}{2}}^{n+\frac{2}{3}}] , \forall i , \forall j . \tag{16.35}$$

Obviously, given the way the equations have been re–ordered, all three sweeps (16.33)–(16.35) can be handled by a single subroutine. Let us denote the subroutine by ONED(...). Then, sweeps (16.33)–(16.35) are dealt with by doing

$$\left. \begin{array}{ll} \text{for the } x \text{ sweep} & \text{CALL ONED}(\rho, u, v, w, p, \Delta t, \Delta x, ...) , \\ \text{for the } y \text{ sweep} & \text{CALL ONED}(\rho, v, u, w, p, \Delta t, \Delta y, ...) , \\ \text{for the } z \text{ sweep} & \text{CALL ONED}(\rho, w, u, v, p, \Delta t, \Delta z, ...) . \end{array} \right\} \tag{16.36}$$

Note that we simply interchange the order of the velocity components in the single subroutine to solve for all the x, y and z sweeps. To implement some of the other splitting schemes such as (16.25) or (16.26), the order in (16.36), as well as the size of the time step, are appropriately changed.

16.3.2 Choice of Time Step Size

The explicit conservative schemes considered here require the computation of a time step Δt to be used in (16.33)–(16.35), such that *stability* of the numerical method is ensured. The choice of Δt depends on (i) the intercell flux used, i.e. the particular method used, (ii) the wave speeds present and (iii) the dimensions of the mesh used. See Sect. 6.3.2 for a full discussion of this topic for one–dimensional problems. To preserve some generality let us assume that the dimensions of cell $I_{i,j,k}$ are $\Delta x_{i,j,k}$, $\Delta y_{i,j,k}$ and $\Delta z_{i,j,k}$. Then, one way of choosing Δt is

$$\Delta t = C_{cfl} \times min_{i,j,k}\Big[\frac{\Delta x_{i,j,k}}{S_{i,j,k}^{n,x}}, \frac{\Delta y_{i,j,k}}{S_{i,j,k}^{n,y}}, \frac{\Delta z_{i,j,k}}{S_{i,j,k}^{n,z}}\Big] . \tag{16.37}$$

Here $S_{i,j,k}^{n,d}$ is the speed of the fastest wave present at time level n travelling in the d direction, with $d = x, y, z$. C_{clf} is the CFL coefficient and is chosen according to the linear stability condition of the particular numerical method in use, which depends on the numerical flux. For the Godunov first–order upwind method $0 < C_{clf} \leq 1$ and in practice one usually takes $C_{clf} \approx 0.9$. Perhaps the most critical point is finding reliable estimates for the speeds $S_{i,j,k}^{n,d}$. The simplest choice is

$$S_{i,j,k}^{n,d} = |d_{i,j,k}^n| + a_{i,j,k}^n \tag{16.38}$$

where $d_{i,j,k}^n$ is the particle velocity component in the d–direction and $a_{i,j,k}^n$ is the sound speed at time level n in cell $I_{i,j,k}$. As remarked in Sect. 6.3.2 of Chap. 6 for one–dimensional problems, choice (16.38) is inadequate for problems with initial conditions involving severe gradients. One may under-estimate the wave speeds, for instance, when the initial conditions are of the shock–tube problem type. In such cases $d_{i,j,k}^n$ might be zero and the only contribution to $S_{i,j,k}^{n,d}$ is the sound speed $a_{i,j,k}^n$, which then results in a gross underestimate for $S_{i,j,k}^{n,d}$. As a consequence a gross overestimate for Δt is computed which leads the method outside its stability range at the beginning of the computations.

16.3.3 The Intercell Flux and the TVD Condition

In discussing the choice of the intercell fluxes in (16.33)–(16.35) we shall consider the case of a single sweep, the x sweep say. The intercell flux $\mathbf{F}_{i+\frac{1}{2},j,k}^n$ in (16.33) may be computed by any of the methods discussed in Chaps. 6 to 12 or the high–order TVD schemes of Chaps. 13 and 14.

The computation of $\mathbf{F}_{i+\frac{1}{2},j,k}^n$ by the Godunov first–order upwind method of Chap. 6 is straightforward. What is new in multidimensional problems is the presence of tangential velocity components. Locally, these are related to shear waves. The particular details of the flux can have a profound effect in the accuracy with which shear waves and vortices are resolved by the method. The Godunov method with the exact Riemann solver of Chap. 4, see Sect. 4.8, gives accurate resolution of shear waves. The approximate Riemann solvers of Chaps. 9, 11, 12 and the HLLC Riemann solver of Chap. 10 will also resolve shear waves accurately. The Flux Vector Splitting schemes of Chap. 8, except for the version of Sect. 8.4.4, and the HLL Riemann solver of Sect. 10.3 of Chap. 10 will smear contact surfaces and shear waves. The FORCE flux of Chap. 7, see also Chap. 14, will smear contacts and shear waves too.

High–order TVD extensions of the previous methods will generally improve the overall accuracy, but if shear waves are unduly smeared by the underlying first–order scheme then their resolution by the high–order extensions will also be unsatisfactory. The implementation of the MUSCL–Hancock scheme to compute $\mathbf{F}^n_{i+\frac{1}{2},j,k}$ is a straightforward application of the one–dimensional approach, see Sect. 14.4 of Chap. 14. The same is true for of the centred schemes FLIC and SLIC presented of Sects. 14.5.2 and 14.5.3 of Chap. 14.

The *limiting* process in enforcing the TVD constraints will now involve the tangential velocity components. For MUSCL–type schemes this is carried out automatically for each equation. The details of the particular Riemann solver used will become important in attempting to resolve contacts and shear waves. Moreover, when applying characteristic limiting, see Sect. 14.4.3, the particular characteristic decomposition used must properly account for the presence of contact and shear waves. For flux–limiter type schemes, such as the WAF and FLIC methods the limiting procedures are different.

The Weighted Average Flux (WAF) method of Sect. 14.3 has intercell flux

$$\mathbf{F}_{i+\frac{1}{2},j,k} = \frac{1}{2}(\mathbf{F}^n_{i,j,k} + \mathbf{F}^n_{i+1,j,k}) - \frac{1}{2}\sum_{k=1}^{5}\text{sign}(c_k)\phi^{(k)}_{i+\frac{1}{2}}\Delta\mathbf{F}^{(k)}_{i+\frac{1}{2}}, \qquad (16.39)$$

where

$$\phi^{(k)}_{i+\frac{1}{2}} = \phi^{(k)}_{i+\frac{1}{2}}(r^{(k)}) \qquad (16.40)$$

is a limiter function for wave k and $r^{(k)} = r^{(k)}(q)$ is the *flow parameter*, computed from ratios of wave jumps of a quantity q, see Sect. 14.3.4. Recall that in the eigenstructure of x–split three–dimensional Euler equations one has one eigenvalue of multiplicity three, that is $\lambda_2 = \lambda_4 = \lambda_5 = u$. Note that we have re–ordered the eigenvalues, consistently with the order of the equations in (16.29). These are associated with three *linear* waves: λ_2 is associated with a contact wave of speed u, λ_4 is associated with a v–shear wave of speed u and λ_5 is associated with a w–shear wave of speed u; $\lambda_1 = u - a$ and $\lambda_3 = u + a$ correspond to the left and right non–linear waves in the solution of the Riemann problem. See Sect. 4.8 for details. For the *limiting* process one needs to compute five limiter functions, one for each wave; they are

$$\left.\begin{array}{ll} \phi^{(1)}_{i+\frac{1}{2}} & \text{is the limiter function for the left non–linear wave}, \\[4pt] \phi^{(2)}_{i+\frac{1}{2}} & \text{is the limiter function for the contact wave}, \\[4pt] \phi^{(3)}_{i+\frac{1}{2}} & \text{is the limiter function for the right non–linear wave}, \\[4pt] \phi^{(4)}_{i+\frac{1}{2}} & \text{is the limiter function for the } v\text{–shear wave}, \\[4pt] \phi^{(5)}_{i+\frac{1}{2}} & \text{is the limiter function for the } w\text{–shear wave}. \end{array}\right\} \quad (16.41)$$

These limiters are then applied to each flux component in (16.39). Recall that the limiter function depends on a flow parameter r, which is computed from a selected quantity q. For the first three limiters one takes $q = \rho$, or the specific internal energy e; for limiter 4 one takes $q = v$ and for limiter 5 one takes $q = w$.

The application of the limiter functions in (16.39) may be viewed as applying the *un–limited* WAF flux to a *dissipated wave structure*, where the original wave paths are displaced. Fig. 16.2 (a) shows the *un–limited* wave structure

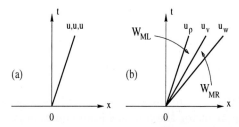

Fig. 16.2. Wave structure for the linear waves (contact and shear waves). (a) un–limited structure before applying the TVD condition, (b) limited structure after applying the TVD condition

for the contact and shear waves; Fig. 16.2. (b) illustrates one possible wave pattern in the dissipated wave structure, where u_ρ, u_v and u_w denote the *new* speeds of the contact, v–shear and w–shear waves respectively. Then the wave pattern of Fig. 14.2. (b) corresponds to

$$\phi^{(2)}_{i+\frac{1}{2}} \le \phi^{(4)}_{i+\frac{1}{2}} \le \phi^{(5)}_{i+\frac{1}{2}}$$

and we call this the 2–4–5 order. The dissipated wave structure results in *two new intermediate states* $\mathbf{W}_{ML}$ and $\mathbf{W}_{MR}$, with corresponding fluxes $\mathbf{F}_{ML}$ and $\mathbf{F}_{MR}$; here $\mathbf{W} = (\rho, u, v, w, p)^T$ denotes the vector of primitive variables. For the two–dimensional Euler equations there are only two possible orders; for the three–dimensional case there are six possible orders. Table 16.1 lists all six possible orders for ρ, v and w. The values for p and u are simply $p_\star$ and $u_\star$ respectively.

A simple procedure is to take the maximum dissipation as the single dissipation to be applied to all three linear waves, that is

	Middle left			Middle right		
Order	ρ	v	w	ρ	v	w
2–4–5	$\rho_{\star R}$	v_L	w_L	$\rho_{\star R}$	v_R	w_L
2–5–4	$\rho_{\star R}$	v_L	w_L	$\rho_{\star R}$	v_R	w_R
4–2–5	$\rho_{\star L}$	v_R	w_L	$\rho_{\star R}$	v_R	w_L
4–5–2	$\rho_{\star L}$	v_R	w_L	$\rho_{\star L}$	v_R	w_R
5–2–4	$\rho_{\star L}$	v_L	w_R	$\rho_{\star R}$	v_L	w_R
5–4–2	$\rho_{\star L}$	v_L	w_R	$\rho_{\star L}$	v_R	w_R

Table 16.1. Six possible cases in dissipated wave structure occuring in limiting the WAF flux for the three dimensional Euler equations

$$\phi_{max}^{(2)} \equiv \phi_{i+\frac{1}{2}}^{(2)} = \phi_{i+\frac{1}{2}}^{(4)} = \phi_{i+\frac{1}{2}}^{(5)} = max[\phi_{i+\frac{1}{2}}^{(2)}, \phi_{i+\frac{1}{2}}^{(4)}, \phi_{i+\frac{1}{2}}^{(5)}] . \qquad (16.42)$$

This will obviously result in excessive dissipation for some of the waves and is not recommended.

Another approach is this. First apply the limiting procedure for waves 1, 2 and 3 as for the one–dimensional case, the first three equations in (16.29). Waves 4 and 5 are treated separately by regarding the equations for the y and z momentum components as scalar equations of the form

$$(\rho\eta)_t + (f_{i+\frac{1}{2}}^1 \eta)_x = 0 , \qquad (16.43)$$

where $\eta = v, w$ and $f_{i+\frac{1}{2}}^1 = \rho u$ is the TVD flux for the mass equation. Then, the TVD flux component for the η momentum equation is

$$(\rho\eta)^{tvd} = [\frac{1}{2}(1 + \phi^\eta)\eta_{i,j,k}^n + \frac{1}{2}(1 - \phi^\eta)\eta_{i+1,j,k}^n] \times f_{i+\frac{1}{2}}^1 , \qquad (16.44)$$

where ϕ^η is a limiter function based on jumps in η across the η–shear wave.

Remark 16.3.1. The attraction of the approach just described is its simplicity as well as its ability to treat, in an entirely similar manner, any passive scalar η. More importantly, one can apply the same procedure for any number of passive scalars, as, for example, in the study of reactive multicomponent flows.

Remark 16.3.2. The equation for total energy per unit mass E, unlike the pure one dimensional case, contains the tangential velocity components and therefore the limiting for the first three equations in (16.29) based on ρ alone is somewhat unsatisfactory.

In the next section we study a methodology to solve multidimensional problems that is distinct from the splitting approach discussed so far in this chapter, namely unsplit finite volume methods. Some of the previous remarks on intercell fluxes and applying the TVD condition will carry over to finite volume methods.

16.4 Unsplit Finite Volume Methods

In this section we present the *unsplit*, finite volume approach for solving multidimensional problems. This is an alternative to dimensional splitting, presented in Sect. 16.2. When finite volume one–dimensional schemes are applied in conjunction with dimensional splitting, there exists a close relationship between both approaches. Both approaches are based on enforcing the integral form of the conservation laws on discrete or finite volumes, as explained in Sect. 16.5 for general non–Cartesian geometries. The difference stems from the fact that in the so–called unsplit finite volume method the solution is advanced by accounting for *all flux contributions in a single step*. In what follows, a finite volume scheme will be understood as an unsplit finite volume scheme. We present two approaches for constructing second–order upwind finite volumes schemes in two and three space dimensions, namely the MUSCL–Hancock and the WAF approaches. These were introduced in Chap. 14 for constructing one–dimensional schemes; in Sects. 16.2 and 16.3 of this chapter these schemes were utilised for solving multidimensional problems via dimensional splitting. There are other approaches for constructing upwind finite volume schemes for multidimensional hyperbolic conservations laws. The reader is referred to the works of Colella [88], LeVeque [208] and Casper and Atkins [60]. The review paper by Vinokur [402] is highly recommended.

16.4.1 Introductory Concepts

Consider a time–dependent two dimensional system of conservation laws

$$\mathbf{U}_t + \mathbf{F}(\mathbf{U})_x + \mathbf{G}(\mathbf{U})_y = \mathbf{0} \ . \tag{16.45}$$

For simplicity assume that the boundaries of the computational domain are aligned with the coordinate directions x and y. Consider a typical finite volume or computational cell $I_{i,j}$ of dimensions $\Delta x \times \Delta y$, as depicted in Fig. 16.3. The cell average $\mathbf{U}_{i,j}^n$ is assigned to *the centre of the cell*, which gives rise to *cell centred methods*. To each intercell boundary there corresponds a numerical flux.

An explicit finite volume scheme to solve (16.45) reads

$$\mathbf{U}_{i,j}^{n+1} = \mathbf{U}_{i,j}^n + \frac{\Delta t}{\Delta x}[\mathbf{F}_{i-\frac{1}{2},j} - \mathbf{F}_{i+\frac{1}{2},j}] + \frac{\Delta t}{\Delta y}[\mathbf{G}_{i,j-\frac{1}{2}} - \mathbf{G}_{i,j+\frac{1}{2}}] \ . \tag{16.46}$$

The cell average $\mathbf{U}_{i,j}^n$ in cell $I_{i,j}$ at time level n is updated to time level $n+1$ via (16.46) in a *single step*, involving flux contributions *from all intercell boundaries*. This conservative formula is the natural extension of one–dimensional conservative formulae, such as (16.33)–(16.35) for example, and is completely determined once the intercell numerical fluxes are specified and a choice of mesh has been made.

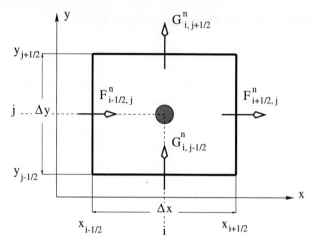

Fig. 16.3. Finite Volume discretisation of Cartesian domain. Typical computing cell $I_{i,j}$ has four intercell boundaries with corresponding intercell fluxes

The simplest, upwind finite volume scheme results from applying the Godunov first–order upwind fluxes across each intercell boundary, in exactly the same way as done for one–dimensional problems, that is

$$\mathbf{F}_{i+\frac{1}{2},j} = \mathbf{F}(\mathbf{U}_{i+\frac{1}{2},j}(0)) ; \quad \mathbf{G}_{i,j+\frac{1}{2}} = \mathbf{G}(\mathbf{U}_{i,j+\frac{1}{2}}(0)) , \tag{16.47}$$

where $\mathbf{U}_{i+\frac{1}{2},j}(x/t)$ is the solution of the Riemann problem

$$\mathbf{U}_t + \mathbf{F}(\mathbf{U})_x = \mathbf{0} , $$
$$\mathbf{U}(x,0) = \left\{ \begin{array}{ll} \mathbf{U}_{i,j}^n & \text{if} \quad x < 0 , \\ \mathbf{U}_{i+1,j}^n & \text{if} \quad x > 0 \end{array} \right\} \tag{16.48}$$

and $\mathbf{U}_{i,j+\frac{1}{2}}(y/t)$ is the solution of the Riemann problem

$$\mathbf{U}_t + \mathbf{G}(\mathbf{U})_y = \mathbf{0} , $$
$$\mathbf{U}(y,0) = \left\{ \begin{array}{ll} \mathbf{U}_{i,j}^n & \text{if} \quad y < 0 , \\ \mathbf{U}_{i,j+1}^n & \text{if} \quad y > 0 . \end{array} \right\} \tag{16.49}$$

For details on one–dimensional fluxes see Chaps. 5 to 14.

Example 16.4.1 (Godunov Finite Volume Method). Consider the linear advection equation

$$u_t + f(u)_x + g(u)_y = 0 ; \quad f(u) = a_1 u , \quad g(u) = a_2 u , \tag{16.50}$$

where a_1 and a_2 are constant velocity components in the x and y directions respectively. By assuming that $a_1 > 0$, $a_2 > 0$, it is easy to verify that the Godunov finite volume scheme (16.46)–(16.47) as applied to (16.50) reads

$$u_{i,j}^{n+1} = u_{i,j}^n + c_1(u_{i-1,j}^n - u_{i,j}^n) + c_2(u_{i,j-1}^n - u_{i,j}^n) , \tag{16.51}$$

where

$$c_1 = \frac{a_1 \Delta t}{\Delta x} \; ; \; c_2 = \frac{a_2 \Delta t}{\Delta y} \qquad (16.52)$$

are the Courant numbers in the x and y directions respectively. The stencil of the scheme is shown in Fig. 16.4. The centre of the stencil is the point (i, j). In addition to the cell average $u_{i,j}^n$ at the centre of the stencil there are only two more data values contributing to the updating. These are $u_{i-1,j}^n$, which is the upwind value in the x–direction, and $u_{i,j-1}^n$, which is the upwind value in the y–direction. Note the scheme reduces identically to the Godunov first–order upwind method, see Sect. 5.3.2 of Chap. 5, if $a_1 = 0$ or $a_2 = 0$. In two dimensions the *upwinding* is not complete; for instance the most obvious upwind value for $a_1 > 0$ and $a_2 > 0$ in a two–dimensional sense is $u_{i-1,j-1}^n$, and this does not contribute to the updating of $u_{i,j}^n$ via (16.51).

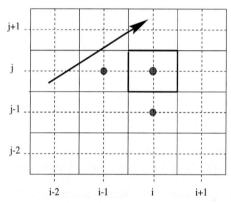

Fig. 16.4. Stencil of the Godunov finite volume scheme for the linear advection equation with positive velocity components a_1 and a_2. The arrow represents the direction of the velocity vector

For three–dimensional time dependent systems of conservation laws

$$\mathbf{U}_t + \mathbf{F}(\mathbf{U})_x + \mathbf{G}(\mathbf{U})_y + \mathbf{H}(\mathbf{U})_z = \mathbf{0} , \qquad (16.53)$$

finite volume schemes read

$$\left. \mathbf{U}_{i,j,k}^{n+1} = \mathbf{U}_{i,j,k}^n \begin{array}{l} + \frac{\Delta t}{\Delta x}[\mathbf{F}_{i-\frac{1}{2},j,k} - \mathbf{F}_{i+\frac{1}{2},j,k}] + \frac{\Delta t}{\Delta y}[\mathbf{G}_{i,j-\frac{1}{2},k} - \mathbf{G}_{i,j+\frac{1}{2},k}] \\ \quad + \frac{\Delta t}{\Delta z}[\mathbf{H}_{i,j,k-\frac{1}{2}} - \mathbf{H}_{i,j,k+\frac{1}{2}}] . \end{array} \right\}$$

$$(16.54)$$

Consider the three–dimensional linear advection equation

$$u_t + f(u)_x + g(u)_y + h(u)_z = 0 \; ; \; f(u) = a_1 u \; , \; g(u) = a_2 u \; , \; h(u) = a_3 u \; , \qquad (16.55)$$

where a_1, a_2 and a_3 are constant velocity components in the x, y and z directions respectively.

Exercise 16.4.1. Assume that $a_1 > 0$, $a_2 > 0$, $a_3 > 0$ and show that the three–dimensional finite volume scheme obtained by straight application of the one–dimensional Godunov flux is

$$u_{i,j,k}^{n+1} = u_{i,j,k}^n + c_1(u_{i-1,j,k}^n - u_{i,j,k}^n) + c_2(u_{i,j-1,k}^n - u_{i,j,k}^n) + c_3(u_{i,j,k-1}^n - u_{i,j,k}^n),$$
(16.56)

where

$$c_1 = \frac{a_1 \Delta t}{\Delta x} \; ; \quad c_2 = \frac{a_2 \Delta t}{\Delta y} \; ; \quad c_3 = \frac{a_3 \Delta t}{\Delta z}$$
(16.57)

are the Courant numbers in the x, y and z directions respectively.

16.4.2 Accuracy and Stability of Multidimensional Schemes

Accuracy Theorems. A useful theorem due to Roe [282] concerning the accuracy of general constant coefficient schemes for the one–dimensional linear advection equation was stated in Sect. 13.2.1 of Chap. 13. We now state the generalisation of this result to two and three dimensional schemes [42].

Two–dimensional schemes such as (16.51) can be written in the general form

$$u_{i,j}^{n+1} = \sum_{\alpha,\beta} b_{\alpha,\beta} u_{i+\alpha,j+\beta}^n,$$
(16.58)

where $b_{\alpha,\beta}$ are the, constant, coefficients of the scheme. Schemes of the form (16.58) are a straight generalisation of the one–dimensional schemes studied in Sect.13.2.1 of Chap. 13. We now state the following result.

Theorem 16.4.1. *The two–dimensional scheme (16.58) to solve (16.50) is m–th–order accurate in space and time if and only if the coefficients satisfy*

$$\sum_{\alpha,\beta} \alpha^q \beta^r b_{\alpha,\beta} = (-c_1)^q (-c_2)^r$$
(16.59)

for all integer pairs (q, r) such that $q \geq 0$, $r \geq 0$ and $q + r \leq m$. Here c_1 and c_2 are the directional Courant numbers (16.52).

Proof. For details of the proof see Billett and Toro [42].

Three–dimensional schemes, such as (16.56), to solve the three–dimensional linear advection equation (16.55) can be written as

$$u_{i,j,k}^{n+1} = \sum_{\alpha,\beta,\gamma} b_{\alpha,\beta,\gamma} u_{i+\alpha,j+\beta,k+\gamma}^n,$$
(16.60)

where $b_{\alpha,\beta,\gamma}$ are the coefficients of the scheme, assumed to be constant. The following result is a generalisation of the previous accuracy theorem to three–dimensional schemes.

Theorem 16.4.2. *The three–dimensional scheme (16.60) to solve (16.55) is mth–order accurate in space and time if and only if the coefficients $b_{\alpha,\beta,\gamma}$ satisfy*

$$\sum_{\alpha,\beta,\gamma} \alpha^q \beta^r \gamma^s b_{\alpha,\beta,\gamma} = (-c_1)^q (-c_2)^r (-c_3)^s \qquad (16.61)$$

for all integer triples (q, r, s) such that $q \geq 0$, $r \geq 0$, $s \geq 0$ and $q + r + s \leq m$. Here c_1, c_2 and c_3 are the directional Courant numbers (16.57).

Proof. For details of the proof see Billett and Toro [42].

Exercise 16.4.2. Apply the two previous accuracy theorems (16.4.1) and (16.4.2) to show that the Godunov finite volume schemes (16.51) and (16.56) to solve the linear advection equations (16.50) and (16.55) are first–order accurate in space and time.

Solution 16.4.1. (Left to the reader).

Computational Test for Linear Stability. There are various methods for analysing the stability of schemes for the linear advection equation; see Hirsch [170], Chaps. 7 to 10, for a comprehensive presentation of techniques. Here we assume the reader to be familiar with the popular von Neumann method.

For a one–dimensional scheme whose coefficients depend only on the Courant number c_1 the von Neumann method derives an algebraic expression

$$A = A(c_1, \theta_1)$$

for the *amplification factor*, where c_1 is the Courant number and θ_1 is the *phase angle* in the range $0 \leq \theta_1 \leq \pi$. For sufficiently simple schemes the stability condition $||A|| \leq 1$ can be used to derive, algebraically, conditions on the Courant number for the scheme to be linearly stable. There are instances, however, where the algebraic task becomes intractable. In this case, drawing a contour plot of $||A||$, the modulus of the complex number A, for a *large number* of values (c_1, θ_1) in a rectangle $[0, c_1^{max}] \times [0, \pi]$, will give an *indication* as to the linear stability region of the scheme. Here c_1^{max} is chosen to be larger than the *expected* stability limit.

For two–dimensional schemes the amplification factor reads

$$A = A(c_1, c_2, \theta_1, \theta_2) .$$

Here c_1, c_2 are the Courant numbers in the x and y directions and θ_1, θ_2 are phase angles in the the x and y directions. Deriving algebraic conditions for the scheme to be stable is now generally much harder than in the one–dimensional case. A *numerical test* can be performed as follows [42]. For a given pair c_1, c_2 of Courant numbers in the rectangle $[0, c_1^{max}] \times [0, c_2^{max}]$ compute $||A||$ for a *large number* M_{ang} of values (θ_1, θ_2) of phase angles in the rectangle $[0, \pi] \times [0, \pi]$. Record the number M_{cfl} for which $||A|| \leq 1$ and define the ratio

$$S = S(c_1, c_2) = \frac{M_{cfl}}{M_{ang}} \ .$$

If $S = 1$ the scheme is regarded as stable for the pair c_1, c_2. A contour plot of $S = S(c_1, c_2)$ for a large number of pairs (c_1, c_2) in the range $[0, c_1^{max}] \times [0, c_2^{max}]$ will give an *indication* of the linear stability of the two–dimensional scheme.

In three space dimensions the amplification factor reads

$$A = A(c_1, c_2, c_3, \theta_1, \theta_2, \theta_3)$$

and one draws contour plots on planes within $[0, c_1^{max}] \times [0, c_2^{max}] \times [0, c_3^{max}]$.

Remark 16.4.1. The above stability tests, useful in practice as they may be, do not constitute proofs of linear stability, and caution is needed when quoting the conclusions of the tests.

Example 16.4.2. Application of the stability test just described to the Godunov finite volume scheme (16.51) *indicates* that the scheme has linear stability condition

$$c_1 + c_2 \leq 1 \ . \tag{16.62}$$

For $c_1 = c_2$ one has $c_1 = c_2 \leq \frac{1}{2}$, which is half the stability range of the one–dimensional Godunov scheme. For the the three–dimensional Godunov finite volume scheme (16.56), application of the above test suggests that the scheme is linearly stable under the condition

$$c_1 + c_2 + c_3 \leq 1 \ . \tag{16.63}$$

For $c_1 = c_2 = c_3$ one has $c_1 = c_2 = c_3 \leq \frac{1}{3}$, which is one third of the stability limit of the one–dimensional, Godunov first–order upwind scheme.

The previous example illustrates the fact that in constructing finite volume schemes by straightforward application of one–dimensional fluxes one, at best, ends up with schemes with a *reduced stability range*. There are cases in which this approach leads to unconditionally unstable finite volume methods, even though the one–dimensional operators are stable for the respective one–dimensional method.

In the next section we construct a finite volume method that is second–order accurate in space and time.

16.5 A MUSCL–Hancock Finite Volume Scheme

Here we extend the second–order MUSCL–Hancock approach, presented in Chap. 14 for one–dimensional systems, to construct unsplit finite volume schemes for multi–dimensional conservation laws. We present the details of the approach for the two–dimensional case and construct a scheme of the form (16.46). Background on the MUSCL–Hancock approach for scalar one–dimensional problems is found in Sect. 13.4.2 of Chap. 13; Sect. 14.4 of Chap. 14 gives details of the approach for one–dimensional non–linear systems. The reader is encouraged to review the details of the one dimensional MUSCL–Hancock scheme before proceeding with this section. As in the one–dimensional case the scheme has the following three steps

(I) **Data Reconstruction and Boundary Extrapolated Values.** The cell averages $\mathbf{U}_{i,j}^n$ are reconstructed, independently, in the x and y directions by selecting respective slope vectors (differences) Δ_i and Δ_j. Boundary extrapolated values are

$$\left.\begin{array}{ll} \mathbf{U}_{i,j}^{-x} = \mathbf{U}_{i,j}^n - \frac{1}{2}\Delta_i \; ; & \mathbf{U}_{i,j}^{+x} = \mathbf{U}_{i,j}^n + \frac{1}{2}\Delta_i \; ; \\[2mm] \mathbf{U}_{i,j}^{-y} = \mathbf{U}_{i,j}^n - \frac{1}{2}\Delta_j \; ; & \mathbf{U}_{i,j}^{+y} = \mathbf{U}_{i,j}^n + \frac{1}{2}\Delta_j \; . \end{array}\right\} \tag{16.64}$$

(II) **Evolution of Boundary Extrapolated Values.**

$$\hat{\mathbf{U}}_{i,j}^l = \mathbf{U}_{i,j}^l + \frac{\Delta t}{2\Delta x}[\mathbf{F}(\mathbf{U}_{i,j}^{-x}) - \mathbf{F}(\mathbf{U}_{i,j}^{+x})] + \frac{\Delta t}{2\Delta y}[\mathbf{G}(\mathbf{U}_{i,j}^{-y}) - \mathbf{G}(\mathbf{U}_{i,j}^{+y})] \tag{16.65}$$

for $l = -x, +x, -y, +y$.

(III) **Solution of Riemann problems.** At each intercell position $(i+\frac{1}{2}, j)$ one solves the x–split one–dimensional Riemann problem with data $\hat{\mathbf{U}}_{i,j}^{+x}$, $\hat{\mathbf{U}}_{i+1,j}^{-x}$ to find a solution $\mathbf{U}_{i+1/2,j}(x/t)$. Similarly, at each interface position $(i, j+\frac{1}{2})$ one solves the y–split one–dimensional Riemann problem with data $\hat{\mathbf{U}}_{i,j}^{+y}$, $\hat{\mathbf{U}}_{i,j+1}^{-y}$ to find a solution $\mathbf{U}_{i,j+1/2}(y/t)$. The corresponding intercell fluxes are found as in the one–dimensional Godunov method, namely

$$\mathbf{F}_{i+\frac{1}{2},j} = \mathbf{F}(\mathbf{U}_{i+\frac{1}{2},j}(0)) \; ; \quad \mathbf{G}_{i,j+\frac{1}{2}} = \mathbf{G}(\mathbf{U}_{i,j+\frac{1}{2}}(0)) \; . \tag{16.66}$$

Example 16.5.1. Here we apply the MUSCL–Hancock finite volume scheme to the linear advection equation (16.50). It is easily shown that the intercell fluxes are

$$\left.\begin{array}{l} f_{i+\frac{1}{2},j} = a_1[u_{i,j}^n + \frac{1}{2}(1 - c_1)\Delta_i - \frac{1}{2}c_2\Delta_j] \; , \\[2mm] g_{i,j+\frac{1}{2}} = a_2[u_{i,j}^n + \frac{1}{2}(1 - c_2)\Delta_j - \frac{1}{2}c_1\Delta_i] \; . \end{array}\right\} \tag{16.67}$$

By taking central difference approximations for the slopes Δ_i, Δ_j and substituting the resulting fluxes into the finite volume formula (16.46) one obtains the scheme

$$u_{i,j}^{n+1} = u_{i,j}^n$$

$$-c_1 \left\{ (u_{i,j}^n - u_{i-1,j}^n) + \tfrac{1}{4}(1-c_1)[(u_{i+1,j}^n - u_{i,j}^n) - (u_{i-1,j}^n - u_{i-2,j}^n)] \right\}$$

$$+\tfrac{1}{4}c_1 c_2[(u_{i,j+1}^n - u_{i,j-1}^n) + (u_{i-1,j+1}^n - u_{i-1,j-1}^n)]$$

$$-c_2 \left\{ (u_{i,j}^n - u_{i,j-1}^n) + \tfrac{1}{4}(1-c_2)[(u_{i,j+1}^n - u_{i,j}^n) - (u_{i,j-1}^n - u_{i,j-2}^n)] \right\}$$

$$+\tfrac{1}{4}c_1 c_2[(u_{i+1,j}^n - u_{i-1,j}^n) + (u_{i+1,j}^n - u_{i-1,j-1}^n)] \, .$$

$$\left. \right\} \tag{16.68}$$

The stencil of the scheme as applied to (16.50) with $a_1 > 0$ and $a_2 > 0$ is shown in Fig. 16.5 and has 10 points.

Exercise 16.5.1. Apply the accuracy theorem (16.4.1), see equation (16.59), to show that scheme (16.68) is second–order accurate in space and time.

Solution 16.5.1. Left to the reader.

Remark 16.5.1. Stability tests according to the method of Sect. 16.4.2 suggest that the MUSCL–Hancock scheme (16.68) has the same stability condition (16.62), as the first–order Godunov finite volume scheme (16.51). The unsplit finite volume MUSCL–Hancock scheme has increased the accuracy of the Godunov finite volume scheme but not its stability range.

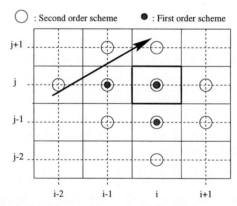

Fig. 16.5. Stencil of the MUSCL–Hancock Finite Volume scheme for the linear advection equation with positive velocity components a_1 and a_2. The arrow represents the direction of the velocity vector. Also shown is the stencil of the first–order version of the scheme

Exercise 16.5.2. Attempt to extend the MUSCL–Hancock approach to construct a three–dimensional finite volume scheme for non–linear systems of conservations laws. Apply the scheme to the three–dimensional linear advection equation (16.55) and show that it is second–order accurate in space and time.

Solution 16.5.2. Left to the reader.

As to the computation of the numerical fluxes, these can be found by using the exact Riemann solver, given in Chap. 6 for the Euler equations, or any of the approximate Riemann solvers presented in Chaps. 9 to 12. One can also use the Flux Vector Splitting schemes of Chap. 8.

In order to *control* spurious oscillations in the two–dimensional schemes, one applies one–dimensional TVD constraints in exactly the same manner as for one–dimensional problems, by replacing the slopes by *limited slopes*; see Chaps. 13 and 14. The resulting schemes are usually referred to, incorrectly, as *two–dimensional* TVD *schemes*. The same remark applies to three–dimensional schemes. We first note that the schemes are *not strictly* TVD as understood for one–dimensional model equations, see Chap. 13; the second order slope limited scheme will still produce spurious oscillations near high gradients. Also note that if the TVD conditions were to be enforced in a two–dimensional sense, then, as proved by Goodman and LeVeque [148], the scheme would be oscillation free but at most first–order accurate.

16.6 WAF–Type Finite Volume Schemes

The Weighted Average Flux (WAF) approach was first introduced [352], [358] to construct schemes to solve one–dimensional hyperbolic systems of conservation laws. The WAF schemes are of the Godunov type and of second order accuracy in space and time. Detailed descriptions of the schemes along with their TVD constraints are presented in Sect. 13.7 of Chap. 13 for scalar problems and in Sects. 14.2–14.3 of Chap. 14 for non–linear systems. By means of dimensional splitting these WAF schemes may be applied to solve multidimensional problems. See Sects. 16.2–16.3 of this chapter about details for the two and three dimensional Euler equations.

The WAF approach may be extended to derive unsplit finite volume schemes for multi–dimensional hyperbolic systems [39], [43], [41]. In the next section we first present some details of the approach as applied to the two and three dimensional linear advection equations. Then we study two of the possible extensions to multidimensional non–linear systems in two and three space dimensions.

16.6.1 Two–Dimensional Linear Advection

Consider the two–dimensional linear advection equation with constant coefficients

$$u_t + f(u)_x + g(u)_y = 0 ; \quad f(u) = a_1 u ; \quad g(u) = a_2 u \qquad (16.69)$$

and finite volume schemes of the form

$$u_{i,j}^{n+1} = u_{i,j}^n + \frac{\Delta t}{\Delta x}[f_{i-\frac{1}{2},j} - f_{i+\frac{1}{2},j}] + \frac{\Delta t}{\Delta y}[g_{i,j-\frac{1}{2}} - g_{i,j+\frac{1}{2}}] \qquad (16.70)$$

to solve (16.69). The objective is to prescribe the intercell fluxes $f_{i+\frac{1}{2},j}$ and $g_{i,j+\frac{1}{2}}$. The WAF approach specifies a flux in the x–direction as follows

$$f_{i+\frac{1}{2},j} = \frac{1}{V(B)} \int_{t_1}^{t_2} \int_{x_1}^{x_2} \int_{y_1}^{y_2} f(u_{i+\frac{1}{2}}(x,y,t)) \, dx \, dy \, dt , \qquad (16.71)$$

where $B \equiv [t_1, t_2] \times [x_1, x_2] \times [y_1, y_2]$ is the domain of integration in t–x–y space, that includes the relevant intercell boundary at $x_{i+\frac{1}{2}}$; $V(B)$ denotes the volume of B. In general $f(u)$ is the flux component perpendicular to the intercell boundary. An analogous definition holds for $g_{i,j+\frac{1}{2}}$. Particular schemes are obtained by specifying (i) the spatial and temporal domain of integration, (ii) the integrand and (iii) integration schemes to evaluate (16.71). Here we make the following choices

$$\left. \begin{array}{llll} t_1 = 0 , & t_2 = \Delta t , & c_1 \leq 1 , & c_2 \leq 1 , \\ x_1 = -\frac{1}{2}\Delta x , & x_2 = \frac{1}{2}\Delta x , & y_1 = 0 , & y_2 = \Delta y . \end{array} \right\} \qquad (16.72)$$

Fig. 16.6 depicts the spatial integration range to compute the intercell flux. The choice of Δt and the conditions on the Courant numbers c_1 and c_2 are

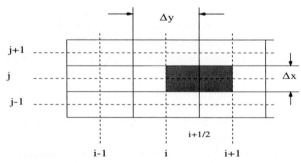

Fig. 16.6. Spacial integration range for evaluating the WAF intercell flux at $i + \frac{1}{2}$

obviously related. The integrand is determined by specifying $u_{i+\frac{1}{2}}(x,y,t)$. Here we assume $u_{i+\frac{1}{2}}(x,y,t)$ to be the solution of *relevant* two–dimensional Riemann problems for (16.69), with initial data $u(x,y,0) = u_{i,j}^n$ for (x,y) in the computing cell $I_{i,j}$, overlapping the spatial range of integration in a time Δt.

For the model equation (16.69) the solution of the two–dimensional Riemann problem is trivial. Exact integration in (16.71) leads to a second–order accurate scheme in space and time, see [39], [43]. Here we only consider schemes arising from approximate evaluation of the integral (16.71).

Assume that $a_1 > 0$ and $a_2 > 0$ in (16.69) and restrict the Courant numbers to satisfy $c_1 \leq \frac{1}{2}$ and $c_2 \leq \frac{1}{2}$; this in turn restricts the number of states that influence $f_{i+\frac{1}{2},j}$ to four, namely $u_{i,j}^n$, $u_{i+1,j}^n$, $u_{i,j-1}^n$ and $u_{i+1,j-1}^n$. These states are laid out in the four quadrants of a rectangle centred at $(x_{i+\frac{1}{2}}, y_{j-\frac{1}{2}})$, and form the initial conditions of a two–dimensional Riemann problem. For different signs of a_1 and a_2 the relevant two–dimensional Riemann problem has different initial conditions.

Having specified the integration domain and the integrand in (16.71), we are left with the task of choosing integration schemes to evaluate the integral.

First–Order Schemes. Apply exact integration in time, the midpoint rule perpendicular to the boundary, and exact integration parallel to the boundary. The integral form of the intercell flux (16.71) simplifies in this case to

$$f_{i+\frac{1}{2},j} = \frac{1}{\Delta t} \frac{1}{\Delta y} \int_0^{\Delta t} \int_{B \bigcap \{x=0\}} f(u_{i+\frac{1}{2}}(0,y,t)) \, dy \, dt \,, \qquad (16.73)$$

where we integrate over the plane $x = 0$, in local coordinates. This gives

$$f_{i+\frac{1}{2},j} = \frac{1}{2}(2 - c_2)f_{i,j} + \frac{1}{2}c_2 f_{i,j-1} \,. \qquad (16.74)$$

The flux $g_{i,j+\frac{1}{2}}$ follows similarly. Substitution of all four intercell fluxes into the finite volume formula (16.70) produces the scheme

$$\left.\begin{array}{r} u_{i,j}^{n+1} = \quad (1 - c_1 - c_2 + c_1 c_2)u_{i,j}^n + c_2(1 - c_1)u_{i,j-1}^n \\ +c_1(1 - c_2)u_{i-1,j}^n + c_1 c_2 u_{i-1,j-1}^n \,. \end{array}\right\} \qquad (16.75)$$

This WAF two–dimensional finite volume scheme turns out to be identical to the *Corner Transport Upwind* (CTU) scheme of Colella [88].

Exercise 16.6.1. Using the accuracy theorem (16.4.1), see (16.59), show that the CTU scheme (16.75) is first–order accurate in space and time.

Fig. 16.7 shows the stencil of two–dimensional first–order upwind scheme (16.75). Compare with the stencil for the Godunov finite volume method shown in Fig. 16.4. The CTU (WAF) finite volume scheme does contain the upwind point $u_{i-1,j-1}^n$ and is the natural two–dimensional extension of the one–dimensional Godunov first–order upwind scheme. Colella proved the scheme (16.75) to be stable under the condition

$$\max\{c_1, c_2\} \leq 1 \,. \qquad (16.76)$$

Thus, the WAF approach is capable of producing first–order two dimensional finite volume schemes with twice the stability limit of the Godunov finite volume scheme (16.51), which is constructed by straightforward application of the Godunov one–dimensional fluxes.

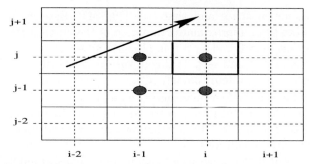

Fig. 16.7. Stencil of the first–order WAF finite volume scheme (identical to the CTU scheme of Colella) for the two–dimensional linear advection equation with positive a_1 and a_2. The arrow indicates the direction of the velocity vector

A Second–Order Scheme. We now consider the midpoint rule in time combined with exact integration in both space directions in (16.71) and assume that $c_1 \leq 1$, $c_2 \leq 1$. This gives the intercell flux

$$f_{i+\frac{1}{2},j} = \left. \begin{array}{l} \tfrac{1}{4}(1+c_1)(2-c_2)f_{i,j} + \tfrac{1}{4}(1-c_1)(2-c_2)f_{i+1,j} \\ +\tfrac{1}{4}(1+c_1)c_2 f_{i,j-1} + \tfrac{1}{4}(1-c_1)c_2 f_{i+1,j-1} \ . \end{array} \right\} \quad (16.77)$$

The flux $g_{i,j+\frac{1}{2}}$ follows similarly. Substitution of all fluxes into the finite volume formula (16.70) gives the scheme in full as

$$u_{i,j}^{n+1} = \left. \begin{array}{l} \left[1 - \tfrac{1}{2}c_1^2(2-c_2) - \tfrac{1}{2}c_2^2(2-c_1)\right] u_{i,j}^n \\ -\tfrac{1}{4}c_1(1-c_1)(2-c_2)u_{i+1,j}^n - \tfrac{1}{4}(2-c_1)c_2(1-c_2)u_{i,j+1}^n \\ -\tfrac{1}{4}c_1 c_2(1-c_1)u_{i+1,j-1}^n - \tfrac{1}{4}c_1 c_2(1-c_2)u_{i-1,j+1}^n \\ +\tfrac{1}{4}\left[c_1(1+c_1)(2-c_2) - 2c_1 c_2^2\right] u_{i-1,j}^n \\ +\tfrac{1}{4}c_1 c_2(2+c_1+c_2)u_{i-1,j-1}^n \ . \end{array} \right\}$$

$$(16.78)$$

When $c_1 = 0$ or $c_2 = 0$, the scheme reduces to the one dimensional Lax–

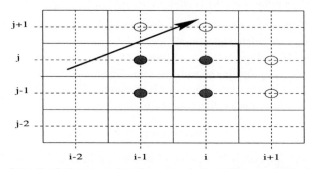

Fig. 16.8. Stencil of the second–order WAF finite volume scheme for the two–dimensional linear advection equation with positive a_1 and a_2. Compare with stencil of first–order scheme of Fig. 16.7

Wendroff scheme, and when $c_1 = c_2 = 1$, it reduces to $u_{i,j}^{n+1} = u_{i-1,j-1}^n$, which reproduces the exact solution under the given initial conditions.

Exercise 16.6.2. Apply the accuracy theorem (16.4.1) to show that scheme (16.78) is second–order accurate in space and time.

Solution 16.6.1. Left to the reader.

The eight–point stencil for the second–order WAF finite volume scheme (16.78) is shown in Fig. 16.8; compare with the ten–point stencil for the second–order MUSCL–Hancock scheme (16.68) shown in Fig. 16.5. Numerical stability tests, as described in Sect. 16.4.2, suggest that scheme (16.78) is stable provided the condition (16.76) holds. The MUSCL–Hancock scheme (16.68) has stability restriction $c_1 + c_2 \leq 1$, which is half that of this WAF finite volume scheme. Both WAF finite volume schemes (16.75) (CTU) and (16.78) are much more stable than the two–dimensional Lax–Wendroff scheme [205], [171].

16.6.2 Three–Dimensional Linear Advection

The WAF approach is now extended to generate finite volume schemes for the linear advection equation in three space dimensions

$$u_t + f(u)_x + g(u)_y + h(u)_z = 0 \; ; \;\; f(u) = a_1 u \, , \;\; g(u) = a_2 u \, , \;\; h(u) = a_3 u \, . \tag{16.79}$$

The finite volume schemes have the form

$$u_{i,j,k}^{n+1} = u_{i,j,k}^n + \frac{\Delta t}{\Delta x}[f_{i-\frac{1}{2},j,k} - f_{i+\frac{1}{2},j,k}] \; + \; \left.\begin{array}{l} \frac{\Delta t}{\Delta y}[g_{i,j-\frac{1}{2},k} - g_{i,j+\frac{1}{2},k}] \\[4pt] + \; \frac{\Delta t}{\Delta z}[h_{i,j,k-\frac{1}{2}} - h_{i,j,k+\frac{1}{2}}] \, , \end{array}\right\} \tag{16.80}$$

where $f_{i+\frac{1}{2},j,k}$, $g_{i,j+\frac{1}{2},k}$ and $h_{i,j,k+\frac{1}{2}}$ are the intercell fluxes in the x, y and z directions respectively. A general WAF intercell flux in three space dimensions reads

$$f_{i+\frac{1}{2},j,k} = \frac{1}{V(B)} \int_{t_1}^{t_2} \int_{x_1}^{x_2} \int_{y_1}^{y_2} \int_{z_1}^{z_2} f(u_{i+\frac{1}{2}}(x,y,z,t)) \; dx \, dy \, dz \, dt \, , \tag{16.81}$$

where $B = [t_1, t_2] \times [x_1, x_2] \times [y_1, y_2] \times [z_1, z_2]$ is the domain of integration in t–x–y–z space; $V(B) = (t_2 - t_1) \times (x_2 - x_1) \times (y_2 - y_1) \times (z_2 - z_1)$ is the volume of B. The fluxes $g_{i,j+\frac{1}{2},k}$ and $h_{i,j,k+\frac{1}{2}}$ are defined similarly.

A First–Order Scheme. A three–dimensional version of the first–order WAF scheme for the two–dimensional linear advection equation is now derived. A choice of parameter values consistent with (16.72) is assumed. By using the midpoint rule in time, the midpoint rule in space perpendicular to the boundary, and exact integration in space in both directions parallel to the boundary, the resulting intercell flux is

$$f_{i+\frac{1}{2},j,k} = \frac{1}{\Delta y \, \Delta z} \int_{B \cap \{x=0\}} f(u_{i+\frac{1}{2}}(0,y,z,\tfrac{1}{2}\Delta t)) \, dy \, dz \,. \qquad (16.82)$$

This can be written explicitly as

$$f_{i+\frac{1}{2},j,k} = \left. \begin{array}{l} \tfrac{1}{4}(2-c_2)(2-c_3)f_{i,j,k} + \tfrac{1}{4}c_2(2-c_3)f_{i,j-1,k} \\[4pt] + \tfrac{1}{4}(2-c_2)c_3 f_{i,j,k-1} + \tfrac{1}{4}c_2 c_3 f_{i,j-1,k-1} \end{array} \right\} \qquad (16.83)$$

The derivation of the fluxes $g_{i,j+\frac{1}{2},k}$ and $h_{i,j,k+\frac{1}{2}}$ follows by symmetry. Substituting these into the finite volume formula (16.80) gives the scheme as

$$u_{i,j,k}^{n+1} = \left. \begin{array}{l} \left[1 - \tfrac{1}{4}c_1(2-c_2)(2-c_3) - \tfrac{1}{4}(2-c_1)(2-c_2)c_3\right] u_{i,j,k}^n \\[4pt] - \left[\tfrac{1}{4}(2-c_1)c_2(2-c_3) - \tfrac{1}{4}(2-c_1)(2-c_2)c_3\right] u_{i,j,k}^n \\[4pt] + c_1 \left[(1-c_2)(1-c_3) - \tfrac{1}{4}c_2 c_3\right] u_{i-1,j,k}^n \\[4pt] + \tfrac{1}{4}(4-3c_1)c_2 c_3 u_{i,j-1,k-1}^n \\[4pt] + c_2 \left[(1-c_1)(1-c_3) - \tfrac{1}{4}c_1 c_3\right] u_{i,j-1,k}^n \\[4pt] + \tfrac{1}{4}c_1(4-3c_2)c_3 u_{i-1,j,k-1}^n \\[4pt] + c_3 \left[(1-c_1)(1-c_2) - \tfrac{1}{4}c_1 c_2\right] u_{i,j,k-1}^n \\[4pt] + \tfrac{1}{4}c_1 c_2(4-3c_3)u_{i-1,j-1,k}^n \\[4pt] + \tfrac{3}{4}c_1 c_2 c_3 u_{i-1,j-1,k-1}^n \,. \end{array} \right\}$$

$$(16.84)$$

By setting $c_1 = c_2 = c_3 = c$ in (16.84) it can be seen that not all the coefficients are positive if $c \geq \frac{2}{3}$; see for example the coefficient of $u_{i-1,j,k}$. A sufficient condition for the scheme to remain *monotone* is

$$\max\{c_1, c_2, c_3\} \leq \frac{2}{3} \,. \qquad (16.85)$$

Exercise 16.6.3. Apply the accuracy theorem (16.4.2) to show that scheme (16.84) is first–order accurate in space and time. This scheme is a natural three–dimensional extension of the one–dimensional Godunov first–order upwind scheme.

A Second Order Scheme. Working from the experience of the two dimensional case, the midpoint rule in time and exact integration in all three space directions will now be used to generate a second–order scheme. The flux in the x–direction becomes

$$f_{i+\frac{1}{2},j,k} = \frac{1}{V(B)} \int_B f(u_{i+\frac{1}{2}}(x,y,z,\tfrac{1}{2}\Delta t)) \, dx \, dy \, dz \,, \qquad (16.86)$$

where $u_{i+\frac{1}{2}}(x,y,z,t)$ is the exact solution to the three–dimensional Riemann problem for (16.79) with initial conditions $\{u_{i,j,k}^n\}$. By assuming $a_1 > 0$, $a_2 > 0$, $a_3 > 0$ in (16.79) and performing the integration we obtain the intercell flux

$$
\begin{aligned}
f_{i+\frac{1}{2},j,k} = \ & \tfrac{1}{8}(1+c_1)(2-c_2)(2-c_3)f_{i,j,k} \\
& +\tfrac{1}{8}(1-c_1)(2-c_2)(2-c_3)f_{i+1,j,k} \\
& +\tfrac{1}{8}(1+c_1)c_2(2-c_3)f_{i,j-1,k} \\
& +\tfrac{1}{8}(1-c_1)c_2(2-c_3)f_{i+1,j-1,k} \\
& +\tfrac{1}{8}(1+c_1)(2-c_2)c_3 f_{i,j,k-1} \\
& +\tfrac{1}{8}(1-c_1)(2-c_2)c_3 f_{i+1,j,k-1} \\
& +\tfrac{1}{8}(1+c_1)c_2 c_3 f_{i,j-1,k-1} \\
& +\tfrac{1}{8}(1-c_1)c_2 c_3 f_{i,j-1,k-1} \ .
\end{aligned}
\tag{16.87}
$$

The fluxes $g_{i,j+\frac{1}{2},k}$ and $h_{i,j,k+\frac{1}{2}}$ follow by symmetry. Substituting all the fluxes into the finite volume formula (16.80) gives the scheme

$$
\begin{aligned}
u_{i,j,k}^{n+1} = \ & u_{i,j,k}^n \\
& -\tfrac{1}{4}\left[c_1^2(2-c_2)(2-c_3)+(2-c_1)c_2^2(2-c_3)+(2-c_1)(2-c_2)c_3^2\right]u_{i,j,k}^n \\
& +\tfrac{1}{8}c_1\left[(1+c_1)(2-c_2)(2-c_3)-2c_2^2(2-c_3)-2(2-c_2)c_3^2\right]u_{i-1,j,k}^n \\
& +\tfrac{1}{8}c_2\left[(2-c_1)(1+c_2)(2-c_3)-2c_1^2(2-c_3)-2(2-c_1)c_3^2\right]u_{i,j-1,k}^n \\
& +\tfrac{1}{8}c_3\left[(2-c_1)(2-c_2)(1+c_3)-2(2-c_1)c_2^2-2c_1^2(2-c_2)\right]u_{i,j,k-1}^n \\
& +\tfrac{1}{8}c_1 c_2\left[(2-c_3)(2+c_1+c_2)-2c_3^2\right]u_{i-1,j-1,k}^n \\
& +\tfrac{1}{8}c_1 c_3\left[(2-c_2)(2+c_1+c_3)-2c_2^2\right]u_{i-1,j,k-1}^n \\
& +\tfrac{1}{8}c_2 c_3\left[(2-c_1)(2+c_2+c_3)-2c_1^2\right]u_{i,j-1,k-1}^n \\
& +\tfrac{1}{8}c_1 c_2 c_3(3+c_1+c_2+c_3)u_{i-1,j-1,k-1}^n \\
& -\tfrac{1}{8}c_1(1-c_1)(2-c_2)(2-c_3)u_{i+1,j,k}^n - \tfrac{1}{8}c_1(1-c_1)c_2(2-c_3)u_{i+1,j-1,k}^n \\
& -\tfrac{1}{8}c_1(1-c_1)(2-c_2)c_3 u_{i+1,j,k-1}^n - \tfrac{1}{8}c_1(1-c_1)c_2 c_3 u_{i+1,j-1,k-1}^n \\
& -\tfrac{1}{8}(2-c_1)c_2(1-c_2)(2-c_3)u_{i,j+1,k}^n - \tfrac{1}{8}c_1 c_2(1-c_2)(2-c_3)u_{i-1,j+1,k}^n \\
& -\tfrac{1}{8}(2-c_1)c_2(1-c_2)c_3 u_{i,j+1,k-1}^n - \tfrac{1}{8}c_1 c_2(1-c_2)c_3 u_{i-1,j+1,k-1}^n \\
& -\tfrac{1}{8}(2-c_1)(2-c_2)c_3(1-c_3)u_{i,j,k+1}^n - \tfrac{1}{8}(2-c_1)c_2 c_3(1-c_3)u_{i,j-1,k+1}^n \\
& -\tfrac{1}{8}c_1(2-c_2)c_3(1-c_3)u_{i-1,j,k+1}^n - \tfrac{1}{8}c_1 c_2 c_3(1-c_3)u_{i-1,j-1,k+1}^n \ .
\end{aligned}
\tag{16.88}
$$

Exercise 16.6.4. Apply the accuracy theorem (16.4.2) to show that scheme (16.88) is second–order accurate in space and time.

Solution 16.6.2. Left to the reader.

Application of the numerical tests for stability according to the method of Sect. 16.4.2 suggests that the three–dimensional scheme (16.88) is stable under the condition

$$
\max\{c_1, c_2, c_3\} \le \frac{2}{3} \ ,
\tag{16.89}
$$

which is also the condition (16.85) for monotonicity for the first–order three dimensional scheme (16.84). The three–dimensional Godunov finite volume scheme (16.56) has stability restriction

$$
\max\{c_1, c_2, c_3\} \le \frac{1}{3} \ ,
\tag{16.90}
$$

which is half that of the WAF second–order finite volume scheme (16.88).

16.6.3 Schemes for Two–Dimensional Nonlinear Systems

The construction of multi–dimensional WAF–type schemes for non–linear systems is based on *re–interpreting* the flux formulae derived for the linear advection equation in multidimensions.

An Average Flux Second–Order Scheme. First we re–write the intercell flux (16.77) as

$$f_{i+\frac{1}{2},j} = \frac{1}{2}(1 + c_1)f(u_{i,j}^{god}) + \frac{1}{2}(1 - c_1)f(u_{i+1,j}^{god}) , \tag{16.91}$$

where

$$\left.\begin{array}{l} u_{i,j}^{god} = u_{i,j}^n + \frac{\frac{1}{2}\Delta t}{\Delta y}\left[g_{i,j-\frac{1}{2}}^{god} - g_{i,j+\frac{1}{2}}^{god}\right] , \\[2mm] u_{i+1,j}^{god} = u_{i+1,j}^n + \frac{\frac{1}{2}\Delta t}{\Delta y}\left[g_{i+1,j-\frac{1}{2}}^{god} - g_{i+1,j+\frac{1}{2}}^{god}\right] \end{array}\right\} \tag{16.92}$$

and $g_{i,j+\frac{1}{2}}^{god} = a_2 u_{i,j}^n$ is the Godunov first–order upwind flux, $a_2 > 0$.

Formulae (16.91) and (16.92) are now *re–interpreted* in terms of two well known *one–dimensional* operators that are valid for non–linear systems. Formulae (16.91) is the Weighted Average Flux scheme that results from solving the one–dimensional Riemann problem for the linear advection equation $u_t + a_1 u_x = 0$, at the intercell position $i + \frac{1}{2}$, with *modified* initial data $(u_{i,j}^{god}, u_{i+1,j}^{god})$. See Sect. 13.3 of Chap. 13 for details on the one dimensional WAF flux for the linear advection equation. The modified data results from applying the Godunov first–order upwind method to the one dimensional advection equation $u_t + a_2 u_y = 0$ for a time $\frac{1}{2}\Delta t$.

We now construct intercell fluxes $\mathbf{F}_{i+\frac{1}{2},j}$ and $\mathbf{G}_{i,j+\frac{1}{2}}$ for the finite volume scheme (16.46) to solve the two–dimensional nonlinear system (16.45) following the WAF approach. This is done by *imitating* (16.91)–(16.92). The intercell fluxes result from applying two one–dimensional operators in succession, namely

$$\mathbf{F}_{i+\frac{1}{2},j} = \mathbf{L}_{x,\frac{1}{2}\Delta t}^{waf}\left(\mathbf{L}_{y,\frac{1}{2}\Delta t}^{god}(\mathbf{U}_{i,j}^n), \mathbf{L}_{y,\frac{1}{2}\Delta t}^{god}(\mathbf{U}_{i+1,j}^n)\right) \tag{16.93}$$

and

$$\mathbf{G}_{i,j+\frac{1}{2}} = \mathbf{L}_{y,\frac{1}{2}\Delta t}^{waf}\left(\mathbf{L}_{x,\frac{1}{2}\Delta t}^{god}(\mathbf{U}_{i,j}^n), \mathbf{L}_{x,\frac{1}{2}\Delta t}^{god}(\mathbf{U}_{i,j+1}^n)\right) . \tag{16.94}$$

Here the operator $\mathbf{L}_{s,dt}^{waf}(\mathbf{Z}_L, \mathbf{Z}_R)$ applies to a pair of states $(\mathbf{Z}_L, \mathbf{Z}_R)$ and is defined as

$$\mathbf{L}_{s,dt}^{waf}(\mathbf{Z}_L, \mathbf{Z}_R) = \frac{1}{\Delta s}\int_{-\frac{1}{2}\Delta s}^{\frac{1}{2}\Delta s} \mathbf{E}\left(\mathbf{Z}_{LR}\left(\frac{s}{dt}\right)\right) ds . \tag{16.95}$$

This is an integral average of the flux component $\mathbf{E}$ in the s–direction normal to the interface between the states $\mathbf{Z}_L$ and $\mathbf{Z}_R$ and gives a WAF–type flux. $\mathbf{Z}_{LR}$ denotes the solution of the Riemann problem with data $(\mathbf{Z}_L, \mathbf{Z}_R)$. In practice one approximates the WAF flux by a summation, namely

$$\mathbf{F}_{LR} = \frac{1}{2}[\mathbf{F}(\mathbf{Z}_L) + \mathbf{F}(\mathbf{Z}_R)] - \frac{1}{2}\sum_{k=1}^{N} \text{sign}(c_{LR}^{(k)})\phi_{LR}^{(k)}\Delta\mathbf{F}_{LR}^{(k)} , \qquad (16.96)$$

where $c_{LR}^{(k)}$ is the Courant number for wave k in the solution of the Riemann problem, $\Delta\mathbf{F}_{LR}^{(k)}$ is the flux jump across wave k and $\phi_{LR}^{(k)} = \phi_{LR}^{(k)}(r_{LR}^{(k)})$ is a one–dimensional limiter function for wave k. For details on the WAF flux see Sect. 14.3 of Chap. 14 and Sect. 16.3.3 of this chapter.

The second operator in (16.93) and (16.94) is the Godunov one dimensional operator

$$\mathbf{L}_{s,dt}^{god}(\mathbf{Z}^n) = \mathbf{Z}^n + \frac{dt}{\Delta s}\left[\mathbf{K}_{l-\frac{1}{2}}^{god} - \mathbf{K}_{l+\frac{1}{2}}^{god}\right] , \qquad (16.97)$$

as applied to a state $\mathbf{Z}^n$ over a time dt in the direction s and $\mathbf{K}_{i+\frac{1}{2}}^{god}$ is the one–dimensional Godunov first–order upwind flux.

Fig. 16.9 depicts the evaluation of the intercell flux $\mathbf{F}_{i+\frac{1}{2},j}$ in (16.93), for which one performs the following two steps:

(I) **Modification of Data:** The initial data on the left and right side of the interface is modified by applying the Godunov first–order upwind scheme in the y–direction (cross direction) for a time $\frac{1}{2}\Delta t$ to it. One obtains the two states $\mathbf{L}_{y,\frac{1}{2}\Delta t}^{god}(\mathbf{U}_{i,j}^n)$ and $\mathbf{L}_{y,\frac{1}{2}\Delta t}^{god}(\mathbf{U}_{i+1,j}^n)$.

(II) **The Riemann Problem:** Solve the Riemann problem in the x–direction with the modified data $(\mathbf{L}_{y,\frac{1}{2}\Delta t}^{god}(\mathbf{U}_{i,j}^n), \mathbf{L}_{y,\frac{1}{2}\Delta t}^{god}(\mathbf{U}_{i+1,j}^n))$ and compute the WAF flux in the usual way; see (16.96). See also (16.39).

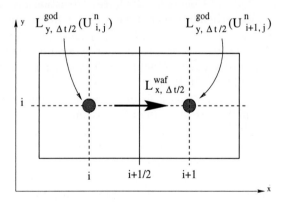

Fig. 16.9. Intercell flux $\mathbf{F}_{i+\frac{1}{2},j}$ is obtained by applying the WAF operator $\mathbf{L}_{x,\frac{1}{2}\Delta t}^{waf}$ in the x–direction to modified data in cells (i,j) and $(i+1,j)$. Modified data is obtained by applying the Godunov operator $\mathbf{L}_{y,\frac{1}{2}\Delta t}^{god}$ in the y–direction to initial states $\mathbf{U}_{i,j}^n$, $\mathbf{U}_{i+1,j}^n$

The evaluation of the intercell flux $\mathbf{G}_{i,j+\frac{1}{2}}$ in (16.94) is entirely analogous. One first modifies the initial data by applying the Godunov operator in the x–direction for a time $\frac{1}{2}\Delta t$ to it and then solves the Riemann problem in the y–direction to compute a WAF flux in the y–direction.

An Average State Scheme. Another extension to non–linear systems is based on the definition of a Weighted Average State (WAS) operator

$$\mathbf{L}_{s,dt}^{was}(\mathbf{Z}_L, \mathbf{Z}_R) = \frac{1}{\Delta s} \int_{-\frac{1}{2}\Delta s}^{\frac{1}{2}\Delta s} \mathbf{Z}_{LR}\left(\frac{s}{dt}\right) ds \qquad (16.98)$$

This is identical to the Weighted Average State version of WAF in one dimension, see Sect. 14.3.2 of Chap. 14. The flux $\mathbf{F}_{i+\frac{1}{2},j}$ is now given by

$$\mathbf{F}_{i+\frac{1}{2},j} = \mathbf{F}\left(\mathbf{L}_{y,\frac{1}{2}\Delta t}^{god}\left(\mathbf{L}_{x,\frac{1}{2}\Delta t}^{was}\left(\mathbf{U}_{i,j}^n, \mathbf{U}_{i+1,j}\right)\right)\right) \qquad (16.99)$$

and the flux $\mathbf{G}_{i,j+\frac{1}{2}}$ is given by

$$\mathbf{G}_{i,j+\frac{1}{2}} = \mathbf{G}\left(\mathbf{L}_{x,\frac{1}{2}\Delta t}^{god}\left(\mathbf{L}_{y,\frac{1}{2}\Delta t}^{was}\left(\mathbf{U}_{i,j}^n, \mathbf{U}_{i,j+1}\right)\right)\right) . \qquad (16.100)$$

First–Order Scheme and the TVD Condition. Recall that the WAF one–dimensional flux with the one dimensional TVD condition includes the Godunov first–order upwind scheme by simply setting the one–dimensional limiter function to an appropriate value, namely $\phi_{LR}^{(k)} = sign(c_{LR}^{(k)}) \times |c_{LR}^{(k)}|$. Therefore, by utilising the WAF operator in its first–order mode in (16.91) and applying the scheme to the two–dimensional linear advection equation (16.69) one obtains the first–order WAF finite volume scheme, which is equivalent to the CTU scheme of Colella for the linear advection equation. As seen earlier this first–order scheme is monotone and has linear stability limit 1. We therefore obtain an extension of the CTU scheme to two–dimensional non–linear systems of hyperbolic conservation laws by simply applying the WAF operator in its first–order mode (Godunov's first order upwind scheme) in (16.93)–(16.94), or (16.99)–(16.100).

The second–order scheme results from the second–order WAF operator in (16.93)–(16.94) or (16.99)–(16.100). In order to control the expected *spurious oscillations* we simply apply the WAF flux together with its TVD constraint. Note that this TVD constraint is applied only in a one–dimensional sense. The resulting scheme *is not* TVD *in a two–dimensional sense*. As a matter of fact spurious oscillations are still present, even for the model two–dimensional equation. In applications to two and three–dimensional nonlinear systems we find that spurious oscillations are small. Billett and Toro [45] have produced some preliminary results on a Total Variation Bounded (TVB) version of the multidimensional WAF schemes.

16.6.4 Schemes for Three–Dimensional Nonlinear Systems

Following the same approach as for two–dimensional non–linear systems, we now construct WAF finite volume schemes of the form (16.54) for solving three– dimensional non–linear systems of hyperbolic conservation laws (16.53). We *re–interpret* the schemes (16.84) and (16.88) derived for the three–dimensional linear advection equation (16.79).

A second–order Weighted Average Flux Scheme has intercell flux in the x–direction given by

$$
\left.
\begin{aligned}
\mathbf{F}_{i+\frac{1}{2},j,k} &= \mathbf{L}^{waf}_{x,\frac{1}{2}\Delta t}(\mathbf{U}^{god}_{i,j,k}, \mathbf{U}^{god}_{i+1,j,k}) , \\[2mm]
\mathbf{U}^{god}_{i,j,k} &= \mathbf{L}^{god}_{y,\frac{1}{2}\Delta t}(\mathbf{L}^{god}_{z,\frac{1}{2}\Delta t}(\mathbf{U}^{n}_{i,j,k})) , \\[2mm]
\mathbf{U}^{god}_{i+1,j,k} &= \mathbf{L}^{god}_{y,\frac{1}{2}\Delta t}(\mathbf{L}^{god}_{z,\frac{1}{2}\Delta t}(\mathbf{U}^{n}_{i+1,j,k})) ,
\end{aligned}
\right\}
\tag{16.101}
$$

numerical flux in the y–direction given by

$$
\left.
\begin{aligned}
\mathbf{G}_{i,j+\frac{1}{2},k} &= \mathbf{L}^{waf}_{y,\frac{1}{2}\Delta t}(\mathbf{U}^{god}_{i,j,k}, \mathbf{U}^{god}_{i,j+1,k}) , \\[2mm]
\mathbf{U}^{god}_{i,j,k} &= \mathbf{L}^{god}_{z,\frac{1}{2}\Delta t}(\mathbf{L}^{god}_{x,\frac{1}{2}\Delta t}(\mathbf{U}^{n}_{i,j,k})) , \\[2mm]
\mathbf{U}^{god}_{i,j+1,j} &= \mathbf{L}^{god}_{z,\frac{1}{2}\Delta t}(\mathbf{L}^{god}_{x,\frac{1}{2}\Delta t}(\mathbf{U}^{n}_{i,j+1,k}))
\end{aligned}
\right\}
\tag{16.102}
$$

and numerical flux in the z–direction given by

$$
\left.
\begin{aligned}
\mathbf{H}_{i,j,k+\frac{1}{2}} &= \mathbf{L}^{waf}_{z,\frac{1}{2}\Delta t}(\mathbf{U}^{god}_{i,j,k}, \mathbf{U}^{god}_{i,j,k+1}) , \\[2mm]
\mathbf{U}^{god}_{i,j,k} &= \mathbf{L}^{god}_{x,\frac{1}{2}\Delta t}(\mathbf{L}^{god}_{y,\frac{1}{2}\Delta t}(\mathbf{U}^{n}_{i,j,k})) , \\[2mm]
\mathbf{U}^{god}_{i,j,k+1} &= \mathbf{L}^{god}_{x,\frac{1}{2}\Delta t}(\mathbf{L}^{god}_{y,\frac{1}{2}\Delta t}(\mathbf{U}^{n}_{i,j,k+1})) .
\end{aligned}
\right\}
\tag{16.103}
$$

Exercise 16.6.5. Show that the intercell fluxes (16.101) reproduces the intercell flux (16.87) for the linear advection equation (16.79) with $a_1 > 0$, $a_2 > 0$, $a_3 > 0$.

Solution 16.6.3. Left to the reader.

Remark 16.6.1. A first–order scheme for three dimensional nonlinear systems is obtained by replacing the WAF flux by the Godunov first order upwind flux in (16.101)–(16.103).

Exercise 16.6.6. Extend the Weighted Average State scheme with fluxes (16.99)–(16.100) to construct a finite volume scheme for three–dimensional non–linear systems. Verify that the corresponding fluxes reproduce (16.87) for the linear advection equation (16.79) with $a_1 > 0$, $a_2 > 0$, $a_3 > 0$.

Solution 16.6.4. Left to the reader.

16.7 Non–Cartesian Geometries

16.7.1 Introduction

A well defined fluid dynamic problem will include (a) a set of partial differential equations (PDEs) or a set of Integral Equations (b) the domain in which the governing equations are to be solved and (c) initial and boundary conditions. The numerical solution of this mathematical problem requires the discretisation of the domain via the generation of a grid or mesh and the discretisation of the equations on the mesh. The grid generation process discretises the continuous domain into a finite number of mesh points (finite difference interpretation) or into a finite number of volumes (finite volume interpretation). So far we have assumed that all domains are Cartesian, that is their boundaries are perfectly aligned with the Cartesian coordinate directions x, y, z. Fig. 16.10 shows two examples of Cartesian domains in two space dimensions. In practice, most domains *are not* Cartesian; Fig. 16.11 shows two simple examples of non–Cartesian domains in two dimensions. Having specified the domain and its boundaries, the process of determining the interior points or finite volumes is known as *grid generation*, or *mesh generation*. The generation of grids for Cartesian domains is exceedingly simple, particularly for a rectangular domain such as that of Fig. 16.10 (a). For non–Cartesian domains such as those of Fig. 16.11, both the generation of a grid and the discretisation of the PDEs is much more involved. The topic of grid generation is a large area of study in its own right and is not pursued here. We assume that a mesh is available. Introductory material on grid generation is found in the textbook by Hoffmann [172]. Comprehensive treatments of the subject are found in the textbook edited by Thompson [347], the book by Thompson, Warsi and Mastin [346] and in the VKI Lectures by Weatherhill [407]; see also [345]. The purpose of this section is to introduce the

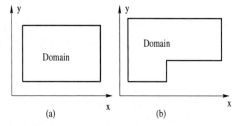

Fig. 16.10. Examples of two–dimensional Cartesian domains in x–y space

reader to approaches for dealing with non–Cartesian domains. In particular, we give details of the finite volume approach, whereby one discretises the PDEs directly in physical space.

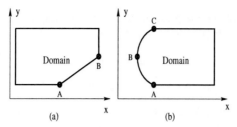

Fig. 16.11. Examples of two–dimensional non–Cartesian domains in x–y space (a) straight segment AB is not aligned with any of the Cartesian coordinates directions (b) arc ABC is not aligned with any of the Cartesian coordinates directions

16.7.2 General Domains and Coordinate Transformation

The simplest approach for dealing with non–Cartesian domains such as those of Fig. 16.11 is to insist on using a Cartesian representation of the boundaries of the domain. In this case those portions of the boundary which are not aligned with any of the coordinate directions will end up with a *staircase* like representation. This simple procedure allows once more the use of Cartesian meshes but the errors introduced at the boundaries are too large and therefore this approach is not recommended. An elegant method that retains the Cartesian mesh in the interior of the domain and treats in a special manner those computing cells that are *cut* by the true boundary is the so–called *Cartesian cut cell* method. Details on this approach are found in [76], [427], [30], [275], [418] and [419].

The most well–known approach for dealing with general domains is based on transforming the domain in *physical space* (t, x, y, z) to a domain in *computational space* (τ, ξ, η, ζ). The coordinate transformation maps the, irregular, domain in physical space to a perfectly regular computational domain, which is then discretised by a perfectly regular mesh as done for Cartesian domains. One defines the transformation via

$$\tau = t \ , \quad \xi = \xi(t, x, y, z) \ , \quad \eta = \eta(t, x, y, z) \ , \quad \zeta = \zeta(t, x, y, z) \ . \quad (16.104)$$

By use of the chain rule, partial derivatives of any quantity $\phi = \phi(t, x, y, z)$ read

$$\left.\begin{aligned}
\phi_t &= \phi_\tau \ + \ \phi_\xi \xi_t + \phi_\eta \eta_t + \phi_\zeta \zeta_t \ , \\
\phi_x &= \phi_\xi \xi_x + \phi_\eta \eta_x + \phi_\zeta \zeta_x \ , \\
\phi_y &= \phi_\xi \xi_y + \phi_\eta \eta_y + \phi_\zeta \zeta_y \ , \\
\phi_z &= \phi_\xi \xi_z + \phi_\eta \eta_z + \phi_\zeta \zeta_z \ ,
\end{aligned}\right\} \quad (16.105)$$

so that PDEs defined in physical space in terms of the (t, x, y, z) coordinates are transformed in computational space in terms of the generalised coordinates (τ, ξ, η, ζ).

Example 16.7.1. Suppose we want to solve the linear advection equation

$$u_t + a_1 u_x + a_2 u_y = 0 \ , \quad (16.106)$$

where a_1 and a_2 are constants, in a non–Cartesian domain. Use of relations (16.105) reduces (16.106) in physical space to

$$u_\tau + \hat{a}_1 u_\xi + \hat{a}_2 u_\eta = 0 \tag{16.107}$$

in computational space. Note that this PDE has variable coefficients

$$\hat{a}_1 = \xi_t + a_1\xi_x + a_2\xi_y \ , \quad \hat{a}_2 = \eta_t + a_1\eta_x + a_2\eta_y \tag{16.108}$$

and is more complicated than the original PDE (16.106), which has constant coefficients. The variable coefficients $\hat{a}_1$ and $\hat{a}_2$ include the transformation derivatives ξ_x, ξ_y, η_x and η_y. These partial derivatives are called *metrics* and may be computed analytically from (16.104), or numerically, as we shall explain.

Next we transform the three–dimensional time dependent Euler equations, or any three–dimensional non–linear system of conservation laws

$$\mathbf{U}_t + \mathbf{F}(\mathbf{U})_x + \mathbf{G}(\mathbf{U})_y + \mathbf{H}(\mathbf{U})_z = \mathbf{0} \ . \tag{16.109}$$

By using relations (16.105) one obtains the transformed system in generalised coordinates

$$\hat{\mathbf{U}}_\tau + \hat{\mathbf{F}}_\xi + \hat{\mathbf{G}}_\eta + \hat{\mathbf{H}}_\zeta = \mathbf{0} \ , \tag{16.110}$$

where

$$\left. \begin{array}{l} \hat{\mathbf{U}} = \frac{1}{J}\mathbf{U} \ , \\ \hat{\mathbf{F}} = \frac{1}{J}[\xi_t\mathbf{U} + \xi_x\mathbf{F} + \xi_y\mathbf{G} + \xi_z\mathbf{H}] \ , \\ \hat{\mathbf{G}} = \frac{1}{J}[\eta_t\mathbf{U} + \eta_x\mathbf{F} + \eta_y\mathbf{G} + \eta_z\mathbf{H}] \ , \\ \hat{\mathbf{H}} = \frac{1}{J}[\zeta_t\mathbf{U} + \zeta_x\mathbf{F} + \zeta_y\mathbf{G} + \zeta_z\mathbf{H}] \ . \end{array} \right\} \tag{16.111}$$

with J to be specified.

Remark 16.7.1. Equations (16.110) are written in conservation form, just as were the original equations (16.109). For details see [172], [5] and [9].

Computation of Metrics. One can relate differentials in the (t, x, y, z) system to differentials in the (τ, ξ, η, ζ) system so that the metrics can be found in terms of derivative of (t, x, y, z) with respect to (τ, ξ, η, ζ) in computational space, namely

$$\mathbf{A} = \mathbf{B}^{-1} \ , \tag{16.112}$$

where

$$\mathbf{A} = \begin{bmatrix} 1 & 0 & 0 & 0 \\ \xi_t & \xi_x & \xi_y & \xi_z \\ \eta_t & \eta_x & \eta_y & \eta_z \\ \zeta_t & \zeta_x & \zeta_y & \zeta_z \end{bmatrix} \ ; \quad \mathbf{B} = \begin{bmatrix} 1 & 0 & 0 & 0 \\ x_\tau & x_\xi & x_\eta & x_\zeta \\ y_\tau & y_\xi & y_\eta & y_\zeta \\ z_\tau & z_\xi & z_\eta & z_\zeta \end{bmatrix} \ . \tag{16.113}$$

See [172] for details. The computation of derivatives in computational space is easily carried out numerically, as this is discretised by perfectly regular meshes. In solving for the components of **A** via (16.112) one requires the computation of the inverse matrix of **B**, which involves the expression

$$J = \frac{1}{det(\mathbf{B})} ,$$

(16.114)

where $det(\mathbf{B})$ is the determinant of **B** and J is called the *Jacobian of the transformation*. Note that this is different from the definition of the Jacobian matrix associated with systems of hyperbolic conservation laws introduced in Chap. 2. Here the Jacobian is a scalar quantity given by

$$J = \frac{1}{x_\xi (y_\eta z_\zeta - y_\zeta z_\eta) - x_\eta (y_\xi z_\zeta - y_\zeta z_\xi) + x_\zeta (y_\xi z_\eta - y_\eta z_\xi)} .$$

(16.115)

In the next section we study the finite volume approach, whereby the equations are directly discretised in physical space.

16.7.3 The Finite Volume Method for Non–Cartesian Domains

The Finite Volume Method is another approach for dealing with general non–Cartesian domains. In this approach the governing conservation laws, expressed in integral form, are discretised *directly* in physical space and one still works with reference to a *Cartesian frame*. Background reading on the finite volume method is found in [170], [402]. Here we present a brief description of the method for two and three dimensional problems. First recall that the integral form of the three–dimensional conservation laws (16.109) is

$$\frac{\mathrm{d}}{\mathrm{d}t} \int \int \int_V \mathbf{U} \, dV + \int \int_A \mathcal{H} \cdot \mathbf{n} \, dA = \mathbf{0} ,$$

(16.116)

where V is a *control volume*, A is the boundary of V, $\mathcal{H} = (\mathbf{F}, \mathbf{G}, \mathbf{H})$ is the tensor of fluxes, **n** is the outward unit vector normal to the surface A, dA is an area element and $\mathcal{H} \cdot \mathbf{n} \, dA$ is the flux component normal to the boundary A. See Sect. 1.5 of Chap. 1 for details on the derivation of the integral form of the equations. The conservation laws (16.116) state that the time–rate of change of **U** inside the volume V depends only on the *total flux* through the surface A, the boundary of the control volume V. In the finite volume approach, the integral form of the conservation laws is enforced on each control volume that results from discretising the domain in physical space into a finite number of finite volumes or computing cells. The conservation equations are thus satisfied at the discrete level, which is a distinguishing property of the Finite Volume Method.

The Two–Dimensional Case. Consider some two–dimensional domain in x–y space and assume this has been discretised into a finite number of computing cells by some grid generation technique. Part of a simple structured mesh is shown in Fig. 16.12, where all computing cells may be identified by the indices i and j, just as in Cartesian meshes assumed previously. Consider now any computing cell from the discretised domain and regard it as the control volume V in the integral form of the equations (16.116). For generality assume that the boundary A of V is the union of N straight segments $A_s A_{s+1}$, where $A_{N+1} \equiv A_1$. Fig. 16.13 shows the special case of a quadrilateral finite volume, $N = 4$. We have assumed that a grid has been generated and that the (x_s, y_s) coordinates of the vertices A_s are known. The *total flux*

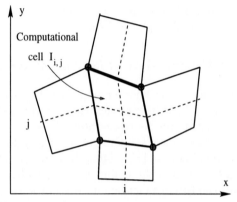

Fig. 16.12. Possible grid configuration for two–dimensional domain in x–y space

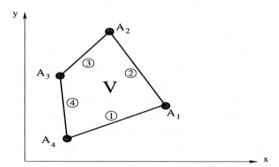

Fig. 16.13. Example of quadrilateral finite volume V in two–dimensional domain in x–y space

through the boundaries in (16.116) may now be written as

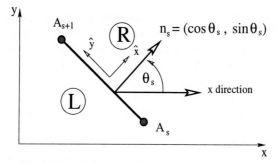

Fig. 16.14. General intercell boundary s of finite volume V in two–dimensional domain in x–y space.

$$\int\int_A \mathcal{H} \cdot \mathbf{n}\,\mathrm{d}A = \sum_{s=1}^{N} \int_{A_s}^{A_{s+1}} \mathcal{H} \cdot \mathbf{n}_s\,\mathrm{d}A , \qquad (16.117)$$

where s denotes the intercell boundary $A_s A_{s+1}$. We consistently adhere to the convention of transversing the boundary so that the interior of the volume always lies on the left hand side of the boundary. Let us now consider any intercell boundary s as shown in Fig. 16.14, where L (left) denotes the interior side of the control volume V and R (right) denotes a state exterior to V and adjacent to side s. The first problem is to determine the direction of the outward unit vector $\mathbf{n}_s$ normal to side s; this is needed to evaluate the integral terms in (16.117). It is convenient to choose the x–direction as the *reference direction* and to define the angle θ_s as the angle formed by the x–direction and the normal vector $\mathbf{n}_s$; see Fig. 16.14. The components of $\mathbf{n}_s$ can be found in terms of the angle θ_s and it is easy to see that

$$\mathbf{n}_s = (cos\theta_s, sin\theta_s) . \qquad (16.118)$$

Then, the total flux in (16.117) is

$$\sum_{s=1}^{N} \int_{A_s}^{A_{s+1}} \mathcal{H} \cdot \mathbf{n}_s\,\mathrm{d}A = \sum_{s=1}^{N} \int_{A_s}^{A_{s+1}} [cos\theta_s \mathbf{F}(\mathbf{U}) + sin\theta_s \mathbf{G}(\mathbf{U})]\,\mathrm{d}A . \quad (16.119)$$

By interpreting the first integral term in equations (16.116) as the time–rate of change of the average of $\mathbf{U}$ inside volume V we may write

$$\frac{\mathrm{d}}{\mathrm{d}t}\int\int\int_V \mathbf{U}\,\mathrm{d}V = |V|\frac{\mathrm{d}}{\mathrm{d}t}\mathbf{U} , \qquad (16.120)$$

where $|V|$ denotes the volume of V, the area of V in the two–dimensional case. Then, by substituting (16.119) and (16.120) into (16.116) we obtain

$$\frac{\mathrm{d}}{\mathrm{d}t}\mathbf{U} = -\frac{1}{|V|}\sum_{s=1}^{N} \int_{A_s}^{A_{s+1}} [cos\theta_s \mathbf{F}(\mathbf{U}) + sin\theta_s \mathbf{G}(\mathbf{U})]\,\mathrm{d}A . \qquad (16.121)$$

This finite volume formula leads to numerical schemes, once approximations to the intercell fluxes for each side s have been made and a time discretisation scheme has been chosen. In devising intercell flux approximations one requires to be more specific as to the locations of the discrete values of $\mathbf{U}$. Various flux approximations are discussed in Chap. 8 of [170]. Here we assume that the discrete values of the variable $\mathbf{U}$ are cell averages within the volumes and assign them to the centres of the cells. This is known as the *cell centred* finite volume method.

One way of constructing intercell flux approximations arises from exploiting the *rotational invariance* of the governing equations. For the Euler equations see Sect. 3.2.1 of Chap. 3. For the two–dimensional case we have

$$cos\theta_s \mathbf{F}(\mathbf{U}) + sin\theta_s \mathbf{G}(\mathbf{U}) = \mathbf{T}_s^{-1}\mathbf{F}(\mathbf{T}_s\mathbf{U}) , \qquad (16.122)$$

where $\mathbf{T}_s \equiv \mathbf{T}(\theta_s)$ is the rotation matrix and $\mathbf{T}_s^{-1}$ is its inverse, namely

$$\mathbf{T}(\theta) = \begin{bmatrix} 1 & 0 & 0 & 0 \\ 0 & \cos\theta & \sin\theta & 0 \\ 0 & -\sin\theta & \cos\theta & 0 \\ 0 & 0 & 0 & 1 \end{bmatrix}, \quad \mathbf{T}^{-1} = \begin{bmatrix} 1 & 0 & 0 & 0 \\ 0 & \cos\theta & -\sin\theta & 0 \\ 0 & \sin\theta & \cos\theta & 0 \\ 0 & 0 & 0 & 1 \end{bmatrix} . \qquad (16.123)$$

We note that $\hat{\mathbf{U}} \equiv \mathbf{T}_s\mathbf{U}$ is the vector of *rotated* conserved variables obtained by applying the rotation matrix $\mathbf{T}_s$ to the original vector of conserved variables $\mathbf{U}$ and is aligned with the new rotated Cartesian frame $(\hat{x}, \hat{y})$ shown in Fig. 16.14. Here the coordinate $\hat{x}$ is the normal direction (normal to the intercell boundary) and $\hat{y}$ is the tangential direction (parallel to the intercell boundary). Then equations (16.121) become

$$\frac{d}{dt}\mathbf{U} = -\frac{1}{|V|} \sum_{s=1}^{N} \int_{A_s}^{A_{s+1}} \mathbf{T}_s^{-1}\mathbf{F}(\mathbf{T}_s\mathbf{U})\, dA . \qquad (16.124)$$

So far, apart from interpretation (16.120), no numerical approximations have been made. In order to define numerical fluxes across the intercell boundaries s we use the equations in the rotated frame $(\hat{x}, \hat{y})$. They become an *augmented* one–dimensional system

$$\hat{\mathbf{U}}_t + \hat{\mathbf{F}}_{\hat{x}} = \mathbf{0} , \qquad (16.125)$$

where $\hat{\mathbf{F}} = \mathbf{F}(\hat{\mathbf{U}})$. This *augmented* one–dimensional system is identical to the split multidimensional systems in Cartesian coordinates considered in Chap. 3, for which numerical fluxes were constructed in Chaps. 6 to 12 and 14.

Each term in the summation (16.124) is now approximated as

$$\int_{A_s}^{A_{s+1}} \mathbf{T}_s^{-1}\mathbf{F}(\mathbf{T}_s\mathbf{U})\, dA \approx \mathcal{L}_s \mathbf{T}_s^{-1}\hat{\mathbf{F}}_s , \qquad (16.126)$$

where $\mathcal{L}_s$ is the length of segment $A_s A_{s+1}$ and $\hat{\mathbf{F}}_s$ is the intercell flux corresponding to the augmented one–dimensional system (16.125). Assuming numerical fluxes $\hat{\mathbf{F}}_s$ have been constructed we then have the *semidiscrete* conservative scheme

$$\frac{\mathrm{d}}{\mathrm{d}t}\mathbf{U} = -\frac{1}{|V|}\sum_{s=1}^{N}\mathcal{L}_s\mathbf{T}_s^{-1}\hat{\mathbf{F}}_s \ , \tag{16.127}$$

where the time discretisation is still open to choice. See Sec. 13.4.7 of Chap. 13.

In a computational setup as depicted in Fig. 16.12 the volume V is a general computing cell labelled $I_{i,j}$. Let us now replace the time derivative by a forward in time approximation; we then obtain the fully discrete scheme

$$\mathbf{U}_{i,j}^{n+1} = \mathbf{U}_{i,j}^{n} - \frac{\Delta t}{|I_{i,j}|}\sum_{s=1}^{N}\mathcal{L}_s\mathbf{T}_s^{-1}\hat{\mathbf{F}}_s \ , \tag{16.128}$$

where $|I_{i,j}|$ is the area of cell $I_{i,j}$.

Exercise 16.7.1. Show that if $I_{i,j}$, $\forall i, j$, is a Cartesian cell of area $\Delta x \times \Delta y$ as shown in Fig. 16.15, then the finite volume formula (16.128) reproduces the finite volume formula

$$\mathbf{U}_{i,j}^{n+1} = \mathbf{U}_{i,j}^{n} + \frac{\Delta t}{\Delta x}[\mathbf{F}_{i-\frac{1}{2},j} - \mathbf{F}_{i+\frac{1}{2},j}] + \frac{\Delta t}{\Delta y}[\mathbf{G}_{i,j-\frac{1}{2}} - \mathbf{G}_{i,j+\frac{1}{2}}] \tag{16.129}$$

studied in Sect. 16.4.

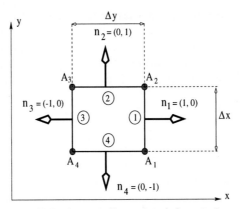

Fig. 16.15. Cartesian cell $I_{i,j}$ in two–dimensional domain in x–y space.

Solution 16.7.1. The outward unit normals for sides 1 to 4 are $\mathbf{n}_1 = (1,0)$, $\mathbf{n}_2 = (0,1)$, $\mathbf{n}_3 = (-1,0)$, and $\mathbf{n}_4 = (0,-1)$. The corresponding angles are $\theta_1 = 0$, $\theta_2 = \frac{1}{2}\pi$, $\theta_3 = \pi$, $\theta_4 = \frac{3}{2}\pi$. The fluxes may be computed from (16.126), or more directly from (16.117); they are

$$\left.\begin{array}{l} \int_{A_1}^{A_2} (\mathbf{F},\mathbf{G}) \cdot \mathbf{n}_1 \, dA = \Delta y \times \mathbf{F}_{12} \; , \\[2mm] \int_{A_2}^{A_3} (\mathbf{F},\mathbf{G}) \cdot \mathbf{n}_2 \, dA = \Delta x \times \mathbf{G}_{23} \; , \\[2mm] \int_{A_3}^{A_4} (\mathbf{F},\mathbf{G}) \cdot \mathbf{n}_3 \, dA = -\Delta y \times \mathbf{F}_{34} \; , \\[2mm] \int_{A_4}^{A_1} (\mathbf{F},\mathbf{G}) \cdot \mathbf{n}_4 \, dA = -\Delta x \times \mathbf{G}_{41} \; . \end{array}\right\} \tag{16.130}$$

By defining $\mathbf{F}_{12} \equiv \mathbf{F}_{i+\frac{1}{2},j}$, $\mathbf{F}_{34} \equiv \mathbf{F}_{i-\frac{1}{2},j}$, $\mathbf{G}_{23} \equiv \mathbf{F}_{i,j+\frac{1}{2}}$, $\mathbf{G}_{41} \equiv \mathbf{G}_{i,j-\frac{1}{2}}$, we obtain the the desired result.

A possible way of constructing intercell numerical fluxes for (16.128) is following the Godunov approach, studied in Chap. 6 as used in conjunction with the exact Riemann solver, and in Chaps. 8 to 12 using approximate Riemann solvers. To obtain a Godunov first–order upwind flux we solve the Riemann problem for (16.125) with initial data

$$\hat{\mathbf{U}}(\hat{x},0) = \begin{cases} \hat{\mathbf{U}}_L = \mathbf{T}_s(\mathbf{U}_L) & \text{if } \hat{x} < 0 \; , \\ \hat{\mathbf{U}}_R = \mathbf{T}_s(\mathbf{U}_R) & \text{if } \hat{x} > 0 \; . \end{cases} \tag{16.131}$$

Here $\mathbf{U}_L$ is the state in the interior of the finite volume V and $\mathbf{U}_R$ is the state outside V separated from $\mathbf{U}_L$ by the intercell boundary $A_s A_{s+1}$. See Fig. 16.14.

Computation of Lengths, Normals and Areas. Here we give formulae for the lengths $\mathcal{L}_s$, the components of the outward unit normals $\mathbf{n}_s$ and the areas of co–planar quadrilateral finite volumes. These quantities are needed in the conservative formula (16.128). Consider the quadrilateral V of Fig. 16.13 and assume the coordinates (x_s, y_s) of the points A_s are known. We first define

$$\Delta x_s \equiv x_{s+1} - x_s \; , \quad \Delta y_s \equiv y_{s+1} - y_s \; . \tag{16.132}$$

Then, the length $\mathcal{L}_s$ of side s is given by

$$\Delta s = \sqrt{\Delta x_s^2 + \Delta y_s^2} \; . \tag{16.133}$$

The components $cos\theta_s$ and $sin\theta_s$ of the outward unit normal $\mathbf{n}_s$ are

$$cos\theta_s = \frac{\Delta y_s}{\Delta s} \; , \quad sin\theta_s = -\frac{\Delta x_s}{\Delta s} \; . \tag{16.134}$$

Finally, the area of the co–planar quadrilateral V is given by

$$|V| = \frac{1}{2}|(x_3 - x_1) \times (y_4 - y_2) - (y_3 - y_1) \times (x_4 - x_2)| \; . \tag{16.135}$$

Examples of Finite Volume Schemes. Consider sufficiently smooth structured grids formed by quadrilateral volumes as shown in Fig. 16.12. Then formula (16.128) contains four flux contributions. The most natural way of updating the solution is by applying (16.128) with all flux contributions included in a single step. We term this the *unsplit finite volume method* and is entirely consistent with the unsplit finite volume schemes for Cartesian meshes studied in Sect. 16.4. One may also reinterpret (16.128) as a *split finite volume scheme*, whereby for instance, flux terms associated with the i index are included in a predictor step and flux terms associated with the j index are included in the corrector step. The practical implementation of the split finite volume schemes is entirely analogous to the split schemes studied in Sects. 16.2 and 16.3.

First–order methods of the Godunov type are easily constructed. The intercell numerical flux depends only on the two rotated states $(\hat{\mathbf{U}}_L, \hat{\mathbf{U}}_R)$, the length $\mathcal{L}_s$ and the normal $\mathbf{n}_s$. The following is a possible algorithm

(I)   For each intercell boundary s rotate the data $(\mathbf{U}_L, \mathbf{U}_R)$ to obtain $\hat{\mathbf{U}}_L = \mathbf{T}_s \mathbf{U}_R$, $\hat{\mathbf{U}}_R = \mathbf{T}_s \mathbf{U}_R$.

(II)   Solve the local Riemann problem with rotated data $(\hat{\mathbf{U}}_L, \hat{\mathbf{U}}_R)$ to obtain the intercell flux $\hat{\mathbf{F}}_s$. Here one uses the exact Riemann solver, see Chaps. 6, or approximate Riemann solvers, see Chaps. 8 to 12.

(III) Rotate $\hat{\mathbf{F}}_s$ back by multiplying it by $\mathbf{T}_s^{-1}$ to obtain the intercell flux $\mathbf{T}_s^{-1}\hat{\mathbf{F}}_s$ for side s.

(IV)   Compute lengths $\mathcal{L}_s$, normals $\mathbf{n}_s$ and areas $|I_{i,j}|$ to complete the intercell flux evaluation.

(V)   Update the solution according to a split or unsplit scheme.

Recall that for Cartesian meshes an unsplit Godunov method has linearised stability limit $\frac{1}{2}$, whereas a split version has limit 1. The WAF–type first order unsplit finite volume scheme of Sect. 16.6.1 also has stability limit 1; this scheme can easily be implemented in non–Cartesian quadrilaterals.

The second–order TVD schemes presented in Chaps. 14, see also Sect. 16.2 and 16.3 of this chapter, can also be implemented for non–Cartesian quadrilateral meshes following the methodology for the Godunov first–order upwind method outlined above. Details on WAF–type methods are found in [321], [322] and [44].

The Three–Dimensional Case. Three–dimensional finite volume schemes are easily constructed. Here one requires the evaluation of the area of of non–coplanar intercell boundaries and as well as volumes. For details see [170]. The rotational invariance property of the three–dimensional Euler equations, proved in Sect. 3.2.2 of Chap. 3, can also be exploited to derive finite volume formulae analogous to (16.128), with corresponding Godunov–type numerical fluxes, see [44] for example.

17. Multidimensional Test Problems

This chapter is concerned with tests for assessing numerical solutions to multidimensional problems. The assessment of the numerical methods to be used in practical computations, prior to their actual application, is of considerable importance and cannot be emphasised enough. There are four classes of test problems that can be used, namely (A) tests with exact solution, (B) tests with reliable numerical solution to equivalent one–dimensional equations obtained under the assumption of symmetry for instance, (C) tests for which other numerical solutions are available and (D) tests for which experimental results are available. In the first three categories of test problems one solves the same or equivalent governing partial differential equations and thus one seeks *complete agreement* in the comparisons. Care is required in class (D) when experimental results are used. The governing PDEs might themselves not be an accurate description of the physical problem being solved. Typical questions to be asked are: will viscosity and heat conduction be important in the problem, will the equation of state be a correct description of the thermodynamics, will turbulence be important, etc. If one can isolate these effects, or account for their influence, then comparison between numerical and experimental results is useful in assessing the performance of the numerical methods. If the numerical solution is reliable, then one would expect the comparison with experimental results to be a way of verifying the validity of the governing equations as a suitable model of the physics. A useful reference here is the report by Albone [2].

Exact solutions for one–dimensional problems can be used to check two and three dimensional programs. For the two–dimensional case one can, for instance, use the exact Riemann solver of Chap. 4 with the initial data states separated by a straight line at an angle to the mesh; this includes the trivial cases of one–dimensional flow along coordinates directions. The same can be done for three–dimensional programs. In this chapter we consider test problems in categories (B), (C) and (D) above.

17.1 Explosions and Implosions

Here we consider test problems for the two and three dimensional Euler equations for ideal gases with $\gamma = 1.4$. The geometry and initial data for the problems are such that cylindrical and spherical symmetry can be enforced. The multidimensional Euler equations may then be simplified to the one–dimensional inhomogeneous system

$$\mathbf{U}_t + \mathbf{F}(\mathbf{U})_r = \mathbf{S}(\mathbf{U}) \;, \tag{17.1}$$

where

$$\mathbf{U} = \begin{bmatrix} \rho \\ \rho u \\ E \end{bmatrix} \;, \quad \mathbf{F} = \begin{bmatrix} \rho u \\ \rho u^2 + p \\ u(E + p) \end{bmatrix} \;, \quad \mathbf{S} = -\frac{\alpha}{r} \begin{bmatrix} \rho u \\ \rho u^2 \\ u(E + p) \end{bmatrix} \;. \tag{17.2}$$

See Sect. 1.6.2 of Chap. 1 for details. Here r is the radial direction, u is the radial velocity and α is a parameter. For $\alpha = 0$ we reproduce plain one–dimensional flow. For $\alpha = 1$ we have cylindrical symmetry and equations (17.1)–(17.2) are equivalent to the two–dimensional Euler equations. For $\alpha = 2$ we have spherical symmetry and (17.1)–(17.2) are equivalent to the three–dimensional Euler equations. The one–dimensional equations (17.1)–(17.2) with a geometric source term may be solved with a reliable one–dimensional method on a very fine mesh to provide very accurate numerical solutions to compare with the numerical solution of the full two and three dimensional Euler equations. These test problems belong to category (B) above.

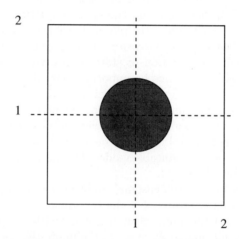

Fig. 17.1. Initial conditions for cylindrical explosion. Dark circular zone has high pressure and high density

17.1.1 Explosion Test in Two–Space Dimensions

The two–dimensional Euler equations, see Chap. 3, are solved on the square domain 2.0×2.0 in the x–y plane. The initial conditions consist of the region inside of a circle of radius $R = 0.4$ centred at $(1, 1)$ and the region outside the circle, see Fig. 17.1. The flow variables take constant values in each of these regions and are joined by a circular discontinuity at time $t = 0$. The two constant states for the two–dimensional Euler equations are chosen to be

$$\left.\begin{array}{llll} \rho_{ins} = 1.0 & , & \rho_{out} = 0.125 \ , \\ u_{ins} = 0.0 & , & u_{out} = 0.0 \ , \\ v_{ins} = 0.0 & , & v_{out} = 0.0 \ , \\ p_{ins} = 1.0 & , & p_{out} = 0.1 \ . \end{array}\right\} \tag{17.3}$$

Subscripts ins and out denote values inside and outside the circle respectively. This test problem is like a two–dimensional extension of the shock–tube problem used for assessing one–dimensional solutions in previous chapters.

We solve the full two–dimensional Euler equations by the WAF method presented in Chap. 14 in conjunction with space splitting, see Sect. 16.2 of Chap. 16; the Riemann solver used is HLLC presented in Chap. 10. The limiter function used is VANLEER, see 13.7 and 13.8 of Chap.13. The CFL coefficient used is $C_{cfl} = 0.9$ and the mesh is of 101×101 computing cells. We remark that in initialising the implosion/explosion problems, we modify the initial data on quadrilateral cells cutting the initial discontinuity, by assigning modified area–weighted values to the appropriate cells at the initial time $t = 0$. This procedure avoids the formation of small amplitude waves created at early times by the staircase configuration of the data; such waves are not spurious, they result from the data as given and the numerical method does its best to resolve them.

Fig. 17.2 shows the density distribution as a function of x and y at the output time $t = 0.25$. The solution exhibits a circular shock wave travelling away from the centre, a circular contact surface travelling in the same direction and a circular rarefaction travelling towards the origin $(1, 1)$. Fig. 17.3 shows the corresponding pressure distribution at the same time. As expected, the pressure is continuous across the contact surface. We remark that as time evolves, a complex wave pattern emerges. Some salient features of the solution are these. The circular shock wave travels outwards, becoming weaker as time evolves. The contact surface follows the shock becoming weaker also; at some point in time the contact comes to rest and then travels inwards. The rarefaction travelling towards the centre reflects, as a rarefaction, and over expands the flow so as to create an inward travelling shock wave; this circular shock wave implodes into the origin, reflects and travels outwards colliding with the contact surface. And so on.

The one dimensional inhomogeneous equations (17.1), (17.2) for the equivalent problem are solved numerically on a domain $[0, 1]$ on the radial line r, with equivalent shock–tube like data. The numerical method used is the Random Choice Method (RCM) described in Chap. 7, together with the splitting techniques of Chap. 15 to account for the geometric source term. We used $M = 1000$ computing cells and a CFL coefficient $C_{cfl} = 0.4$. Recall that the RCM resolves discontinuities as true discontinuities, the only errors

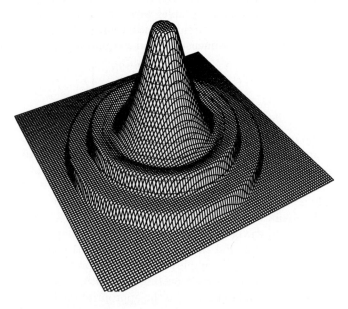

Fig. 17.2. Cylindrical explosion. Density distribution at time $t = 0.25$. Solution exhibits circular shock, circular contact and circular rarefaction

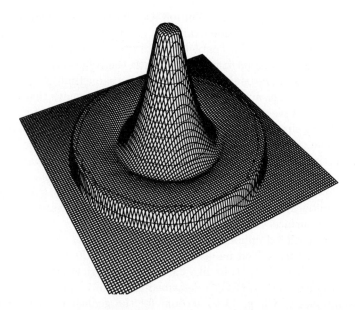

Fig. 17.3. Cylindrical explosion. Pressure distribution at time $t = 0.25$. Solution exhibits circular shock and circular rarefaction. Compare with Fig. 17.2

being those of the position of the waves. Under these circumstances we may regard the RCM solution as the *exact solution*.

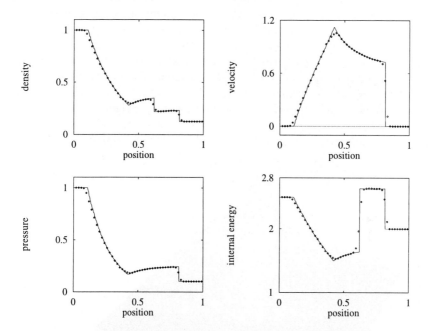

Fig. 17.4. Comparison between the two–dimensional WAF solution (symbol) and the one–dimensional radial RCM solution (line) at time=0.25

Fig. 17.4 shows a comparison between the one–dimensional radial solution (line) and the two–dimensional solution (symbols) along the radial line that is coincident with the x–axis. Comparisons along other radial directions give virtually identical results. This is confirmed by the very symmetric character of the numerical solution shown in Figs. 17.2 and 17.3. The shock wave is resolved with two mesh points and the contact surface is resolved with two to three mesh points. It is worth remarking that, at least for this test problem, the resolution of discontinuities that travel in *all directions* is essentially the same as that achieved in one–dimensional problems; see results of Chap. 14.

17.1.2 Implosion Test in Two Dimensions

The initial data of the type (17.3) may be reversed so as to create an implosion problem, with *shock focussing* taking place as part of the solution. This can also be used to test programs and methods.

17.1.3 Explosion Test in Three Space Dimensions

The three–dimensional Euler equations, see Chap. 3, are solved on a cube $2.0 \times 2.0 \times 2.0$ in x–y–z space. The initial conditions consist of the region inside of a sphere with radius $R = 0.4$ centred at $(1, 1, 1)$ and a region outside the sphere. The flow variables take constant values in each of these regions and are joined by a spherical discontinuity at time $t = 0$. The data values are the same as those for the two–dimensional case in (17.3), with the w velocity component initialised as $w_{ins} = w_{out} = 0$. This test problem is like a three–dimensional extension of the shock–tube problem.

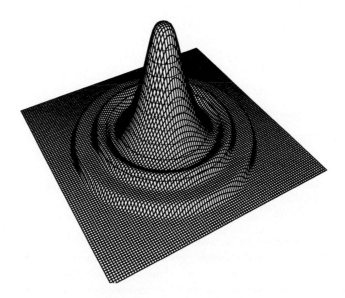

Fig. 17.5. Spherical explosion. Density distribution on plane $z = 0$ at time $t = 0.25$. Solution exhibits a spherical shock, a spherical contact and a spherical rarefaction

Fig. 17.5 shows the density distribution as a function of x and y on the plane $z = 0$ at the output time $t = 0.25$. The solution exhibits a spherical shock wave, a spherical contact surface travelling in the same direction and a spherical rarefaction travelling towards the origin $(1, 1, 1)$. Fig. 17.6 shows the corresponding pressure distribution; the pressure is continuous across the contact surface.

The one–dimensional equations (17.1)–(17.2) for the equivalent problem are solved on a domain $[0, 1]$ on the radial line r with equivalent shock–tube like data by the Random Choice Method, with $M = 1000$ computing cells and a CFL coefficient of $C_{cfl} = 0.4$.

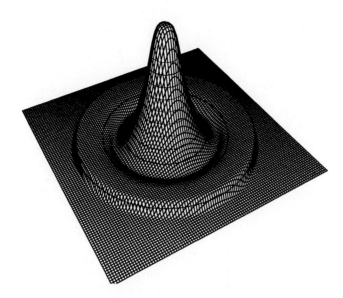

Fig. 17.6. Spherical explosion. Pressure distribution on plane $z = 0$ at time $t = 0.25$. Solution exhibits a spherical shock and a spherical rarefaction. Compare with Fig. 17.5

Fig. 17.7 shows a comparison between the one–dimensional radial solution (line) and the three–dimensional solution (symbols) along the radial line that is coincident with the x–axis. Agreement is good and, as for the two–dimensional equations, the resolution of discontinuities in three space dimensions is as good as in one–dimensional problems.

17.2 Shock Wave Reflection from Wedges

The subject of shock waves is large and embodies a substantial international community of scientists. *The International Symposium on Shock Waves* is a series of bi–annual meetings devoted exclusively to shock waves. A good source of information are the proceedings of these meetings. See for example Bershader and Hanson [35], Grönig [153], Takayama [339] and Brun and Dumitrescu [57].

Shock wave phenomena of considerable physical interest are those arising from the reflection of a shock wave from a wedge placed at an angle to the incident shock wave direction; see the sketch of Fig. 17.8. At the initial time a shock wave of shock Mach number M_S, perfectly perpendicular to the x–direction is placed at a position x_S. The shock travels in the x–direction and encounters a wedge that makes an angle ϕ with the shock direction. The

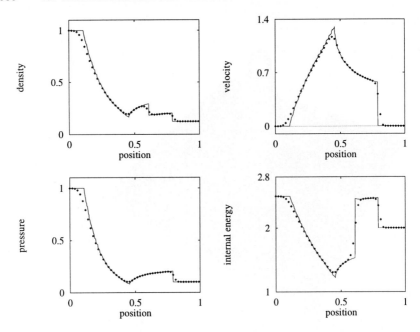

Fig. 17.7. Spherical explosion. Comparison between the three–dimensional WAF solution (symbol) and the one–dimensional radial RCM solution (line) at time=0.25

shock wave reflection pattern that emerges may fall into one of four possible categories. These are called: (a) regular reflection, (b) single Mach reflection, (c) complex Mach reflection and (d) double Mach reflection. The occurrence of a particular type of reflection depends on the shock Mach number M_S and the wedge angle ϕ. A comprehensive study of these wave patterns is found in the book by Ben–Dor [25]. Informative references, amongst many others, are those of Heilig [166], Heilig and Reichenbach [166], Dewey [108] and Takayama [338].

Two very useful reference are the UTIAS report by Glaz, Colella, Glass and Deschambault [139] and the paper by the same authors [140]. Many experimental and numerical results for this type of problems are presented and discussed in these works.

17.2.1 Mach Number $M_S = 1.7$ and $\phi = 25$ Degrees

Fig. 17.9 shows experimental and computational results for the case of an incident shock of shock Mach number $M_S = 1.7$ and a wedge angle of $\phi = 25$ degrees. Fig. 17.9 (a) shows the experimental result (Courtesy of Professor K. Takayama, Shock Wave Research Center, Sendai, Japan). Figs. 17.9 (b) and 17.9 (d) show computed results. These results were obtained on a rectangular domain of 25.0×16.5 on the x–y plane, with the apex of the wedge placed at

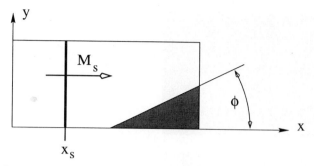

Fig. 17.8. Sketch of initial conditions for shock wave reflection from wedge

$x = 4.69$ and the initial shock placed at $x = 4.0$. The conditions ahead of the shock are ambient conditions, with $\rho_a = 1.225 \ kg/m^3$, $p_a = 1.01325 \times 10^5$ Pa, $u_a = 0$, $v_a = 0$. Given conditions ahead of the shock and the shock Mach number M_S, the state behind the shock is obtained from the *Rankine–Hugoniot Conditions*, see Sect. 3.1.3 of Chap. 3. *Start up errors* appear in the numerical solution, which may affect the fine details of the solution. A common practice [140], [169] is to run the problem until the *numerical shock* has been established and then reset the initial conditions connecting the states ahead and behind the shock, not through the Rankine–Hugoniot Conditions, but through the numerical shock. For this purpose one may simply use a one–dimensional code to find the numerical shock profile and give this as the initial conditions for the problem, or one may run the two–dimensional code and propagate the shock for some distance and then reset the state behind the shock. In the computations shown, we use the latter option and allow the shock to travel a distance of 90 % of that between its initial position and that of the apex of the wedge.

The numerical method used here is the WAF method with dimensional splitting, see Sect. 16.2 and 16.3 of Chap. 16. The Riemann solver used is HLLC of Sect. 10.4 of Chap. 10. The limiter function used is the VANLEER limiter; see Sects. 13.7 and 13.8 of Chap. 13, see also Chap. 14. The CFL coefficient is $C_{cfl} = 0.9$ and the maximum wave speed is estimated as in Sect. 16.3.2 of Chap. 16.

Fig. 17.9 (b) shows the numerical solution obtained with a regular curvilinear mesh of 800×528. Fig. 17.9 (d) shows the numerical solution obtained with a scheme that combines the CHIMERA approach [14], [325], [26], [68], [69], an Adaptive Mesh Refinement (AMR) approach [28], [31], [29], [274], [80], [275] and a Cartesian mesh approach. The base grid used is 50×33 with two further levels of refinement, each with a refinement factor of four. Fig. 17.9 (c) shows the structure of the mesh at the output time. Details of the CHIMERA–AMR–Cartesian approach used are found in Boden [46] and Boden and Toro [47].

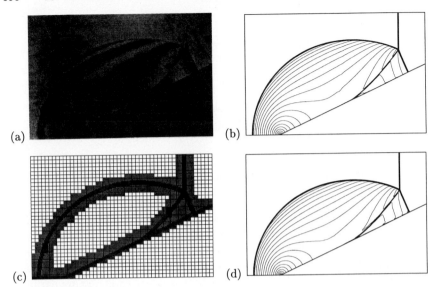

Fig. 17.9. Shock reflection problem for $M_S = 1.7$ and wedge angle of $\phi = 25$ degrees. (a) experimental result (b) density contours from fine regular mesh solution (c) AMR–CHIMERA grid structure at output time, (d) density contours from AMR–CHIMERA solution

The wave pattern of the solution corresponds to *single Mach reflection*. There are three shocks meeting at the triple point, namely, part of the incident shock, the reflected shock and the Mach stem, which joins the triple point with the wedge surface. From the triple point there emerges a slip surface that joins the wedge surface at a sharp angle. The qualitative agreement between the experimental result and the computations is excellent. The reader is warned however, that for large shock Mach numbers, when both real gas and dissipative effects are important, the agreement deteriorates, unless of course the Euler equations are replaced with the Navier–Stokes equations and the ideal gas equation of state is replaced with a suitable real–gas equation of state. See the paper by Glaz et. al. [139], [140] on these issues.

Generally, the overall wave structure of this problem can be computed well with almost any modern shock–capturing method. A delicate feature is the *slip surface*, across which discontinuities in density and velocity occur. Godunov–type upwind methods will generally be more accurate than other upwind methods, such as the Flux Vector Splitting Method of Chap. 8. The particular Riemann solver used in Godunov–type methods can, however, have a profound influence on the solution. For instance, the HLL Riemann solver of Chap. 10, the structure of which does not account for contact and shear waves, will smear the slip surface to unacceptable levels, as illustrated in the next section. Other Riemann solvers may be sensitive to extreme conditions.

17.2.2 Mach Number $M_S = 1.2$ and $\phi = 30$ Degrees

Computed and experimental results for the case of a shock Mach number $M_S = 1.2$ and wedge angle of $\phi = 30$ degrees are shown in Figs. 17.10 to 17.12. The experimental result was kindly given to us by Professor K. Takayama, Shock Wave Research Center, Sendai, Japan. This case also corresponds to *single Mach reflection* but the incident shock is weak. Weak shocks, and generally weak flow features, are difficult to compute accurately. The results of Fig. 17.10 were obtained with the WAF method in conjunction with the HLLC approximate Riemann solver of Sect. 10.4 of Chap. 10. All computational details are as for the previous section, but the mesh used is a regular curvilinear mesh of 420×240 cells. Overall the agreement between the computed and experimental results is satisfactory. The results of Figs. 17.11 and 17.12 were obtained (courtesy of Dr T. Saito, Cray Japan) from the Piecewise Linear Method (PLM) of Colella [87]; see Sect. 13.4.3 of Chap.13. Fig. 17.11 shows the results of the PLM with the HLLC Riemann solver; these look very similar to those of the WAF method, see Fig. 17.10. The results of Fig. 17.12 were obtained with the PLM and the HLL approximate Riemann solver of Sect. 10.3 of Chap. 10. This approximate Riemann solver assumes a two–wave structure for the solution of the Riemann problem and thus the contact surface is neglected. As a consequence the slip surface is smeared very significantly. Results obtained with the WAF method and HLL are similar to those of Fig. 17.12 and are thus omitted.

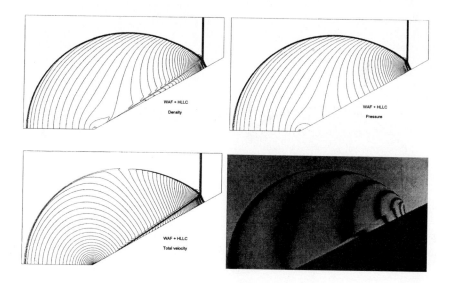

Fig. 17.10. Numerical and experimental results for shock Mach number $M_S = 1.2$ and wedge andgle $\phi = 30$ degrees. WAF method with HLLC Riemann solver

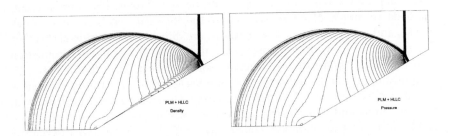

Fig. 17.11. Numerical and experimental results for shock Mach number $M_S = 1.2$ and wedge andgle $\phi = 30$ degrees. PLM with HLLC Riemann solver

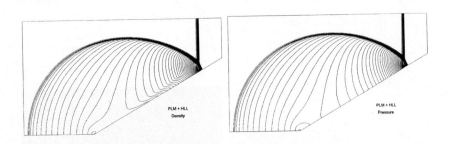

Fig. 17.12. Numerical and experimental results for shock Mach number $M_S = 1.2$ and wedge andgle $\phi = 30$ degrees. PLM with HLL Riemann solver

18. Concluding Remarks

18.1 Summary

The objective of this book is to present, in a more or less coherent manner, a class of shock–capturing numerical methods for solving hyperbolic conservation laws. The methodology has been presented via simple model equations. For non–linear systems we have chosen the time–dependent compressible Euler equations as the representative system of hyperbolic conservations laws. Our original intention was to include the application of the methods to three other systems of practical interest, namely the steady supersonic Euler equations, the shallow water equations and the artificial compressibility equations associated with the incompressible Navier–Stokes equations, but we ran out of space.

Several techniques have been presented in complete detail. These, fall broadly into two categories (a) Upwind TVD methods and (b) Centred TVD methods. Within the class of upwind methods we have Godunov–type methods with the Riemann solvers of Chaps. 4, 9, 10, 11 and 12, and Flux Vector Splitting Methods of Chap. 8 with three different splittings. The Random Choice Method of Chap. 7 is somewhat special, in that it uses the exact solution of the Riemann problem, see Chap. 4, and gives infinite resolution of discontinuities but does not extend to general systems with more than two independent variables. Godunov–type methods are more sophisticated than Flux Vector Splitting Methods. For shock waves, both classes of upwind schemes give comparable performance, with Godunov methods having a slight advantage. Contact surfaces and shear waves are the class of problems for which the difference between Godunov–type methods and FVS methods is significant; this is most evident for the case of slowly–moving contact discontinuities and has a bearing on the accuracy of computations for more complex systems, such as the Navier–Stokes equations, in resolving shear layers for instance. FVS methods are simpler and more efficient than Godunov–type methods, but care is required here. Explicit FVS schemes can have a more restrictive stability condition than that of Godunov–type methods. Centred TVD methods are the simplest of all methods presented.

We have studied two centred TVD methods. For shock waves these give results of comparable quality to those obtained from the upwind methods used with the most diffusive of limiter functions. For contact and shear waves these centred methods are similar to FVS methods and to Godunov–type methods used with two–wave Riemann solvers, such as HLL. The centred methods are considerably simpler than any of the upwind methods and are well suited for use by scientists and engineers who do not want to get involved with sophisticated numerical methods for their applications. Of the two centred TVD methods given we recommend SLIC for practical applications. For serious applications we strongly recommend the use of Godunov–type methods with appropriate Riemann solvers.

There are important topics at the level of numerical methods that have not been included. One example is the Piecewise Parabolic Method (PPM) of Colella and Woodward [91], [413]. This is an excellent method of the Godunov type and can be easily implemented following the basic material presented here. We have not included high–order methods such as the ENO/UNO schemes. See the works of Harten and Osher [165]; Harten, Engquist, Osher and Chakravarthy [162]; Harten [160], [161]; Shu and Osher [312], [309]; Casper and Atkins [60]; Liu, Osher and Chan [222] and Jiang and Shu [184]. We have not included implicit methods. Useful references here are Harten [159]; Yee [424] and specially the VKI Lectures [422]. We have not attempted to include current research areas such as multi–dimensional upwinding. The reader will find the following references useful: Roe [285]; Deconinck, Powell, Roe and Struijs [105]; van Leer [396]; Deconinck, Roe and Struijs [106]; García–Navarro, Hubbard and Priestley [129] and Baines and Hubbard [15].

18.2 Extensions and Applications

The methods of this book can be applied to a large variety of practical problems. The recent paper by Roe [289] contains some examples of very ambitious applications of upwind methods. In what follows we mention some specific areas and give a partial list of references that the reader may find useful.

18.2.1 Shallow Water Equations

The shock–capturing methods presented in this book are directly applicable to the two–dimensional time dependent shallow water equations. The derivation of these equations is found in Chap. 1. General background reading is found in the book by Stoker [328]. Useful references are Marshall and Méndez [230]; Glaister [136]; Toro [357]; Watson, Peregrine and Toro [406]; Alcrudo and García–Navarro [3]; Bermúdez and Vázquez [34]; García–Navarro, Hubbard and Priestley [129]; Bermúdez et. al. [33]; Fraccarollo and Toro [127]; and the forthcoming textbook by Toro [370].

18.2.2 Steady Supersonic Euler Equations

Under the assumption of steady and supersonic flow, the Euler equations derived in Chap. 1 produce a non–linear system that is hyperbolic in the flow direction. The Riemann problem can be solved exactly and approximately and all the methods developed in this book can be applied in a *space marching* approach. Useful references are Marshall and Plohr [231]; Roe [281]; Pandolfi [257]; Dawes [104]; Toro and Chou [377] and Toro and Chakraborty [376].

18.2.3 The Incompressible Navier–Stokes Equations

The steady incompressible Navier–Stokes equations may be formulated via the artificial compressibility approach of Chorin [72] to produce the Artificial Compressibility Equations. Details of the derivation of these equations are given in Chap. 1. The inviscid part of the equations form a hyperbolic system and one can therefore apply all the methods presented in this book. The approach can also be extended to solve the time–dependent incompressible Navier–Stokes equations. Useful references are the following: Marx [236], [237]; Drikakis [109]; Drikakis [110]; Weinan and Shu [410]; Turkel and Arnone [384]; Hänel and Sharma [157]; Tamamidis, Zhang, Assanis [342]; Toro [366] and Drikakis *et. al.* [111].

18.2.4 The Compressible Navier–Stokes Equations

The compressible Navier–Stokes equation, see Chap. 1, are the obvious next extension of the methods presented in this book. Generally, viscous and heat conduction terms are discretised using central difference approximations, and in principle, any of the methods studied here may be applied to solve the Navier–Stokes equations. A very useful reference here are the VKI Lectures by Hänel [156]. See also, for example, the papers by van Leer, Thomas and Roe [397]; Brown [56]; Toro and Brown [375]; Drikakis and Tsangaris [112] and Allmaras [4].

A word of caution is in order, when applying TVD methods to solve the Navier–Stokes equations. TVD methods are effectively a smart way of introducing numerical viscosity to control spurious oscillations near large gradients. As pointed out in Chap. 13, Sect. 13.9.2, in the presence of physical viscosity, the standard TVD condition is strictly speaking incorrect, in that it does not acknowledge the existence of physical viscosity. The net result is excessive artificial viscosity. For most problems, satisfactory results are reported in the literature, but for special problems concerned with fine details of the flow physics, one would expect the limitations of standard TVD methods for viscous problems to become apparent. Standard TVD methods are techniques that have been constructed for inviscid problems. For viscous problems, new TVD methods are needed. See [356] for possible extensions of TVD methods for viscous problems.

18.2.5 Compressible Materials

The Euler equations can be used to model wave propagation in general compressible media, gases being the most obvious example. For problems under extreme thermodynamic conditions, real gas effects become important and use of real gas equations of state must be considered. For Godunov–type schemes this requires the solution of the Riemann problem for a general equation of state. The media may be some material other than a gas, a liquid for instance, which at high pressure is compressible. See Ivings, Causon and Toro [180] for details on Riemann solvers for compressible liquids. The review paper by Menikoff and Plohr [240] is highly recommended. See also the papers by Colella and Glaz [89]; Glaister [137], [138] and Saurel, Larini and Loraud [298].

18.2.6 Detonation Waves

Detonation waves in gases and solids is an extensive area of application of the numerical methods of this book. General references on detonation waves are, amongst many others, Oppenheim and Soloukhin [249]; Fickett and Davis [124]; Williams [412]; Clarke [79], [78]; Edwards, Thomas and Williams [116]; Gelfand, Frolov and Nettleton [132] and Gilbert [133]. Useful references on numerical computation of detonation waves are Taki and Fujiwara [340]; Oran and Boris [250], [251]; Kailasanath, Oran, Boris and Young [189]; Colella, Majda and Roytburd [90]; Bourlioux, Majda and Roytburd [53]; Bourlioux and Majda [52]; Singh [313]; Singh and Clarke [314]; Clarke et. al. [80] and Nikiforakis and Clarke [247].

18.2.7 Multiphase Flows

Continuum models for multiphase reactive flows involving a compressible gas phase (combustion products) and granular reactive solid particles consist of partial differential equations of similar form to the basic equations of gas dynamics. There is set of PDEs for each phase, which are coupled through the variable porosity and the source terms. Each phase has its own velocity vector and note that this is not like in multicomponent flow, where a single velocity vector is needed, for the carrier fluid. The most popular of these multiphase flow models are of mixed hyperbolic–elliptic type, which leads to mathematical and therefore numerical difficulties. See Stewart and Wendroff for a review on the subject of mathematical models and numerics for multiphase flows [327]. Attempts have been made to extend Godunov–type methods to solve problems of this kind. See Toro [354]; Toro et. al. [371]; Sainsaulieu [292]; Saurel, Daniel and Loraud [296]; Saurel, Forestier, Veyret and Loraud [297]; Toumi and Kumbaro [383]; Toumi [381]; Toumi and Caruge [382]; Cortes, Debussche and Toumi [96].

18.2.8 Magnetohydrodynamics (MHD)

A very complex system of hyperbolic conservation laws are the equations of Magneto Hydrodynamics or MHD equations. Upwind methods have been applied to these equations. See the papers by Brio and Wu [54]; van Puten [398]; Dai and Woodward [100], [101]; Tanaka [343]; Barmin, Kulikovskiy and Pogorelov [17] and Falle and Komissarov [122]. See also the recent related paper by Wegmann [409] and references therein.

18.2.9 Relativistic Gas Dynamics

In several branches of science, such as Astrophysics, Cosmology and Nuclear Physics, the equations of Relativistic Gas Dynamics offer a suitable mathematical model. The exact solution of the Riemann problem for Relativistic Gas Dynamics was first reported by Martí and Müller [233]. They extended the shock–tube problem solution of Thompson [348] and highlighted the importance of their work in the context of numerical relativistic hydrodynamics via Godunov–type methods. See also the papers by Schneider, Katscher, Rischke, Waldhauser, Maruhn and Munz [300]; Plohr and Sharp [264]; Marquina, Martí, Ibañez, Miralles and Donat [229]; Martí et. al. [232], [233], [234] and Falle and Komissarov [121]. The reader interested in the simulation of astrophysical flows should consult the recent book by LeVeque, Mihalas, Dorfi and Müller [210].

18.2.10 Waves in Solids

The techniques presented in this textbook may also be applied to model wave propagation in solids. Useful references are, amongst many others, Kim and Ballmann [194]; Lin and Ballmann [215], [218], [216], [217].

18.3 NUMERICA

Most numerical methods presented in this book have been illustrated via practical numerical examples. Use has been made of codes from *NUMERICA* [369], which is a library of source codes for teaching, research and applications. The library is based on the contents of the present work and the forthcoming textbook [370]. The source codes of the library can be useful as *learning and teaching* tools. First, they can be used in a *self–study situation* as another way of reading the details of a numerical method and second, they can be used as a method of demonstrating, in a *teaching situation*, the main steps of a numerical technique and a possible way of coding the method. The programs can also be utilised as *development programs*. First, they can most easily be converted into practical tools for *real* applications. They can also form the bases for much more ambitious applications, such as the ones listed in Sect. 18.2.

For details on *NUMERICA* the reader should contact the author directly: Professor E. F. Toro, Department of Computing and Mathematics, Manchester Metropolitan University, Chester Street, Manchester, M1 5GD, United

Kingdom, email: E.F.Toro@doc.mmu.ac.uk. Alternatively, the reader may obtain information from the publishers, email: inquiries@numeritek.com, website: www.numeritek.com

References

1. R. Abgrall. How to Prevent Pressure Oscillations in Multicomponent Flow Calculations: A Quasiconservative Approach. *J. Comput. Phys.*, 125:150–160, 1996.
2. C. M. Albone. Report on the AGCFM Working Party on Software Quality Assurance. Technical Report AERO 2105, Royal Aircraft Establishment, Farnborough, UK, 1987.
3. A. Alcrudo and P. García-Navarro. A High Resolution Godunov–Type Scheme in Finite Volumes for the 2D Shallow Water Equations. *Int. J. Num. Meth. Fluids*, 16:489–505, 1993.
4. S. R. Allmaras. Contamination of Laminar Boundary Layers by Artificial Dissipation in Navier–Stokes Solutions. In *Numerical Methods for Fluid Dynamics, 4*, pages 443–449. Oxford University Press, 1993.
5. D. A. Anderson, J. C. Tannehill, and R. H. Pletcher. *Computational Fluid Mechanics and Heat Transfer*. Hemisphere Publishing Corporation, 1984.
6. D. S. Anderson and J. J. Gottlieb. On Random Numbers for the Random Choice Method. Technical Report 258, UTIAS, University of Toronto, Canada, 1987.
7. J. D. Anderson. *Hypersonic and High Temperature Gas Dynamics*. Mc Graw–Hill, 1989.
8. J. D. Anderson. *Modern Compressible Flow*. Mc Graw-Hill, 1990.
9. J. L. Anderson, S. Preiser, and E. L. Rubin. Conservation Form of the Equations of Hydrodynamics in Curvilinear Coordinate Systems. *J. Comput. Phys.*, 2:279–287, 1968.
10. W. Anderson, J. L. Thomas, and B. van Leer. Comparison of Finite Volume Flux Vector Splittings for the Euler Equations. *AIAA Journal*, 24:1453–1460, 1986.
11. W. K. Anderson, J. L. Thomas, and L. Rumsey. Extension and Application of Flux–Vector Splitting to Calculations on Dynamic Meshes. *AIAA Journal*, 27:673–674, 1989.
12. T. M. Apostol. *Mathematical Analysis*. Addison–Wesley Publishing Company, 1974.
13. M. Arora and P. L. Roe. On Postshock Oscillations Due to Shock Capturing Schemes in Unsteady Flows. *J. Comput. Phys.*, 130:25–40, 1997.
14. E. H. Atta and J. Vadyak. A Grid Overlapping Scheme for Flowfield Computations About Multicomponent Configurations. *AIAA Journal*, 21(9):1271–1277, 1983.
15. M. J. Baines and M. E. Hubbard. Multidimensional Upwinding with Grid Adaptation. In *Numerical Methods for Wave Propagation. Toro, E. F. and Clarke, J. F. (Editors)*, pages 33–54. Kluwer Academic Publishers, 1998.

16. M. J. Baines and P. K. Sweby. A Flux Limiter Based on Fromm's Scheme. Technical Report 9/84, Department of Mathematics, University of Reading, UK, 1984.

17. A. A. Barmin, A. G. Kulikovskiy, and N. V. Pogorelov. Shock–Capturing Approach and Nonevolutionary Solutions in Magnetohydrodynamics. *J. Comput. Phys.*, 126:77–90, 1996.

18. P. Batten, N. Clarke, C. Lambert, and D. Causon. On the Choice of Wave Speeds for the HLLC Riemann Solver. *SIAM J. Sci. and Stat. Comp.*, 18:1553–1570, 1997.

19. P. Batten, M. A. Leschziner, and U. C. Goldberg. Average–state jacobians and implicit methods for compressible viscous and turbulent flows flows. *J. Comput. Phys.*, 137:38–78, 1997.

20. A. Bayliss and E. Turkel. Radiation–Boundary Conditions for Wave–Like Equations. *Comm. Pure and Appl. Maths.*, 33:708–725, 1980.

21. E. Becker. *Gas Dynamics*. Academic Press, 1968.

22. M. Ben-Artzi. Application of the Generalised Riemann Problem Method to 1–D Compressible Flows with Interfaces. *J. Comput. Phys.*, 65:170–178, 1986.

23. M. Ben-Artzi and J. Falcovitz. A Second Order Godunov-Type Scheme for Compressible Fluid Dynamics. *J. Comput. Phys.*, 55:1–32, 1984.

24. M. Ben-Artzi and J. Falcovitz. A High Resolution Upwind Scheme for Quasi 1–D Flows. In *Numerical Methods for the Euler Equations of Fluid Dynamics*, pages 66–83. INRIA, SIAM, 1985.

25. G. Ben-Dor. *Shock Wave Reflection Phenomena*. Springer-Verlag, 1992.

26. J. A. Benek, J. L. Steger, and F. C. Dougherty. A Flexible Grid Embedding Technique with Application to the Euler Equations. AIAA Paper 83–1944, 1983.

27. F. Benkhaldoun and A. Chalabi. Characteristic Based Scheme for Conservation Laws with Source Terms. Technical Report 95.15, Centre National de la Recherche Scientifique, Université Paul Sabatier, UFR-MIG, 118 route de Narbone, 31062, Toulouse Cedex, France, 1995.

28. M. J. Berger. *Adaptive Mesh Refinement for Hyperbolic Partial Differential Equations*. PhD thesis, Dept. of Computer Science, Stanford University, California, 1982.

29. M. J. Berger and P. Colella. Local Adaptive Mesh Refinement for Shock Hydrodynamics. *J. Comput. Phys.*, 82:64–84, 1989.

30. M. J. Berger and R. J. LeVeque. Stable Boundary Conditions for Cartesian Grid Calculations. Technical Report ICASE 90–37, NASA Langley Research Center, USA, 1990.

31. M. J. Berger and J. Oliger. Adaptive Mesh Refinement for Hyperbolic Partial Differential Equations. *J. Comput. Phys.*, 53:484–512, 1984.

32. A. C. Berkenbosch, E. F. Kaasschieter, and J. H. M. ten Thije Boonkkamp. The Numerical Computation of One–dimensional Detonation Waves. Technical Report RANA 94–21, Eindhoven University of Technology, Department of Mathematics and Computer Science, The Netherlands, 1994.

33. L. Bermúdez, A. Dervieux, J. A. Desideri, and M. E. Vázquez. Upwind Schemes for the Two–Dimensional Shallow Water Equations with Variable Depth Using Unstructured Meshes. Technical Report 2738, INRIA, Rocquencourt, 78153 Le Chesnay, France, 1995.

34. L. Bermúdez and M. E. Vázquez. Upwind Methods for Hyperbolic Conservation Laws with Source Terms. *Computers and Fluids*, 23:1049–1071, 1994.

35. H. Bershader and R. Hanson (Editors). *Shock Waves and Shock Tubes (Proc. 15th Int. Symp. on Shock Waves and Shock Tubes, Berkeley, California, USA 1985)*. Stanford University Press, 1986.

36. M. Berzins, P. M. Dew, and R. M. Fuzerland. Developing Software for Time–Dependent Problems Using the Method of Lines and Differential Algebraic Integrators. *Applied Numerical Mathematics*, 5:375–397, 1989.

37. M. Berzins and J. M. Ware. Solving Convection and Convection Reaction Problems using M.O.L. *Applied Numerical Mathematics*, 20:83–89, 1996.

38. S. J. Billett. Numerical Aspects of Convection Diffusion Problems. MSc. Thesis, College of Aeronautics, Cranfield Institute of Technology, UK, 1991.

39. S. J. Billett. A Class of Upwind Methods for Conservation Laws. Ph.D thesis, College of Aeronautics, Cranfield University, UK, 1994.

40. S. J. Billett and E. F. Toro. Restoring Monotonicity of Slowly–Moving Shocks Computed with Godunov–Type Schemes. *CoA Report 9207, Cranfield Institute of Technology, UK*, 1992.

41. S. J. Billett and E. F. Toro. Implementing a Three–Dimensional Finite Volume WAF–Type Scheme for the Euler Equations. In *Proc. Third ECCOMAS Computational Fluid Dynamics Conference, 9–13 September 1996, Paris, France*, pages 732–738. Wiley, 1996.

42. S. J. Billett and E. F. Toro. On the Accuracy and Stability of Explicit Schemes for Multidimensional Linear Homogeneous Advection Equations. *J. Comp. Phys.*, 131:247–250, 1997.

43. S. J. Billett and E. F. Toro. WAF–Type Schemes for Multidimensional Hyperbolic Conservation Laws. *J. Comp. Phys.*, 130:1–24, 1997.

44. S. J. Billett and E. F. Toro. Unsplit WAF–Type Schemes for Three Dimensional Hyperbolic Conservation Laws. In *Numerical Methods for Wave Propagation. Toro, E. F. and Clarke, J. F. (Editors)*, pages 75–124. Kluwer Academic Publishers, 1998.

45. S. J. Billett and E. F. Toro. TVB Limiter Functions for Multidimensional Schemes. In *Proc. 6th Intern. Symposium on Computational Fluid Dynamics, Lake Tahoe, Nevada, USA*, pages 117–122, September 4-8, 1995.

46. E. P. Boden. *An Adaptive Gridding Technique for Conservation Laws on Complex Domains* . PhD thesis, Department of Aerospace Science, Cranfield University, UK, 1997.

47. E. P. Boden and E. F. Toro. A Combined Chimera–AMR Technique for Hyperbolic PDEs. In *Proc. Fifth Annual Conference of the CFD Society of Canada*, pages 5.13–5.18. University of Victoria, Canada, 1997.

48. D. L. Book, J. P. Boris, and K. Hain. Flux Corrected Transport II: Generalizations of the Method. *J. Comput. Phys.*, 18:248–283, 1975.

49. J. P. Boris and D. L. Book. Flux–Corrected Transport. I. SHASTA, A Fluid Transport Algorithm that Works. *J. Comput. Phys.*, 11:38–69, 1973.

50. J. P. Boris and D. L. Book. Flux–Corrected Transport III: Minimal–Error FCT Algorithms. *J. Comput. Phys.*, 20:397–431, 1976.

51. J. P. Boris and E. S. Oran. *Numerical Simulation of Reactive Flow*. Elsevier, Amsterdam–New York, 1987.

52. A. Bourlioux and A. J. Majda. Theoretical and Numerical Structure for Unstable Two–Dimensional Detonations. *Combustion and Flame*, 90:211–229, 1992.

53. A. Bourlioux, A. J. Majda, and V. Roytburd. Theoretical and Numerical Structure for Unstable One–Dimensional Detonations. *SIAM J. Appl. Math.*, 51:303–343, 1991.

54. M. Brio and C. C. Wu. An Upwind Differencing Scheme for the Equations of Ideal Magnetohydrodynamics. *J. Comput. Phys.*, 75:400–422, 1988.

55. M. Brown. *Differential Equations and their Applications*. Springer–Verlag, 1975.

56. R. E. Brown. Numerical Solution of the 2D Unsteady Navier–Stokes Equations Using Viscous–Convective Operator Splitting. MSc. Thesis, Department of Aerospace Science, Cranfield University, UK, 1990.

57. R. Brun and L. Z. Dumitrescu (Editors). *Shock Waves @ Marseille (Proc. 19th Int. Symp. on Shock Waves, Marseille, France, 1993), Four Volumes.* Springer–Verlag, 1995.

58. M. Cáceres. Development of a Third–Order Accurate Scheme of the MUSCL Type for the Time–Dependent One–Dimensional Euler Equations. MSc. Thesis, Department of Aerospace Science, Cranfield University, UK, 1993.

59. Carasso, Raviart, and Serre. *Proc. First International Conference on Hyperbolic Problems.* Springer, 1986.

60. J. Casper and H. L. Atkins. A Finite Volume High–Order ENO Scheme for Two-Dimensional Hyperbolic Systems. *J. Comput. Phys.*, 106:62–76, 1993.

61. V. Casulli and E. F Toro. Preliminary Results on a Semi-Lagrangian Version of the WAF Method. Technical Report 9018, College of Aeronautics, Cranfield Institute of Technology, UK, 1990.

62. S. R. Chakravarthy, D. A. Anderson, and M. D. Salas. The Split–Coefficient Matrix Method for Hyperbolic Systems of Gas Dynamics Equations. *AIAA Paper 80–0268, AIAA 19th Aerospace Science Meeting*, 1980.

63. A. Chalabi. Stable Upwind Schemes for Conservation Laws with Source Term. *IMA J. Numer. Anal.*, 12:217–241, 1992.

64. A. Chalabi. On Convergence of Numerical Schemes for Hyperbolic Conservation Laws with Stiff Source Terms. Technical Report 95.22, Centre National de la Recherche Scientifique, Université Paul Sabatier, UFR-MIG, 118 route de Narbone, 31062, Toulouse Cedex, France, 1995.

65. A. Chalabi. An Error Bound for the Polygonal Approximation of Hyperbolic Conservation Laws with Source Terms. *Computers and Mathematics with Applications*, 32:59–63, 1996.

66. A. Chalabi. On Convergence of Numerical Schemes for Hyperbolic Conservation Laws with Stiff Source Terms. *Math. of Comput.*, 66:527–545, 1997.

67. D. N. Cheney. Upwinding Convective and Viscous Terms via a Modified GRP Approach. MSc. Thesis, Department of Aerospace Science, Cranfield University, UK, 1994.

68. G. Chesshire and W. D. Henshaw. Composite Overlapping Meshes for the Solution of Partial Differential Equations. *J. Comput. Phys.*, 90:1–64, 1990.

69. G. Chesshire and W. D. Henshaw. A Scheme for Conservative Interpolation on Overlapping Grids. *SIAM J. Sci. Stat.*, 15(4):819–845, 1994.

70. B. L. Chessor, Jr. Evaluation of Computed Steady and Unsteady Quasi–One–Dimensional Viscous/Inviscid Interacting Internal Flow Fields Through Comparison with Two–Dimensional Navier–Stokes Solutions. MSc. Thesis, 1992, Mississippi State University, USA, 1992.

71. T. Chiang and Hoffmann K. Determination of computational time step for chemically reacting flows. AIAA 20th Fluid Dynamics, Plasma Dynamics and Lasers Conference, Buffalo, New York, June 12–14, USA, 1989, Paper AIAA 89–1855, 1989.

72. A. J. Chorin. A Numerical Method for Solving Viscous Flow Problems. *J. Comput. Phys.*, 2:12–26, 1967.

73. A. J. Chorin. Random Choice Solutions of Hyperbolic Systems. *J. Comput. Phys.*, 22:517–533, 1976.

74. A. J. Chorin. Random Choice Methods with Applications to Reacting Gas Flow. *J. Comput. Phys.*, 25:253–272, 1977.

75. A. J. Chorin and J. E. Marsden. *A Mathematical Introduction to Fluid Mechanics.* Springer–Verlag, 1993.

76. D. K. Clarke, M. D. Salas, and H. A. Hassan. Euler Calculations for Multiele-
    ment Airfoils Using Cartesian Grids. *AIAA J.*, 24:353–, 1987.

77. J. F. Clarke. Compressible Flow Produced by Distributed Sources of Mass: An
    Exact Solution. Technical Report CoA Report 8710, College of Aeronautics,
    Cranfield Institute of Technology, Cranfield, UK, 1987.

78. J. F. Clarke. Chemical Reactions in High Speed Flows. *Phil. Trans. Roy. Soc.
    London*, A 335:161–199, 1989.

79. J. F. Clarke. Fast Flames, Waves and Detonations. *Prog. Energy Comb. Sci.*,
    15:241–271, 1989.

80. J. F. Clarke, S. Karni, J. J. Quirk, L. G. Simmons, P. L. Roe, and E. F. Toro.
    Numerical Computation of Two–Dimensional, Unsteady Detonation Waves in
    High Energy Solids. *J. Comput. Phys.*, 106:215–233, 1993.

81. J. F. Clarke and M. McChesney. *Dynamics of Relaxing Gases*. Butterworths,
    1976.

82. J. F. Clarke and E. F. Toro. Gas Flows Generated by Solid–Propellant Burning.
    In R. Glowinski, B. Larrouturou, and R. Teman, editors, *Proc. Intern. Confer.
    on Numerical Simulation of Combustion Phenomena*, number 241 in Lecture
    Notes in Physics, pages 192–205, INRIA, Sophia Antipolis, France, May 21–24
    1985.

83. B. Cockburn and C. Shu. Nonlinearly Stable Compact Schemes for Shock
    Calculations. *SIAM J. Numer. Anal.*, 31(3):607–627, 1994.

84. E. Coddington and N. Levinson. *Theory of Ordinary Differential Equations*.
    McGraw–Hill, New York, 1955.

85. P. Colella. *An Analysis of the Effect of Operator Splitting and of the Sampling
    Procedure on the Accuracy of Glimm's Method*. PhD thesis, Department of
    Mathematics, University of California, USA, 1978.

86. P. Colella. Glimm's Method for Gas Dynamics. *SIAM J. Sci. Stat. Comput.*,
    3(1):76–110, 1982.

87. P. Colella. A Direct Eulerian MUSCL Scheme for Gas Dynamics. *SIAM J. Sci.
    Stat. Comput.*, 6:104–117, 1985.

88. P. Colella. Multidimensional Upwind Methods for Hyperbolic Conservation
    Laws. *J. Comput. Phys.*, 87:171–200, 1990.

89. P. Colella and H. H. Glaz. Efficient Solution Algorithms for the Riemann
    Problem for Real Gases. *J. Comput. Phys.*, 59:264–289, 1985.

90. P. Colella, A. Majda, and V. Roytburd. Theoretical and Numerical Structure
    for Reacting Shock Waves. *SIAM J. Sci. Stat. Comput.*, 7:1059–1080, 1986.

91. P. Colella and P. R. Woodward. The Piecewise Parabolic Method (PPM) for
    Gas Dynamical Simulation. *J. Comput. Phys.*, 54:174–201, 1984.

92. P. Concus and W. Proskurowski. Numerical Solution of a Nonlinear Hyperbolic
    Equation by the Random Choice Method. *J. Comput. Phys.*, 30:153–166, 1979.

93. S. D. Conte and C. de Boor. *Elementary Numerical Analyis*. McGraw–Hill
    Kogakusha Ltd., 1980.

94. J. M. Corberán and M. Ll. Gascón. TVD Schemes for the Calculation of Flow
    in Pipes of Variable Cross–Section. *Mathl. Comput. Modelling*, 21:85–92, 1995.

95. J. Corner. *Theory of the Interior Ballistics of Guns*. Wiley, 1950.

96. J. Cortes, A. Debussche, and I. Toumi. A Density Perturbation Method to
    Study the Eigenstructure of Two–Phase Flow Equation Systems. *J. Comput.
    Phys.*, 147:1–22, 1998.

97. R. Courant and K. O. Friedrichs. *Supersonic Flow and Shock Waves*. Springer-
    Verlag, 1985.

98. R. Courant, E. Isaacson, and M. Rees. On the Solution of Nonlinear Hyperbolic
    Differential Equations by Finite Differences. *Comm. Pure. Appl. Math.*, 5:243–
    255, 1952.

99. M. G. Crandall and A. Majda. Monotone Difference Approximations for Scalar Conservation Laws. *Mathematics of Computations*, 34:1–34, 1980.

100. W. Dai and P. R. Woodward. An Approximate Riemann Solver for Ideal Magnetohydrodynamics. *J. Comput. Phys.*, 111:354–372, 1994.

101. W. Dai and P. R. Woodward. Extension of the Piecewise Parabolic Method to Multidimensional Ideal Magnetohydrodynamics. *J. Comput. Phys.*, 115:485–, 1994.

102. S. F. Davis. TVD Finite Difference Schemes and Artificial Viscosity. Technical Report ICASE 84–20, NASA Langley Research Center, Hampton, USA, 1984.

103. S. F. Davis. Simplified Second–Order Godunov–Type Methods. *SIAM J. Sci. Stat. Comput.*, 9:445–473, 1988.

104. A. S. Dawes. *Natural Coordinates and High Speeed Flows. A Numerical Method for Reactive Gases*. PhD thesis, College of Aeronautics, Cranfield Institute of Technology, UK, 1992.

105. H. Deconinck, K. G. Powell, P. L. Roe, and R. Struijs. Multi-dimensional Schemes for Scalar Advection. *AIAA*, 91(1532), 1991.

106. H. Deconinck, P. L. Roe, and R. Struijs. Multi–dimensional Generalisation of Roe's Flux Difference Splitter for the Euler Equations. *Computers and Fluids*, 22:215–222, 1993.

107. T. DeNeff and G. Moretti. Shock Fitting for Everyone. *Computers and Fluids*, 8:327–334, 1980.

108. J. M. Dewey. Mach Reflection Research–Paradox and Progress. In *Shock Waves (Proc. 18th Int. Symp. on Shock Waves, Sendai, Japan, 1991. Takayama, Editor), Vol. I*, pages 113–120. Springer–Verlag, 1992.

109. D. Drikakis. A Characteristic–Based Method for Incompressible Flow. *Int. J. Numer. Meth. Fluids*, 19:667–685, 1994.

110. D. Drikakis. A Parallel Multiblock Characteristic–Based Method for Three–Dimensional Incompressible Flows. *Advances in Engineering Software*, 26:111–119, 1996.

111. D. Drikakis, O. P. Iliev, and D. P. Vassileva. A Nonlinear Multigrid Method for the Three–Dimensional Incompressible Navier–Stokes Equations. *J. Comput. Phys.*, 146:301–321, 1998.

112. D. Drikakis and S. Tsangaris. On the Solution of the Compressible Navier–Stokes Equations Using Improved Flux Vector Splitting Methods. *Appl. Math. Modelling*, 17:282–297, 1993.

113. F. Dubois and G. Mehlman. A Non–Parameterized Entropy Fix for Roe's Method. *AIAA Journal*, 31 (1):199–200, 1993.

114. J. K. Dukowicz. A General, Non–Iterative Riemann Solver for Godunov's Method. *J. Comput. Phys.*, 61:119–137, 1985.

115. P. Dutt. A Riemann Solver Based on a Global Existence Proof for the Riemann Problem. Technical Report ICASE 86–3, NASA Langley Research Center, USA, 1986.

116. D. H. Edwards, G. O. Thomas, and T. L. Williams. Initiation of Detonation by Steady Planar Incident Shock Waves. *Combustion and Flame*, 43:187–198, 1981.

117. B. Einfeldt. On Godunov–Type Methods for Gas Dynamics. *SIAM J. Numer. Anal.*, 25(2):294–318, 1988.

118. B. Einfeldt, C. D. Munz, P. L. Roe, and B. Sjögreen. On Godunov–Type Methods near Low Densities. *J. Comput. Phys.*, 92:273–295, 1991.

119. B. Engquist and B. Gustafsson. *Proc. Third International Conference on Hyperbolic Problems, Vol. I and II*. Chartwell–Bratt, 1991.

120. B. Engquist and S. Osher. One Sided Difference Approximations for Nonlinear Conservation Laws. *Math. Comp.*, 36(154):321–351, 1981.

121. S. A. E. G. Falle and S. S. Komissarov. An Upwind Numerical Scheme for Relativistic Hydrodynamics with a General Equation of State. *Monthly Notices of the Royal Astronomical Society*, 278:586–602, 1996.

122. S. A. E. G. Falle and S. S. Komissarov. A Multidimensional Upwind Scheme for Magnetohydrodynamics. *Monthly Notices of the Royal Astronomical Society*, 297:265–277, 1997.

123. M. Fey, R. Jeltsch, and S Müller. The Influence of a Source Term, an Example: Chemically Reacting Hypersonic Flow. Technical Report 92–06, SAM, Eidgenössische Technische Hochschule (ETH), Zürich, Switzerland, 1992.

124. W. Fickett and W. Davis. *Detonation*. University of California Press, 1979.

125. C. A. J. Fletcher. *Computational Techniques for Fluid Dynamics, Vols. I and II*. Springer–Verlag, 1988.

126. L. Fraccarollo and E. F. Toro. A Shock-Capturing Method for Two Dimensional Dam–Break Problems. *Proceedings of the Fifth International Symposium in Computational Fluid Dynamics, Sendai, Japan*, 1993.

127. L. Fraccarollo and E. F. Toro. Experimental and Numerical Assessment of the Shallow Water Model for Two–Dimensional Dam–Break Type problems. *Journal of Hydraulic Research*, 33:843–864, 1995.

128. J. E. Fromm. A Method for Reducing Dispersion in Convective Difference Schemes. *J. Comput. Phys.*, 3:176–189, 1968.

129. P. Garcia-Navarro, M. E. Hubbard, and A. Priestley. Genuinely Multidimensional Upwinding for the 2D Shallow Water Equations. *J. Comput. Phys.*, 121:79–93, 1995.

130. P. H. Gaskell and A. K. C. Lau. Curvature Compensated Convective Transport: SMART, a New Boundedness Preserving Transport Algorithm. *Int. J. Numer. Meth. Fluids*, 8:617–641, 1988.

131. C. W. Gear. *Numerical Initial–Value Problems in Ordinary Differential Equations*. Prentice–Hall, Englewood Cliffs, NJ, 1971.

132. B. E. Gelfand, S. M. Frolov, and M. A. Nettleton. Gaseous Detonations–A Selective Review. *Prog. Energy Comb. Sci.*, 17:327–371, 1991.

133. R. B. Gilbert and R. A. Strehlow. Theory of Detonation Initiation Behind Reflected Shock Waves. *AIAA Journal*, 4:1777–1783, 1966.

134. M. B. Giles. Non-Reflecting Boundary Conditions for Euler Equation Calculations. *AIAA J.*, 28:2050–2058, 1990.

135. L. Giraud and G. Manzini. Parallel Implementations of 2D Explicit Euler Solvers. *J. Comput. Phys.*, 123:111–118, 1996.

136. P. Glaister. Difference Schemes for the Shallow Water Equations. Technical Report 9/87, Department of Mathematics, University of Reading, England, 1987.

137. P. Glaister. An Approximate Linearised Riemann Solver for the Euler Equations of Gas Dynamics. *J. Comput. Phys.*, 74:382–408, 1988.

138. P. Glaister. An Approximate Linearised Riemann Solver for the Three–Dimensional Euler Equations for Real Gases Using Operator Splitting. *J. Comput. Phys.*, 77:361–383, 1988.

139. H. M. Glaz, P. Colella, I. I. Glass, and R. L. Deschambault. A Detailed Numerical, Graphical and Experimental Study of Oblique Shock Wave Reflections. Technical Report 285, Institute for Aerospace Science, University of Toronto (UTIAS), 1985.

140. H. M. Glaz, P. Colella, I. I. Glass, and R. L. Deschambault. A Numerical Study of Oblique Shock–Wave Reflections with Experimenatl Comparisons. *Proc. Roy. Soc. London*, A 398:117–140, 1985.

141. J. Glimm. Solution in the Large for Nonlinear Hyperbolic Systems of Equations. *Comm. Pure. Appl. Math.*, 18:697–715, 1965.

142. J. Glimm, M. J. Graham, J. W Grove, and B. J. Plohr. *Proc. Fifth International Conference on Hyperbolic Problems*. World Scientific, 1996.

143. J. Glimm, G. Marshall, and B. Plohr. A Generalized Riemann Problem for Quasi-One-Dimensional Gas Flows. *Advances in Applied Mathematics*, 5:1–30, 1984.

144. E. Godlewski and P. A. Raviart. *Numerical Approximation of Hyperbolic Systems of Conservation Laws*. Applied Mathematical Sciences, Vol. 118, Springer–Verlag, New York, 1996.

145. S. K. Godunov. A Finite Difference Method for the Computation of Discontinuous Solutions of the Equations of Fluid Dynamics. *Mat. Sb.*, 47:357–393, 1959.

146. S. K. Godunov, A. V. Zabrodin, and G. P. Prokopov. –. *J. Comp. Math. Phys. USSR 1 (1962)*, pages 1187–, 1962.

147. S. K. (Ed.) Godunov. *Numerical Solution of Multi-Dimensional Problems in Gas Dynamics*. Nauka Press, Moscow, 1976.

148. J. B. Goodman and R. J. LeVeque. On the Accuracy of Stable Schemes for 2D Scalar Conservation Laws. *Math. Comp.*, 45(21):15–21, 1985.

149. J. J. Gottlieb. Staggered and Non–Staggered Grids with Variable Node Spacing and Local Timestepping for Random Choice Method. *J. Comput. Phys.*, 78:160–177, 1988.

150. J. J. Gottlieb and C. P. T. Groth. Assessment of Riemann Solvers for Unsteady One–Dimensional Inviscid Flows of Perfect Gases. *J. Comput. Phys.*, 78:437–458, 1988.

151. S. Gottlieb and C. W. Shu. Total Variation Diminishing Runge–Kutta Schemes. Technical Report ICASE 96–50, NASA Langley Research Center, Hampton, USA, 1996.

152. D. F. Griffiths, A. M. Stuart, and H. C. Yee. Numerical Wave Propagation in an Advection Equation with a Nonlinear Source Term. *SIAM J. Numer. Anal.*, 29:1244–1260, 1992.

153. H. Grönig (Editor). *Shock Tubes and Waves (Proc. 16th Int. Symp. on Shock Tubes and Waves, Aachen Germany 1987)*. VCH Verlagsgessellschaft mbH, 1988.

154. R. Haberman. *Mathematical Models*. SIAM, 1998.

155. J. M. Hammersley and D. C. Handscombe. *Monte Carlo Methods*. Chapman and Hall, 1964.

156. D. Hänel. On the Numerical Solution of the Navier–Stokes Equations for Compressible Flow. VKI Lecture Series 1989–04 on Computational Fluid Dynamics, 1989.

157. D. Hänel and S. D. Sharma. Simulation of Hydrodynamical Free–Surface Flow. In *Proc. Third ECCOMAS Computational Fluid Dynamics Conference, 9–13 September 1996, Paris, France*, pages 1019–1024. Wiley, 1996.

158. A. Harten. High Resolution Schemes for Hyperbolic Conservation Laws. *J. Comput. Phys.*, 49:357–393, 1983.

159. A. Harten. On a Class of High Resolution Total Variation Stable Finite Difference Schemes. *SIAM J. Numer. Anal.*, 21(1):1–23, 1984.

160. A. Harten. Preliminary Results on the Extension of ENO Schemes to Two–Dimensional Problems. In *International Conference on Non–Linear Hyperbolic Problems*, pages 23–51. Springer–Verlag, Berlin, 1987.

161. A. Harten. ENO Schemes with Subcell Resolution. *J. Comput. Phys.*, 83:148–184, 1989.

162. A. Harten, B. Engquist, S. Osher, and S. R. Chakravarthy. Uniformly High Order Accuracy Essentially Non–oscillatory Schemes III. *J. Comput. Phys.*, 71:231–303, 1987.

163. A. Harten and J. M. Hyman. Self Adjusting Grid Methods for One–Dimensional Hyperbolic Conservation Laws. *J. Comput. Phys.*, 50:235–269, 1983.

164. A. Harten, P. D. Lax, and B. van Leer. On Upstream Differencing and Godunov–Type Schemes for Hyperbolic Conservation Laws. *SIAM Review*, 25(1):35–61, 1983.

165. A. Harten and S. Osher. Uniformly High–Order Accurate Nonoscillatory Schemes I. *SIAM J. Numer. Anal.*, 24(2):279–309, 1987.

166. W. Heilig and H. Reichenbach. High Speed Interferometric Study of Unstationary Shok Wave Processes. Technical report, Fraunhofer Institut für Kurzzeitdynamik, Ernst–Mach–Institut, Freiburg im Breisgau, Germany, 1982.

167. P. W. Hemker and S. P. Spekreijse. Multiple Grid and Osher's Scheme for the Efficient Solution of the Steady Euler Equations. *Applied Numerical Mathematics*, 2:475–493, 1986.

168. F. B. Hildebrand. *Introduction to Numerical Analysis*. Tata McGraw–Hill Publishing Co. Limited, New Delhi, 1974.

169. R. Hillier. Numerical Modelling of Shock Wave Diffraction. In *Shock Waves @ Marseille, Vol. 4 (Brun and Dumitrescu, Editors)*, pages 17–26. Springer–Verlag, 1995.

170. C. Hirsch. *Numerical Computation of Internal and External Flows, Vol. I: Fundamentals of Numerical Discretization*. Wiley, 1988.

171. C. Hirsch. *Numerical Computation of Internal and External Flows, Vol. II: Computational Methods for Inviscid and Viscous Flows*. Wiley, 1990.

172. K. A. Hoffmann. *Computational Fluid Dynamics for Engineers*. Engineering Education Systems, Austin, Texas, USA, 1989.

173. M. Holt. *Numerical Methods in Fluid Dynamics*. Springer–Verlag, 1984.

174. H. Honma and I. I. Glass. Random Choice Solutions for Weak Spherical Shock–Wave Transitions of N–Waves in Air with Vibrational Excitation. Technical Report 253, UTIAS, University of Toronto, Canada, 1983.

175. H. Honma, M. Wada, and K. Inomata. Random Choice Solutions for Two–Dimensional and Axi–Symetric Supersonic Flows. Technical Report S. P. Number 2, The Institute of Space and Astronautical Science, 1984.

176. L. Hörmander. *Lectures on Nonlinear Hyperbolic Differential Equations, Mathématiques et Applications 26*. Springer–Verlag, 1997.

177. T. Y. Hou and LeFloch P. Why Non–Conservative Schemes Converge to the Wrong Solutions: Error Analysis. *Math. of Comput.*, 62:497–530, 1994.

178. E. L. Ince and I. N. Sneddon. *The Solution of Ordinary Differential Equations*. Longman Scientific and Technical, 1987.

179. M. J. Ivings, D. M. Causon, and E. F. Toro. On Hybrid High–Resolution Upwind Methods for Multicomponent Flows. *ZAMM*, 77, Issue 9:645–668, 1997.

180. M. J. Ivings, D. M. Causon, and E. F. Toro. On Riemann Solvers for Compressible Liquids. Technical Report 97–4, Department of Mathematics and Physics, Manchester Metropolitan University, UK, 1997.

181. A. Jameson and P. D. Lax. Conditions for the Construction of Multi–Point Total Variation Diminishing Difference Schemes. Technical Report MAE 1650, Department of Mechanical and Aerospace Engineering, University of Princeton, USA, 1984.

182. A. Jameson, W. Schmidt, and E. Turkel. Numerical Solution of the Euler Equations by Finite Volume Methods using Runge-Kutta Stepping Schemes. Technical Report 81–1259, AIAA, 1981.

183. A. Jeffrey. *Quasilinear Hyperbolic Systems and Waves*. Pitman, 1976.

184. G. S. Jiang and C. W. Shu. Efficient Implementation of Weigthed ENO Schemes. Technical Report ICASE 95–73, NASA Langley Research Center, Hampton, USA, 1995.

185. F. John. *Partial Differential Equations.* Springer–Verlag, 1982.

186. L. W. Johnson and R. D. Riess. *Numerical Analysis.* Addison–Wesley Publishing Company, 1982.

187. P. Jorgenson and E. Turkel. Central–Difference TVD and TVB Schemes for Time–Dependent and Steady State Problems. Technical Report ICOMP–91–27; AIAA–92–0053, NASA Lewis Research Center, Cleveland, Ohio, USA, 1992.

188. D. Kahaner, C. Moler, and S. Nash. *Numerical Methods and Software.* Prentice Hall, Englewood Cliffs, New Jersey, 1989.

189. K. Kailasanath, E. S. Oran, J. P. Boris, and T. R. Young. Determination of the Detonation Cell Size and the Role of Transverse Waves in Two–Dimensional Detonations. *Combustion and Flame,* 61:199–209, 1985.

190. S. Karni. Accelerated Convergence to Steady State by Gradual Far–Field Damping. *AIAA J.,* 30:1220–1228, 1992.

191. S. Karni. Multicomponent Flow Calculations Using a Consistent Primitive Algorithm. *J. Comput. Phys.,* 112(1):31–43, 1994.

192. S. Karni. Hybrid Multifluid Algorithms. Technical Report 95–001, Courant Mathematics and Computing Laboratory, 1995.

193. S. Karni and S. Čanić. Computations of Slowly Moving Shocks. *J. Comput. Phys.,* 136:132–139, 1997.

194. K. S. Kim and J. Ballmann. Numerische Simulation Mechanischer Wellen in Geschichteten Elastischen Körpern. *Z. Angew. Math. Mech.,* 70:204–206, 1990.

195. B. Koren and S. P. Spekreijse. Multigrid and Defect Correction for the Efficient Solution of the Steady Euler Equations. In *Notes on Numerical Fluid Mechanics, Vol. 17,* pages 87–100. Vieweg, 1987.

196. D. Kröner. *Numerical Schemes for Conservation Laws.* Wiley Teubner, 1997.

197. A. Lafon and H. C. Yee. Dynamical Approach Study of Spurious Steady–State Numerical Solutions of Nonlinear Differential Equations, Part III. The Effects of Nonlinear Source Terms in Reaction–Convection Equations. *Intern. J. Computational Fluid Dynamics,* 6:1–36, 1996.

198. A. Lafon and H. C. Yee. Dynamical Approach Study of Spurious Steady–State Numerical Solutions of Nonlinear Differential Equations, Part IV. Stability vs. Methods of Discretizing Nonlinear Source Terms in Reaction–Convection Equations. *Intern. J. Computational Fluid Dynamics,* 6:89–123, 1996.

199. J. D. Lambert. *Computational Methods in Ordinary Differential Equations.* John Wiley and Sons, 1973.

200. L. D. Landau and E. M. Lifshitz. *Fluid Mechanics.* Pergamon Press, 1982.

201. P. D. Lax. Weak Solutions of Nonlinear Hyperbolic Equations and Their Numerical Computation. *Comm. Pure. Appl. Math.,* VII:159–193, 1954.

202. P. D. Lax. The Formation and Decay of Shock Waves. *American Mathematical Monthly,* 79:227–241, 1972.

203. P. D. Lax. *Hyperbolic Systems of Conservation Laws and the Mathematical Theory of Shock Waves.* Society for Industrial and Applied Mathematics, Philadelphia, 1990.

204. P. D. Lax and B. Wendroff. Systems of Conservation Laws. *Comm. Pure Appl. Math.,* 13:217–237, 1960.

205. P. D. Lax and B. Wendroff. Difference Schemes for Hyperbolic Equations with High Order Accuracy. *Comm. Pure Appl. Math.,* XVII:381–393, 1964.

206. B. P. Leonard. Simple High–Accuracy Resolution Program for Convective Modelling of Discontinuities. *Int. J. Numer. Methods in Fluids,* 8:1291–1318, 1988.

207. R. J. LeVeque. *Numerical Methods for Conservation Laws*. Birkhäuser Verlag, 1992.
208. R. J. LeVeque. Simplified Multidimensional Flux Limiter Methods. In M. J. Baines and K. W. Morton, editors, *Numerical Methods in Fluid Dynamics 4: Proceedings of the 1992 International Conference on Numerical Methods in Fluids*, pages 173–189, Reading, 1993.
209. R. J. LeVeque. Balancing Source Terms and Flux Gradients in High–Resolution Godunov Methods. *J. Comput. Phys.*, 146:346–365, 1998.
210. R. J. LeVeque, D. Mihalas, E. A. Dorfi, and E. Müller. *Computational Methods for Astrophysical Flow*. Springer–Verlag, 1998.
211. R. J. LeVeque and K. M. Shyue. Two–Dimensional Front Tracking Based on High Resolution Wave Propagation Methods. *J. Comput. Phys.*, 123:354–368, 1996.
212. R. J. LeVeque and H. C. Yee. A Study of Numerical Methods for Hyperbolic Conservation Laws with Stiff Source Terms. *J. Comput. Phys.*, 86:187–210, 1990.
213. K. M. Li and M. Holt. Numerical Solutions to Water Waves Generated by Shallow Underwater Explosions. *Phys. Fluids*, 24:816–824, 1981.
214. NAG Library. NAG Library Routines, Mark 18, Routines D03PWF and D03PXF, 1996.
215. X. Lin and J. Ballmann. A Riemann Solver and a Second–Order Godunov Method for Elastic–Plastic Wave Propagation in Solids. *Int. J. Impact Engng.*, 13(3):463–478, 1993.
216. X. Lin and Ballmann J. Numerical Method for Elastic–Plastic Waves in Cracked Solids, Part 1: Anti–Plane Shear Wave. *Archive of Applied Mechanics*, 63:261–282, 1993.
217. X. Lin and Ballmann J. Numerical Method for Elastic–Plastic Waves in Cracked Solids, Part 2: Plain Strain Problem. *Archive of Applied Mechanics*, 63:283–295, 1993.
218. X. Lin and Ballmann J. Reconsideration of Chen's Problem by Finite Difference Method. *Engineering Fracture and Mechanics*, 44(5):735–739, 1993.
219. W. B. Lindquist. *Current Progress in Hyperbolic Systems: Riemann Problems and Computations*. American Mathematical Society, 1989.
220. M. S. Liou. Recent Progress and Applications of AUSM+. In *Sixteenth International Conference on Numerical Methods in Fluid Dynamics, Lecture Notes in Physics 515*, pages 302–307. Springer–Verlag, 1998.
221. M. S. Liou and C. J. Steffen. A New Flux Splitting Scheme. *J. Comput. Phys.*, 107:23–39, 1993.
222. X. D. Liu, S. Osher, and T. Chan. Weighted Essentially Non–oscillatory Schemes. *J. Comput. Phys.*, 115:200–212, 1994.
223. Xu-Dong Liu and P. D. Lax. Positive Schemes for Solving Multi–Dimensional Hyperbolic Conservation Laws. Technical Report 95–003, Courant Mathematics and Computing Laboratory, New York University, USA, 1995.
224. C. Y. Loh, M. S. Liou, and W. H. Hui. An Investigation of Random Choice Method for Three Dimensional Steady Supersonic Flow. In *First Asian Computational Fluid Dynamics Conference*, pages 963–971. The Hong Kong University of Science and Technology, 1995.
225. J. Lorenz and H. J. Schroll. Stiff Well–Posedness for Hyperbolic Systems with Large Relaxation Terms. Technical Report 125, Institut für Geometrie un Praktische Mathematik, RWTH Aachen, Germany, 1996.
226. A. Majda and S. Osher. Numerical Viscosity and Entropy Condition. *Comm. Pure and Applied Mathematics*, 32:797–838, 1979.

227. J. C. Mandal and S. M. Desphande. Kinetic Flux Vector Splitting for the Euler Equations. *Computers and Fluids*, 23:447–, 1994.

228. M. J. Maron and R. J. Lopez. *Numerical Analysis*. Wadsworth, 1991.

229. A. Marquina, J. M. Martí, J. M. Ibañez, Miralles J. A., and R. Donat. Ultrarelativistic Hydrodynamics: High–Resolution Shock–Capturing Methods. *Astron. Astrophys.*, 258:566–, 1992.

230. E. Marshall and R. Méndez. Computational Aspects of the Random Choice Method for Shallow Water Equations. *J. Comput. Phys.*, 39:1–21, 1981.

231. G. Marshall and B. Plohr. A Random Choice Method for Two–Dimensional Steady Supersonic Shock Wave Diffraction Problems. *J. Comput. Phys.*, 56:410–427, 1984.

232. J. M. Martí, J. M. Ibañez, and J. A. Miralles. Numerical Relativistic Hydrodynamics: Local Characteristic Approach. *Phys. Rev.*, 43:3794–, 1991.

233. J. M. Martí and E. Müller. The Analytical Solution of the Riemann Problem for Relativistic Hydrodynamics. *Journal of Fluid Mechanics*, 258:317–333, 1994.

234. J. M. Martí, E. Müller, J. A. Font, J. M. Ibañez, and A. Marquina. Morphology and Dynamics of Relativistic Jets. Technical Report MPA 895, Max–Planck–Institut für Astrophysik, Garching, Germany, 1995.

235. Y. P. Marx. A Practical Implementation of Turbulence Models for the Computation of Three–Dimensional Separated Flows. *Int. J. Numer. Meth. Fluids*, 13:775–796, 1991.

236. Y. P. Marx. Evaluation of the Artificial Compressibility Method for the Solution of the Incompressible Navier–Stokes Equations. In *Notes on Numerical Fluid Mechanics, Vol. 35*, pages 201–210. Vieweg, 1991.

237. Y. P. Marx. Time Integration Schemes for the Unsteady Incompressible Navier–Stokes Equations. *J. Comput. Phys.*, 112:182–209, 1994.

238. J. H. Mathews. *Numerical Methods*. Prentice–Hall International, Inc., 1987.

239. C. Y. McNeil. *Efficient Upwind Algorithms for Solution of the Euler and Navier–Stokes Equations*. PhD thesis, Department of Aerospace Science, Cranfield University, UK, 1995.

240. R. Menikoff and B. J. Plohr. The Riemann Problem for Fluid Flow of Real Materials. *Reviews of Modern Physics*, 61:75–130, 1989.

241. A. R. Mitchell and D. F. Griffiths. *The Finite Difference Method in Partial Differential Equations*. John Wiley and Sons, 1980.

242. G. Moretti. The λ–scheme. *Computers and Fluids*, 7:191–205, 1979.

243. G. Moretti. Computation of Flows with Strong Shocks. *Ann. Rev. Fluid Mech.*, 19:313–337, 1987.

244. G. Moretti and G. Bleich. A Time–Dependent Computational Method for Blunt–Body Flows. *AIAA J.*, 4:2136–2141, 1966.

245. C. D. Munz. A Tracking Method for Gas Flow into Vacuum Based on the Vacuum Riemann Problem. *Mathematical Methods in Applied Sciences*, 17:597–612, 1994.

246. C. D. Munz, R. Schneider, and O. Gerlinger. The Numerical Approximation of a Free Gas–Vacuum Interface. In *Numerical Methods for the Navier–Stokes Equations*, pages 181–190. Vieweg–Verlag, 1994.

247. N. Nikiforakis and J. F. Clarke. Quasi–Steady Structures in the Two Dimensional Initiation of Detonations. *Proc. Roy. Soc. Lond. A*, 452:2023–2042, 1996.

248. H. Olivier and H. Grönig. The Random Choice Method Applied to Two-Dimensional Shock Focusing and Diffraction. *J. Comput. Phys.*, 63:85–106, 1986.

249. A. K. Oppenheim and R. I. Soloukhin. Experiments in Gas Dynamics of Explosions. *Annual Review of Fluid Mechanics*, 5:31–58, 1973.

250. E. S. Oran and J. P. Boris. Detailed Modelling of Combustion Systems. *Prog. Energy. Comb. Sci.*, 7:1–72, 1981.

251. E. S. Oran and J. P. Boris. *Numerical Simulation of Reactive Flow*. Elsevier, New York, 1987.

252. S. Osher. Riemann Solvers, the Entropy Condition and Difference Approximations. *SIAM J. Numer. Anal.*, 21(2):217–235, 1984.

253. S. Osher and S. Chakravarthy. Upwind Schemes and Boundary Conditions with Applications to Euler Equations in General Geometries. *J. Comput. Phys.*, 50:447–481, 1983.

254. S. Osher and S. Chakravarthy. High Resoultion Schemes and the Entropy Condition. *SIAM J. Numer. Anal.*, 21(5):955–984, 1984.

255. S. Osher and F. Solomon. Upwind Difference Schemes for Hyperbolic Conservation Laws. *Math. Comp.*, 38,158:339–374, 1982.

256. Liu T. P. and J. A. Smoller. On the Vacuum State for the Isentropic Gas Dynamics Equations. *Advances in Applied Mathematics*, 1:345–359, 1980.

257. M. Pandolfi. Computation of Steady Supersonic Flows by a Flux–Difference Splitting Method. *Computers and Fluids*, 10:37–46, 1985.

258. R. B. Pember. Numerical Methods for Hyperbolic Conservation Laws with Stiff Relaxation I: Spurious Solutions. *SIAM J. Appl. Math.*, 53:1293–1330, 1993.

259. R. B. Pember. Numerical Methods for Hyperbolic Conservation Laws with Stiff Relaxation II: Higher–Order Godunov Methods. *SIAM J. Sci. Comput.*, 14:824–859, 1993.

260. L. L. Pennisi. *Elements of Complex Analysis*. Holt, Rinehart and Winston, 1963.

261. B. Perthame. Boltzmann Type Schemes for Gas Dynamics and the Entropy Property. *SIAM J. Numer. Anal.*, 27:1405–1421, 1990.

262. B. Perthame. Second–Order Boltzman Schemes for Compressible Euler Equations in one and two Space Dimensions. *SIAM J. Numer. Anal.*, 29:1992, 1992.

263. J. Pike. Riemann Solvers for Perfect and Near–Perfect Gases. *AIAA Journal*, 31(10):1801–1808, 1993.

264. B. J. Plohr and D. H. Sharp. Riemann Problems and their Application to Ultra–Relativistic Heavy Ion Collisions. In *The VIII International Conference on Mathematical Physics (Mebkhout and Sénéor, editors)*, pages 708–713. World Scientific Pub. Co., Singapore, 1987.

265. K. H. Prendergast and K. Xu. Numerical Hydrodynamics from Gas–Kinetic Theory. *J. Comput. Phys.*, 109:53–66, 1993.

266. D. I. Pullin. Direct Simulation Methods for Compressible Inviscid Ideal Gas Flow. *J. Comput. Phys.*, 34:231–244, 1980.

267. N. Qin. A Comparative Study of Two Upwind Schemes as Applied to Navier–Stokes Solutions for Resolving Boundary Layers in Hypersonic Viscous Flow. Technical Report GU Aero Report 9120, Department of Aerospace Engineering, University of Glasgow, UK, 1992.

268. N. Qin and G. W. Foster. Study of Flow Interactions due to a Supersonic Lateral Jet Using High Resolution Navier–Stokes Solutions. In *AIAA–95–2151 Paper, 27th AIAA Fluid Dynamics Conference, June 1995, San Diego, USA*, 1995.

269. N. Qin and A. Redlich. Massively Separated Flows due to Transverse Sonic Jet in Hypersonic Laminar Stream. In *AIAA–96–1934 Paper, 27th AIAA Fluid Dynamics Conference, June 1996, New Orleans, USA*, 1996.

270. N. Qin and B. E. Richards. Sparse Quasi–Newton Method for High Resolution Schemes. In *Notes on Numerical Fluid Mechanics, Vol. 20*, pages 310–317. Vieweg, 1988.

271. N. Qin and B. E. Richards. Sparse Quasi–Newton Method for Navier–Stokes Solution. In *Notes on Numerical Fluid Mechanics, Vol. 29*, pages 474–483. Vieweg, 1990.

272. N. Qin, K. W. Scriba, and B. E. Richards. Shock–Shock, Shock–Vortex Interaction and Aerodynamic Heating In Hypersonic Corner Flow. *The Aeronautical Journal*, pages 152–160, 1991.

273. N. Qin, X. Xu, and B. E. Richards. Newton–Like Methods for Fast High Resolution Simulation of Hypersonic Viscous Flows. *Computing Systems in Engineering*, 3:429–435, 1992.

274. J. J. Quirk. An Adaptive Grid Algorithm for Computational Shock Hydrodynamics. PhD Thesis, College of Aeronautics, Cranfield Institute of Technology, UK, 1991.

275. J. J. Quirk. An Alternative to Unstructured Grids for Computing Gas Dynamic Flows Around Arbitrarily Complex Two Dimensional Bodies. *Computers and Fluids*, 23(1):125–142, 1994.

276. R. D. Richtmyer and K. W. Morton. *Difference Methods for Initial Value Problems*. Interscience-Wiley, New York, 1967.

277. N. H. Risebro and A. Tveito. Front Tracking Applied to Nonstrictly Hyperbolic Systems of Conservation Laws. *SIAM J. Sci. Stat. Comput.*, 12:1401–1419, 1991.

278. A. Rizzi. Numerical Implementation of Solid-Body Boundary Conditions for the Euler Equations. *ZAMM*, 58:301–304, 1978.

279. P. J. Roache. *Computational Fluid Dynamics*. Hermosa Publishers, 1982.

280. T. W. Roberts. The Behavior of Flux Difference Splitting Schemes near Slowly Moving Shock Waves. *J. Comput. Phys.*, 90:141–160, 1990.

281. P. L. Roe. Approximate Riemann Solvers, Parameter Vectors, and Difference Schemes. *J. Comput. Phys.*, 43:357–372, 1981.

282. P. L. Roe. Numerical Algorithms for the Linear Wave Equation. Technical Report 81047, Royal Aircraft Establishment, Bedford, UK, 1981.

283. P. L. Roe. Some Contributions to the Modelling of Discontinuous Flows. In *Proceedings of the SIAM/AMS Seminar*, San Diego, 1983.

284. P. L. Roe. Generalized Formulation of TVD Lax–Wendroff Schemes. Technical Report ICASE 84–53, NASA Langley Research Center, Hampton, USA, 1984.

285. P. L. Roe. Discrete Models for the Numerical Analysis of Time–Dependent Multidimensional Gas Dynamics. *J. Comput. Phys.*, 63:458–476, 1986.

286. P. L. Roe. Upwind Differencing Schemes for Hyperbolic Conservation Laws with Source Terms. In *Proc. First International Conference on Hyperbolic Problems, Carasso, Raviart and Serre (Editors)*, pages 41–51. Springer, 1986.

287. P. L. Roe. Remote Boundary Conditions for Unsteady Multidimensional Aerodynamic Computations. *Computers and Fluids*, 17:221–231, 1989.

288. P. L. Roe. Sonic Flux Formulae. *SIAM J. Sci. Stat. Comput.*, 13(2):611–630, 1992.

289. P. L. Roe. The Harten Memorial Lecture–New Applications of Upwinding. In *Numerical Methods for Wave Propagation. Toro, E. F. and Clarke, J. F. (Editors)*, pages 1–31. Kluwer Academic Publishers, 1998.

290. P. L. Roe and J. Pike. Efficient Construction and Utilisation of Approximate Riemann Solutions. In *Computing Methods in Applied Science and Engineering*. North–Holland, 1984.

291. V. V. Rusanov. Calculation of Interaction of Non–Steady Shock Waves with Obstacles. *J. Comput. Math. Phys. USSR*, 1:267–279, 1961.

292. L. Sainsaulieu. Finite Volume Approximation of Two Phase–Fluid Flows Based on an Approximate Roe–Type Riemann Solver. *J. Comput. Phys.*, 121:1–28, 1995.

293. T. Saito and I. I. Glass. Application of Random Choice to Problems in Shock and Detonation Wave Dynamics. Technical Report UTIAS 240, Institute for Aerospace Studies, University of Toronto, 1979.

294. D. A. Sánchez. *Ordinary Differential Equations and Stability Theory.* Dover Publications, Inc., 1968.

295. R. H. Sanders and K. H. Prendergast. On the Origin the Three Kiloparsec Arm. *Astrophys. J.*, 188:489–500, 1974.

296. R. Saurel, E. Daniel, and J. C. Loraud. Two–Phase Flows: Second Order Schemes and Boundary Conditions. *AIAA Journal*, 32(6):1214–1221, 1994.

297. R. Saurel, A. Forestier, D. Veyret, and J.-C. Loraud. A Finite Volume Scheme for Two–Phase Compressible Flows. *Int. J. Numer. Meth. Fluids*, 18:803–819, 1994.

298. R. Saurel, M. Larini, and J. C. Loraud. Exact and Approximate Riemann Solvers for Real Gases. *J. Comput. Phys.*, 112:126–137, 1994.

299. M. Schleicher. Ein Einfaches und Effizientes Verfahren zur Loesung des Riemann–Problems. *Z. Flugwiss. Weltraumforsch.*, 17:265–269, 1993.

300. V. Schneider, U. Katscher, D. H. Rischke, B. Waldhauser, J. A. Maruhn, and C. D. Munz. New Algorithms for Ultra–Relativistic Numerical Hydrodynamics. *J. Comput. Phys.*, 105:92–, 1993.

301. H. J. Schroll and A. Tveito. A System of Conservation Laws with a Relaxation Term. Technical Report 103, Institut für Geometrie un Praktische Mathematik, RWTH Aachen, Germany, 1994.

302. H. J. Schroll, A. Tveito, and R. Winther. An Error Bound for a Finite Difference Scheme Applied to a Stiff System of Conservation Laws. Technical Report 1994-3, Department of Informatics, University of Oslo, Norway, 1994.

303. H. J. Schroll and R. Winther. Finite Difference Schemes for Scalar Conservation Laws with Source Terms. Technical Report 107, Institut für Geometrie un Praktische Mathematik, RWTH Aachen, Germany, 1994.

304. F. W. Sears and G. L. Salinger. *Thermodynamics, Kinetic Theory and Statistical Thermodynamics.* Addison–Wesley, 1986.

305. C. C. L. Sells. . Technical Report RAE TR 80065, Royal Aircraft Establishment, Bedford, UK, 1980.

306. L. F. Shampine. *Numerical Solution of Ordinary Differential Equations.* Chapman and Hall, London and New York, 1994.

307. J. S. Shang. A Fractional–Step Method for Solving 3D, Time–Domain Maxwell Equations. *J. Comput. Phys.*, 118:109–119, 1995.

308. Z. C. Shi and J. J. Gottlieb. Random Choice Method for Two Dimensional Planar and Axisymmetric Steady Supersonic Flows. Technical Report 297, UTIAS, University of Toronto, Canada, 1985.

309. C. Shu and S. Osher. Efficient Implementation of Essentially Non–oscillatory Shock–Capturing Schemes II. *J. Comput. Phys.*, 83:32–78, 1988.

310. C. W. Shu. TVB Boundary Treatment for Numerical Solutions of Conservation Laws. *Math. Comp.*, 49(179):123–134, 1987.

311. C. W. Shu. TVB Uniformly High–Order Schemes for Conservation Laws. *Math. Comput.*, 49(179):105–121, 1987.

312. C. W. Shu and S. Osher. Efficient Implementation of Essentially Non-oscillatory Shock–Capturing Schemes. *J. Comput. Phys.*, 77:439–471, 1988.

313. G. Singh. *A Numerical Study of the Evolution of Piston Driven Detonation.* PhD thesis, Department of Arospace Science, Cranfield University, UK, 1990.

314. G. Singh and J. F. Clarke. Transient Phenomena in the Initiation of a Mechanically Driven Plane Detonation. *Proc. Roy. Soc. London A*, 438:23–46, 1992.

315. G. D. Smith. *Numerical Solution of Partial Differential Equations*. Clarendon Press, Oxford, 1978.
316. J. Smoller. *Shock Waves and Reaction–Diffusion Equations*. Springer–Verlag, 1994.
317. G. A. Sod. A Numerical Study of a Converging Cylindrical Shock. *J. Fluid Mechanics*, 83:785–794, 1977.
318. G. A. Sod. A Survey of Several Finite Difference Methods for Systems of Nonlinear Hyperbolic Conservation Laws. *J. Comput. Phys.*, 27:1–31, 1978.
319. G. A. Sod. *Numerical Methods in Fluid Dynamics*. Cambridge University Press, 1985.
320. G. A. Sod. A Random Choice Method for the Stefan Problem. *Computers Math. Applic.*, 19:1–9, 1990.
321. W. Speares. A Finite Volume Approach to the Weighted Average Flux Method. MSc thesis, College of Aeronautics, Cranfield Institute of Technology, UK, 1991.
322. W. Speares and E. F. Toro. A High-Resolution Algorithm for Time–Dependent Shock Dominated Problems with Adaptive Mesh Refinement. *Z. Flugwiss. Weltraumforsch. (Journal of Flight Sciences and Space Research)*, 19:267–281, 1995.
323. S. P. Spekreijse. Multigrid solution of monotone second–order discretizations of hyperbolic conservation laws. *Math. Comp.*, 49(179):135–155, 1987.
324. S. P. Spekreijse. *Multigrid Solution of the Steady Euler Equations*. PhD thesis, Centre for Mathematics and Computer Science, Amsterdam, The Netherlands, 1988.
325. J. L. Steger, F. C. Dougherty, and J. A. Benek. A Chimera Grid Scheme. In *Advances in Grid Generation, ASME FED, Vol. 5*, 1983.
326. J. L. Steger and R. F. Warming. Flux Vector Splitting of the Inviscid Gasdynamic Equations with Applications to Finite Difference Methods. *J. Comput. Phys.*, 40:263–293, 1981.
327. H. B. Stewart and B. Wendroff. Two–Phase Flow: Models and Methods. *J. Comput. Phys.*, 56:363–409, 1984.
328. J. J. Stoker. *Water Waves. The Mathematical Theory with Applications*. John Wiley and Sons, 1992.
329. G. Strang. On the Construction and Comparison of Difference Schemes. *SIAM J. Numer. Anal.*, 5(3):506–517, 1968.
330. R. C. Swanson and E. Turkel. On Central–Difference and Upwind Methods. Technical Report ICASE 90–44, NASA Langley Research Center, Hampton, Virginia, USA, 1990.
331. B. K. Swartz and B. Wendroff. AZTEC: A Front Tracking Code Based on Godunov's Method. *Applied Numerical Mathematics*, 2:385–397, 1986.
332. P. K. Sweby. *Shock Capturing Schemes*. PhD thesis, Department of Mathematics, University of Reading, UK, 1982.
333. P. K. Sweby. High Resolution Schemes Using Flux Limiters for Hyperbolic Conservation Laws. *SIAM J. Numer. Anal.*, 21:995–1011, 1984.
334. P. K. Sweby. TVD Schemes for Inhomogeneous Conservation Laws. In *Notes on Numerical Fluid Mechanics, Vol. 24, Non–Linear Hyperbolic Equations–Theory, Computation Methods and Applications*, pages 599–607. Vieweg, 1989.
335. P. K. Sweby and M. J. Baines. On Convergence of Roe's Scheme for the General Non–linear Wave Equation. *J. Comput. Phys.*, 56(1):135–148, 1984.
336. E. Tadmor. Convenient Total Variation Diminishing Conditions for Non–Linear Difference Schemes. *SIAM J. Numer. Anal.*, 25:1002–1014, 1988.
337. Y. Takano. An Application of the Random Choice Method to a Reactive Gas with Many Chemical Species. *J. Comput. Phys.*, 67:173–187, 1986.

338. K. Takayama. Optical Flow Visualization of Shock Wave Phenomena (Paul Vieille Memorial Lecture). In *Shock Waves @ Marseille, Vol. 4 (Brun and Dumitrescu, Editors)*, pages 7–16. Springer–Verlag, 1995.

339. K. Takayama (Editor). *Shock Waves (Proc. 18th Int. Symp. on Shock Waves, Sendai, Japan, 1991), Two Volumes.* Springer–Verlag, 1992.

340. S. Taki and T. Fujiwara. Numerical Analysis of Two–Dimensional Nonsteady Detonations. *AIAA Journal*, 16:73–77, 1978.

341. C. K. W. Tam and J. C. Webb. Dispersion–Relation–Preserving Finite Difference Schemes for Computational Acoustics. *J. Comput. Phys.*, 107:262–281, 1993.

342. P. Tamamidis, G. Zhang, and D. N. Assanis. Comparison of Pressure–Based and Artificial Compressibility Methods for Solving 3D Steady Incompressible Viscous Flows. *J. Comput. Phys.*, 124(1):1–13, 1996.

343. T. Tanaka. Finite Volume TVD Scheme on an Unstructured Grid System for Three–Dimensional MHD Simulation of Inhomogeneous Systems Including Strong Background Potential Fields. *J. Comput. Phys.*, 111:381–, 1994.

344. J. W. Thomas. *Numerical Partial Differential Equations, Texts in Applied Mathematics 22.* Springer–Verlag, 1998.

345. J. F. Thompson, F. C. Thames, and C. W. Mastin. Automatic Numerical Grid Generation of Body–Fitted Curvilinear Coordinate System for Field Containing any Number of Arbitrary Two–Dimensional Bodies. *J. Comput. Phys.*, 15:299–319, 1974.

346. J. F. Thompson, J. F. Warsi, and C. W. Mastin. *Grid Generation: Foundations and Applications.* North–Holland, 1982.

347. J. F. (Editor) Thompson. *Numerical Grid Generation.* North–Holland, 1982.

348. K. W. Thompson. The Special Relativistic Shock Tube. *J. Fluid Mechanics*, 171:365–, 1986.

349. K. W. Thompson. Time Dependent Boundary Conditions for Hyperbolic Systems. *J. Comput. Phys.*, 68:1–24, 1987.

350. E. F. Toro. A New Numerical Technique for Quasi-Linear Hyperbolic Systems of Conservation Laws. Technical Report CoA-8608, College of Aeronautics, Cranfield Institute of Technology, UK, 1986.

351. E. F. Toro. A Fast Riemann Solver with Constant Covolume Applied to the Random Choice Method. *Int. J. Numer. Meth. Fluids*, 9:1145–1164, 1989.

352. E. F. Toro. A Weighted Average Flux Method for Hyperbolic Conservation Laws. *Proc. Roy. Soc. London*, A423:401–418, 1989.

353. E. F. Toro. Random Choice Based Hybrid Schemes for one and Two–Dimensional Gas Dynamics. In *Non–linear Hyperbolic Equations – Theory, Computation Methods and Applications. Notes on Numerical Fluid Mechanics*, pages 630–639. Vieweg, Braunschweig, 1989.

354. E. F. Toro. Riemann–Problem Based Techniques for Computing Reactive Two–Phase Flows. In Dervieux and Larrouturrou, editors, *Proc. Third. Intern. Confer. on Numerical Combustion*, number 351 in Lecture Notes in Physics, pages 472–481, Antibes, France, May 1989.

355. E. F. Toro. A Linearised Riemann Solver for the Time–Dependent Euler Equations of Gas Dynamics. *Proc. Roy. Soc. London*, A434:683–693, 1991.

356. E. F Toro. Some Aspects of Shock Capturing Methods for Gas Dynamics. Technical Report COA–9112, College of Aeronautics, Cranfield Institute of Technology, UK, 1991.

357. E. F. Toro. Riemann Problems and the WAF Method for Solving Two–Dimensional Shallow Water Equations. *Phil. Trans. Roy. Soc. London*, A338:43–68, 1992.

358. E. F. Toro. The Weighted Average Flux Method Applied to the Time–Dependent Euler Equations. *Phil. Trans. Roy. Soc. London*, A341:499–530, 1992.

359. E. F. Toro. Viscous Flux Limiters. In J. B. Vos, A. Rizzi, and I. L. Ryhming, editors, *Notes on Numerical Fluid Dynamics, Vol. 35*, pages 592–600. Vieweg, 1992.

360. E. F. Toro. Defects of Conservative Approaches and Adaptive Primitive–Conservative Schemes for Computing Solutions to Hyperbolic Conservation Laws. Technical Report MMU 9401, Department of Mathematics and Physics, Manchester Metropolitan University, UK, 1994.

361. E. F. Toro. Direct Riemann Solvers for the Time–Dependent Euler Equations. *Shock Waves*, 5:75–80, 1995.

362. E. F. Toro. MUSCL–Type Primitive Variable Schemes. Technical Report MMU–9501, Department of Mathematics and Physics, Manchester Metropolitan University, UK, 1995.

363. E. F. Toro. On Adaptive Primitive–Conservative Schemes for Conservation Laws. In M. M. Hafez, editor, *Sixth International Symposium on Computational Fluid Dynamics: A Collection of Technical Papers*, volume 3, pages 1288–1293, Lake Tahoe, Nevada, USA, September 4–8, 1995.

364. E. F Toro. Some IVPs for Which Conservative Methods Fail Miserably. In M. M. Hafez, editor, *Sixth International Symposium on Computational Fluid Dynamics: A Collection of Technical Papers*, volume 3, pages 1294–1299, Lake Tahoe, Nevada, USA, September 4-8, 1995.

365. E. F. Toro. On Glimm–Related Schemes for Conservation Laws. Technical Report MMU–9602, Department of Mathematics and Physics, Manchester Metropolitan University, UK, 1996.

366. E. F. Toro. Exact and Approximate Riemann Solvers for the Artificial Compressibility Equations. Technical Report 97–02, Department of Mathematics and Physics, Manchester Metropolitan University, UK, 1997.

367. E. F. Toro. On Two Glimm–Related Schemes for Hyperbolic Conservation Laws. In *Proc. Fifth Annual Conference of the CFD Society of Canada*, pages 3.49–3.54. University of Victoria, Canada, 1997.

368. E. F. Toro. Primitive, Conservative and Adaptive Schemes for Hyperbolic Conservation Laws. In *Numerical Methods for Wave Propagation. Toro, E. F. and Clarke, J. F. (Editors)*, pages 323–385. Kluwer Academic Publishers, 1998.

369. E. F. Toro. *NUMERICA*, A Library of Source Codes for Teaching, Research and Applications. Numeritek Ltd., www.numeritek.com, 1999.

370. E. F. Toro. *Front–Capturing Methods for Free–Surface Shallow Flows*. John Wiley and Sons, 2000 (to appear).

371. E. F. Toro, S. J. Billet, and E. P. Boden. Advanced Modelling Techniques for Propulsion Systems. In *Proceedings of Conference on Energetic Materials and Propulsion Technology, Salsbury, Australia, April 18–19, 1996*, pages –. –, 1996.

372. E. F. Toro and S. J. Billett. Centred TVD Schemes for Hyperbolic Conservation Laws. Technical Report MMU–9603, Department of Mathematics and Physics, Manchester Metropolitan University, UK, 1996.

373. E. F. Toro and S. J. Billett. A Unified Riemann–Problem Based Extension of the Warming–Beam and Lax–Wendroff Schemes. *IMA J. Numer. Anal.*, 17:61–102, 1997.

374. E. F. Toro and S. J. Billett. Centred TVD Schemes for Hyperbolic Conservation Laws. *IMA J. Numerical Analysis*, to appear, 1999.

375. E. F. Toro and R. E. Brown. The WAF Method and Splitting Procedures for Viscous, Shocked Flows. In *Proceedings of the 18th International Symposium on*

Shock Waves, Tohoku University, Sendai, Japan (K. Takayama, Editor), pages 1119–1126. Springer–Verlag, 1992.

376. E. F. Toro and A. Chakraborty. Development of an Approximate Riemann Solver for the Steady Supersonic Euler Equations. *The Aeronautical Journal*, 98:325–339, 1994.

377. E. F. Toro and C. C. Chou. A Linearised Riemann Solver for the Steady Supersonic Euler Equations. *Int. J. Numer. Meth. Fluids*, 16:173–186, 1993.

378. E. F. Toro and P. L. Roe. A Hybridised High–Order Random Choice Method for Quasi-Linear Hyperbolic Systems. In Grönig, editor, *Proc. 16th Intern. Symp. on Shock Tubes and Waves*, pages 701–708, Aachen, Germany, July 1987.

379. E. F. Toro and P. L. Roe. A Hybrid Scheme for the Euler Equations Using the Random Choice and Roe's Methods. In *Numerical Methods for Fluid Dynamics III. The Institute of Mathematics and its Applications Conference Series, New Series No. 17, Morton and Baines (Editors)*, pages 391–402. Oxford University Press, New York, 1988.

380. E. F. Toro, M. Spruce, and W. Speares. Restoration of the Contact Surface in the HLL–Riemann Solver. *Shock Waves*, 4:25–34, 1994.

381. I Toumi. An Upwind Numerical Method for Two–Fluid Two–Phase Flow Models. *Nuclear Science and Engineering*, 123:147–168, 1996.

382. I. Toumi and D. Caruge. An Implicit Second–Order Numerical Method for Three–Dimensional Two–Phase Flow Calculations. *Nuclear Science and Engineering*, 130:1–13, 1998.

383. I. Toumi and A. Kumbaro. An Approximate Linearized Riemann Solver for a Two–Fluid Model. *J. Comput. Phys.*, 124:286–300, 1996.

384. E. Turkel and A. Arnone. Pseudo–Compressibility Methods for the Incompressible Flow Equations. Technical Report ICASE 93–66, NASA Langley Research Center, USA, 1993.

385. A. Tveito and R. Winther. *Introduction to Partial Differential Equations*. Springer–Verlag, 1998.

386. G. D. van Albada, B. van Leer, and W. W. Roberts. A Comparative Study of Computational Methods in Cosmic Gas Dynamics. *Astron. Astrophysics*, 108:76–84, 1982.

387. B. van Leer. Towards the Ultimate Conservative Difference Scheme I. The Quest for Monotonicity. *Lecture Notes in Physics*, 18:163–168, 1973.

388. B. van Leer. Towards the Ultimate Conservative Difference Scheme II. Monotonicity and Conservation Combined in a Second Order Scheme. *J. Comput. Phys.*, 14:361–370, 1974.

389. B. van Leer. MUSCL, A New Approach to Numerical Gas Dynamics. In *Computing in Plasma Physics and Astrophysics*, Max–Planck–Institut für Plasma Physik, Garchung, Germany, April 1976.

390. B. van Leer. Towards the Ultimate Conservative Difference Scheme III. Upstream–Centered Finite Difference Schemes for Ideal Compressible Flow. *J. Comput. Phys.*, 23:263–275, 1977.

391. B. van Leer. Towards the Ultimate Conservative Difference Scheme IV. A New Approach to Numerical Convection. *J. Comput. Phys.*, 23:276–299, 1977.

392. B. van Leer. Towards the Ultimate Conservative Difference Scheme V. A Second Order Sequel to Godunov's Method. *J. Comput. Phys.*, 32:101–136, 1979.

393. B. van Leer. Flux–Vector Splitting for the Euler Equations. Technical Report ICASE 82–30, NASA Langley Research Center, USA, 1982.

394. B. van Leer. Flux Vector Splitting for the Euler Equations. In *Proceedings of the 8th International Conference on Numerical Methods in Fluid Dynamics*, pages 507–512. Springer–Verlag, 1982.

395. B. van Leer. On the Relation Between the Upwind–Differencing Schemes of Godunov, Enguist-Osher and Roe. *SIAM J. Sci. Stat. Comput.*, 5(1):1–20, 1985.

396. B. van Leer. Progress in Multi–Dimensional Upwind Differencing. Technical Report CR-189708/ICASE 92–43, NASA, Sept. 1992.

397. B. van Leer, J. L. Thomas, and P. L. Roe. A comparison of numerical flux formulas for the euler and navier–stokes equations. AIAA 8th Computational Fluid Dynamics Conference, June 9–11, 1987, Honolulu, Hawaii. Paper AIAA–87–1104–CP, 1987.

398. M. H. P. M van Puten. A Numerical Implementation of MHD in Divergence Form. *J. Comput. Phys.*, 105:339–353, 1993.

399. M. O. Varner, W. R. Martindale, W. J. Phares, and J. C. Adams. Large Perturbation Flow Field Analysis and Simulation for Supersonic Inlets. Technical Report NASA CR–174676, NASA, USA, 1984.

400. M. E. Vázquez-Cendón. *Estudio de Esquemas Descentrados para su Aplicación a las Leyes de Conservación Hiperbólicas con Términos de Fuente.* PhD thesis, Departamento de Matemáticas Aplicadas, Universidad de Santiago de Compostela, España, 1994.

401. M. E. Vázquez-Cendón. Improved Treatment of Source Terms in Upwind Schemes for the Shallow Water Equations in Channels with Irregular Geometry. *J. Comput. Phys.*, 148:497–526, 1999.

402. M. Vinokur. An Analysis of Finite–Difference and Finite– Volume Formulations of Conservation Laws. *J. Comput. Phys.*, 81:1–52, 1989.

403. S. Y. Wang, P. B. Buttler, and H. Krier. Non–Ideal Equations of State for Combusting and Detonating Explosives. *Prog. Energy Combust. Sci.*, 11:311–331, 1985.

404. R. F. Warming and R. W. Beam. Upwind Second Order Difference Schemes with Applications in Aerodynamic Flows. *AIAA Journal*, 24:1241–1249, 1976.

405. Z. U. A. Warsi. *Fluid Dynamics: Theoretical and Computational Approaches.* CRC Press Inc., 1992.

406. G. Watson, D. H. Peregrine, and E. F. Toro. Numerical Solution of the Shallow Water Equations on a Beach Using the Weighted Average Flux Method. In *Computational Fluid Dynamics 1992, Vol. 1*, pages 495–502. Elsevier, 1992.

407. N. P. Weatherill. Numerical Grid Generation. Lecture series 1990–06, von Karman Institute for Fluid Dynamics, 1990.

408. N. P. Weatherill. An Upwind Kinetic Flux Vector Splitting Method on General Mesh Topologies. *Intern. J. Numer. Meth. Engineering*, 37:623–643, 1994.

409. R. Wegmann. An Upwind Difference Scheme for the Double–Adiabatic Equations. *J. Comput. Phys.*, 131:199–215, 1997.

410. E. Weinan and C. W. Shu. A Numerical Resolution Study of High–Order Essentially Non–Oscillatory Schemes Applied to Incompressible Flow. Technical Report ICASE 92–39, NASA Langley Research Center, USA, 1992.

411. G. B. Whitam. *Linear and Non–linear Waves.* John Wiley and Sons, 1974.

412. F. A. Williams. *Combustion Theory.* Benjamin Cummings Publishing Company, Inc Merlo Park, California, 1985.

413. P. Woodward and P. Colella. The Numerical Simulation of Two–Dimensional Fluid Flow with Strong Shocks. *J. Comput. Phys.*, 54:115–173, 1984.

414. K. Xu. Gas kinetic schemes for unsteady compressible flow simulations. VKI Fluid Dynamics Lecture Series, 1998–3, 1998.

415. K. Xu, L. Martinelli, and A. Jameson. Gas–Kinetic Finite Volume Methods, Flux–Vector Splitting and Artificial Diffusion. *J. Comput. Phys.*, 120:48–65, 1995.

416. K. Xu and K. H. Prendergast. Numerical Navier–Stokes Solutions from Gas–Kinetic Theory. *J. Comput. Phys.*, 114:9–17, 1994.

417. N. N. Yanenko. *The Method of Fractional Steps.* Springer–Verlag, New York, 1971.

418. G. Yang, D. M. Causon, D. M. Ingram, R. Saunders, and P. Batten. A Cartesian Cut Cell Method for Compressible Flows Part A: Static Body Problems. *Aeronautical Journal*, 101:47–56, 1997.

419. G. Yang, D. M. Causon, D. M. Ingram, R. Saunders, and P. Batten. A Cartesian Cut Cell Method for Compressible Flows Part B: Moving Body Problems. *Aeronautical Journal*, 101:57–65, 1997.

420. J. Y. Yang, M. H. Chen, I. N. Tsai, and J. W. Chang. A Kinetic Beam Scheme for Relativistic Gas Dynamics. *J. Comput. Phys.*, 136:19–40, 1997.

421. H. C. Yee. Construction of Explicit and Implicit Symmetric TVD Schemes and their Applications. *J. Comput. Phys.*, 68:151–179, 1987.

422. H. C. Yee. A class of high–resolution explicit and implicit shock–capturing methods. Von Karman Institute of Fluid Dynamics, Lecture Series 1989–04, 1989.

423. H. C. Yee and P. K. Sweby. Non–Linear Dynamics and Numerical Uncertainties in CFD. Technical Report TM 110398, NASA Ames Research Center, Moffett Field, USA, 1996.

424. H. C. Yee, R. F. Warming, and A. Harten. Implicit Total Variation Diminishing (TVD) Schemes for Steady State Problems. *J. Comput. Phys.*, 57:327–360, 1985.

425. E. C. Zachmanoglou and D. W. Thoe. *Introduction to Partial Differential Equations.* Dover Publications, Inc. New York, 1986.

426. S. T. Zalesak. Fully Multidimensional Flux–Corrected Transport Algorithms for Fluids. *J. Comput. Phys.*, 31:335–362, 1979.

427. D. D. Zeeuw and K. G. Powell. An Adaptively Refined Cartesian Mesh Solver for the Euler Equations. *J. Comput. Phys.*, 104:56–68, 1993.

428. Y. B. Zeldovich and Y. P. Raizer. *Physics of Shock Waves and High Temperature Hydrodynamic Phenomena.* Academic Press, 1966.

429. G-C. Zha and E. Bilgen. Numerical Solution of Euler equations by a New Flux Vector Splitting Scheme. *Int. J. Numer. Meth. Fluids*, 17:115–144, 1993.

430. T. Zhang and Y. Zheng. Conjecture on the Structure of Solutions of the Riemann Problem for Two Dimensional Gas Dynamics. *SIAM J. Math. Anal.*, 21(3):593–630, 1990.

Index

Printing: Mercedes-Druck, Berlin
Binding: Stürtz AG, Würzburg